Proceedings

MW01700362

Geometric Modeling and Processing 2000

Theory and Applications

Edited by

Ralph Martin and Wenping Wang

Front Cover:

Graphics designer: Bin Chan
3D model credit: Platinum Pictures Multimedia, Inc.

Proceedings

Geometric Modeling and Processing 2000

Theory and Applications

April 10-12, 2000

Hong Kong, China

Organized by

The University of Hong Kong

In cooperation with

Hong Kong Institution of Engineers
Control, Automation and Instrumentation Division
IEEE Computer Society, Hong Kong Chapter
SIAM Activity Group on Geometric Design
The Institute of Mathematics and Its Application, UK
The Geometric Modeling Society, UK
The Society of Manufacturing Engineers, Hong Kong Chapter

Sponsored by

The K.C. Wong Foundation
Spatial Technology Ltd.

IEEE COMPUTER SOCIETY

Los Alamitos, California

Washington • Brussels • Tokyo

IEEE Computer Society Order Number PR00562
ISBN 0-7695-0562-7
ISBN 0-7695-0564-3 (microfiche)
Library of Congress Number 00-100381

Additional copies may be ordered from:

IEEE Computer Society
Customer Service Center
10662 Los Vaqueros Circle
P.O. Box 3014
Los Alamitos, CA 90720-1314
Tel: + 1-714-821-8380
Fax: + 1-714-821-4641
E-mail: cs.books@computer.org

IEEE Service Center
445 Hoes Lane
P.O. Box 1331
Piscataway, NJ 08855-1331
Tel: + 1-732-981-0060
Fax: + 1-732-981-9667
mis.custserv@computer.org

IEEE Computer Society
Asia/Pacific Office
Watanabe Building,
1-4-2 Minami-Aoyama
Minato-ku, Tokyo 107-0062 JAPAN
Tel: + 81-3-3408-3118
Fax: + 81-3-3408-3553
tokyo.ofc@computer.org

Editorial production by Danielle C. Young

Cover art production by Joseph Daigle/Studio Productions

Printed in the United States of America by The Printing House

IEEE
COMPUTER
SOCIETY

Table of Contents

Geometric Modeling and Processing 2000 (GMP 2000)

Session 8: Mesh Processing and Subdivision Surfaces II

Session 9: Minkowski Roots, Parametrization, and Implicitization

Session 10: Computation in Geometric Modeling

Session 11: Geometric Processing and Tolerancing

Preface

This book contains the proceedings of Geometric Modeling and Processing 2000, the first in a biennial international conference series on solid modeling, shape representation, and geometric computation.

Modeling and processing of geometric shape is of importance in many disciplines, such as computer graphics, computer aided design, computer vision, medical imaging, and scientific computation.

The topics addressed by the papers in these proceedings fall into several main areas. The largest number address various issues in curve and surface modeling, from long-standing issues such as interpolation and approximation, to ideas of growing popularity such as polyhedral and subdivision methods, and finally on to more specific areas such as blending, deformations, implicitization, and parametrized design. Solid modeling is the second largest category of papers, covering issues such as input of models from sketches and from scanned data, subdivision modeling, and trimmed patches. A third category of papers addresses various application issues, some in computer graphics — polygon mesh refinement, for example, and others elsewhere — modeling the draping of cloth, mould generation, and watermarking of polygon meshes for security.

Of the papers presented in these proceedings, ten are invited contributions authored by leading international experts. Twenty-nine further papers were selected from those submitted after rigorous refereeing by an international program committee. Contributors to this volume come from America, Asia, and Europe, representing eighteen countries (or regions). Overall, the organizers were very pleased with the generally high quality of submissions received, and feel that this bodes well for the planned future conferences in this series, to be held in Japan in 2002 and Korea in 2004.

The Conference Organizing Committee under the chairmanship of S.H. Lo is to be thanked for its work in helping to enable the conference. Thanks go to all reviewers who have contributed with their time and expertise by reviewing and recommending papers. We thank in particular Myung-Soo Kim for his efforts and ideas that led to this conference. We also wish to thank all the cooperating organizations for their assistance. We thank the K.C. Wong Foundation for its generous financial support, and Spatial Technology for donating the Best Presentation Award. Finally, we wish to thank all the participants at the conference for taking part.

Sooi Thor Tan, Conference Chair

Ralph Martin, Program Co-Chair

Wenping Wang, Program Co-Chair

Conference Committee

Conference Chair

Sooi Thor Tan

Organizing Committee Chair

Sai Huen Lo

Organizing Committee

Kit Wah Chan
Yonghua Chen
Ian Gibson
Wing Suen Sze
Tak Nam Wong

Program Committee Co-Chairs

Ralph Martin
Wenping Wang

Program Committee

Karim Abdel-Malek
Kunwoo Lee
Hans-Peter Seidel
Chandrajit Bajaj
Weiyin Ma
Yoshihisa Shinagawa
Brian Barsky
Takashi Maekawa
Vadim Shapiro
Wim Bronsvoort
Jai Menon
Yohanes Stefanus
Stephen Cameron
Ahmad Nasri
Kokichi Sugihara
Fuhua (Frank) Cheng
Alex Pasko
Hiromasa Suzuki
Eng Wee Chionh
Nick Patrikalakis
Chiew Lan Tai
B.K. Choi
Qunsheng Peng
Demetri Terzopoulos
Wendelin Degen
Michael Pratt
Michael Wagner
Gershon Elber
Hong Qin
Jiaye Wang
Gerald Farin
Kaihuai Qin
Joe Warren
Ronald Goldman
Ari Rappoport
Tony Woo
Bert Juettler
Alyn Rockwood
Brian Wyvill
Myung-Soo Kim
Malcolm Sabin
Geoff Wyvill

Additional Reviewers

Yonghua Chen

Sai Huen Lo

Vijay Srinivasan

Yi-King Choi

Chung Lun Li

Wing Suen Sze

Rida Farouki

Oleg Okunev

Sooi Thor Tan

Jieqing Feng

Helmut Pottmann

Tamas Varady

Ian Gibson

Srinivas Raghothama

Tak Nam Wong

Kin-Chuen Hui

Takis Sakkalis

Guoxin Yu

Session 1

Model Acquisition and Reverse Engineering

Reverse Engineering B-rep models from Multiple Point Clouds

Tamás Várady, Pál Benkő
Computer and Automation Research Institute,
Hungarian Academy of Sciences
{varady,benko}@sztaki.hu

Abstract

The procedure of reconstructing conventional engineering objects from multiple-view, 3D point clouds is described. Emphasis is put on producing accurate and topologically consistent B-rep models, ready to be used in computer aided design and manufacture. After describing the basic phases of reverse engineering, related algorithmic difficulties are analyzed.

Three particular areas are discussed in more detail: segmenting point data into regions, creating translational and rotational surfaces with smooth, constrained profiles, and finally adding blends. The results of the algorithms are illustrated using a mechanical engineering object, which is a well-known benchmark in the RE community.

1. Introduction

Reconstructing objects has been a central problem in computer vision for a long time [6, 7, 19]. Due to recent technological advances in laser scanning, now it is possible to produce multiple 3D point clouds with high density and high accuracy, and in this way it has become a realistic expectation to generate accurate models with correct topology which can be directly transferred to and utilised by CAD/CAM systems. There is a wide diversity of reverse engineering methods for converting measured data into geometric models [34]. Methods differ in the quality of measurements, the quantity of points, the geometric characteristics of the objects in question, the amount of user interaction required, and the final representation used for the model.

About twenty years ago, the two main areas of geometric modelling development addressed (i) geometrically demanding free-form surface models or (ii) topologically complex solid models with simple surface elements. These techniques were eventually integrated into commercial CAD/CAM systems. The current situation in reverse engineering is somewhat similar. For free-form shapes, the emphasis is put on developing techniques by means of which the user can manually segment point clouds. Segmentation is followed by fitting high-quality surfaces [21, 22] and the result is typically a collection of trimmed parametric surface patches. For conventional engineering objects, which are bounded mainly by simple implicit surfaces, the main efforts are directed towards building accurate and consistent CAD models with little or no user interaction. These approaches are based on certain assumptions concerning the regularity of the objects which make it possible to automate the process.

It is important to point out that, although there are many valuable contributions in computer graphics and computer vision in object reconstruction, the majority of these cannot directly be applied for CAD/CAM purposes in mechanical engineering. Some of these approaches produce fair, triangulated approximations (see the pioneering work of Hoppe et al. [20]), while others prefer to represent the boundary of objects by special uniform surface representations such as A-patches or superquadric primitives (see for example [2, 28]). But these representations do not conform with standard CAD representations, which have been widely used for data exchange and manufacturing for a long time.

Approaches exist which use standard free-form surfaces, such as B-splines, to reconstruct conventional engineering objects — see [12, 16, 23, 33]. Unfortunately these free-form models cannot reproduce perfectly the original design as would be required by most CAD/CAM applications. As an example, take the well-known benchmark object in Figure 1, which — according to the authors' classification — is a conventional engineering object. (This object was selected a few years ago by Hoschek: see [21].) From a functional point of view, this part is the union of three components. LEFT has rotational symmetry, MIDDLE is a part of a blended cone, and RIGHT is bounded by two planar faces and a cone in between, and a sequence of extruded side faces. The intersection edges have been blended by constant radius rolling ball blends.

The difficulty of using free-form surface models is the following. If we use multiple surfaces, it is hard to find

0-7695-0562-7/00 $10.00 © 2000 IEEE

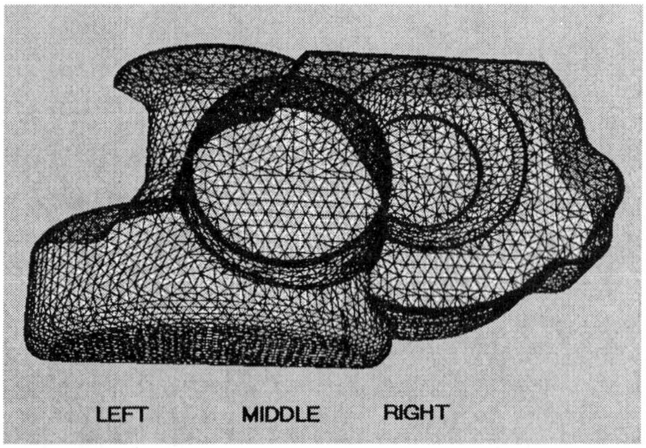

Figure 1. Triangulated and decimated data points

good segmenting curves, and to assure smoothness and watertightness between the the adjacent elements. To fit a single B-spline is a challenging task in surface approximation due to the sudden variations in curvature. We may approximate the data points within tolerance, but B-spline control points do not exist which properly reproduce the regularity of the surfaces with appropriate curvature distribution.

The above arguments justify dedicated approaches for conventional engineering objects. We set up the following requirements. Simple analytic surfaces must be accurately represented. Translational and rotational symmetries must be recognised and recorded explicitly. If two adjacent faces are tangent continuous, smoothness must be guaranteed. Blends need to be treated as special transition surfaces dependent on other surfaces. Since surface fitting to measured data sets will definitely lead to inaccuracies, we need to improve our models by recognising the most frequent engineering constraints such as perpendicularity, parallelism, concentricity etc. and enforcing these on the model.

This paper deals with the reverse engineering of conventional objects in the above sense. The majority of previous publications mostly handled single view range data and only simple surfaces (for example [9, 15, 18, 30]). Here we assume a more general data model and extend the class of bounding surfaces. The point cloud is ungridded coming from dense, multiple view measurements, the objects are bounded by relatively large *primary surfaces* of analytic surface type: such as planes, natural quadrics and tori; we also permit general translational and rotational surfaces. *Blends* may also be present and are assumed to be small transition surfaces.

As will be shown, due to the regularity and dependency of the bounding surfaces it is possible to interpret and compute the underlying topological structure of many objects in an automatic manner. In this sense, translational and rotational surfaces with free-form profiles can also be included. However, this concept does not seem appropriate for complex free-form surface models where some sort of internal segmentation is required.

After surveying the whole reconstruction process and identifying the successive algorithmic steps in Section 2, three particular topics are chosen and discussed in more detail: segmenting point clouds, reconstructing translational and rotational surfaces, and adding blends (Sections 3–6).

2. Basic phases of reconstructing solid models

The reverse engineering procedure can be decomposed into the following consecutive steps:

1. Data capture
2. Merging multiple point clouds
3. Triangulation/decimation
4. Segmentation
5. Surface fitting for simple surfaces
6. Reconstructing translational and rotational surfaces
7. Building an adjacency graph
8. Beautifying, constrained fitting
9. B-rep model creation and further beautifying
10. Adding blends

For *data capture* typically laser scanners are used, which produce measured data sets with high density and reasonable accuracy. A single point cloud is generally projectable onto a plane or a cylinder, but is not necessarily gridded. Unfortunately, even these data sets are not free from outlying data points; measurements can be unstable in the neighbourhood of silhouette curves and concave edges, and there are always missing portions due to occlusion. This means that discontinuities in depth do occur, and this may hold even for merged point clouds too.

A single view represents only a part of an object. In order to get a complete or at least a multiple-view model, point clouds need to be *merged*, i.e. appropriate transformations need to be found by means of which the different data sets are moved into a common coordinate system. This is often called *registration*, which has been and still is an important problem in computer vision [5, 11, 27]. Iterative closest point methods are based only on point data. Most ICP algorithms assume that the user has already roughly positioned the point clouds into a common coordinate system before the actual iterative process is started. To obtain reliable results, large overlaps are necessary.

Other methods use calibrating balls or special markers to find the correspondence between the individual point sets. High-precision rotary tables are often used, since transformations can robustly be computed; however, in this case the bottom face and other downward pointing faces are not visible.

Basically, there are two different approaches to merging. If it is done at *point cloud level*, a more complex data structure is obtained with large, united point regions, but these may carry additional inaccuracies and uneven, irregular point distributions. If we create single view geometric models, we have to deal with discontinuities, but simpler models are obtained. These may contain various geometric features which help merging at *model level*. The single view models models must be consistent and accurate to be able to unite them using Boolean operations in the final stage. The pros and cons of point cloud merging and model merging are discussed in detail in [3].

Even a single point cloud may contain several tens or hundreds of thousands of data points. In order to determine the boundary surface of our objects, first we have to determine the neighbourhood and connectivity of the data points. This is a difficult problem, particularly if we have a merged data set with noise, uneven point distributions, discontinuities, holes and disjoint parts. There are various techniques to build a consistent triangulation, which differ in their basic assumptions, their generality, and the computational efficiency of their solutions — see for example [2, 13, 20, 25]. We cannot go into details here, but just assume that a 3D *triangulation* of the point cloud has already been created. For efficiency reasons, the original triangulation often needs to be *decimated*, forming a so-called *carrier surface*. The carrier surface reflects the original topology of the point cloud; each data point is either a vertex of the triangulation or associated with the closest triangle of the carrier. The carrier surface plays an important role in speeding up further computations.

One of the most crucial elements of reverse engineering is *segmentation*, i.e. identifying point regions which are disjoint subsets and mark out connected portions within the whole point cloud sharing some common property. These are considered to be pre-images of single faces (or possibly multiple faces) of the B-rep model. For each face a single surface is used to approximate the related points within a given tolerance. Various methods for *fitting* simple analytic surfaces have been described in [7, 14, 15, 17, 29, 36]. The authors' view is that the use of geometric, or geometrically faithful, distances is more satisfactory for surface reconstruction than algebraic distances. For example, instead of an approximating quadric surface, we prefer to compute the best possible circular cylinder in this explicit geometric form.

In general, segmentation is a complex, iterative process [34]. Often, iterative *region growing* techniques are applied [6, 28, 33]. In the current context we prefer another solution called *direct segmentation* [36], which is feasible for regular objects and matches our basic assumptions.

Direct segmentation produces disjoint smooth regions, each of which is approximated by a simple analytic or swept surface. Having identified these, a so-called *region adjacency graph* needs to be built, which reflects the complete topology of the final B-rep model. The nodes of the adjacency graph are the regions; if two regions are found to be adjacent (ignoring small blends for the time being), an adjacency edge is created in the graph, which of course corresponds to a real 3D edge. Taking a given region and its adjacent nodes, we can determine the edges and edge loops of a face. These edges are computed either by surface-surface intersection, or they are silhouette edges (*jump* edges in computer vision), which need to be approximated based on the discontinuous border of the triangulation. Since surfaces fitted to large regions are numerically stable, the intersection curves are also smooth and fair.

Surface-surface intersections work well for *sharp* edges, but are not robust for *smooth* edges, which need to be treated in a different way. We detect *multiple face smooth regions*, which cannot be approximated by a single surface and further segmentation is necessary. Then smoothness constraints are enforced and eventually a sheet object with several faces is generated. Thus the smooth edges are directly computed without intersection calculations. An important subset of multiple regions comprises translational or rotational surfaces (see Section 4). Here we fit a sweeping profile in a way that the consecutive elements are constrained to be smooth. For general multiple regions, *constrained fitting* is applied [4], where several surfaces with already detected surface types and parameter estimates are fitted simultaneously.

Smooth edges which are the boundaries of *blending surfaces* occur in a different way. Blends have been temporarily removed during direct segmentation and are reconstructed only in the very last phase of model building [24]. The blending surfaces need to be tangential to the primary surfaces and are generated accordingly: see Section 5.

A crucial step is to *beautify* B-rep models. When noisy measured data are used, the generated model is likely to be imperfect. The exclusion of very small edges and facets, filling of little holes, ensuring orthogonality, parallelism, concentricity, etc. are all important requirements for real CAD/CAM applications. The latter types of beautification can also be solved by constrained fitting. As discussed earlier, for composite regions smoothness constraints are added and the surface elements are simultaneously fitted. Similarly, for example, if two planes were found to be nearly perpendicular, the points of the two related surfaces can be refitted under a perpendicularity constraint, which will lead to a better model. There are many papers on constrained geometric modelling [8], but its relevance to reverse engineering has been recognised only recently — see [4, 37].

The actual construction of B-rep models is generally performed using a solid modelling kernel. Two basic approaches are known. *Prismatic* models are based on gluing together swept volumes of the individual faces acquired from different views, and thus closed volumes are always obtained. In *stitched* models, the individual faces are just glued together along their common edges, which lead to composite shell objects. More on these alternatives can be found in [3].

Having concluded the overview of the reconstruction process, we continue with specific solutions which can be plugged into the above general framework.

3. Direct segmentation

In this section we describe the basic concept of direct segmentation, but lack of space prevents us for giving full details, for which see [4, 36]. Our purpose is to subdivide the point data into disjoint subsets, which are smooth and to which one (or more) surfaces can locally be fitted. As noted earlier, the fundamental structure of the object is determined by primary surfaces. Along the intersection of two primary surfaces there is either a sharp edge or a small blend. Alternatively the surfaces may be smoothly connected.

As will be shown, in the first half of direct segmentation we attempt to separate regions by removing triangles which are in the vicinity of sharp edges or small blends. Thus, disjoint regions are obtained, which can be classified either as *simple*, i.e. these can be approximated by a single analytic surface, or as *multiple*, which need to be further subdivided by smooth edges.

An important feature of the current algorithm is that, for each data point we estimate a surface *normal vector* and a value which characterises the *planarity* of the point neighbourhood. Our experience shows that the best normal vector estimates are based on fitting a second order algebraic surface to surrounding points in the neighbourhood. This means that first we translate all the points in the neighbourhood so that the given point $\mathbf{p}$ goes to the origin (the image point set is denoted by P'), then we look for a symmetric matrix A and unit vector $\mathbf{d}$ which minimise

$$\sum_{\mathbf{x}\in P'} \left(\mathbf{x}^T A\mathbf{x} + \langle \mathbf{d}, \mathbf{x}\rangle\right)^2.$$

This leads to a three dimensional eigenvalue problem and the normal we seek is just $\mathbf{d}$. ($\langle\cdot,\cdot\rangle$ denotes the scalar product of two vectors.)

Figure 2. Planarity filtering — simple and smooth, multiple regions

The planarity is characterised by taking a point and the related least-square plane fitted to its neighbours, and computing the normalised errors from it over a small disk: for relatively smooth primary surfaces this value will be small, while for highly curved parts this value will be greater. There are other alternatives to identify highly curved parts using the triangulation; for example, the angular variation of the normals was suggested in [23], or curvature estimates can be computed [1].

After choosing an appropriate threshold value, planarity filtering will remove those triangles which are classified as highly curved. For example, in Figure 2, we have filtered out the small blends between the components LEFT, MIDDLE and RIGHT, and the small blends between the side faces and the top and bottom planar faces. The larger blends for the rotational and translational regions were kept. We obtained two smooth, multiple regions.

Note that our example object is a variant of the original benchmark of Hoschek, which did not contain sufficiently dense data for our purposes, making it impossible to reconstruct blends reliably. Our variant was redesigned and sampled in such a way that we can demonstrate important features of our algorithm.

Given a region, we do not immediately know whether it is simple or multiple. A sequence of hypotheses is set to decide whether the given point set can be approximated by some simple surface. If we fail, we proceed to try more complex surface types. Finally multiple regions are decomposed. We always proceed in order of increasing complexity. Fortunately, simpler surfaces (i) occur more frequently in mechanical engineering, and (ii) can be detected in a more reliable and efficient manner. The decision procedure thus follows the simplified scheme below:

1. Test whether the region is a plane or a sphere; if so: determine parameters. OTHERWISE

2. Test for translational symmetry; if so:

   (a) Test whether it is a cylinder; if so: determine parameters, ELSE

   (b) Test whether the sweeping contour can be decomposed into a smooth sequence of straight line segments and circular arcs; if so: multiple region, ELSE

   (c) Compute a free-form profile. OTHERWISE

3. Test whether it is a cone; if so: determine parameters. OTHERWISE

4. Test for rotational symmetry; if so:

   (a) Test, whether it is a torus, if so: determine parameters, ELSE

   (b) Test whether the sweeping contour can be decomposed into a smooth sequence of straight line segments and circular arcs if so: multiple region, ELSE

   (c) Compute a free-form profile. OTHERWISE

5. Decompose non-regular, multiple smooth regions

   (a) Detect planes

   (b) Detect translational subregions

   (c) Search for independent axes of rotation, and for each axis detect rotational subregions

(For 5b and 5c proceed as in 2 and 4.)

The above tests are based on consecutive least-squares procedures for the geometric parameters of these simple surfaces — see [36]. The scheme shows the significance of detecting translational and rotational symmetries, which are based on the surface normal estimates associated with each point.

3.1. Determining the Translational Direction

The normal vectors of a translational surface must be perpendicular to a common translational direction. We may use the deviation of these angles from $\pi/2$, but the simplest is to minimise the cosine of the angles, which is just the scalar product of the given normal vectors and the unknown direction. The translational characteristic of a shape can be visualised with the help of the Gaussian sphere: the normal vectors of a translational surface lie on a great circle, that is, in a plane which goes through the origin and is perpendicular to the translational direction.

Formally: if the estimated normal vectors of the region are denoted by $\{\mathbf{n}_i\}$, then we look for the unit vector $\mathbf{d}$ which minimises

$$\sum \langle \mathbf{n}_i, \mathbf{d} \rangle^2.$$

This is a well-known three dimensional eigenvalue problem.

3.2. Determining the Rotational Axis

The normal lines of a rotational surface must intersect a common axis (in a projective sense, i.e. the lines may also be parallel). Ideally we would like to minimise the sum of the squared angles γ_i between each normal line l_i and the plane which contains the axis and the related point $\mathbf{p}_i$. This leads to a non-linear minimization. However, Pottmann suggested an interesting alternative solution [32], which proves to be valuable and efficient in practice. The problem is slightly modified: instead of a rotational surface, we look for a helical one, as described below.

So-called Plücker coordinates of lines are used. A line through point $\mathbf{p}$ in direction $\mathbf{d}$ is represented by a sextuple $(\mathbf{d}, \mathbf{d} \times \mathbf{p})$ (the latter, the *momentum vector* is denoted by $\bar{\mathbf{d}}$). This is independent of the choice of $\mathbf{p}$ as can be seen. This is a homogeneous representation, i.e. $(\alpha \mathbf{l}, \alpha \bar{\mathbf{l}})$ represents the same line as $(\mathbf{l}, \bar{\mathbf{l}})$ for non zero α. For $(\mathbf{l}, \bar{\mathbf{l}})$ to be a proper line, it must satisfy $\langle \mathbf{l}, \bar{\mathbf{l}} \rangle = 0$.

Let us calculate the angle γ: first we need the normal of the plane containing the axis (with given Plücker coordinates $(\mathbf{a}, \bar{\mathbf{a}})$) and a data point $\mathbf{p}$. If $\mathbf{c}$ is a point on the axis, then the normal of the plane is

$$(\mathbf{p} - \mathbf{c}) \times \mathbf{a} = \mathbf{p} \times \mathbf{a} + \mathbf{a} \times \mathbf{c} = \mathbf{p} \times \mathbf{a} + \bar{\mathbf{a}},$$

but must be normalised. As the angle of the normal and the direction $\mathbf{d}$ of the line and γ are conjugate we obtain

$$\begin{aligned} \sin \gamma &= \left\langle \mathbf{d}, \frac{\mathbf{p} \times \mathbf{a} + \bar{\mathbf{a}}}{\|\mathbf{p} \times \mathbf{a} + \bar{\mathbf{a}}\|} \right\rangle = \frac{\langle \mathbf{d} \times \mathbf{p}, \mathbf{a} \rangle + \langle \mathbf{d}, \bar{\mathbf{a}} \rangle}{\|\mathbf{p} \times \mathbf{a} + \bar{\mathbf{a}}\|} \\ &= \frac{\langle \bar{\mathbf{d}}, \mathbf{a} \rangle + \langle \mathbf{d}, \bar{\mathbf{a}} \rangle}{\|\mathbf{p} \times \mathbf{a} + \bar{\mathbf{a}}\|}. \end{aligned}$$

The denominator is just the distance of the point from the axis. For simplicity's sake, we ignore it, and then the distance of the two lines multiplied by the sine of their angle is obtained. One can argue that this is a fair measure: we want these lines to either intersect (zero distance) or be parallel (zero sine). Thus we have a linear formula with two constraints: $\|\mathbf{a}\| = 1$ and $\langle \mathbf{a}, \bar{\mathbf{a}} \rangle = 0$. Simplifying again, we may omit the latter. This is Pottmann and Randrup's idea, so we actually minimise $\sum (\langle \bar{\mathbf{d}}, \mathbf{a} \rangle + \langle \mathbf{d}, \bar{\mathbf{a}} \rangle)^2$. For details see [32].

This method is sufficiently accurate to compute the rotational axis in most cases. If not, it can be used as an initial guess for iteratively improving the above mentioned quantities to obtain a more accurate solution.

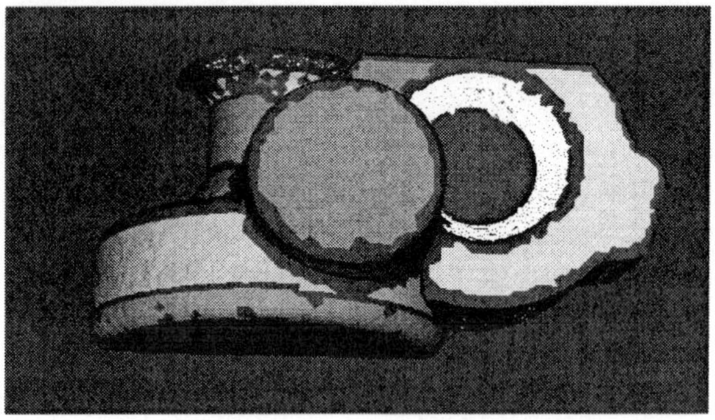

Figure 3. Segmented object, simple translational and rotational surfaces detected

3.3. Multiple regions

In the general scheme of direct segmentation, we may identify smooth multiple regions. These are defined by sweeping a smooth profile, which consists of tangential line-segments-and-circular-arcs or free-form profiles. Details of reconstructing conventional profiles are given in the next section.

For many engineering objects, after the first four phases of direct segmentation no unresolved regions will be left. If there are still multiple regions to be dealt with (we are at step 5), a useful tool called *dimensionality filtering* needs to be applied. Taking the normal vectors of a multiple region mapped onto the Gaussian sphere, the normals of any planar region will be mapped into a point cluster; those of a translational or conical region onto a circular arc. Classifying the normal vector distributions on the Gaussian sphere as 0- or 1-dimensional, the planar and translational surfaces can be separated and detected — see details in [36].

The only problem left is when more than one rotational surface exists within a smooth subregion. In this case, we may identify all likely rotational axes using an approach suggested by Lukács in [29]. The basic idea is that any four normal lines determine two axes, which intersect all four lines, and can be computed easily by a second degree equation. After clustering the guess axes, we detect rotational surfaces locally. This step concludes the direct segmentation procedure.

Our segmented object can be seen in Figure 3. The regions are coloured differently and the detected simple surfaces have already been fitted. The multiple regions, the rotational axis of LEFT and the sweeping direction of the side face of RIGHT have also been computed and decomposed according to their smooth profile curves, i.e. we generate a sequence of connected cylinders, cones, and tori for LEFT, and connected cylinders for RIGHT, respectively.

4. Straight-circular Profiles

We continue with computing straight-circular profiles; the related smooth, multiple region has already been detected.

4.1. Guiding Polygon

The *guiding polygon* is a key element in profile fitting. Approaches which are based only on point data may find it difficult to approximate the point set after projection. Due to the inaccuracies of the translational direction or the rotational axis we may obtain a point cloud which is 'thick', and thus it is hard to order the points, a step which is generally needed for the different approximating methods. An approach to thin point clouds was reported recently in [26]. We propose to work over the triangulation, which has the advantage that not only the point information, but their *connectivity* can be transformed into the plane of the profile.

Once we know the translational direction or the rotational axis, we can create a polygon which approximates the profile by intersecting the triangulation with a plane. The problem is that there may be planes which do not contain the *entire* profile due to intersections with other surfaces — it may even be the case that *all* planes have this unfortunate property. This is why we have to apply a gradual approach to build up the complete polygon. We start at an arbitrary triangle — possibly in the middle of the point set. We intersect it with the projection plane through its barycentre and extend the polygon by taking the adjacent triangles along that plane. Of course, this may be just a part of the guiding polygon and it may need to be extended. After building the initial polygon we check the remaining triangles. Each triangle is projected (or rotated) into the profile plane and its image is projected onto the polygon. If this image is not fully covered by the current polygon, the polygon will be extended accordingly. In order to build up the polygon correctly, we do not traverse the triangles in an arbitrary order: we always take a triangle adjacent to a previously processed one.

If the distance between the two endpoints of the polygon becomes too small (say, smaller than the average distance of adjacent vertices), we close the polygon and stop the building process.

4.2. Constrained Profiles

Of course, every profile can be approximated by a free-form curve, but many engineering applications require an explicit straight-circular arc representation of the profile where applicable. While for free-form curves the guiding polygon provides a sensible parametrisation of the projected points, here it helps the segmentation of the profile into circular arcs and straight line segments to apply a 'recover & select' [28] procedure.

Circles are represented in Pratt's form [31], which includes lines as special cases and not as singularities. After segmentation, circles whose radii are too great are replaced by straight lines. Then the points are grouped with the help of the guiding polygon and either a circle or a straight line is fitted to each group, such that each is tangent to its neighbouring elements. There are two strategies to achieve this goal. One of them is to fit the elements sequentially and always satisfying the smoothness constraint with the previous element. This strategy may lead to a contradiction in the case of a closed profile; moreover, the errors are accumulated: the first object to be fitted will be quite exact, but as we proceed we obtain worse and worse approximations. We prefer an alternative strategy which computes the elements simultaneously, i.e. approximates the points and satisfies the tangency constraints.

The equation of a circle written in Pratt's form is the following:

$$A(x^2 + y^2) + Bx + Cy + D = 0,$$

where $B^2 + C^2 - 4AD = 1$. With the given normalisation condition, the expression on the left hand side is an approximation of the signed geometric distance of the point (x, y) from the circle, and consequently it is appropriate for a least squares minimisation. Taken in itself, this corresponds to a generalised eigenvector problem. As one can see, if $A = 0$, we obtain the equation of a line. The constraint expresses that a unit normal vector is used in the equation.

For constrained fitting we need equations which express the tangency between two circles or a circle and a line.

For two circles:

$$(A_1B_2 - A_2B_1)^2 + (A_1C_2 - A_2C_1)^2 - (A_2 \pm A_1)^2 = 0.$$

For a line and a circle:

$$B_lB_c + C_lC_c - 2D_lA_c \pm 1 = 0.$$

Based on these, a large system is set up to be minimised at the zeros of another system of equations. One may apply the general method of Lagrangian multipliers, or its special variant, constrained fitting, as described in details in a companion paper [4]. This solution requires, of course, an iterative process starting from initial values. These are computed during segmentation of the guiding polygon by independent unconstrained fits.

Returning to our test object, Figure 4 shows the segmented profile curve of the LEFT component. In Figure 5 the profile of RIGHT is shown. The construction elements, i.e. the full circles, can be seen in Figures 6 and 7. Having the profile curves, we can create the exact surface elements for the multiple face, which can directly be used in the final model building phase.

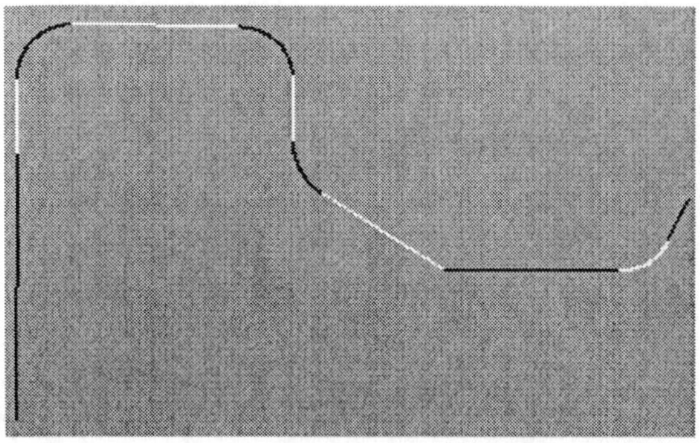

Figure 4. Reconstructed rotational profile

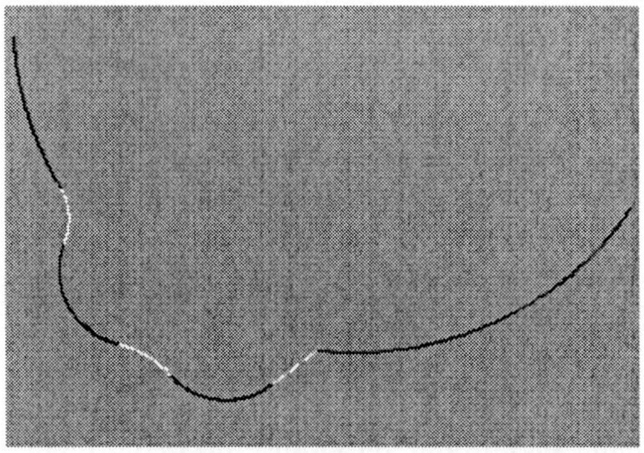

Figure 5. Reconstructed translational profile

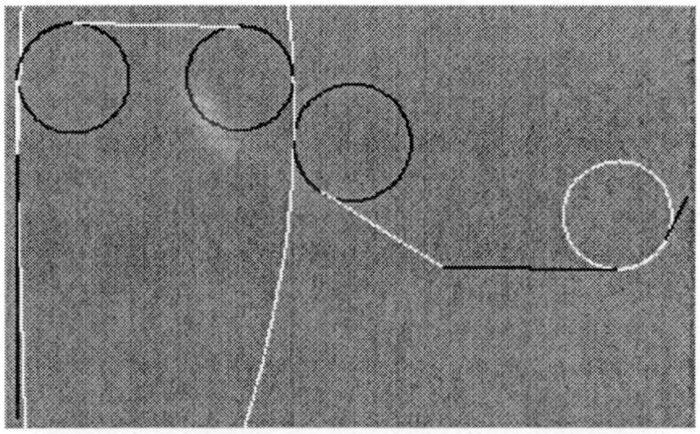

Figure 6. Circles of the rotational profile

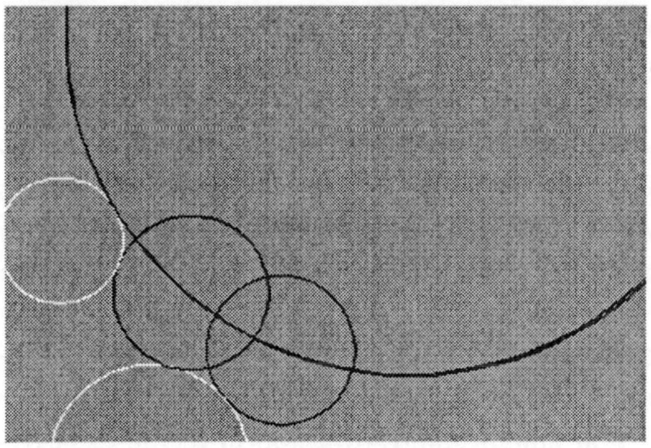

Figure 7. Circles of the translational profile

Figure 8. Test object without small blends

Figure 9. Test object with small blends

5. Blending surfaces

In the early planarity filtering phase we filtered out the data points which belong to blending surfaces. Now, having already reconstructed the primary surfaces, we can reconstruct the blends. It is assumed that these were generated by constant radius rolling ball blending. We must determine the best approximation to the appropriate radius and just forward this information to the underlying solid modeller to actually incorporate these into the B-rep model. Based on the region neighbourhood graph and the triangulation, we ignore the point regions around vertices, and take the disjoint point sets which represent single edge blends between two given primary faces. Note that in this context the primary face can possibly be a smooth, multiple face: this makes no difference from the point of view of blending.

There are different approaches to estimating blend radii; these are thoroughly analyzed in [24]. *Iterative spine* methods are based on the principle that the spine curve of a rolling ball blend is computed by intersecting offsetted primary surfaces. Having an initial radius estimate, the spine and the corresponding blend can be computed. After calculating the distances of the data points from this estimated blend, the radius can be corrected and a new spine computed; this is done iteratively until all the points are within tolerance. This method is particularly efficient in special cases when the intersection of the primary surfaces is a simple, analytic curve, but it also works for surfaces of arbitrary type.

Another approach is to compute a so-called *maximum ball* for each data point such that the ball contains the point and touches the two adjacent primary surfaces. Having these radius estimations for each point, we compute an average for the whole blend. This method is also efficient and numerically stable, and it has the advantage that it can be generalised to reproduce varying radius rolling ball blends.

We can attach blend information to the edges, and the modelling kernel incorporates the edge blends into the B-rep model. We do not reconstruct vertex blends in the current development phase, since in the case of small blends we only have a few data points. Vertex blends are considered to be by-products which automatically connect edge blends and are constructed by the solid modeller. Setback vertex blends provide a reasonable transition surface in the case of arbitrary edge blend configurations [35].

Figure 8 shows our test object with sharp edges. The blended object can be seen in Figure 9.

6. Implementation and experimental details

A prototype system for solid model reconstruction is under development at the Geometric Modelling Laboratory of the Computer and Automation Research Institute. A REPLICA laser scanner system (3D Scanners, London, UK) is used for data acquisition. The program is written in C++ and uses the VTK graphics library; it runs under LINUX and Windows NT. The ACIS solid modelling kernel [10] is used for solid model creation and blending.

The original data set of our simulated test object shown in the Figures contained 18970 points. The size of the bounding box was $100 \times 50 \times 40$. The noise added to the sampled data points was 0.05. In Figure 1 the point set was decimated to 8238 points with 16432 triangles. The estimated rotational axes and the translational direction were (0.0028, 0.9999, -0.0008) and (0.0232, 0.0052, 0.9997); compare these with the exact vectors of (0, 1, 0) and (0, 0, 1). The radius of the large cylinder of LEFT was computed as 30.1478 (cf. 30); that of RIGHT was computed as 26.876 (cf. 27). The radii of the profile elements were reconstructed with reasonable accuracy, for example, for the rotational profile the computed radii are 4.02 (cf. 4), 3.84 (cf. 4), 4.31 (cf. 4), 3.97 (cf. 4). The blend reconstruction also gave reasonable values, for example, for the top faces of MIDDLE and RIGHT we obtained 2.03 (cf. 2) and 2.01 (cf. 2); along their intersection 1.01 (cf. 1).

7. Conclusion

A concept for reverse engineering conventional solid objects has been presented, which can be directly used in mechanical engineering CAD/CAM. The complexity of the procedure was illustrated by describing the consecutive algorithmic steps and their links. The importance of reproducing the regularity of the objects — simple surfaces, translational and rotational symmetries, constraints — was strongly emphasized. The current prototype implementation shows that for reasonably dense and accurate measured data, solid reconstruction is possible with minimal user interaction by setting only certain threshold and tolerance values which control the various steps of the algorithm. Current and future efforts are mainly directed towards reliably segmenting complex multiple regions, constrained fitting and model beautification.

8. Acknowledgments

The solid reconstruction project started in 1995 within the framework of a EU supported COPERNICUS grant (RECCAD 1068). The authors highly appreciate Ralph Martin's (Cardiff University, Wales) cooperation in this research and the contribution of the GML development team, namely Géza Kós, Gábor Lukács, László Andor and Péter Salvi. This research was also supported by the National Science Foundation of the Hungarian Academy of Sciences (OTKA, No. 26203).

References

[1] L. Alboul and R. van Damme, "Polyhedral metrics is surface reconstruction: tight triangulations", In: *The Mathematics of Surfaces VII*, Eds: T. Goodman, R. R. Martin, Information Geometers Ltd, 1997, pp. 309–336

[2] C. Bajaj, F. Bernardini, J. Chen and D. Schikore, "Automatic Reconstruction of 3D CAD models", In: *Proc. Int. Conf. on Theory and Practice of Geometric Modeling II*, Eds: W. Strasser, R. Klein, R. Rau, Blaubeuren, Germany, October 1996

[3] P. Benkő, G. Lukács, R. R. Martin and T. Várady, "Algorithms to Create B-rep Models from Measured Point Clouds", *Geometric Modelling Studies*, GML1998/6, Computer and Automation Research Institute, Budapest, 1998

[4] P. Benkő, L. Andor, G. Kós and T. Várady, "Constrained Fitting in Reverse Engineering", *Geometric Modelling Studies*, GML1999/4, Computer and Automation Research Institute, Budapest, 1999

[5] R. Bergevin, M. Soucy, H. Gagnon and D. Laurendeau, "Towards a General Multi-view Registration Technique", IEEE PAMI, Vol 18, No 5, 1996, pp. 540–547

[6] P. J. Besl and R. C. Jain, "Segmentation through Variable-order Surface Fitting", IEEE PAMI, Vol 10, No 2, 1988, pp. 167–192

[7] R. M. Bolle and B. C. Vemuri, "On Three-dimensional Surface Reconstruction Methods", IEEE PAMI, Vol 13, No 1, 1991, pp. 1–13

[8] X. Chen and C. M. Hoffmann, "On Editability of Feature-based Design", *Computer Aided Design*, Vol 27, No 12, 1995, pp. 905–914

[9] P. N. Chivate and A. G. Jablokow, "Solid-model generation from measured point data", *Computer Aided Design*, Vol 25, No 9, 1992, pp. 587–600

[10] J. Corney: *3D Modeling using the ACIS Kernel and Toolkit*, John Wiley and Sons, Chichester, 1997

[11] C. Dorai, G. Wang, A. K. Jain and C. Mercer, "Registration and Integration of Multiple Object Views for 3D Model Construction", IEEE PAMI, Vol 20, No 1, 1998, pp. 83–89

[12] M. Eck and H. Hoppe, "Automatic Reconstruction of B-Spline Surfaces of Arbitrary Topological Type", *Computer Graphics* (SIGGRAPH 96), Vol 30, 1996

[13] H. Edelsbrunner and E. Mücke, "Three-dimensional Alpha Shapes", ACM TOG, Vol 13, No 1, 1994, pp. 43–72

[14] O. D. Faugeras and M. Hebert, "The Representation, Recognition and Locating of 3D Objects", *International Journal of Robotics Research*, Vol 5, No 3, 1986, pp. 27–52

[15] A. W. Fitzgibbon, D. W. Eggert and R. B. Fisher, "High-level CAD Model Acquisition from Range Images", *Computer-Aided Design*, Vol 29, No 4, 1997, pp. 321–330

[16] G. Greiner and K. Hormann, "Interpolating and Approximating Scattered 3D Data with Hierarchical Tensor Product B-Splines", In: *Surface Fitting and Multiresolution Methods*, Eds: A. Le Méhauté, C. Rabut, L. L. Schumaker, Vanderbilt University Press, 1997, pp. 163–172

[17] M. Hebert and J. Ponce, "A New Method for Segmenting 3D Scenes into Primitives", *Procs. 6th Intl. Conf. on Pattern recognition*, (München), 1982, pp. 836–838

[18] A. Hoover, D. Goldgof and K. W. Bowyer, "Extracting a Valid Boundary Representation from a Segmented Range Image", IEEE PAMI, Vol 17, No 9, 1995, pp. 920–924

[19] A. Hoover et al., "An Experimental Comparison of Range Image Segmentation Algorithms", IEEE PAMI, Vol 18, No 7, 1996, pp. 673–689

[20] H. Hoppe, T. DeRose, T. Duchamp, J. McDonald and W. Stuetzle, "Surface Reconstruction from Unorganised Points", *Computer Graphics* (SIGGRAPH 92), Vol 26, 1992, pp. 71–78

[21] Reverse Engineering, Eds.: J. Hoschek, W. Dankwort, B. G. Teubner, Stuttgart, 1996

[22] Advanced Course on FAIRSHAPE, Eds.: J. Hoschek, P. Kaklis, B. G. Teubner, Stuttgart, 1996

[23] J. Hoschek, U. Dietz and W. Wilke, "A Geometric Concept of Reverse Engineering of Shape: Approximation and Feature Lines", In: *Mathematical Methods for Curves and Surfaces II*, Eds: M. Dæhlen, T. Lyche, L. L. Schumaker, Vanderbilt University Press, 1998, pp. 253–262

[24] G. Kós, R. R. Martin and T. Várady, "Recovery of Blend Surfaces in Reverse Engineering", *Computer Aided Geometric Design*, Vol 17, 2000, pp. 127–160

[25] G. Kós, "An algorithm to triangulate surfaces in 3D using unorganised point clouds", *Geometric Modelling Studies*, GML1999/3, Computer and Automation Research Institute, Budapest, 1999 (presented at the Dagstuhl Geometric Modelling Seminar, May 1999, submitted to Computing)

[26] I. K. Lee, "Curve Reconstruction from Unorganised Points", Institut für Geometrie, Technical Report No. 55, Technische Universität Wien, 1998

[27] P. Krsek, T. Pajdla, V. Hlaváč and R. R. Martin, "Range image registration driven by hierarchy of surface differential features", *Proc. 22nd Workshop of the Austrian Association for Pattern Recognition*, May 1998, pp. 175–183

[28] A. Leonardis, A. Jaklič and F. Solina, "Superquadrics for Segmenting and Modeling Range Data", IEEE PAMI Vol 19, No 11, 1997, pp. 1289–1295

[29] G. Lukács, A. D. Marshall and R. R. Martin, "Faithful Least-squares Fitting of Spheres, Cylinders, Cones and Tori for Reliable Segmentation", In: *Computer Vision-ECCV 98*, Lecture Notes in Computer Science, Eds: H. Burkhardt and B. Neumann, Springer-Verlag, 1998

[30] B. Parvin and G. Medioni, "B-rep from Unregistered Multiple Range Images", *Proc. Int. Conf. Robotics and Automation*, Nice, 1992, pp. 1602–1607

[31] V. Pratt, "Direct Least-squares Fitting of Algebraic Surfaces", *Computer Graphics* (SIGGRAPH 87), Vol 21, No 4, 1987, pp. 145–152

[32] H. Pottmann and T. Randrup, "Rotational and Helical Surface Approximation for Reverse Engineering", *Computing*, 1998, pp. 307–322

[33] N. S. Sapidis and P. J. Besl, "Direct Construction of Polynomial Surfaces from Dense Range Images through Region Growing", ACM TGO, Vol 14, No 2, 1995, pp. 171–200

[34] T. Várady, R. R. Martin and J. Cox, "Reverse Engineering of Geometric Models — an Introduction", *Computer-Aided Design*, Vol 29, No 4, 1997, pp. 255–268

[35] T. Várady and C. M. Hoffmann, "Vertex Blending: Problems and Solutions", In: *Mathematical Methods for Curves and Surfaces II*, Eds: M. Dæhlen, T. Lyche, L. L. Schumaker, Vanderbilt University Press, 1998, pp. 501–528

[36] T. Várady, P. Benkő and G. Kós, "Reverse Engineering Regular Objects: Simple Segmentation and Surface Fitting Procedures", *Int. Journal of Shape Modeling*, Vol 4 (1998), pp. 127–141

[37] N. Werghi, R. Fisher, C. Robertson and A. Ashbrook, "Object reconstruction by incorporating geometric constraints in reverse engineering", *Computer-Aided Design*, Vol 31, 1999, pp. 363–399

A System for Constructing Boundary Representation Solid Models from a Two-Dimensional Sketch

P. A. C. Varley, R. R. Martin

Dept. of Computer Science
Cardiff University
Cardiff, Wales, UK
{Peter.Varley, Ralph.Martin}@cs.cf.ac.uk

Abstract

This paper describes a system which converts a two-dimensional sketch with hidden lines removed of a single polyhedral object into a boundary representation solid model.

The system improves on a previous system by Grimstead by analysing the sketch for symmetry elements and regularities. It uses the results of this analysis to produce a more plausible topological completion of the hidden parts of the object, and to enforce exact constraints on the geometry of the boundary-representation model.

1. Introduction

This paper describes a system which converts a two-dimensional sketch with hidden lines removed of a single polyhedral object with no cracks or shadows into a boundary representation solid model. It reports on the current capabilities of the system, which is still under development. Until we have established a set of methods which cover our input domain, it would be premature to attempt to abstract any general principles; in places, our approach is based on analysing and reconstructing various special cases.

Textbooks on engineering drawing (e.g. [6]) emphasise the importance of freehand sketching in the design process. Studies such as Jenkins [14] have shown that such a system would be of considerable interest to engineers and architects, whose current working practices involve freehand sketching on paper and subsequent manual input to a CAD package.

Although the theoretical impossibility of perfect conversion of 2D to 3D is both obvious and well-known, there exists an expanding set of objects for which conversion can be achieved in practice. Some assumptions concerning the sketch are required — removing some of the restrictions stated below will be the subject of future research. We assume that the sketch is of a single manifold polyhedral object. Topologically, we currently assume that all vertices in the object are trihedral and that no face contains more than one loop of edges. Geometrically, we assume that all faces are planar, and that the object is in a general position, with no faces lying perpendicular to the view direction. Lastly, we assume that the object has been sketched from the "most interesting viewpoint", i.e. that there is nothing at the rear of the object which could not reasonably be inferred from the visible part of the object.

In terminology, we use *region*, *line* and *junction* to denote *atoms* of a two-dimensional *sketch*, and *face*, *edge* and *vertex* to denote atoms of the three-dimensional *object* which has been sketched. Most, but not all, junctions are also vertices; visible vertices are junctions. All lines are edges or parts of edges; visible edges and visible parts of partially-obscured edges are lines. Regions are visible faces or the visible parts of partially-obscured faces. A junction where two lines meet is *biconnected*; a junction of three lines is *triconnected*.

We also distinguish *recognition*, where the computer chooses one of a finite set of candidate objects, from *interpretation*, where the computer creates a new object from an infinite set of constructible objects. We are interested only in the latter.

The problem is to convert the sketch into a boundary representation solid model of the most plausible 3D interpretation of the sketch, and to do so in a reasonable time (a second or less on a powerful personal computer is "reasonable").

Several variations on the type of input are possible: the sketch could show shaded faces or simply lines representing edges, and in the latter case the sketch could have hidden lines drawn, or only the visible lines drawn as in the case of our system.

A method requiring a complete wireframe object has been developed by Lipson and Shpitalni [22, 23]. It is tolerant of freehand sketching errors, and allows flat and cylindrical faces. It assumes the object is in a general position. Many of the ideas contained in it are of interest. In particular, they provide a list of useful compliance functions: face planarity; line parallelism; line verticality; isometry; corner orthogonality; skewed facial orthogonality; skewed facial symmetry; line orthogonality; minimum standard deviation of angles; face perpendicularity; regularity of end faces of prisms; line collinearity; planarity of skewed chains. These compliance functions are weighted according to their degrees of accuracy in freehand sketches. The system produces depth values by performing a least-squares fit against these. The authors have examined several optimisation methods and chose cyclic application of Brent's method [3] to each vertex in turn. They stress the need for reasonable preliminary estimates in order to minimise optimisation time.

The disadvantage of using sketches with hidden lines visible, that of ambiguity (Necker reversal), is in principle insoluble, although frequently the symmetry of the drawn object will make the two potential interpretations the same. There are other disadvantages of using wire-frame sketches rather than sketches without hidden lines: it is more natural to draw only that which can be seen, and it is quicker and easier to draw a smaller rather than a larger number of lines.

Grimstead [8] provides a prototype system based on input of sketches made with hidden lines removed. This comprises five stages: incremental line labelling; two-dimensional drawing tidying; conversion to 3D using a linear system; recovery of hidden parts; and three-dimensional drawing tidying. It is described in more detail below. Grimstead's system makes the same assumptions as those listed above. In order to make the system interactive, further limitations were also introduced; these are discussed below.

A further advantage of interpreting sketches with only visible lines is that this makes it easier to incorporate work deriving from attempts to recognise real-world objects. Naturally, if the two-dimensional data is obtained from a photograph of a real object rather than from a sketch, hidden lines will not be visible. This process is inevitably more complex, as it cannot enforce assumptions on the real world in the way that a sketching interface can enforce assumptions on its user. It must be able to handle multiple objects, only some of which are of interest; it must be able to convert shading indicating differently-oriented faces into lines representing edges; it must be able to cope with noisy input. Nevertheless, it has one advantage which analysis of sketch input does not: it can assume that the input is valid, that it genuinely represents a real object, and that the projection is "drawn" correctly.

One such recognition system by Barrow and Tenenbaum [1] obtains line drawings by preprocessing greyscale pictures into region boundaries. A line may represent either a true edge between adjoining faces, or a depth discontinuity where one face obscures another. Their objective is to produce depth information on a per-pixel basis. The system can handle semicircles, cylinders and spheres, using Chakravarty's junction catalogue [4]. However, if the input sketch contains these, the iterative algorithms used converge too slowly to be useful for an interactive system. More interestingly for us are the special cases used for polyhedra. They use an iterative optimisation scheme which adjusts the Z coordinates of vertices in order to optimise a single global measure of regularity, preferring one of three measures: sum of squares of angles between faces, sum of squares of cosines of angles between faces; and sum of squares of ($2\pi -$ sum of angles at a vertex). All produce similar results.

Similar methods are also used in analysis of aerial photographs, although as Mayer's survey [28] shows, this field has subsequently become more specialised. Early work analysed single pictures by modelling buildings as firstly rectangular prisms [11] and later by grouping rectangular regions and parallel lines [29]. Artefacts such as skewed symmetry [15] are equally applicable to such pictures and to interpretation of sketches. This is a method for calculating the normal of a face given two axes on the face which are believed to be perpendicular in the 3D world. Given α and β, the angles of the two axes chosen such that the angle between them is obtuse, and expressed in the convention that anticlockwise is positive and zero is to the right along the horizontal axis, it can be shown that the face normal $[P, Q, R]$ is given by $P/R = \rho \cos \lambda$ and $Q/R = \rho \sin \lambda$ where $\lambda = (\alpha + \beta)/2$ and $\rho = \sqrt{-\cos(\alpha - \beta)}/\cos((\alpha - \beta)/2)$. The derivation assumes a correct orthographic projection. We know of no sensitivity analysis of the effect of inaccuracies in the projection, as would be present in a sketch, on the results.

More distant fields also produce results of interest. One such is the detection of motion of an object in an image, or of egomotion (movement of the viewpoint with respect to the image). Useful mathematical results can be found in Kanatani [17].

Interpretations of sketches generally involve some method of labelling lines. The most common line labelling method is that using the Clowes-Huffman [5, 13] catalogue. This assumes trihedral vertices, but can in principle be extended to curved surfaces using Chakravarty's catalogue [4] or Turner's [35]. The terminology used in this paper follows Chakravarty: junctions are L-junctions, T-junctions, W-junctions or Y-junctions, and lines are arrows (occluding), concave or convex. When it is necessary to indicate particular types of junction, these are extended by adding the suffices a for an incoming arrow, b for an outgoing arrow, c for a convex line, and d for a concave line, reading clockwise; thus the legal trihedral junction labels are *Yccc, Yddd, Yabd, Wbca, Wdcd, Wcdc, Lac, Lab, Lcb, Lbd, Lba, Lda, Tbaa, Tbab, Tbac, Tbad.*

Various algorithms for line labelling have been proposed. In general, these are based on constraint propagation [17, 40]; an attempt to improve on these using genetic algorithms was unsuccessful [33]. An alternative method of labelling lines proposed by Kanatani [17] assumes not only that vertices are trihedral but that the object is orthogonal. It not only labels lines according to their concavity/convexity but also uses their orientation to classify them by axis as i, j or k lines. It is more successful in rejecting impossible objects, and more capable of providing useful information for later stages of the recognition process. The assumption of object orthogonality is however a major limitation.

A method for constructing the most plausible interpretation of a sketch, as determined by human perception, must of necessity be heuristic rather than mathematical in some places. Draper[7], in advocating sidedness reasoning, makes the interesting statement that such an approach is "more intuitively correct" than gradient and dual space algorithms. We are of the opinion that "intuitive correctness", although difficult to make precise, should also be an important criterion for assessing our method.

We consider that in a quick, natural sketch input system the advantages of drawing only the visible lines outweigh the disadvantages, so we take Grimstead's method as our starting point. We extend his approach by using symmetry and regularity relationships inferred from the sketch (a) to produce topologically more plausible completions of the hidden parts of objects, and (b) to enforce exact constraints on the geometry of the boundary-representation model. This paper gives an overview of our approach, and gives some examples for which it improves on Grimstead's approach; more technical detail is given in related papers [37, 38, 39].

Section 2 describes Grimstead's system. Section 3 outlines our extended system. Some results are shown in section 4, and section 5 lists possible future work.

The Figures are selected from our wide repertoire of test sketches as the ones which best illustrate particular points (rather than being the only ones we can solve!). These test sketches, currently numbering over 130, are culled from various sources, including [8, 34], all of the solid objects (but not the paper objects or wire-frames) from [21] and all of the planar objects from [43]. We continue looking for additional sketches of solid objects with planar faces and would welcome input from readers.

2. Grimstead's System

In order to infer depth information for the Y-, L- and W-junctions in a 2D sketch (we call this the *frontal geometry*), the following stages are common to all approaches summarised above: labelling of junctions and lines in the two-dimensional sketch; identifying loops or regions in the two-dimensional sketch; identifying artefacts of interest in the two-dimensional sketch; deducing constraints from the artefacts; and producing depth information such that the constraints are satisfied while the resulting shape fits the original sketch as well as possible (the order of these stages may differ). In addition, to produce a complete and valid solid model, both the topology and the geometrical positions of any hidden vertices, edges and faces must be deduced. Grimstead's system included a rudimentary method for doing this.

2.1. Incremental Line Labelling

In Grimstead's system, this stage incorporated both line labelling and region identification. Line labelling used Waltz's algorithm [40], and the Clowes-Huffman catalogue for trihedral vertices. Line-labelling both validated the sketch, and identified the faces and the edges which bound them. Little use was made of the line or junction labels in subsequent stages of the system.

2.2. Two-Dimensional Drawing Tidying

In Grimstead's system, this stage identified only one regularity, 2D line parallelism, and adjusted the sketch to improve this. The purpose was to enhance this regularity in order to make other artefacts which depend on parallelism easier to detect.

The algorithm used for this process allocated angles of 2D lines to buckets, and merged nearby buckets until it was evident which lines should be parallel. This bucketing process [8] is not very robust. A simplistic interpretation produces an unreliable implementation which can introduce problems with some types of sketch.

To illustrate this, consider the sketch in Figure 1. It is evident that edges A, B and C should be parallel, and any reasonable bucketing process will detect this. It is also evident to an intelligent observer that edges D, E and F should be parallel; however, depending on the quality of the sketch, D and E may be closer to G than to F, and an algorithmic process may therefore decide to make D, E and G parallel instead. It is also quite possible that F and G may be within the allowed angle below which buckets are merged, and made parallel; this is clearly nonsense geometrically. It is not clear how these problems should be resolved.

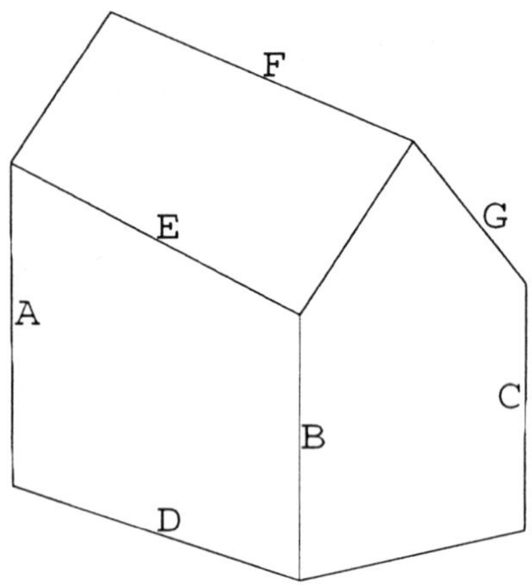

Figure 1. House

Even if the system correctly determines that edges D, E and F should be parallel, problems can occur. Suppose that lines A, B and C are well-drawn, being both parallel and of matching lengths, but that line F is misdrawn. A least-squares fit, requiring D, E and F to be parallel while trying to keep A, B and C parallel too and to preserve existing vertex positions, will change the orientations of lines D and E, and therefore change the lengths and possibly the orientations of A and B, while C remains unaffected.

Regularity may be lost, not gained, by two-dimensional tidying.

2.3. Generation of Frontal Geometry

Grimstead generated the frontal geometry of the object using a linear system. Let each face be of the form $Px + Qy + Rz + C = 0$, and each vertex (X, Y, Z). Constraining a vertex to lie on a face, and assuming $R = 1$ (no face is parallel to the projection direction), leads to a linear equation with the unknowns P, Q, C and Z. Such equations were generated for each vertex-face pairing.

Equations $m(Z_{n_1} - Z_{n_2}) = n(Z_{m_1} - Z_{m_2})$ (where m and n are the lengths of the two lines M and N, and $Z_{m_1}, Z_{m_2}, Z_{n_1}, Z_{n_2}$ are the depths of the start and end points of the lines) were generated for each pair of 2D parallel lines identified.

Equations $P = \rho \cos \lambda$ and $Q = \rho \sin \lambda$ were generated for each skewed symmetry artefact detected, using a variant of the equations described previously.

The system was then solved to give equations for each face using a weighted least-squares algorithm [2]. The original weighting of each equation was somewhat arbitrary. The solution process was iterative, with the equation with the largest residual error (the one which disagreed most with the consensus) being dropped at each subsequent iteration. The performance was assessed, once per iteration, using right-angle fit and minimum standard deviation of angles. Eventually, a self-consistent set of equations remained, and the output of the linear system gave P, Q and C values for each face.

The linear system approach as a whole suffered from several disadvantages:

- It was analytically opaque, in that if it did not work as well as was desired, modifications were made to weightings of equations, rather than to algorithms. The resultant weightings thus appear arbitrary, being the values which worked best for the test sketches used.

- It did not distinguish between constraints which *must* be met and equations for which a best-fit approach is adequate. Even the constraint that ensures that vertices lie on faces was only approximately satisfied, and had to be enforced at a later stage. Others, such as making two faces parallel or orthogonal, were also only approximately enforced.

- The variables being optimised were of different kinds, thus creating doubt as to whether the fit really was 'best'.

- The approach influenced unduly the choice of artefacts which were used in creating the fit, in that only those constraints which could be written as equations linear in P, Q, Z and C were admitted. For example, the assumption of corner orthogonality, first suggested by Mackworth [25] and used successfully in several systems, does not translate into a linear equation and so was not used.

- The skewed symmetry equations for P and Q were decoupled. The iterative least-squares fit might drop one and retain the other when rejecting inconsistent equations. A value of P or Q on its own is meaningless; both are needed for either to make sense.

- A hidden weighting was also given to skewed symmetry estimates of 'front-on', rather than 'side-on', faces, through the representation of the face normal as $[P, Q, 1]$ rather than $[P, Q, R]$. However, this is justifiable on the grounds that it is easier for the user to draw 'front-on' faces, so they will be more accurate.

Despite these disadvantages, the approach also had one major advantage: it ran sufficiently quickly to be interactive for sketches such as Figure 2.

2.4. Recovery of Hidden Parts

Grimstead's system assumed that every hidden face in the object met an occluding edge. The topological reconstruction of the hidden part of the object thus produced the simplest possible object, not necessarily the most plausible.

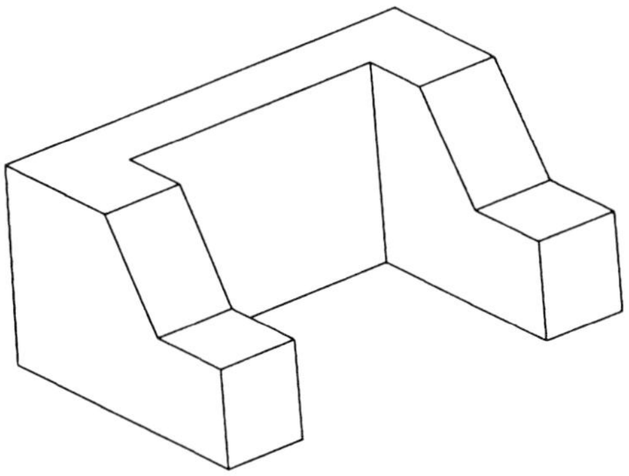

Figure 2. Grimstead's Bracket

Where three or more points on a hidden face existed in the 3D conversion of the visible sketch, the equation of the hidden face could be obtained directly. Where only two points existed, the equation was deduced on the basis of topological information.

Edges were extended through T-junctions, and new edges created behind the visible object from visible L-junctions. Groups of three such edges were then tested for consistency — if the three edges were allocated to three faces, an additional hidden vertex was created where the three faces intersected, and the three edges were removed from the set of edges requiring completion.

Grimstead's algorithm for recovery of hidden parts was successful in the test cases he used. However, Grimstead acknowledges that knowledge of the algorithm makes it simple to find sketches for which the algorithm does not work correctly. Figure 3 is one such — T-junction A would be joined to junctions B (correctly) and C (incorrectly).

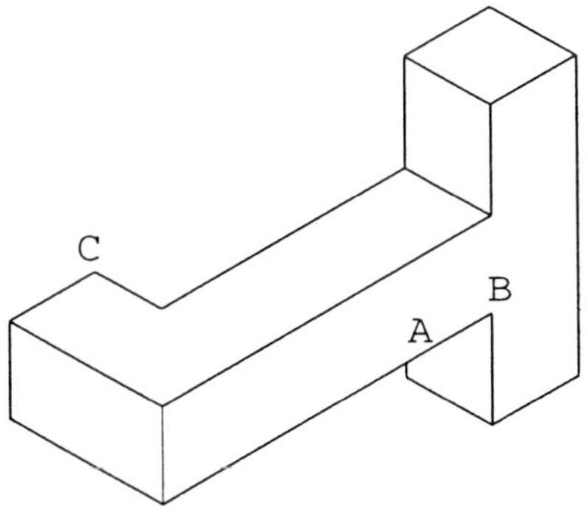

Figure 3. J Block

2.5. Three-Dimensional Drawing Tidying

Grimstead's final three-dimensional tidying process used the face equations generated above to give the 3D positions of each vertex by intersecting the appropriate faces, the values of X and Y from the sketch and Z from the linear system being discarded. This ensured a self-consistent boundary representation model.

One disadvantage with this approach was that the equations of hidden faces were derived from positions of visible vertices, themselves derived from a least-squares fit of various constraints, not all of which were fully-satisfied. Numerical errors were propagated at each stage, and using these error-prone values to calculate the positions of hidden vertices further compounded the error.

As already noted, the approach to geometrical completion of the object was that of a best fit, frequently producing edges which were nearly, but not quite, parallel, and corners which were nearly, but not quite, perpendicular. Regularity constraints such as parallelism of faces were not strictly enforced on the final solid.

The assumption that all hidden faces have at least two visible vertices was not always plausible — it is simple to construct objects where a more complex but symmetrical reconstruction is psychologically preferable to the minimal reconstruction given by Grimstead's system. However, the minimal approach may enable the system to produce an interpretation, albeit an undesired one, of a sketch where a more complex system may fail altogether (for example, the dodecahedron in Figure 4).

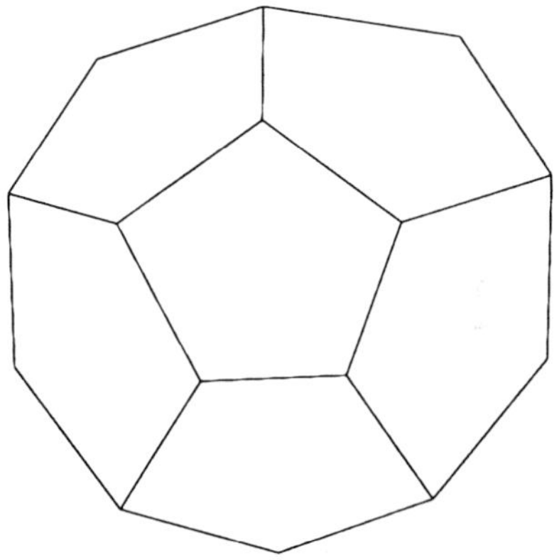

Figure 4. A Platonic Solid (Dodecahedron)

3. A System using Symmetries and Regularities

We consider the most serious defects in Grimstead's system to be that the topological completion of hidden parts can be implausible and that regularities implied by the sketch are not strictly enforced. Lockwood [24] describes the aesthetic appeal of symmetry, and Martin and Dutta [26] give various technical reasons why the possession of symmetry may be beneficial in designing shapes, and consider various approaches for making almost-symmetric shapes symmetric. It is therefore plausible that sketches are intended to be symmetrical in some way, so we use implied symmetry and regularities in the sketch as the means to overcome these defects.

In many cases, there may be several plausible alternatives for reconstruction. We attempt to assign a *figure of merit* to each. This is a numerical value intended to represent the plausibility of a particular choice; higher values represent more plausible choices, and of a set of alternatives, the one with the highest figure of merit will be considered first by our algorithm.

Following Kettle [18], *symmetry* is characterised by the fact that it is possible to carry out operations which, whilst interchanging the positions of some or all of the atoms, give arrangements of atoms which are indistinguishable from the initial arrangement. Rotation, reflection and inversion are thus symmetry operations. Rotation axes and mirror planes are termed *symmetry elements*. Other artefacts which give clues to the structure of an object, such as parallelism or extrusion, which do not meet the strict definition of symmetry are here termed *regularities*. Identification of a symmetry element or regularity in an object produces one or more *constraints*, which limit the possible positions of the vertices.

Grimstead's system does not use mirror or rotational symmetry in object completion. When mirror and rotational symmetry artefacts are used, these affect the whole object, not just the visible part. For this reason, we advocate that the topological structure of the model including the hidden parts should be deduced *before* finalising the geometry of faces and vertices.

For convenience of implementation, our sketch interpretation system is therefore subdivided into four stages: labelling, symmetry identification, topological 3D completion and geometric finishing.

3.1. Labelling

This stage comprises a Clowes-Huffman line-labeller and identification of the regions of the sketch, including the 'background region', and follows the algorithm outlined by Kanatani [17]. The output is a list of vertices, edges and faces. For each vertex, the x and y coordinates, the junction type, and the incident edges and faces are listed. For each edge, the edge type, vertices and faces are listed. For each face, the vertices and edges are listed.

3.2. Symmetry Identification

We concluded above that two-dimensional tidying provides little benefit and some potential harm. It is the identification of symmetry artefacts which is of importance, not the small movements of junction x and y coordinates to improve them. Therefore, the two-dimensional tidying stage has been replaced by a stage where various symmetry artefacts and other global and local regularities in the sketch are identified.

Symmetry and regularity are used in two ways in later stages: in assisting the process of recovery of hidden parts, and in producing constraints to be met by the face normals and positions and vertex positions.

Figures of merit are assessed for potential symmetries and regularities, and also for hypotheses drawn from these artefacts at various stages of processing. The numerical values used are still the subject of experimentation [37]. They are intended to reflect the probability that an apparent artefact is a genuine feature of the object the user intended to sketch. When the system has a choice of actions, the one with the highest figure of merit will be tried first.

In general, figures of merit take into account the following four factors:

- the probability of a particular artefact occurring — some artefacts, such as face parallelism, are common, whereas others, such as C_5 rotational symmetry, are comparatively rare.
- a measure of how well a particular artefact fits in with what is already known about the object — regular pentagons may be quite rare, but become more plausible if several pentagons are already visible in the sketch.
- the probability that the artefact corresponds to the sketcher's intentions — where two lines are 1° apart, they are probably intended to be parallel, whereas if they are 10° apart, the intention is less certain.
- the complexity of the artefact — if an artefact is hard to draw (see, for example, the sixty-sided semi-regular solid in Figure 5), then anything approximating to it is probably intended to be that artefact.

Note that in some cases, such as partially-occluded faces, it will not be possible to determine whether or not the completed face has a particular symmetry or regularity. It is convenient to assume in such cases that the artefact is present, but to assign it a low figure of merit. An alternative approach might consider artefacts to have three states — completely present and thus likely; partially present and thus possible; and demonstrably absent on the basis conflicting evidence. We have not pursued this, as it is not clear that this is a sufficiently-detailed classification. For example, an artefact present locally, but contradicted by evidence elsewhere in the sketch, might still be of use in local reconstruction of hidden parts, and should neither be rejected immediately nor accepted unequivocally.

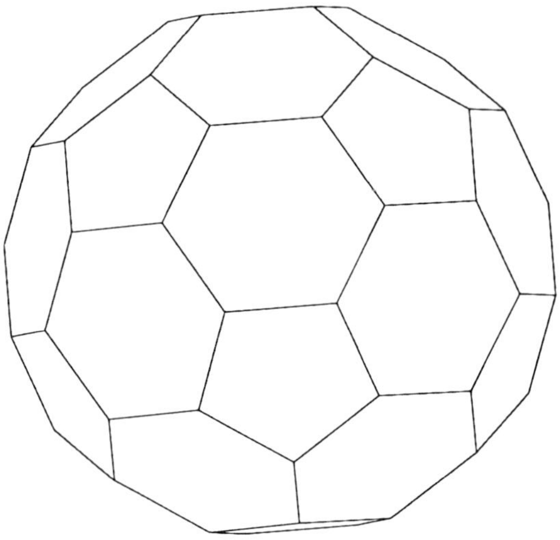

Figure 5. An Archimedean Solid (Truncated Icosahedron)

3.2.1 Bucketing

A bucketing process similar to Grimstead's is used to identify groups of parallel lines. In our system, bucketing requires lines which are to be grouped as parallel to be no more than 15° apart, and each group to be at least 15° from any other group. The angles are arbitrary, but sufficient to permit at least six groups of parallel lines in any sketch (Figure 6 has six lines, none of them parallel to one another), and up to twelve in well-drawn sketches.

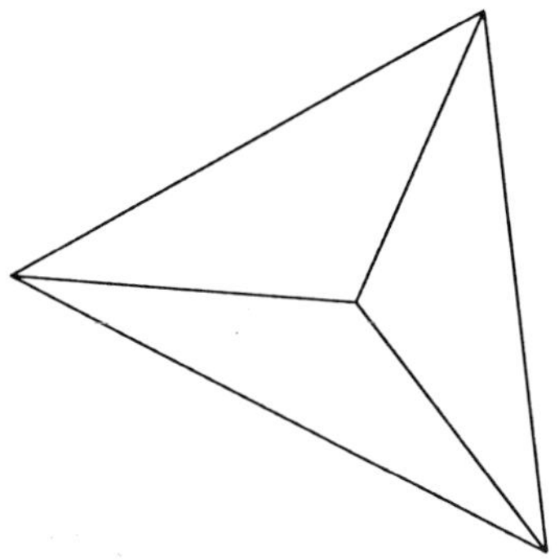

Figure 6. Tetrahedron

There are various possible outcomes of this:

- It may not be possible to allocate lines to distinct buckets in this manner (Figure 5 cannot be drawn in a way which meets the requirements above), in which case the sketch will be treated as having no parallel lines, but the system does not know whether or not this is intended.
- The bucketing process may produce an unacceptable result — two lines meeting at a junction are identified as being parallel. In this case also, bucketing is considered to have failed.
- Bucketing may succeed but each bucket contains one line (Figure 6 is a sketch where this occurs), in which case it is known that the lines are intentionally not parallel.
- Bucketing may succeed and identify candidate parallel lines.

An adaptive bucketing process could be developed to interpret badly-drawn sketches, as discussed in [37].

3.2.2 Frontal Geometry

Grimstead's system does not use mirror or rotational symmetry in object completion. If these are added, it becomes necessary to take account of the entire object when calculating spatial positions of vertices, not just of the visible part. For this reason, precise calculation of spatial positions is deferred until after the topology of the hidden part of the object has been deduced.

It is still necessary to have a topologically-correct completion of the frontal part of the object with added depth information. Accurate geometry is not required, but it is desirable that the depth-ordering of the visible vertices is correct, and necessary that adjacent vertices are ordered correctly (for example, skewed symmetry produces multiple solutions and selection of the correct one is based on knowledge of this ordering).

It can be observed in Figure 7 that lines from a $Wbca$ junction (A or C) to an Lba junction (B or D) are travelling away from the viewer. Similarly, lines from mostly convex junctions to mostly concave junctions are away from the viewer. Frequently, the longer the line, the further away from the viewer it travels.

Thus to achieve a correct ordering of adjacent vertices, we create a linear system with the number of equations given by the number of visible lines in the sketch and the number of unknowns given by the number of visible vertices in the sketch. The unknowns are the depth values; the equations are of the form $W * (Z_a - Z_b) = W * L * K$, with Z_a and Z_b being the depths of the near and far end points of the line, L being the length of the line in the two-dimensional sketch, K being a constant which depends on the junction types of the start and end points of the line and

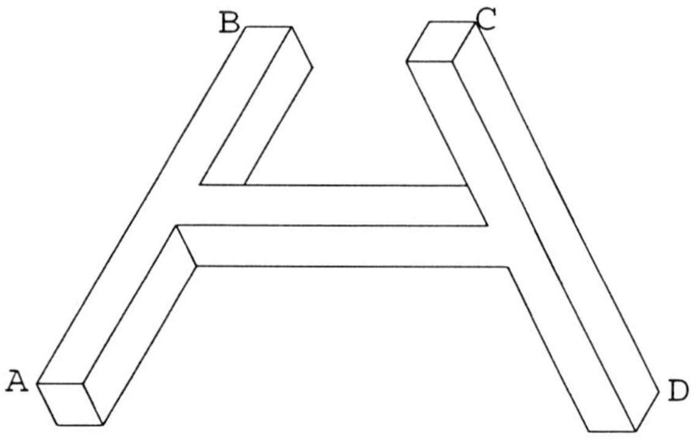

Figure 7. H Block

W (range 1..10) is a subjective measure of the confidence in the correctness of the value of K. Vertex 0 is constrained to an arbitrary fixed depth. The algorithm is described in more detail, and the K and W values tabulated, elsewhere [37].

For a simplistic method, this works surprisingly well in practice for most junction types, although it is found that Lcb junctions are frequently mispositioned. It fails entirely for sketches which are not graph-complete — if it is not possible to find a path from any edge to any other edge without making use of T-junctions, the resulting linear system cannot be solved. Various refinements have been tried, without significant improvement. One which has been adopted is to add further equations to the linear system to encourage lines in the same 2D bucket to be parallel in 3D. It has not been found desirable to follow this by forcing all visible vertices to lie on planar faces — this improves accuracy slightly for well-drawn sketches at the expense of exaggerating inaccuracies in poorly-drawn sketches.

3.2.3 Local Symmetries and Regularities

While it is not possible to determine the symmetry of an entire object from a sketch which only shows part of the object, it is possible to gain clues to the structure of the object by considering the symmetry of each region. Each region is usually a skewed view of all or part of the corresponding 3D face. Line lengths may not be preserved, and junction angles will almost certainly be distorted, but parallel lines remain parallel within sketching tolerance.

For two arrangements of atoms to be indistinguishable, the atoms occupying the corresponding positions in the two arrangements must be *compatible*. Two vertices are compatible if, in the final object, they have the same numbers of concave and convex edges. Two corners of a face are compatible if, when viewing a clockwise circumnavigation of the face from outside the object, they are both left turns or both right turns. Two edges are compatible if, in the final object, they are both convex or both concave (occluding edges will become convex); additionally, it must be possible to group their end points into two pairs of compatible vertices. Two faces are compatible in a particular orientation if they have the same number of edges, their edges when paired are compatible, their vertices when paired are compatible, and their corners when paired are compatible.

To identify rotational symmetry, it is noted that any axis of rotational symmetry must pass through either a vertex possessing C_3 symmetry (vertices are trihedral), an edge possessing C_2 symmetry about its mid-point, or a face possessing any rotational symmetry about its centre. It is therefore sufficient to identify such axes, which can be used later as "seeds" for potential global symmetry.

Triconnected junctions are candidates for local C_3 symmetry if their faces are compatible in the orientations which pair the junction with itself. Not all types of junctions need be analysed: for example, $Wcdc$ and $Yabd$ junctions cannot have C_3 symmetry as they have one concave and two convex edges in 3D. Of the biconnected junctions, Lba junctions are provisionally assumed to be candidates for C_3 symmetry as there is only one visible face.

The mid-points of non-occluding edges are candidates for local C_2 symmetry if the faces are compatible in the orientations which pair the edge with itself. At the time of writing, no artefacts are produced for the mid-points of occluding edges — this will be investigated in the future.

Candidate local rotational symmetry operations $C_2, C_3, C_4, C_5, C_6, \ldots$ for each region can be identified (our implementation stops at C_6). Firstly, a particular symmetry operation C_n is only a candidate if the region has $n, 2n, 3n, \ldots$ junctions. The candidate operation should be rejected if the face in its original orientation is not compatible with itself in its rotated orientation – for example, C_4 is only a candidate symmetry operation for an octagonal face if the junctions are either all convex or alternately convex and concave. Even-numbered candidate symmetry operations $C_2, C_4, C_6, \ldots$ should also be rejected if opposing lines are not parallel to within sketching tolerance. For C_5 symmetry operations, each set of five vertices $V_i, V_{n+i}, V_{2n+i}, V_{3n+i}, V_{4n+i}$ must be a regular pentagon; the undrawn line joining vertices V_{n+i} to V_{4n+i} must be parallel to the base of the pentagon, the line joining V_{2n+i} to V_{3n+i}, and so on for each edge taken as base in turn. Similar rules are used for C_6 symmetry.

Candidate mirror symmetry operations can again be localised to a seed, which in this case must be either an edge or a face with mirror symmetry. Face-based mirror planes can be further subdivided into vertex-to-vertex planes and edge-to-edge planes in faces with an even number of sides, and vertex-to-edge planes in faces with an odd number of sides.

A face is a candidate for local mirror symmetry if all vertices (other than any which are in the symmetry plane) can be paired sequentially across the mirror axis, and lines joining these pairs are approximately parallel with one another and centred on the mirror axis. The angle of the axis, and the angle of the lines joining sequential pairs, are recorded (they are the input angles required by skewed symmetry) — see Figure 8, where line A is the candidate mirror line and

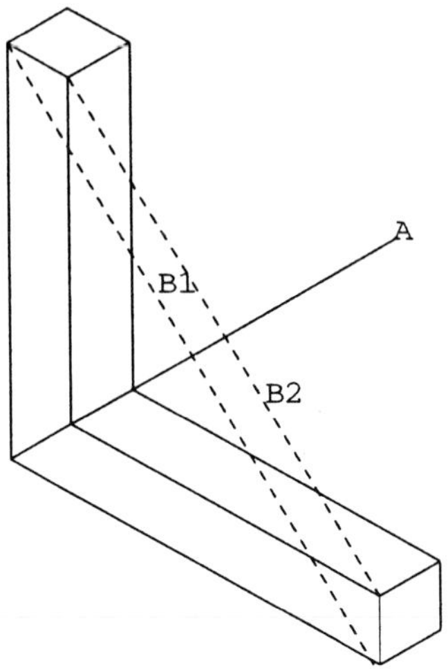

Figure 8. L Block

lines $B1$ and $B2$ join pairs of vertices. It has been found that generating "partial" mirror planes, where the face and its reflection are not compatible, can nevertheless be useful for the purposes of recovery of hidden parts — see Figure 9, where a reconstruction based on the partial mirror sym-

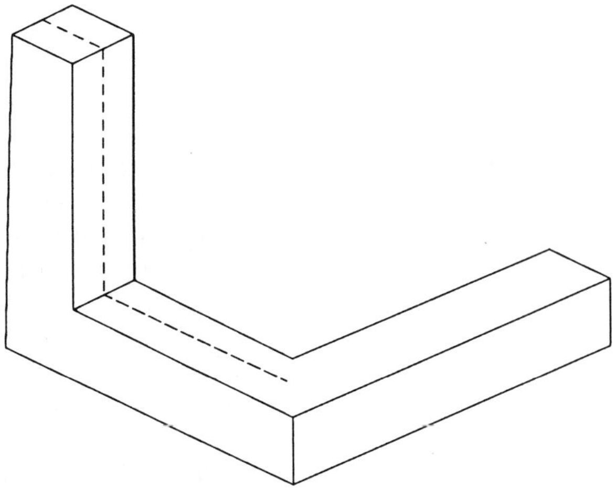

Figure 9. Z Block

metry marked in dotted lines gives a topologically-correct completion.

Candidate mirror planes can be chained: where a face mirror plane terminates at an edge mid-point, it can be chained with another face mirror plane terminating at the same edge as in Figure 9. Where a face mirror plane terminates at a vertex, it can be chained with an edge mirror plane including that vertex, and vice versa, as in Figure 8.

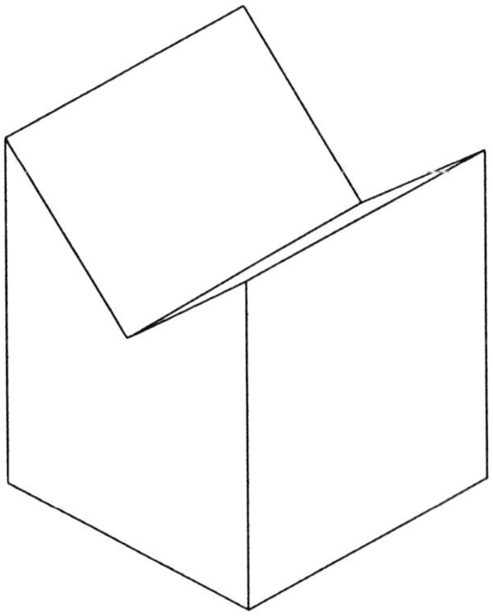

Figure 10. An Extrusion

3.2.4 Global Regularities and Symmetry Groups

An attempt is made to classify the sketch into one of several categories. A successful classification speeds up the later stages of processing considerably. Since an object may fall into more than one of the categories, a figure of merit is assigned for each classification.

Global symmetry and regularities are hypothesised on the basis of an agglomeration of local artefacts, described above, or on the basis of grouping of lines in the sketch into sets of parallel lines.

Line parallelism has been used in other systems [8, 14, 22]. We extend this idea to consider face parallelism and coplanarity. These regularities can be hypothesised either as a result of identification of a regularity which would imply them, or on the basis of having similar estimated numerical values for face constants P, Q and C. At this stage, we are concerned mostly with parallelism, and we identify various special cases.

The idea of face parallelism can further be extended to the hypothesis that the object is a right extrusion, with two identical and parallel faces joined by parallelograms. This is possible if no more than one visible face has other than four sides, all vertices are either on this special face or on edges which terminate on this face, and each junction label indicates that the vertex will have no more than one concave edge. If these requirements are met, the figure of merit for this hypothesis is based on the degree of parallelism of the edges which do not bound the special end face.

Right frusta such as Figure 11 can be identified by similar criteria, although in this case, the lines between the vis-

'nd the back face should meet at a single .. being parallel. The figure of merit for this . depends on how close to a single point these come, and is reduced if the object is likely to be an extrusion (this is in order to prevent the system miscategorising an extrusion as a frustum where the nearly-parallel lines meet a long way away).

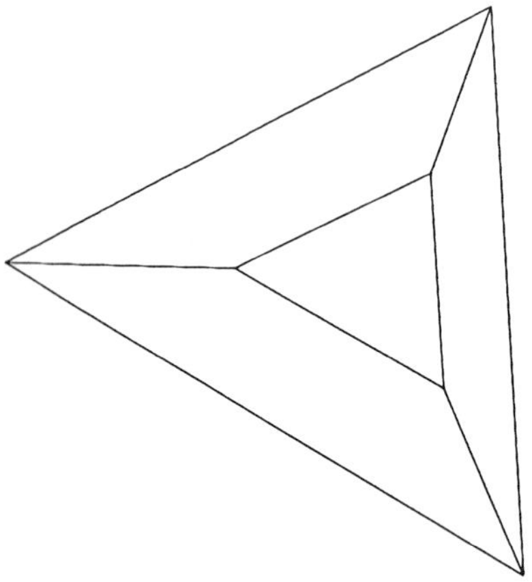

Figure 11. A Frustum

Orthogonality has been used in other systems. Lipson and Shpitalni [23] suggest that if a histogram of line angles is plotted and there are only three peaks, these probably correspond to the three perpendicular axes and the object is axis-aligned. In our system, if the bucketing process successfully identifies exactly three groups of line angles, the sketch contains no fully-visible odd-sided faces, and the fully-visible even-sided faces have four more right turns than left turns, the sketch is marked as being likely to be axis-aligned. The figure of merit of the axis-aligned hypotheses decreases as the peaks of the angle histogram become less sharp. Kanade [15] observes that the object can only be aligned to orthogonal axes if all three of the angles at Y-junctions are obtuse — since this is true of accurate projections but not necessarily of sketches, if the sketch contains Y-junctions with acute angles, the figure of merit for axis-alignment is reduced, but the hypothesis is not rejected entirely.

This concept can be extended to a category of semi-orthogonal objects such as the one shown in Figure 2. The sets of three line directions which meet at junctions are listed, and the number of occurrences (or potential occurrences for L-junctions) are counted. If one set occurs more than any other, this may indicate that these lines correspond to orthogonal axes. The figure of merit for this is calculated as for axis-aligned objects, but is reduced by a factor based on the proportion of non-occurrences of the chosen set of groups of line orientations.

In practice, many semi-orthogonal objects have a single mirror plane which is axis-aligned. Since this combination provides enough information to deduce the face normals of many hidden faces, it is detected as a special case. The figure of merit is based on that for semi-orthogonality, as above, the length of the mirror chain, and the quality of the pairing produced by propagating mirror pairings through the sketch. It would be simple to extend this further to a combination of semi-orthogonality with either a C_2 or C_4 rotation, which would also provide enough information to deduce the face normals of many hidden faces, but which appear to be less common in engineering practice.

Similar to the concept of an axis-aligned object is one where the mirror planes and rotation axes define three mutually-orthogonal axes — see for example Figure 12.

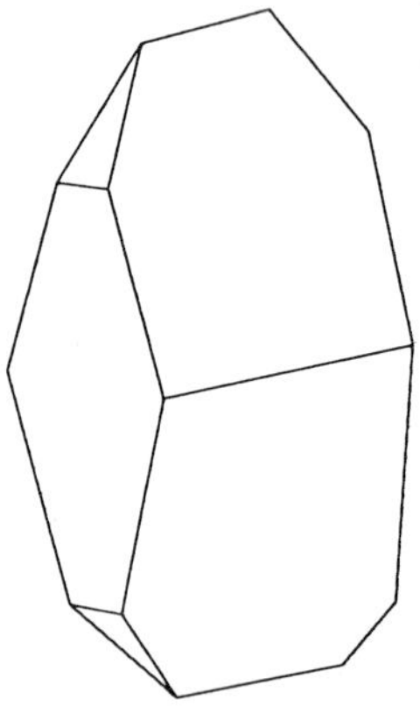

Figure 12. Tapered Wedge

Although the Platonic and Archimedean solids such as Figures 4 and 5 are useful test cases for general-case rotations, we prefer in practice to treat them as a special category. This is identified by comparing the numbers of sides (a, b, c) of the faces meeting at each vertex. If all vertices with three visible faces match (a, b, c), all vertices with two visible faces match (a, b), (a, c) or (b, c), and all vertices with one face match (a), (b) or (c), then the solid may be uniform. Additionally, all edges must be convex (or occluding).

As the system is extended, other special categories could be added — for example, if the entire sketch is consistent with C_n symmetry, $n > 4$. However, a survey of one engineering graphics textbook [43] yielded no examples in this category which could not equally well be classed as extrusions.

Identification of objects which can best be processed by splitting them into two smaller objects (for example, Figure 13) will be the subject of future work. We believe that the presence of junctions of two concave and one convex edge, such as $Wdcd$, will be a useful clue in identifying such objects.

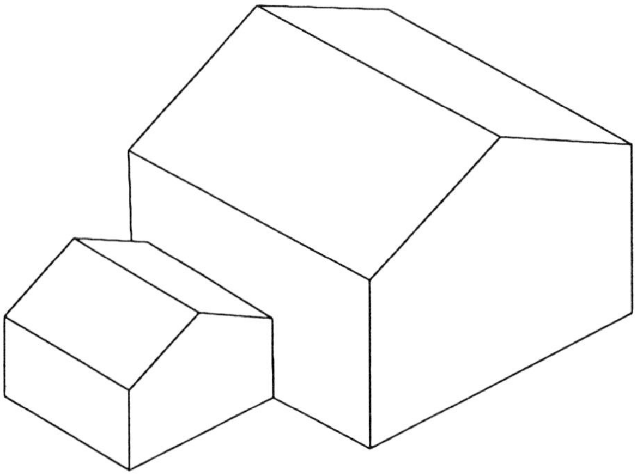

Figure 13. House with Extension

3.2.5 Outputs

To the outputs of the labelling process, the artefact identification process adds: for each vertex, a provisional Z coordinate; for each edge, the mean angle of the bucket to which the line has been allocated; for each face, a list of the possible rotational symmetries, and an indication for each vertex as to whether it is convex or concave; a list of the mirror planes found in the object; and a list of the special-case configurations which the object matches. The presence of vertex Z coordinates and edge bucket angles are dependent, respectively, on estimation of frontal geometry and on bucketing.

3.3. Topological 3D Completion

We now attempt to recover the topology of the hidden parts of the object to complete the topology of a consistent and plausible 3D object. A greedy algorithm is used for this. While the back of the object remains incomplete, hypotheses are made on the basis of the current state of the object concerning possible *moves* towards completion. The various moves are assessed, and the highest-rated is chosen. Various alternative approaches using different ideas as to what constitutes a move (e.g. adding a vertex and enough edges to join it to the existing object, or adding an n-sided face with some of its vertices and edges being existing parts of the object) are described elsewhere [38] and their respective merits are assessed.

In any object which contains exactly two biconnected junctions and these junctions are not joined by an existing edge, it is reasonable to add a single hidden edge connecting them to complete the object. In trihedral objects, this can be considered to take priority over any other possible move. It is not such a good assumption for non-trihedral objects, as Figure 14 shows.

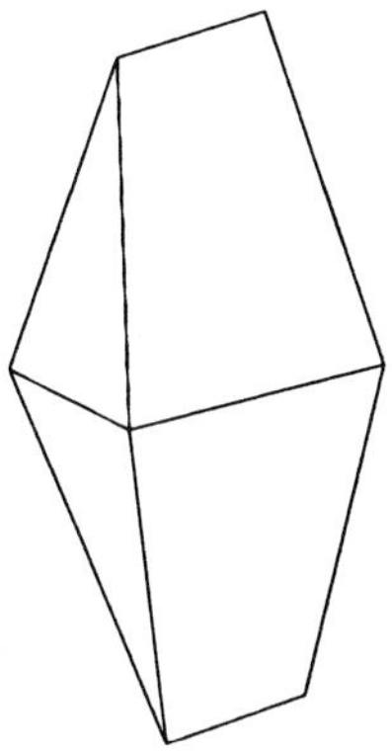

Figure 14. Wedge

Similarly, in any object which contains exactly one biconnected junction and exactly one uncompleted T junction, and the biconnected junction is located approximately on a continuation of the line defining the T junction, it is reasonable to extend this line to the biconnected junction to complete the object.

If an object contains no remaining biconnected junctions or T junctions, it is reasonable to conclude that all vertices and edges have been found.

In any of these three cases, any additional faces required to make all vertices trifacial and all edges bifacial are generated by testing for unused loops of half-edges in the object. If this completes the object (we have no test case where it fails) the object is deemed to be topologically complete.

If none of these near-complete cases applies, moves are selected on the basis of generating candidates by the following methods and picking the one with the best figure of merit. Whenever a new hidden vertex is hypothesised, a prediction is made concerning its position. Since it is found that some hypotheses are better than others at predicting vertex positions accurately, this prediction has its own figure of merit. For trustworthy hypotheses, this may be equal to the figure of merit of the hypothesis as a whole; for hypotheses such as mirror chains, which are good at predicting topology but poor at geometry, it may be considerably lower. Whenever equivalent hypotheses are merged, the predicted positions are also merged taking account of this latter figure of merit.

Where a chain of one or more mirror planes has been identified, an attempt is made to create a list of correspondences between the current partially-complete object and the mirror-image which would be generated by reflection. Each vertex, edge or face in the mirror chain is matched to its equivalent after the reflection operation, and then an attempt is made to propagate the matches through the rest of the object (the propagation method is described in detail in [38]).

Atoms in the mirror-image which have no correspondence to atoms in the original are assumed to correspond to extra atoms which will be required, and the appropriate hy-

·rating them can be made.

·ased on chains of mirrors are allocated fig- ·ıt based on the figure of merit of the chain — ·creases with the number of mirror planes in the chain ·ıd also increases if the chain starts or ends at an object boundary. It is decreased by a factor based on the number of new vertices required by attempting to match the object with its mirror-image – the more new vertices are required, the greater the decrease.

Hypotheses derived from rotational symmetries use a similar algorithm, incorporating the same pairing mechanism, but using the rotating single face, combination of edge and two faces, or combination of vertex and three faces, as its seed.

If the object is potentially axis-aligned, lines are extended from incomplete vertices in the same manner as in Grimstead's system. Vertices are hypothesised where these lines intersect. The figure of merit of any individual hypothesised hidden vertex is increased if three of the proposed lines cross in approximately the same place. Depth information is generated by assuming that the hypothesised lines are parallel in 3D to visible lines in the same bucket.

Additionally, if two such lines are almost collinear, it is probable that they are really the same edge, and the two vertices should be connected. The figure of merit for this is high, decreasing as the angle between them increases and the distance between them increases.

Similar hypotheses, though with a lower figure of merit, are produced for semi-orthogonal objects.

If an object is potentially an extrusion, hypotheses can be made concerning a back end face of the object (it is identical to and parallel to the front end face, and includes all vertices which are not part of the front end face) and the sides (they are parallelograms, and include two vertices from the front end face and the corresponding two vertices from the back end face).

Similar hypotheses can be made if an object is potentially a frustum. It is topologically equivalent to an extrusion but with tapered sides.

No additional hypotheses are generated for other special cases such as uniform (Platonic and Archimedean) objects, as it has been found in the cases we have tested that the mirror and rotation hypotheses described above are adequate for completing the object.

Hypotheses can be made for T-junction completion. A true vertex must exist somewhere further along the line past where the T-junction intersects the occluding line. It is plausible that this vertex is directly connected to the next real vertex (vertices) obtained by following the occluded region (regions) for arrow (concave, convex) T-junctions around along the obscuring line; the figure of merit is based on the proportion of known faces in the object which would have the same number of sides as this method would predict. It is also worth generating the other hypothesis used by Grimstead: that for arrow T-junctions, the true vertex is connected to the next incomplete junction in the other direction around the object boundary; this may or may not be true in practice, so the figure of merit is low.

Both hypotheses are correct for Figure 15 — the former hypothesis correctly predicts the edge between A and B, and the latter hypothesis correctly predicts the edge between A and C. This is not the case for Figure 3, where the former hypothesis correctly predicts the edge between A and B but the latter hypothesis incorrectly predicts an edge between A and C.

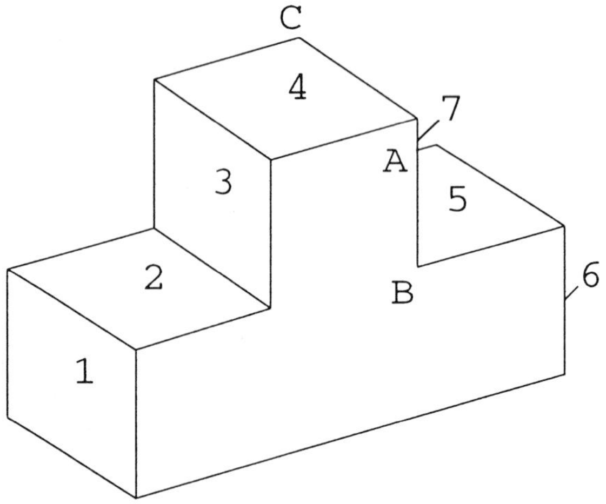

Figure 15. T Block

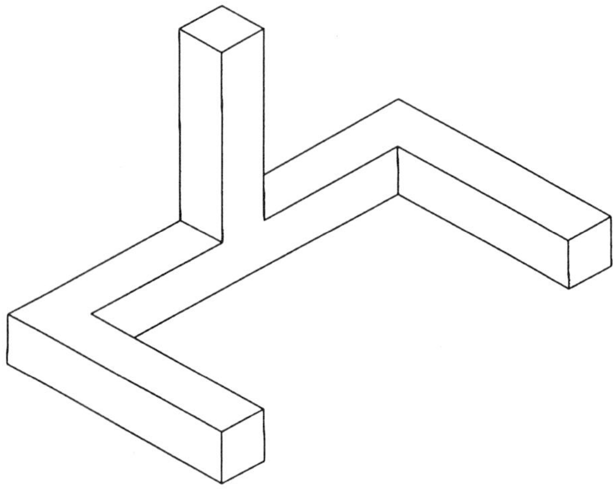

Figure 16. Double Z Block

Two local configurations of vertices lead to useful hypotheses. Firstly, given any two biconnected vertices A and B, if they are separated by a single triconnected vertex C then adding a new vertex D to form a parallelogram $ACBD$ is plausible. The figure of merit is based on a fixed value for this type of move multiplied by the proportion of quadrilateral faces in the partially-completed object. It is reduced if the object is not believed to be orthogonal or semi-orthogonal. By way of illustration, this hypothesis would generate the back face of Figure 2 and the two back faces of Figure 17.

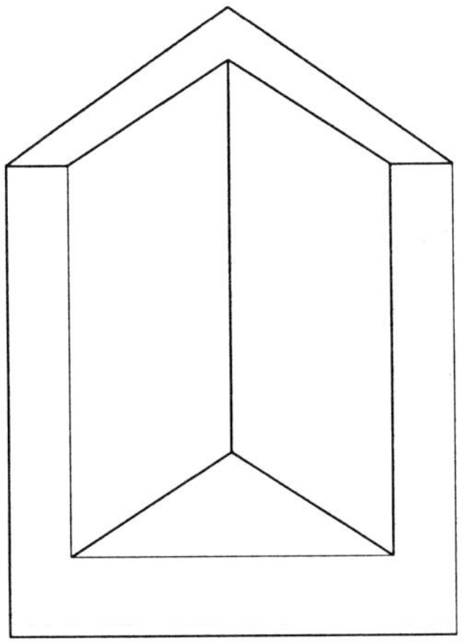

Figure 17. Angle Bracket

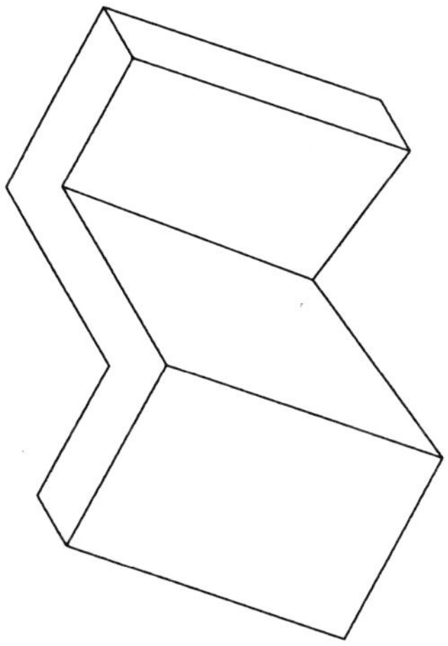

Figure 18. S Block

If two biconnected vertices A and B are separated by two triconnected vertices C and E, then adding an edge to join A and B to form the quadrilateral $ACEB$ is plausible. The figure of merit is based on a fixed value for this type of completion, the proportion of quadrilateral faces in the partially-completed object and the actual degree of parallelism of the supposedly parallel lines AC and EB. It is further reduced if the object is not believed to be orthogonal or semi-orthogonal. By way of illustration, this hypothesis would complete the front, partially-obscured face of Figure 2.

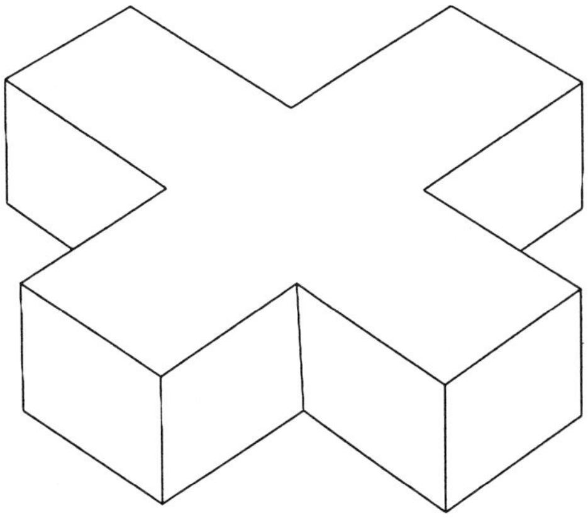

Figure 19. X Block

Finally, if there are exactly three biconnected junctions remaining as in Figures 8 and 17, it can be hypothesised that there exists a single hidden vertex connected to all three.

Where two or more methods suggest the same move, or moves which appear to represent the same thing, the hypotheses are merged and the figure of merit reinforced. An example might be that two methods suggest creating a face by adding a new vertex, and completing a partially-occluded face by extending an existing line through a T-junction.

A hypothesis which places a new, supposedly hidden vertex in a *visible* location has its figure of merit reduced, in order to allow for the fact that the geometry of the partial object is at best provisional, and thus locations derived from it inaccurate, without rejecting better hypotheses which place the new vertex in a hidden location.

Figures of merit are also reduced if the hypothesised move conflicts with beliefs about the object as a whole, or for other reasons which reduce their plausibility [38].

A backtracking mechanism has been implemented so that the chosen move may be rejected either immediately, if after making the move the object is invalid, or after following all branches of the resulting tree of subsequent moves, if none of these results in a valid object. The former has been found useful, but the latter is only sporadically useful: in practice, a valid but poor move leads more often to a valid but unexpected object than to an invalid object.

A move is rejected entirely if, while the object remains incomplete, the resulting tree of subsequent moves is empty. It can also be rejected entirely if, after making the move, the object becomes invalid or too complex [38].

3.3.1 Outputs

To the outputs of the regularity recognition process, the topology completion process adds: a number of extra vertices; for each vertex, a depth estimate; a number of extra edges; a number of extra faces; and the connectivity information to provide a boundary-representation topological structure.

The topology completion process handles most of our test cases in interactive time, producing topologically valid solids; in a few cases, these solids are not geometrically realisable in 3D. We conjecture that those cases which do not complete in interactive time would eventually also produce topologically valid solids.

3.4. Geometric Finishing

This stage of processing takes a topologically-correct object and a group of symmetry and regularity artefacts, and aims to produce geometric information about the geometry of vertices and faces which satisfies as many of the plausible constraints as possible. Grimstead's three-dimensional tidying stage is thus extended to force the face equations of the object to conform to symmetry constraints identified in previous stages of the system, rather than merely find a best fit. The output of this stage is a complete list of geometric data for vertices, edges and faces which together with the previous topological information determines a boundary representation solid model.

The stage is subdivided into five substages: estimation of face normals; generation of constraints from symmetries and regularities; enforcing constraints on face normals; optimising face distances while enforcing constraints; and intersecting faces to obtain final vertex positions. For some special cases of objects, the face normals can be generated directly. It is still necessary to generate constraints from symmetries for use in the distance optimisation substage.

Six types of constraint are identified:

- a parallelism constraint requires two faces M and N to be parallel.
- a two-way perpendicularity constraint requires two faces M and N to be perpendicular.
- a three-way perpendicularity constraint requires three faces M, N and O to be mutually perpendicular.
- an angular constraint of angle ρ requires two faces M and N to be at a defined, non-perpendicular angle.
- a rotation constraint through angle ρ about the face normal through the centre of O requires three faces M, N and O to be oriented such that the rotation moves M to the position occupied by N.
- a mirror constraint requires two faces M and N and a mirror chain C to be related such that reflection through the mirror plane defined by C moves M to the position occupied by N and vice versa; additionally, all of the faces C_1, C_2, ... in the mirror chain must have their face normals in the mirror plane.

For the purposes of constraints, mirror planes are treated like faces: they have normals and distances, and can be constrained using the constraint types listed.

Constraints are allocated a figure of merit. The aim is to accommodate constraints in descending order of merit, with constraints being accepted if there are enough degrees of freedom left in the object to accommodate them, or if they agree with the geometry produced by previously accepted constraints.

In addition to the method described here, alternatives have been tried; these are discussed elsewhere [39].

3.4.1 Estimation of Face Normals

Vertex coordinates generated during the topological completion stage are used to provide preliminary estimates of face normals. For all but triangular faces (which can be solved explicitly) we use a linear system [2] where the variables are P, Q and C for the face (setting $R = 1$), and equations place each vertex on the face (with weightings which give priority to visible vertices). Adopting the convention that face normals are directed to look towards the face from outside the object, normals of visible (including partially-occluded) faces are provisionally assigned positive z-components, and normals of hidden faces are assigned negative z-components (this has the advantage of being quick, and thus repeatable: a more rigorous determination of the direction of face normals, deduced from convexity/concavity of edges, is performed once only at a later substage).

3.4.2 Special Categories of Object

For some of our more complex test cases, the general-case optimisation process described below was too slow to be considered interactive. There are also theoretical doubts about the general-case algorithm for face normals. We therefore use the object categorisation performed during artefact identification to take short cuts through the process for some categories of object.

If the object is believed to be a right extrusion (i.e. this category has a higher figure of merit than any other) the end caps are made parallel and the sides are made perpendicular to the end caps. If the front end face is believed to have mirror or rotational symmetry, the face normals of the sides are adjusted to preserve this symmetry. Since there are no further constraints on face distances, no further constraints are generated for extrusions, and the general-case process is bypassed entirely. Similar methods are also used for frusta.

For axis-aligned objects, three perpendicular axes are formed, close to the average values of the appropriate face normals, and each face normal is constrained to the appropriate axis. The general-case face normal process is bypassed, but mirror and rotation constraints are still generated for use during the distance optimisation substage — this is required, for example, in order to ensure that the sides of the T-block (Figure 15, faces 2 and 5) are the same height and length.

For objects which are semi-orthogonal with an axis-aligned mirror plane, three perpendicular axes are formed, as for axis-aligned objects. The mirror plane is used to identify which of the hidden faces ought to be axis-aligned. The face normal of each face which should be axis-aligned is constrained to the appropriate axis. No part of the general-case process described below is bypassed. Optimisation of face normals takes place as for the general case, in order that relationships between non-axis-aligned faces can be established, but will be quicker by virtue of the knowledge already gained concerning parallelism and perpendicularity of axis-aligned faces and the relationships between non-axis-aligned faces and the mirror plane.

Since the uniform solids are finite in number and easily-identified, they are most easily treated as individual special cases. In practice, we adjust face normals iteratively by adding to each face normal the value which would be obtained by rotating adjacent faces by the appropriate angle (the general case is bypassed entirely). This provides a test-bed for procedures also used in the general case and a template for handling non-uniform objects with multiple rotational symmetry.

Special case processing of objects with three axis-aligned sets of mirror planes, and non-uniform objects with clear higher-order rotational symmetry, is the subject of ongoing work. Similar methods to the above will be used.

3.4.3 Generation of Constraints

Constraints, firstly on the face normals and then on the face positions, are generated from the symmetry artefacts and allocated a figure of merit, both for the general case and for special categories which do not bypass the general case entirely.

Each rotational symmetry artefact produces constraints. The existing object is matched with the object which would be generated by applying the rotation about the rotating face O. For each face N in the rotated object which matches a face M in the existing object, a constraint is produced. This is normally a rotation constraint, except for the special case where $M = N$: $M = N = O$ is trivial and is ignored, and otherwise, M must be parallel to O, so a parallelism constraint is produced.

Hypotheses based on rotations are allocated figures of merit based on a fixed probability for each type of rotation, with C_4 and C_6 being given higher probabilities. The assessment is decreased by a factor based on the number of unmatched vertices left after attempting to match the object with its rotated equivalent — the more unmatched vertices which remain, the greater the decrease.

Each mirror chain generates constraints. The existing object is matched with the object generated by reflection through the mirror plane C. For each face N in the reflected object which matches a face M in the existing object, constraints are produced. One will normally be a mirror constraint, except for the special case where $M = N$, where M must be perpendicular to the mirror plane C and a two-way perpendicularity constraint is produced. Additionally, the mirror plane C is perpendicular to each face in the chain, C_1, C_2, ..., and two-way perpendicularity constraints are generated for these. As with rotations, the figures of merit are based on the quality of the match between object and reflection.

If an object is potentially axis-aligned, but this is not the most favoured interpretation, four perpendicularity constraints (three two-way and one three-way) are still generated for the faces meeting at each vertex. Their figures of merit are based on the orientation of the face normals involved and the overall figure of merit for the hypothesis that the object is axis-aligned.

In a similar manner, if the object is potentially semi-orthogonal, perpendicularity constraints are generated for those vertices at which the three presumed orthogonal axes meet. The figure of merit is calculated in a similar manner.

If the object is potentially an extrusion or a frustum, but these are not the favoured interpretations, a parallelism constraint is generated between the front and back end faces.

If bucketing is successful, a parallelism constraint is generated for any pair of faces which use edges from the same set of buckets.

If any two faces are approximately perpendicular or approximately parallel in the topological completion, it can be hypothesised that they are exactly parallel in the true object. Similarly, if the faces form an angle of $30°$, $45°$ or $60°$, or an angle the tangent of which is approximately a ratio of two small integers, it can be hypothesised that this is intentional and this angle should be retained in the true object.

3.4.4 Enforcing Constraints on Face Normals

In the general case, our aim is to find and enforce the best set of geometric constraints which can be satisfied simultaneously by the topology. This is a knapsack problem, exact solution of which is infeasible, so we use a greedy algorithm. Constraints are enforced in descending order of merit until they specify a unique object:

- the constraint with the highest figure of merit is always accepted and enforced
- for each other constraint, in descending order of merit:

- if logical reasoning using known information about the face normals can show that the constraint is necessarily valid, it is accepted;
- if logical reasoning can show that the constraint is necessarily invalid, it is rejected;
- if the constraint is already satisfied numerically by the object, it is accepted;
- if enough degrees of freedom remain in the affected faces, the constraint is accepted;
- if the object has enough degrees of freedom elsewhere, an attempt is made to adjust the existing face normals to accommodate the new constraint as well as all previous accepted constraints; if this succeeds, the new face normals are stored and the constraint is accepted; otherwise, the previous face normals are restored;
- otherwise, the constraint is rejected.

Logically, the relationship between two faces is considered to be one of three: they are either parallel, perpendicular, or at some other angle. A set representing the three possible relationships is stored for each pair of faces. Initially, all three relationships are possible, except that faces sharing an edge cannot be parallel. If the object is axis-aligned no pair of faces can be other than parallel or perpendicular. When new information is added, it is propagated to all faces parallel to the faces being updated. Interrogating the status of relationships (known to be true, possible but uncertain, or known to be false) is straightforward on the whole; the most complex inference made is that two faces are necessarily parallel if that is still a possible relationship and they are both perpendicular to the same two mutually-perpendicular faces.

The logical reasoning sets an upper limit to the number of orientational degrees of freedom of a face [39]. To find the available number of degrees of freedom, this must then be reduced to allow for previously-accepted mirror, rotational and angular constraints; this is a non-trivial problem to which we know of no perfect solution.

Methods of calculating the number of degrees of freedom after satisfying a number of geometric constraints have been the subject of several studies. Sugihara [34] lists a number of these. More recently, Kramer [19] has shown the difficulty of allowing for geometrical coincidences in two dimensions. Latham [20] has methods for resolving this issue which appear to improve on existing algorithms, but these are based on symbolic reasoning and are therefore unsuitable for current interactive systems. Whitely's [42] extension to 3D of a method which is successful in 2D is unsatisfactory in that it is intolerant of sketching inaccuracies and that it is limited to triangular and quadrilateral faces.

Rather than attempt to improve further, we recognise that since perfect conversion of sketches to objects is formally impossible, and that we hope for most sketches to bypass the general case, an algorithm which works more often than not for the remainder is acceptable. Therefore, we use a practically effective algorithm which would be hard to justify theoretically [39].

Different methods of adjusting face normals have been tested [39]. Some are quicker than others, but none is quick enough to be considered interactive for objects which depend mostly on numerical, rather than logical, reasoning.

3.4.5 Optimising Face Distances

The objective of this substage is to provide values of face distances for each face which fit the faces of the object as closely as possible to the visible vertices while enforcing accepted constraints. The starting-point is an n-dimensional iterative optimisation [30, 31] where n is the number of faces in the object and the variables being optimised are the face distances. The objective function being optimised is the sum of the squares of the 2D distances between the predicted x and y coordinates of each vertex and its actual position in the original sketch.

Mirror constraints and C_2 rotation constraints affect this in two ways. Firstly, if the mirror plane or rotation axis is perpendicular to the two faces M and N, and M and N are parallel, then they must also be coplanar. The distance for N is then dropped from the optimisation (the number of variables is decremented) and set equal to that for M on each iteration (see faces 2 and 5 in Figure 15). Secondly, if faces M and N are parallel, and two other paired faces M' and N' are also parallel to one another and to M, then the distances obey the equation $C_M - C_{M'} = C_{N'} - C_N$. The distance for N' is then dropped from the optimisation (the number of variables is again decremented) and set equal to $C_M + C_N - C_{M'}$ on each iteration (see faces 1 and 3, and 6 and 7, in Figure 15).

In addition, mirror planes must bisect the faces they lie on, producing constraints on the distances of faces meeting the edges of the mirroring face. This is the subject of ongoing work.

The use of rotational symmetry in adjusting face distances is the subject of ongoing work. This is needed for sketches such as the dodecahedron, which has a completely hidden face for which visible vertex positions provide no information.

The assumption of axis alignment, and the perpendicularity constraints it produces, have no effect on distances.

3.4.6 Intersecting Faces

This matches Grimstead's three-dimensional tidying process — the x, y and z coordinates of each vertex are finally recalculated from the equations of the three faces on which it lies.

4. Results

This section gives some examples of objects generated from sketches, and makes general observations about the merits of our approach.

4.1. Experimental Results

Table 1 collates the results obtained using all of the trihedral sketches illustrated in the Figures in this paper and a few other examples. They are grouped by category, and within a category are in order of increasing complexity. A tick $\surd$ indicates that results are as expected, and a cross $\times$ that they are not, in which case see the referenced note. Timings are in milliseconds and were obtained on a Sun Ultra 10 running Unix.

Sketch (Figure no.)	Topology	Geometry	Time (ms)
Cube (–)	$\surd$	$\surd$	30
L Block (8)	$\surd$	$\surd$	30
T Block (15)	$\surd$	$\surd$	60
J Block (3)	$\surd$	$\surd$	150
Z Block (9)	$\surd$	$\surd$ (A)	70
S Block (18)	$\surd$	$\surd$ (B)	70
X Block (19)	$\surd$	(C)	570
double Z Block (16)	$\times$ (A)	N/A	N/A
H Block (7)	$\surd$	$\surd$ (D)	560
Notched Extrusion (10)	$\surd$	$\surd$	10
Hexagonal Prism (–)	$\surd$	$\surd$	50
House (1)	$\surd$	(E)	40
Triangular Frustum (11)	$\surd$	$\surd$ (F)	10
Hexagonal Frustum (–)	$\surd$	$\surd$ (F)	20
Tetrahedron (6)	$\surd$	$\surd$	10
Dodecahedron (4)	$\surd$	$\surd$ (G)	370
Truncated Tetrahedron (–)	$\surd$	$\surd$ (H)	50
Truncated Octahedron (–)	$\surd$	$\surd$ (G)	390
Truncated Icosahedron (5)	$\surd$	$\times$ (G)	28220
Angle Bracket (17)	$\surd$	$\surd$ (B)	130
Grimstead's Bracket (2)	$\surd$	$\surd$ (I)	2530
Tapered Wedge (12)	(B)	N/A	490
House with Extension (13)	$\times$ (C)	N/A	N/A

Table 1. Results of Sketch Interpretation

4.1.1 Topology

Our system reconstructs the correct topology for the majority of the Figures.

(A) At an early stage of topological reconstruction of the double-Z-block, the hypothesis of a rectangular base to the near wing of the block has a high figure of merit, and is accepted. The expected object can be produced using different tunings for figures of merit.

(B) The tapered wedge produces a topologically-valid but unexpected interpretation based on regular triangles and hexagons.

(C) The absence of symmetry in the sketch makes the weightings of figures of merit for hypotheses arbitrary. Incorrect hypotheses are accepted.

4.1.2 Geometry

Our system reconstructs the expected geometry in over half of the Figures. In some cases the expected geometry is not ideal, since the symmetry operation which would produce the ideal geometry is not one we detect. Where unexpected geometry is added to correct topology, this is usually because a higher degree of symmetry has been imposed in preference to a lower-degree but intuitively more correct symmetry.

(A) The two ends of the Z-block may not be the same length — although the object as a whole has a C_2 rotation, this is centred on an occluding edge so is not detected.

(B) The two end sections of the S-block are the same width, but the centre section may be a different width — no symmetry relation exists between them. The same is true of the width of the U-shaped face of the Angle Bracket.

(C) The C_4 rotational symmetry is enforced in preference to the axis-aligned mirror symmetry planes, so the resulting geometry is not as expected.

(D) S_n symmetries are not detected, so the two end-pieces of the H-block may be of different lengths.

(E) The sketch of a house is converted to a pentagonal prism (C_5 rotations are considered hard to draw, so the figure of merit for the C_5 rotation is higher than that for sides being parallel).

(F) The height of the frustum is arbitrary since the sketch contains no clues.

(G) The rear face of the dodecahedron is too large — its position is not constrained by any visible vertices. Similarly, the rear square of the truncated octahedron is too small. The problem is amplified in the truncated icosahedron, where several faces in the topological reconstruction contain no visible vertices. Treating uniform objects as special cases would enable correct interpretation of such sketches, but would not cure the problem in the general case.

(H) The triangular faces of the truncated tetrahedron are different sizes, reflecting inaccuracies in the original sketch.

(I) The angle between the axis-aligned and angled faces is to some extent arbitrary, and not necessarily the one which would be obtained by 'de-skewing' the end face.

4.2. Conclusions

The set of sketches which can be interpreted plausibly improves on Grimstead's method. In the majority of cases, the goal of producing the expected interpretation is achieved. We believe that a further improvement can be achieved by refining the figures of merit.

On the whole, the system meets the goal of interactive response times. The exceptions occur when a poor choice of move is made during the early stages of topological reconstruction. This can make the process very slow — of the order of minutes — before backtracking brings the system back to a more sensible choice.

We have added one more assumption concerning the sketch to our initial set — that it is from a viewpoint from which the visible junctions are graph-connected. An alternative method of generating provisional depth information would remove this restriction.

Objects which meet one of the special-case categories are in general categorised correctly. More commonly, objects will meet the requirements of several of the special-case categories, and the category to which they are allotted can be arbitrary. This can also vary depending on how well-drawn the sketch is — different versions of Figure 1 are categorised as a frustum, a semi-axis-aligned sketch with a mirror plane, or an extrusion.

The two non-axis-aligned Platonic solids, the tetrahedron and dodecahedron, are reconstructed correctly, although the positioning of faces containing no visible vertices is suspect. Successful conversion of the Archimedean solids has also been achieved but is not interactive for the most complex of these and the geometry is poor. Treating them as special cases throughout would lead to perfect conversion in interactive time.

We have no extrusion test case which fails. However, sketches of hexagonal prisms are categorised as extrusions (which they are) rather than as uniform solids (which is also true, and gives more information).

Intentionally axis-aligned objects can be detected (rather than assumed, as in [17]) with a reasonable degree of confidence. They are converted to exactly axis-aligned boundary representation models (rather than approximately axis-aligned, as in [8]) interactively. We have no axis-aligned test case which consistently fails, but in some cases correct topological interpretation depends on the numerical values of the figures of merit assigned to various hypotheses — the double-Z-block is one such. Poorly-drawn versions of the T-block and other similar objects intended to be axis-aligned are generally categorised as extrusions, and are reconstructed correctly via that route.

Simple semi-axis-aligned objects with mirror planes are interpreted correctly, but more slowly and less reliably than the other special categories. Where the number of non-axis-aligned faces is low, it is still quick enough to be considered interactive.

General-case objects meeting no special categories are sometimes interpreted successfully, with sketches containing ten or twelve junctions being interpreted in interactive time. The time taken rises sharply with the number of junctions. Reconstruction of topology is more stable than reconstruction of geometry. In some cases, the system can add symmetry where it was not intended, further distorting the geometry.

5. Future Work

Further investigation into the numerical values of figures of merit is planned, with the aim of correctly converting those of our test cases which are currently failing.

Thereafter, it is planned to attempt to extend the range of sketches which can be interpreted by allowing non-trihedral vertices. It is also planned to attempt to extend the range of permitted sketches by allowing sketches of objects with hole loops, and to extend the scope of reconstruction by partitioning difficult sketches into smaller, categorisable subsketches.

Further in the future, a system which can deal with simple curved objects may be attempted.

Acknowledgements

The authors are grateful to Unigraphics Solutions Inc. for providing Parasolid to the authors for use in this research.

We also wish to thank the referees, whose observations have helped us to clarify the paper.

References

[1] H.G.Barrow and J.M.Tenenbaum, *Interpreting Line Drawings as Three-Dimensional Surfaces,* Artificial Intelligence **17**, 75–116, 1981.

[2] L.Bauer, *Elimination with Weighted Row Combinations for Solving Linear Equations and Least Squares Problems,* Contribution I/9 in ed. J.H.Wilkinson and C.Reinsch, Handbook for Automatic Computation **II**, Linear Algebra, Springer-Verlag, 1971.

[3] R.P.Brent, *Algorithms for Minimization without Derivatives,* Prentice-Hall, 1973.

[4] I.Chakravarty, *A Generalised Line and Junction Labelling Scheme with Applications to Scene Analysis,* IEEE Trans. Pattern Analysis, Machine Intelligence **1**(2), 202–205, 1979.

[5] M.B.Clowes, *On Seeing Things,* Artificial Intelligence, **2**, 79–116, 1970.

[6] H.T.Davey and R.J.Wilkins, *Engineering Drawing,* Macdonald and Co, 1952.

[7] S.W.Draper, *The Use of Gradient and Dual Space in Line-Drawing Interpretation,* Artificial Intelligence **17**, 461–508, 1981.

[8] I.J.Grimstead, *Interactive Sketch Input of Boundary Representation Solid Models,* PhD Thesis, Cardiff University, October 1997.

[9] I.J.Grimstead and R.R.Martin, *Incremental Line Labelling for Sketch Input of Solid Models,* Computer Graphics Forum, **15**(2), 155–166, 1996.

[10] I.J.Grimstead and R.R.Martin, *Creating Solid Models from Single 2D Sketches,* in ed. C.Hoffmann and J.Rossignac, Proceedings of Third Symposium on Solid Modelling and Applications, 323–337, ACM Press, 1995.

[11] M.Herman and T.Kanade, *The 3D MOSAIC Scene Understanding System: Incremental Reconstruction of 3D Scenes from Complex Images,* in Proceedings of the Image Understanding Workshop, 137–148, 1984.

[12] E.Horowitz and S.Sahni, *Fundamentals of Computer Algorithms,* Computer Science Press Inc, 1979.

[13] D.A.Huffman, *Impossible Objects as Nonsense Sentences,* Machine Intelligence **6**, 295–323, New York American Elsevier, 1971.

[14] D.L.Jenkins, *The Automatic Interpretation of Two-Dimensional Freehand Sketches,* PhD Thesis, Cardiff University Computer Science Department, May 1992.

[15] T.Kanade, *Recovery of the Three-Dimensional Shape of an Object from a Single View,* Artificial Intelligence **17**, 409–460, 1981.

[16] K.Kanatani *Reconstruction of consistent shape from inconsistent data: Optimisation of $2\frac{1}{2}D$ sketches.* International Journal of Computer Vision **3**(4), 261–292, 1989.

[17] K.Kanatani, *Group-Theoretical Methods in Image Understanding,* Springer Series in Information Sciences No 20, Springer-Verlag, 1990.

[18] S.F.A.Kettle, *Symmetry and Structure,* Wiley, 1985.

[19] G.A.Kramer, *Solving Geometric Constraint Systems,* MIT Press, 1992.

[20] R.S.Latham, *Combinatorial Algorithms for the Analysis and Satisfaction of Geometric Constraints,* PhD Thesis, Department of Computer Science, Brunel University, December 1996.

[21] Y.G.Leclerc and M.A.Fischler, *An Optimization-based Approach to the Interpretation of Single Line Drawings as 3D Wire Frames,* International Journal of Computer Vision, **9**(2) 113–136, 1992.

[22] H.Lipson and M.Shpitalni, *Optimization-based Reconstruction of a 3D Object from a Single Freehand Line Drawing,* Computer-Aided Design **28**, 651–663, 1996.

[23] H.Lipson and M.Shpitalni, *An Interface for 3D Conceptual Design Based on Freehand Sketching,* in ed M.J.Pratt, R.D.Sriram, M.J.Wozny, Product Modeling for Computer Integrated Design and Manufacture, Chapman and Hall, 1997.

[24] E.H.Lockwood, *Geometric Symmetry,* Cambridge University Press, 1978.

[25] A.K.Mackworth, *Model-driven Interpretation in Intelligent Vision Systems,* Perception **5**, 349–370, 1987.

[26] R.R.Martin and D.Dutta, *Tools for Asymmetry Rectification in Shape Design,* Journal of Systems Engineering **6**, 98–112, 1996.

[27] D.Marr, *Representing Visual Information,* in A.Hanson and E.M.Riseman, Computer Vision Systems, Academic Press NY, 1978.

[28] H.Mayer, *Automatic Object Extraction from Aerial Imagery - A Survey Focusing on Buildings,* Computer Vision and Image Understanding **74**(2), 138–149, 1999.

[29] R.Mohan and R.Nevatia, *Using Perceptual Organisation to Extract 3D Structures,* IEEE Transactions on Pattern Analysis and Machine Intelligence **11**(11), 1121–1139, 1989.

[30] J.A.Nelder and R.Mead, *Computer Journal,* **7**, 308–313, 1965.

[31] W.H.Press, S.A.Teukolsky, W.T.Vettering and B.P.Flannery, *Numerical Recipes in C, The Art of Scientific Computing,* Second Edition, Cambridge University Press, 1994.

[32] Y.Shirai, *Analysing Intensity Arrays Using Knowledge About Scenes,* in The Psychology of Computer Vision, ed P.H.Winston, McGraw-Hill, New York, 1975

[33] L.A.E.Stevens, *Genetic Algorithm to Optimise Line Labelling,* MSc dissertation, Cardiff University, 1994.

[34] K.Sugihara, *Machine Interpretation of Line Drawings,* MIT Press, 1986.

[35] K.Turner, *Computer Perception of Curved Objects using a Television Camera,* PhD Dissertation, Edinburgh University, November 1974.

[36] T.Vetter and T.Poggio, *Symmetric 3D objects are an easy case for 2D object recognition,* in ed C.W.Tyler, Human Symmetry Perception, 349–359, 1996.

[37] P.A.C.Varley and R.R.Martin, *Constructing Boundary Representation Solid Models from a Two-Dimensional Sketch — Frontal Geometry and Sketch Categorisation* , 1st Korea-UK Joint Workshop on Geometric Modeling and Computer Graphics, April 2000, to appear.

[38] P.A.C.Varley and R.R.Martin, *Constructing Boundary Representation Solid Models from a Two-Dimensional Sketch — Topology of Hidden Parts,* 1st Korea-UK Joint Workshop on Geometric Modeling and Computer Graphics, April 2000, to appear.

[39] P.A.C.Varley and R.R.Martin, *Constructing Boundary Representation Solid Models from a Two-Dimensional Sketch — Geometric Finishing,* 1st Korea-UK Joint Workshop on Geometric Modeling and Computer Graphics, April 2000, to appear.

[40] D.M.Waltz, *Generating Semantic Descriptions from Drawings of Scenes with Shadows,* Tech Report AI-TR-271, M.I.T., Cambridge USA, 1972.

[41] C.E.Weatherburn, *Elementary Vector Analysis,* Bell, 1921.

[42] W.Whitely, *Matroid and Rigid Structures,* in ed. N.White, *Matroid Applications,* Cambridge University Press, 1991.

[43] H.W.Yankee, *Engineering Graphics,* Prindle, Weber and Schmidt, 1985.

Rapid 3D Model Acquisition from Images of Small Objects

Wee Kheng Leow, Zhiyong Huang, Yong Zhang, and Rudy Setiono
School of Computing, National University of Singapore
Lower Kent Ridge Road, Singapore 119260
leowwk, huangzy, zhangyon, rudys@comp.nus.edu.sg

Abstract

This paper describes a system for rapid acquisition of 3D models of objects for use in applications such as CAD and VR. The system consists of an image capturing platform, which comprises a PC-controlled turntable and CCD camera, and associated computer vision and graphics algorithms for 3D model acquisition. An algorithm for accurately recovering 3D points from image sequences is described. Experiments performed on synthetic and real data show that the algorithm is accurate in recovering the coordinates of 3D points and is robust against noise in 2D feature location and 3D object rotation. Results of applying the system to acquiring 3D models of real objects are illustrated.

1. Introduction

3D models of objects and scenes are essential components of CAD and VR systems. However, the acquisition of the geometry of 3D models is a long-standing difficult problem in computer graphics. In the past, forward methods such as CSG, B-rep, NURBS, and sweeping have been successful in modeling regular shapes such as mechanical and architectural parts. These methods allow a user to directly and interactively define the shape of an object. On the other hand, natural objects and scenes have irregular shapes and complex surface structures. It is very difficult to use forward methods to interactively define the details of the complex shapes and surfaces.

With the availability of vision equipment and the maturity of computer vision technology, there is now a trend to reverse engineer the acquisition process by recovering the geometric and topological information from measurements of real scenes and objects. Examples of this approach include the acquisition of human faces from photographs [3, 10], human cornea from specular reflection patterns [11], and sculpture models from range images [4]. While some of these methods are designed to acquire specific object models, others (e.g., [4]) are applicable to modeling general, natural objects.

There are two main approaches to the rapid acquisition of the models of natural objects and scenes: structured light and multiple views. The first approach projects a structured light pattern, usually a stripe of laser beam, onto the surface of an object and uses a camera to capture the surface contour as revealed by the reflected laser light. The contours along different surface patches are then combined to compute the 3D geometry of the object. Such laser scanning products are currently available in two versions: hand-held and automated. The hand-held version (e.g., [16]) requires the user to manually move the laser stripe to various parts of the object's surface. It, therefore, takes a lot of manual work to capture all the surface details of an object. The automated version (e.g., [18, 20]) uses a motorized system to move the laser in a manner similar to a 2D image scanner. 3D laser scanners typically produce accurate results but are expensive, especially the automated version and the color scanners.

The second approach uses one or more cameras to capture multiple views of an object and uses the multiple views to recover 3D coordinates of the feature points on the object's surface. The single-camera system (e.g., [14, 15, 17]) either moves the camera or the object so as to capture various views of the object. The multiple-camera system (e.g., [19]) typically employs static cameras to simultaneously capture multiple views of the object. However, it would require many cameras to adequately capture various facets of a complex object, and would therefore be more complex and expensive than the single-camera system.

This article describes an inexpensive 3D model acquisition system that is affordable to the home users. It consists of an image capturing platform, which comprises a PC-controlled turntable and a CCD camera, and associated computer vision and graphics algorithms for 3D model reconstruction. In the following sections, a comparative review of related single-camera systems is presented in Section 2. Next, our system for rapid acquisition of 3D models is described Section 3. Results of applying the system to acquiring 3D models of real objects are illustrated in Section 4. In addition, Section 4 also reports experiments conducted to measure the system's reconstruction accuracy and robustness against noise. Finally, Section 5 summarizes and concludes our research work.

2. Related Works

The systems proposed by Matsumoto et al. [14], Fitzgibbon et al. [9], Mehren and Rodehorst [15], and Szeliski and co-workers [13, 21] are most similar to ours. They also use a turntable to capture multiple views of an object. In particular, the methods of Matsumoto et al. and Fitzgibbon et al. reconstruct 3D model from image silhouette. These methods have the disadvantage that concave parts of the object may not appear on the silhouette and, thus, cannot be recovered. In contrast, our method recovers 3D structures based on the feature points on the surface of the object and it does not have the limitation of the silhouette method.

Mehren and Rodehorst's method adopts an affine camera model which is less accurate then the perspective model used in our system. The methods of Szeliski and co-workers adopt approximations of the perspective camera models and recover a quasi-Euclidean structure of the object. Our method, on the other hand, recovers the true 3D model of the object.

Whereas the above systems focus on acquiring 3D models of objects, the systems of Teller [22] and Debevec et al. [6] aim at acquiring buildings and architectural structures from multiple perspective views and aerial images. Teller's system consists of a camera mounted on a pan-tilt head which is, in turn, mounted on a mobile platform. The platform also includes instrumentation for maintaining estimates of global position (GPS), heading information (IMU), and for dead-reckoning (mechanical wheel encoders). These devices are controlled by a PC which is also mounted on the mobile platform. Debevec et al. apply a hybrid approach consisting of image-based and geometry-based modeling. The approach consists of two components. The first component is a photogrammetric modeling method which facilitates the recovery of the basic geometry from scene images. The second component is a model-based stereo algorithm which recovers depth from widely-spaced image pairs. Human intervention is required to interactively indicate to the system block structures, feature correspondences, and un-occluded texture regions in the scene. In contrast, our system is automatic and does not require human intervention. It is also simpler than Teller's system because our goal is to acquire 3D models of small objects instead of buildings.

Figure 1. The acquisition hardware platform consisting of a PC, a PC-controlled turntable, and a CCD camera.

Figure 3. Object with known corner points used for calibrating the camera.

3. 3D Model Acquisition System

3.1. The Acquisition Framework

Our 3D model acquisition system consists of three hardware components (Fig. 1): a PC, a PC-controlled turntable, and a CCD camera. The turntable is rotated by a high-precision stepper-motor with a smallest step size of $0.036°$ and a maximum error of 1% (i.e., $0.01°$ for $1°$ rotation). Therefore, the amount of rotation can be controlled by the PC accurately.

The software framework of the system is illustrated in Fig. 2. It consists of four main modules: Camera Calibration, Feature Tracking, 3D Recovery, and Triangulation. The first module, Camera Calibration, is performed only once. A calibration object (Fig. 3) is placed on the turntable and its image is captured by the CCD camera. A refined version of the Beaudet's corner detector [2] is applied to accurately locate the 2D coordinates of the corners in the image. The known 3D positions of the corners and their corresponding 2D image coordinates are input into Tsai's calibration algorithm [24] which computes the intrinsic and extrinsic parameters of the camera.

After calibration, the object to be acquired is placed on

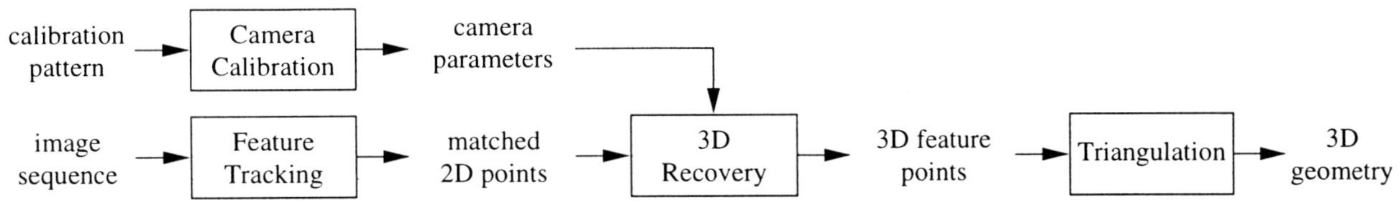

Figure 2. The acquisition platform: software functional modules.

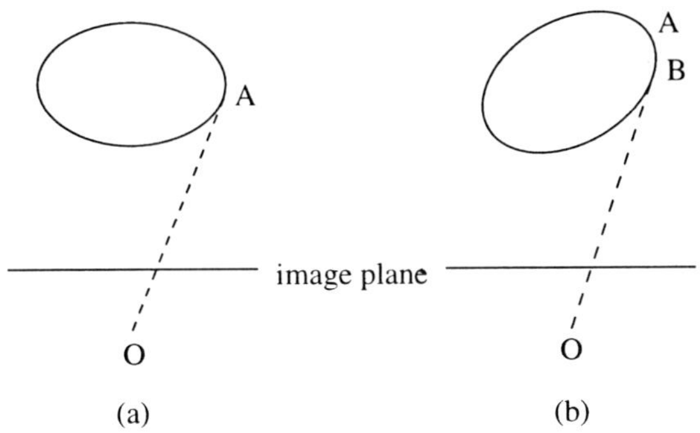

Figure 4. The dashed lines denote the lines of projection from the object's boundary through the image plane (viewed from the top) to the focal point O. As the object is rotated, its boundary point changes from point A to point B.

the turntable and rotated through 360° to capture different views of the object. The current implementation of the system rotates the object at a step size of 1.8° thus capturing a total of 200 images. Next, the Kanade-Lucas-Tomasi (KLT) tracker [23] is applied to the 200 images to match the corresponding feature points in consecutive images. It is observed that as the object is rotated, smooth curve surfaces of the object present different feature points on the 2D object boundaries at different time (Fig. 4). The feature tracking algorithm may misinterpret the boundary points in consecutive images as arising from the same 3D feature point. To overcome this problem, the original KLT tracker is modified to ignore feature points that are located near object boundaries.

The coordinates of the matched 2D feature points and their corresponding rotation angles, together with the camera's parameters, are input into the 3D Recovery module to recover the 3D coordinates of the feature points. An algorithm, called the *Ray Intersecting* (RI) algorithm, is developed to accurately compute the 3D coordinates. Details about RI will be described in Section 3.2.

In the final stage, the 3D coordinates of feature points are input to the Triangulation module to reconstruct the surfaces of the object. Currently, the 3D Alpha Shapes algorithm introduced by Edelsbrunner and Mucke [8] is used. The algorithm outputs a 3D model represented by a polygon mesh. Polygonization from scattered points is an important problem in computer graphics and CAGD. Well known polygonization methods include crust algorithm [1], zero-set algorithm [12], and Alpha Shapes algorithm [7, 8]. In the Alpha Shapes method, a simplex (edge, triangle, or tetrahedron) is included in an Alpha Shape if it contains some circumspheres with no interior sample points. A circumsphere is a sphere of radius α whose surface touches at least three sample points.

We adapted the implementation of Alpha Shapes algorithm from CGAL library [5]. It works well for most cases. However, if the feature points' sampling rate varies a lot on different parts of an object, it is impossible to have a unique α value that polygonizes the whole model. This situation can happen because the 2D feature points tracked by KLT algorithm may not be evenly distributed. We are now working on a new polygonization algorithm that can adapt to different sampling rates on different parts of the object.

3.2. Ray Intersecting Algorithm

The Ray Intersecting (RI) algorithm is a variation of the triangulation method. It recovers 3D coordinates of feature points as follows. A 3D point W at world coordinate $\mathbf{X} = (X, Y, Z)^\top$ is projected to a 2D point w at coordinates $\mathbf{x} = (x, y)^\top$ in the image plane through a perspective transformation:

$$\begin{aligned} x &= \frac{f\xi(\mathbf{R}_1 \cdot \mathbf{X} + D_X) + s\,(\mathbf{R}_2 \cdot \mathbf{X} + D_Y)}{\mathbf{R}_3 \cdot \mathbf{X} + D_z} + o_x \\ y &= \frac{f(\mathbf{R}_2 \cdot \mathbf{X} + D_y)}{\mathbf{R}_3 \cdot \mathbf{X} + D_z} + o_y \end{aligned} \tag{1}$$

where f is the focal length, ξ the aspect ratio, s the skew, (o_x, o_y) the principal point, $\mathbf{R}$ the rotation matrix, and $\mathbf{D}$ the translation matrix of the camera. By rearranging terms in Eq. 1, we obtain a parametric equation that describes the projection line from a 3D world point to a 2D image point:

$$\begin{aligned} X_0 &= (B_1 D_2 - D_1 B_2)/(A_1 B_2 - B_1 A_2) \\ Y_0 &= (D_1 A_2 - A_1 D_2)/(A_1 B_2 - B_1 A_2) \\ u &= (B_1 C_2 - C_1 B_2)\,/(A_1 B_2 - B_1 A_2) \\ v &= (C_1 A_2 - A_1 C_2)\,/(A_1 B_2 - B_1 A_2) \end{aligned} \tag{2}$$

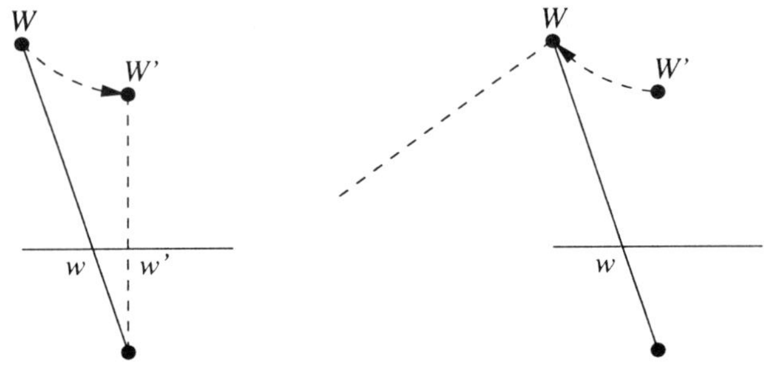

Figure 5. Ray Intersecting method. (a) World point W casts a projection line (solid line) onto the image point w on the image plane. As W rotates to its new position W', it casts a new projection line (dashed line) onto the image plane. (b) By rotating the new projection line backward, it now intersects the original line at the world point W.

where

$$\begin{aligned} A_1 &= f\xi R_{11} + sR_{21} + (o_x - x)R_{31} \\ B_1 &= f\xi R_{12} + sR_{22} + (o_x - x)R_{32} \\ C_1 &= f\xi R_{13} + sR_{23} + (o_x - x)R_{33} \\ D_1 &= f\xi D_X + sD_Y + (o_x - x)D_Z \\ A_2 &= fR_{21} + (o_y - y)R_{31} \\ B_2 &= fR_{22} + (o_y - y)R_{32} \\ C_2 &= fR_{23} + (o_y - y)R_{33} \\ D_2 &= fD_Y + (o_y - y)D_Z \,. \end{aligned} \tag{3}$$

Equation 2 can be written in the vector form

$$\mathbf{X} = \mathbf{X}_0 + t\mathbf{u} \tag{4}$$

where $\mathbf{X}_0 = (X_0, Y_0, 0)^\top$ and $\mathbf{u} = (u, v, 1)^\top$.

The world coordinate system is fixed onto the turntable such that the turntable rotates about the Y-axis. As the world point W is rotated about the Y-axis, it projects several lines

$$\mathbf{X} = \mathbf{X}_{0i} + t_i\mathbf{u}_i \tag{5}$$

onto the image plane at various rotation angles θ_i, where $\mathbf{X}_{0i} = (X_{0i}, Y_{0i}, 0)^\top$ and $\mathbf{u}_i = (u_i, v_i, 1)^\top$. By rotating the new projection lines backward about the Y-axis by the corresponding angles, the new world points would now coincide with the original world point W (Fig. 5).

The backward rotated projection lines are given by

$$\mathbf{X} = \mathbf{M}_i(\mathbf{X}_{0i} + t_i\mathbf{u}_i) \tag{6}$$

where

$$\mathbf{M}_i = \begin{bmatrix} \cos\theta_i & 0 & \sin\theta_i \\ 0 & 1 & 0 \\ -\sin\theta_i & 0 & \cos\theta_i \end{bmatrix} . \tag{7}$$

A closed-form solution of the 3D world point can now be obtained by solving for the intersection of any two backward rotated projection lines, for example, line 0 (Eq. 4) and line i (Eq. 6). For notational convenience, denote $\mathbf{X}'_0 = \mathbf{M}_i\mathbf{X}_{0i}$ and $\mathbf{u}' = \mathbf{M}_i\mathbf{u}_i$. Then, the 3D world point is given by Eq. 4 at the following t:

$$t = \frac{(X'_0 - X_0)v' + (Y_0 - Y'_0)u'}{uv' - u'v} \,. \tag{8}$$

In practice, due to the limited resolution of the camera's digitizer and the presence of noise, the backward rotated projection lines may not intersect at exactly one point. A good approximation of the intersection can be obtained by finding the point $\mathbf{Q}$ nearest to the lines. That is, find the $\mathbf{Q}$ that minimizes the total distance to the lines

$$D_T = \sum_i \left\| (\mathbf{Q} - \mathbf{X}_{0i}) - \frac{(\mathbf{Q} - \mathbf{X}_{0i}) \cdot \mathbf{u}_i}{\|\mathbf{u}_i\|^2} \mathbf{u}_i \right\| \tag{9}$$

This minimization can be performed efficiently and reliably given a good initial guess of the actual solution, which can be obtained from the intersection of any two projection lines using the closed-form solution (Eq. 8).

Note that, in contrast to most existing algorithms, RI recovers each 3D point independently of other 3D points. Consequently, if an object has mostly simple surfaces and a small number of complex, convoluted surfaces, RI's error in localizing the 3D points on complex surfaces does not affect its accuracy in recovering those on simple surfaces. Thus, RI is robust against localized acquisition inaccuracy.

4. Experiments and Results

Two sets of experiments were conducted. The first set of experiments measured the accuracy and robustness of the RI algorithm using both synthetic data and a real object. The second set of experiments applied the system to acquire the 3D models of three real objects: a calibration cube, a duck and a mouse.

4.1. Accuracy and Robustness Measurement

To measure the accuracy and robustness of the RI algorithm, twenty synthetic world points were randomly selected from a sphere of radius 200mm. These world points were projected onto the image plane through a perspective camera model to generate the synthetic 2D image points. The values of the camera parameters were set according to a real calibrated camera. In particular, the focal length was set at 8mm and the effective resolution was set at 100 pixels/mm instead of infinity. In contrast, published tests on synthetic data typically assume infinite camera resolution. The world points were rotated about the Y-axis through an

Table 1. Reconstruction error in the presence of noise in image pixel location. Noise is measured in pixels and mean error in mm.

noise	0	1	2	3	4	5
error	0.38	0.73	1.33	1.89	2.65	3.13

Table 2. Reconstruction error due to inaccuracy in rotation. Noise is measured in degrees of rotation and mean error in mm.

noise	0.0	0.1	0.2	0.5	1.0	2.0
error	0.38	0.33	0.35	0.47	0.73	1.31

angle of 90° in steps of 10° to generate the multiple views required for the reconstruction. Moreover, random noise was added to the coordinates of the 2D image points and to the rotation angles of the object.

In the test, both calibration and reconstruction errors were measured. Calibration error was computed as the mean distance from a recovered 3D point to the projection line (Eq. 4) through its corresponding 2D image point. The reconstruction error was calculated as the mean absolute error between a recovered 3D point and the corresponding actual 3D point. Tables 1 and 2 show that, in the absence of noise, RI has a mean error of 0.38mm, which is 0.19% of the size of the object. The test results also show that, despite the addition of noise, RI's reconstruction error does not increase significantly. Therefore, RI is a robust algorithm for recovering the 3D coordinates of feature points.

An experiment was also conducted to measure the accuracy of RI and Tsai's calibration algorithm in combination. In this test, the real calibration object (Fig. 3) was used. The object has a size of about $100 \times 100 \times 100$ mm^3. 3D coordinates of 72 corner points on the cube were measured manually. These coordinates were used to calibrate a real camera using Tsai's algorithm. Next, the cube was rotated and the images at different views were captured. 3D coordinates of the corner points were computed using the RI algorithm. This experiment was performed three times and the mean error was averaged over the three tests. Test results in Table 3 show that in recovering the 3D coordinates of the corner points, RI and Tsai's algorithm in combination have a mean error of about 1% of the object's size. This amount of error is about 6 times larger than that of RI alone under noise-free condition (compare with Table 1). The main source of the error is attributed to calibration error, which accounts for about 70% of the error. If calibration error is removed, then the reconstruction error amounts to 0.34% of the object's size, which is only 1.8 times that of RI under noise-free condition. This experiment again shows that RI is robust against noise in recovering 3D coordinates of real objects. More accurate calibration, such as precise measurement of the actual 3D points, would improve the overall reconstruction accuracy.

Table 3. Mean errors in calibrating and recovering the 3D coordinates of corner points on a real calibration object.

error source	error	object size	% object size
calibration	0.77	100	0.77%
calibration & reconstruction	1.11	100	1.11%
reconstruction (noise-free)	0.38	200	0.19%

4.2. Model Acquisition

Figure 3 shows an image of the calibration object. The corners of the squares on the object's surfaces are extracted and their 3D coordinates are recovered. Fig. 6 shows three perspective views of the squares acquired. As expected, the squares in the acquired model lie on flat surfaces.

Figure 7 shows two different views of a duck figurine. Figure 8 illustrates four perspective views of the acquired duck model in polygon mesh. These figures show that the general shape of the duck is correctly acquired. In particular, the duck's head, bill, neck, body, and tail are correctly acquired.

Figure 9 illustrates two views of a mouse figurine and Fig. 10 illustrates three perspective views of the acquired mouse model in polygon mesh. The general shape of the mouse is well acquired. The mouse's ears, body, forelimbs, and hindlimbs are apparent in the perspective views.

Most of the computation time was spent in tracking 2D features and recovering 3D coordinates of feature points. For the duck model, tracking of the 9921 feature points in 200 images took about 2.5 hours. For the mouse model, tracking of the 3848 feature points in 200 images took about 1 hour. The RI algorithm took about 1/10 the amount of time to compute.

5. Conclusions

3D model acquisition from images of existing scenes and objects is a promising geometric modeling approach. This reverse engineering approach, which can be fully automated, is complementary to forward approaches such as

Figure 6. Perspective views of the acquired cube's faces.

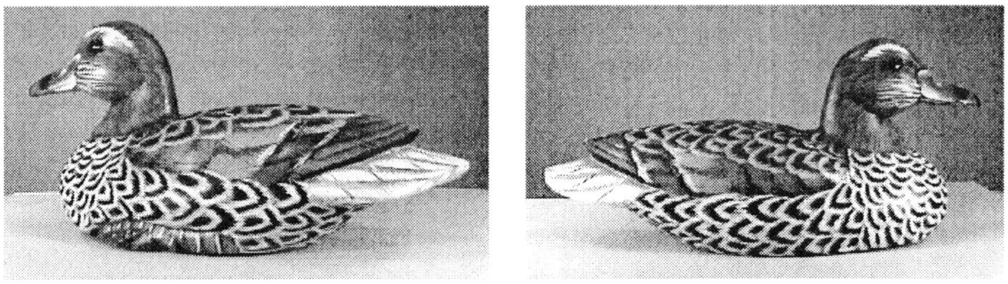

Figure 7. Two views of a duck figurine.

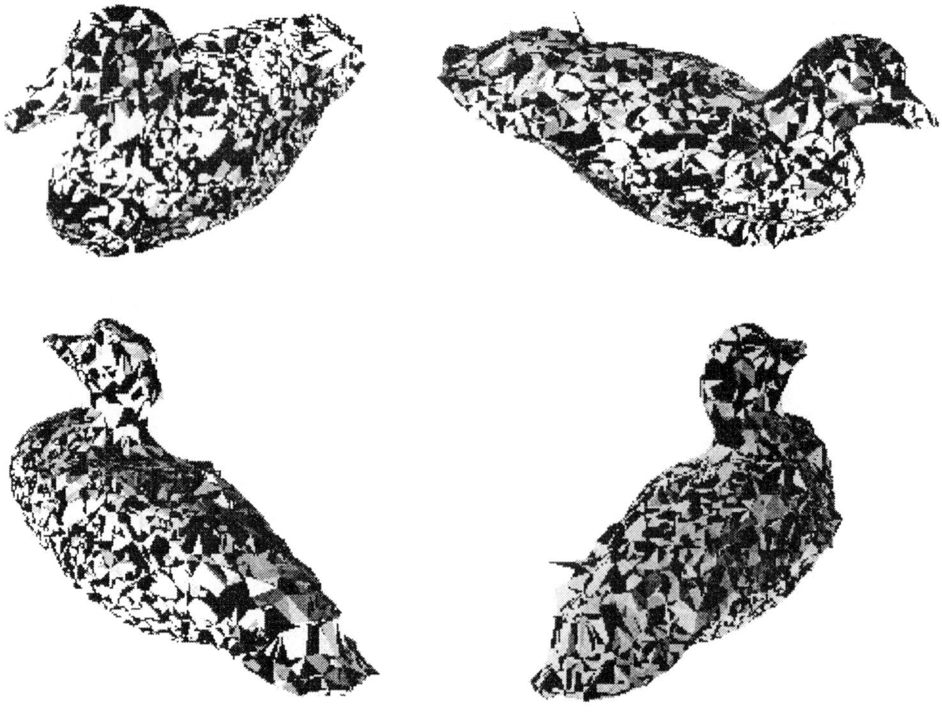

Figure 8. Four perspective views of the acquired model of the duck.

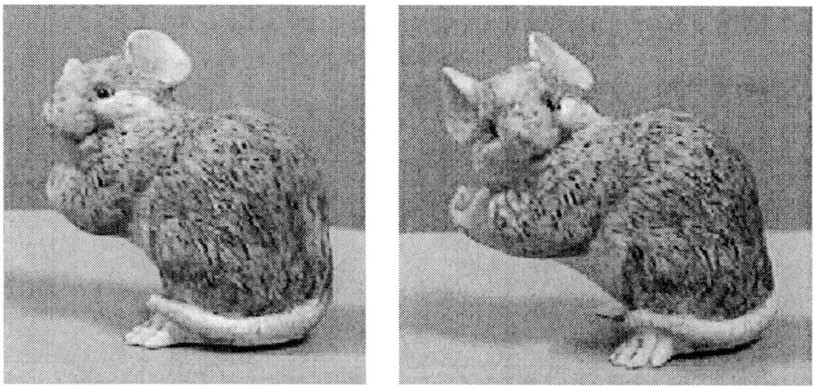

Figure 9. Two views of a mouse figurine.

(a) (b) (c)

Figure 10. Three perspective views of the acquired model of the mouse: (a) front, (b) front-top, (c) back.

CSG, B-rep, NURBS, and sweeping. With the availability of inexpensive equipments such as digital camera and computer-controlled turntable, 3D scanning technology is becoming more affordable to home users. In the near future, we expect automated 3D scanners to be widely available to home users, just like 2D image scanners.

In this paper, a rapid 3D model acquisition framework is presented. The framework consists of an image capturing platform and associated algorithms of computer vision and graphics. Experimental results show that it is appropriate for acquiring the geometric models of small objects. Moreover, the algorithm for recovering 3D feature points from 2D image points is accurate and robust against noise in pixel location and object rotation. Further research is being performed to develop better polygonization algorithms for reconstructing complex surfaces such as convoluted surfaces, discontinuous surfaces, and surfaces on a thin volume.

Acknowledgment

This research is supported by NUS Academic Research Fund RP3982704. We also like to thank Indriyati Atmosukarto for her help in touching up the displayed images.

References

[1] N. Amenta, M. Bern, and M. Kamvysselis. A new Voronoi-based surface reconstruction algorithm. In *Proc. ACM SIGGRAPH 98*, pages 415–421, 1998.

[2] P. R. Beaudet. Rotational invariant image operators. In *Proc. 4th Int. Conf. Pattern Recognition*, pages 579–583, 1978.

[3] V. Blanz and T. Vetter. A morphable model for the synthesis of 3D faces. In *Proc. ACM SIGGRAPH 99*, pages 187-194, 1999.

[4] B. Curless and M. Levoy. A volumetric method for building complex models from range images. In *Proc. ACM SIGGRAPH 96*, pages 303-312, 1996.

[5] S. Schirra, R. Veltkamp, and M. Yvinec eds. *CGAL release 2.0 Reference Manual.* http://www.cs.uu.nl/CGAL/, 1999.

[6] P. E. Debevec, C. J. Taylor, and J. Malik. Modeling and rendering architecture from photographs: A hybrid geometry- and image-based approach. In *Proc. of ACM SIGGRAPH 96*, pages 11–20, 1996.

[7] H. Edelsbrunner, D. G. Kirkpatrick, and and R. Seidel. On the shape of a set of points in the plane. In *IEEE Trans. on Information Theory*, 29, pages 551–559, 1983.

[8] H. Edelsbrunner and E. P. Mucke. Three-dimensional Alpha Shapes. In *ACM Trans. on Graphics*, 13, pages 43–72, 1994.

[9] A. W. Fitzgibbon, G. Cross, and A. Zisserman. Automatic 3D model construction from turn-table sequences. In L. Van Gool and R. Koch, eds., *Structure and Motion from Multiple Images in Large-Scale Environments*. Springer, 1998.

[10] B. Guenter, C. Grimm, D. Wood, H. Malvar, and F. Pighin. Making faces. In *Proc. ACM SIGGRAPH 98*, pages 55-66, 1998.

[11] M. A. Halstead, B. A. Barsky, S. A. Klein, and R. B. Mandell. Reconstructing curved surfaces from specular reflection patterns using spline surface fitting of normals. In *Proc. ACM SIGGRAPH 96*, pages 335-342, 1996.

[12] H. Hoppe, T. DeRose, T. Duchamp, J. McDonald, and W. Stuetzle. Surface reconstruction from unorganized points. In *Proc. ACM SIGGRAPH 92*, pages 71–78, 1992.

[13] S. B. Kang. Quasi-Euclidean recovery from unknown but complete orbital motion. Technical Report CRL 97/10, Cambridge Research Laboratory, 1997.

[14] Y. Matsumoto, H. Terasaki, K. Sugimoto, and T. Arakawa. A portable three-dimensional digitizer. In *Proc. of Int. Conf. on Recent Advances in 3-D Digital Imaging and Modeling*, pages 197–204, 1997.

[15] D. Mehren and V. Rodehorst. Gestaltsanalyse komplexer Objekte bei kontrollierter Bewegung, Diplomarbeit, TU Berlin, Fachbereich Informatik, 1994.

[16] ModelMaker. 3D Scanners Ltd, UK. http://www.3dscanners.com.

[17] ReCaSs. Fraunhofer Institute for Computer Graphics (IGD), Germany. http://www.igd.fhg.de/projects/index.html.

[18] REPLICA. 3D Scanners Ltd, UK. http://www.3dscanners.com.

[19] H. Saito, S. Baba, M. Kimura, S. Vedula, and T. Kanade. In *Proc. 2nd Int. Conf. on 3-D Digital Imaging and Modeling*, 1999.

[20] Synchronized Laser Scanners. National Research Council, Canada. http://www.vit.iit.nrc.ca/.

[21] R. Szeliski and S. B. Kang. Recovering 3D shape and motion from image streams using non-linear least squares. *Journal of Visual Communication and Image Representation*, 5(1):10–28, 1994.

[22] S. Teller. Toward Urban Model Acquisition from Geo-Located Images. In *Proc. of Pacific Graphics'98*, pages 45–52, 1998.

[23] C. Tomasi and T. Kanade. Shape and motion from image streams under orthography: a factorization method. *IJCV*, pages 137–154, 9(2), 1992.

[24] R. Y. Tsai. A versatile camera calibration technique for high-accuracy 3D machine vision metrology using off-the-shelf TV cameras and lenses. *IEEE Trans. on Pattern Analysis and Machine Intelligence*, 3(4):323–344, 1987.

Session 2

Analysis and Applications of Geometric Models

Geometric Processing for Analysis

Armstrong, CG; McKeag, RM; Ou, H and Price, MA
The Queen's University of Belfast
Ashby Building
Stranmillis Road
Belfast BT9 5AH
N Ireland

e-mail: c.armstrong@qub.ac.uk

Abstract

Processing geometric models for finite element analysis is a major factor in the time required to develop and optimise a design. Efficient analysis may require simplification of the geometric model by dimensional reduction (replacing thin sheets or slender bars of material with equivalent line or surface elements) or detail suppression (the removal of small features below the scale of interest in the analysis). There are many situations where detailed local models of full dimension could be combined with global models containing reduced dimensional elements, or in which the abstract models used in preliminary design could be used to generate an initial version of the 3D solid model. Furthermore, the most appropriate representation of an object is dependent on the analysis type. These requirements provide significant challenges for future generations of CAE software in feature recognition and the modelling and representation of geometric objects.

1. Introduction

This paper aims to survey some of the challenges involved in processing geometric design models for the analysis of phenomena such as stress, vibration or heat flow.

Processing of geometric models for analysis is currently a major impediment to the rapid development and optimisation of product designs. The development of an appropriate analysis model can take so long that it is difficult to keep design and analysis in sync, even though the real benefit of analysis occurs when its results are available to guide the evolution of the design.

Why is geometric processing for analysis necessary? Preparing an analysis model typically involves *Mesh Generation*: the subdivision of an arbitrary geometry into a collection of finite elements of simple shape. Assuming the form of the solution within each element allows a system of equations to be assembled and solved in terms of variables at the nodes to which the elements are connected, giving a piecewise approximation to the global behaviour. Note that representing complex features of the geometry or analysis solution usually requires that the element size should vary over the domain. Mesh generation with simplex elements (2D triangles, 3D tetrahedra) is at an advanced state of development [11, 25], but much less progress has been made, for example, in mesh generation for 3D hexahedra [3, 17, 18, 24].

In order to make an efficient analysis, more radical transformations of the geometry may be necessary. Specialised elements for the analysis of thin sheets of material or long slender parts have been used from the earliest days of computer aided analysis. In a thin sheet, these elements replace the 3D continuum with an equivalent surface plus a thickness attribute. A long slender part is represented as an equivalent line with a cross-section attribute. This process is usually known as *Dimensional Reduction*.

Another frequent problem is that the geometric model contains features which are below the scale of interest in the analysis. This problem of *Detail Suppression* is of course dependent on the accuracy and type of analysis required.

2. Analysis Modelling Requirements

The first requirement in analysis modelling is usually to acquire a geometric representation of the domain to be analysed. Given the differences in internal representation, algorithms and tolerances among different modellers, the

0-7695-0562-7/00 $10.00

transfer of model data between systems is an extremely difficult problem which is the focus of much commercial activity [10, 23]. An alternative approach is to define an API through which applications such as analysis interrogate [13, 20] and modify the geometric model. The Djinn API [1], proposed as a mechanism for researchers to access modelling technology, defines an API which is independent of the underlying representation (B-rep, CSG, facetted, octree etc.) and provides a coherent interface to multi-dimensional, cellular, non-manifold models.

Whilst no complete implementations of Djinn exist, the ideas have been very influential in the development of analysis modelling technology at QUB. Analysis models can range from simple 1D models of beam behaviour or fluid flow in pipes to complex 3D geometry with different regions of different materials. In Djinn these would be represented as separate cells in the material. Using the Djinn approach, analysis models can contain cracks and features of different dimension, such as 1D steel reinforcing rods in a 3D concrete matrix. Djinn introduced the notion of partitioning operators, where one object is used as a 'pastry cutter' to imprint its structure on another object. This has many applications in analysis modelling - an area of a face over which a pressure is to be applied can be isolated in this way. Some "structured" mesh generation techniques demand that the geometry be partitioned into sub-domains of simple shape which are then filled with either mesh templates or regular arrays of elements, [14]. There are corresponding departitioning operations, in which a cell that separates two touching cells is removed from the model and the point-sets of the touching cells are united. These serve as a useful description for the operations needed for detail suppression, described below.

3. Medial Axis Transform

The Medial Axis of a planar domain may be regarded as the locus of the centre of all discs of maximal diameter which are contained within the object, Figure 1(a). In conjunction with the radius function of the inscribed disc, which describes the local distance to the boundary curves, it is a complete description of the object geometry. It was proposed by Blum [5] for the characterisation of biological shape, but has since attracted interest in a number of areas including mesh generation [18, 22].

This alternative representation provides a lower-dimensional skeleton of the object. The touching points of the inscribed disc centred on the Medial Axis identify parts of the object boundary which are in geometric proximity even if they are not topologically adjacent, Figure 1(a). It facilitates the identification of relative sizes and aspect ratios for identifying features of structural significance such as long slender parts, thin sheets and small features embedded in a large amount of material.

The 3D Medial Surface is the locus of the centre of all maximal inscribed spheres contained within the object.

Whilst the computation of exact medial geometry and topology is a challenging task, a useful approximation to the MAT can be obtained from a Delaunay triangulation of points distributed over the object boundary, Figure 1(b). The Medial Axis of the convex hull of the object can be calculated in the same way.

4. Detail Suppression

A generic requirement in finite element modelling is to create an analysis model that is significantly simpler that the original geometric model supplied. The geometric model may contain small features below the scale of interest in the analysis which should be suppressed during the process of mesh generation. If the geometric model was constructed in such a way that these small features are identifiable and removable, then obviously they can be suppressed at the level of the feature hierarchy and mesh generation can be carried out directly on a modified geometric model. However different analyses may have different requirements for feature suppression and it is unlikely that all the appropriate idealisations can be identified *a priori* during model construction. Unintentional small features of size down to the level of the modeller precision can be introduced during construction by nearly degenerate geometry.

At a high level, the operations to remove these features can be viewed in terms of the Djinn partitioning and departitioning operations. In Figure 2(a), departitioning has been used to merge or unite the pointsets of the faces *abed* and *bcfe*. Edges *ab* and *bc* have been merged, as have edges *de* and *ef*. The edge meshing algorithm knows that it is not required to place an element node at the original model vertices *e* and *b*, whilst the surface meshing algorithm knows that it is not required to align element edges with the original model edge *be*. Similarly, in Figure 2(b), the edge blend has been partitioned into two and then each half merged with the neighbouring face by departitioning. The effect on the topology of the mesh generated is an implicit collapse or dimensional reduction of the blend face, since the target element size given to the meshing algorithm is much greater than the size of the blend in the merged face. This approach to guiding the mesh generation algorithm about the topology of the mesh required has been called "logical modelling" [2] or "virtual topology" [21]. No geometric modification to the original model is needed, and the virtual topology is primarily used to define the mesh topology required. The mesh geometry may be created by projecting an initial nodal position onto the original geometric model. It is also possible to construct the required mesh by meshing the geometry including the details and then removing features such as short edges at the mesh level [7]. Marcum [15] has recently described an approach to computing a single parametric surface approximating a collection of original surfaces, which obviously simplifies the subsequent meshing. Each of these approaches has merits and further research is clearly required, though it is suggested that changes to the original geometric model should be avoided, since these may have unpredictable consequences for other applications.

The automatic identification of small features is perhaps the most difficult problem in detail suppression. It will be argued below that for linear elasticity problems, geometric measures based on the Medial Axis Transform can be used for *a priori* identification of small features. However for other problems purely geometric measures are unlikely to be sufficient.

The medial axis radius provides a measure to the local thickness of material at its touching points on the boundary. If this thickness is large compared to the size of the geometric feature on the boundary, then that feature will have only a localised effect on the localised behaviour of the structure. The most complex boundary structure which can be removed in a single step is an inner loop of edges. From the medial axis or the equivalent Delaunay triangulation, the convex hull of the hole (associated with the Short-cut edge in Figure 3) can be found. A loop aspect ratio, defined as the total length around the hole divided by the minimum disc diameter touching the hole, can be used to identify those loops which are embedded in a substantial amount of material. Recursive application of this procedure allows the cluster of small holes in the centre of Figure 4 to be identified as a single feature and removed.

Similar techniques for identifying and computing the relative size of features such as protrusions and notches can be developed. In Figure 5 the boundary of the notch feature can be identified from the touching points of the exterior medial axis *v2* and *v3*. Points *a* and *b* delimit where the medial axis touches any of the boundary elements defining the notch, so between these limits the depth of the notch (defined as the distance from the notch root to the convex hull between *v2* and *v3*) should be small compared to the minimum size of the discs touching between *a* and *b*. Using these ideas, it is possible to identify notches within notches, Figure 5(c), in the presence of compound features such as Figure 5(d).

At a later stage in the design process, it may be necessary to investigate the behaviour of the component in the vicinity of details. There are several approaches to this:

- ***total re-analysis of the domain including the detail:*** superficially this is an attractive option, but as larger and larger systems are analysed the need for intelligent identification and treatment of details becomes even more pressing. A given increase in computing power can be more productively employed in evaluating more design alternatives or tackling more challenging problems such as simulating manufacturing processes;
- ***submodelling:*** where a local model of the detail is analysed in isolation, Figure 6, using the results from the global model to impose appropriate displacements on the boundary of the submodel;
- ***substructuring:*** where the equations describing the behaviour of the local model of Figure 6 are condensed out so that the global model perceives only the boundary nodes of the feature - it behaves like a single "superelement".

Substructuring has the advantage over submodelling that the true stiffness of the detail is transmitted to the global model. If the stiffness of a submodel is significantly different from that of the coarse global model of the same feature, the analysis of global behaviour may be incorrect.

5. Dimensional Reduction

Dimensional Reduction is the process whereby the behaviour of the continuum in an analysis model is represented by special-purpose finite elements which have a lower manifold dimension. Figure 1(c) shows the stress contours in a 2D continuum when a bending moment is applied at one end. It can be seen that the stresses vary linearly over the cross-section everywhere except around the change in section. In a beam element, it is **assumed** that stress varies linearly over the cross-section, so it is only necessary to discover by analysis how the stresses (arising from bending moment, shear force and axial force) vary along the **length** of the beam. This variation is represented as a variation along a 1D line by beam elements, Figure 1(d). Similar considerations apply to other phenomena, such as laminar flow of Newtonian fluid.

This special behaviour of elastic material in long slender members (beam theory) and thin sheets (shell theory for curved sheets, plate theory for flat sheets) has been extensively developed in the classical literature and provides valuable insights into structural behaviour. It can be rigorously justified using Perturbation Methods, with which it can be shown that introducing a small parameter representing the aspect ratio into the non-dimensionalised governing equations of elasticity provides the required specialised equations for beams, plates or shells [12]. The approach is illustrated for the simplest case of 2D beam theory in McCune *et al.* [16]. It is important to note that these simplifications only apply if both the geometry and the loading vary slowly over the beam length or plate/shell surface. Round the points of load application and the change in section in Figure 1(c), complex stress distributions are evident. Similar effects are seen around discontinuities in material properties.

The Medial Axis Transform has obvious applications in identifying features of interest for Dimensional Reduction [9]. In 2D, the length of the region associated with a given branch of the Medial Axis can be compared with the behaviour of the radius function along the branch. If the length of the region is large compared to the maximum inscribed circle radius, then it is safe to assume that beam theory can be used, at least away from the ends of the slender area, Figure 1(c). Various definitions of aspect ratio are possible, Figure 7. The length of the region could be associated with the medial axis length l_2, or the distances l_1 and l_3 traversed by the touching point of the inscribed disc as it moves along the medial axis branch. A conservative estimate of aspect ratio when looking for slender areas of large aspect ratio is l_{min}/r_{max}, which correctly identifies features like the ends of beams as not "slender". Figure 8 shows regions identified by the MAT as having an aspect ratio greater than the critical value shown in the diagram.

Any model containing long slender regions or thin sheets will always include some discontinuities in geometry, loading or material where the simplified theories of beam, plate or shell are not applicable. These do not invalidate a model based on using dimensionally reduced elements - if the aim is to estimate the overall deflection of a structure, then inaccurate estimates of the stress and deflection in short, stiff, chunky areas of the structure will have relatively little effect. Analyses to determine the natural frequency of vibration use reduced-dimensional elements almost exclusively, since very large (~ 2 orders of magnitude) reduction in computing times to find the first few eigenvalues of the system are possible. However dimensionally-reduced elements give locally accurate results only in slender regions - they will not give accurate estimates of the stress concentrations arising due to changes in section of loading. If the discontinuities form a substantial part of the whole domain, then of course the achievable global accuracy is reduced.

The treatment of discontinuities in geometry, loading or material in a reduced-dimensional analysis is currently based largely on heuristics. Figure 1 shows three slender regions meeting at a joint. There is no point which represents the intersection of the mid-line of the three slender arms of the body. Whilst the Medial Axis of the domain is well-defined, Figure 1(a), it contains features corresponding to the inscribed disc rolling around the concavities which are of small aspect ratio and which cannot be accurately represented by beam elements. If beam elements are used, their length must be much greater than the largest dimension of the cross-section. If plate or shell elements are used, their edges must be much longer than the sheet thickness. For areas where the aspect ratio is not large, such as around the connection in Figure 1(c), the character of the elasticity solution will be different from that assumed in the reduced-dimension elements.

In 2D, several treatments of these areas of small aspect ratio are possible:

- they can be treated as rigid point masses. This ensures that a vibration analysis, or a stress analysis in which self-weight is significant, will have the correct mass.
- adjacent 1D regions can be extrapolated to their point of closest approach, Figure 1(d). This gives a better estimate of the stiffness, and a point mass can be associated with the resulting node so that the mass and moment of inertia are correctly evaluated.
- as a mixed dimensional model, where the vicinity of the discontinuity is modelled in full, with the slender parts only modelled by beams. In the next section it will be argued that this approach can produce answers of guaranteed accuracy and is not necessarily much more expensive.

A number of authors have questioned the use of the MAT for feature identification because small perturbations to the geometry of the domain can cause large changes to the topology and geometry of the Medial Axis. For the purposes of Dimensional Reduction, this deficiency is not important since dimensionally reduced elements are only truly applicable where the aspect ratio is large and the region is clearly slender. If there are short segments of the medial axis which might collapse under a small perturbation to the geometry they should not be represented in the reduced-dimensional analysis anyway.

Limited functionality for creating dimensionally-reduced sheets from thin sections in solids exists in some current CAE modellers [19], but the general problem of identifying thin sections between surfaces of arbitrary geometry has not been solved. The proper analysis of chunky connections between thin sheets has not been explored in any depth. A further unsolved problem is the automatic identification of long slender sections in a general 3D solid, though some preliminary ideas have been investigated [8].

It is worth noting that others [6] are investigating the use of highly anisotropic meshes of full dimension to analyse object with slender parts. This is an attractive option, since it builds on meshing technology developed for the solution of problems where highly oriented and stretched meshes are required and avoids the use of Dimensional Reduction altogether. Potential problems include some increase in computing cost, numerical stability and difficulties with generating slender elements in thin, curved sheets. For the moment, a great deal of proven technology and useful experience exists for reduced-dimensional elements which practitioners are reluctant to abandon.

6. Mixed Dimensional Modelling

In the model of Figure 1(c), the solution to the governing equations of elasticity in the vicinity of the connection between the arms is fundamentally two dimensional - there is no general basis for deriving a reasonable estimate of the stresses from the corresponding 1D beam model. An obvious solution therefore is to use 1D elements for the long slender arms but 2D elements for the connection. The difficulty with this approach is that the nodal deformations in the elements of different dimension need to be connected at the interface between dimensions. Note that all the nodes on the 2D side of the interface need to deform in a manner compatible with the displacement and rotation of the node on the 1D side of the interface. Traditional techniques for coupling together elements of different dimension usually introduced spurious stresses at the interface, which negated the benefit of making the idealisation.

A solution to this problem has recently been proposed [16]. Whilst this is not the right forum to discuss the technical details, the basic idea is to analyse first how the stresses vary over cross-section of a body which is known to be long and slender in response to unit force or moment and then equate the work done on both sides of the interface between dimensions. This ensures that the stresses on the 2D side of the interface match (at least in a least squares sense) those implied by the reduced dimension assumption.

These ideas generalise into coupling plates to solids (2D-3D), beams to plates (1D-2D) or even direct beam to solid connections (1D-3D), Figure 9. Note that there are no disturbances to the stresses around the interface. Whilst the coupling procedures can be generated automatically, selection of the slender regions and appropriate partitions in general 3D solids still requires manual intervention.

Figure 10 shows an offshore structure where the partitioning is simple and obvious but has substantial performance benefits. The simplest representation of the structure would be as a collection of 1D beam elements as shown in Figure 10(a). These give an accurate description of the behaviour everywhere except in the vicinity of the connections where several slender tubes meet. Using a mixed dimensional model, Figure 10(b), the 3D variation in stress round the stress concentrations in the connection are captured using 3D elements, which are then coupled to the 1D beams at a distance greater than one tube diameter away from the stress concentration. Using the substructuring techniques described in the section on Detail Suppression above, the behaviour of this complex 3D connection can be abstracted in terms of the displacements and rotations of the four nodes connected to the beams. The entire 3D model of the connection is reduced to an equivalent spring with four point connections, Figure 10(c). Adding these spring elements to the beam model causes a negligible increase in the cost and complexity, but the structural stiffness of the connection is properly represented. Furthermore, once the global model has been analysed, the local stresses and deformations in the 3D model of the connection can be evaluated from the known displacements and rotations where the local model is connected to the global.

Similar abstraction of the failure behaviour of substructures may be useful. In the substructure of Figure 11, only two loads F_Y and F_T are significant. In this case, the failure of the material at a given nodal point in the 3D substructure occurs when

$$\alpha_1 F_Y^2 + \alpha_2 F_Y F_T + \alpha_3 F_T^2 = \sigma_Y^2$$

where σ_Y is the yield stress of the material and α_1, α_2, α_3 are parameters which can be determined for each finite element node by three initial analyses of the substructure. Determining whether the yielding load in the substructure has been exceeded at any node amounts to determining whether the point (F_T, F_Y) lies outside the domain formed by the intersection of the ellipses for all nodal points, Figure 12, a problem which will be familiar to CSG modellers. In the general case, the structure might contain $O(10^5)$ nodes (ellipses) and $O(10^2)$ loads (dimensions), so efficient methods of eliminating redundant primitives would be needed. In Figure 10 failure always occurs first in one of only six nodes, so only six ellipses are necessary.

The representation of mixed dimensional analysis models is an interesting issue. The obvious solution is to have a mixed dimensional geometric model. An alternative approach which has some advantages is a partitioned model, Figure 10(d), where different cells have attributes which tell the mesh generation application what dimension of element to use and the orientation of the reduced dimension elements, e.g. which pair of edges or faces the elements should lie between. This is particularly simple if the model has been created by sweeping a cross-section profile along an axis to form a beam or by thickening a sheet, see the section on Dimensional Addition below. The internal partitions between domains of different dimension define the cross-sections at which coupling must be generated. Note that in the mixed-dimensional analysis model of the offshore structure the 1D beam elements and the 3D solid meshes do not touch, since the beam elements are positioned on the tube centreline.

7. Dimensional Addition

In some industries, preliminary design evaluation is done on highly abstract models. Typically, 1D beam elements and 2D shell elements are employed to evaluate the global behaviour of the structure and to investigate different layouts and concepts. Figure 13(a) shows a beam model for the chassis of a truck trailer. At this level of detail it is easy to change the number of transverse elements, which one of the available standard sections they will be made from, whether different sections need to be used for the longitudinal members in different areas, etc. Most importantly, the analysis model is small enough that rapid solutions are possible, many design concepts can be explored and optimisation of the design with respect to, for example, the section properties is feasible.

This abstract layout is then used as an informal guide to the construction of detailed 3D models of the individual components. In aerospace engineering, there is usually a requirement that the final component designs be appropriately idealised so that modified properties can be put back into the abstract global model, which is then used to check that the load paths through the structure have not changed. If the loading on some components has changed, redesign at a very late stage in product development may be necessary. There is an obvious need to maintain alternative representations of the geometry in the product data model and to ensure consistency between representations.

Whilst sweep operations for creating geometry have been used since the earliest modelling systems, the paradigm of creating solids by *Dimensional Addition* from the abstract beam/shell models used in preliminary design is not well supported in current CAE systems, nor extensively employed in industrial practice. The 1D beams in Figure 13(a) have associated properties such as cross-sectional area, moments of inertia, offsets of the centroid, all of which are computed from the cross-section; the analysis model is implicitly the sweep of the cross-section along the beam axis. Figure 13(b) shows the result of applying sweeps of the cross-sectional profiles to the 1D elements. Taper or draft angle would imply that successive elements have different average properties.

This simple procedure allows the user of the beam element model to confirm that the sections are oriented correctly. It could also be used for direct visualisation of the stress results, which cannot easily be displayed on the line model. It provides the basis for the solid model to be used for subsequent detailed design. Detailed design often requires detailing joints and connections - in the example of Figure 13 there are a number of approaches to designing the joint between the longitudinal and transverse beams based on welding, mechanical fasteners, etc. This detailed design is facilitated if there is a separate cell in the solid representing the connection and its immediate neighbourhood.

Partitioning the solid into separate cells representing each slender member and each connection allows design changes in a given part to be reflected into appropriate changes in the behaviour of that cell in the global model. Thus the global effect of local changes to the design can be assessed. Depending on the analysis requirements, the joints where the stress distribution is locally complex could be treated in the analysis as either a simple node, or as a local 3D substructure in a mixed dimensional global model, Figure 10(c).

"Fleshing Skeletons" as a technique for geometric construction and modification has recently been explored by Blanding [4].

8. Conclusions

Processing geometric models for analysis poses significant challenges to both the geometric modelling and analysis communities. Geometric models are often too detailed or expensive to analyse and it is very difficult to keep models at different levels of abstraction coherent.

Dimensional reduction and detail suppression are useful paradigms for model abstraction. Operations to construct dimensionally reduced meshes which ignore small features are feasible.

Mixed dimensional modelling allows the most efficient element type to be used in each part of the analysis domain.

The most difficult unresolved problems relate to the automatic identification of features which are significant in an analysis context. For linear elastic stress analysis, measures of proximity and aspect ratio are useful. These can most naturally be found using the Medial Axis Transform.

Dimensional addition is a potentially useful paradigm for design by structural feature in fields such as transport and aerospace, where preliminary design is done on highly abstract models.

9. Acknowledgements

The authors would like to thank the current and former members of the Finite Element Modelling Group at Queen's University for their contributions to this work.

10. References

[1] Armstrong, C.G., Bowyer, A., Cameron, S., Comey, J., Jared, G.E.M., Martin, R.R., Middleditch, A.E., Salmon, J., Sabin, M.A. and Woodwark, J.R. "Djinn, a Geometric Interface for Solid Modelling", Information Geometers Ltd, to appear, 2000.

[2] Armstrong, C.G., Robinson, D.J., McKeag, R.M., Li, T.S., Bridgett, S.J., Donaghy, R.J. and McGleenan, C.A. "Medials for Meshing and More", *Proc 4th Int Meshing Roundtable*, pp. 277-288, 1995.

[3] Blacker, T.D. "The Cooper Tool", *Proc 5th Int Meshing Roundtable*, pp. 13-29, 1996.

[4] Blanding, R., Brooking, C., Ganter, M. and Storti, D. "A Skeletal-based Solid Editor", *Proc 5th Symposium on Solid Modelling and Applications*, W.F. Bronsvoort and D.C. Anderson (eds), ACM Press, pp. 141-150, 1999.

[5] Blum, H. "A Transformation for Extracting New Descriptions of Shape", *Models for the Perception of Speech and Visual Form*, W. Wathen-Dunn (ed), MIT Press, pp. 362-380, 1967.

[6] Coupez, T. "Adaptive Meshing for Forming Processes", *Numerical Grid Generation in Computational Field Simulation*, M. Cross, B.K. Soni, J.F. Thompson, J. Hauser and P.R. Eiseman (eds), Int Soc Grid Generation, pp. 3-19, 1998.

[7] Dey, S., Shephard, M.S. and Georges, M.K. "Elimination of the Adverse Effects of Small Model Features by Local Modifications of Automatically Generated Meshes", *Eng with Computers*, Vol 13, Issue 3, pp. 134-152, 1997.

[8] Donaghy, R.J. "Dimensional Reduction of Stress Analysis Models", *PhD Thesis*, The Queen's University of Belfast, 1998.

[9] Donaghy; R.J., Armstrong, C.G. and Price, M.A. "Dimensional Reduction of Surface Models for Analysis", *Engineering with Computers*, accepted for publication, 1999.

[10] FEGS Ltd, http://www.fegs.co.uk.

[11] George, P.L. and Borouchaki, H. *Delaunay Triangulation and Meshing: Application to Finite Element*, HERMES, Paris 1998.

[12] Goodier, J.N. "On the Problems of the Beam and Plate in the Theory of Elasticity", *Trans Royal Society of Canada*, Vol 32, Sect III, pp. 65-88, 1938.

[13] Haimes, R. and Follen, G.J. "Computer Analysis Programming Interface", *Numerical Grid Generation in Computational Field Simulations*, pp. 663-672, 1998.

[14] Li, T.S., Armstrong, C.G. and McKeag, R.M. "Quad Mesh Generation for K-Sided Faces and Hex Mesh Generation for Trivalent Polyhedra", *Finite Elements in Analysis and Design*, Vol 26, pp. 279-301, 1997.

[15] Marcum, D.L. and Gaither, J.A. "Unstructured Surface Grid Generation using Global Mapping and Physical Space Approximation", *Proc 8th Int Meshing Roundtable*, pp. 397-406, 1999.

[16] McCune, R.W., Armstrong, C.G. and Robinson, D.J. "Mixed Dimensional Coupling in Finite Element Models", *Int J Numerical Methods in Engineering*, to appear, 1999.

[17] Owen, S.J. "Constrained Triangulation: Application to Hex-Dominant Mesh Generation", *Proc 8th Int Meshing Roundtable*, pp. 31-41, 1999.

[18] Price, M.A. and Armstrong, C.G. "Hexahedral Mesh Generation by Medial Surface Subdivision: Part II - Solids with Flat and Concave Edges", *Int J Numerical Methods in Engineering*, Vol 40, pp. 111-136, 1997.

[19] Rezayat, M. "Midsurface Abstraction from 3D Solid Models: General Theory and Applications", *Computer-Aided Design*, Vol 28, pp. 905-915, 1996.

[20] Shah, J.J., Dedhia, H., Pherwani, V. and Solkhan, S. "Dynamic Interfacing of Applications to Geometric Modelling Services via Modeler Neutral Protocol", *Computer-Aided Design*, Vol 29, pp. 811-824, 1997.

[21] Sheffer, A., Blacker, T.D., Clements, J. and Bercovier, M. "Virtual Topology Construction and Application", *Geometric Modelling Theory and Practice*, Springer-Verlag, pp. 247-259, 1997.

[22] Sheffer, A., Etzion, M., Rappoport, A. and Bercovier, M. "Hexahedral Mesh Generation using the Embedded Voronoi Graph", *Proc 7th Int Meshing Roundtable*, pp. 347-364, 1998.

[23] SPATIAL Technology Ltd, http://www.spatial.com and http://3dmodelserver.com/.

[24] Tautges, T., Blacker, T. and Mitchell, S. "The Whisker Weaving Algorithm: A Connectivity-based Method for Constructing All-hexahedral Finite Element Meshes", *Int J Numerical Methods in Engineering*, Vol 39, pp. 3327-3349, 1996.

[25] Thompson, J.F., Soni, B.K. and Weatherill, N.P. (eds). *Handbook of Grid Generation*, CRC Press, 1998.

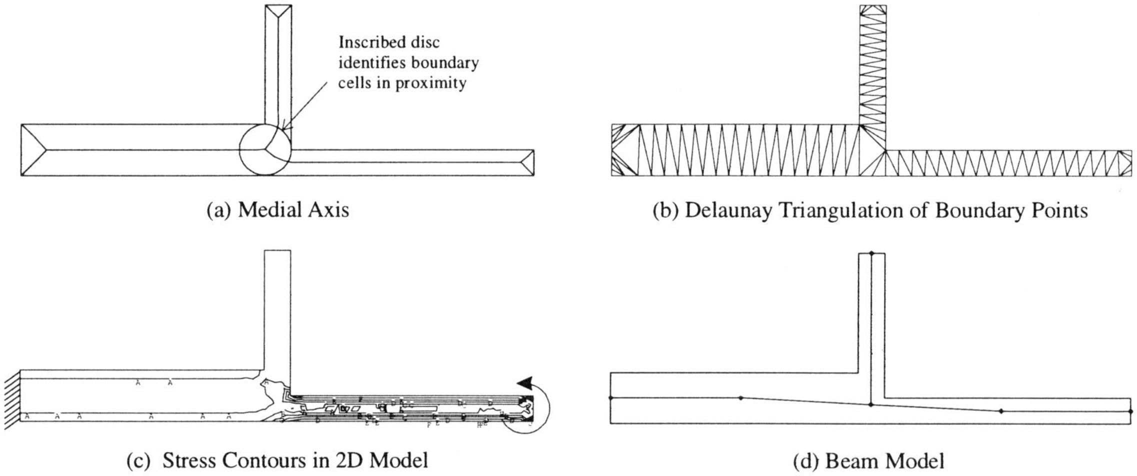

Figure 1. Various representations of a structure containing slender parts.

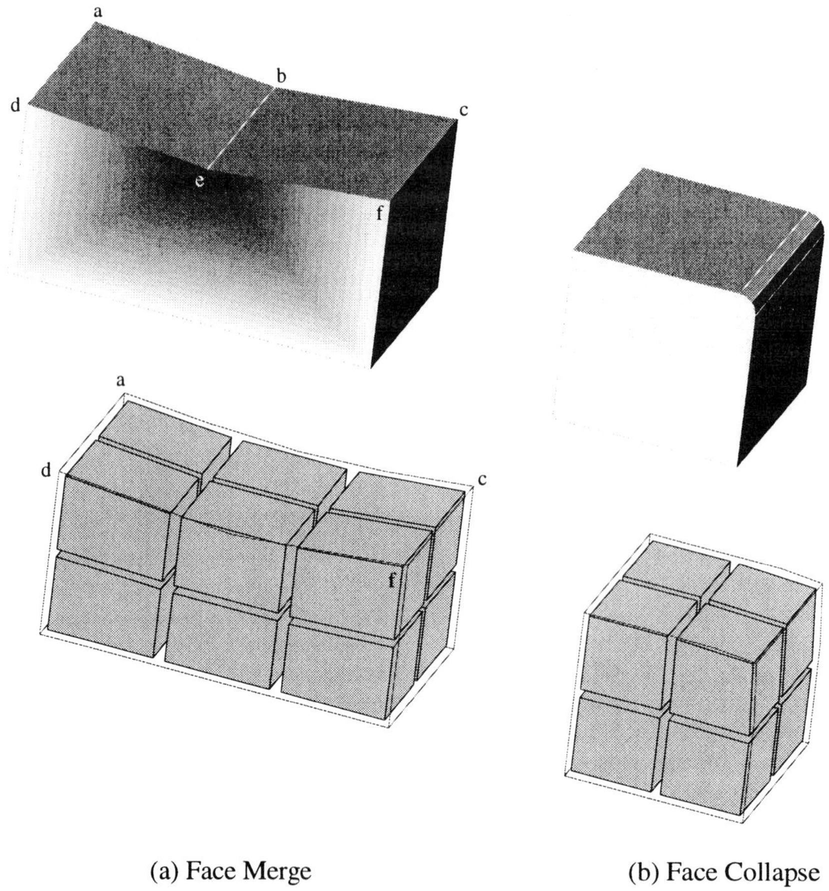

Figure 2. (De-)Partitioning operations for detail supression during meshing.

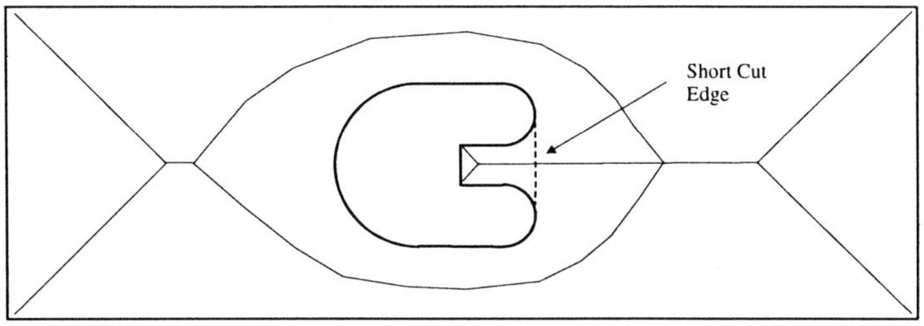

Figure 3. Using the MAT to find the convex hull of an interior hole.

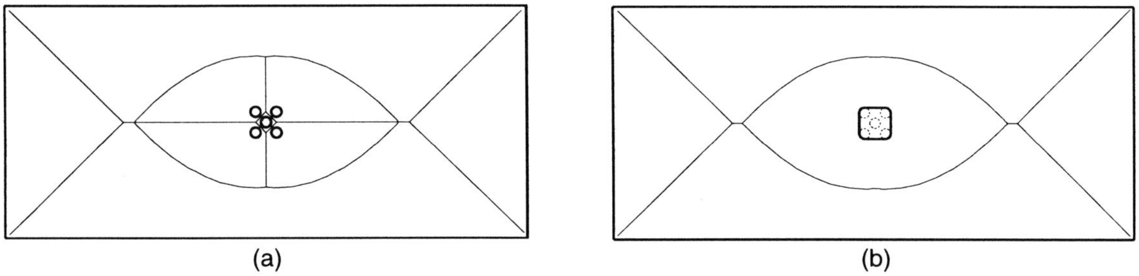

Figure 4. Cluster of small holes with (a) loop around each individual hole, and (b) with a loop around the entire group. This outermost loop can be used to indicate removal of the entire group.

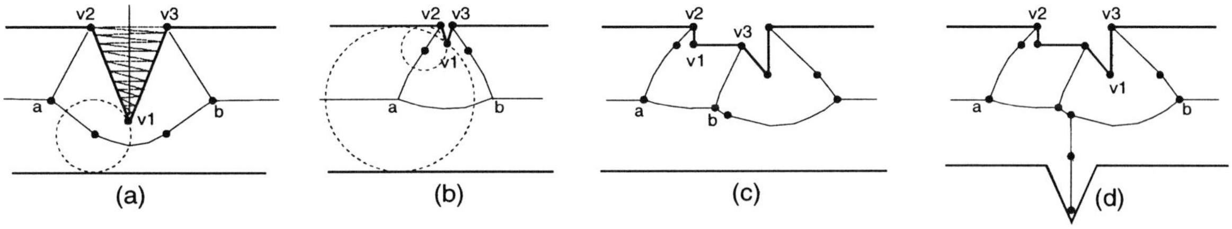

Figure 5. Notch aspect ratios based on the internal and external MATs.

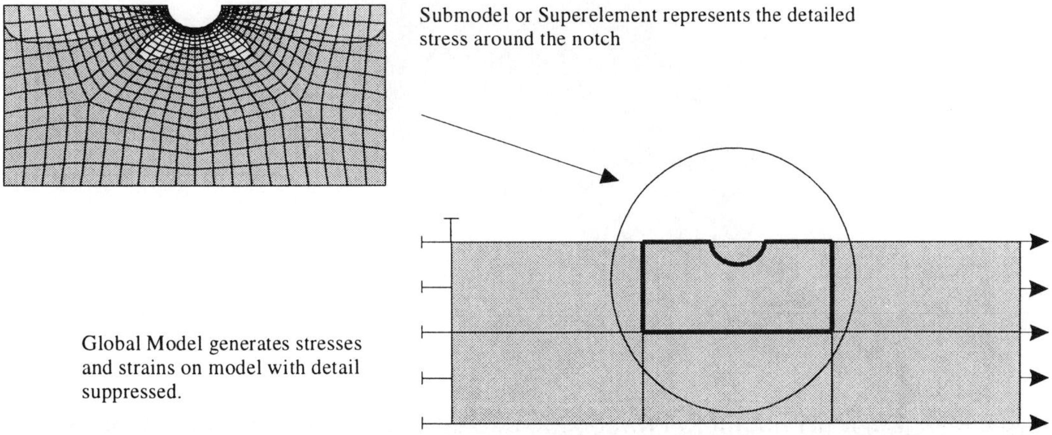

Figure 6. Local analysis of detail by submodelling or substructuring.

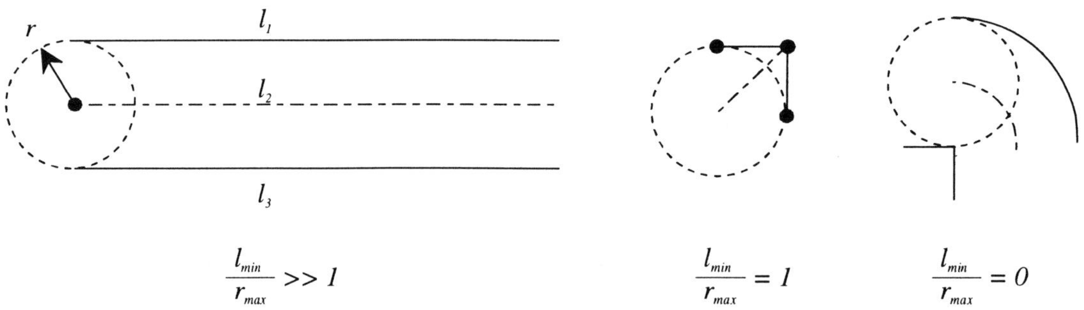

Figure 7. Aspect ratios for dimensional reduction.

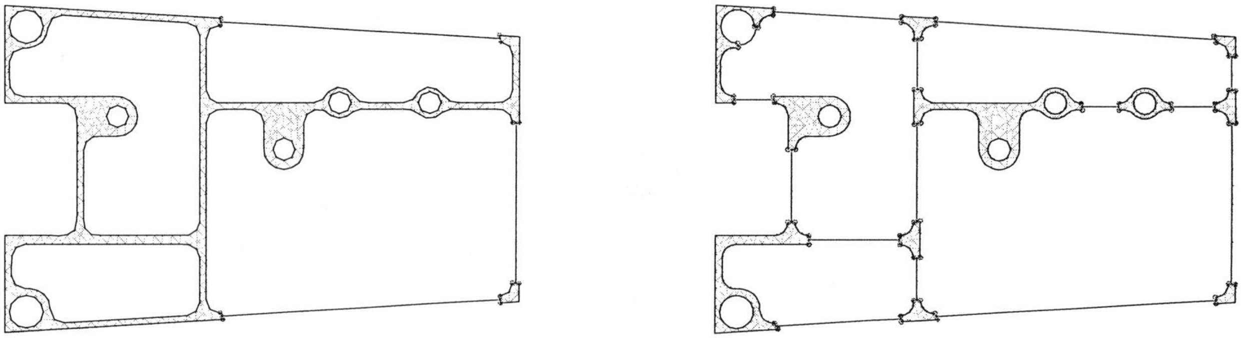

(a) Aspect ratio >= 20

(b) Aspect ratio >= 2

Figure 8. Using the MAT to identify long slender regions.

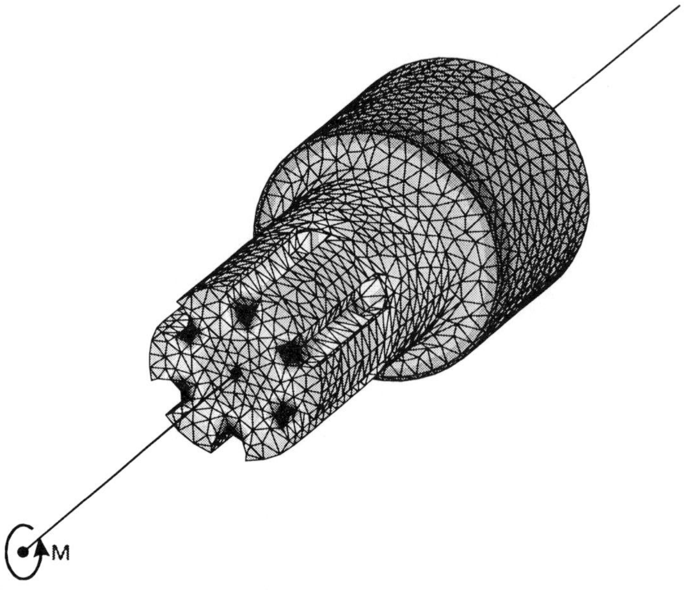

Figure 9. Coupling 3D model of stress concentration at joint to 1D shaft models.

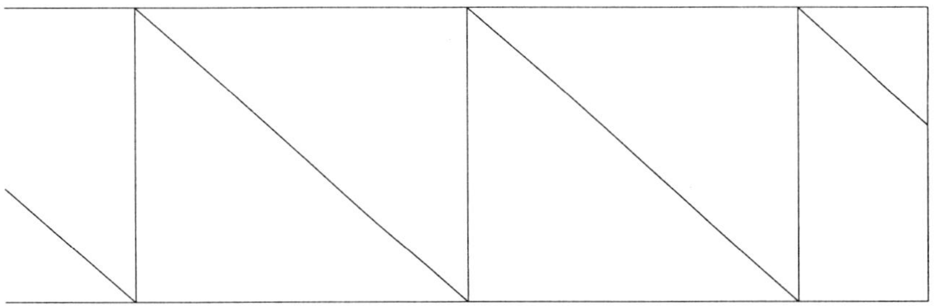

(a) 1D Beam Model

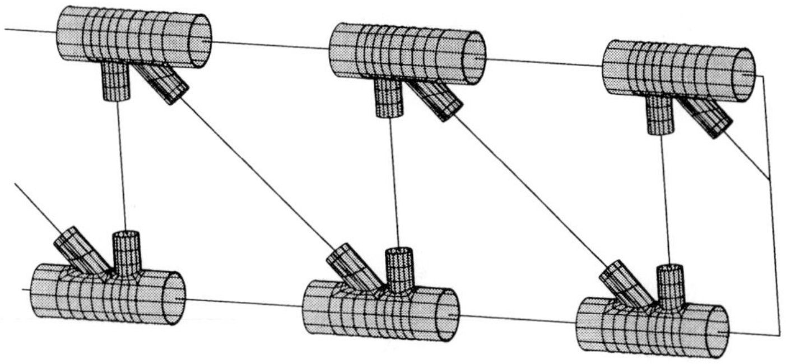

(b) Mixed Dimensional Model: 3D Analysis of Connections, 1D Analysis of Tubes

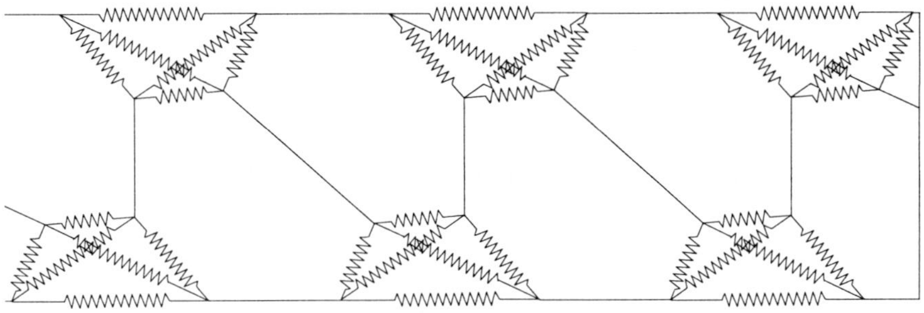

(c) Mixed Dimensional Model with Substructuring

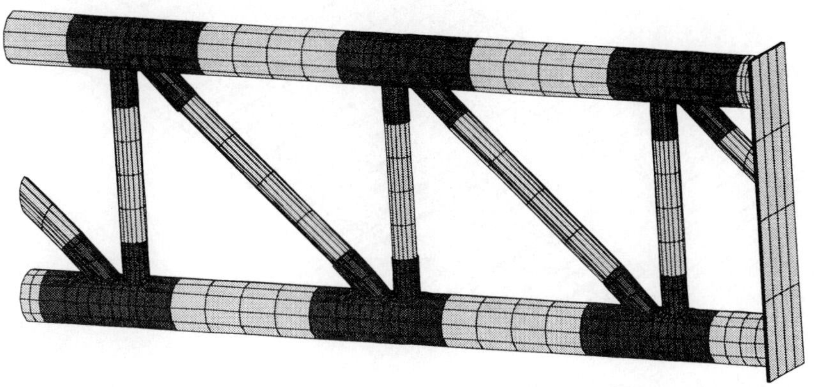

(d) 3D Model Partitioned into Regions for 3D and 1D Analysis

Figure 10. Various models of offshore structure.

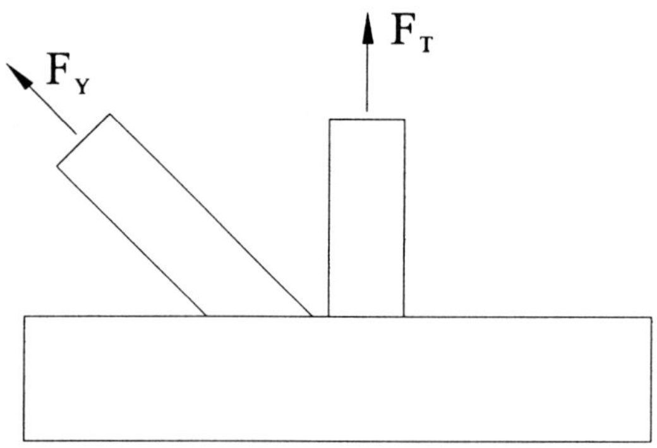

Figure 11. Loads on joint substructure.

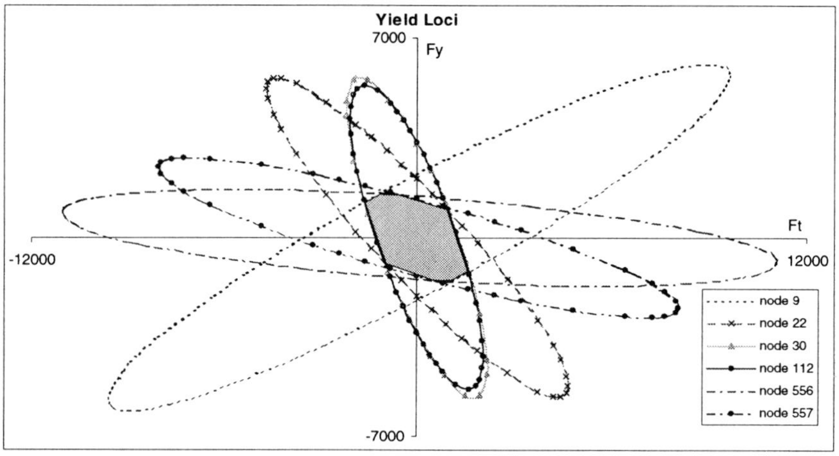

Figure 12. Yield locus of structure as intersection of yield loci of nodes.

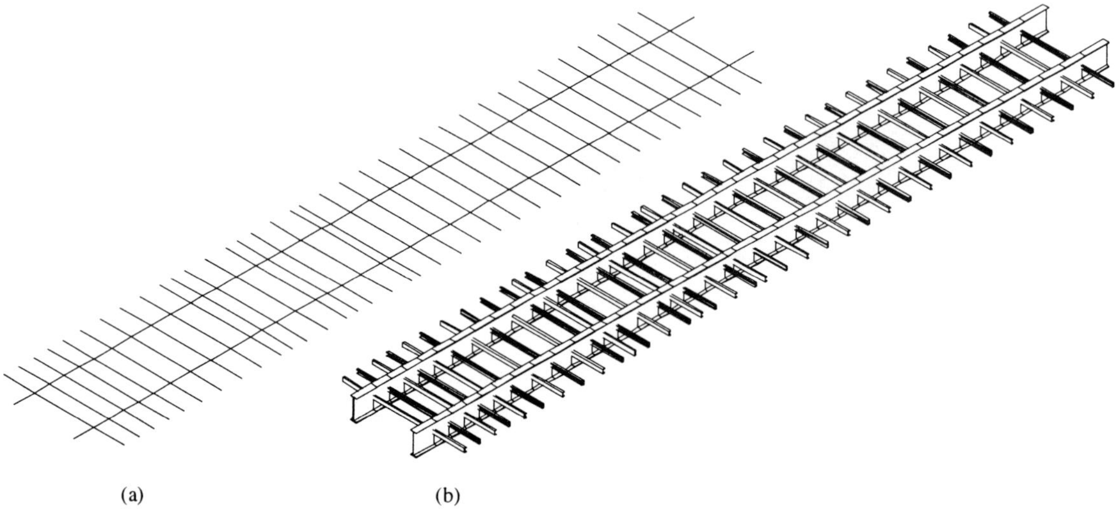

Figure 13. 1D analysis model of truck trailer with 3D model generated by dimensional addition.

A Feature-Based Approach to Automatic Injection Mold Generation

Zhenyong ZHOU *Shuming GAO*
Zhengchao GU *Jiaoying SHI*
State Key Laboratory of CAD&CG
Zhejiang University
Hangzhou, P.R.China, 310027
Email: zyzhou@cad.zju.edu.cn
smgao@cad.zju.edu.cn
Zhengchao.Gu@sdrc.com
Jyshi@cad.zju.edu.cn

Abstract

This paper presents a feature-based approach to automatic generation of injection mold. In the approach, all the molding features of a molded part are recognized first using a universal hint-based feature recognition algorithm, and the optimal parting direction is determined based on the feature model. Then, all the faces related with the parting line are classified into three groups according to the determined parting direction, and two kinds of parting lines are automatically determined by extracting the largest profile of each face group. Finally, the parting surface is generated based on the optimal parting line, thereafter, the mold core and mold cavity are automatically set up by splitting the mold box of the molded part with the generated parting surface. By using recognized molding features, the approach makes it possible to identify all undercuts, to determine internal parting lines and internal parting surfaces, to generate the mold core and cavity of a complex molded part with undercuts as well.

1. Introduction

Injection mold has been widely used in modern industry. As product life cycle is becoming shorter and shorter, there is a strong demand for reducing lead time of mold design and manufacturing. In recent years, research on automatic mold design is paid more and more attentions in modern mold industry. From the viewpoint of concurrent engineering methodology, the mold design and manufacturing factors should be considered while molded parts are designed, so that the complicated mold structure can be avoided. The mold parting direction, parting line and parting surface are among those key factors. Therefore, the automatic determination of the parting direction and parting line and automatic generation of the parting surface, mold core and cavity are the important parts of research on automatic mold design.

A few basic molding terminologies are introduced first [1]. A basic mold consists of two plates, called cavity and core, that form an impression into which molten material is injected. The cavity determines the external shape of a molded part and the core fixes the part's internal shape. The parting direction is a pair of opposite directions, along which the core and cavity are separated. Undercuts are "sideways" recesses or protrusions of the molded part that prevent the removal of the molded part from the mold along the parting direction. There are two kinds of undercuts, external and internal. An external undercut is formed by the external shape of a molded part, preventing removal of the molded part from the cavity, while an internal undercut is formed by the internal shape of a part,

0-7695-0562-7/00 $10.00 © 2000 IEEE

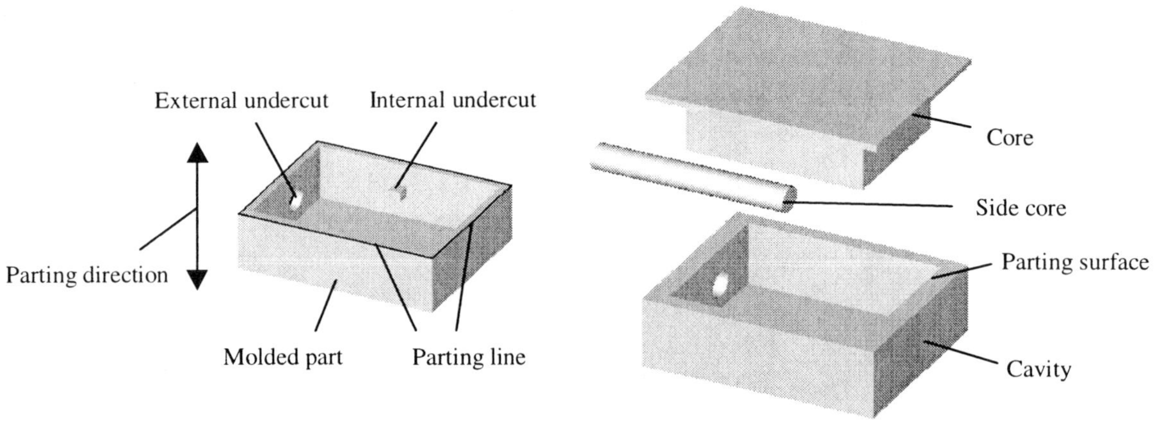

Figure 1. Some basic molding terminologies

preventing removal of the molded part from the core along the parting direction. Side core is used to mold the external undercut, and split core is adopted to mold the internal undercut. Parting line refers to the boundary where a molded part and its mold touch each other. Parting surface is the touch surface between the mold core and the mold cavity of a molded part when the mold is closed. Figure 1 illustrates some basic molding terminologies.

In this paper, we present a feature-based approach to automatic injection mold generation. Our objective is to use the feature model to enable the approach to identify all undercuts, to determine internal parting lines and internal parting surface, and to generate the mold core and the cavity of a complex molded part with undercuts. In our approach, the feature model of a molded part is derived from the solid model of the part by automatic feature recognition.

2. Literature Review

Some research work on automatic generation of injection mold has been conducted in recent years, which mainly focuses on automatic determination of parting direction and parting line and automatic generation of parting surface.

Hui[2] presented an approach which was based on blockage concept to determine the main parting direction, side core direction and split core direction. A subdivision technique was developed to evaluate the geometry of an undercut. The approach was applied to polyhedral solids. When curved surfaces are contained in a molded part, it is difficult for the method to give the partition space. Chen et al. [3] introduced the concept of visibility to determine the parting direction. In their approach, the surface visibility is obtained by the Gaussian mapping and the visibility map. If a surface is not completely visible from a direction, an undercut may exist in that direction. The parting direction is chosen based on the criterion of minimizing undercuts. The method did not take into account internal undercuts, because the visibility map of an internal undercut is zero. Machizuki [4] recognized undercut features using the difference between the part and its convex hull. Nee et al. [5] [6] gave a new classification of undercuts according to external loops and internal loops of a molded part. The parting directions of each possible undercut are determined based on the boundary information of the part, and the optimal parting direction is chosen according to the number of possible undercuts and their corresponding undercut volumes. The method primarily used geometric and topological relations of a part to detect undercuts. They also presented a method for automatically determining parting line by classifying faces into groups and extracting the largest profile of each face group. Urabe et al. [7] selected three principal coordinate directions as the candidate parting directions, calculated the number of undercuts, the projected area, the number of cone surfaces of the molded part along each candidate direction and used them as major mold factors to determine the principal parting direction. This approach was limited to simple 3D parts that don't contain complicated features.

Tan [8] [9] [10] proposed two methods for generating parting line. In the first method, a parting direction of a molded part is selected first. Then, all the surfaces of the part are triangulated and the obtained triangles are classified into visible faces, invisible faces and degenerate faces. The edges on the 1part separating the visible faces from the invisible faces are considered as the tentative paring lines. The parting line is finally set up based on the tentative parting lines. Their second method is to locate the parting line by recursively slicing the molded part. It can apply to the molded part with free-form surfaces. However, there are some problems with its reliability and accuracy since the slicing plane is discrete. Besides the two methods, Tan et al. also presented an algorithm of generating 3D parting surface.

Ravi and Srinivasan [11] proposed nine criteria for computer-aided parting line and parting surface design. Serrar [12] classified the parting surface into several patches including external parting patches and internal parting patches. The external parting patches are divided into land patch, non-contact patch and clamping patch further. The algorithms of generating these patches were proposed. The parting surface was generated with these patches. In the method, generating internal parting surface is difficult and it may result in wrong parting surface when a molded part contains some complex feature such as snap fit.

3. Approach Overview

In this section, we overviews the feature-based approach to automatic injection mold generation, meanwhile explain why we use feature model and automatic feature recognition to support automatic mold generation.

It is observed that determining and dealing with undercuts of a molded part is a key but difficult task in automatic mold generation. Since undercuts are exactly some kinds of molding features, in our approach, the feature model of a molded part is adopted to solve undercut problem. To set up the feature model of a molded part, there are two ways: one is feature-based design, another is feature recognition. Considering that the standard model of a molded part is still B-rep up to now and the automatic feature recognition technique is becoming more and more mature, we think it is advisable to directly use automatic feature recognition to set up the feature model of a molded part. In our approach, the feature model of a molded part is also used for supporting determination of internal paring line, supporting automatic generation of mold core and cavity of a molded part with undercuts, and so on, which can not be effectively solved by current approaches.

The flow chart of the feature-based approach to automatic mold generation is shown in Figure 2.

4. Universal Hint Based Feature Recognition

In our approach, the feature model of a molded part is derived from the part's solid model by automatic molding feature recognition. Considering there may exist convex features, concave features as well as interacting features in a molded part, we extend the universal hint based automatic machining feature recognition approach [13] and use it to automatically recognize molding features of molded parts. The original approach is extended in two aspects. First, several kinds of new features, such as protrusion, rib and snap-fit, are added into the feature library. Second, algorithms for recognizing general rib and snap-fit features are developed. The detailed algorithm will be presented in another paper. In Figure 3, shading area shows a recognized slot feature.

5. Determination of the Optimal Parting Direction

After feature recognition, the feature model of the molded part is built. Based on the molding feature model of the molded part, we first determine all candidate and feasible parting directions of each feature, then determine all candidate parting directions of the molded part, finally choose the optimal parting

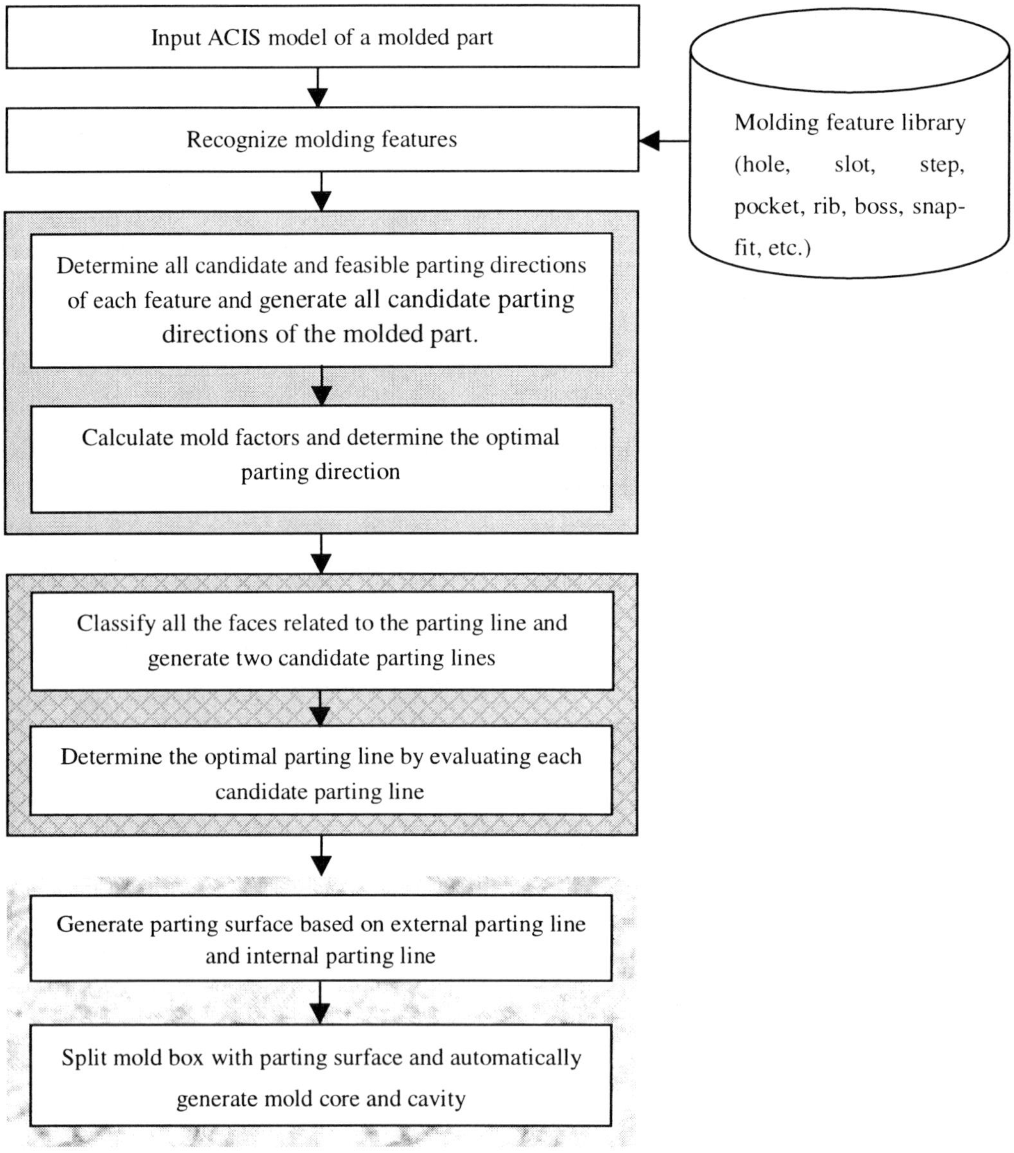

Figure 2. Flow chart of the feature-based approach to automatic mold generation

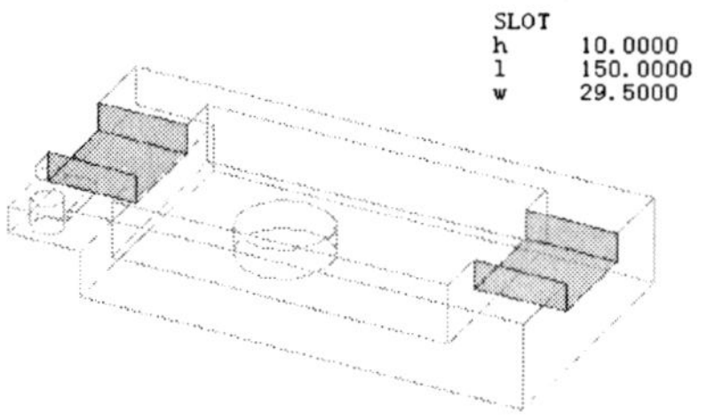

Figure 3. An example of automatic feature recognition

direction through evaluating all candidate directions of the molded part.

5.1 Determination of Candidate and Feasible Parting Directions for Each Kind of Molding Feature

Considering the lay out of the mold ejection mechanism and the convenience for mold machining,

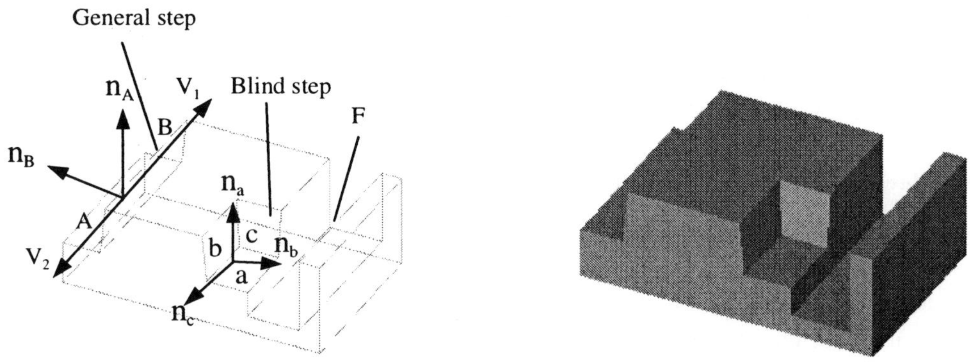

(a) (b)

Figure 4. Determination of CPD for step feature

practical parting directions are usually chosen from certain faces' normals of a molded part. If along a direction, a feature volume can be moved and has no interference with the molded part, then this direction is a feasible parting direction of the feature. A feature may or may not have feasible parting directions (called FPD below), and a feature's FPD can be infinite (when it contains a range of directions) or a null depending on the dihedral between its constituent faces. The candidate parting directions (called CPD below) are chosen from FPDs, which are a number of representative FPDs for mold factor evaluation. A feature's CPD can be determined by faces' normals of the feature and their correlation depending on the feature type. As an example, the analysis and determination of the step's CPDs and FPDs are described in the following.

Step can be classified into general step and blind step (see Figure 4a). The number of CPDs of general step may be 2 or 4. Taking Figure 4 as an example, where $\mathbf{n}_A$ and $\mathbf{n}_B$ are the normals of face A and face B respectively and $\mathbf{V}_1 = \mathbf{n}_B \times \mathbf{n}_A$, $\mathbf{V}_2 = \mathbf{n}_A \times \mathbf{n}_B$. If the dihedral between face A and face B is an obtuse or right angle, $\mathbf{V}_1$, $\mathbf{V}_2$, $\mathbf{n}_A$ and $\mathbf{n}_B$ are all CPDs. However, if the dihedral is an acute angle, only $\mathbf{V}_1$ and $\mathbf{V}_2$ are CPDs. The FPDs for general step is a range of directions: $\{\mathbf{V} \mid \mathbf{V} \cdot \mathbf{n}_A \geqslant 0.0 \cap \mathbf{V} \cdot \mathbf{n}_B \geqslant 0.0\}$. A blind step's CPDs are determined based on the dihedral angles between every two of its three constituent faces, say a, b and c in Figure 4. Three cases exist. If all the dihedrals are right or obtuse angles, three CPDs exist and they are the normals of three constituent faces of the blind step. If only one dihedral between two faces, see a and b in Figure 4, is an acute angle, there are three CPDs too. One is the normal of face c. The other two are the cross products of normal of face c and normals of face a and face b respectively. The CPDs point to the outside of the part. If more than two acute dihedrals exist, the three CPDs are the cross products of the normals of every two constituent faces and point to the outside of the part. The FPDs of a blind step is a range of directions: $\{\mathbf{V} \mid \mathbf{V} \cdot \mathbf{n}_A \geqslant 0.0 \cap \mathbf{V} \cdot \mathbf{n}_B \geqslant 0.0 \cap \mathbf{V} \cdot \mathbf{n}_C \geqslant 0.0\}$.

It is observed that the primary candidate parting directions may be invalid due to that there is a blockage face in that direction. For example, since face b is blocked by face F in Figure 4, n_b is not a valid candidate parting direction of the part. In order to determine whether these primary CPDs are real CPDs, blockage test is used. The blockage test consists of two aspects. First certain feature face is swept along the primary CPD for a certain distance to create a temporary solid. Then the intersection operation is performed between the created solids and the molded part. If the intersection is non-empty, it indicates the feature has blockage face and the tested primary CPD is not a real CPD, hence it is deleted from the primary CPDs. Otherwise, the feature has no blockage face in this direction and the tested primary CPD becomes a real CPD. The test results are recorded and they are also used in undercut determination later.

The determination of CPDs and FPDs of other kinds of features is similar to that of step feature, so it will not be described here.

5.2 Determination of the Optimal Parting Direction

In this section, we discuss how to determine the optimal parting direction (OPD) of a molded part based on the CPDs and FPDs of the part's molding features. It consists of several phases. All the CPDs of a molded part are determined first. Then each CPD of the molded part is evaluated using an object evaluation function. Finally the CPD with maximum evaluation value is selected as the OPD of the molded part.

The primary CPDs of a molded part consist of all distinguished real CPDs of all molding features contained in the part and all normals of the convex faces of the part. In the primary CPDs of the molded part, two completely reversed CPDs are merged as one real CPD of the molded part (hereafter called RCPD). For each RCPD, we set up a list to record all the features that their real CPDs are the same as or completely reversed to this RCPD. Observe that one feature may be recorded in several different RCPDs' list, since a feature can have several real CPDs. In this way, a set of RCPDs together with their related feature lists of the molded part are generated and denoted as: RCPD1 = {dir 1, feature 1, feature 2, ···, feature n}, RCPD2 = {dir 2, feature 1, feature 4, ···, feature m}, ···, RCPDk = {dir k, feature 3, ···, feature l}.

The object evaluation function for determining the optimal parting direction used in our approach is as follows: SMF = $\sum$weight_factor$_i$ * mold_factor$_i$ (i = 1 ··· 4), where SMF is the total evaluation value, mold_factor refers to a specific factor calculated to assess the RCPD using a certain molding rule or criterion, weight_factor is the weight value for a specific mold factor. The assignment of a weight_factor is based on how much its corresponding mold_factor affects mold cost, quality and productivity. In this paper, four kinds of mold factors are selected to assess RCPDs. They are number of undercuts, projected area, boxed area and height of a molded part. The specific methods for calculating each kind of mold factor are described below.

1) The number of undercuts

The number of undercuts refers to the number of features that obstruct to separate the mold halves along a certain candidate direction. For a certain RCPD, all possible undercuts in this direction consist of all the features of the part that are not in the RCPD's feature list. Each possible undercut feature is further checked to determine whether it is a real undercut or not as follows: First check if the FPD of the feature contains the RCPD.

①If it is true, the blockage test along the RCPD is taken to check the existence of blockage faces of the feature. If there is no blockage faces, the feature is not an undercut with respect to the RCPD. Otherwise the feature is a real undercut.

② If the FPD of the feature does not contain the RCPD, the feature is certainly an undercut.

When the feature is verified as an undercut, it can be further determined as an internal or external undercut by checking whether or not it has blockage faces along the feature's CPDs. If there is any blockage face, the feature is an internal undercut. Otherwise it is an external undercut. The side core direction or the split core direction of the undercut feature is simply its CPD. In this way, the number of real undercut features is obtained. Since reducing the number of undercuts can greatly decrease the process time and lower tooling and manufacturing cost, the direction with fewer undercuts indicates a strong preference to be the optimal parting direction.

2) The projected area and the boxed area

The projected area is the bounding area of silhouette formed by projecting the molded part to the plane perpendicular to a candidate direction, while the boxed area refers to the sum of the areas of all faces of the part's bounding box. The direction with larger projected area and boxed area facilitates removal of the part from the mold and decreases the needed draft angle and cutting depth in mold manufacturing, therefore it has high preference to be the optimal parting direction.

3) The height of the molded part

The height of a molded part is the maximum projection length of the part in the candidate direction. The height of a molded part affects the productivity. The direction with a smaller height of the part is given

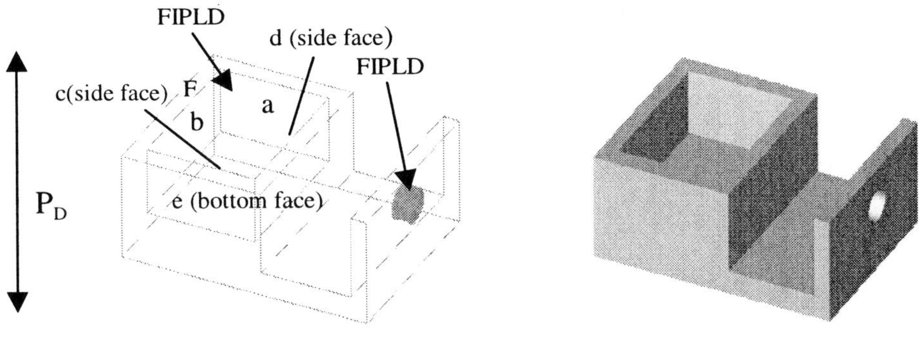

Figure 5. Illustration of FIPLD

preference to be the optimal parting direction.

6. Determination of the Optimal Parting Line

After the optimal parting direction is determined, we start to determine the parting line in that direction. The algorithm of automatically determining parting line includes the following five steps.

6.1 Elimination of the Faces Irrelated with the Parting Line Determination

In order to increase the efficiency of the algorithm, we first eliminate all the faces irrelated with the parting line determination (hereafter called FIPLD) from the face set of the molded part. We observe that all the faces of undercut features as well as blind hole, blind pocket, internal protrusion and rib belong to FIPLD. This is because the projection of these faces on the plane perpendicular to the parting direction is always within the profile of a certain molded part face, therefore making no contribution to the parting line determination. Taking Figure 5 as an example, the projection of the blind pocket feature's faces a, b, c, d and e on the plane perpendicular to P_D is within the profile of face F and they obviously have no use for determining the parting line.

6.2 Classification of the Faces Related with the Parting Line Determination

The faces related with the parting line determination (hereafter called FRPLD) consist of all the faces of a molded part except its FIPLD. It is observed that the parting line of a molded part must be the boundary separating the visible faces from the invisible faces of the molded part. Hence, to determine the parting line of a molded part, we first classify all FRPLD and find out all the visible faces, invisible faces and degenerate faces in the parting direction. The specific classification method is simple. Visible face is the face that the dot product of its normal and the parting direction is greater than zero. Invisible face is the face that the dot product of its normal and the parting direction is less than zero. Degenerate face refers to the face that is neither visible face nor invisible face. Degenerate face is a transition from visible face to invisible face. It is the degenerate face that makes the parting line non-unique. In order to eliminate those obviously unreasonable parting lines, we reclassify the degenerate faces. The reclassification rule is as follows. If there is any shared edge between a degenerate face and some visible face and there is no shared edge between the degenerate face and all invisible faces, the degenerate face is reclassified into visible face group. If there is any shared edge between a degenerate face and some invisible face and there is no shared edge between the degenerate face and all visible faces, the degenerate face is reclassified into invisible face group. Otherwise, the degenerate face keeps unchanged. Note that, for a curved surface, maybe just a part of the surface is visible. Hence, it is necessary to generate its silhouette first. If its silhouette exists, the curved surface should

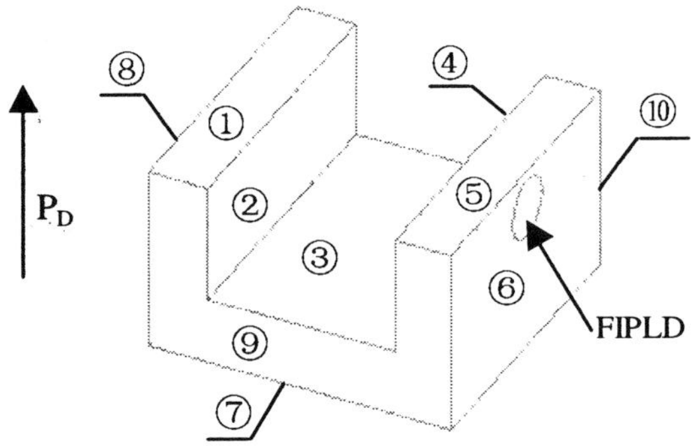

Figure 6. An example of the classification of FRPLD

Table 1

Type	Initial classification result	Reclassification result
Visible face	①、③、⑤	①、③、⑤、②、④
Degenerate face	⑧、②、④、⑥、⑨、⑩	⑧、⑥、⑨、⑩
Invisible face	⑦	⑦

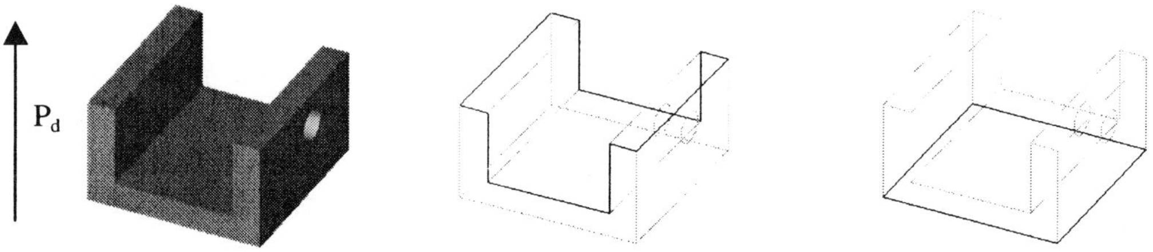

(a) Molded part (b) Profile of visible face group (c) Profile of invisible face group

Figure 7. Illustration of two kinds of external parting lines

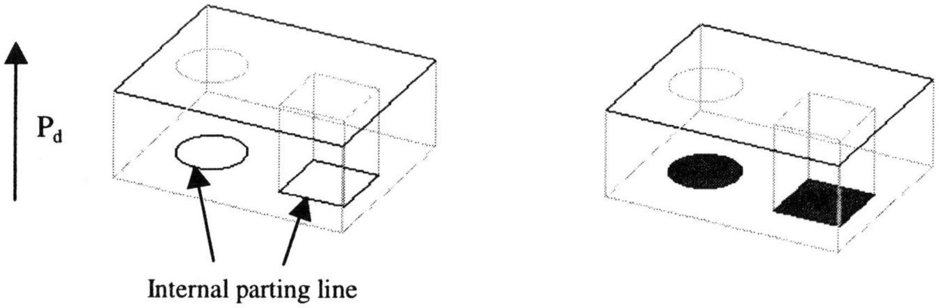

(a) (b)

Figure 8. An example of internal parting line

be partitioned into a visible part and an invisible part along its silhouette.

The classification result of the FRPLD of the molded part shown in Figure 6 is listed in Table 1.

6.3 Determination of the External Parting Line

Observing that all the visible faces are connected, all the invisible faces are connected, and their profiles respectively form parting lines, therefore we create two kinds of candidate parting lines by generating the profiles of visible face group and invisible face group respectively. The profiles of visible face group and invisible face group are obtained by merging the

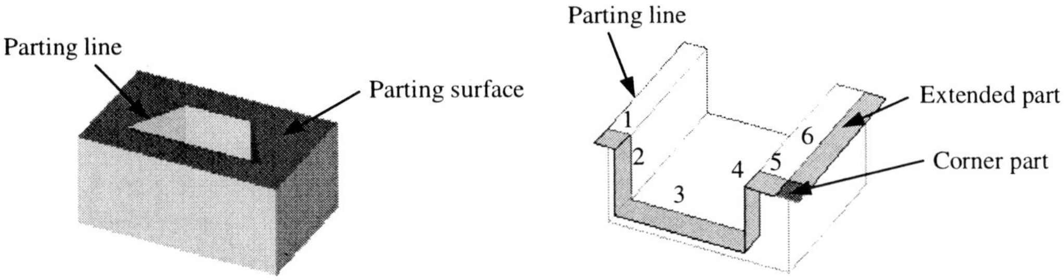

(a) For a planar parting line

(b) For a non-planar parting line

Figure 9. Generation of the external parting surface

boundary of adjacent faces. That is to get the common external loop of the adjacent faces by eliminating their shared edges and linking other edges together. Figure 7 shows a molded part and its two kinds of external parting lines.

6.4 Determination of the Internal Parting Line

The internal parting line is just related with some special kinds of features such as through hole and through pocket. That is, when a molded part contains some through hole and/or some through pocket which are not undercuts, there exist internal parting lines. The specific method for determining internal parting line is simple. For a through hole, the intersection line of the through hole and the molded part's bottom face is the internal parting line. Similarly, the internal parting line of a through pocket can be obtained. An example of an internal parting line is shown in Figure 8.

6.5 Determination of the Optimal Parting Line

The optimal parting line is finally determined by comparing the two candidate parting lines using an objective function. According to the injection mold design experience, three factors are selected as the evaluation parameters in our method. They are parting line complexity, draw depth and machining complexity.

The specific objective function is as follows:

$$F_{obj} = F1\,(\text{parting line complexity}) * w1 + F2\,(\text{draw depth}) * w2 + F3\,(\text{machining complexity}) * w3$$

where w1, w2 and w3 are weight factors. The assignment of these weight factors is based on how the corresponding evaluation factor affects the mold cost, quality and productivity. The calculation methods of F1, F2 and F3 are similar to Weinstein's [14]. However, we take into account the number of joint faces between the mold core and the mold cavity in calculating machining complexity. The number of joint faces is important because the manufacture precision of joint faces is much higher than other faces and it greatly affects the mold cost.

7. Automatic Generation of Parting Surfaces, Mold Core and Mold Cavity

After the parting line is determined, the parting surface is generated automatically based on the parting line and then the mold core and the mold cavity are constructed using the parting surface in an automatic way.

7.1 Generation of the Parting Surface

The parting surface is divided into external parting surface and internal parting surface. The former is generated by the external parting line while the latter is constructed using the internal parting line.

Our algorithm for generating the external parting surface is divided into two cases so as to improve the algorithm efficiency. If the external parting line is planar, its generation is relatively simple. Here assume that mold box of the part has been created. In this case, the face whose boundary consists of the external parting line and the intersection lines between the mold box and the plane where the external parting line locates is exactly the required external parting surface. For another case, i.e. non-planar external parting line, we first divided the parting line into a number of groups so that all lines in the same group are the boundary of the same degenerate face. For example, line 1-5 in Figure 9(b) form a group, while line 6 forms another group. Then, each group of parting lines is extended along the normal of the corresponding degenerate face to create the partial parting surface. At last, the complete external parting surface comprises all the partial surfaces as well as the corresponding corner parts. A corner part exists between two adjacent partial surfaces. An instance of the corner part is depicted in Figure 9(b). Also, the whole generation process of the external parting surface is illustrated by Figure 9.

In contrast to the external parting surface generation, the creation of the internal parting surface is simple. The internal parting surface is simply the face whose boundary comprises internal parting loop, see Figure 8(b) as an example.

7.2 Automatic Generation of Mold Core and Cavity

The mold core and mold cavity are automatically generated by splitting mold box using the parting surface. It consists of four steps.

1) Create the mold that is 10% larger than the enclosing box of the molded part.
2) Put the molded part into the mold box and center the mold box with the part.
3) Perform subtraction operation between the mold box and the part.
4) Generate the mold core and cavity by splitting the mold box subtracted by the part using the parting surface.

It should be pointed out that when the molded part contains external undercuts, after splitting the mold box by parting surface, the mold core and mold cavity are still connected. In our approach, this problem is solved with the help of the recognized external undercut features. First, for each external undercut, we construct a side core with the volume of the undercut feature. Then subtract the mold box, which has been subtracted by the part and split by the parting surface, with the side cores. After that, the mold core is separated from the mold cavity.

8. Implementation

The approach presented in this paper has been implemented with Visual C++ based on the geometrical solid modeling kernel ACIS5.0. Figure 11 shows a simplified mouse shell part as well as its optimal parting direction, optimal parting line and mold core and cavity automatically generated by our approach. The evaluation results used to determine the optimal parting direction and optimal parting line of the test part are given in Figure 10.

9. Discussions and Conclusions

This paper presents a feature-based approach to automatic mold generation. The proposed approach has following characteristics:

1) Introduce automatic feature recognition into automatic mold generation. The feature model set up by recognizing all molding features from a molded part's solid model is adopted to effectively support automatic mold generation.
2) By utilizing molding features, the approach can identify all undercuts, construct internal parting line, generate internal parting surface, and create the mold core and cavity of a complex molded part with undercuts.
3) Increase the efficiency of parting line determination algorithm by effectively excluding all the faces irrelated with the parting line.
4) Employ proper objective functions to determine the optimal parting direction and optimal parting line.

Current approach can deal with molded parts with planes and quadratic surfaces, and just generate two kinds of parting lines. Our future work will focus on enabling the approach to handle molded parts with free form surfaces and to generate other kinds of practical parting lines.

*****The Determination of Optimal Parting Direction*****

Parting direction weight value:

ProjAreaWgt = 0.3, UcNumWgt = 0.5, BoxAreaWgt = 0.1, HeightWgt = 0.1

Evaluation value:

GrpNo	dir.x	dir.y	dir.z	ProjArea	UcNum	BoxArea	Height	SMF
0	0.00	1.00	0.00	4668.40	5	16130.02	50.00	0.687
1	1.00	0.00	0.00	2837.97	5	16130.02	72.50	0.547
2	0.00	0.00	-1.00	3638.48	1	16130.02	36.25	0.934
3	-0.41	0.00	0.91	4478.81	5	24502.55	62.63	0.626

Selected group No: 2

Main parting direction: (0.00)i + (0.00)j + (1.00)k

External undercut: Through_pocket **Side core direction:** (0.00)i + (1.00)j +(0.00)k

*****The Determination of Optimal Parting Line*****

Parting line weight value:

F1Wgt = 0.4, F2Wgt = 0.2, F3Wgt = 0.4

Evaluation value:

GrpNo	F1	F2	F3	AssVal
0	10.04	36.25	0.88	0.68
1	4.94	36.25	0.63	1.00

Selected group No: 1 (as shown in Figure 11(c))

Figure 10. Evaluation results

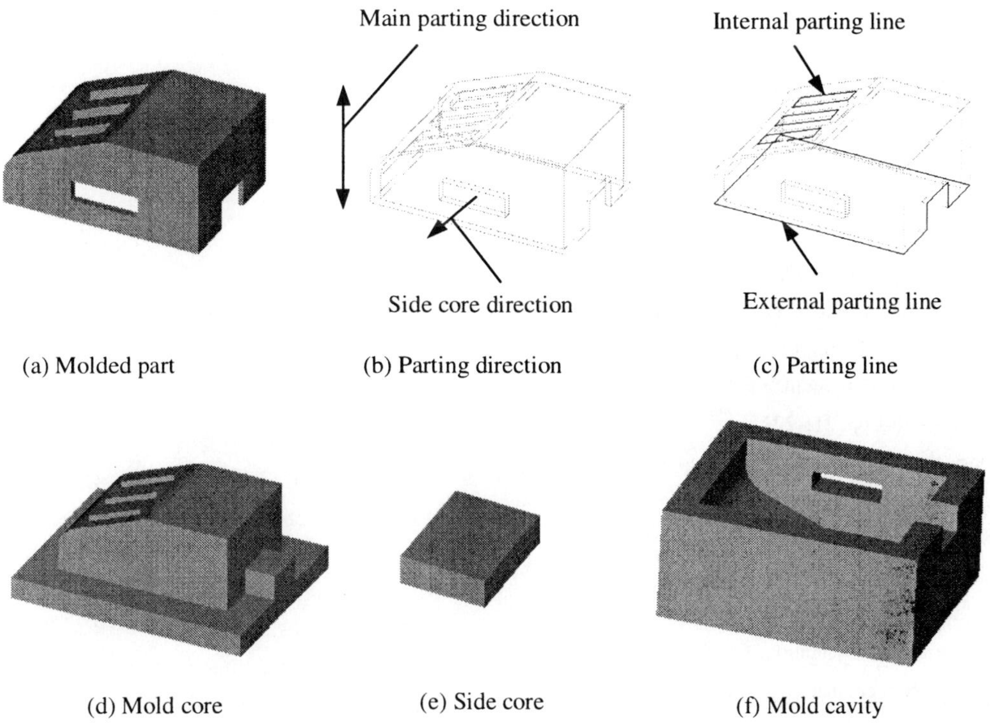

Figure 11. Automatic mold generation of a simplified mouse shell part

Acknowledgements

The research is supported by National Natural Science Foundation of China and 863 High Technology Foundation of China. We are also grateful to K.C.Wong Foundation for its supporting one author to attend GMP2000 Conference in Hong Kong.

References

[1] R. G. W. Pye, Injection mould design: A textbook for the novice and a design manual for the thermoplastics industry (4th ed), John Wiley & Sons, New York, 1989.

[2] K. C. Hui, Geometric aspects of the mouldability of parts, Computer-Aided Design, 1997, 29(3), pp. 197-208.

[3] L. L. Chen, S. Y. Chou, and T. C. Woo, Parting directions for mould and die design, Computer-Aided Design, 1993, 25(12), pp. 762-768.

[4] T. Mochizuki, and N. Yuhara, Methods of extracting potential undercut and determining optimum withdrawal direction for mold designing, Int. J. Japan Soc. Prec. Eng., 1992, 26, pp. 68-73.

[5] A. Y. C. Nee, M. W. Fu, J. Y. H. Fuh, etc., Determination of optimal parting directions in plastic injection mold design, Annals of the CIRP, 1997, 46(1), pp. 429-432.

[6] A. Y. C. Nee, M. W. Fu, et al., Automatic determination of 3-D parting lines and surfaces in plastic injection mould design, Annals of the CIRP, 1998, 47(1), pp. 95-98.

[7] K. Urabe and P. K. Wright, Parting direction and parting planes for the CAD/CAM of plastic injection molds, Proceedings of DETC'97, 1997, DETC97DFM4336. Sept. 14-17, Sacramento, California, USA.

[8] S. T. Tan, M. F. Yuen, et al., Parting lines and parting surfaces of injection moulded parts, Proc Instn Mech Engrs, 1990, 204, pp. 211-221.

[9] T. Wong, S. T. Tan, W. S. Sze, Computer-aided parting line generation for free-form solid models, International Conference on Manufacturing Automation, April 28-30, 1997, Hong Kong, pp. 1097-1105.

[10] S. T. Tan, et al., A method for generation of parting surface for injection moulds, Conference on computer aided production engineering, 1988, Edinburgh, pp. 401-408.

[11] B. Ravi, M. N. Srinivasan, Decision criteria for computer-aided parting surface design, Computer-Aided Design, 1990, Jan/Feb, 22(1), pp. 11-18.

[12] M. Serrar, G. A. Gabriele, Automatic generation of parting surfaces and mold halves, Proceedings of the Computers in Engineering Conference and the Engineering Database Symposium, ASME, 1995.

[13] S. Gao, and J. J. Shah, Automatic recognition of interacting machining features based on minimal condition subgraph. Computer-Aided Design, 1998, 30(9), pp. 727-739.

[14] M. Weinstein, and S. Manoochehri, Optimum parting line design of molded and cast parts for manufacturability, ournal of Manufacturing Systems, 1997, 16(1), pp. 1-12.

Effect of Fabric Properties on Cloth Draping Modeling

C. K. Au
Z. Wu
M. M. F. Yuen
Department of Mechanical Engineering
Hong Kong University of Science and Technology
chikit.au@ugsolutions.com

Abstract

Cloth draping is of particular interest in the field of garment engineering. Most cloth draping simulation stresses the visual effects of garment design. However, the accuracy of the draping effect should not be overlooked, particularly for garment manufacturing. The accuracy of the draping behaviour is largely affected by the fabric properties. This paper discusses the issue of incorporating the fabric properties into the cloth model, so that its raping behaviour can be realistically simulated. These fabric properties are expressed in terms of a set of measurable mechanical properties. The draping behaviour of the cloth is derived based on elastic theory. Dynamic draping simulation is presented with illustrative examples.

1. Introduction

Cloth modelling is of particular interest in the field of garment design and manufacturing. Much research concentrates on the physical based modelling of the cloth object. Feynman[1] modelled the shape of the cloth in an equilibrium state by energy minimization. Terzopoulos[2] proposed a deformable surface to model a piece of cloth based on Lagrange's equation of motion. The derivation concentrated on the balance of forces: inertial force, damping force, other external forces and the internal elastic and bending force acting on cloth. Provot[3] proposed a mass-spring system to animate the cloth shape under constraints. The cloth was represented by an array of point mass related by a set of spring constants.

In fact, both the energy minimization method and force equilibrium method originate from the same principle. Celniker and Gossard[4] pinpointed the fact that the energy stored in a deformable model was the sum of the energy due to stretching and bending, and the work done by the external forces. Minimizing this energy function yielded a set of differential equations governing various forces under equilibrium conditions.

Recently, cloth modelling tends to focus on two major areas:

1. Incorporating fabric property into the cloth model to give an accurate simulation[5,6].
2. Improving the computational efficiency to give a fast simulation[7,8].

Most cloth draping simulation stresses the visual effects of garment design. However, the accuracy of the draping effect should not be overlooked, particular for garment manufacturing. The accuracy of the draping behaviour is largely affected by the fabric properties. This paper discusses how the fabric properties account for the draping behaviour of the cloth object. The fabric properties are expressed in terms of a set of mechanical properties and are incorporated into the formulation though the strain energies. As a result, the draping effect can be realistically simulated. Several draping examples for various cloth materials are presented.

2. Strain energy density due to deformation

The analysis concentrates on the geometric changes that take place in a piece of cloth during its motion from one configuration to another. Assuming that a piece of cloth is a deformable surface defined in a parametric space $\mathbf{\Omega}_u^2$ such that every point in the surface is represented by a vector (u_1, u_2). The configuration of the cloth is described by the vector $\mathbf{s}$ in a material space $\mathbf{\Omega}_s^2$ given as $\mathbf{s} = (s_1, s_2)$ where $s_i = s_i(u_1, u_2), \forall i = 1,2$. At time instant t, the configuration of the cloth in a spatial space $\mathbf{\Omega}_x^3$ is expressed as $\mathbf{x} = (x_1, x_2, x_3)$ with $x_j = x_j(u_1, u_2, t)$, $\forall j = 1,2,3$.

0-7695-0562-7/00 $10.00

The deformations in these two spaces are mainly due to stretching, shearing, bending and twisting. The strain energy densities of the cloth are expressed in terms of the changes of the metric tensor $\overline{G}_{\alpha\beta}$ and curvature tensor $\overline{B}_{\alpha\beta}$:

$$U_G = \frac{1}{8}(\overline{G}_{\alpha\beta} - \overline{G}^0_{\alpha\beta})C_1^{\alpha\beta\delta\gamma}(\overline{G}_{\delta\gamma} - \overline{G}^0_{\delta\gamma}) \quad (1)$$

$$U_B = \frac{1}{2}(\overline{B}_{\alpha\beta} - \overline{B}^0_{\alpha\beta})C_2^{\alpha\beta\delta\gamma}(\overline{B}_{\delta\gamma} - \overline{B}^0_{\delta\gamma}) \quad (2)$$

where $C_1^{\alpha\beta\delta\gamma}$ and $C_2^{\alpha\beta\delta\gamma}$ are the material modulus tensors describing the metric and curvature behaviours of the cloth, U_G is the strain energy density due to stretching and shearing, U_B is the strain energy density due to bending and twisting, $\overline{G}^0_{\alpha\beta}$ and $\overline{B}^0_{\alpha\beta}$ are the initial values of $\overline{G}_{\alpha\beta}$ and $\overline{B}_{\alpha\beta}$.

The metric tensor $\overline{G}_{\alpha\beta}$ and the curvature tensor $\overline{B}_{\alpha\beta}$ (which are defined in the physical space Ω^2_S) are related to the first and second fundamental quantities $G_{\alpha\beta}$ and $B_{\alpha\beta}$ (which are defined in the parametric space Ω^2_U) of the cloth as

$$\overline{G}_{\alpha\beta} = T^{\alpha\beta\delta\gamma}G_{\delta\gamma} \quad (3)$$

$$\overline{B}_{\alpha\beta} = T^{\alpha\beta\delta\gamma}B_{\delta\gamma} \quad (4)$$

where $T^{\alpha\beta\delta\gamma} = \dfrac{\partial u_\delta}{\partial s_\alpha}\dfrac{\partial u_\gamma}{\partial s_\beta}$ is a fourth order tensor.

3. Fabric properties

Considering the warp and weft directions of the fabric as the co-ordinate axes of the space Ω^2_S yields the symmetric material modulus tensors. As a result, only five components, C_r^{1111}, C_r^{1122}, C_r^{1212}, C_r^{2211} and C_r^{2222} (r=1, 2) are non-zero. Comparing with Hook's law, these components can be expressed in terms of a set of measurable mechanical properties.

$$\left.\begin{aligned} C_1^{1111} &= \frac{D_1}{1-\upsilon_1\upsilon_2} \\ C_1^{1122} = C_1^{2211} &= \frac{\upsilon_2 D_1}{1-\upsilon_1\upsilon_2} \\ C_1^{2222} &= \frac{D_2}{1-\upsilon_1\upsilon_2} \\ C_1^{1212} &= D_s \end{aligned}\right\} \quad (5)$$

$$\left.\begin{aligned} C_2^{1111} &= \frac{H_1}{1-\mu_1\mu_2} \\ C_2^{1122} = C_2^{2211} &= \frac{\mu_2 H_1}{1-\mu_1\mu_2} \\ C_2^{2222} &= \frac{H_2}{1-\mu_1\mu_2} \\ C_2^{1212} &= H_s \end{aligned}\right\} \quad (6)$$

where D_1, D_2 are the tensile modulus in the principal axes, D_s is the shear modulus, ν_1, ν_2 are Poisson's ratio, H_1, H_2 are the bending modulus in the principal axes, H_s is the twisting modulus, μ_1, μ_2 are physical quantities analogous to Poisson's ratios.

The mechanical properties tensile modulus D_1, D_2, shear modulus D_s and bending modulus H_1, H_2 can be measured directly. Other parameters such as Poisson ratio and twisting modulus are obtained from the tensile, shear and bending modulus by using the following equations. If the cloth is stretched and bent respectively in a direction making an angle θ with the one of the principal axes direction, the membrane strain ε_θ and bending strain κ_θ are related to the in-plane stress σ^θ and out-of-plane stress τ^θ as $D_\theta = \dfrac{\sigma^\theta}{\varepsilon_\theta}$ and $H_\theta = \dfrac{\tau^\theta}{\kappa_\theta}$. By equating the strains ε_θ, κ_θ with (ε_{11}, ε_{22}, ε_{12}), (κ_{11}, κ_{22}, κ_{12}) and stresses σ^θ, τ^θ with (σ^{11}, σ^{22}, σ^{12}), (τ^{11}, τ^{22}, τ^{12}) respectively, the tensile modulus and the bending modulus in this direction are given by

$$\frac{1}{D_\theta} = \frac{1}{D_1}\cos^4\theta + \left(\frac{1}{D_s} - \frac{\nu_1}{D_1} - \frac{\nu_2}{D_2}\right)\cos^2\theta\sin^2\theta + \frac{1}{D_2}\sin^4\theta \quad (7)$$

$$\frac{1}{H_\theta} = \frac{1}{H_1}\cos^4\theta + \left(\frac{1}{H_s} - \frac{\mu_1}{H_1} - \frac{\mu_2}{H_2}\right)\cos^2\theta\sin^2\theta + \frac{1}{H_2}\sin^4\theta \quad (8)$$

The "Poisson ratios" of bending μ_1 and μ_2 are usually small and can be neglected. Taking the angle $\theta = 45^0$, the tensile Poisson ratio and the twisting modulus are written as

$$\nu_1 = \frac{1}{2}D_1\left(\frac{1}{D_1} + \frac{1}{D_2} + \frac{1}{D_s} - \frac{4}{D_{45^0}}\right),\ \nu_2 = \frac{D_2}{D_1}\nu_1 \quad (9)$$

$$H_s = \left(\frac{4}{H_{45^0}} - \frac{1}{H_1} - \frac{1}{H_2}\right)^{-1} \quad (10)$$

The detail derivation of these equations can be found in reference 9.

4. Discretization

The unit square domain of the parameter **u** is discretized as a mesh of m×n grid of nodes. The in-plane and out-of-plane energy for nodal point $\mathbf{x}_{ij}$ can be expressed as

$$U_G^{ij} = \frac{1}{8}\left(\overline{G}_{\alpha\beta}^{ij} - \overline{G}^{0ij}_{\alpha\beta}\right) C_1^{\alpha\beta\delta\gamma}\left(\overline{G}_{\delta\gamma}^{ij} - \overline{G}^{0ij}_{\delta\gamma}\right) \quad (11)$$

$$U_B^{ij} = \frac{1}{2}\left(\overline{B}_{\alpha\beta}^{ij} - \overline{B}^{0ij}_{\alpha\beta}\right) C_2^{\alpha\beta\delta\gamma}\left(\overline{B}_{\delta\gamma}^{ij} - \overline{B}^{0ij}_{\delta\gamma}\right), \quad \text{for } 1 \le i \le m$$

and $1<j<n$ (12)

where the superscript ij denotes the quantities at the nodal point $\mathbf{x}_{ij}$. The unit normal vector at $\mathbf{x}_{ij}$ is given as

$$\mathbf{n}_{ij} = \frac{(\mathbf{x}_{(i+1)j} - \mathbf{x}_{ij}) \times (\mathbf{x}_{i(j+1)} - \mathbf{x}_{ij})}{\left|(\mathbf{x}_{(i+1)j} - \mathbf{x}_{ij}) \times (\mathbf{x}_{i(j+1)} - \mathbf{x}_{ij})\right|} \quad (13)$$

The strain energy densities due to deformation are then rewritten as

$$U_G = \sum_{i=1}^{m}\sum_{j=1}^{n} U_G^{ij} \quad (14)$$

$$U_B = \sum_{i=1}^{m}\sum_{j=1}^{n} U_B^{ij} \quad (15)$$

The internal forces acting on the cloth can be categorized into in-plane forces and out-of-plane forces. Stretching and shearing are examples of in-plane forces while out-of-plane forces include bending and twisting. Let $\mathbf{f}_G^{ij} = \frac{\partial U_G}{\partial \mathbf{x}_{ij}}$ and $\mathbf{f}_B^{ij} = \frac{\partial U_B}{\partial \mathbf{x}_{ij}}$ be the in-plane and out-of-plane forces acting on the nodal point $\mathbf{x}_{ij}$ respectively, then these forces are

$$\mathbf{f}_G^{ij} = \frac{\partial U_G^{(i-1)j}}{\partial \mathbf{x}_{ij}} + \frac{\partial U_G^{ij}}{\partial \mathbf{x}_{ij}} + \frac{\partial U_G^{i(j-1)}}{\partial \mathbf{x}_{ij}} \quad (16)$$

$$\mathbf{f}_B^{ij} = \frac{\partial U_B^{(i-1)j}}{\partial \mathbf{x}_{ij}} + \frac{\partial U_B^{(i-1)(j-1)}}{\partial \mathbf{x}_{ij}} + \frac{\partial U_B^{i(j+1)}}{\partial \mathbf{x}_{ij}} + \frac{\partial U_B^{ij}}{\partial \mathbf{x}_{ij}} + \frac{\partial U_B^{i(j-1)}}{\partial \mathbf{x}_{ij}} + \frac{\partial U_B^{(i+1)j}}{\partial \mathbf{x}_{ij}} \quad (17)$$

with $U_G^{ij} = U_B^{ij} = 0$ if $i < 0$ or $i > m$ and $j < 0$ and $j > n$.

Balancing the forces acting on nodal point $\mathbf{x}_{ij}$ yields

$$\mathbf{f}_G^{ij} + \mathbf{f}_B^{ij} = \begin{cases} \rho_{ij}\mathbf{g} & \text{for static draping} \\ \rho_{ij}(\mathbf{g} - \ddot{\mathbf{x}}_{ij}) - \gamma_{ij}\dot{\mathbf{x}}_{ij} & \text{for dynamic draping} \end{cases} \quad (18)$$

where ρ_{ij} is the mass density of nodal point $\mathbf{x}_{ij}$, γ_{ij} is the damping coefficient and $\mathbf{g}$ is the gravitational acceleration. The static draping equation can be considered as a special case of the dynamic draping equation with $\ddot{\mathbf{x}}_{ij} = \dot{\mathbf{x}}_{ij} = 0$.

Assembling equation (18) for each of the grid points gives

$$\mathbf{M}\ddot{\mathbf{X}} + \mathbf{R}\dot{\mathbf{X}} + \mathbf{F}_G + \mathbf{F}_B = \mathbf{F} \quad (19)$$

where $\mathbf{M}$ is the mass matrix. It is a diagonal matrix with ρ_{ij} variables as the components. $\mathbf{R}$ is the damping matrix similar to $\mathbf{M}$ with γ_{ij} as the components. $\mathbf{F}_G$, $\mathbf{F}_B$, $\mathbf{X}$ and $\mathbf{F}$ are designated as the m×n column vectors corresponding to $\mathbf{f}_G^{ij}$, $\mathbf{f}_B^{ij}$, $\mathbf{x}_{ij}$ and $\rho_{ij}\mathbf{g}$ respectively. This equation can be solved with the initial conditions $\mathbf{X}_0 = \mathbf{X}_{-\Delta t}$ and $\dot{\mathbf{X}}_0 = 0$.

5. Collision detection and response

There are two types of collisions in modelling the draping effect of the cloth over an object:

self–cloth collision and

cloth–object collision.

Both cloth and object are reorganized into two triangular meshes respectively. Collision is detected between the nodal points of the cloth and each of the elements in the meshes. As shown in figure 1, $\mathbf{n}_{ij}$ is the unit normal vector at $\mathbf{x}_{ij}$. A vector $\mathbf{v}=\mathbf{q}-\mathbf{x}_{ij}$ is defined for checking the collision. The signed value d is calculated from the inner product $\mathbf{v}$ and $\mathbf{n}_{ij}$ such that $d=(\mathbf{v}, \mathbf{n}_{ij})$. The point $\mathbf{q}$ is on the triangle if d is zero while collision between the point and the triangle occurs when d is negative.

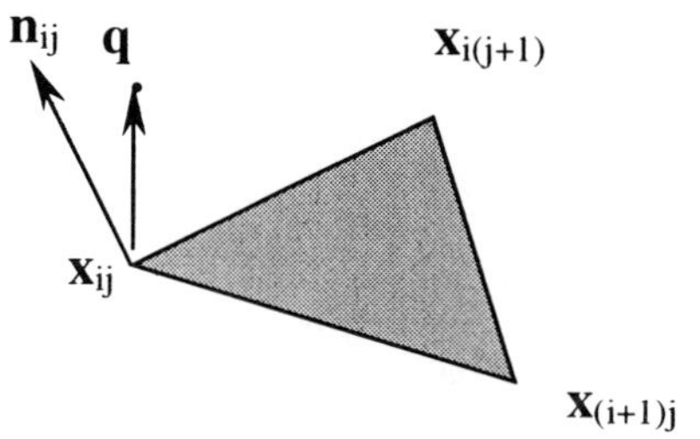

Figure 1: Collision detection

Force-based[2] method and displacement-based[10] method are two commonly used collision responses for cloth draping simulation. In this discussion, the displacement-based method is used for the cloth-object collision response, while the force-based method is employed for self-collision response. These responses are implemented by imposing constraints on the nodal points of the cloth at a specific time interval during the solving of equation (18). These constraints refer to the specific positions in space. They are categorized into:

1. user-defined displacement constraints such as the pasting point between the object and the cloth. These can be considered as the boundary conditions of the cloth at time instant t,
2. contact constraints such as the slipping of the cloth either over itself or the object.

Refer to reference 2 and 10 for detail description of collision detection. The pseudo code of calculating the state of the nodal point is listed as follows:

```
/*Read initial condition*/
input current_state                    /* x and ẋ of a nodal point*/
initialize time to zero

/*Initialize the types of collisions*/
initialize collision_type to null

/*Flag to indicate the convergence of the nodal point state*/
initialize convergence_flag to false

/*Loop if X does not converge*/
loop while convergence_flag=false
    increment time

    /*Compute the next X*/
    compute next_state

    /*X converge*/
    if next_state = current_state
        set convergence_flag to true
    else
        /*Set collision_type by collision detection function*/
        collision_type←collision_detection

        /*Pasting point exists between cloth and object*/
        /*Update the next state by user-defined displacement constraint*/
        if collision_type=pasting_point
            update next_state

        /*Self collision is detected*/
        /*Update next state by force-based method*/
        else if collision_type=self_collision
            update F
            compute next_state

        /*Cloth object collision is detected*/
        /*Update next state by displacement-based method*/
        else
            update next_state
        endif
    current_state←next_state
    output current_state
    endif
end loop
stop
```

6. Examples

Three examples are raised to illustrate the derivations. The work was performed on a Pentium 200, using C++ and OpenGL with Windows NT environment.

Example 1: Table cloth draping simulation

The content of the table cloth is 60% wool and 40% polyester. The material's mechanical properties measured by the Kawabata Evaluation System are:

Mass Density:	ρ=0.125 kg/m2,
Tensile Modulus:	D_1=3657.0 N/m,
	D_2=3267.0 N/m,
Shear Modulus:	D_s=36.8 N/m,
Bending Modulus:	H_1=17.6 μNm,
	H_2=12.7 μNm.
The constants:	
Poisson's ratios:	ν_1=0.125 and ν_2=0.177,
Twisting modulus:	H_s= 2.7 μNm,

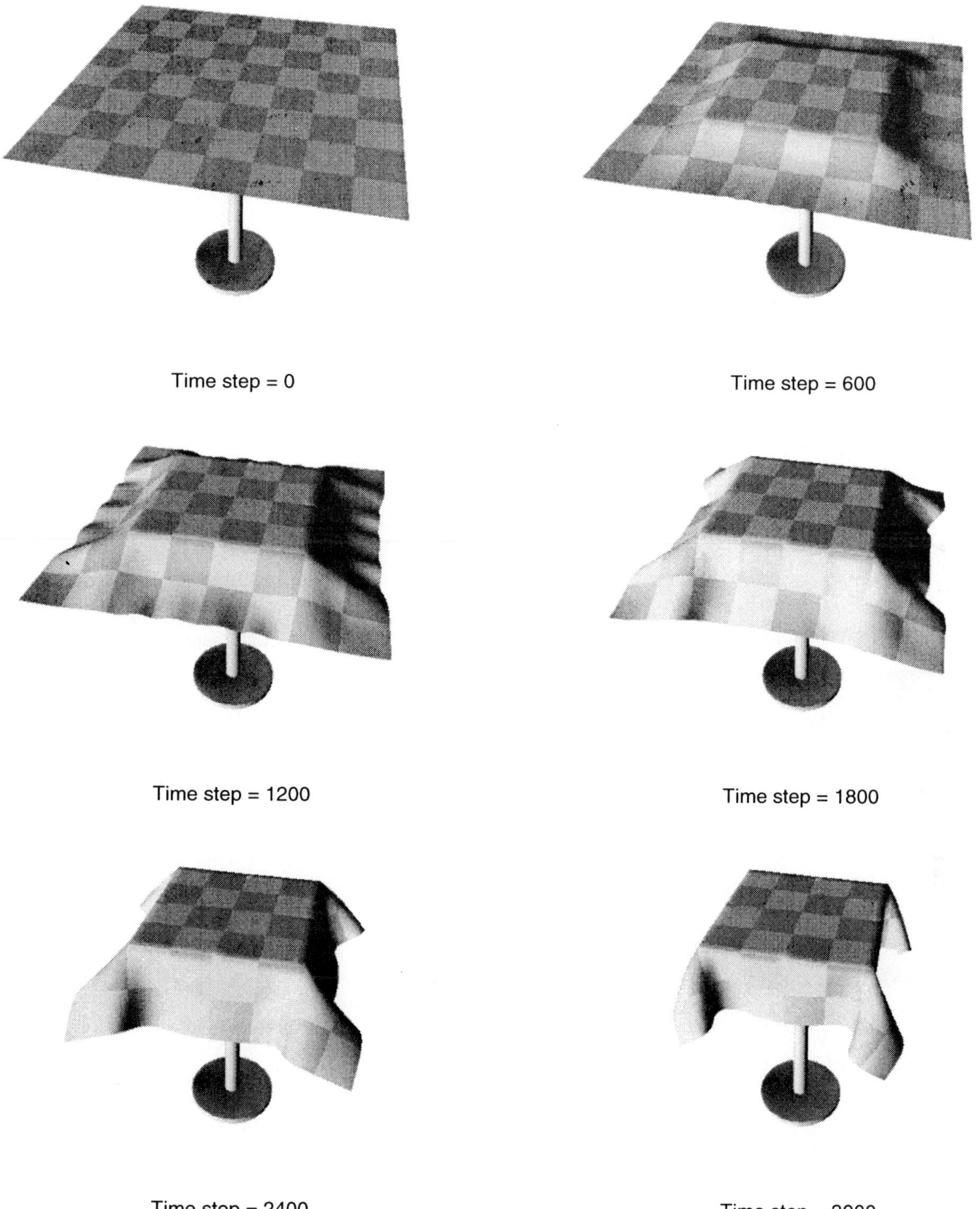

Time step = 0 | Time step = 600 | Time step = 1200 | Time step = 1800 | Time step = 2400 | Time step = 3000

Figure 2: Tablecloth Simulation

are calculated by equation (7) and (8).

The size of the table cloth pattern is 1m×1m and the table size is 0.5m×0.5m. Due to symmetry, only a quarter of the cloth is analyzed using a 21×21 mesh. The processing cpu time is about 14 minus. The simulation process is shown in Figure 2.

Example 2: Table cloth draping with various fabric properties

The same simulation in example 1 is performed by using a cloth with varying fabric properties. These properties are listed in table 1. Each simulation is then compared with the appearance of real table cloth situations as shown in figure 3.

Table 1: Material Properties of cotton and rayon

	ρ kg/m2	D_1 N/m	D_2 N/m	D_s N/m	H_1 μNm	H_2 μNm	H_s μNm
Cotton	0.208	3475	2865	191	17.6	12.7	2.7
Rayon	0.129	1847	3644	15.2	4.4	2.3	0.71

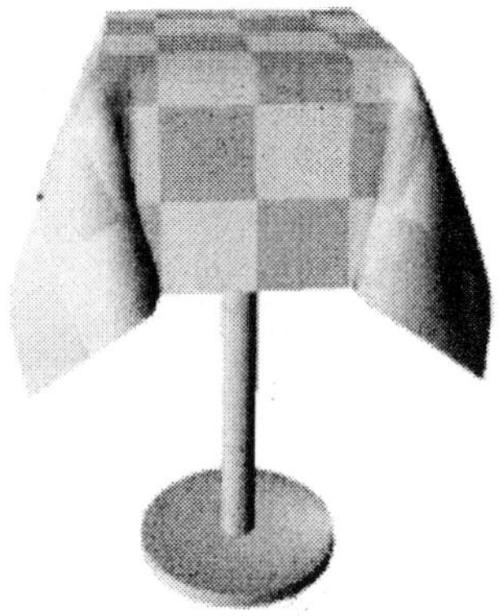

simulated result

real image

(a) Cotton

real image

simulated result

(b) Rayon

Figure 3: Tablecloth Simulation with Various Fabric Properties

Example 3: Dresses with various fabric properties
Three typical fabric properties, cotton, polyester and silk, are listed in table 2. A 96×96 mesh is used to model the dress pattern. The results are simulated with a mannequin as shown in figure 4.

Table 2: Material Properties of cotton, polyester and silk

Material Sample	ρ kg/m2	D_1 N/m	D_2 N/m	D_s N/m	H_1 μNm	H_2 μNm	H_s μNm
Cotton	0.208	3475	2865	191	87.7	61.4	11.42
Polyester	0.212	3071	1823	46.7	10.9	9.0	2.31
Silk	0.076	1388	827	12.4	0.81	0.76	0.089

Figure 4: Dresses Simulation with Various Fabric Properties

7. Discussion

An approach to insert various measurable fabric properties into the deformable model is proposed in this paper. It is developed based on the macroscopic view of the fabric properties of the whole piece of cloth. The fabric is an anisotropic material. Figure 5 shows the tensile modulus of cloth.Its property is considered in three different directions: the two principal directions and an axis of 45^0 from one of the principal axes. Fabric properties are measured directly in these three directions by the Kawabata Evaluation system and calculated from equation (7) and (8). Comparing with Breen's[6] model, which microscopically considered the fabric properties of each thread of the cloth, the proposed model requires much shorter computation time. As a result, cloth with various materials such as silk, cotton etc can be modelled efficiently and realistic cloth animation can be created.

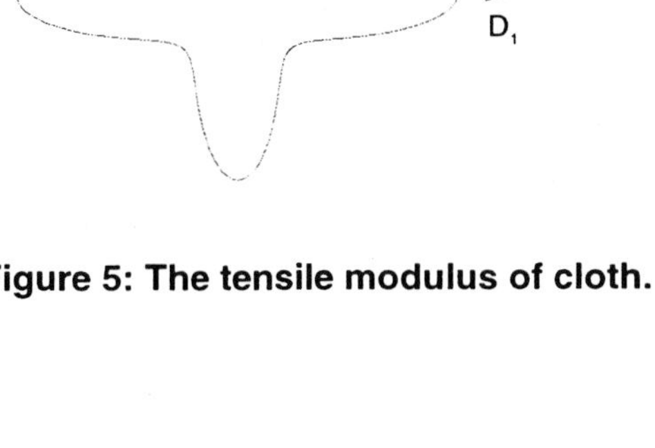

Figure 5: The tensile modulus of cloth.

In equation (19), $\mathbf{F}_G$ and $\mathbf{F}_B$ are considered as the in-plane force and out-of-plane force acting on the cloth. The force $\mathbf{F}_G$ can be written in the form of $\mathbf{F}_G=\mathbf{K}_G{\cdot}\mathbf{X}$, but not $\mathbf{F}_B$ since the unit normal vector $\mathbf{n}$ is involved. The characteristics of large bending deformation and small deformation in the membrane direction indicates that cloth material has a high coefficient of stiffness for the in-plane forces. In fact, the force $\mathbf{F}_G$ is most responsible for the stiffness of the equation (19).

Therefore, the dynamic equation (19) is re-arranged as

$$\mathbf{M}\ddot{\mathbf{X}} + \mathbf{R}\dot{\mathbf{X}} + \mathbf{K}_G\mathbf{X} = \mathbf{F} - \mathbf{F}_B \qquad (20)$$

where $\mathbf{K}_G\mathbf{X} = \mathbf{F}_G$ and is solved by conjugate gradient iteration. This implies that the force $\mathbf{F}_B$ of the previous time step is used to evaluate the deformation of the current time step. The error is negligible since the force $\mathbf{F}_B$ is small compared with the other force terms in the equation.

8. Conclusion

A cloth draping simulation based on cloth fabric properties is presented. The fabric properties are expressed in term of a set of measurable cloth mechanical properties and are fully incorporated into the deformable model based on elastic deformation theory. These properties control the draped effect intuitively. The prediction for the draping behaviour of cloths with different fabric properties is helpful to garment design and manufacturing. The proposed model generates a satisfactory visualisation with comparative small computational time.

9. Reference

1. C. Feynman, Modeling the appearance of cloth. Master Thesis, Dept. of EECS, Massachusetts Inst. of Technology, Cambridge, 1986

2. D. Teropoulos and K. Fleischer, Deformable models, Visual Computer 4, 1988, pp.306-331
3. X. Provot, Deformation constraints in a mass-spring model to describe rigid cloth behavior, Proceedings of Graphics Interface, 1995, pp.147-154.
4. G. Celniker and D. Gossard, Deformable curve and surface finite elements for free-form shape design. SIGGRAPH '91, Computer Graphics Proceedings, Annual Conference Series, 1991, pp.257-266.
5. B. Wang, Z. Wu, Q. P. Sun and M. M. F. Yuen, A deformation model of thin flexible surfaces. The Sixth International Conferences in Central Europe on computer Graphics and Visualization' 98. Plzen-Bory, 1998, pp.440-447
6. D. E. Breen, D. H. House and P. H. Getto, A physically-based particle model of woven cloth, Visual Computer, 8, 1992 pp.264-277.
7. D. Baraff and D. Witkin, Large steps in cloth simulation. SIGGRAPH '98, Computer Graphics Proceedings, Annual Conference Series, 1998, pp.43-54.
8. S. T. Tan, T. N. Wong, Y. F. Zhao and W. J. Chen, A constrained finite element method for modelling cloth deformation, Visual Computer, 15(2), 1999, pp.90-99.
9. Z. Wu Z, Material-based cloth draping simulation, Ph.D. Dissertation, Hong Kong University of Science and Technology, 1999.
10. Jen-Duo Liu, Ming-Tat Ko and Ruei-Chuan Chang (1996). Collision avoidance in cloth animation, Visual Computer, 12, 1996, pp.234-243.

Session 3

Model Representations and Parameterized Design

A Representation Independent Geometric Modeling Kernel

A. E. Middleditch* C. M. P. Reade** A. J. Gomes*
* *Centre for Geometric Modelling and Design, Brunel University, Uxbridge Middlesex UB8 3PH, UK*
alan.middleditch@brunel.ac.uk abel.gomes@brunel.ac.uk
** *Kingston University, Kingston Hill, Kingston, Surrey KT2 7LB, UK*
chris.reade@kingston.ac.uk

Abstract

This paper is concerned with "open kernel" geometric modelling systems. It uses industrial needs to identify requirements for the geometric objects of a proposed interface. The necessary mixed-dimension cellular objects are formally defined as the fundamental data abstractions for that interface. Such definitions are imperative for the construction of robust functions and the integrity of data transfer. As an example, formal specifications are provided for "set-like" object combination functions.

Acknowledgements

The Djinn API design decisions that are discussed or implicit in this paper reflect a consensus of the Djinn project team: Cecil Armstrong, Adrian Bowyer, Stephen Cameron, Jonathan Corney, Graham Jared, Ralph Martin, Alan Middleditch, Malcolm Sabin, Jonathan Salmon, and John Woodwark. The Djinn project was funded by the Engineering and Physical Sciences Research Council (EPSRC) of the UK.

Keywords: *Geometric modelling, solid modelling, computer aided design, geometric kernel, semianalytic sets, stratification, continuity, orientation, set-operations.*

1 Introduction

Many computer aided design (CAD), computer aided manufacturing (CAM) and other industrial applications exploit geometric modelling kernel software. Unfortunately, current kernels have incompatible interfaces and provide many system specific operations. They are unsuitable as "open kernels" primarily due to their interface's lack of representation independence, i.e. independence from their representation of geometric structure (e.g. CSG or B-Rep), and independence from their curve and surface formulations. In contrast, the Djinn application programming interface (API) to a geometric modelling kernel is representation-independent [1].

This paper develops formal definitions of the topology and geometry of the main data abstractions of the Djinn API. It deliberately avoids discussion of implementation issues in order to focus on these data abstractions. The lack of such abstractions leads to imprecise function descriptions in terms of the data representations and implementation language that are the cause of many problems with current geometric modelling kernels.

Traditional geometric modellers have functions to combine representations of solid objects that implement set-operations on solid point-sets. These functions are unsuitable for current industrial applications that require objects with cellular structure and cells of various dimension. Recent systems provide the necessary object structure and "set-like" combination functions, but the lack of a theoretical foundation leads to ill-defined objects and unpredictable function results. This paper provides such a foundation for Djinn object structure and "set-like" combinations.

2 Requirements

The geometric computations and objects relevant to CAD, CAM and other industrial applications that involve physical objects are discussed in [1]. This section provides a summary of the geometric requirements of such applications. Subsequent sections show how these requirements translate into API requirements.

Objects of mixed dimension ... Traditional geometric modellers were concerned with single dimension objects embedded in 2D- or 3D-space, e.g. drafting systems (1D-arcs in 2D-space) and 3D-design systems (2D-surfaces or 3D-solids in 3D-space). Such systems do not meet the needs of the many applications that require the ability to

0-7695-0562-7/00 $10.00 © 2000 IEEE

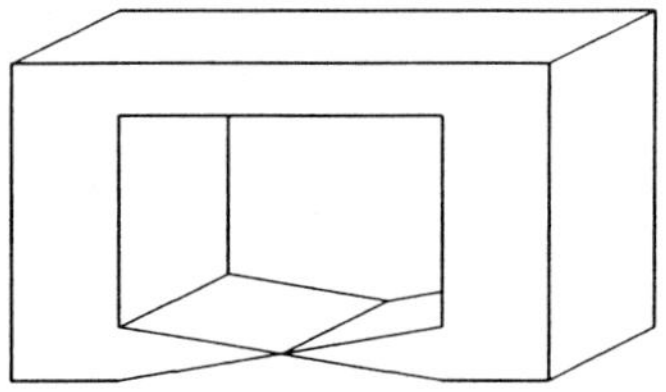

Figure 1. Non-manifold object

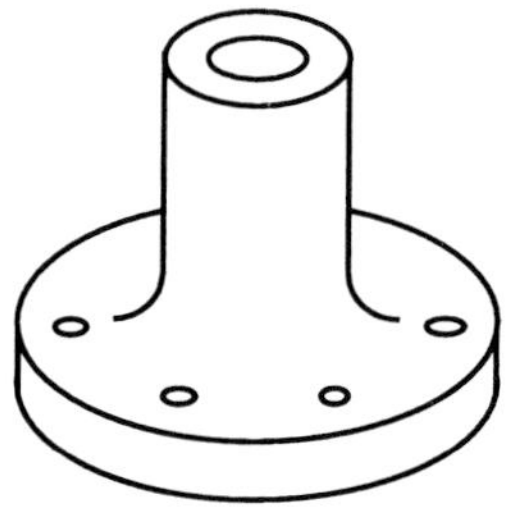

Figure 2. Bearing

model mixed-dimensional components in various spatial dimensions. For example:-

- A car body modelled as a surface rather than a solid must allow the subtraction of a solid to create the hole for a head-lamp.
- An end mill traverses a *curve* to create a *solid* slot and that surface must be derived from representations of the solid cutter and the required *solid* void.
- Parts of a solid component may be structurally analysed as a surface shell with a thickness *property*, and a bolt hole may be treated as a line segment with loads and constraints.

Non-manifold objects ... Some applications require non-manifold objects other than those of mixed dimension, e.g. solids with an edge incident to more than two faces (fig. 1). For example, a manifold 3D-region may be machined using a cutter centre-line that traverses a non-manifold region.

Object boundaries ... Parts of an object's boundary must be separately accessible in order to associate application data such as surface treatment or loads and constraints for analysis. Object boundaries are typically accessed via subdivision into faces, edges and vertices but, although a boundary is unique, its face and edge subdivisions are not [11]. The intuitive definition bounds a face by tangent discontinuities, but applications such as finite element analysis sometimes require a face that is a part of such an *intuitive* face and sometimes require adjacent intuitive faces to be considered as a single face.

Cellular Objects ... Many applications require the ability to partition an object model into a set of disjoint pieces to represent, for example, component assemblies, objects composed of different materials, finite element meshes and cracks.

Features ... Mechanical components, buildings and other objects are typically designed in terms of features and such features may be designed in terms of other features, thus creating a feature hierarchy. For example, the bearing of figure 2 might be defined in terms of a simple bearing with a flange containing a bolt hole circle. Such features may overlap, e.g. the flanges of two adjacent bearings.

Process plans for machining are sequences of operations that remove material from a work-piece. The component can be modelled as the initial work-piece together with a set of external void features. Such features also may overlap, e.g. some machine tool fixtures have orthogonal T-slots that cross each other.

Contrary to opinions in much of the feature literature, such design and manufacturing features are not features of the computer representation, e.g. not collections of solid boundary elements. They have the same geometric properties as the objects of which they are features.

Constraints ... Constraint based design systems enable geometry to be specified by relationships between bounding geometric entities rather than co-ordinate values. Sets of such relationships often relate an entity to more than one other entity. Thus, unlike features, constraint relationships constitute a graph [5].

* * * * *

In conclusion, design, analysis and manufacturing applications require geometric objects consisting of an aggregation of disjoint pieces that may be isolated points, arcs, faces and solids.

3 Djinn object geometry and topology

It is sometimes convenient to construct geometric models from intermediate models that are unbounded. Infinite points also can be generated by projective transformations that are necessary to generate perspective images and object distortions. However, since projective spaces lack a distance metric and are not orientable, *Djinn geometry is defined in Euclidean space using Cartesian co-ordinates.* Projective transformations are supported, but they are restricted to avoid points at infinity.

3.1 Shape

Early set-theoretic (CSG) geometric modelling systems apply set-operations to *solid* point-sets whilst early boundary (B-Rep) modellers operate on the closed boundaries of solids by exploiting topological concepts such as Euler operations. *Djinn objects are defined using point-sets* without losing access to bounding faces and edges (also defined as point-sets).

The class of point-set used in geometric modelling must be restricted to *sufficiently tame* to avoid extreme cases such as space filling curves. This has led to the prevalence of semi-algebraic sets in CAD systems, i.e. sets defined by polynomial and rational polynomial functions. Unfortunately, such sets do not provide sufficient geometric coverage to model many common engineering objects such as screw threads.

The semialgebraic sets are a sub-class of the semianalytic sets, i.e. subsets of $\mathbb{R}^n$ which are locally a finite union of subsets defined by a finite number of analytic equalities and inequalities. This definition involves analytic functions defined in a neighbourhood of each point of $\mathbb{R}^n$ because, unlike algebraic functions, analytic functions need not be defined at all points of $\mathbb{R}^n$. *Djinn object point-sets are semianalytic* because:-

- semianalytic sets are sufficiently general for free-form curves and surfaces including screw-threads and a large class of physical artifacts.
- semianalytic sets are closed under (locally-finite) union, finite intersection and complement, i.e. they form a total Boolean algebra [10].
- semianalytic sets are closed under (topological) closure, interior, exterior, frontier and boundary operations [10].
- semianalytic sets admit Whitney C^r-stratification ($r > 0$) [10] and the subsequently defined strong relative topological stratification.
- semianalytic sets are triangulable [3, 10].

The existence of triangulations enables orientability to be established and the ability to Whitney stratify facilitates the manipulation of discontinuities and singularities and provides a well defined cellular structure.

3.2 Cells

Assemblies and objects with cellular structure are represented by *Djinn objects consisting of a set of separately identifiable disjoint cells*. Hierarchies of overlapping features can be represented by subsets of an object's cells, and cracks can be represented by cells with specific attributes.

Traditional geometric modelling systems prohibit non-solid geometry such as isolated edges and faces, and cracks in solids by the use of *regular closed sets*, i.e. sets equal to the closure of their interior [4, 9]. But there is a requirement for such entities and also for separately identifiable frontiers with associated (possibly non-geometric) properties. *Djinn accommodates such objects and non-manifold objects with cells that are manifolds of various dimensions*, i.e.

- 3D-solids without their bounding faces, edges or vertices (3-manifolds),
- 2D-surface patches without their bounding edges or vertices (2-manifolds),
- 1D-arcs without their end-points (1-manifolds), and
- 0D-points (0-manifolds).

The missing edges and vertices constitute frontiers; the *frontier* of a point-set X is the set of boundary points of X that are not in X, i.e. $frn(X) = clo(X) - X$. It is usually a point-set one dimension lower, but degenerate frontiers of lower dimension are possible. For example, an unbounded conical surface in $\mathbb{R}^3$ without its apex has that apex as its 0-dimensional frontier.

In order for Djinn objects to be closed under "set-like" combination, *a Djinn cell can have an infinite number of connected components*. For example, an unbounded sine curve and the X-axis are valid Djinn cells whose set-intersection is an infinite set of isolated points which can constitute a single Djinn cell. However, a Djinn object is at least *locally finite*, i.e. each point has a neighbourhood that meets only finitely many connected components of cells and each cell has frontier cells of lower dimension. This prevents, for example, a line from being treated as an infinite collection of 0-manifolds (points).

Some point-sets have an empty frontier, e.g. a spherical surface or a plane. In traditional boundary modelling systems such entities are used as construction geometry and solid objects are represented as topological structures of faces, edges and vertices. This sometimes demands artificial frontiers, e.g. the representation of a spherical face with two vertices and an edge. *Djinn abandons the distinction between constructional and topological entities and supports curves, surfaces and solids that can have empty frontiers*.

Since applications such as finite element analysis sometimes require slope discontinuities within edges and faces of solids, *Djinn imposes no restriction on the degree of a cell's derivative discontinuity*. They can contain cusps and disjoint pieces. For example, a cube can have one, six or any other number of faces.

3.3 Object structure

In order to provide each cell with an explicit frontier and simplify navigation of the cellular structure, *Djinn objects satisfy the frontier condition*, i.e. the frontier of each cell is a subset of the object's cells. The frontier condition on cells

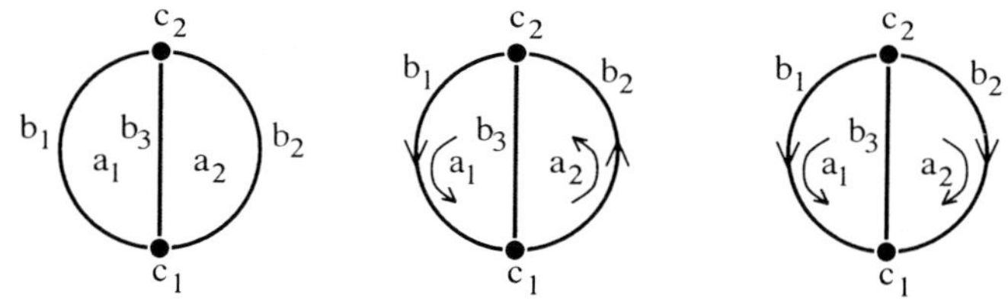

Figure 3. Cell orientation

is stronger than the frontier condition on their connected components. For example, if the 2D-regions a_1 and a_2 of figure 3 are separate cells, the former requires the edges b_1 and b_2 to be separate cells whilst the latter does not .

Both a Whitney stratification (of a closed set) and a CW-complex [13] is locally finite and satisfies the frontier condition. Since the point-set of a Djinn object is semianalytic, it can be Whitney stratified and further decomposed into a CW-complex of topological cells. However, the Djinn notion of cell need not be identified with such strata nor topological cells in order to satisfy application requirements. It suffices that *A Djinn object is a finite C^0-stratification of a closed set whose strata are Djinn cells that satisfy the strong relative topological condition*:

- A **strong relative topological** stratification of X is a stratification in which each stratum M contains points of the same topological type with respect to $M \cup M'$ for each stratum M'. Points have the same topological type with respect to a topological space if they have homeomorphic neighbourhood intersections with that space.

The strong relative topological stratification drops the smoothness constraint of Whitney but retains the topological constraint. The need for the topological constraint is evident from a conical surface in $\mathbb{R}^3$ that is stratified into a line passing through its apex, and the rest of the surface. This is not a strong relative topological stratification because the apex has a different topological type (with respect to the surface stratum) from that of all other points on the line. There is a unique minimal stratification of the cone which satisfies the strong relative topological condition by separating the apex from the remainder of the surface. Thus, since a Djinn cell is not necessarily smooth, *a Djinn object is not necessarily a Whitney stratification*, and since a Djinn cell is not necessarily bounded nor homeomorphic to an open ball, *a Djinn object is not necessarily a cell-complex* (fig. 4).

3.4 Orientation

Some applications require computations that involve the integration of a differential form over a point-set, but such forms are defined *only on orientable* smooth manifolds ([8], p.164). Therefore, the point-set of *each Djinn cell is oriented*. This eliminates the representation of a Möbius band as a single cell, but it can be represented as a Djinn object if the continuous loop is broken by an edge cell. A connected Djinn cell has one of two possible orientations, whilst a cell with n connected components has one of 2^n different orientations.

Each oriented Djinn cell induces usually one but possibly two orientations on each component of a frontier cell one dimension lower. In the former case, the induced orientation may (fig. 3b) or may not (fig. 3c) be equal to that of the frontier cell(s). This relative orientation must be accessible at the Djinn interface. For example, when a pressure load is integrated over a face, that face must be oriented

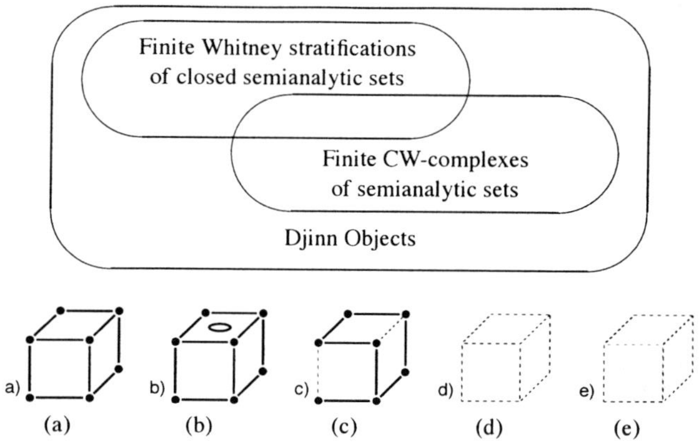

	a	b	c	d	e
Whitney stratification	yes	yes	no	no	no
CW-complex	yes	no	yes	no	no
Djinn object	yes	yes	yes	yes	no

a) A solid cube stratified into its interior, separate vertices, edges, and faces.

b) A cylindrical hole in the stratification of (a) such that its cylindrical face and its two edges with the top and bottom faces constitute new cells. This decomposition is not a CW-complex because the cylindrical face is not homeomorphic to $\mathbb{R}^2$ and the circular edges are not homeomorphic to $\mathbb{R}^1$.

c) The stratification of (a), modified by coalescing three pairs of adjacent faces with their common edge. This is not a Whitney stratification because the faces are not C^1.

d) A solid cube stratified into a single surface cell and an interior cell. This is not a CW-complex because the face is not homeomorphic to $\mathbb{R}^2$ and it is not a Whitney stratification because the face is not C^1.

e) A solid cube stratified into its interior and an infinite number of surface points. This is not a CW-complex because it is not closure finite. It is neither a Whitney stratification nor a Djinn object because the points do not constitute a locally finite set.

Figure 4. Stratifications, complexes and cells

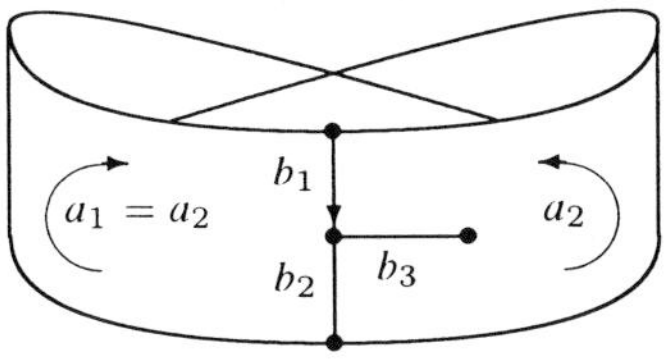

Figure 5.

consistently with the solid it bounds or the result adjusted to compensate. Meaningful relative orientation for multi-component cells demands that *each Djinn object has consistently oriented cells*, i.e. each cell a has *uniform relative orientation* with respect to each cell b in its frontier that is one dimension lower. Cells a and b have uniform relative orientation if and only if the orientation of each component of b is that induced by a or every orientation is the opposite of that induced by a. Unfortunately, a minimal strong relative topological stratification may require further decomposition before it admits consistent orientations. For example, components b_1, b_2 and b_3 of figure 5 could be part of the same stratum in a strong relative topological stratification, but b_1 and b_2 have the same orientation induced from both sides whilst b_3 has opposite orientations induced.

3.5 Definitions

A Djinn cell is an orientable semi-analytic (topological) C^0-submanifold of Euclidean space, i.e. $\mathbb{R}^n$ $(-1 \le d \le n)$. It is not necessarily connected nor smooth, but semi-analyticity ensures that it is at least piecewise smooth. It can have infinitely many connected components, may be unbounded and need not be homeomorphic to a ball (of any dimension).

Cell geometry is specified by the following Z-schema for a type *Cell* (see appendix for an explanation of our use of Z-notation). In this schema, **Orientations**(C, d, S) denotes the set of possible orientations of the set C of a d-manifold's connected components in a space S, and **Orientable**(C, d, S) is true if each component has two possible orientations. **Semianalyticsets**$(\mathbb{R}^n)$ denotes all the semianalytic subsets of $\mathbb{R}^n$, **Submanifolds**(d, S) denotes the set of (topological) submanifolds of S of dimension d and **ConnectedComps**(X, S) denotes the set of all connected components of X in a space S. Thus, the cell is specified by selector functions that provide its spatial dimension (*SDim*), manifold dimension (*MDim*), point-set (*PointSet*), orientation (*Orient*), connected components (*Components*) and those components which are oriented:-

Cell

$SDim \in \{0 \ldots 3\}$	[spatial dimension]
$MDim \in \{-1 \ldots SDim\}$	[manifold dimension]
$PointSet \in \mathbb{P}(\mathrm{R}^{SDim})$	
$Components \in \mathbb{P}(\mathbb{P}(\mathrm{R}^{SDim}))$	[connected components]
$OComponents \in \mathbb{P}(\mathbb{P}(\mathrm{R}^{SDim}))$	[oriented components]
$Orient \in$ **Orientations**$(OComponents, MDim, \mathrm{R}^{SDim})$.	[component orientations]
$PointSet \in$ **Semianalyticsets**(R^{SDim})	[cell is semianalytic]
$PointSet \in$ **Submanifolds**$(MDim, \mathrm{R}^{SDim})$	[cell is MDim-submanifold]
$(PointSet = ?) \Leftrightarrow (MDim = -1)$	[convention]
$PointSet = \bigcup Components$	
$Components =$ **ConnectedComps**$(PointSet, \mathrm{R}^{SDim})$	
Orientable$(Components, MDim, \mathrm{R}^{SDim})$	[cell is orientable]
$OComponents \subseteq Components$	

A Djinn object is a finite collection of consistently oriented disjoint Djinn cells that constitute a strong relative topological stratification of a closed semianalytic subset of $\mathbb{R}^n$. It is locally finite and satisfies the frontier condition.

A Djinn object is specified by the following Z-schema using **StrongRT**(X, R) to denote the set of all strong relative topological stratifications of a subset X of a space R and denoting a consistently oriented stratification in $\mathbb{R}^n$ by the predicate **ConsOrientation**$(S, \mathbb{R}^n)$ which is true if and only if the set of strata S are completely and consistently oriented. Thus, Djinn objects form a type *Object* that has selector functions to provide an object's spatial dimension (*SDim*), its point-set (*PointSet*) and its cells (*Cells*). The cells are **DCells**, i.e. the previously specified cells restricted to be completely oriented and non-empty.

Object

$SDim \in \{0 \ldots 3\}$	[spatial dimension]
$Cells \in \mathbb{F}$ DCell	[finite set of non-empty completely oriented cells]
$PointSet \in \mathbb{P}(\mathrm{R}^{SDim})$	
$PointSet = \bigcup_{c \in Cells} c.PointSet$	
$PointSet = clo(PointSet)$	[closed point-set]
$\forall c \in Cells \bullet c.SDim = SDim$	[common cell spatial dimension]
$\forall a, b \in Cells \bullet a \neq b \Rightarrow a.PointSet \cap b.PointSet = ?$	[cells are disjoint]
$\{c.PointSet \mid c \in Cells\} \in$ **StrongRT**$(PointSet, \mathrm{R}^{SDim})$	[cells form a strong relative topological stratification]
ConsOrientation$(\{\langle c.PointSet, c.Orient\rangle \mid c \in Cells\}, \mathrm{R}^{SDim})$	[the stratification is consistently oriented]

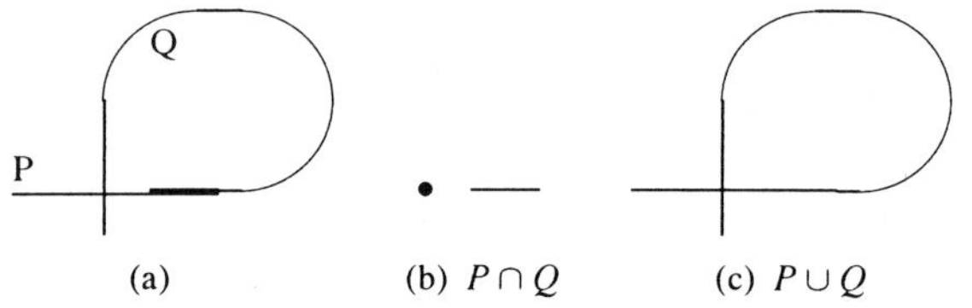

Figure 6. Set-combinations of manifolds

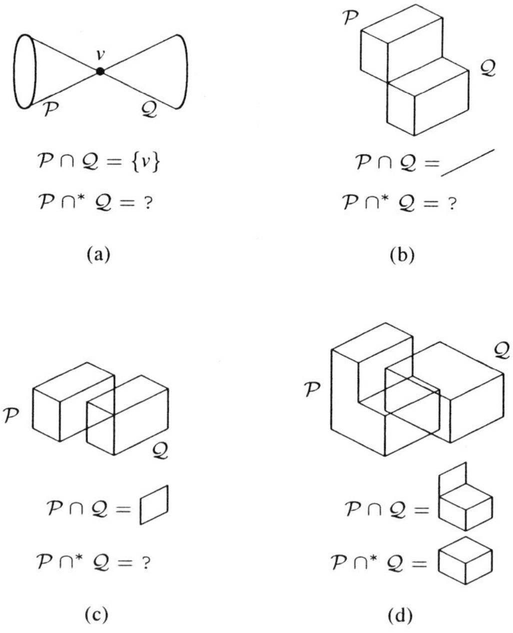

Figure 7. Regularised set-operations

4 Set-operations

4.1 Conventional set operations

The conventional set-combination of Djinn object point-sets does not accommodate their cellular structure. "Set-like" combinations of objects that preserve cellular structure require constituent operators to combine individual cells. The set-combination of manifold cells must consist of manifolds, but the conventional set-combination of two manifolds is not necessarily a manifold.

Figure 6a shows a horizontal line P and a curve Q that intersects P in a co-incident segment (depicted thick) and an isolated point (fig. 6b). The result of the set-intersection (fig. 6b) or the set-union (fig. 6c) of manifolds P and Q is not a manifold; $P \cap Q$ is not dimensionally homogeneous and $P \cup Q$ is not locally homeomorphic to a ball of the same dimension everywhere. Thus, further stratification is necessary to create manifold cells.

If a and b are point-sets in $\mathbb{R}^n$ of dimension i and j respectively:-

- Every dimension in $0 \ldots min(i,j)$ can arise for manifolds in $a \cap b$ when i and j are less than n. For example, two spherical surfaces in $\mathbb{R}^3$ can be coincident (a 2-manifold intersection), intersect in a (1-manifold) circle, or touch at a (0-manifold) point. More irregular surfaces can have intersections of all three dimensions simultaneously.
- When $i = n$ or $j = n$, $a \cap b$ is either empty, or contains a single manifold of dimension $min(i,j)$.

4.2 Regularised set-operations

Early solid modellers supported only homogeneously 3-dimensional objects, ensuring homogeneity by regularising conventional set-combinations of closed point-sets. These operators apply interior (int) and closure (clo) to the conventional set-intersection in order to remove unwanted edges and faces in the intersection of touching solid objects [4, 9]. For example, the regularised set-intersection $(clo(int(P \cap Q)))$ of homogeneously 3-dimensional closed point-sets P and Q in figures 7a, 7b and 7c is the empty set because their conventional set-intersection is dimension deficient; its dimension is less than that of the embedding space. The regularised set-intersection of figure 7d is dimension complete because regularisation removes the dimension deficient dangling face.

In contrast to early geometric modelling systems, current systems that admit cells of different dimensions usually represent cells by point-sets that are relatively open. Cracks in the union of touching point-sets are eliminated by the open regularised set-operators:-

$$P \cup^* Q = int(clo(P \cup Q))$$

$$P \cap^* Q = int(clo(P \cap Q))$$

$$P -^* Q = int(clo(P - Q))$$

Since both open and closed regularisation always reduce dimension deficient point-sets to the empty point-set, they are incompatible with the requirement to handle point-sets of different dimension.

5 "Set-like" combination of Djinn objects

Object combination should not affect cellular structure in regions where the point-sets of the operands do not overlap. In a region of overlap, possible alternatives retain all the original cell frontiers, no cell frontiers, or frontiers from only one operand. For example, a rectangular object subdivided by a diagonal cell and one subdivided into four rectangular cells (fig. 8a), may be united to create the unions of

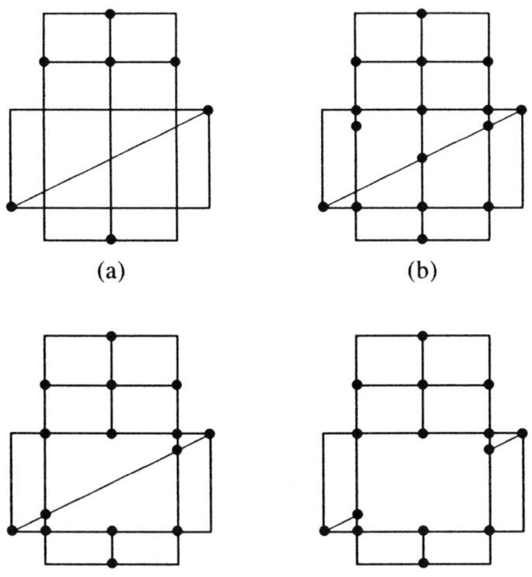

Figure 8. Possible cellular object unions

figures 8b, 8c and 8d. Since applications may or may not require the cracks and surfaces eliminated by the regularised set-operators, *the Djinn operators retain all cell frontiers*, possibly subdivided. An application can obtain the alternatives and/or discard degenerate intersections, using Djinn de-partitioning and cell elimination functions respectively, thus achieving a "regularisation" appropriate to the application's requirements.

5.1 Orientation inheritance

Djinn "set-like" object intersection is defined in section 5.3 in terms of the pairwise intersection of cells and subsequent stratification. Each cell created by such an intersection may inherit its orientation from the (one or two) "parent" cells from which it derives. Its point-set is a subset of each parent and no other cell, but its dimension can be lower than that of either parent.

Each component of a cell intersection with the same dimension as exactly one parent cell, inherits its orientation from the relevant component of that parent. Orientation is inherited also if the dimension is the same as that of both parents, and the relevant component orientations agree. If they do not agree an oriented result requires cell intersection to be non-symmetric or the orientation to be arbitrary.

If the dimension of a component of a cell intersection is more than one lower than that of either parent, it is not possible to deterministically compute an orientation from the parents' orientations. If it is exactly one dimension lower, deterministic computation may or may not be possible. For example, the orientation of an edge at the intersection of two surfaces can be computed from the cross product of their surface normals, but this product does not exist if the surfaces are tangential, and the surface normal is not defined at a tangent discontinuity. In general, the latter problem cannot be resolved using the normal of an adjacent regular region because oppositely directed cross products may result from regions on opposite sides of the discontinuities.

A Djinn object has completely oriented cells that are consistently oriented, but all components of cells that result from "set-like" combination may not inherit their orientation. In the subsequent definitions, inherited orientations are denoted by the relation **InheritOrient** between a cell set and two objects; **InheritOrient** holds if the (possibly partial) orientation of each cell is inherited from the objects.

5.2 Orientation Consistency

Consider a cell set that results from cell intersection and subsequent stratification to separate manifolds. Before it can be used to constitute a Djinn object, it must have all its cell components oriented without violating uniform relative orientation between cells. Such completions are denoted by the relation **Completion**; C' is a completion of C if it is a set of completely oriented cells that satisfy the consistent orientation condition, and there is a bijection (1:1 correspondence) f between the partially oriented **Cells** C and the completely oriented **DCells** C' such that corresponding cells have the same point-set and, if they are both oriented, the same orientation.

Completions are not necessarily unique and may not even exist (see the example in figure 11, section 5.5). When no completion exists, consistent orientation can be achieved by subdividing multi-component cells. The set of possible results is denoted by the **Splits** relation; C' is a split of C if it has the same point-set and the components of each cell in C' are components of a cell in C.

Relational composition of splits and completions yields the relation **CSplits** for all possible completed splits; $(C, C') \in$ **CSplits** whenever C' is the completion of a split of C. It is preferable to minimise the subdivision performed to obtain a completion, i.e. to avoid subdividing already completed sets. Such minimal **CSplits** are denoted by the relation **ConsOrients** in subsequent definitions.

5.3 Intersection

Djinn objects can not be intersected by pairwise intersecting their cells and separately stratifying the result of each cell intersection to obtain manifolds, because the union of the separate stratifications may violate the frontier condition. For example, consider the angled line in $\mathbb{R}^3$ (fig. 9b):-

$$b_2 = \{\langle 0, y, 0\rangle \mid y \leq 0\} \cup \{\langle -y, y, 0\rangle \mid y > 0\}$$

and the surface which is the XY plane without this line:-

$$b_1 = \{\langle x, y, 0\rangle\} - b_2$$

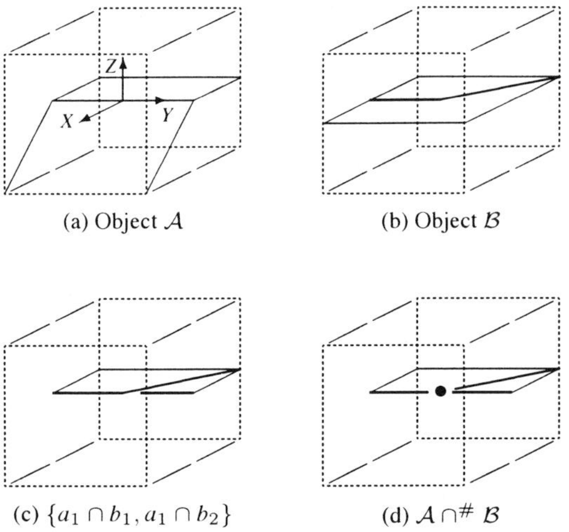

(a) Object $\mathcal{A}$ (b) Object $\mathcal{B}$

(c) $\{a_1 \cap b_1, a_1 \cap b_2\}$ (d) $\mathcal{A} \cap^{\#} \mathcal{B}$

Figure 9. Object intersection

Let $\mathcal{B}$ (fig. 9b) be an object that consists of two cells with point-sets b_1 and b_2 respectively, and let $\mathcal{A}$ (fig. 9a) consist of a single cell with a point-set a_1 that constitutes an angled plane:-

$$a_1 = \{\langle x, y, 0\rangle \mid x \leq 0\} \cup \{\langle x, y, -x\rangle \mid x > 0\}$$

Now consider the intersection of $\mathcal{A}$ and $\mathcal{B}$ computed by the pairwise intersection of cells. The intersection of b_2 and a_1 simply yields b_2, but the intersection of b_1 and a_1 yields half the XY-plane without the angled line b_2. This 2D-region must be stratified (fig. 9c) into two point-sets, half the plane without the Y-axis (s) and half the Y-axis (e):-

$$s = \{\langle x, y, 0\rangle \mid x < 0\} - \{\langle -y, y, 0\rangle \mid y > 0\}$$

$$e = \{\langle 0, y, 0\rangle \mid y > 0\}$$

The frontier of line e in figure 9c is the origin but it is not a separate stratum, and the collection $\{b_2, s, e\}$ is not a strong relative topological stratification because it violates the frontier condition. Further stratification is necessary to subdivide the angled line b_2 at the origin (fig. 9d).

The definition of object intersection uses **RTCellSet** to denote a cell set that constitutes a strong relative topological stratification, and **minStrongRT**$(C, \mathbb{R}^n))$ to denote the minimal stratification of the set $\bigcup C$ that contains a partition of each point-set C. Thus, object intersection is defined via a single such minimal stratification of cell point-set intersections such that each cell component inherits its orientation whenever possible (by satisfying **InheritOrient**) and the remaining cell components are subdivided if necessary and assigned orientations to consistently orient the result (by satisfying **ConsOrients**). The (not necessarily unique) result is defined by the indeterminate Djinn object intersection:-

$\cap^{\#} \in$ Object $\times$ Object $\rightarrow\!\!\!\!\!+$ Object

---

$\forall \mathcal{X}, \mathcal{Y}, \mathcal{Z} \in$ Object $\bullet$ $(\mathcal{Z} = \mathcal{X} \cap^{\#} \mathcal{Y}) \Rightarrow$
  $(\mathcal{X}.SDim = \mathcal{Y}.SDim) \wedge (\mathcal{Z}.SDim = \mathcal{X}.SDim)$
    [invariant spatial dimension]
  $\exists C \in$ **RTCellSet** $\bullet$
    $\{c.Pointset \mid c \in C\} =$ **minStrongRT**$(D, \mathbb{R}^{\mathcal{Z}.SDim})$,
    $D = \{a \cap b \mid a \in \mathcal{X}.Cells, b \in \mathcal{Y}.Cells\}$.
    [stratification of pairwise cell intersections]
    $(C, \mathcal{X}, \mathcal{Y}) \in$ **InheritOrient**
    $(C, \mathcal{Z}.Cells) \in$ **ConsOrients**

Since an object's cells are disjoint, it is clear from this definition that the point-set of any set-intersection of objects is the intersection of the point-sets of its arguments.

5.4 Difference

One Djinn object can be subtracted from another by subtracting all its cells from each cell of the latter. Cells remain disjoint after such subtraction, but cells that result from non-trivial subtractions do not satisfy the frontier condition because they do not unite to form a closed set. It is shown in [6] that *the complement of a Djinn object's point-set is a valid cell point-set* and that a complementary object can be created by amalgamating the complement cell with the object's boundary cells. Therefore, Djinn object difference $(-^{\#})$ can be defined in terms of object complement $(\neg)$ and intersection, validity following from the validity of those operations:-

$-^{\#} \in$ Object $\times$ Object $\rightarrow\!\!\!\!\!+$ $\mathbb{P}$ Object

---

$\forall \mathcal{X}, \mathcal{Y} \in$ Object $\bullet$
  $(\mathcal{X}.SDim = \mathcal{Y}.SDim) \Rightarrow (\mathcal{X} -^{\#} \mathcal{Y} = \mathcal{X} \cap^{\#} (\neg \mathcal{Y}))$

Since Djinn object difference completes the frontier of the naive subtraction with the relevant cells of the second operand, object difference leaves an imprint of the subtracted cell's frontier cells. Figure 10d shows an object $\mathcal{A}$ consisting of a 2D cell and its boundary cell, and a 2D object $\mathcal{C}$ partitioned into two 2D cells together with the necessary edge and vertex cells. When $\mathcal{C}$ is subtracted from $\mathcal{A}$, edges and vertices of $\mathcal{C}$ leave an imprint on the frontier of the result (fig. 10e). This enables a surface to be subtracted from a solid to create internal cells, possibly representing a crack, (figs. 10a and 10b).

5.5 Union

The union of two Djinn objects consists of

- cells created by the pairwise set-intersection of cells from each operand, subsequently stratified if necessary,

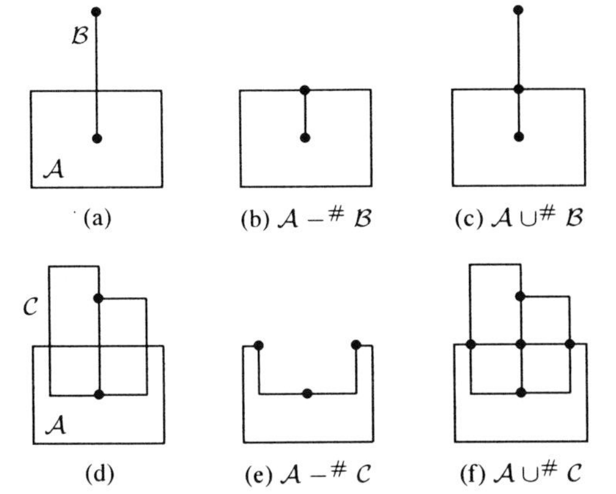

Figure 10. Object difference and union

- those parts of the first operand's cells outside the point-set of the second operand, and
- those parts of the second operand's cells outside the point-set of the first operand.

For example, consider the rectangular object $\mathcal{A}$ and the line object $\mathcal{B}$ in figure 10a, each with a single interior and frontier cell. Their union is the 7-celled object of (fig. 10c). The union of the rectangular object $\mathcal{A}$ of figure 10d and the object $\mathcal{C}$ with two 2D, three 1D and two 0D-cells is the object of figure 10f with five 2D, nine 1D and five 0D-cells.

Figure 11 shows the union of two rectangular surface objects in $\mathbb{R}^3$, each with a single interior and a single frontier edge cell, and one surface almost forming a Möbius strip. The intersection of these surfaces contains a 0D-cell of four points, two 1D-cells each with 2 components, and a 2D-cell with two components. Only one of the 2D-components inherits its parents' orientation; the other has no inherited orientation because the parents' orientations are different. Each component of the 1D-cell that is interior to the Möbius strip inherits its orientation from the edge cell of the other object, but these components have opposing orientations with respect to that part of the latter object outside the Möbius strip. These components can be consistently oriented only if they are separated into 2 distinct cells. This in turn demands the separation of the 0D points into distinct cells in order to restore the strong relative topological stratification.

The subtraction of an object from a cell either yields a single manifold or it is empty, i.e. a **Cell** but not necessarily a **DCell**. This operation naturally extends to the subtraction ($-^c$) of an object from a set of cells whilst excluding empty cells. The point-set of the result is not necessarily closed but the cells do constitute a strong relative topological stratification. Thus, the set of cells in a Djinn object union consists of the cells of their intersection $\cap^\#$ and the

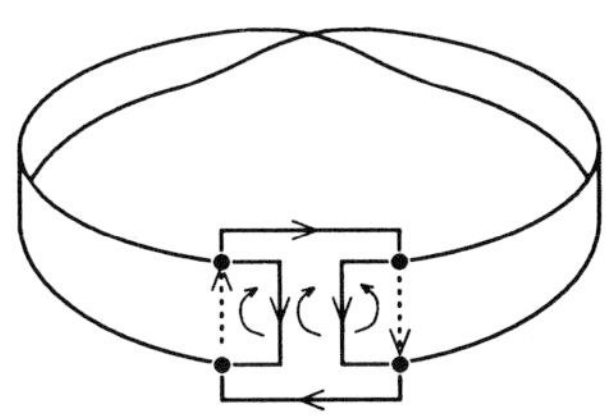

Figure 11.

cells of the two possible differences $-^c$, subjected to a minimal splitting completion and the assignment of consistent relative orientations:-

$$\cup^\# \in \text{Object} \times \text{Object} \nrightarrow \text{Object}$$

$$\begin{aligned}
&\forall \mathcal{X}, \mathcal{Y}, \mathcal{Z} \in \text{Object} \bullet (\mathcal{Z} = \mathcal{X} \cup^\# \mathcal{Y}) \Rightarrow \\
&\quad (\mathcal{X}.SDim = \mathcal{Y}.SDim) \wedge (\mathcal{Z}.SDim = \mathcal{X}.SDim) \\
&\quad \exists C \in \mathbf{RTCellSet} \bullet \\
&\qquad \{c.Pointset \mid c \in C\} = \mathbf{minStrongRT}(D, \mathrm{R}^{\mathcal{Z}.SDim}), \\
&\qquad D = \{a \cap b \mid a \in \mathcal{X}.Cells, b \in \mathcal{Y}.Cells\}. \\
&\quad \exists C' \in \mathbf{RTCellSets} \bullet \\
&\qquad C' = C \cup (\mathcal{X}.Cells -^c \mathcal{Y}) \cup (\mathcal{Y}.Cells -^c \mathcal{X}) \\
&\qquad (C', \mathcal{X}, \mathcal{Y}) \in \mathbf{InheritOrient} \\
&\qquad (C', \mathcal{Z}.Cells) \in \mathbf{ConsOrients}
\end{aligned}$$

6 Summary and conclusions

This paper summarises the important mathematical background necessary for the formalisation of a general class of object for geometry based applications in order to provide a firm foundation for such objects. Arrows in figure 12 indicate how the proposed object abstraction satisfies industrial requirements, and show implications between object properties. When two arrows are directed at a property, that property is due to one of the two causes, but if the arrows are combined with an &, both causes are necessary. For example, semianalytic sets are chosen as the basis for Djinn geometric models because such sets have the requisite geometric coverage and they can be stratified.

Formal definitions in terms of point-sets are provided for the cells and objects of the Djinn API [1]. Thus, the point-set of a Djinn cell is an orientable semianalytic manifold and a Djinn object is a generalisation of a finite Whitney stratification and a finite CW-complex. It is a finite stratification of a semianalytic point-set whose strata are cells that satisfy the frontier condition and have uniformly related frontier orientations. These definitions do not restrict the Djinn API to a particular modelling paradigm like other recent APIs (e.g. [2]).

The Djinn restriction to objects that satisfy the frontier condition enables sensible navigation of cellular structure via frontier relations. Its restriction to objects whose cells taken pairwise have the same relative orientation for all components, is necessary for numerical integrations that involve orientation. The paper shows that these restrictions also lead to the need for the strong relative stratification introduced in [7], and further refinement to enable consistent relative orientations between the connected components of adjacent cells. These complexities significantly complicate "set-like" combination, demanding the clear and unambiguous definitions of this paper to prevent incompatible implementations.

A Appendix - Specification

This paper uses Z-notation (see e.g. [12]) to specify operations on mathematical models of Djinn objects. The first part of object and cell specifications lists typed selector functions that provide defining properties, and the second part (below the middle line) provides constraints on those properties. Comments are incorporated to enable the specifications to be understood without familiarity with Z-notation.

$\mathbb{P}(X)$ denotes the power set of a set X, i.e. the set of all sets of elements of X, and $\mathbb{F}(X)$ denotes a power set that only includes finite collections. The scope of a universal ($\forall$) or existential ($\exists$) quantifier is from the following $\bullet$ to the beginning of the first line whose indentation is less than or equal to that of the quantifier. $\exists\, 1x$ is used for "there exists a unique x".

For clarity, we use some minor self explanatory notational deviations from Z (e.g. the use of $\in$ instead of :). Also, since a complete Z-specification is unnecessary for the purposes of this paper, formal definition of all mathematical concepts from first principles is replaced with functions that represent concepts that are well known or described in this paper.

References

[1] C. Armstrong, A. Bowyer, S. Cameron, J. Corney, G. Jared, R. Martin, A. Middleditch, M. Sabin, J. Salmon, and J. Woodwark. *Djinn: a geometric interface for solid modelling*. Information Geometers Ltd., Winchester, UK (to be published), 2000.

[2] EDS. Overview of parasolid. Parasolid Business Unit, March 1995.

[3] H. Hironaka. Subanalytic sets. In Y. Kusunoki, S. Mizohata, M. Nagata, H. Toda, M. Yamaguti, and H. Yoshizawa, editors, *Number Theory, Algebraic Geometry and Commutative Algebra*, pages 453–493. Kinokuniya Book-Store Co., Ltd., Tokyo, Japan, 1973.

[4] A. Middleditch. Some notes on solid object boundary evaluation. Technical Report PCL/CAE/74:3, Polytechnic of Central London, September 1974.

[5] A. Middleditch and C. Reade. A kernel for geometric features. In C Hoffmann and W Bronsvort, editors, *Proceedings: Fourth Symposium on Solid Modeling and Applications*, pages 131–140. ACM Press, May 1997.

[6] A. Middleditch, C. Reade, and A. Gomes. Rationale and formal definitions for the objects of the Djinn API to a geometric modelling kernel. *(Submitted for publication)*, 1998.

[7] A. Middleditch, C. Reade, and A. Gomes. Theoretical foundations for computer aided design systems. *(Submitted for publication)*, 1998.

[8] M. Nakahara. *Geometry, Topology and Physics*. Graduate Student Series in Physics. Adam Hilger, IOP Publishing Ltd, 1990.

[9] A. Requicha. Representations of rigid solids: Theory, methods, and systems. *ACM Computing Surveys*, 12(4):437–464, December 1980.

[10] M. Shiota. *Geometry of Subanalytic and Semialgebraic Sets*. Progress in Mathematics, Vol. 150. Birkhauser, Boston, USA, 1997.

[11] C. Silva. Alternative definitions of faces in boundary representations of solid objects. Production Automation Project 36, Uni. of Rochester, November 1981.

[12] J. Spivey. *The Z Notation*. Prentice-Hall Europe, 1992.

[13] J. Whitehead. Combinatorial homotopy. i. *Bulletin American Mathematical Society*, 55:213–245, 1949.

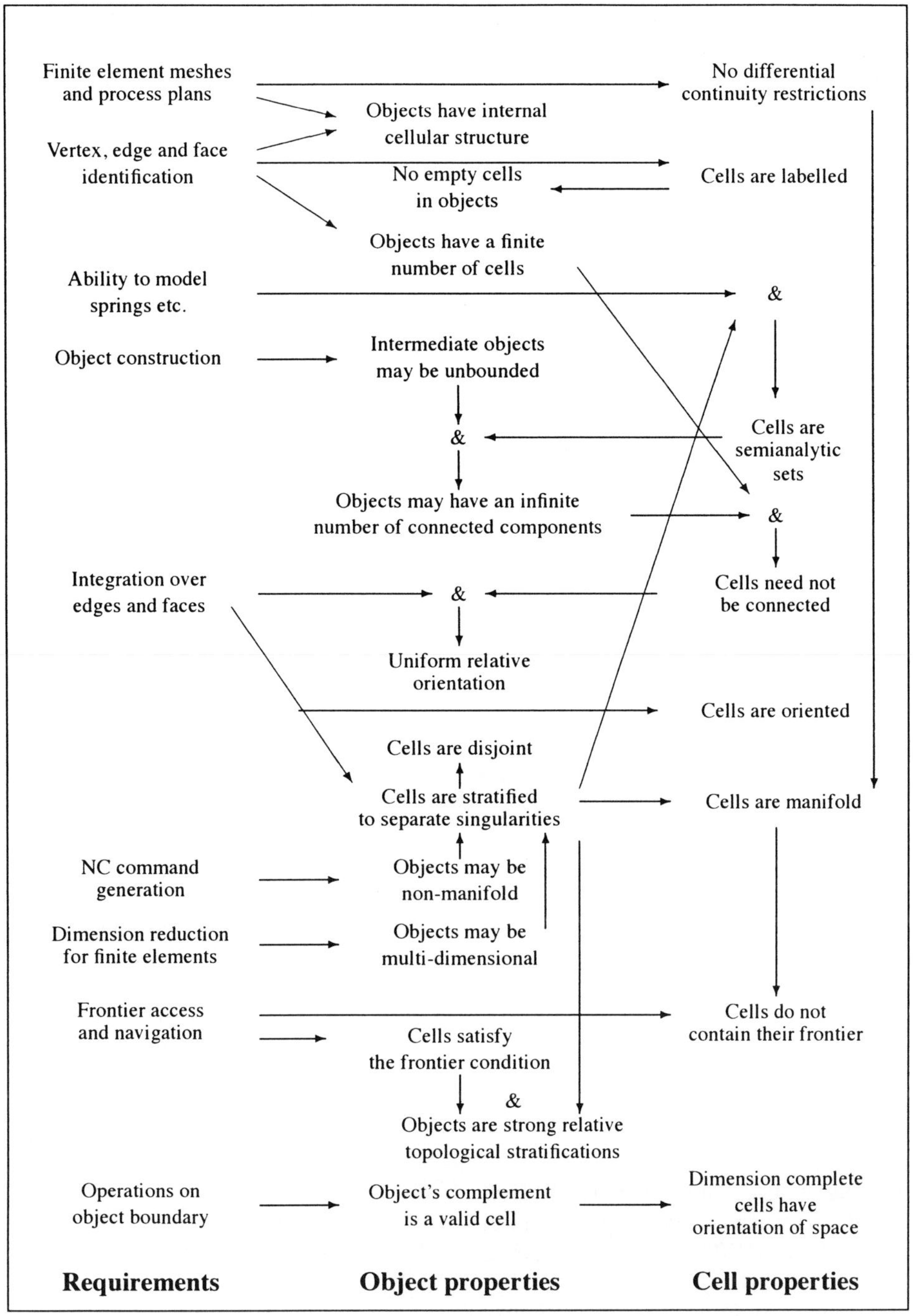

Figure 12. Modelling requirements and some implications

Dimension-Driven Parameterized Design of Free Form Objects

K.M. Yu
Department of Manufacturing Engineering
The Hong Kong Polytechnic University
Hong Kong
mfkmyu@polyu.edu.hk

W.M. Zhu
Department of Manufacturing Engineering
The Hong Kong Polytechnic University
Hong Kong
mfwmzhu@inet.polyu.edu.hk

Abstract

The paper develops a theoretical basis for dimension-driven parameterized design of free form objects. In traditional parametric Computer-aided Design (CAD) system, research is focused on the representation of dimension parameters and constraint satisfaction algorithms. Dimension-driven free form object parameterized design is to obtain a new free form object similar in shape to the old one but with different set of dimension parameter instantiation. In other words, a new member of the master free form object's family will be designed. However, research on dimension-driven parameterized design of free form objects is an unattempted area. Also, there is no commercial CAD system that can deal with dimension-driven free form objects successfully. In this research, we exploit a new concept of similar shape applicable to 3D rational curves and surfaces. Dimension-driven variation of similar free form objects is then explained and illustrated with an example.

1. Introduction

In traditional parametric Computer-aided Design (CAD) system, research is focused on the representation of dimension parameters [1,2,3,4]. The parametric design functionality allows users to instantiate their designs by supplying parameter values to implicitly or explicitly identified attributes of the design. It is a technique that can assist a designer in developing new geometry that is related to but distinct from those previously designed. Fowler has given a good survey in this area recently [5]. Dimension-driven parameterized design is a kind of advanced and flexible design method in current CAD system. Parden described a dimension-driven parametric design system [6], where a dimension is considered to express an implicit relationship between a set of construction lines. He presented a number of applications, such as families of parts, design changes, mechanisms, manufacturing dimensions and tolerances and other downstream systems, which can be tackled very efficiently. Figure 1 explains the application of design change. The original drawing is in the top of the figure. Changes of the parameters are then made as shown in the middle figure. The whole drawing will be adjusted automatically, and at the same time preserving all explicit and implicit geometric constraints. The final result is shown in the bottom of Figure 1. The theory of dimension-driven parameterized design for non-free form geometry has been studied for many years [7]. Some excellent CAD systems with the parametric functions have won great successes in the last decade, for example, T-FLEX CAD [8], Pro/Engineer, and CADDS5. They have brought convenience to users on changing geometric models by modifying user-defined parameters.

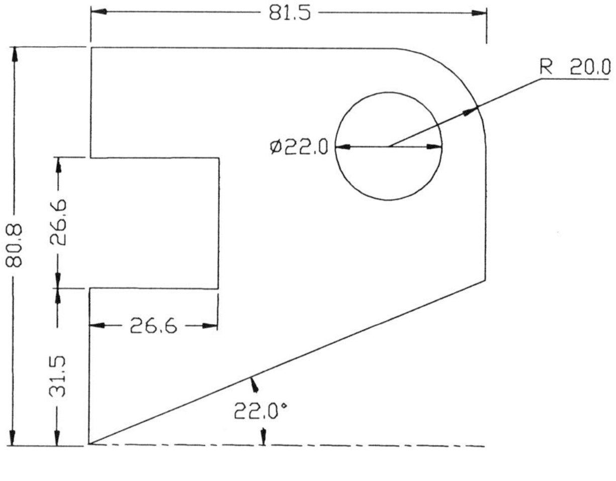

0-7695-0562-7/00 $10.00 © 2000 IEEE

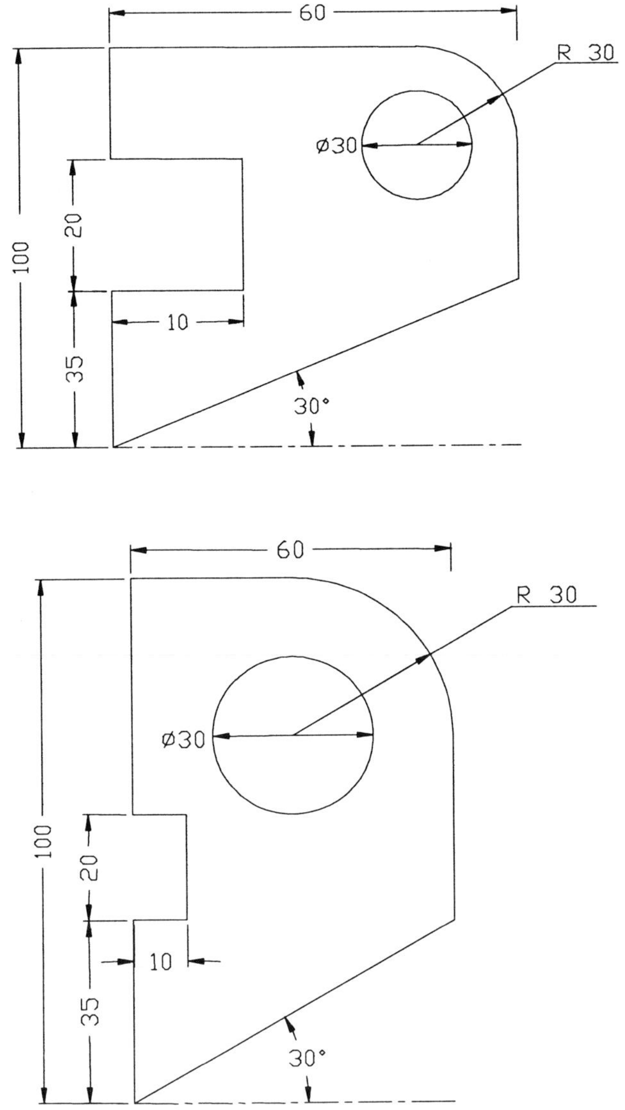

Figure 1. Dimension-based parameterized design for non-free form geometry

Dimension-driven free form object parameterized design is to obtain a new free form object that is similar to the old one but with different set of dimension parameter values. In other words, a new member of the old free form object's family will be designed. An example of the design is shown in Figure 2. The left apple is the original design drawn from free form curves and with height and width dimensions being parameterized. When the dimension values are changed, a new apple will be obtained on the right. The two apples should be similar because they are from the same family. This technology is very useful in serious design. However, there is no commercial CAD system that can deal with dimension-driven free form objects successfully.

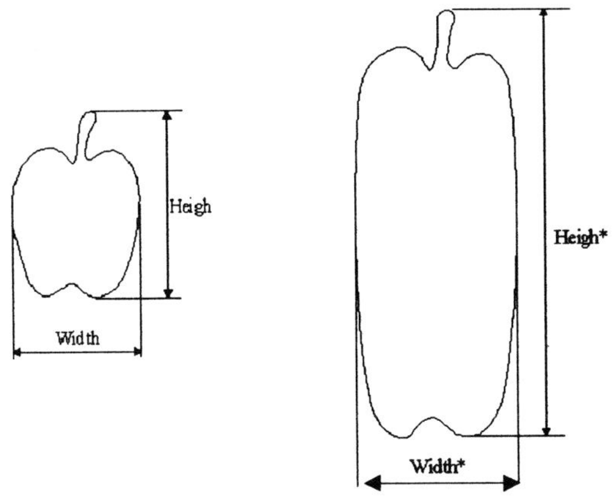

Figure 2. Dimension-based parameterized design for free form object

In dimension-driven free form object research, the first problem to be solved is a proper definition of shape. This includes how to describe shapes for free form objects, how to define similar free form objects and how to preserve shapes in variational free form geometry. Some works associated with shapes of free form objects were done in recent years. Su and Jones studied a shape-preservation theorem [9,10]. Kantorowitz extended the theorem to cases of hyperconvex and looped polygon [11]. Goodman researched on shape preserving properties on the basis of generalized Ball [12]. Piegl discussed the use of shape handles to modify the shape of rational B-spline curves [13]. Joris presented a method to design B-spline surface on the basis of data points [14]. Although these works covered shapes of free form objects, research on dimension-driven parameterized design of free form objects is still an unattempted area. The paper thus proposes a theoretical base for dimension-driven parameterized design of free form objects.

We first revise the definition of similar polygonal shape and investigate differentiation rules of varying degrees of similarity. The qualitative classification of shape angle is presented in detail. Next, similar 2D free form curves are defined in relation to similar polygons. Our study is then extended to 3D free form curves and free form surfaces. After that, dimension-driven variation of similar free form objects is explained and an example is given. Finally, conclusions and future work of this research are presented.

2. Similar Polygonal Shape

2.1. Definition of Similar Polygons

We define two kinds of similarity for polygons. One is *metric similar* and the other is *shape similar*.

The definition of metric similar is derived from that of similar triangles. In Euclidean geometry, two triangles that have equal corresponding angles or proportional corresponding side lengths are considered similar. Since polygons can always be triangulated, they can be considered similar if all the corresponding triangles in the tessellations are similar. These similar polygons are called metric similar, that is, they have equal corresponding angles and proportional corresponding side lengths. An example of two metric similar pentagons is shown in Figure 3. The definition of metric similar is, however, too restrictive. Instead, the quantitative shape control restriction is relaxed. Shape similar only constrains the shape qualitatively. As mentioned before, two objects that are metric similar will have equal corresponding angles. In shape similar, the corresponding angles only need to be of the same type, says both acute or both convex. Obviously, metric similar can be considered as a special case of shape similar. The definition of shape similar for polygons is based on their corresponding angles and ratios of their corresponding side lengths. Note that corresponding angle equality and corresponding side length equality are independent properties. Even for the simple triangle case, it can be proved by using the sine law that equal corresponding angles imply equal ratios of corresponding side lengths. But in general, equal ratios of corresponding side lengths do not imply equal corresponding angles. For two polygons with the same kinds and same numbers of sides, they are classified into different similarities as enunciated in Table 1. In Table 1, α_i, α_i' and r_i are the original angles, the corresponding angles, and ratios of corresponding side lengths respectively.

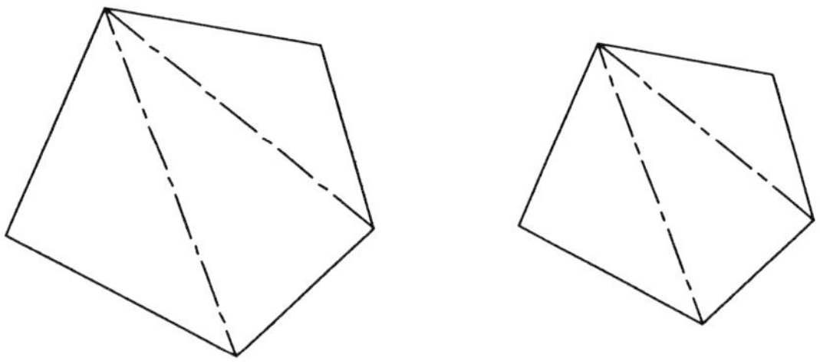

Figure 3. Metric similar objects

Table 1. Types of similar polygons

TYPE	ANGLE	RATIOS
I	$\alpha_i = \alpha_i'$	$r_i = const.$
II	$\exists\alpha_i \neq \alpha_i'$	$r_i = const.$
III	$\alpha_i \neq \alpha_i'$	$r_i = const.$
IV	$\alpha_i = \alpha_i'$	$\exists r_i \neq const.$
V	$\exists\alpha_i \neq \alpha_i'$	$\exists r_i \neq const.$
VI	$\alpha_i \neq \alpha_i'$	$\exists r_i \neq const.$
VII	$\alpha_i = \alpha_i'$	$r_i \neq const.$
VIII	$\exists\alpha_i \neq \alpha_i'$	$r_i \neq const.$
IX	$\alpha_i \neq \alpha_i'$	$r_i \neq const.$

2.2. The Classification of Angle

In shape similarity, an angle has two pieces of information -- the quantitative value and the qualitative shape. Depending on the design intent, we can have two classification schemes for the angle shape. In the first scheme, the acuteness scheme, the angle is categorized by its range of value. See Table 2. The angle α_i is within the range of [0°, 180°) since $\alpha_i = 180° - \cos^{-1}(\mathbf{t}_i \bullet \mathbf{t}_{i+1})$, where $\mathbf{t}_i$ represents the directional vector of a side, and the terminating end of $\mathbf{t}_i$ will always be the starting end of $\mathbf{t}_{i+1}$. See Figure 4. In dimension-driven variation, we need to check whether those modified angles will still be within their previous categories, i.e., acute remains acute and obtuse remains obtuse. The second angle shape classification scheme, convexity scheme, is less subtle. In this scheme, an angle is classified into convex, concave or straight line, as shown in Table 3. In Figure 4, $\mathbf{n}$ is the unit normal vector of the 2D polygon. The direction sense of $\mathbf{n}$ is determined by that of $\mathbf{t}_i$'s. The angle shape of two neighbouring sides can be expressed as:

$$(\mathbf{n} \times \mathbf{t}_i) \cdot \mathbf{t}_{i+1} \qquad (1)$$

Again in dimension-driven variation, we need to check whether convex angle remains convex, concave angle remains concave, and straight-line angle remains straight-line. The above discussion is not limited to closed polygons but is also applicable to open polylines.

Table 2. Acuteness scheme

Angle Values	Angle Shapes
$\alpha_i = 0°$	Straight Line
$0° < \alpha_i < 90°$	Acute
$\alpha_i = 90°$	Right Angle
$90° < \alpha_i < 180°$	Obtuse

Table 3. Convexity scheme

$(\mathbf{n} \times \mathbf{t}_i) \bullet \mathbf{t}_{i+1}$	Shape
< 0	Concave
$= 0$	Straight Line
> 0	Convex

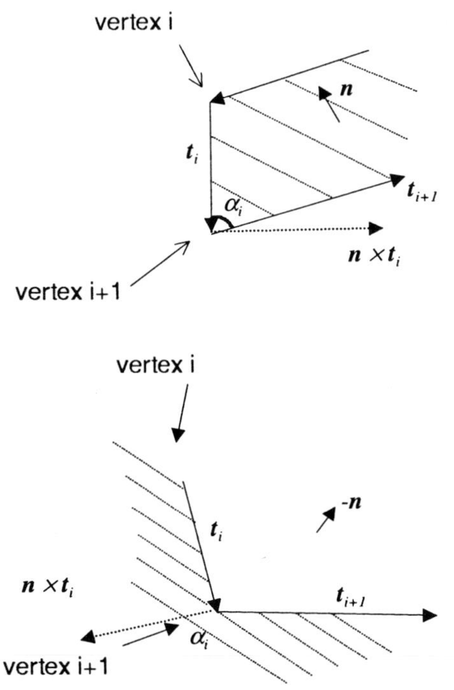

Figure 4. Convex and concave check

3. Similar 2D Free Form Curves

3.1. The Definition of Similar 2D Free Form Curves

The main purpose of this research is to study shape-preserving similar transform of free form objects. Free form curve representations being commonly used in contemporary CAD systems are Bezier, B-spline or the more general NURBS curves. Unlike similar polygons, free form curves cannot be defined by angles and side length ratios. A simple way to obtain similar free form curve is by metric similarity transform as depicted in Figure 5 [15]. The mathematical expression for definition of metric similar free form curves is:

$$\mathbf{P}'(t) = S \cdot \mathbf{P}(t) \qquad (2)$$

where **P**(t) and S are a free form curve and its metric similarity transform matrix respectively. The transformation image **P′**(t) is metric similar to **P**(t) (see Figure 6). Since metric similar is not flexible enough to the designers, we will develop definition for shape similar free form curves. In differential geometry, the shape of a free form curve is characterized by its *curvature*. In essence, the curvature κ quantifies the deviation of the free form curve from a straight line [16],

$$\mathbf{t}' = \kappa \mathbf{n} \qquad (3)$$

$$\kappa \mathbf{b} = \frac{\dot{\mathbf{P}} \times \ddot{\mathbf{P}}}{\dot{s}^3} \qquad (4)$$

where **t** is the tangent vector, **n** is the principal normal vector, **b** is the binormal vector and *s* is the arc length of the curve respectively. Shape similar free form curves are defined as free form curves with the same sequence of curvature types, i.e., positive, negative or zero. The sequence is a discrete distribution curve where absolute curvature value is not required. Note that the sign of the curvature is determined by the principal normal vector **n** that is parallel to both **P″** (arc length differentiation) and $\ddot{\mathbf{P}}$ (parametric differentiation). However, whether the curvature will degenerate into zero or not is determined by both $\dot{\mathbf{P}}$ and $\ddot{\mathbf{P}}$. Examples shown in Figure 7 are three shape similar free form curves having the same sequence of curvature type +,-,+,-,+. The examples are for open 2D free form curves. For closed ones, the curvature type sequence should be checked cyclically. Figure 8 depicts two closed free form curves. If their sequences of curvature type from A and B are checked, the results are +,-,+,-,+,-,+,-,+,- and -,+,-,+,-,+,-,+,-,+ respectively. However when cyclicity is considered, the two curves are shape similar with the same sequence of curvature type. Definitely, curvature type sequence is an intrinsic property independent of the starting point.

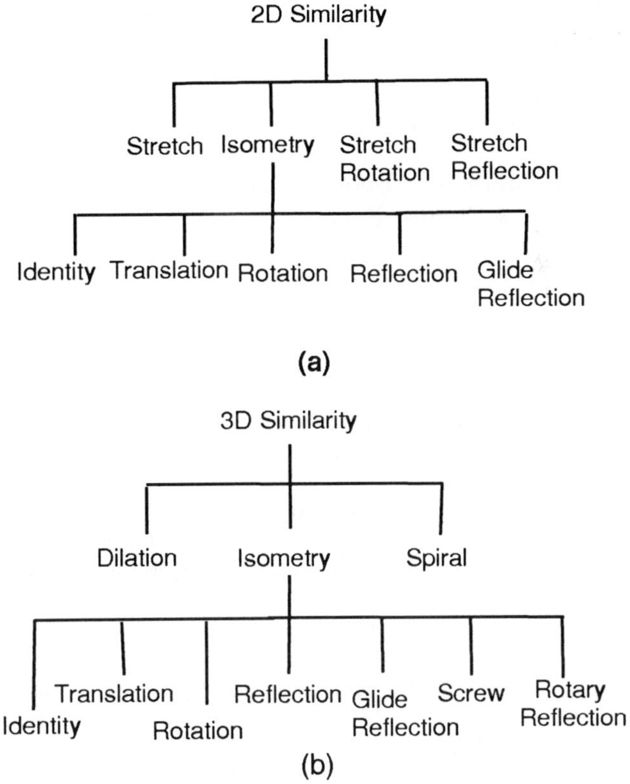

Figure 5. Metric similar transformation

Figure 6. Similar 2D free form curves by uniform scaling

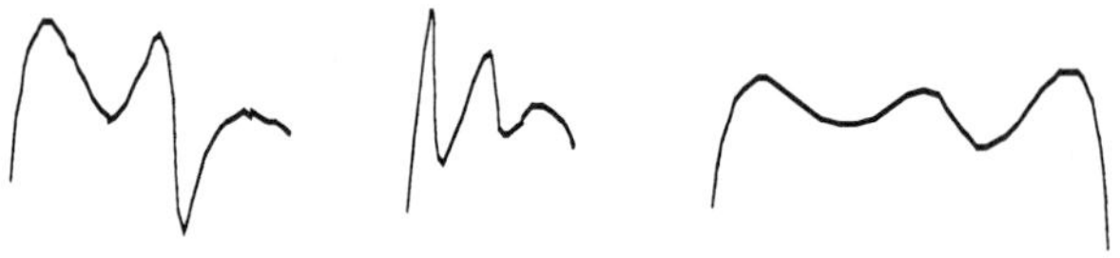

Figure 7. Shape similar 2D free form curves

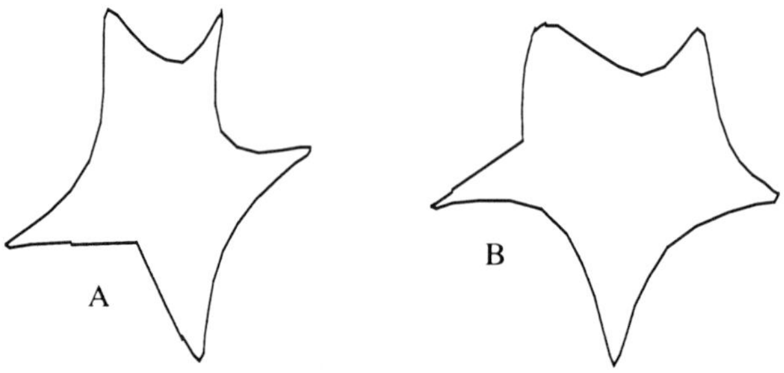

Figure 8. Closed 2D free form curves

3.2. Relationship Between Similar Free Form Curves and Similar Polygons

From curve modeling, a free form curve can be determined uniquely by its control vertices and blending functions. For NURBS curve, the formulation is

$$\mathbf{P}(t, \mathbf{V}) = \sum_{i=0}^{k} \mathbf{V}_i R_{i,k}(t) \tag{5}$$

where $\mathbf{V}_i$ are control vertices, $\mathbf{R}_{i,k}(t)$ are rational blending functions, and $\mathbf{V}$ is the whole control polygon. When transformation of control polygon is made, the variational position vector is,

$$\Delta\mathbf{P}(t, \Delta\mathbf{V}) = \sum_{i=0}^{k} \Delta\mathbf{V}_i R_{i,k}(t) \tag{6}$$

So the expression for the new NURBS curve becomes,

$$\mathbf{P}'(t, \mathbf{V}) = \mathbf{P}(t, \mathbf{V}) + \Delta\mathbf{P}(t, \Delta\mathbf{V}) = \sum_{i=0}^{k} \left(\mathbf{V}_i + \Delta\mathbf{V}_i\right) R_{i,k}(t) \tag{7}$$

Here, the shape of the new free form curve is determined by its control polygon. From Kantorowitz's shape preservation theorem [11] and properties of NURBS curves [17], a free form curve, in general, follows the shape of its control polygon. Therefore, we can conclude that similar control polygons result in similar free form curves (see Figure 9). For shape similar transform, we need only to maintain that the control vertex labels in the control polygon are invariant, i.e., convex remains convex or acute remains acute. The result also applies to free form curves with repeated control vertices or linear segments [17, 18]. However, one exceptional case should be avoided. A 2D self-intersecting control polygon can generate a self-intersecting or a non-self-intersecting free form curve depending on its parameters [11,19].

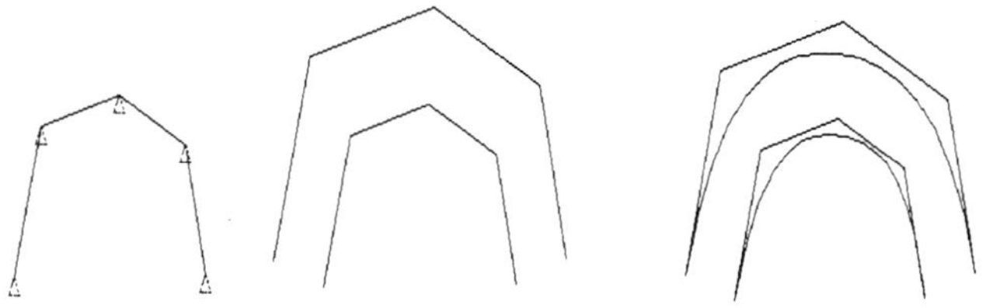

Figure 9. Similar free form objects from similar control polygons

4. Similar 3D Free Form Curves

4.1. The Definition of Similar 3D Free Form Curves

The result for 2D free form curves also applies to 3D cases. However from differential geometry, a 3D curve needs to be characterized by its *torsion* in addition to its curvature [16]. In essence, the torsion τ quantifies the deviation of the free form curve from a plane [16],

$$\mathbf{n}' = \tau\mathbf{b} - \kappa\mathbf{t} \tag{8}$$

$$\tau = \frac{\dot{\mathbf{P}} \cdot (\ddot{\mathbf{P}} \times \dddot{\mathbf{P}})}{\dot{s}^6 \kappa^2} \tag{9}$$

so that the torsion of a curve is positive when the curve normal $\mathbf{n}$ turns out of the osculating plane into the positive direction of the binormal vector $\mathbf{b}$. Sequence of torsion type (positive, negative or zero) along the 3D curve can thus be identified. Hence, 3D free form curves are shape similar if they have the same sequence of curvature and torsion types.

4.2. The Description of Shapes for 3D Free Form Curves

In the first step of shape description, the curvature label sequence similar to 2D-curve case is found. In the second step, torsion label sequence of the curve is considered. The control polygon will be used to describe the torsion. As shown in Figure 10, control vertices $\mathbf{V}_{i-1}$ will be connected to $\mathbf{V}_{i+1}$. The virtual linkings along the control polygons result in a ribbon of triangles. The folds or included angles between adjacent triangles will determine the torsion of the curve.

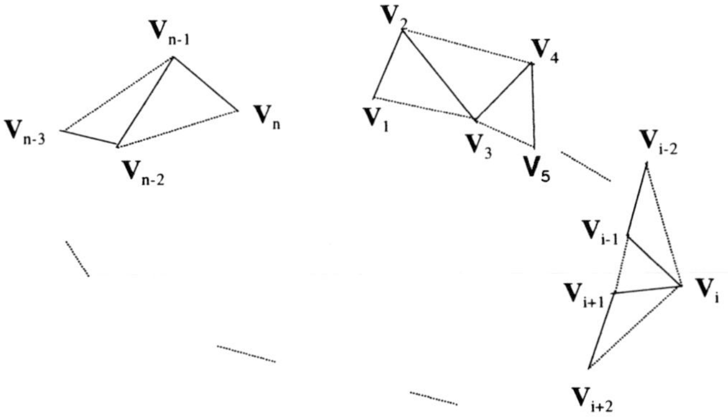

(a) Open

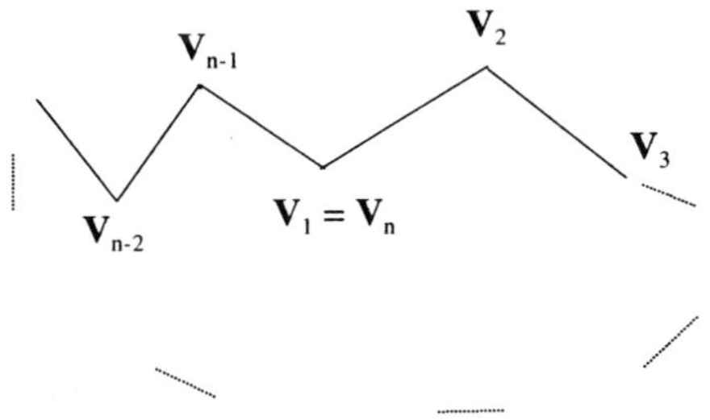

(b) Close

Figure 10. Ribbon of triangles

The included angles in the ribbon can be labelled in either the convexity scheme or the acuteness scheme. This time, the labels are applied to edges of the control polygons. Since there is no included angle for the first and last edges of the open ribbon, they have null labels. For the convexity scheme, the angle is computed from [20],

$$(\mathbf{n}_{i-1} \times \mathbf{V}_i\mathbf{V}_{i+1}) \cdot \mathbf{n}_i \qquad (10)$$

where
$$\mathbf{n}_{i-1} = \frac{\mathbf{V}_i\mathbf{V}_{i+1} \times \mathbf{V}_i\mathbf{V}_{i-1}}{\|\mathbf{V}_i\mathbf{V}_{i+1} \times \mathbf{V}_i\mathbf{V}_{i-1}\|} \qquad (11)$$

$$\mathbf{n}_i = \frac{\mathbf{V}_i\mathbf{V}_{i+1} \times \mathbf{V}_i\mathbf{V}_{i+2}}{\|\mathbf{V}_i\mathbf{V}_{i+1} \times \mathbf{V}_i\mathbf{V}_{i+2}\|} \qquad (12)$$

See Figure 11. For the acuteness scheme, the angle is calculated from $\cos^{-1}(\mathbf{n}_{i-1} \bullet \mathbf{n}_i)$.

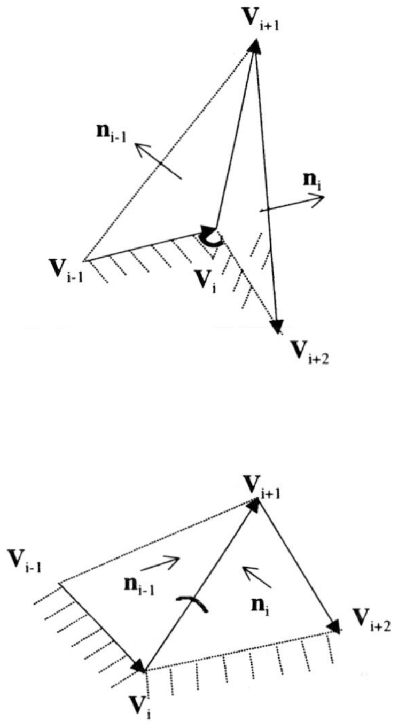

Figure 11. Angle subtended between two triangles

After finding the curvature and torsion label sequences of the curve, the shape similarity of the 3D free form curves can be checked. For closed 3D curves, cyclicity should also be considered.

5. Similar Free Form Surfaces

5.1. The Definition of Similar Free Form Surfaces

From differential geometry, the oscillation of a smooth surface can be characterized by its principal curvatures (see Table 4). Free form surfaces are similar if they have the same distribution of principal curvatures. For metric similar surfaces, they have identical principal curvature values. For shape similar, they have matched principal curvature labels.

Table 4. Types of surface shape composed of rectangular patches

Type	Principal Curvatures	Figure
Flat	Both zero.	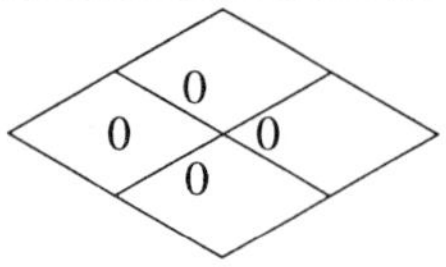
Convex Elliptic (Bump)	Both positive.	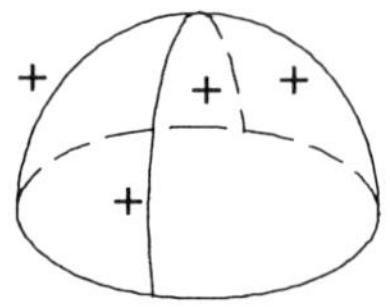
Concave Elliptic (Hollow)	Both negative	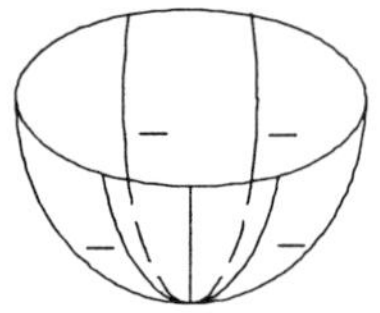
Hyperbolic (Saddle)	One is positive, the other is negative.	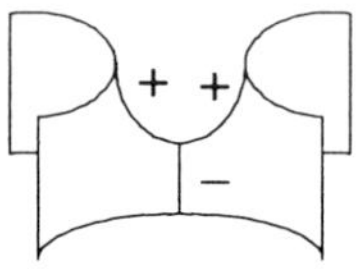
Convex Parabolic	One is positive, the other is zero.	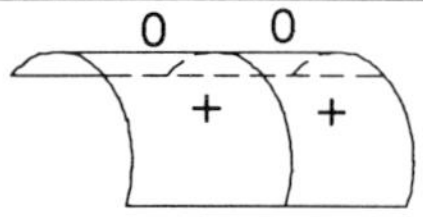
Convex Parabolic	One is negative, the other is zero.	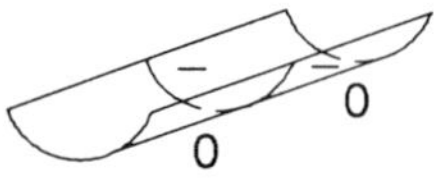

5.2. Free Form Surfaces and Control Points

From surface modeling, a free form surface can be determined uniquely by its control net and blending functions. For NURBS surface, the formulation is

$$\mathbf{P}(s,t,\mathbf{V}) = \sum_{i=0}^{n}\sum_{j=0}^{m} \mathbf{V}_{i,j} R_{i,j}(s,t) \tag{13}$$

where $\mathbf{V}_{i,j}$ form a control net and $\mathbf{R}_{i,j}(s, t)$ are the biparametric rational B-spline surface blending functions. Similar to the curve case, a new surface equation due to control net variation is,

$$\mathbf{P}'(s,t,\mathbf{V}) = \mathbf{P}(s,t,\mathbf{V}) + \Delta\mathbf{P}(s,t,\Delta\mathbf{V}) = \sum_{i=0}^{n}\sum_{j=0}^{m} \left(\mathbf{V}_{i,j} + \Delta\mathbf{V}_{i,j}\right) R_{i,j}(s,t) \tag{14}$$

5.3. Shape description of Free Form Surface

From properties of NURBS surfaces [17], a free form surface follows, in general, the shape of its control net. Therefore, we can conclude that similar control nets result in similar free form surface.

For shape similar transform, we need only to maintain that the labels for control vertex and control edge are invariant, i.e., convex remains convex or acute remains acute.

When topological rectangular control net is used for a composite of biparametic surface patch, a labelling scheme analogous to principal curvatures, can be employed to control vertices and edges (see Figure 12(c)). If the control net is not composed of quadrilateral, says triangular or polygonal patches as shown in Figure 13, the same scheme can also be used by labelling all edges incident to each control vertex. Figure 12 shows a control vertex and its incident edges. Each pair of adjacent edges defines a triangular area and a normal vector, e.g., $\mathbf{n}_{1,2}=\mathbf{t}_1\times\mathbf{t}_2$. The triangular area may belong to an interior or exterior portion of the control net. A boundary or corner control vertex will have one exterior triangular area while an interior control vertex has none. In Figure 12(b), $\mathbf{t}_1\times\mathbf{t}_2$ subtends one triangle while $\mathbf{t}_2\times\mathbf{t}_1$ subtends another triangle.

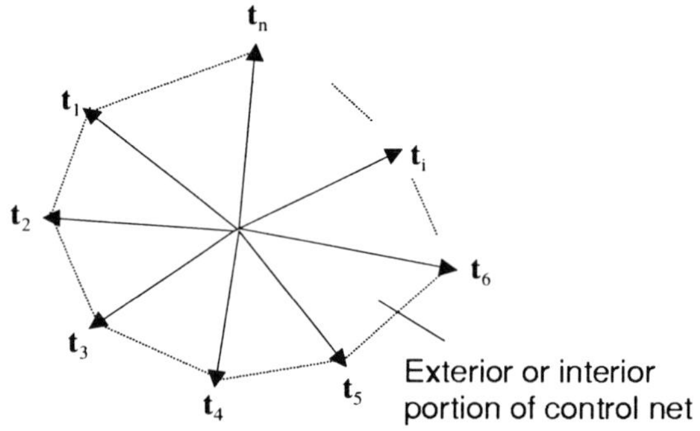

(a)

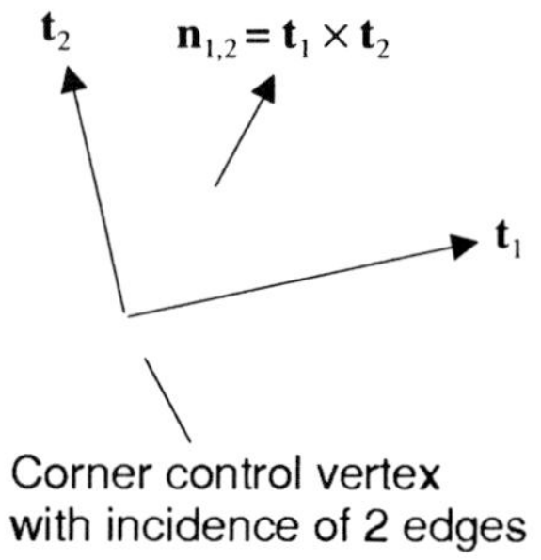

(b)

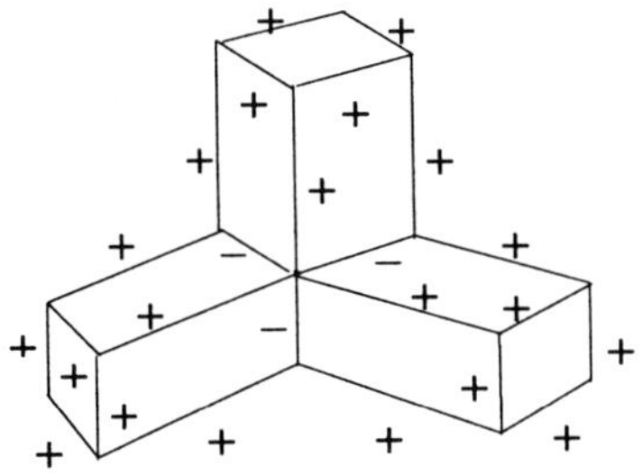

(c)

Figure 12. Control net labelling

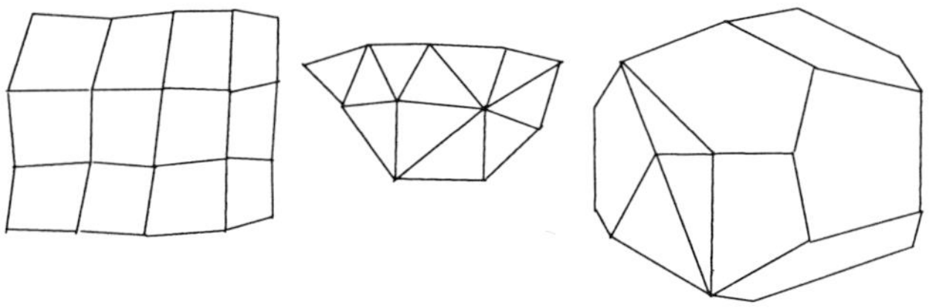

Figure 13. Control net for rectangular and non-rectangular patches

Edges' labels can then be found by using normal vectors of adjacent triangular areas. Convexity or acuteness edge label is obtained in the same way as described in the 3D curve cases. In fact, the triangles partitioning each control vertex in the 3D surface control net can be considered as a degenerate case of the triangle ribbon for a closed 3D curve (see Figure 14).

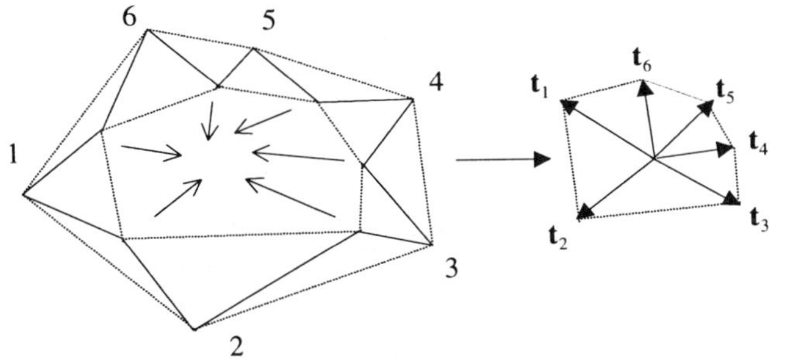

Figure 14. Ribbon in 3D curve degenerates into a star of spokes in 3D surface

For control nets of open surfaces, the acuteness label for boundary control edges is anti-parallel. See Table 5 where β_i is the included angle. For convexity labelling scheme, additional check for anti-parallelism is required to differentiate boundary edge from flatness since $\alpha_i = 0$ in both cases. The above result also applies to free form surfaces with repeated control vertices or linear segments [17, 18].

Table 5. Included angle of control edges

$\beta_i = \cos^{-1}(\mathbf{n}_i \bullet \mathbf{n}_{i+1})$	Label
$\beta_i = 0°$	Planar ($\mathbf{n}_i = \mathbf{n}_{i+1}$)
$0° < \beta_i < 90°$	Acute
$\beta_i = 90°$	Right Angle
$0° < \beta_i < 180°$	Obtuse
$\beta_i = 180°$	Boundary ($\mathbf{n}_i = -\mathbf{n}_{i+1}$)

6. Dimension-Driven Variation of Similar Free Form Objects

In dimension-driven parameterized design of free form objects, a new free form object is obtained by substituting a new set of dimension values to the old free form object. The dimensions we give to the free form object are only critical dimensions characterizing key design intent (e.g., shape or function), while maintaining the shape to be similar. In other words, the relaxed shape similar is more desirable than the restrictive metric similar here. Unlike fixed form geometry, the dimensions are not sufficient to define all geometric information of the free form object. Since a free form curve is an oscillatory curve and sine curve is a simple analytical oscillatory curve, the dimensioning scheme of a sine curve is adapted to suit the more general free form curves. Sine curve is characterized by the amplitude and period dimensions. In extending to free form curve, peak-to-peak and peak-to-base linear dimensions are analogous to the amplitude dimension (see Figure 15). Similarly, peak-to-peak linear and angular dimensions can be employed to define the period dimension of the free form curve. In the same way, amplitude and period dimensions can be used to define free form surface oscillations. For instance, the height and width dimensions of the apple in Figure 2 are critical to be satisfied accurately as specified by the designer.

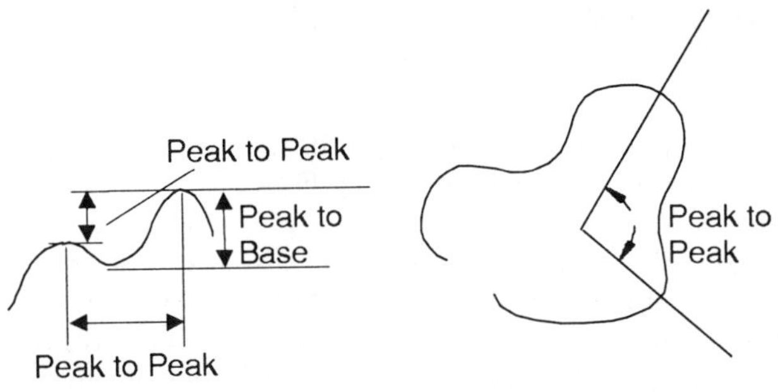

Figure 15. Linear and angular dimensions

Figure 16 shows the flow chart for dimension-driven parameterized design of free form objects. It follows the steps of a typical dimension-driven variational geometry procedure, namely, sketch, dimension, instantiate and regenerate. The free form shape can be input via its control polygon or by data points fitting. There is a hidden step that detects and extracts all implicit constraints in the input shape. A new step pertinent to free form parameterized design is to fix the similarity type (i.e., convexity or acuteness labelling) and the variation range permissible. Since dimensions can be treated as metric constraints, a constraint satisfaction algorithm, numerical or geometrical reasoning follows. If successful, a new shape similar in shape but different in size will be constructed.

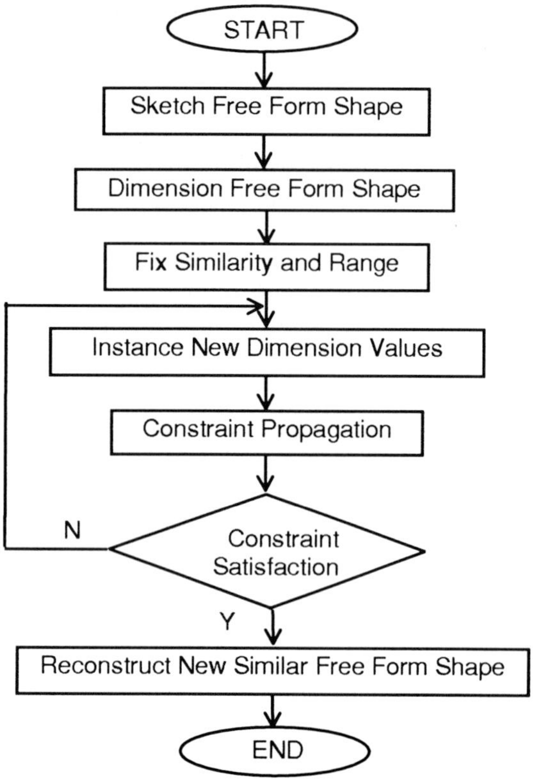

Figure 16. Dimension-based variation of similar free form objects

An example of facial profiles shown in Figure 17 demonstrates the procedure [21]. Figure 17(a) is the initial shape. We first input the control polygon and its facial profile is created. Next, some dimension parameters, the lengths of segments DE and EF, the values of angles D, E and F are specified. In this example, convexities or acuteness of the angles are constrained but not the side lengths. New dimension values are then assigned (see Table 6). A new control polygon is generated after successful constraint satisfaction. A shape similar facial profile will be output as shown in Figure 17(b). Figure 17(c) is another shape similar facial profile but constrained by the acuteness scheme.

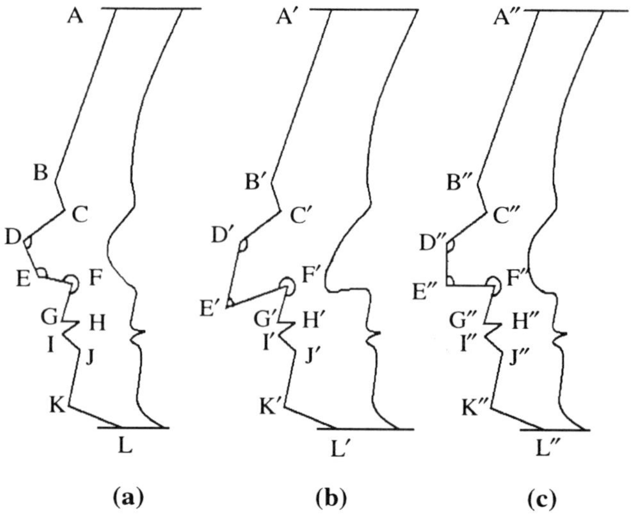

Figure 17. An example of dimension-based variation of similar free form objects

Table 6. Variational new dimension values assigned of the example in Figure 17

(a)		
Angle	Original	Original
∠D	Convex	90°
∠E	Convex	Obtuse
∠F	Concave	Obtuse

(b)		
Angle	Fix	Vary
∠D′	Convex	Obtuse
∠E′	Convex	Acute
∠F′	Concave	Obtuse

(c)		
Angle	Fix	Vary
∠D″	Convex	Obtuse
∠E″	Convex	90°
∠F″	Concave	Obtuse

7. Discussion and Further Work

This paper has developed a theoretical basis for dimension-driven parameterized design of free form objects. However for practical implementation and future development, some other points have to be considered.

First, self-intersecting curves are not catered because self-intersecting control polygons may generate non-self-intersecting curves [11,19]. Shape similar variation of the control polygon will thus not preserve self-intersection.

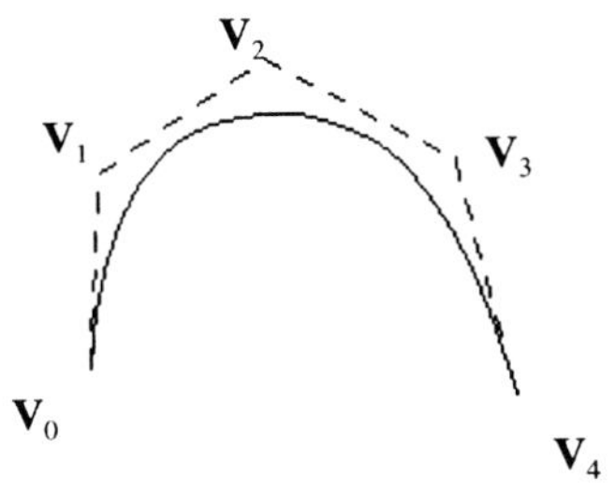

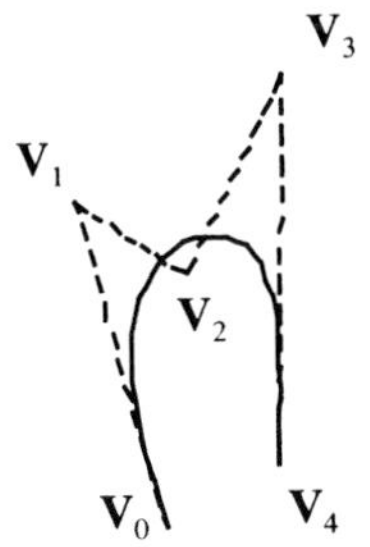

Figure 18. Similar free form curve from non-similar control polygons

Second, vertex and edge labelling scheme discussed are very general. Non-smooth curve with slope and curvature discontinuities due to repeated control vertices [18] can be handled. Other shape parameters, like slope and curvature continuities [15] due to linear segments can also be tackled.

Third, similar control polygons can generate similar free form shapes but similar free form shape can be created from more than one control polygons. Farin illustrated that there exist convex curves that have a nonconvex control polygon [22]. So there exists non-similar control polygons (some convex, some non-convex) that can generate similar convex curves (see Figure 18). Another way to obtain similar free form curves from non-similar control polygons is from polygons composed of different number of entities. A Bezier curve with n control vertices can be approximated by another Bezier curve with $n-1$ vertices through degree reduction [22]. The two control polygons are obviously not similar. An example is shown in Figure 19.

Fourth, in addition to designing customer products and engineering parts with free form features, dimension-driven free form variation is useful in other areas. For instance, garment design, modeling of human organs, and reverse engineering are potential applications.

Finally, a prototype for 2D free form curves has been implemented [21], extension to 3D curves and surfaces will be done in the future.

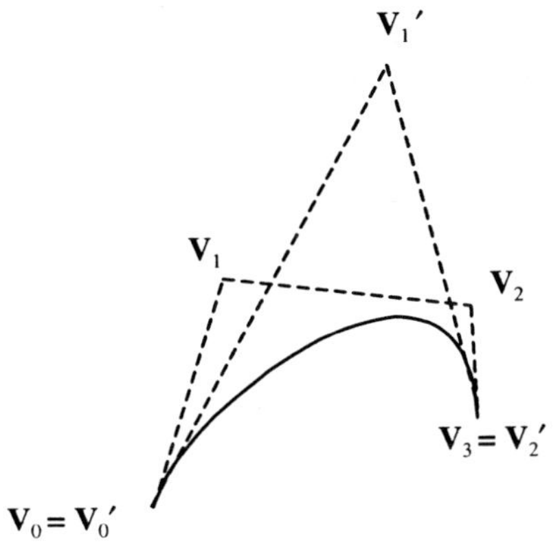

Figure 19. Same free form curve from different control polygons

8. Conclusions

This paper has redefined similar free form objects and demonstrated a potential way to achieve dimension-driven parameterized free form objects. The theory is suitable to characterize shapes of 2D and 3D NURBS curves and surfaces. Control vertex-based similar transformation of free form objects is developed that preserves the shape qualitatively.

Acknowledgements

The work described in this paper was supported by a grant from The Hong Kong Polytechnic University (Project No. G-S087).

References

[1] P. Koparkar, "Designing parametric blends: surface model and geometric correspondence", *The Visual Computer*, v7, 1994, pp.39-58.

[2] S. Gopalsamy, and T.S. Reddy, "Compact representation for planar sections of parametric surfaces", *IFIP Transactions B: Computer Applications in Technology*, B-9, 1993, pp.279-286.

[3] B. Ravani, and J.W. Wang, "Computer aided geometric design of line constructs", *Journal of Mechanical Design*, 133(4), 1991, pp.363-371.

[4] G.C. Onwubolu, "Modeling sculptured solids in computer-aided design", *Computers in Industry*, 17(1), 1991, pp.67-75.

[5] J.E. Fowler, "Variant design for mechanical artifacts: A state-of-the-art survey", *Engineering with Computers*, 12(1), 1996, pp.1-15.

[6] G. Parden, and R.G. Newell, "A dimension based parametric design system", *CAD'84*, 1984, pp.252-259.

[7] K.M. Yu, *Dimensioning and Tolerancing in Geometric Modeling*, Ph.D. thesis, University of Hong Kong, 1990.

[8] *T-FLEX CAD User Guide*, Top Systems Ltd., Russia, 1994

[9] B.Q. Su, and D.Y. Liu, *Computational Geometry, Curve and Surface Modeling*, Academic Press, 1989.

[10] A.K. Jones, "Shape control of curves and surfaces through constrained optimization" in G.E. Farin (ed.) *Geometric Modeling: Algorithms and New Trends*, SIAM, 1987, pp.265-279.

[11] E., Kantorowitz, and Y. Schechner, "Managing the shape of planar splines by their control polygons", *Computer Aided Design*, 25(6), 1993, pp.355-364.

[12] T.N.T. Goodman, and H.B. Said, "Shape preserving properties of the gerenalised Ball basis", *Computer Aided Geometric Design*, v8, 1991, pp.115-121.

[13] L. Piegl, "Modifying the shape of rational B-splines. Part 1: curves", *Computer Aided Design*, 21(8), 1989, pp.509-518.

[14] J.S.M. Vergeest, "Surface fitting for interactive shape design", *Computers in Industry*, v13, 1989, pp.1-13.

[15] G.E. Martin, *Transformation Geometry: an Introduction to Symmetry*, Springer, 1982.

[16] I.D. Faux, and M.J. Pratt, *Computational Geometry for Design and Manufacture*, Ellis Horwood, 1979.

[17] D.F. Rogers, and J.A. Adams, *Mathematical Elements for Computer Graphics*, McGraw-Hill, 1989.

[18] J. Rooney, and P. Steadman, *Principles of Computer-Aided Design*, Pitman, 1987.

[19] D.S. Meek, and D.J. Walton, "Shape determination of planar uniform cubic B-spline segments", *Computer Aided Design*, 22(7), 1990, pp.434-441.

[20] H. Chiyokura, *Solid Modeling with Design Base Theory and Implementation*, Addison-Wesley, 1988.

[21] Y.C. Yip, *Parametric Based Object Modeling*, Department of Mechanical Engineering, Hong Kong Polytechnic University, BEng (Hon) Final Year Project Report, 1996.

[22] G. Farin, *Curves and Surfaces for Computer Aided Geometric Design*, Academic Press, 3rd Ed, 1993.

On Families of Objects and Their Semantics

Rafael Bidarra and Willem F. Bronsvoort
Faculty of Information Technology and Systems
Delft University of Technology
Zuidplantsoen 4, NL-2628 BZ Delft, The Netherlands
email: (Bidarra/Bronsvoort)@cs.tudelft.nl

Abstract

An approach is presented to define a family of objects in parametric modelling. It avoids an inherent problem of boundary re-evaluation in current parametric modelling systems. Even more important, it offers an effective way to define the properties, or semantics, of a family of objects.
A family of objects is defined by a prototype feature model, including several types of constraints to precisely specify its semantics. A family membership test is used to check whether an instance corresponding to a particular set of parameter values belongs to a family. The underlying geometric representation is a non-manifold cellular model.

Keywords: parametric modelling, families of objects, semantics, feature modelling, cellular model

1. Introduction

In a parametric modelling system, once an object has been modelled, a user can easily create variants of that object by changing its parameters. This approach has a number of obvious advantages: it can be helpful in fine tuning the shape of an object, several objects that differ only in some details can be modelled with much less effort than with traditional solid modelling techniques, and libraries of standard parts can be built (Anderl and Mendgen 1995).

The set of possible variants of an object is called a *class* or a *family of objects*. So a family of objects represents a set of objects with "similar" shape. The user can create an *instance* from such a family by specifying values for the parameters that characterise the family. Constraints play an important role in the definition of parametric objects; these specify certain properties for the objects, for example that some parameter value has to be within some prescribed range. Each instance has to satisfy all constraints specified for the family.

Most current parametric modelling systems use a dual representation for an instance of a family of objects: on the one hand a model history and the set of values for the parameters, on the other hand a boundary representation of the resulting shape.

Shapiro and Vossler (1995) posed an interesting question about parametric modelling: what is a good definition for a parametric family of solids? Such a definition should determine which objects are a member of a family, and which objects are not. It has been shown that, in current parametric modelling systems, such a definition is hampered by certain peculiarities of boundary representations (Raghothama and Shapiro 1998). Even more important, specification mechanisms currently available for parametric objects, in particular the available constraint types, are insufficient to adequately maintain the design intent of a family in all its instances.

Therefore, in this paper it is argued that it would be better to use an alternative definition and representation for a parametric family of objects, in which the desirable properties for its instances are more precisely specified, and which avoids the pitfalls of the boundary representation. A family of objects here is defined by a semantic feature model (Bidarra and Bronsvoort 2000). Such a model contains features and constraints that explicitly define the semantics of the family of objects. A *family membership test* can be used to determine whether a given set of parameter values specifies a valid instance of the family. This test greatly benefits from the underlying non-manifold geometric representation, a cellular model.

In Section 2, the problems with families of objects in current parametric modelling systems will be illustrated. In Section 3, the new mechanism for defining families of objects will be introduced. In Section 4, the family membership test will be described. In Section 5, some results and conclusions will be given.

0-7695-0562-7/00 $10.00 © 2000 IEEE

2. Families of objects in current parametric modelling systems

In most current parametric modelling systems, a user can create a model of an object, sometimes called a *prototype*, and then create variants of that object by specifying different parameter values. The parameters may be dimensions of basic shapes that constitute the object, or user-defined parameters that specify other dimensions of the combined shape. If the modelling system is a parametric feature modelling system, the basic shapes are feature shapes. The set of possible variants is designated a *family of objects*.

In the prototype, several constraints can occur, for example to align two faces of two basic shapes of the object, or to restrict the range of some parameter. The sequence of modelling operations used to create the prototype, including all its basic shapes and constraints, is called the *model history* of the object. This is in fact a procedural representation, used to re-compute the boundary representation of the object whenever parameter values are modified. Basically, this comes down to a stepwise boundary re-evaluation of all basic shapes in the model with their new parameter values.

Several researchers, for example Chen and Hoffmann (1995), Lequette (1997) and Raghothama and Shapiro (1998), have observed that unexpected, and even improper, models can result from such a boundary re-evaluation process in commercial modelling systems. Shapiro and Vossler (1995) related this problem to the question on how to define families of objects in parametric modelling. The main issue in the latter is to determine which objects belong to a family and which do not. It may happen that, even if all constraints defined for the prototype are satisfied for some set of parameter values, the corresponding instance of the object does not belong to the family of objects as intended by the user.

An example of getting an unexpected model is given in Figure 1, inspired by examples in the references given above. The model consists of a block, a protrusion and a slot. The protrusion has been attached to the block so that their coplanar top faces are merged into one face, f_1, see Figure 1.a. The through slot, which intersects both the block and the protrusion, has been attached to face f_1, causing it to be split into two faces, f_{2a} and f_{2b}, see Figure 1.b. By modifying a positioning parameter, the protrusion is displaced downwards. When the model history is re-executed, after this parameter modification, the user will probably expect the model of Figure 1.c, but, in at least one current commercial system, he obtains the model of Figure 1.d instead. Although this model is a valid model, it is apparently not according to the user intent. It is therefore questionable whether it should be a member of the

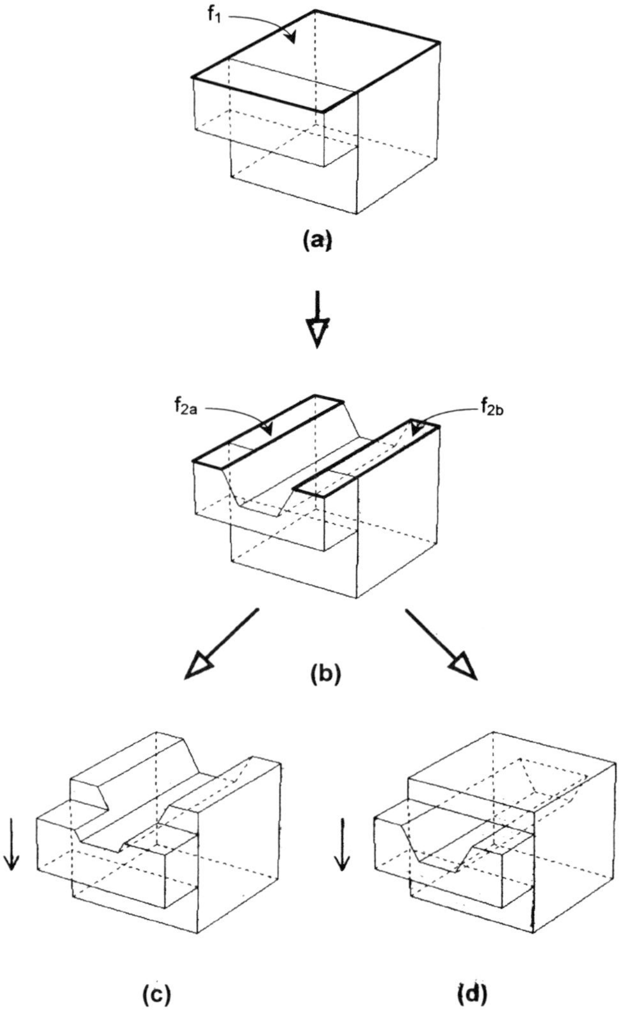

Figure 1 – Obtaining an unexpected model after boundary re-evaluation

family of objects represented by the prototype model in Figure 1.b.

However, boundary re-evaluation can even behave worse, as the resulting boundary model may become disconnected. An example of this, inspired by an example in Raghothama and Shapiro (1998), is given in Figure 2. The model in Figure 2.a has a cylindrical through slot, positioned at a distance a from the indicated edge, causing the top face of the block to be split into faces f_1 and f_2. A rib is then attached to face f_2, at a distance b of the slot edge indicated in Figure 2.b. The slot position is now modified, making the parameter value a equal to the radius of the slot, by which face f_1 in Figure 2.a collapses into an edge. According to the positioning constraints specified, the user would most likely expect the model in Figure 2.c as the result of this parameter modification. However, some current parametric modelling systems produce the model

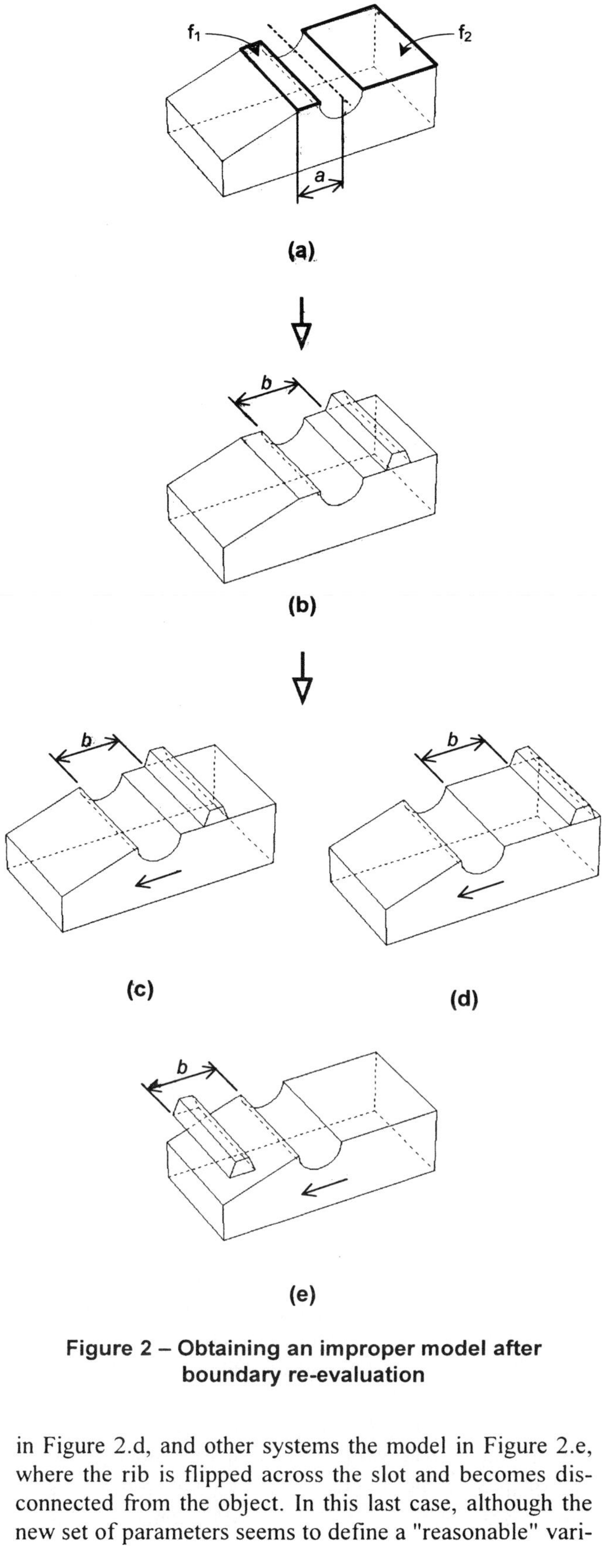

Figure 2 – Obtaining an improper model after boundary re-evaluation

in Figure 2.d, and other systems the model in Figure 2.e, where the rib is flipped across the slot and becomes disconnected from the object. In this last case, although the new set of parameters seems to define a "reasonable" variant of the original object, the modelling system produces an improper solid. So the only possible conclusion is that the variant should not be in the family of objects defined by the prototype, because it is definitely not in line with the user intent.

The underlying problem in both examples is the so-called *persistent naming problem*. At each modelling step in the model history, references can be made to topologic entities in the intermediate boundary representation, which is the result of all previous modelling steps. However, these entities may have been split, merged or deleted as a result of previous modelling operations, depending on the parameter values. As a result, references can be made to entities of intermediate boundary representations that are no longer there. In the example of Figure 1, face f_1 resulted from merging the coplanar block and protrusion top faces. When these are no longer coplanar, one of them is selected by the system as the reference for attaching the through slot. In the example of Figure 2, the slot edge used to position the rib (Figure 2.b) is merged with another edge, and the system selects the merged edge (Figure 2.c), another edge (Figure 2.d), or the merged edge but the other side of it (Figure 2.e), to position the rib.

Several schemes have been proposed to solve the persistent naming problem (Kripac 1995, Capoyleas et al 1996, Lequette 1997), so that the boundary re-evaluation can at least be consistently executed, and in a deterministic way. However, the fundamental problem remains that the resulting model is not always as expected by the user: most users probably expect the model of Figure 1.c, but some users may nevertheless expect the model of Figure 1.d. Depending on how the persistent naming scheme works, the user will get one of them, but this one may not be according to his intent. One might also say that the semantics of the family of objects is not well defined.

Raghothama and Shapiro (1998) have introduced *boundary representation deformations* as a first step to come to a sounder definition of a parametric family of objects. Basically it is argued that as long as there is a continuous boundary deformation from the prototype model to the new instance possible, the new instance is considered to be a member of the family. However, as Raghothama and Shapiro also admit, continuous deformations seem to be too restrictive, and operations like splitting and merging of model entities have to be dealt with, in order to allow topological changes such as the elimination of holes, but this has not been elaborated. A major problem here seems to be that such operations can only be generically allowed or disallowed, e.g. for all holes in an object or for no holes at all. And, again, which objects belong to a family is determined, at least partly, by how the modelling system works, in particular which operations are available to allow discontinuous deformations.

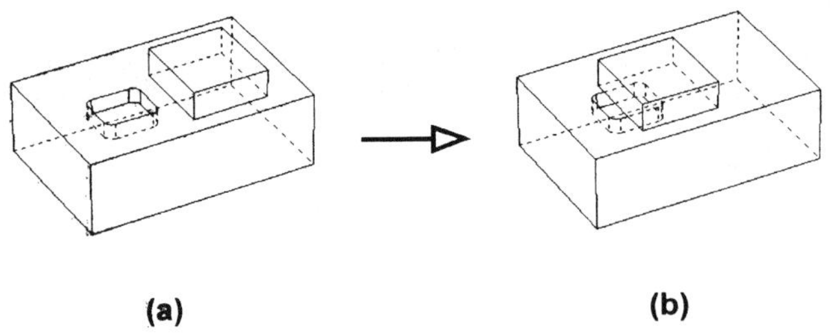

Figure 3 - Changing semantics with a parameter modification

An even more important shortcoming in current parametric modelling systems, which makes it difficult to come to a sound definition of a family of objects, is that currently available constraint types are not always expressive enough to adequately specify the design intent of a prototype object, and thus of the corresponding family of objects.

As an example, Figure 3.a shows a prototype model consisting of a block and, on its top face, a protrusion and a blind pocket. By modifying a positioning parameter for the protrusion, the variant given in Figure 3.b can be created. In this model, the pocket is partially closed by the protrusion, which might well be against the design intent of the prototype, in which case the variant should not be in the corresponding family of objects.

With currently available constraint types, the variant of Figure 3.b can only be excluded by setting distance constraints, for example, between the side faces of the pocket and of the protrusion. However, if there would be more than one protrusion on top of the block, separate distance constraints would have to be specified for each protrusion and the pocket to guarantee that the latter would never be closed for any variant.

For a somewhat complicated model, it is obviously unfeasible to specify the full design intent by such mutual constraints between pairs of faces. Another solution to accept or reject a variant would be to have the user inspect the resulting shape and decide. But that can be rather awkward, and might even be undesirable, in particular if the family of objects has been defined as a standard.

What is needed in the example of Figure 3 is a type of constraint that specifies that the pocket should remain open in all cases, i.e. that the top face of its shape should never be *on* the boundary of a variant. More in general, a set of constraint types is needed to explicitly specify which topological variations are allowed. Such a set of constraints, in combination with the more traditional geometric and algebraic constraints, can effectively specify the design intent, or *semantics*, of a prototype object, and therefore also of the corresponding family of objects. In addition, an automatic family membership test is needed that accepts a variant as a member of a family if and only if all its specified constraints are satisfied.

In this paper, the use of such a set of constraints is proposed to enhance the specification of a family of objects. All variants that satisfy the constraints specified for the prototype are said to belong to the corresponding family of objects. More specifically, a prototype object is defined by a *semantic feature model*, which consists of a set of features and a set of constraints; this is elaborated in Section 3. All objects are represented by a feature graph and a non-manifold cellular model, not only avoiding the problems of evaluation with a boundary representation, but also enabling a family membership test; this is elaborated in Section 4.

3. Defining semantics for a family of objects

This section presents the framework within which semantics is effectively incorporated in the definition of a family of objects. First, a variety of constraint types is defined, each of which can capture specific aspects of design intent. Next, these are integrated in the definition of a feature class. Finally, a formal definition for a semantic feature model is presented, providing the basis for a comprehensive definition of a family of objects.

3.1. Constraints

Constraints offer a convenient means to specify properties which the user requires in a feature model. Generally speaking, these properties may be classified as *geometric* or *topologic*.

Geometric properties

Two types of constraints can be used to specify geometric properties in and among features of a model: *geometric constraints* and *algebraic constraints*.

- **Geometric constraints**: specify geometric relations (e.g. parallelism, perpendicularity or distance) between feature entities;
- **Algebraic constraints**: specify expressions (equalities or inequalities) among feature parameters.

Geometric constraints can be used to specify geometric properties both within a feature instance or between entities of different feature instances. An example of the former is the requirement that the sides faces of a slot should be parallel, and an example of the latter is that the axes of two holes should be at a specified distance.

An essential property of geometric constraints is that they operate on entities of features in the model, e.g. their faces or axes, rather than on topologic entities of its geometric representation. This is crucial because, as described

Table 1 – Classification of interactions

Interaction type	Description
Splitting	Splits the boundary of a feature into two (or more) disconnected subsets.
Disconnection	Causes the volume of an additive feature (or part of it) to become disconnected from the model.
Boundary clearance	Causes (partial) obstruction of a closure face of a subtractive feature.
Volume clearance	Causes partial obstruction of the volume of a subtractive feature.
Closure	Causes some subtractive feature volume(s) to become a closed void inside the model.
Absorption	Causes a feature to cease completely its contribution to the model shape.
Geometric	Causes a mismatch between a nominal parameter value and the actual feature geometry.
Transmutation	Causes a feature instance to exhibit the shape imprint characteristic of another feature class.
Topologic	Corresponds to the violation of a boundary constraint in a given feature.

in the previous section, topologic entities can have an ephemeral existence in the geometric model, and are therefore unsuitable for being referenced in a persistent manner. References to feature entities, instead, remain valid as long as the respective feature instances remain in the model. By using them, the persistent naming problem outlined in the previous section is avoided.

Algebraic constraints can be used to specify parameter relations within a feature instance, e.g. an equality between the width and the length of a square-section passage feature, or among parameters of different features, e.g. requiring the depth of a pocket to be twice the diameter of a hole. Another use of algebraic constraints is for the specification of the allowable range of values for a feature parameter, by means of inequality relations.

Topologic properties

Topologic properties can be specified on features in a model by using two types of constraints: *boundary constraints* and *interaction constraints*. Both types of constraints, generally designated as *topologic constraints*, specify which topologic variants of a feature instance are allowed:

Boundary constraints: specify the extent to which its feature faces should be *on* the model boundary;

Interaction constraints: specify that a particular feature interaction type is not allowed for the feature instance.

Boundary constraints are of two types: *onBoundary*, which means the feature face should be present on the model boundary, and *notOnBoundary*, which means the feature face should not be present on the model boundary. Furthermore, both types of boundary constraints are parameterised, stating whether the presence or absence on the model boundary is *completely* or only *partly* required. An example of this is a blind hole class for which the top face has a *notOnBoundary(completely)* constraint, the side face has an *onBoundary(partly)* constraint, and the bottom face has an *onBoundary(completely)* constraint.

Several functional aspects of feature semantics, however, cannot be fully specified by boundary constraints alone. Those are better described in terms of the topology of the feature volume and/or of the feature boundary as a whole, and therefore require a higher-level specification, not directly based on feature entities. Such functional requirements can be violated by *feature interactions*, which are modifications of the shape aspects represented by a feature that affect its functional meaning. Examples of this are a *transmutation* interaction of a blind hole into a through hole, or an *absorption* interaction that causes a feature to cease completely its contribution to the model shape. A non-exhaustive classification of feature interactions can be found in (Bidarra 1999). For completeness, it is briefly summarised in Table 1. Topologic interaction, corresponding to the violation of a boundary constraint, is by definition always disallowed.

3.2. Definition of a feature class

Features can be defined as "representations of shape aspects of a product that are mappable to a generic shape and are functionally significant for some product life-cycle phase". In other words, each feature has a well-defined meaning, expressed through its geometric and topologic properties. A *feature class* is a structured description of all such properties, defining a template for all instances of a given feature type. These properties include the *validity conditions* that all feature instances of that type should satisfy. These conditions, as well as the feature shape and its parameters, are specified using the constraint types presented in the previous subsection.

The basis of a feature class is a parameterised shape. For a simple feature, this is a *basic shape*, e.g. a cylinder for a hole. A basic shape encapsulates a set of geometric constraints that relate its parameters to the corresponding shape entities. For a compound feature, the shape is a combination of several, possibly overlapping, basic shapes, e.g. two cylinders for a stepped hole.

The geometry of a feature, designated the feature's *shape extent*, accounts for the bounded region of space comprised by its volumetric shape. Moreover, its boundary is decomposed into functionally meaningful subsets, the *shape faces*, each one labelled with its own generic name, to be used in modelling operations. For example, a cylinder shape has a *top*, a *bottom* and a *side* face.

A feature class comprises also the notion of *feature nature*, indicating whether its feature instances represent material added to or removed from the model (respectively *additive* and *subtractive* natures).

Creation of an instance of a given feature class typically requires assigning numeric values to a set of scalar parameters $P_F = \{p_1, p_2, \ldots, p_f\}$ –so-called *feature parameters*– for determining its shape, and specifying some references to entities of other features for positioning it.

3.3. Definition of a semantic feature model

The two previous subsections described the specification of feature semantics by means of constraints. The use of such constraints should now be distinguished, according to whether they are specified *within* a feature class, thus holding for all its feature instances, or they are specified by the designer *among* (or *on* a) feature instance(s). The former are called *feature constraints*, as they are members of a feature class, whereas the latter are called *model constraints*. A model constraint instance, just like a feature instance, may require assigning numeric values to its scalar *constraint parameters* $P_C = \{p_1, p_2, \ldots, p_c\}$, as well as specification of any feature entities/parameters which it refers to.

Considering *n* features in a model, together with its *k* model constraints, we can define the set of *model parameters*, P_M, as

$$P_M = \{p_1, p_2, \ldots, p_m\} = \bigcup_{i=1}^{n} P_{F_i} \cup \bigcup_{i=1}^{k} P_{C_i}$$

Each sequence, $\boldsymbol{p}$, of m numeric values assigned to the model parameters p_j, j=1,...,m, defines a point in the parameter space $\mathbb{R}^m$. Such a sequence fully determines the shape and position of all features, and thus also the set of points of E^3 that is represented by the solid model.

A *semantic feature model* can then be defined according to the following:

Definition: a semantic feature model is a triple

$\mathcal{M} = \langle F, C, \boldsymbol{p} \rangle$

where

F is a set of interrelated feature instances, each one with its own set of feature constraints;

C is a set of model constraints applied on features of F; and

p *is a sequence of numeric values for the model parameters that fully determines the geometry of all feature instances of F.*

Typically, not all points in $\mathbb{R}^m$ correspond to a meaningful object representation; for example, some parameters in P_M, in particular feature dimensions, will be limited to a non-negative range of values. In general, the semantic feature model is said to be *valid* if and only if the validity conditions of all features in the model and all model constraints are satisfied; otherwise, it is an *invalid* model. In other words, a valid semantic feature model satisfies all its constraints or, formally, its sequence $\boldsymbol{p}$ of parameter values is such that the feature constraints of all features $f_j \in F$, and all model constraints in C, are satisfied.

3.4. Definition of a family of objects

Considering a semantic feature model $\mathcal{M}$ as the prototype model for a family of objects, $\mathcal{F}$, we propose the following definition:

Definition: the family of objects defined by the semantic feature model $\mathcal{M} = \langle F, C, \boldsymbol{p} \rangle$ *is the set of models*

$$\mathcal{F} = \{ \mathcal{M}' : \mathcal{M}' = \langle F, C, \boldsymbol{p}' \rangle \text{ is a valid model, } \boldsymbol{p}' \in \mathbb{R}^m \}$$

All instances of a family of objects have the same features and the same model constraints as the prototype. Essential in the definition is that *all* constraints must be satisfied for every member of the family.

The dimension of the parameter space of a family of objects does not necessarily have to be the same as that of the prototype. For example, a family parameter, say *throughHoleDiameter*, might be mapped to the *diameter* parameters of two through holes in the prototype. More generally, the specification of the family of objects may include a system of equations to map *n family parameters* to the *m* model parameters necessary to fully determine all features in an instance of the family. Such a system can be represented in the family specification by means of alge-

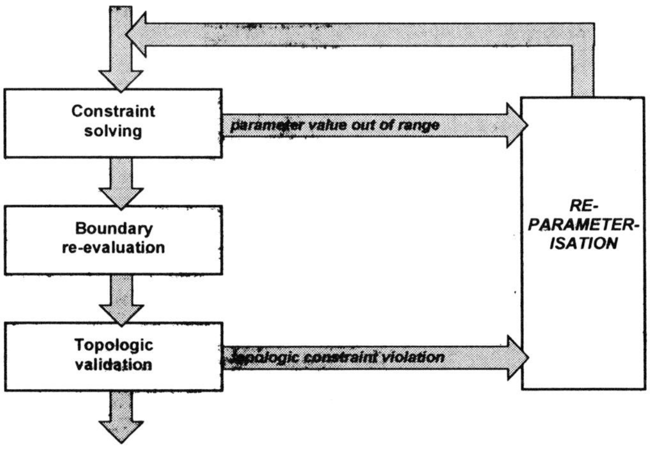

Figure 4 - Phases of the membership test

braic constraints.

Family parameters provide a convenient interface through which variants may easily be instantiated. In the remainder of the paper, family parameters will be designated simply as parameters, without further reference to any possible mapping between parameter spaces.

4. Family membership test

As mentioned in Section 2, a family membership test is required to determine whether a given sequence of parameter values $\boldsymbol{p}'$ specifies a valid instance of the family. According to the definitions in the previous section, checking membership comes down to assessing the validity of the resulting feature model.

A model validity maintenance process within the context of building a semantic feature model was described in detail by Bidarra and Bronsvoort (1999a). It has been implemented in the SPIFF system, a prototype multiple-view feature modelling system, developed at Delft University of Technology (Bronsvoort et al. 1997). In this system, a semantic feature model has a two-level structure that clearly distinguishes *modelling entities*, i.e. the entities on which all modelling operations are performed, from *entities in the evaluated geometric model*. The former are kept in the first level of the model –the so-called *Feature Dependency Graph*–, which contains all feature and (model) constraint instances, interrelated by *dependency relations*. The second level contains the evaluated geometric representation of the product in the so-called *Cellular Model*. Its entities are kept internal, being only required to "reflect" the geometry that results from the modelling operations performed on the first level.

We now concentrate on validity checking in the family membership test. This can be split into three phases, see Figure 4: (i) *constraint solving*, (ii) *boundary re-evaluation* and (iii) *topologic validation*, which will be subsequently described. In the end, we briefly explain how the user is assisted in re-parameterisation of an invalid instance to make it a member of the family.

4.1. Constraint solving

The first phase of the membership test is the internal geometric and algebraic constraint solving process. Its goal is to update the model parameters, i.e. dimensions, position and orientation of all features in the model, according to the new parameter values $\boldsymbol{p}'$ specified.

This task is performed by a Constraint Manager, which invokes two dedicated constraint solvers: a geometric constraint solver based on extended 3D degrees of freedom analysis (Kramer 1992), and a SkyBlue algebraic constraint solver (Sanella 1992). The iterative co-operation of these solvers, under the control of the Constraint Manager, is described by Dohmen (1997).

In this phase, the family membership test fails whenever some model parameter gets assigned a value out of its range of allowable values (specified with algebraic constraints, see Subsection 3.1). Such algebraic constraint violations are passed to the re-parameterisation mechanism for further processing, as depicted in Figure 4.

4.2. Boundary re-evaluation

When this second phase is reached, each feature in the model has all its parameters successfully updated. In particular, all feature shape extents have their dimensions, position and orientation fully determined. The Cellular Model may thus be updated, so that the effects of the operation are also reflected in the evaluated geometric model.

In this subsection, the main characteristics of the Cellular Model are first summarised, and then the two stages of the boundary re-evaluation process are shortly described: *re-evaluation of the Cellular Model* and its *history-independent interpretation.*

The Cellular Model

The Cellular Model is a *non-manifold* geometric representation of the semantic feature model, integrating the contributions from all its features. It has been described in detail in (Bidarra et al. 1998).

The Cellular Model represents an object's geometry as a connected set of volumetric quasi-disjoint *cells* of arbitrary shape, in such a way that each cell lies either entirely *inside* a shape extent or entirely *outside* it. The cells represent the point sets of the shape extents of all features in the model. Each shape extent is, thus, represented in the Cellular Model by a connected subset of cells.

Furthermore, the cellular decomposition is interaction-driven, i.e. for any two overlapping shape extents, some of

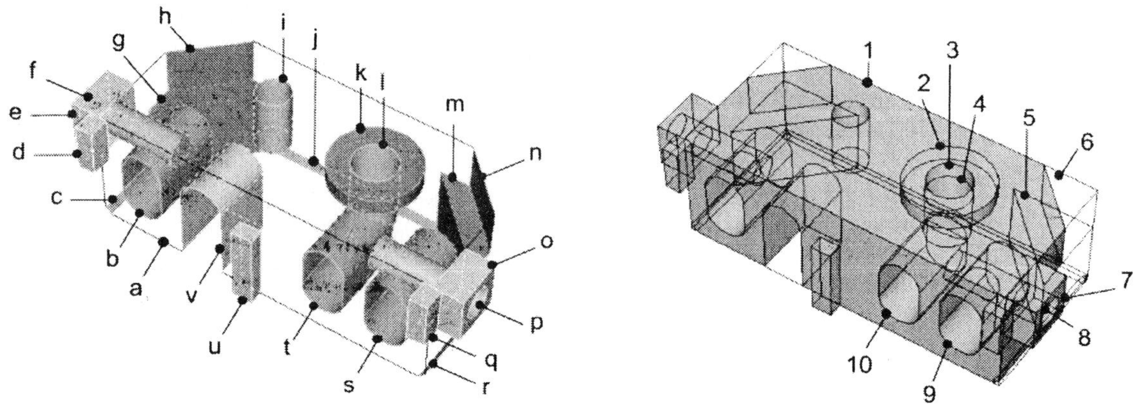

(a) object

(b) Cellular Model

a block
b pocketLeft
c chamferLeft
d ribLeft
e protrusionLeft
f blindHoleLeft
g cornerBotLeft
h cornerTopLeft
i blindHoleBack
j chamferBack
k cylindricalProtrusion
l blindHoleTop
m cornerBotRight
n cornerTopRight
o protrusionRight
p blindHoleRight
q ribRight
r chamferRight
s pocketRight1
t pocketRight2
u ribFront
v blindSlot

(c) features

1 <block>
2 <cylindricalProtrusion>
3 <cylindricalProtrusion, blindHoleTop>
4 <block, blindHoleTop>
5 <block, cornerBotRight>
6 <block, cornerTopRight>
7 <protrusionRight>
8 <protrusionRight, blindHoleRight>
9 <block, pocketRight1>
10 <block, pocketRight2>
.
.
.

(d) some cell owner lists

Figure 5 – Cell owner lists in the Cellular Model

their cells lie in both shape extents (and are called *interaction cells*), whereas the remaining cells lie in either of them. As a consequence of this, two cells can never volumetrically overlap. They may, however, be adjacent, in which case there is an interior *cell face* separating them.

As described in Subsection 3.2, the boundary of a feature's shape extent is decomposed into functionally meaningful subsets, the shape faces, each one labelled with its own generic name. Each shape face is represented by a connected set of cell faces. In order to be able to search and analyse features and their faces in the Cellular Model, each cell has an attribute –called *owner list*– indicating which shape extents it belongs to, see Figure 5. Similarly, each cell face has also an owner list, indicating which shape faces it belongs to.

Just like for features, the *nature of a cell* expresses whether its volume represents "material" of the object or not. Similarly, the *nature of a cell face* expresses whether it lies on the boundary of the object or not. Cell and cell face natures are determined after Cellular Model re-evaluation, in the interpretation stage.

The Cellular Model, including its attribute mechanism to maintain and propagate the owner lists of cells and cell faces, was implemented using the Cellular Topology husk of the Acis Geometric Modeller (Spatial 1999).

Re-evaluation of the Cellular Model

In contrast with most parametric systems, which use two non-associative set operations (union and difference) to evaluate the geometric model, in our approach *only one set operation* is used to evaluate the Cellular Model: it is computed by performing the *non-regular cellular union* of the shape extents of all features. Because it is a union operation, the order in which the shape extents are processed is now irrelevant for the final Cellular Model obtained. By these non-regular cellular operations, the cellular decomposition described above is computed. Essential in this process is the correct propagation of the owner lists of each cell and cell face when these are further decomposed.

Typically, when some model parameters are modified, only a few feature shapes are actually geometrically affected (i.e., in their dimensions, position and/or orientation). Consequently, *only* the shape imprint of those features needs to be updated. For this, their shapes are *removed* from the Cellular Model and then *re-added* with their new parameter values.

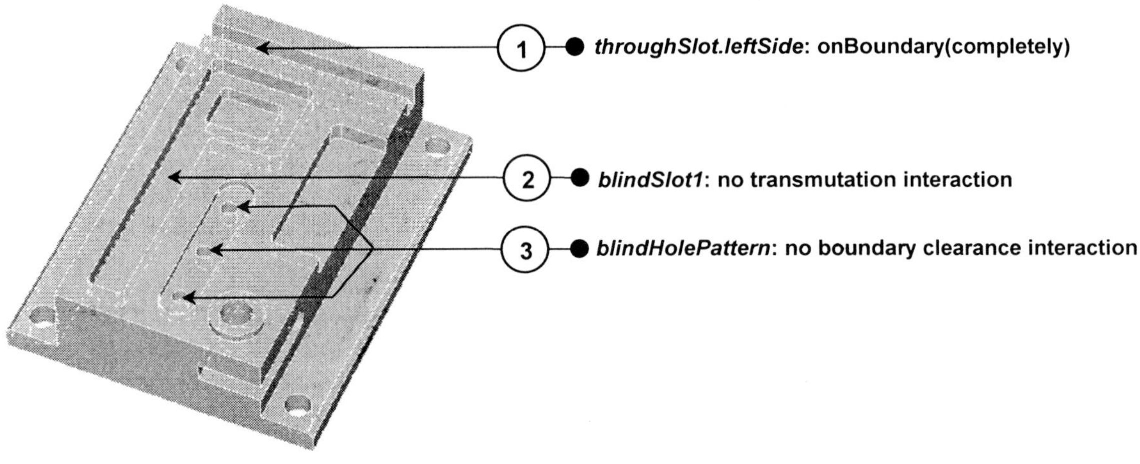

Figure 6 – Prototype model and three of its topologic constraints

History-independent interpretation

Interpretation of the re-evaluated Cellular Model consists of determining whether the point set represented by each cell does belong to (or represent "material" of) the object, i.e. determining the nature of that cell. Interpretation of the Cellular Model requires deciding *which* of the features in its owner list "prevails", either as additive or as subtractive.

For this, precedence relations among features are determined, based on dependencies between features and on possible overlap between independent features. Once all precedence relations have been established, a *global ordering* of the set *F* of all features in the model can be easily performed by a classical topological sorting algorithm. The features in the resulting sorted sequence are then assigned unique, increasing precedence numbers. Eventually, every cell owner list (a subset of *F*) is sorted according to these precedence numbers, and each cell's nature becomes therefore automatically determined: it is the nature of the last feature in its owner list, i.e. the feature with the highest precedence number. This scheme always yields an interpretation of the Cellular Model that is unambiguously determined, independent of the model history. For details on the precedence criteria and sorting algorithms involved in Cellular Model interpretation, as well as for a discussion on the advantages of history-independent boundary evaluation, the reader is referred to (Bidarra and Bronsvoort 1999b; Bidarra and Bronsvoort 2000).

4.3. Topologic validation

Once the model boundary has been re-evaluated, the shape imprint of all modified features exhibits their updated geometry and topology, and detection of violations in topologic constraints can take place. In this third phase, the model is considered invalid, and thus the family membership test fails, if any boundary or interaction constraint is violated for some feature.

For checking each boundary constraint and each interaction constraint, see Table 1, topologic entities of the Cellular Model are queried and analysed, using the feature information maintained in their owner lists. Details on the interaction detection algorithms can be found in (Bidarra 1999). Eventually, the set of constraint violations, if any, is analysed, and their causes are identified and passed for further processing to the re-parameterisation mechanism, see Figure 4.

The re-parameterisation process includes reporting to the user the constraint violations detected, documenting their scope and causes, and, whenever possible, providing corrective hints. In case of a parameter that has been assigned a value out of its specified range, this is straightforward. In case of a topologic constraint violation, re-parameterisation hints are presented to the user in terms of the relevant parameter values causing the violation.

5. Results and conclusions

We have introduced a new approach to define a family of objects, and to determine whether a particular instance belongs to the family. A family of objects is defined by creating a prototype semantic feature model (Bidarra and Bronsvoort 2000), in which the semantics of the family can be fully specified by several types of constraints, in a relatively easy way. Given a set of parameter values, the family membership test can be used to determine whether the corresponding instance belongs to the family.

In Figures 6-8, we show an example of its application

in the SPIFF modelling system. In Figure 6 the geometry of a prototype model is shown, together with three of its topologic constraints relevant for the example. In Figure 7 three instances are shown that do belong to the family defined by the prototype, and in Figure 8 three instances that do not belong to the family.

The first advantage of the approach is that the instance corresponding to a set of parameter values always has a shape that may be expected on the basis of the prototype and the parameter values. Unexpected results, which regularly occur in current parametric modelling systems, in particular because of the persistent naming problem in boundary re-evaluation, do not occur. The reason for this is that all references during modelling are made to persistent feature entities instead of non-persistent entities in a boundary representation.

The second advantage of the approach is illustrated by the example given above. If one considers the shape of the prototype in Figure 6 and of the instances in Figures 7 and 8 only, it is not obvious that the instances in Figure 7 should indeed belong to the family defined by the prototype, and the instances in Figure 8 not. In our approach, the membership of these instances is determined by the constraints in the definition of the prototype of the family. More generally, the design intent, or semantics, of a family of objects can be specified much more precisely than in current parametric modelling systems, in particular with the topologic constraints.

Regarding the implementation, it has been shown that a cellular model can be very valuable to represent objects in the context of defining families of objects, as has also been suggested by Shapiro and Vossler (1995) and Raghothama and Shapiro (1998). The cellular model is in particular exploited by the family membership test; for example, the model offers even the possibility to give, in case an instance is not a member of the family, hints to the user to re-parameterise the instance in a way to make it a member of the family. Bidarra et al. (1998) already showed that a cellular model can be very useful in several other contexts too, e.g. multiple-view feature modelling. So although cellular models are not used in current modelling systems, they might well become more popular in the future.

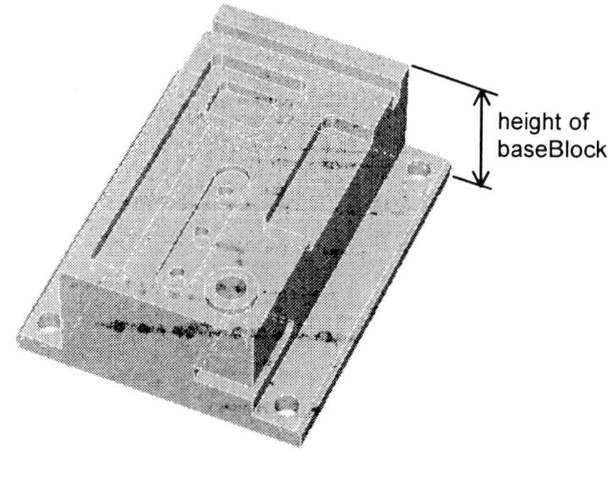

(a) height of *baseBlock* increased

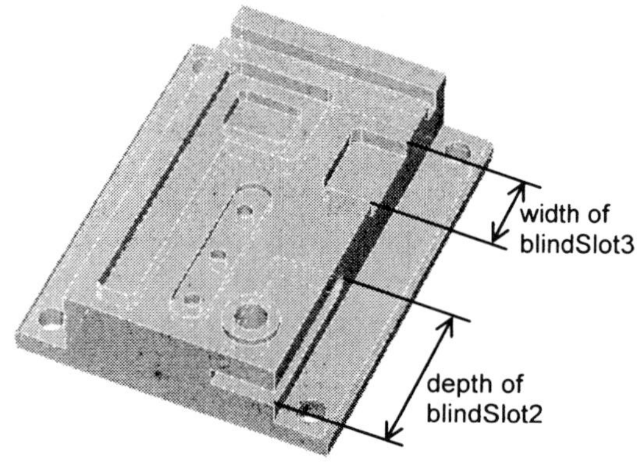

(b) depth of *blindSlot2* increased; width of *blindSlot3* decreased

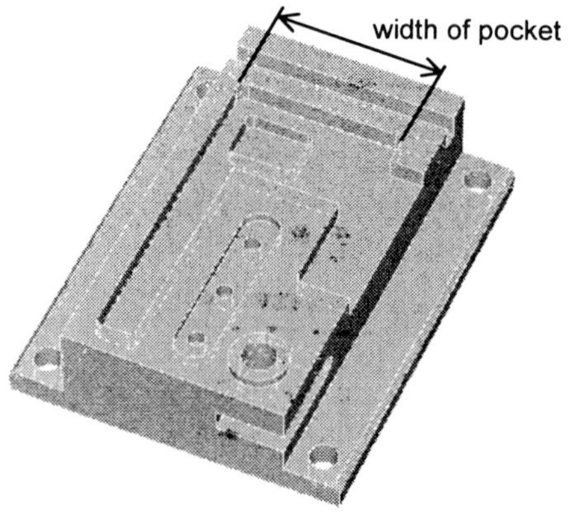

(c) width of *pocket* increased

Figure 7 – Examples of members

References

Anderl, R. and Mendgen, R. (1995) Parametric design and its impact on solid modeling applications. In: *Proceedings Solid Modeling '95 – Third Symposium on Solid Modeling and Applications, 17–19 May, Salt Lake City, UT, USA*, Hoffmann, C.M. and Rossignac, J.R. (Eds.), ACM Press, New York, pp. 1–12

Bidarra, R. (1999) Validity maintenance in semantic feature modeling. PhD Thesis, Delft University of Technology, The Netherlands

Bidarra, R. and Bronsvoort, W.F. (1999a) Validity maintenance of semantic feature models. In: *Proceedings of Solid Modeling '99 – Fifth Symposium on Solid Modeling and Applications, 9–11 June, Ann Arbor, MI, USA*, Bronsvoort, W.F. and Anderson, D.C. (Eds.), ACM Press, New York, pp. 85–96

Bidarra, R. and Bronsvoort, W.F. (1999b) History-independent boundary evaluation for feature modeling. In: *CD-ROM Proceedings of the 1999 ASME Design Engineering Technical Conferences, 12–15 September, Las Vegas, NV, USA*, ASME, New York

Bidarra, R. and Bronsvoort, W.F. (2000) Semantic feature modelling. To be published in: *Computer Aided-Design* **32**(3)

Bidarra, R., de Kraker, K.J. and Bronsvoort, W.F. (1998) Representation and management of feature information in a cellular model. *Computer-Aided Design* **30**(4): 301–313

Bronsvoort, W.F., Bidarra, R., Dohmen, M., van Holland, W. and de Kraker, K.J. (1997) Multiple-view feature modelling and conversion. In: *Geometric Modeling: Theory and Practice – The State of the Art,* Strasser, W., Klein, R. and Rau, R. (Eds.), Springer, Berlin, pp. 159–174

Capoyleas, V., Chen, X. and Hoffmann, C.M. (1996) Generic naming in generative, constraint-based design. *Computer-Aided Design* **28**(1): 17–26

Chen, X. and Hoffmann, C.M. (1995) On editability of feature based design. *Computer-Aided Design* **27**(12): 905–914

Dohmen, M. (1997) Constraint-based feature validation. PhD Thesis, Delft University of Technology, The Netherlands

Kramer, G.A. (1992) Solving geometric constraint systems: a case study in kinematics. The MIT Press, Cambridge, MA

Kripac, J. (1995) A mechanism for persistently naming topological entities in history-based parametric solid models. In: *Proceedings Solid Modeling '95 – Third Symposium on Solid Modeling and Applications, 17–19 May, Salt Lake City, UT, USA,* Hoffmann, C.M. and Rossignac, J.R. (Eds.), ACM Press, New York, pp. 21–30. Also in: *Computer-Aided Design* 1997; **29**(2): 113–122

Lequette, R. (1997) Considerations on topological naming. In: *Product Modeling for Computer Integrated Design and Manufacturing – Proceedings TC5/WG5.2 International Workshop on Geometric Modeling in Computer Aided Design, 19–23 May 1996, Airlie, VA, USA*, Pratt, M., Sriram, R.D. and Wozny, M.J. (Eds.), Chapman & Hall, London, pp. 394–403

Raghothama, S. and Shapiro, V. (1998) Boundary representation deformation in parametric solid modeling. *ACM Transactions on Graphics* **17**(4): 259–286

Sanella, M. (1992) The SkyBlue constraint solver. Technical Report 92-07-02, University of Washington, WA, USA

Shapiro, V. and Vossler, D.L. (1995) What is a parametric family of solids? In: *Proceedings of Solid Modeling '95 – Third Symposium on Solid Modeling and Applications, 17–19 May, Salt Lake City, UT, USA,* Hoffmann, C.M. and Rossignac, J.R. (Eds.), ACM Press, New York, pp. 43–54

Spatial (1999) Acis 3D Modeling Kernel, Version 5.3, Spatial Technology Inc., Boulder, CO, USA

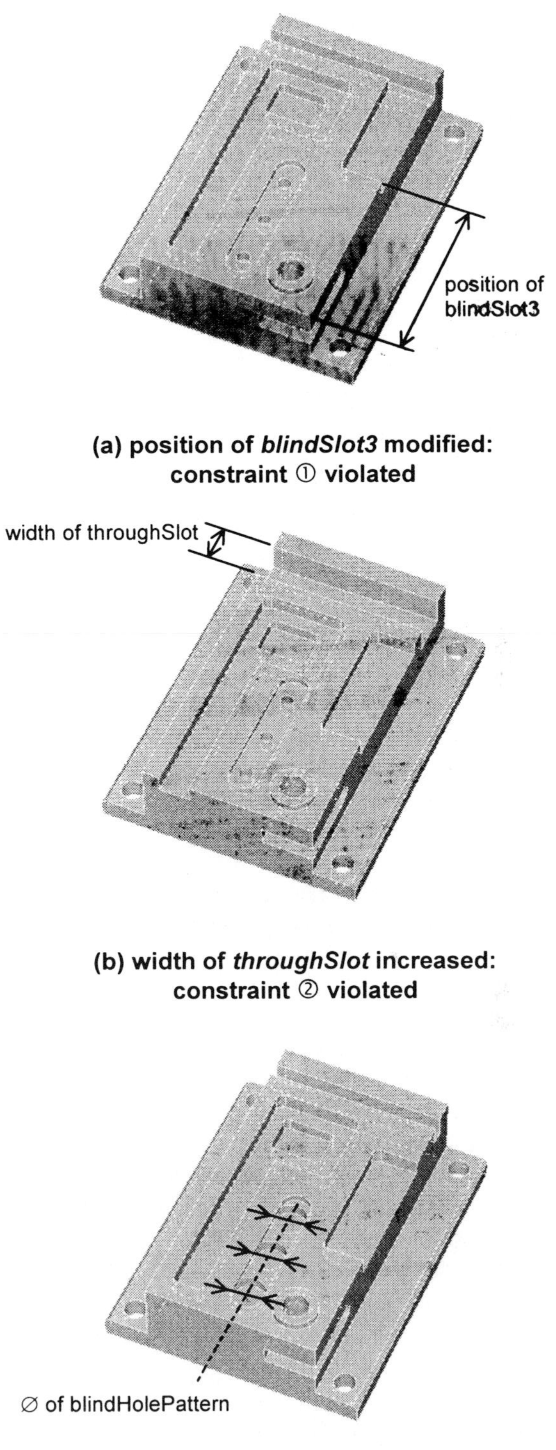

(a) position of *blindSlot3* modified: constraint ① violated

(b) width of *throughSlot* increased: constraint ② violated

(c) diameter of *blindHolePattern* increased: constraint ③ violated

Figure 8 – Examples of non-members

Session 4

Proximity Query and Voronoi Regions

Proximity Queries for Modeling and Manufacturing

Stephen Cameron and Joe Pitt-Francis
Oxford University Computing Laboratory
Parks Road, Oxford OX1 3QD, UK
{cameron,jmpf}@comlab.ox.ac.uk

Abstract

Collision detection and distance calculation are two examples of proximity queries between modelled objects. They are interesting both in their own right, and as useful tools to solve other problems (e.g., robot path planning). They also form useful abstractions away from the underlying representation of a geometric model. This paper will consider the solution and use of such problems in geometric modeling and robotics, give examples from our own work, and discuss the possible future of the technology.

1. Introduction

There are many problems whose solution depends critically on the shapes of the objects involved, for example: constructing an assembly, removing a part for maintenance, VLSI layout, stacking or orienting parts, robot planning, pipe routing, simulation for virtual reality and computer games, and moving furniture. Such *spatial reasoning* problems have proved difficult to solve by computer; partly because humans solve some of these problems on a daily basis without really knowing how! There have been many attempts to produce algorithmic solutions to these problems, some of which have been largely successful and are in widespread use (e.g., VLSI layout), and some of which have not been widely used (e.g., motion planning). It is instructive to examine the reasons why this is true of motion planning, despite two decades of research in this area.

- The theory behind path planners is mathematical in nature and can get quite complex. This has important consequences not only when trying to persuade a potential user that the techniques are useful, but also in trying to persuade the user that the program code is reliable (i.e., no bugs!) and can be maintained in the field.
- Motion planning has a reputation of being too slow to be useful, especially for examples in which the geometry is a realistic representation of standard industrial parts.

The mismatch between research and practice seems even greater to many potential users because path planning is something that we do every day; say when crossing the street! The user would like to think of path planning in fuzzy terms (e.g., pass that obstacle on the left; approach this point slowly; go between these objects) whereas most path planning algorithms focus on the details (e.g., we can now move between 114–893mm forwards). We characterise this by saying that the user would like to think at a *strategic level* and not be bothered too much by the *geometric level*, and although there are execptions — sometimes you really need to know that you have exactly 23.7mm to go before hitting an obstacle — many other spatial reasoning problems can be separated in a similar manner.

Recently it has become apparent that many problems in robotics can be solved with the help of an efficient solution to the problem of computing the distance between two modelled objects; in particular, the interference detection problem can be solved by looking for when this distance becomes zero, and the evolution of distance over simulated time admits solutions to the collision detection (i.e., continuous intersection detection) problem and the path planning problem [6, 24]. There are also problems in other general areas — such as simulation, assisted tele-operation, virtual reality, and computer games — whose solution can be greatly simplified by the availability of a fast method for computing distances.

Given our experience in these areas we propose that *proximity measures* form a useful bridge between complex geometric calculations and strategic-level reasoning. In the rest of this paper we will descrie some of our experiments in this field; the design of our path planning system that uses distances; and ways of computing the distance between objects. We also discuss a few more general proximity measures.

2. Path planning

Producing paths for robotic devices that avoid collisions has been a goal of robotics research for almost 3 decades, and a serious subject of academic study for over 10 years. For the purpose of this paper we simply refer the reader to [14], [3], or to Latombe's encyclopaedic textbook [18], for overviews on path planning methods, and proceed to state the issues when it comes to planning for large, complex environments.

Most path planners either perform their computations in *real space* (i.e., normal Euclidean space), or in *configuration space* [20]. The latter considers the actions of the robot within a space which has a coordinate direction for each degree-of-freedom (dof) of the robot, and a point in that space corresponds to the robot being moved so that those freedoms[1] take specific values. Thus the complete configuration space for a typical 5 actuator industrial manipulator would be 5 dimensional. The one-to-one correspondence between configuration space and the actions of a robot made it an early favourite for writers of path planners, but the way that real world obstacles map into configuration space obstacles is extremely complex for useful robots, and so algorithms that construct explicit maps of configuration space have fallen out of favour. An alternative is to sample the configuration space at some resolution, and this is the basis of many useful planners. In any case, the number of possible paths is (infinitely!) large, and there is a secondary question as to whether an optimal solution is sought, or whether a bounded search is used. One successful technique for performing a bounded search is the so-called *potential field* method, in which the robot is attracted to the goal configuration under the influence of a simulated vector field that also contained repulsive terms that attempted to keep the robot away from obstacles [17]. Potential field techniques turn out to be ideal for robots that could move freely ('free-flying robots'), and have been successfully used also for manipulator robots (e.g., [1, 21]).

Recent motion planners tend to be based on either: (a) taking a sample of configuration space, in order to reduce the size of the search space; (b) lazy evaluation of configuration space; or (c) potential field techniques. Configuration space sampling turns out to be equivalent to *collision detection*, and many of the potential field techniques can computed by *distance calculation*. Distance calculation turns out to be sometimes sufficient for lazy evaluation of configuration space, and by testing whether the distance is zero we can also perform collision detection. Thus we argue that a useful path planning tool can be developed based on distance calculation, and this observation forms the basis for our path planning shell, OxSim. A (simple) example of a path being planned by the system is shown in Fig. 1; the target workcell more complex than this, being made up of thousands of polygons.

[1]For example, positions for a free-roaming robot, or joint angles for a manipulator

2.1. OxSim overview

A functional overview of OxSim is show in Fig. 2. A geometric modeller is used to produce polyhedral models of parts, and convert them into a suitable input format, and OxSim is compiled with an appropriate local path planner. Then commands are sent to OxSim to, essentially, run a simulation of the situation, guided by the local planner and by explicit calls to the distance computation module.

A key part of OxSim is thus the computation of the (Euclidean) distance between links of the robot and the environment. It is not always necessary for this measurement to be accurate when objects are far apart, but the distance measure will be called many times as part of an inner loop of the planner. We have used a variation on the algorithm of Gilbert, Johnson and Keerthi (GJK) for convex polyhedra [11, 4], which is discussed in more detail in §.

For complex environments we may have many hundreds or thousands of component parts, like out convex polyhedra, and finding the smallest distance between any pair in a naive fashion would require a quadratic number of component-component computations to be completed. We can (and do) greatly reduce this cost by using the following heuristic scheme [24]. If we construct a simple bounding shape (such as a sphere) around each piece then we can quickly compute the minimum and maximum distances between the bounding shapes, and a guess as to the distance between the pieces. The latter number can be computed in a number of ways — thus the term 'heuristic' — but a simple method would be to take the distance between the centres of the bounding shapes. Then if we express the minimum bound, maximum bound, guess distance, and actual distance between two objects A and B as $\min(A, B)$, $\max(A, B)$, $h(A, B)$, and $d(A, B)$ we can proceed as follows:

1. Construct a list $\mathcal{L}$ of object pairs $[\langle A_i, B_i \rangle]$, ordered by guess distance (i.e., $h(A_i, B_i) \le h(A_j, B_j)$ for $i \le j$), and set an upper bound u on the minumum distance between the composite objects by
$$u = \min_i \max(A_i, B_i).$$
2. Now, while $\mathcal{L}$ is non-empty:
3. Remove the pair $\langle A_j, B_j \rangle$ from the top of $\mathcal{L}$.
4. If $\min(A_j, B_j) \ge u$, jump back to step 3.
5. Compute $D = d(A_j, B_j)$. If $D < u$ then this pair achieves a smaller minimum distance than we have seen before. Record the pair, set $u = D$.

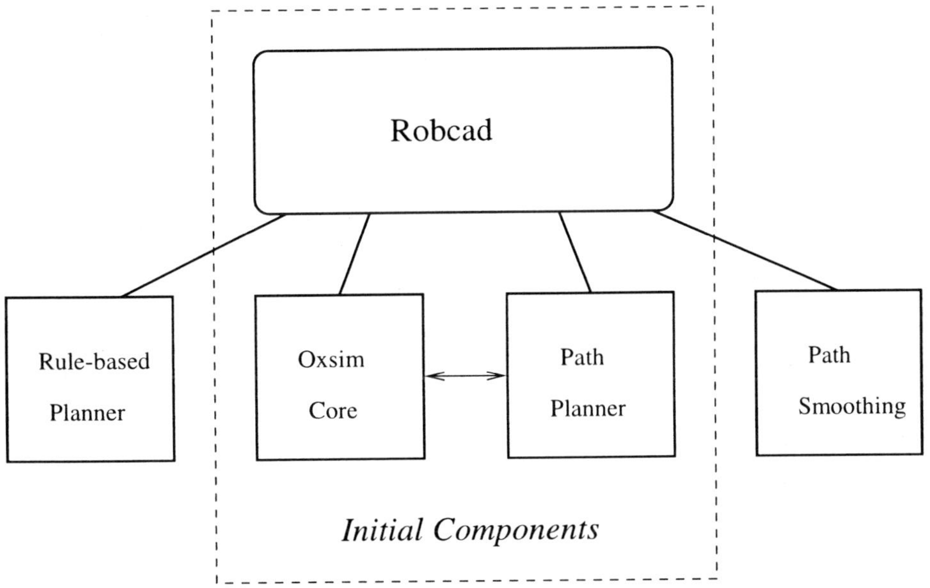

Figure 3. Major Components of the Large Scale Demonstrator

6. Return to step 3.

If the heuristic used to compute h is a good one then this algorithm should find the true minimum distance quickly, and avoid computing d (which is a relatively expensive operation) for many pairs of objects.

Variations of this algorithm are used in most distance algorithms. We have found that a useful one is to construct a binary tree of bounds, with each internal node bounding the bounds of its child nodes. Then we can use this extra infomation to make top-down traversals of the trees of bounds to prune out more potential pairs in many cases (although the theoretical worst-case behaviour is just as bad as the easier algorithm).

2.2. Large-scale demonstrator

In order to demonstrate that OXSIM and other planning techniques can be built up into really useful planning programs we have embarked on the construction of a large-scale demonstrator of the technology. The ROBCAD robot simulation system was chosen as the host system because it is relatively straightforward to add modules made up of user-code to the system, and because it was already in use with one of our industrial collaborators (British Aerospace). OXSIM does not tackle the whole of the planning process, in that it requires a rough plan as input, and its output may require smoothing to take into account the robot dynamics and low-level controller. Rather than trying to produce the rough input plan automatically, the system, at least in its first incarnation, will expect the user to suggest a number of via points for the tip of the robot. This means of working fits in well with what is easily provided by the existing ROBCAD framework, which has tools for defining robot positions and chaining these up into motion profiles. However rather than then relying on ROBCAD's visualisation and analysis tools to verify the path, the intention is that a rough path be passed on to OXSIM for refinement. Furthermore, if the path coming from OXSIM appears acceptable to the user, it can be passed on to further stages that can tweak the path to produce something that is easier for the robot to execute, given knowledge of the robot's dynamics and controller. Thus rather than trying to produce a fully automated solution, the user is presented with a number of advanced tools, as in Fig. 3.

We actually regard this method of working as *more* useful than a fully automated solution, at least for the time being. For the foreseeable future no planning program can hope to take account of all the considerations that a human process planner can; for example, certain paths may be disallowed for issues of safety, and some preferred because of the likely flow pattern of materials on a production line. We are therefore trying to produce tools that the process planner can use *and control*; it is, after all, their job that is on the line if something goes wrong.

3. Computing distance

In this section we give some more detail on how we can efficiently compute the distance between the pieces. Work in this area has concentrated on computing the distance between convex polyhedral objects. In retrospect, the reason for this can be summarised by two points:

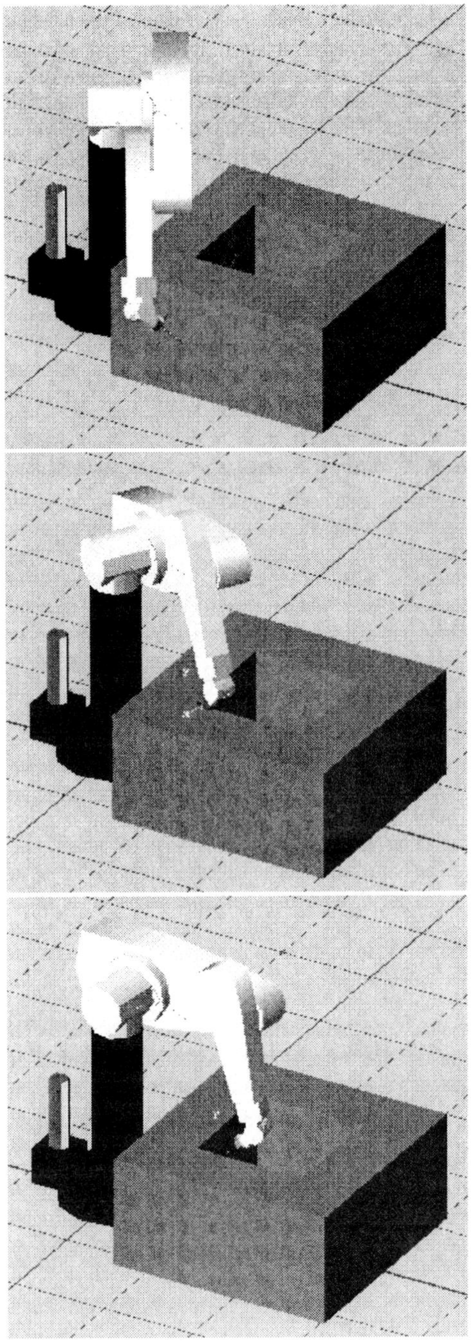

Figure 1. Planning A Simple Robot Motion

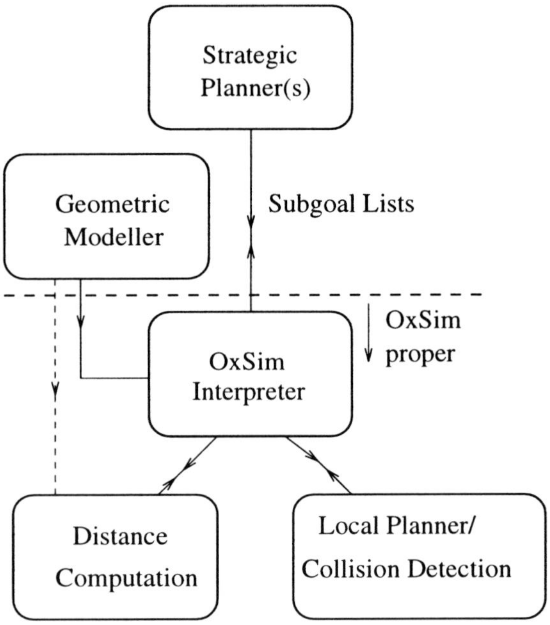

Figure 2. Major Components of OxSim

- polyhedral models are easier to deal with than more general models, are always supported by three-dimensional modeling systems, and can be made sufficiently accurate for most tasks; and
- restricting the objects to be convex admits to very fast solutions to the problems.

In turn, the basic reason why convex objects admit fast solutions to the problem is that the distance problem between convex bodies is itself a 'convex problem' in the numerical optimisation sense. Essentially this means that iterative solutions to the problem can always be formulated that

(a) given a proposed solution to the problem, in the form of a pair of points from the two objects which are thought to possibly realise the minimun distance between them (so-called *witness points*), we can check whether these points are indeed witness points by *local operations*, that is, by scanning only those parts of each objects that are neighbours of the proposed witness points; and

(b) if the proposed witness points are not found to be true witness points, we can obtain a better solution by some form of gradient descent operation.

The three best-known algorithms for computing distance both work in this way. The algorithm of Gilbert, Johnson and Keerthi (GJK, [11]) and the algorithms Lin-Canny (LC, [19]) and V-Clip ([22]) all have outer loops that is finding a sequence of pairs of proposed witness points which are then improved by gradient descent [4]. As the objects considered are polyhedra then the outer loops must converge in a finite number of steps to the true solution, and this number of steps can be bounded given some measures of the complexity of the polyhedra. The main difference between the algorithms is in the way that the proposed witness points are selected: for GJK they are taken from points within or on the surface of the polyhedra, whereas for LC and V-Clip only surface points are considered.

For many application areas the distance problem is not called just once for a particular pair of objects; rather, we track the evolution of the distance over a number of time-steps. For this *tracking problem* we can use the temporal coherence between calls to obtain even faster solutions to the problem. That is, we can use the fact that the function that describes the distance between two objects undergoing continuous motions is itself a continuous function in order to cache information from one call to the next. In the case of LC (and V-Clip, which is related) they cached the witness points (plus some other information) so that at the next call they can use property (a) above to test whether the same witness points are still solutions, and if not to quickly find new witness points amongst the neighbours (by property (b)). Using this structure LC exhibits 'almost constant time' behaviour, and we have shown [5] that the simple modification to GJK described in the next sub-section (which we call *enhanced GJK*, or EGJK) gives this algorithm a very similar time complexity[2].

3.1. Support mapping

Both LC and GJK need to check whether the proposed witness points are true witness points, and in both cases this is done by checking the *support mapping*. This is illustrated in Fig. 4 for two convex shapes in the plane A and B. Given proposed witness points $\mathbf{a}$ and $\mathbf{b}$ on A and B we can check whether they are true witness points by considering the direction $\Delta = \mathbf{b} - \mathbf{a}$, and looking for maximal values of the *support function* h, defined as

$$h_X(\eta) = \max_{\mathbf{x} \in X} \mathbf{x} \cdot \eta$$

where X is a set of points, η is a direction, and $\cdot$ is the scalar product function[3]. Then $\mathbf{a}$ and $\mathbf{b}$ are witness points if and only if

$$h_A(\Delta) = \mathbf{a} \cdot \Delta \text{ and } h_B(-\Delta) = \mathbf{b} \cdot (-\Delta) \qquad (1)$$

[2]Specifically, GJK takes around 100 floating-point multiplies per iteration, and for LC and V-Clip the figure is nearer 50. However the core of GJK is rather smaller than for the other two algorithms, resulting in fairly comparable timings. As a rule of thumb, EGJK takes around 10μS per iteration on a 400Mhz Pentium processor.

[3]The max in the definition of h would cause problems if X is not a topologically closed set. This can be avoided by using a sup operation instead, but for modeling purposes it is usual to only deal with closed sets anyway.

which on the figure corresponds to checking whether A lies to the left of the line through $\mathbf{a}$ and perpendicular to Δ (shown dotted), and with a similar relationship holding for $\mathbf{b}$ and B.

By convexity h_X can be computed by hill-climbing over the surface of X until a maximal point is reached. Noting that in the general case this maxima may not be unique, we can then define the support mapping s_X to be any mapping that, given a direction η, returns a point in X satisfying $h_X(\eta) = s_X(\eta) \cdot \eta$. The support mapping is thus a critical part of both LC and GJK. For polyhedra the support mapping can always return vertices of the polyhedra, and simple hill-climbing can be implemented by the following scheme.

1. Given the support direction η and a current vertex $\mathbf{v}$, compute (or recall) $\mathbf{v} \cdot \eta$.

2. For every neighbouring vertex $\mathbf{v}_i$ of $\mathbf{v}$ compute $\mathbf{v}_i \cdot \eta$. If some $\mathbf{v}_j$ has $\mathbf{v}_j \cdot \eta > \mathbf{v} \cdot \eta$, set the current vertex to $\mathbf{v}_j$ and return to step 1[4].

3. Otherwise, return $\mathbf{v}$ as the result.

The support mapping also has other uses in geometry and robotics; for example, in Chung's interference detection routine [9].

In general parts will be provided to the system that are not convex. Although in theory an automated solution to the problem of splitting an arbitrary polyhedra into convex pieces exists [8], in practise it is very difficult to do so without hitting degenerate cases that 'blow up' the implementation. However using some heuristics and the ACIS modeling system we have had some success, as illustrated in Fig. 5.

3.2. Penetration distance

Most people will have a pretty good idea of what the minimum distance between two non-overlapping objects is, but what about objects in collision? Although such a configuration is physically impossible, simulations may often drive virtual objects into collisions, and having some idea of how 'deep' the collision is is often a good way of winding back the simulation to the moment of collision. This observation was the rationale behind the definition of *minimum translational distance* (MTD) [7] which has been used in several (dynamic) collision detection and path planning systems. MTD is defined in terms of the smallest translation required to take the two objects into a contact-only situation, with penetrations regarded as giving a negative value, and the non-penetration case being synonymous with the usual Euclidean distance.

[4][13] discusses possible choices for $\mathbf{v}_j$.

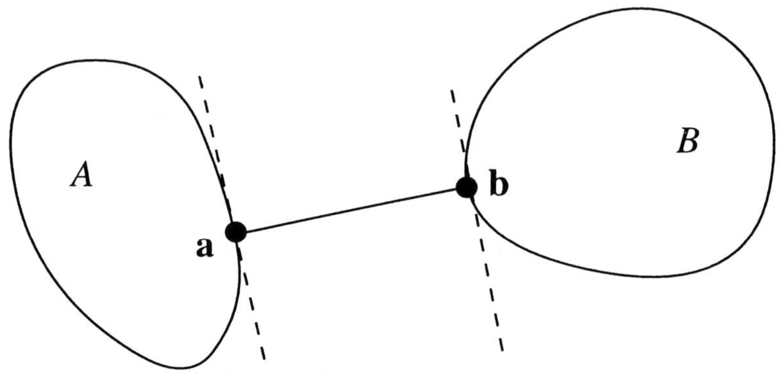

Figure 4. Illustration of the Support Function

3.3. Distances for non-polyhedral objects

Not all objects look like polyhedra to a sufficient level of detail. For convex objects the GJK algorithm has a natural extension, namely the Gilbert-Foo algorithm [10, 26], which uses the same outer loop structure as GJK but with a different, numerical, termination procedure. But again, not all objects can be decomposed into a finite number of convex pieces of whatever shape — as an example, consider a block with a circular hole drilled in it. For such shapes methods have been developed which treat the computation of the minimum distance as a general numerical optimisation problem, although by adding some domain-specific knowledge significant speedups can be achieved, and we can also compute penetration distances [25].

3.4. The Gilbert-Foo algorithm

GJK has a quite natural extension to curved objects, as described by Gilbert and Foo [10]. GJK worked by keeping track of between 1 and 4 pairs of vertices of the two objects, which between them define a pair of three-dimensional simplices[5] which are subsets of the two objects, and the action of GJK determines simplices that contain the witness points for the two objects. At each step the support mapping is used to determine whether the proposed witness points are true witness points, and if not new vertices are discovered that are added to the vertex pairs set to describe new simplices.

Gilbert-Foo works in exactly the same way, except that now the simplices are described by pairs of arbitrary surface points of the two objects, as the support mapping need not always return object vertices[6]. In this case the distances between the simplices will still converge, but is not guaranteed in a finite number of steps. Thus another termination criteria for the outer loop is required, to return an answer once we are 'close enough'. This is not difficult to provide, although further work is required to try to optimise the criteria for common surface types.

Given the termination criteria then, we note that the *only* question we have to answer for a particular type of (convex) surface is the support mapping.

[5]That is, points, lines, triangles or tetrahedra.

[6]Consider, for example, a sphere, which has no natural vertices.

3.5. Non-Euclidean distances

For planning purposes what we really want to estimate when using 'distance' is an answer to the question 'How much *effort* is required to go that way'? Sometmes 'effort' equates to time, sometimes to energy, and often to some curious mixture! For example, when planning the route for a rechargeable robot truck we should ideally take into account the time it would take to recharge, as well as the time to travel.

For these reasons amongst others it is often quite sufficient to return some non-Euclidean measure of proximity. Simple such measures include the L^1 and L^∞ measures, or some weighted measure of distance and turning angle.

However we know little about the efficient computation of such measures. One exception is Ong's 'growth distance function' [12], which has the added advantage of returning a sensible measure even when objects inter-penetrate.

3.6. Computing distances for NURBS

NURBS and NURBS patches have become a de facto standard in industry for representing and designing models. They include as special cases Beziér cases, all non-rational B-Splines, conics and straight edges, so obviously cover a large number of required shapes. Also, they are pleasingly smooth internally and it is straightforward to ensure C^n continuity at joins. Perhaps most usefully, they are also closed under all translation, rotation, scaling and projection transformations.

We can write down an analytical solution for the support function for a NURBS quite simply; for NURBS of degree p we need to be able to find roots of non-rational polynomials of degree $2p - 1$. In two dimensions this technique

Figure 5. Two objects automatically decomposed into convex pieces

works quite well, as illustrated in Fig. 6 [26]. In three dimensions the technique works, but is less practical by itself as (a) many parts are difficult to describe as a partition of convex pieces, and (b) the interfaces to modeling systems for NURBS are not always as well implemented by the underlying modeller as they could be. An alternative approach that also shows promise is to use sub-division schemes to calculate the distance to any required accuracy [15, 16].

4. Conclusions

In this paper we have expounded on the principles of *problem decomposition* and *the separation of geometry and planning* that allow (we claim) the development of planners that are useful outside of a research environment. We have also experimented with several alternative planners, and are now re-engineering OXSIM as part of an EPSRC funded project at Oxford in order to test whether the OXSIM philosophy will scale to make solutions of industrial-sized problems possible (and easy to maintain). This project will deal with the planning of robot motions within large structures with relatively small clearances. The size of these geometric structures are large (tens to hundreds of thousands of curved polygons), which provides a considerable challenge to our ideas. Further work at Oxford is also looking at alternative means of computing distances [23], use of sensed distance data to construct planners that work at run-time, and at incorporating knowledge of the robot controller within the planner [2].

Acknowledgements

Thanks are due to EPSRC grant GR/L32408 and to British Aerospace for support. I am also indebted to the other members of the research team — Roy Featherstone, Joe Pitt-Francis, Federico Thomas, Colin Turnbull, and Yonghang Bu — for making this idea a reality.

References

[1] J. Barraquand and J. Latombe. Robot motion planning: A distributed representation approach. *Int. J. Robotics Res.*, 10(6):628–649, 1991.

[2] Y. Bu and S. Cameron. Active motion planning and collision avoidance for redundant manipulators. In *Proc. IEEE Int. Symp. Assembly and Task Planning*, pages 13–18, Marina Del Ray (Los Angeles), Aug. 1997.

[3] S. Cameron. Obstacle avoidance and path planning. *Industrial Robot*, 21(5):9–14, 1994.

[4] S. Cameron. A comparison of two fast algorithms for computing the distance between convex polyhedra. *IEEE Trans. Robotics & Automation*, 13(6):915–920, Dec. 1997.

[5] S. Cameron. Enhancing GJK: computing minimum and penetration distances between convex polyhedra. In *IEEE Int. Conf. Robotics & Automation*, pages 3112–3117, Albuquerque, Apr. 1997.

[6] S. A. Cameron. A study of the clash detection problem in robotics. In *IEEE Int. Conf. Robotics & Automation*, pages 488–493, St. Louis, Mar. 1985.

[7] S. A. Cameron and R. K. Culley. Determining the minimum translational distance between two convex polyhedra. In *IEEE Int. Conf. Robotics & Automation*, pages 591–596, San Francisco, Apr. 1986.

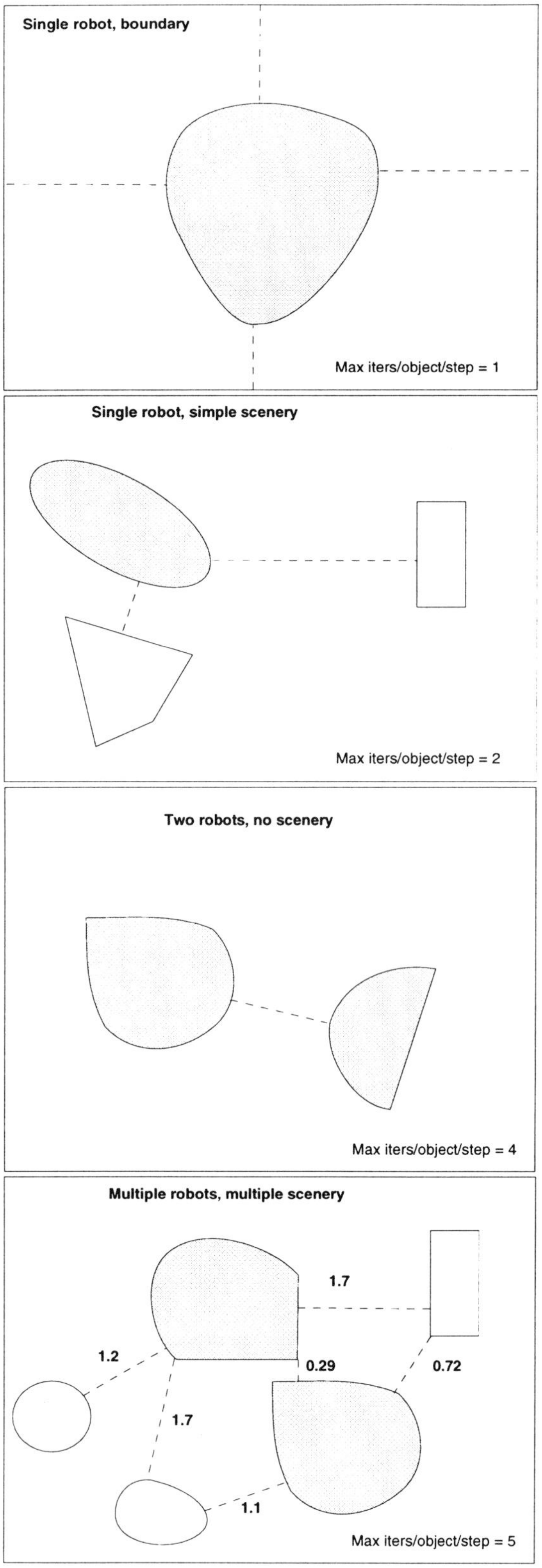

Figure 6. Two-dimensional examples of distance calculations.

[8] B. Chazelle. Convex partitions of polyhedra. *SIAM J. Computing*, 13(3):488–507, 1984.

[9] K. Chung and W. Wang. Quick collision detection of polytopes in virtual environments. In *ACM Symposium on Virtual Reality Software and Technology*, Hong Kong, July 1996.

[10] E. G. Gilbert and C.-K. Foo. Computing the distance between general convex objects in three-dimensional space. *IEEE Trans. Robotics & Automation*, 6(1):53–61, Feb. 1990.

[11] E. G. Gilbert, D. W. Johnson, and S. S. Keerthi. A fast procedure for computing the distance between complex objects in three-dimensional space. *IEEE Trans. Robotics & Automation*, 4(2):193–203, Apr. 1988.

[12] E. G. Gilbert and C. J. Ong. New distances for the separation and penetration of objects. In *IEEE Int. Conf. Robotics & Automation*, pages 579–586, San Diego, May 1994.

[13] E. G. Gilbert and C. J. Ong. The Gilbert-Johnson-Keerthi distance algorithm: A fast version for incremental motions. In *IEEE Int. Conf. Robotics & Automation*, Albuquerque, Apr. 1997.

[14] Y. K. Hwang and N. Ahuja. Gross motion planning—a survey. *ACM Computing Surveys*, 24(3):219–291, Sept. 1992.

[15] D. E. Johnson and E. Cohen. A framework for efficient minimum distance computations. In *IEEE Int. Conf. Robotics & Automation*, pages 3678–3684, Leuven, May 1998.

[16] D. E. Johnson and E. Cohen. Bound coherence for minimum distance computations. In *IEEE Int. Conf. Robotics & Automation*, pages 1843–1848, Detroit, May 1999.

[17] O. Khatib. A unified approach for motion and force control of robot manipulators: The operational space formulation. *IEEE J. Robotics & Automation*, pages 43–53, Feb. 1987.

[18] J.-C. Latombe. *Robot Motion Planning*. Kluwer Academic Publishers, 1991.

[19] M. Lin and J. Canny. A fast algorithm for incremental distance calculation. In *IEEE Int. Conf. Robotics & Automation*, pages 1008–1014, Sacremento, Apr. 1991.

[20] T. Lozano-Pérez. Spatial planning—a configuration space approach. *IEEE Trans. Comput.*, C-32(2):108–120, Feb. 1983.

[21] A. W. McLean and S. A. Cameron. The virtual springs method: Path planning and collision avoidance for redundant manipulators. *Int. J. Robotics Res.*, 15(4):300–319, Aug. 1996.

[22] B. Mirtich. V-clip: Fast and robust polyhedral collision detection. Technical Report TR-97-05, Mitsubishi Electric Research Laboratory, Cambridge MA (www.merl.com), July 1997.

[23] J. Pitt-Francis and R. Featherstone. Automatic generation of sphere hierarchies from cad data. In *IEEE Int. Conf. Robotics & Automation*, Leuven, May 1998.

[24] C. Qin, S. Cameron, and A. McLean. Towards efficient motion planning for manipulators with complex geometry. In *Proc. IEEE Int. Symp. Assembly and Task Planning*, pages 207–212, Pittsburgh, Aug. 1995.

[25] F. Thomas, C. Turnbull, and S. Cameron. Computing signed distances between free-form objects. Submitted Int. Conf. Rob. Aut. 2000, Sept. 1999.

[26] C. Turnbull and S. Cameron. Computing distances between NURBS-defined convex objects. In *IEEE Int. Conf. Robotics & Automation*, pages 3685–3690, Leuven, May 1998.

Distance Computation between Non-convex Polyhedra at Short Range Based on Discrete Voronoi Regions

Katsuaki Kawachi and Hiromasa Suzuki
Department of Precision Machinery Engineering, The University of Tokyo
7-3-1 Hongo, Bunkyo-ku, Tokyo 113-8656, Japan
{kawachi, suzuki}@cim.pe.u-tokyo.ac.jp

Abstract

An algorithm for calculating the minimum distance between non-convex polyhedra is described. A polyhedron is represented by a set of triangles. In calculating the distance between two polyhedra, it is important to search efficiently the pair of the triangles which gives the pair of closest points.

In our algorithm discrete Voronoi regions are prepared as voxels around a non-convex polyhedron. Each voxel is given the list of triangles which have possibility to be the closest to the points in the voxel. When a triangle on the other object is intersecting a voxel, the closest triangles can be efficiently searched from this list on the voxel.

The algorithm has been implemented, and the results of distance computations show that it can calculate the minimum distance between non-convex polyhedra composed of a thousand of triangles at interactive rates.

1 Introduction

Computing distance between objects is an important problem in various field including computer graphics animation and simulation. In order to prevent the moving objects from interference, it is needed to detect if the objects are colliding or not. In addition to this interference detection, it is useful to calculate the minimum distance between the objects for predicting the instance of collision in dynamics simulation.

The algorithms of distance computation can be classified by their targets as follows.

- Non-polyhedral Objects
  - CSG models
  - Curved surfaces
  - Clouds of points
- Polyhedral Objects *without* topological information
  - Sets of Polygons
- Polyhedral Objects *with* topological information
  - Convex polyhedra
  - Non-convex polyhedra

In this paper we present an algorithm for polyhedral objects. The object consists of a set of triangles and is not necessarily needed to have topological information.

2 Related Work

For distance computation between convex polyhedral models specified by a boundary representation, there are two major algorithms: closest-feature algorithms [7][8] and simplex-based algorithms [2][1].

The Lin-Canny closest feature algorithm [7] defines Voronoi region on its features (faces, edges, and vertices). If two features are closest, each feature is inside the Voronoi region of the other. In dynamics simulation the algorithm can efficiently find the closest features by tracking the pair of the closest features. V-Clip [8] employs the improved closest feature algorithm which can handle the penetration between objects and calculate the distance robustly. In these closest feature algorithms the method of searching the closest feature pair utilizes Voronoi regions defined around convex polyhedra. Unfortunately these algorithms are based on convex polyhedra and cannot be directly applied to non-convex ones.

In this paper, we deal with the non-convex problem. We utilize Voronoi regions for picking up the closest triangles between non-convex polyhedra. Because it is difficult to calculate accurate structure of the Voronoi regions on a non-convex polyhedron, they are defined by the discrete approximation method using voxels [6].

3 Discrete Voronoi Regions

In this section we describe the definition of the discrete Voronoi regions and the algorithm of distance computation between non-convex polyhedra.

3.1 Cost of Distance Computation

The total cost of interference detection between two polyhedra can be formulated as the following equation [3]:

$$T = N_v \times C_v + N_p \times C_p, \quad (1)$$

where

T : total cost function for interference detection
N_v : number of bounding volume pair overlap tests
C_v : cost of testing a pair of bounding volumes for overlap
N_p : number of feature pairs tested for interference
C_p : cost of testing a pair of features for interference

As C_p is expensive in general, bounding volume with smaller C_v can save T by decreasing the number N_p. Typical algorithms for defining bounding volumes include axis-aligned boxes, octrees [10], sphere trees [9], OBBTree [3], swept sphere volumes [5] etc. Avobe all, the PQP library utilizing swept sphere volumes can perform efficient distance calulation by transforming distance computation to intersection detection.

In calculating the minimum distance between non-convex polyhedra the simplest algorithm tests the all pairs of features in them and picks up the pair with the smallest distance. In the same way of collision detection, because the cost of distance computation between a pair of features is expensive, the total cost can be saved by reducing the number of pairs of features with some bounding volume.

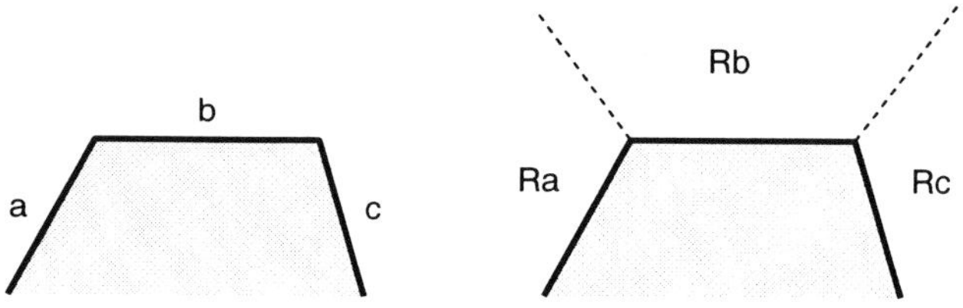

Figure 1. Voronoi Regions Defined around Convex Polygon

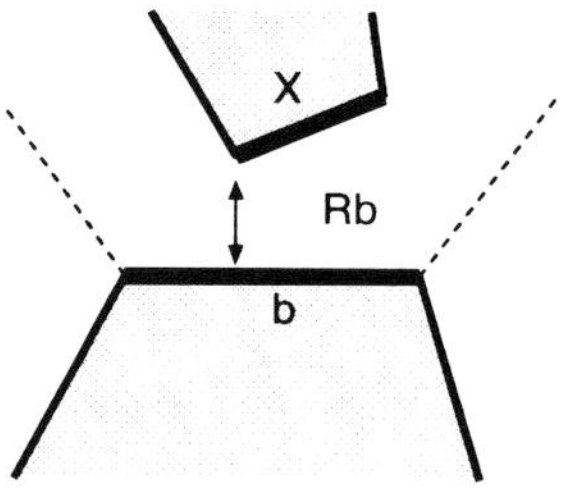

Figure 2. Closest Feature in Voronoi Regions

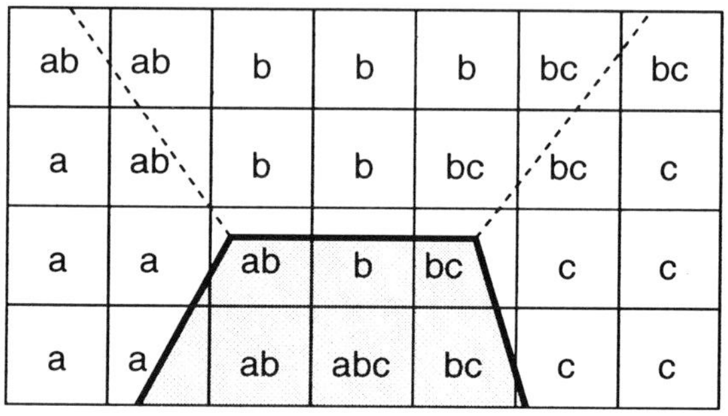

Figure 3. CFL (Closest Feature List) on Voxels

3.2 Voxels around Objects

A Voronoi region can be defined for each feature on a polyhedron:

Definition A Voronoi region associated with a feature is a set of points closer to that feature than any other.

Figure 1 shows a two dimensional illustration of the Voronoi regions defined around a convex object. If a feature X on another polygon is included in region R_b (Figure 2), it is closest to the edge b, and it is needless to calculate its distance against the other features a, c. The regions are easily defined on a convex polyhedron, and I-Collide [7] and V-Clip [8] utilize the Voronoi regions for calculating the distance between convex polyhedra.

If we can define Voronoi regions on the features of a non-convex polyhedron, a feature X of the other polyhedron can be closest to the feature Y whose Voronoi region includes X. Because it is difficult to calculate accurate structure of the Voronoi regions of a non-convex polyhedron, we discretely define Voronoi regions by voxelizing the space around the polyhedron and do not explicitly calculate the structure of Voronoi regions. Each voxel is given the list of the features which have possibility to be closest to the voxel (Figure 3). This list is called "Closest Feature List" (CFL). If the feature X has interference with the voxel V, we need to calculate the distances between X and the features in CFL on the voxel V.

3.3 Representation of Non-convex Polyhedra

In this paper we present an algorithm for calculating distance between objects with the following properties:

- A polyhedron is represented by a set of triangles. In the closest feature algorithms, a polyhedron is constituted by three features (faces, edges, and vertices). In our algorithm edges and vertices are not regarded as independent features but as a part of a triangle. By unifying them to triangles, we can decrease the number of distance calculations.
- A triangle does not necessarily have topological information about neighboring triangles across the edges. If topology is known, it is utilized for culling the voxels which cannot include the triangle on their CFLs.
- Triangles in a polyhedron may have interference each other.

These properties are useful for dealing with polyhedra with boundaries, messy polygon models etc. As the algorithm does not calculate the Voronoi regions explicitly, triangles in a polyhedron can be freely arranged. The algorithm does not distinguish the front and back of a triangle. So the distance calculated by the algorithm is always zero or more than zero.

3.4 Closest Triangles to Voxel

In creating CFL on voxels, we utilize the minimum and maximum values of the distance between a point in a voxel and a triangle.

If a point x and triangles are given, one triangle which is closest to the point x can be chosen by the relative position of x to the triangles. If x exists inside a voxel V, the closest triangle varies as x moves in V. CFL on a voxel can be created by picking up all the closest triangles for the all points in voxel V. If the triangle Y on the other polyhedron is interfering with V, the triangles in the list on V have possibility to be closest to Y.

Given a point x and a triangle A, we can calculate their minimum distance $d_A(x)$. In voxel V there exists the points $x_A^{\min}$ and $x_A^{\max}$ which give the minimum and maximum value of $d_A(x)$ (Figure 4). By this definition, for all points x in the voxel V we can write

$$d_A(x_A^{\min}) \leq d_A(x) \leq d_A(x_A^{\max}),\ \forall x \in V. \quad (2)$$

In the same way, the minimum distance $d_B(x)$ can be defined for another triangle B, and we get the points $x_B^{\min}$ and $x_B^{\max}$ satisfying the following condition:

$$d_B(x_B^{\min}) \leq d_B(x) \leq d_B(x_B^{\max}),\ \forall x \in V. \quad (3)$$

If the points $x_A^{\max}$ and $x_B^{\min}$ satisfies $d_A(x_A^{\max}) < d_B(x_B^{\min})$, the definitions (2) and (3) leads:

$$d_A(x) < d_B(y),\ \forall x \in V,\ \forall y \in V, \quad (4)$$

that is, any points in V are given smaller minimum distance by the triangle A than B. Consequently the triangle B cannot be closest to any points in V.

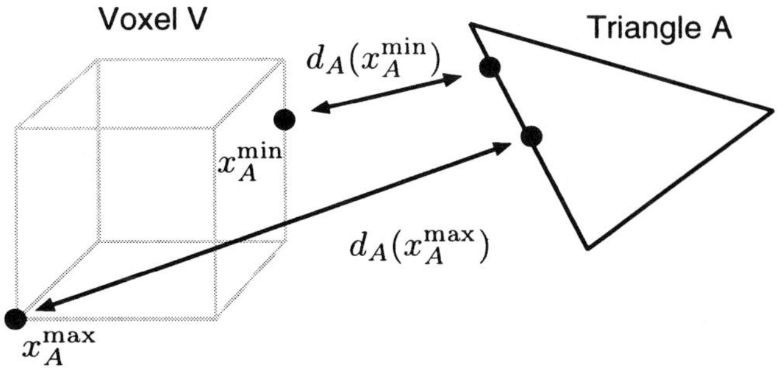

Figure 4. Minimum Distance between Triangle and Point in Voxel

In picking up all the closest triangles for the voxel V, the minimum value $d_i^{\min}$ and maximum value $d_i^{\max}$ of the distance d_i between V and the triangle i are calculated for all triangles at first. Then the triangles are sorted by their minimum values of distance. If some triangles have the same minimum value, they are sorted again by the maximum values of distance. By this procedure we can get the triangle k with the smallest $d_k^{\min}$ and $d_k^{\max}$. By the condition (4), the triangle i satisfying $d_k^{\max} < d_i^{\min}$ cannot be closest to the voxel V and is not included in CFL on V.

In our algorithm voxels are defined on the local coordinate system of a non-convex polyhedron. The CFLs are calculated only once as a preprocess of distance computations. They are kept with the geometric information of the polyhedron and are reused when the relative orientation of the objects is changed, because the CFL calculation generally takes much time. For instance, the CFLs on $30 \times 30 \times 30$ voxels around a polyhedron with 1314 triangles requires about 40 minutes to compute. The CFL calculation can be much accelerated if we utilize the computation method of Voronoi region using graphics hardware [4].

3.5 Calculating Distance with CFL

With CFLs we compute the distance between non-convex polyhedra A and B with the following four steps.

At the first step, the voxels on A which have interference with the triangles in B are picked. The interference test is done efficiently by defining a BSP tree of voxels on A and an OBBTree [3] on the triangles of B.

At the second step, the closest feature tests are applied for the intersecting triangles and voxels found in the first step. The Lin-Canny closest feature algorithm utilizes the necessary and sufficient condition that each feature is inside the

Voronoi region of the other. This condition can be rewritten for the discrete Voronoi regions with CFLs as follows:

$$\left\{\begin{array}{l}\text{Feature } X \text{ intersects voxel } V_Y \\ \text{(the CFL of } V_Y \text{ includes feature } Y) \\ \quad \textit{and} \\ \text{Feature } Y \text{ intersects voxel } V_X \\ \text{(the CFL of } V_X \text{ includes feature } X)\end{array}\right. \quad (5)$$

Because the condition(5) is not sufficient but necessary for non-convex polyhedra, if a voxel V on A intersecting a triangle T in B are found at the first step, they can not be closest and are ignored at the second step when the condition(5) is not satisfied.

In order to check the condition, the voxels on B which have interference with the triangles in A are picked at the second step. If a triangle in the CFL of V has interference with the voxel with the CFL including T are found, the condition is satisfied.

At the third step, the picked voxels are sorted by their minimum values of distance $d_k^{\min}$, and at the final step, for each voxel in the order of $d_k^{\min}$, the minimum distance between the triangles in the CFL and the triangle interfering the voxel is calculated.

By sorting the voxels at the third step, we can efficiently eliminate the excess voxels. If in the way of final step we have a voxel k with distance $d_k^{\min}$ which is greater than the minimum distance given by the previous voxels, we can cut off the distance computations for the following voxels, which cannot be the closest to the polyhedron any more.

In the implementation of our algorithm, voxels are defined in the region which is twice as large as an object's bounding box. If the object is out of this region, the distance is approximated by that of the convex hull of the original polyhedra.

3.6 Voxel Culling

If topological informations between triangles are available, the voxels where the triangle can be closest can be culled by the approximated Voronoi region on the triangle. The voxels which have no interference with the region cannot be closest and do not need to be calculated their distances to the triangle.

The Voronoi region of the triangle is approximately defined by the planes containing the edges on the triangle. Each plane is parallel to the average of the normal vector of the triangle and that of the neighboring triangle.

When the polyhedron is convex, the approximated region on a triangle is identical to the accurate Voronoi region. When non-convex, it is not necessarily identical, but it includes the accurate Voronoi region, so the approximation by neighboring triangles does not make an incomplete CFL.

As shown in Figure 5, the approximated Voronoi region for a triangle is partitioned with planes on edges (plotted with broken lines). The voxels interfering with the region can be the closest to the triangle.

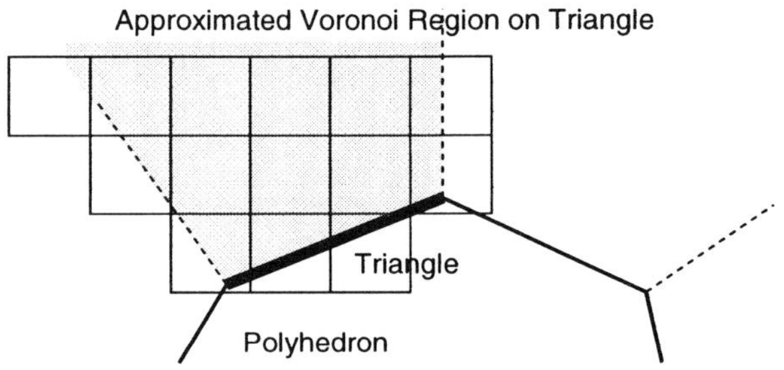

Figure 5. Voxel Culled by Approximated Voronoi Region

4 Results

Based on the algorithm of the discrete Voronoi regions we have implemented a program for calculating the minimum distance between non-convex polyhedra.

4.1 Length of CFL

Table 1 shows the number of voxels and the average length of CFL defined on a bunny model of 1314 triangles (Figure 6). The average length of CFL per voxel is affected by the size of the voxel. If the size of a voxel is increased, more triangles per voxel are listed. It can be said that smaller voxels make the length of CFL shorter and that the efficiency of distance calculation can be improved, but we have found that the length of CFL do not decreases very rapidly when the size of voxels is smaller than the size of triangles. The change of triangles picked for distance calculation in some voxel sizes is shown in Figure 7 as dark triangles. The thick line between two objects indicates the pair of closest points.

CFL on voxels are calculated on each polyhedron as a preprocess of distance calculation. For instance, a dark triangle on the back of the bunny is given in CFL on the black voxels (Figure 8).

Table 1. Number of Voxels and Length of CFL

Number of Voxels	Average Length of CFL
$10 \times 10 \times 10$	29.08
$15 \times 15 \times 15$	11.92
$20 \times 20 \times 20$	10.90
$30 \times 30 \times 30$	6.35

Figure 6. Non-convex Polyhedron with 1314 Triangles

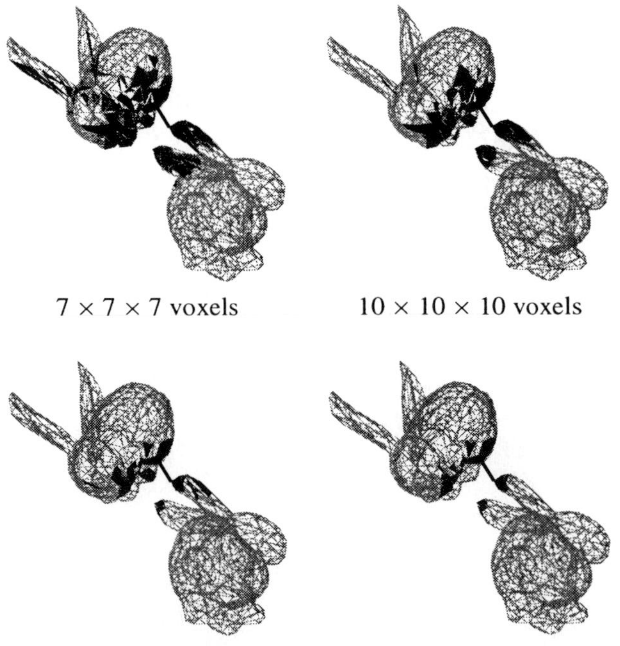

7 × 7 × 7 voxels　　10 × 10 × 10 voxels

20 × 20 × 20 voxels　　30 × 30 × 30 voxels

Figure 7. Picked Triangles and Size of Voxels

4.2 Distance Calculation with CFL

In calculating the distance with 30 × 30 × 30 voxels, the total number of distance calculations between a pair of triangles is diminished to 172 on average, which is about 1/10000 compared with the number of all combinations of triangles. The average of CPU time consumed in distance computation is 0.06 second with MIPS R10000 250MHz. The triangles which are picked from CFL on voxels and calculated their distance are shown in dark color (Figure 9). The thick line between two objects indicates the pair of closest points.

The algorithm enables us fast computation of the distance between non-convex polyhedra which consist of a thousand of triangles, and we utilize this algorithm for our interactive rigid body dynamics simulator named "pVRML", which can simulate the dynamics of physically-based models defined in 3D scene of VRML1.0.

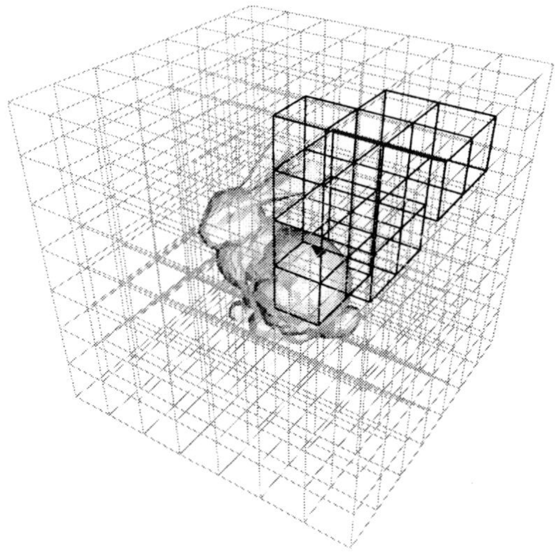

Figure 8. Voxels Associated with a Triangle

5 Conclusion and Future Work

We have described an the discrete Voronoi regions for calculating the distance between non-convex polyhedra. By defining the minimum and maximum values of distances between voxel and triangles, we can efficiently eliminate the voxels which cannot be closest. The algorithm enables us to calculate the distance between non-convex polyhedra made of a thousand of triangles at interactive rates, and we utilize it for an interactive dynamics simulator.

It is found that the average length of CFL is affected by the size of the voxel. In order to achieve more efficient distance computation algorithm, a method for deciding the optimal size of voxels for an polyhedron is needed.

If the polyhedra are interfering, our algorithm answers zero as the distance between them. In terms of dynamics simulation, it can be useful to extend the algorithm to treat penetration as minus distance in estimating the instance of collision.

Table 2 shows the comparison of the distance computation cost with the PQP library [5]. Our implementation is about 3.4 times as slow as PQP. The polyhedron in Figure 6 is used for this test, and CFLs are defined on 30 × 30 × 30 voxels. This result is mainly caused by the large N_v, i.e. the number of intersection tests and closest feature tests between triangles and voxels performed in the first, second and third step described in the section 3.5, and these tests cost more than 70% of all computation time. By constructing a hierarchical structure of triangles in CFLs of voxels, they can be more efficient and it is likely to be possible to reduce N_v as small as that of PQP .

Table 2. Performance of Distance Computation

Algorithm	Av. Computation Time	Av. N_p	Av. N_v
PQP	0.018 sec.	105.4	1781
CFL	0.061 sec.	171.6	7970

References

[1] S. Cameron. Enhanceing GJK: Computing minimum penetration distances between convex polyhedra. In *IEEE International Conference on Robotics and Automation*, April 1997.

[2] E. G. Gilbert, D. W. Johnson, and S. S. Keerthi. A fast procedure for computing the distance between complex objects in three-dimensional space. *IEEE Journal of Robotics and Automation*, 4(2):193–203, April 1988.

[3] S. Gottschalk, M. C. Lin, and D. Manocha. OBBTree: A hierarchical structure for rapid interference detection. In *Computer Graphics Proceedings, Annual Conference Series*, pages 171–180. ACM SIGGRAPH, 1996.

[4] K. E. H. III, T. Culver, J. Keyser, M. Lin, and D. Manocha. Fast computation of generalized voronoi diagrams using graphics hardware. In *Computer Graphics Proceedings, Annual Conference Series*, pages 277–285. ACM SIGGRAPH, 1999.

[5] E. Larsen, S. Gottschalk, M. Lin, and D. Manocha. Fast proximity queries with swept sphere volumes. Technical Report TR99-018, Department of Computer Science, University of North Carolina, Chapel Hill, 1999.

[6] D. Lavender, A. Bowyer, J. Davenport, A. Wallis, and J. Woodwark. Voronoi diagrams of set-theoretic solid models. *IEEE Computer Graphics and Applications*, 12(5):69–77, September 1992.

[7] M. C. Lin and J. F. Canny. Efficient algorithm for incremental distance computation. In *IEEE Conference on Robotics and Automation*, 1991.

[8] B. Mirtich. V-Clip: Fast and robust polyhedral collision detection. Technical Report TR-97-05, MERL, July 1997.

[9] J. Pitt-Francis and R. Featherstone. Automatic generation of sphere hierarchies from cad data. In *Proceedings of IEEE International Conference on Robotics and Automation*, 1998.

[10] H. Samet. *Spatial Data Structures: Quadtree, Octree and Other Hierarchial Methods*. Addison Wesley, 1989.

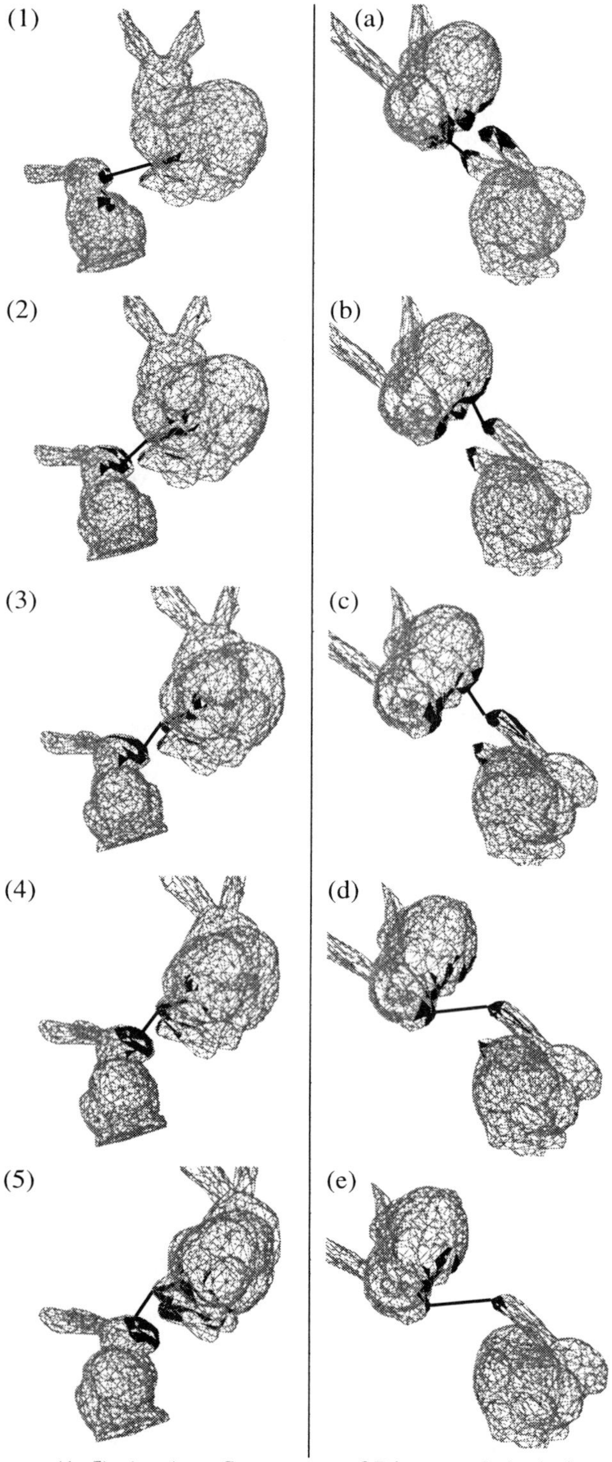

(1–5), (a–e): Sequences of Distance Calculation
Dark Triangles: Picked Triangles in Calculation
Thick Line: Pair of Closest Points

Figure 9. Closest Points between Two Moving Non-convex Polyhedra

A Sequence of Generalized Coordinate Systems Based on Voronoi Diagrams and Its Application to Interpolation

Hisamoto Hiyoshi and Kokichi Sugihara
Department of Mathematical Engineering and Information Physics
University of Tokyo
Hongo 7-3-1, Bunkyo-ku, Tokyo 113-8656, Japan
hiyoshi@simplex.t.u-tokyo.ac.jp
sugihara@simplex.t.u-tokyo.ac.jp

Abstract

This paper presents a general framework for constructing a variety of multi-dimensional interpolants based on Voronoi diagrams. This framework includes previously known methods such as Sibson's interpolant and Laplace's interpolant; moreover it contains infinitely many new interpolants. Computational experiments suggest that the smoothness can be improved by the proposed generalization. In addition, this framework also includes the piecewise linear interpolant over the Delaunay triangulation, which is a finite-element interpolant. This fact suggests that already established techniques in the finite element method might be brought into the research of the Voronoi-based approach. Hence this framework gives a new and promising direction of research on the interpolation based on Voronoi diagrams.

1. Introduction

Interpolation is an extremely important technique to solve various problems in engineering such as differential equations and geometric modeling. The finite element method is one of the most practical approaches to the interpolation problem, and is well-established today. However, this method has disadvantage in the sense that the interpolant is not continuous in the displacements of the data sites. To overcome this disadvantage, there is an alternative approach, which utilizes Voronoi diagrams. In this paper, we call this approach the *Voronoi-based approach*. See, for example, [1, 5, 7] for the Voronoi diagram theory.

The interpolation problem is formulated in the following way. Let $P_1, \ldots, P_n$ be fixed points in $\mathbf{R}^d$, called *data sites*. Suppose that $z_1, \ldots, z_n$ are given reals. The (exact) interpolation problem is to find a *good* function $\tilde{\varphi}$ such that

$$\tilde{\varphi}(P_i) = z_i$$

for $i = 1, \ldots, n$. The meaning of the word "good" depends on the context.

Thiessen first applied Voronoi diagrams to the interpolation problem [11]. Let P be the target point the value on which is to be estimated. In his method, the Voronoi diagram for the data sites is constructed. If the point P belongs to the Voronoi region of P_i, the value at P is estimated at the value at the point P_i. By the definition, Thiessen's interpolant is a piecewise constant function.

Recently, Sibson found another interpolation method [8, 9]. In his method, the Voronoi diagram for $\{P_1, \ldots, P_n, P\}$ is constructed. If the Voronoi regions of P and P_i are adjacent via a $(d-1)$-dimensional facet, we call P_i a *neighbor* of P. The critical fact he found is that the position vector of P can be expressed as a convex combination of the position vectors of the neighbors of P with the coefficients computed from the second-order Voronoi diagram [8]. Hence we can interpret the coefficients of this convex combination as the coordinates of P. Sibson constructed C^0 and C^1 interpolants based on this coordinate system. Sibson's interpolation method was further researched by Farin [2] and Piper [6]. Farin proposed another C^1 interpolant based on the Bernstein polynomials [2].

On the other hands, Hiyoshi and Sugihara found another coordinate system, and proposed a C^0 interpolant [3, 4, 10]. We call this system Laplace's coordinate system, because it is achieved from discretization of Laplace's equation.

Compared with the finite element interpolation, the Voronoi-based approach has a lot of virtues [9]. For example, Voronoi-based interpolants behaves continuously when the data sites move. However, the Voronoi-based approach is less flexible in some points: for example, it is difficult to construct even C^2 interpolants by the Voronoi-based ap-

0-7695-0562-7/00 $10.00 © 2000 IEEE

proach, as explained in Section 2.4.

In this paper, we generalize the coordinate systems in order to make the Voronoi-based approach more flexible, concentrating on the case when $d = 2$. Indeed we construct a large class of Voronoi-based interpolants, which contains both Sibson's and Laplace's interpolants, and which contains infinitely many other interpolants. Computational experiments suggest that the smoothness is improved by the proposed generalization. Hence, this generalization gives a new and promising direction of research on the Voronoi-based interpolation. In addition, another special case of the generalization leads the piecewise linear interpolant over the Delaunay triangulation. This fact suggests that some already established techniques in the finite element method might be brought into the research of the Voronoi-based approach.

Section 2 reviews Sibson's and Laplace's coordinate systems. Section 3 gives another proof to Sibson's result, which provides the basic idea of our generalization. Section 4 proposes the generalization. Section 5 gives Computational experiments. Section 6 concludes our research.

2. Previous works

In this section, we review the previous works briefly.

2.1. Notations

First, let us introduce some notations.

Assume that a finite number of points $P_1, \ldots, P_n \in \mathbf{R}^2$ are given. P_i's are called the *generators*. Let $V(P_i)$ be the set of all the points $Q \in \mathbf{R}^2$ such that $\mathrm{d}(Q, P_i) < \mathrm{d}(Q, P_j)$ for $j \neq i$, where $d(P, Q)$ denotes the Euclidean distance between two points P and Q. We call $V(P_i)$ the *Voronoi region* of the generator P_i. By the definition, the boundary of each $V(P_i)$ is a convex polygon. The Euclidean plane $\mathbf{R}^2$ is almost partitioned into $V(P_1), \ldots, V(P_n)$, that is, the measure of the set of all the points that do not belong to any $V(P_i)$ is zero. The collection of all $V(P_i)$'s is denoted by $\mathcal{V}(P_1, \ldots, P_n)$ and is called the *Voronoi diagram* for the generator set $\{P_1, \ldots, P_n\}$. When two Voronoi regions $V(P_i)$ and $V(P_j)$ are adjacent via some open line segment, this line segment is called the *Voronoi edge* between P_i and P_j, and is denoted by $E(P_i, P_j)$. If $E(P_i, P_j)$ exists, P_j is called a *neighbor* of P_i (and vice versa).

For $i \neq j$, let $V(P_i, P_j)$ be the set of all the points $Q \in \mathbf{R}^2$ such that $\mathrm{d}(Q, P_i) < \mathrm{d}(Q, P_j) < \mathrm{d}(Q, P_k)$ for $k \neq i, j$. $V(P_i, P_j)$ is called the *second-order Voronoi region* of the ordered pair (P_i, P_j). By the definition, each $V(P_i)$ is almost partitioned into $V(P_i, P_j)$, $i \neq j$. The collection of all $V(P_i, P_j)$'s is called the *second-order Voronoi diagram* for the generator set $\{P_1, \ldots, P_n\}$.

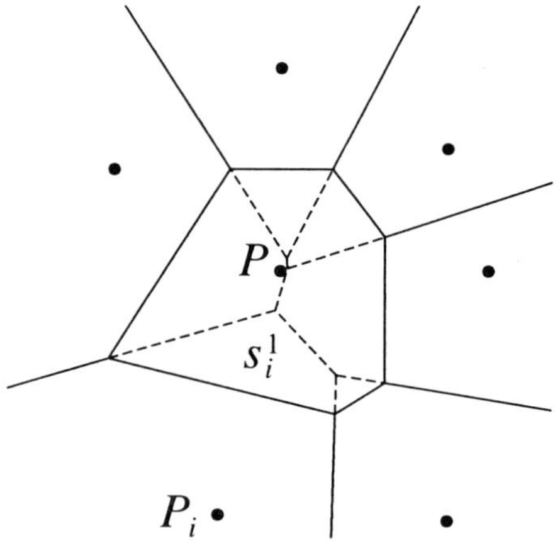

Figure 1. The Voronoi region $V(P)$ is partitioned into the subregions $V(P, P_i)$.

See, for example, [1, 5, 7] for the detail of the Voronoi diagram theory.

2.2. Sibson's coordinates

Now let us describe Sibson's interpolation method [8, 9] briefly.

Let $P_1, \ldots, P_n \in \mathbf{R}^2$ be the given data sites, and let $z_1, \ldots, z_n \in \mathbf{R}$ be the data value associated with $P_1, \ldots, P_n$, respectively. Assume that we want to evaluate the value on the target point P. Here, we require that P is an inner point of the convex hull of $P_1, \ldots, P_n$, but does not equal to any of $P_1, \ldots, P_n$.

First, the Voronoi diagram $\mathcal{V}(P_1, \ldots, P_n, P)$ is constructed. Each $V(P_i)$ is not necessarily bounded, but $V(P)$ is always bounded because P is an inner point of the convex hull of $P_1, \ldots, P_n$. Now partition $V(P)$ into $V(P, P_i)$'s, and let $s_i^1(P)$ denote the measure of $V(P, P_i)$. Note that $s_i^1(P) > 0$ if and only if P_i is a neighbor of P.

Figure 1 shows this situation. In this figure, the Voronoi region $V(P)$ is partitioned into the subregions $V(P, P_i)$. The boundaries of this partition are drawn by the dashed lines.

Let $\boldsymbol{p}_1, \ldots, \boldsymbol{p}_n$ and $\boldsymbol{p}$ denote the position vectors of $P_1, \ldots, P_n$ and P, respectively. Sibson [8] proved the following identity:

$$\sum_{i=1}^{n} s_i^1(P)\boldsymbol{p} = \sum_{i=1}^{n} s_i^1(P)\boldsymbol{p}_i. \tag{1}$$

We will give another proof of this identity in Section 3, which gives the basic idea of the generalization we propose.

Now define that

$$\hat{s}_i^1(P) = s_i^1(P) \Big/ \sum_{j=1}^{n} s_j^1(P).$$

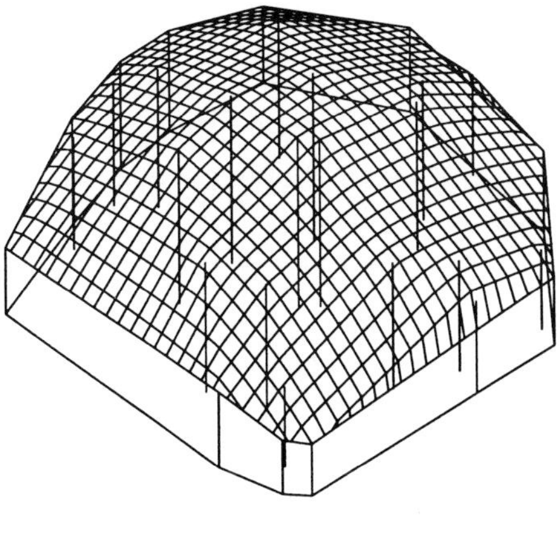

Figure 2. An example of a surface obtained from Sibson's interpolant.

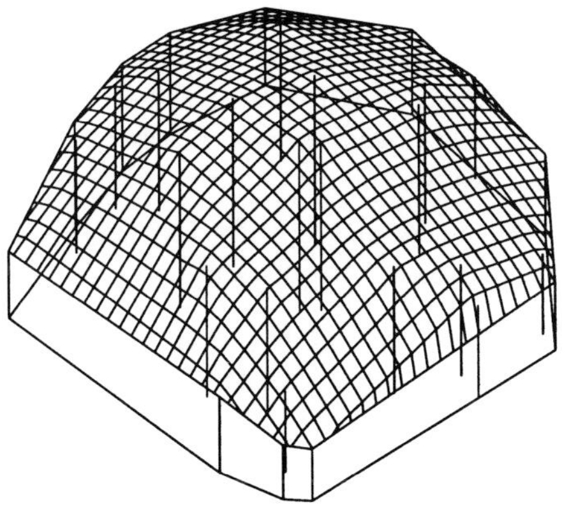

Figure 3. An example of a surface obtained by Laplace's interpolant.

It follows from the definition that

$$\sum_{i=1}^{n} \hat{s}_i^1(P) = 1, \quad 0 \le \hat{s}_i^1(P) \le 1.$$

Thus $\boldsymbol{p}$ is expressed as the following convex combination:

$$\boldsymbol{p} = \sum_{i=1}^{n} \hat{s}_i^1(P)\boldsymbol{p}_i.$$

We call $\hat{s}_i^1(P)$'s *Sibson's coordinates* of the point P. We also call $s_i^1(P)$'s Sibson's coordinates when there is no confusion. In fact, $(s_1^1, \ldots, s_n^1)$ and $(\hat{s}_1^1, \ldots, \hat{s}_n^1)$ denote the same point in the $(n-1)$-dimensional real projective space. Note that $\hat{s}_j(P) \to \delta_{ij}$ as P approaches P_i. Therefore we define

$$\hat{s}_j^1(P) = \delta_{ij}$$

when P coincides with P_i.

From Sibson's coordinates, Sibson constructed the following C^0 interpolant:

$$\tilde{\varphi}^1(P) = \sum_{i=1}^{n} z_i \hat{s}_i^1(P).$$

Figure 2 shows an example of a surface obtained from the above interpolant.

2.3. Laplace's coordinates

Recently, Hiyoshi and Sugihara found another coordinate system [3, 4, 10]. In their method, the Voronoi diagram $\mathcal{V}'(P_1, \ldots, P_n, P)$ is also constructed. For $i = 1, \ldots, n$, let us define that

$$s_i^0(P) = l_i(P)/r_i(P),$$

where $l_i(P)$ is the length of the Voronoi edge $E(P, P_i)$ if any, and $r_i(P)$ is the Euclidean distance between P and P_i. Note that $s_i^0(P) > 0$ if and only if P_i is a neighbor of P. Then, the following identity follows, as proved in Section 3.1.

$$\sum_{i=1}^{n} s_i^0(P)\boldsymbol{p} = \sum_{i=1}^{n} s_i^0(P)\boldsymbol{p}_i. \quad (2)$$

Now let us define that

$$\hat{s}_i^0(P) = s_i^0(P) / \sum_{j=1}^{n} s_j^0(P).$$

We call $s_j^0(P)$'s *Laplace's coordinates*, because the coordinates are obtained from the discretization of Laplace's equation. Then, $\boldsymbol{p}$ can be expressed as another convex combination:

$$\boldsymbol{p} = \sum_{i=1}^{n} \hat{s}_i^0(P)\boldsymbol{p}_i.$$

It is clear that $\hat{s}_j^0(P) \to \delta_{ij}$ as P approaches P_i.

Hiyoshi and Sugihara proposed the following interpolant from Laplace's coordinates [3, 4, 10]:

$$\tilde{\varphi}^0(P) = \sum_{i=1}^{n} z_i \hat{s}_i^0(P).$$

Figure 3 shows an example of a surface obtained by $\tilde{\varphi}^0$.

2.4. Continuity

As for continuity of Laplace's and Sibson's interpolants, the following theorem holds:

Theorem 1 *Let k be 0 or 1. Then, $\tilde{\varphi}^k(P)$ is C^0 but not necessarily C^1 if P coincides with some data site; $\tilde{\varphi}^k(P)$ is C^k but not necessarily C^{k+1} if P lies on the Delaunay circles of the data sites but does not coincide with any data site; $\tilde{\varphi}^k(P)$ is C^∞ otherwise.*

Proof. See Hiyoshi-Sugihara [3] for Laplace's interpolant. See Piper [6] for Sibson's interpolant. □

The above theorem suggests that the points where the smoothness of $\tilde{\varphi}^k(P)$ is lost can be classified into the following two cases:

1. the data sites, and
2. the points on the Delaunay circles of the data sites excluding the data sites.

Independently, Sibson [9] and Farin [2] remove the nonsmoothness of the case 1, and constructed globally C^1 interpolants. However, a device for removing the nonsmoothness of the case 2 has not been presented. The framework presented in Section 4 seems to remove the nonsmoothness of the case 2. This conjecture will be encouraged by the computational experiments in Section 5.

3. Another proof of Sibson's identity

In this section, we give another proof of Sibson's identity. Observing this proof closely leads to the generalization proposed in Section 4.

3.1. Proof of the identity for Laplace's coordinates

At first, let us give a proof of the identity (2), which comes from Hiyoshi and Sugihara [4]. For this purpose, the following lemma is used as a fundamental tool.

Lemma 2 *For any region $V \subset \mathbf{R}^d$, the following equation holds:*

$$\int_{Q\in\partial V} \boldsymbol{n} \mathrm{d}S = \mathbf{0}, \qquad (3)$$

where ∂V denotes the boundary of V, $\boldsymbol{n}$ denotes the unit outer normal vector to ∂V at Q, and $\mathrm{d}S$ denotes the infinitesimal surface element at Q.

Proof. It follows from Gauss's divergence theorem that

$$\int_{Q\in V} \nabla \cdot \boldsymbol{A} \mathrm{d}Q = \int_{Q\in\partial V} \boldsymbol{A} \cdot \boldsymbol{n} \mathrm{d}S$$

for any vector field A. Substituting $\boldsymbol{A}$ by $\boldsymbol{c}\varphi$ for an arbitrary constant vector $\boldsymbol{c}$ and an arbitrary scalar field φ, we get

$$\boldsymbol{c} \cdot \int_{Q\in V} \nabla\varphi \mathrm{d}Q = \boldsymbol{c} \cdot \int_{Q\in\partial V} \varphi\boldsymbol{n} \mathrm{d}S.$$

Since $\boldsymbol{c}$ is arbitrary, we get

$$\int_{Q\in V} \nabla\varphi \mathrm{d}Q = \int_{Q\in\partial V} \varphi\boldsymbol{n} \mathrm{d}S.$$

Setting φ to be a constant in the above equation, we get

$$\int_{Q\in\partial V} \boldsymbol{n} \mathrm{d}S = \mathbf{0}.$$

□

Let $P_{\iota(1)}, \ldots, P_{\iota(k)}$ be the neighbors of the target point P. In general, the boundary of the Voronoi region of some generator Q is a (possibly unbounded) polygon each edge of which is a part of the perpendicular bisector of the line segment $\overline{QQ'}$ with another generator Q'. Therefore, the unit outer normal vector to the Voronoi edge $E(P, P_{\iota(i)})$ is denoted by $(1/r_{\iota(i)})\overrightarrow{PP_{\iota(i)}}$. From this fact and Lemma 2, we get

$$\sum_{i=1}^{k} \frac{l_{\iota(i)}}{r_{\iota(i)}} \overrightarrow{PP_{\iota(i)}} = \mathbf{0}.$$

Note that $l_i = 0$ if P_i is not a neighbor of P. Hence we get

$$\sum_{i=1}^{n} \frac{l_i}{r_i} \boldsymbol{p} = \sum_{i=1}^{n} \frac{l_i}{r_i} \boldsymbol{p}_i,$$

which proves (2).

3.2. Proof of Sibson's identity

Observing the proof we gave in Section 3.1 closely, we notice that the crucial fact is that for each neighbor P_i of P, the Voronoi edge $E(P, P_i)$ is perpendicular to the vector $\overrightarrow{PP_i}$. Therefore, if we find a polygon each of whose edges is perpendicular to the vector $\overrightarrow{PP_i}$ with some neighbor P_i, we can obtain another coordinate system from that polygon.

At first, we describe a procedure for constructing such polygons. For this purpose, the Voronoi diagram for the generator set $\{P_1, \ldots, P_n, P\}$ is constructed. We denote the neighbors of P by $P_{\iota(1)}, \ldots, P_{\iota(k)}$ in the counterclockwise order. Without loss of generality, we assume that $P_{\iota(1)}$ is the nearest generator to P (choose arbitrary one if not unique). Let $S_i \in \overline{V(P, P_{\iota(i)})}$ be the point furthest from the line containing $E(P, P_{\iota(i)})$, and let h_i be the distance of S_i from the line containing $E(P, P_{\iota(i)})$.

Let us pay attention to $V(P, P_{\iota(1)})$. Since $P_{\iota(1)}$ is the nearest generator to P, $\overline{V(P, P_{\iota(1)})}$ contains the target point P. We start constructing the polygon by stroking its first edge inside $V(P, P_{\iota(1)})$. The following procedure outputs the vertices of a desired polygon.

1. Let L_1 be the line perpendicular to $\overrightarrow{PP_{\iota(1)}}$ such that $L_1 \cap \overline{V(P, P_{\iota(1)})}$ has a positive area.

2. Let $\{Q_1, Q_2\} = \partial V(P, P_{\iota(1)}) \cap L_1$ such that S_1 is located on the left of $\overrightarrow{Q_1 Q_2}$.

3. Set $i \leftarrow 2$ and $j \leftarrow 2$.

4. Let L_i be the line that is perpendicular to $\overrightarrow{PP_{\iota(i)}}$ and that contains Q_i.

5. If $L_i \cap V(P, P_{\iota(i)})$ is empty, go to 7.

6. Set $j \leftarrow j+1$. Let Q_j be the other endpoint of the line segment $L_i \cap V(P, P_{\iota(i)})$.

7. Set $i \leftarrow i+1$.

8. If $i \leq k$, go to 4. Otherwise output $Q_1, \ldots, Q_{j-1}$ and terminate.

For $i = 1, \ldots, k$, let v_i denote the distance of the edge of the obtained polygon inside $V(P, P_{\iota(i)})$, if any, from the Voronoi edge $E(P, P_{\iota(i)})$, and let $u_i = h_i - v_i$. We denote by $C(u_1)$ the obtained polygon since some edge of the obtained polygon necessarily lies inside $V(P, P_{\iota(1)})$. Note that $C(u_1)$ tends to the boundary of $V(P)$ as $u_1 \to h_1$. In fact, the above procedure was obtained by mimicking the incremental algorithm for constructing Voronoi diagrams.

By setting $\partial V = C(u_1)$ in Lemma 2, we get the following identity:

$$\sum_{i=1}^{n} s_i^0(u_1; P)\boldsymbol{p} = \sum_{i=1}^{n} s_i^0(u_1; P)\boldsymbol{p}_i, \tag{4}$$

where

$$s_i^0(u_1; P) = l_i(u_1; P)/r_i(P),$$
$$l_i(u_1; P) = (\text{the length of } C(u_1) \cap V(P_i)).$$

We call $s_i^0(u_1; P)$'s *Laplace's coordinates with parameter* u_1. Note that $s_i^0(u_1; P) \to s_i^0(P)$ as $u_1 \to h_1$. Hence we define $s_i^0(h_1; P) = s_i^0(P)$.

For the purpose of proving Sibson's identity, we require another fact. Let $e_i(u_1)$ denote the line segment $V(P_{\iota(i)}) \cap C(u_1)$.

Lemma 3 *For two polygons $C(u_1)$ and $C(u_1 + \mathrm{d}u_1)$, assume that the edges $e_i(u_1)$ and $e_i(u_1 + \mathrm{d}u_1)$ exist. Let $\mathrm{d}u_i$ be the width between $e_i(u_1)$ and $e_i(u_1 + \mathrm{d}u_1)$. Then,*

$$r_{\iota(1)}\mathrm{d}u_1 = r_{\iota(i)}\mathrm{d}u_i.$$

Proof. Let b denote the perpendicular bisector of the line segment $\overline{P_{\iota(1)}P_{\iota(i)}}$. Then, b contains the Voronoi edge $E(P_{\iota(1)}, P_{\iota(i)})$. Let φ denote the angle generated by the lines b and $\overline{PP_{\iota(1)}}$, and let ψ denote the angle generated by the lines b and $\overline{PP_{\iota(i)}}$. Figure 4 shows these notations.

It is clear from the definitions that

$$\frac{\mathrm{d}u_1}{\cos\varphi} = \frac{\mathrm{d}u_i}{\cos\psi}.$$

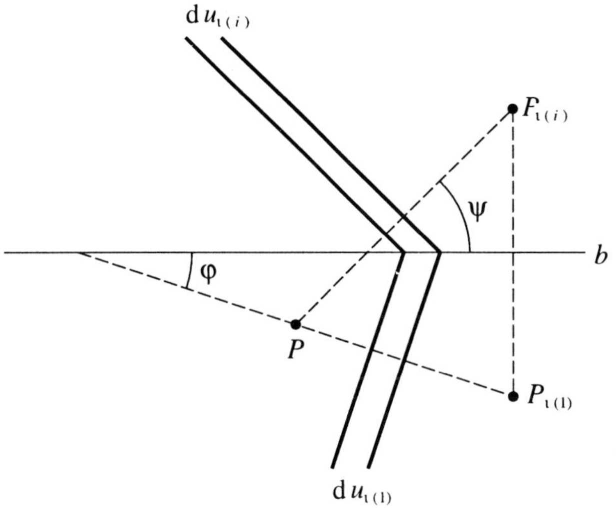

Figure 4. Edges of the polygons $C(u_1)$ and $C(u_1 + \mathrm{d}u_1)$.

Since $b \perp \overline{P_{\iota(1)}P_{\iota(i)}}$, the following equations hold:

$$\angle PP_{\iota(1)}P_{\iota(i)} = 90^\circ - \varphi, \quad \angle PP_{\iota(i)}P_{\iota(1)} = 90^\circ - \psi.$$

On the other hand, applying the sine theorem to the triangle $PP_{\iota(1)}P_{\iota(i)}$ yields the equation

$$\frac{r_{\iota(i)}}{\sin\angle PP_{\iota(1)}P_{\iota(i)}} = \frac{r_{\iota(1)}}{\sin\angle PP_{\iota(i)}P_{\iota(1)}}.$$

From the above equations, we get

$$r_{\iota(1)}\mathrm{d}u_1 = r_{\iota(i)}\mathrm{d}u_i.$$

□

At last, we give another proof of Sibson's identity. Lemma 2 guarantees (4) for any $0 < u_1 < h_1$. Operating $r_{\iota(1)} \int_0^{h_1} \mathrm{d}u_1$ on (4), we get

$$r_{\iota(1)} \sum_{i=1}^{n} \int_0^{h_1} s_i^0(u_1; P)\mathrm{d}u_1\boldsymbol{p} = r_{\iota(1)} \sum_{i=1}^{n} \int_0^{h_1} s_i^0(u_1; P)\mathrm{d}u_1\boldsymbol{p}_i.$$

From Lemma 3, we get

$$\sum_{i=1}^{k} \int_0^{h_i} l_{\iota(i)}(u_1(u_i); P)\mathrm{d}u_i\boldsymbol{p} = \sum_{i=1}^{k} \int_0^{h_i} l_{\iota(i)}(u_1(u_i); P)\mathrm{d}u_i\boldsymbol{p}_{\iota(i)}.$$

Since

$$s_{\iota(i)}^1(P) = \int_0^{h_i} l_{\iota(i)}(u_1(u_i); P)\mathrm{d}u_i,$$

we get

$$\sum_{i=1}^{n} s_i^1(P)\boldsymbol{p} = \sum_{i=1}^{n} s_i^1(P)\boldsymbol{p}_i,$$

the above equation completes the proof.

4. Generalization of coordinate systems

Observing the proof given in Section 3.2 carefully, we notice that Sibson's coordinate system is not the only one that we can obtain; other coordinate systems can be obtained by modifying the expression slightly. The following are typical modifications:

1. Integration in subinterval. Any subinterval (a, b) of $(0, h_1)$ is available as the integration interval.

2. Integration in arbitrary interval. Any interval (a, b) with $0 < a < b$ is available when we extend $C(u_1)$ naturally for $u_1 > h_1$. $C(u_1)$ with $u_1 > h_1$ lies outside the Voronoi region $V(P)$.

3. Weighted integration. In the integration step, the weight function $w(u_1)$ can be multiplied.

4. Multiple integration. We can construct other coordinate systems by integrating previously obtained coordinates repeatedly.

4.1. Standard coordinates

Although a lot of coordinate systems can be constructed by combining the above modifications, the authors guess the following subclass of $s_i^k(P)$ is especially important. For each $k = 1, 2, \ldots$, the coordinate $s_i^k(P)$ is defined as follows:

- If P_i is not a neighbor of P, $s_i^k(P) = 0$.
- If P_i is a neighbor of P, there exists j such that $\iota(j) = i$. Then, $s_i^k(P)$ is calculated in the following recursion:

$$s_i^k(u_1(u_j); P) = r_i \int_0^{u_j} s_i^{k-1}(u_1(u_j'); P) du_j',$$
$$s_i^k(P) = s_i^k(h_1; P).$$

We get the following generalized identity:

$$\sum_{i=1}^{n} s_i^k(P) \boldsymbol{p} = \sum_{i=1}^{n} s_i^k(P) \boldsymbol{p}_i. \tag{5}$$

Let $\hat{s}_i^k(P)$ denote the normalization of $s_i^k(P)$:

$$\hat{s}_i^k(P) = s_i^k(P) / \sum_{j=1}^{n} s_j^k(P).$$

Rewriting (5), we get

$$\boldsymbol{p} = \sum_{i=1}^{n} \hat{s}_i^k(P) \boldsymbol{p}_i,$$

$$\sum_{i=1}^{n} \hat{s}_i^k(P) = 1, \quad 0 \leq \hat{s}_i^k(P) \leq 1.$$

Thus we can express $\boldsymbol{p}$ as a convex combination of $\boldsymbol{p}_1, \ldots, \boldsymbol{p}_n$. We call $\hat{s}_i^k(P)$ the *order-k standard coordinates.* Laplace's coordinate system and Sibson's coordinate system are the order-0 and order-1 coordinate systems, respectively. The corresponding generalized interpolants are given by

$$\tilde{\varphi}^k(P) = \sum_{i=1}^{n} z_i \hat{s}_i^k(P). \tag{6}$$

4.2. Piecewise linear interpolant over the Delaunay triangulation

The coordinates $\hat{s}_i^0(u_1; P)$ become Laplace's coordinates when $u_1 = h_1$, and become Sibson's coordinates when integrated over the interval $(0, h_1)$. In this subsection, we consider the case when $u_1 = 0$.

Assume that u_1 is sufficiently small, and that each quadruple of $\{P_1, \ldots, P_n\}$ are not cocircular. Then, the polygon $C(u_1)$ surrounds some Voronoi vertex in the Voronoi diagram $\mathcal{V}(P_1, \ldots, P_n)$ (in the exceptional case, $C(u_1)$ surrounds some Voronoi edge; this occurs when P lies on some line segment $\overline{P_iP_j}$ where P_i and P_j are neighbors). Since the degree of each Voronoi vertex is three, $C(u_1)$ is a triangle. Hence $\sigma_i^0(u_1; P)$, $\sigma_j^0(u_1; P)$ and $\sigma_k^0(u_1; P)$ are the barycentric coordinates of P with respect to the associated three data sites where the Voronoi vertex surrounded by $C(u_1)$ is generated by P_i, P_j and P_k. The crucial fact is that the set of all the points P such that $C(u_1)$ surrounds some Voronoi vertex is the Delaunay triangle associated with that Voronoi vertex (see, e.g., Okabe et al. [5], for Delaunay triangulation). Thus the limit of

$$\tilde{\varphi}^0(u_1; P) = \sum_{i=1}^{n} z_i \hat{s}_i(u_1; P)$$

when $u_1 \to 0$ is the piecewise linear interpolant over the Delaunay triangulation. The above result is true even when $P_1, \ldots, P_n$ are not located in general positions.

5. Computational experiments

In this section, we present some computational results for the order-k standard interpolant.

Figure 5 shows the data sites used in these experiments. In this figure, there are eleven data sites $P_1, \ldots, P_{11}$. The data value 1 is given to P_1, while the data value 0 is given to the other data sites. Substituting the data value in the interpolant (6), we get

$$\varphi^k(P) = \hat{s}_1^k(P).$$

Thus the obtained function is a k-th order standard coordinate of P.

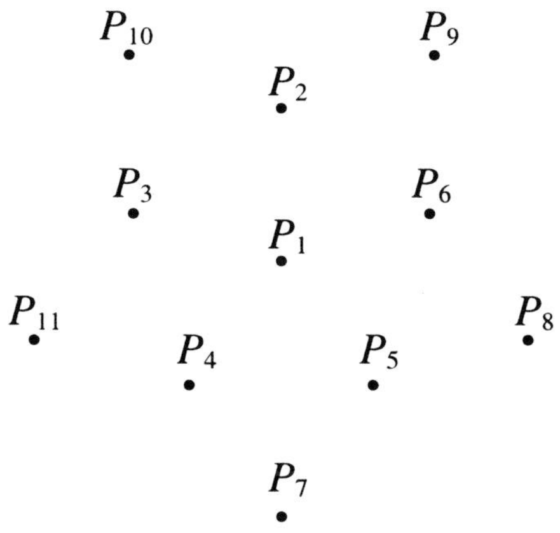

Figure 5. The data sites used in the experiments.

Figure 6 describes the surfaces of the interpolants $\tilde{\varphi}^0$, $\tilde{\varphi}^1$ and $\tilde{\varphi}^2$. Although our eyes can hardly see the difference among the figures, the following discussion clarifies their difference.

Figure 7 describes the cross sections of the interpolants by a vertical plane. If we see the cross section of $\tilde{\varphi}^0$, we notice that some non-smooth points, as explained in Section 2.4. The slopes of these cross sections are drawn in Figure 8. The differentiation was done by the numerical manner. Figure 8 tells that $\tilde{\varphi}^1$ loses C^2-continuity, whereas $\tilde{\varphi}^2$ still has C^2-continuity. Therefore these computational experiments suggests that the obtained surface becomes smoother as k increases. We conjecture that $\tilde{\varphi}^k$ has C^k-continuity except at the data sites.

6. Concluding remarks

This paper generalized Sibson's and Laplace's coordinate systems, and plotted a general framework for producing new Voronoi-based interpolants. The authors guess that the standard coordinates are especially important because the computational experiments suggest that their continuity is better than the previously known Voronoi-based interpolants. In addition, the proposed framework also produces the piecewise linear interpolants over the Delaunay triangulation, so already established techniques in the finite element method might be brought into the research of the Voronoi-based interpolation. Thus this framework implies a promising direction of the research of the Voronoi-based approach.

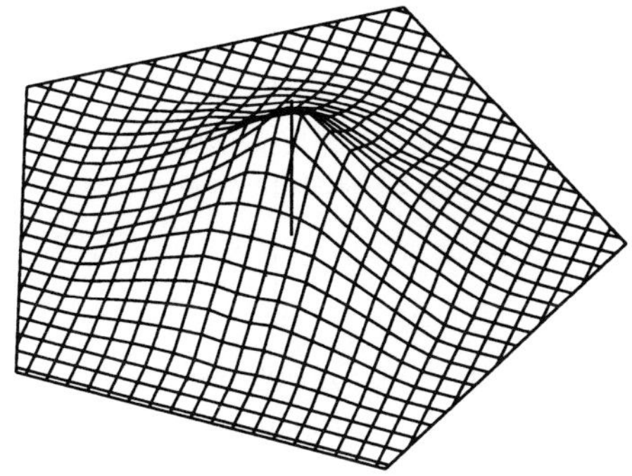

(a) A zeroth order standard coordinate function.

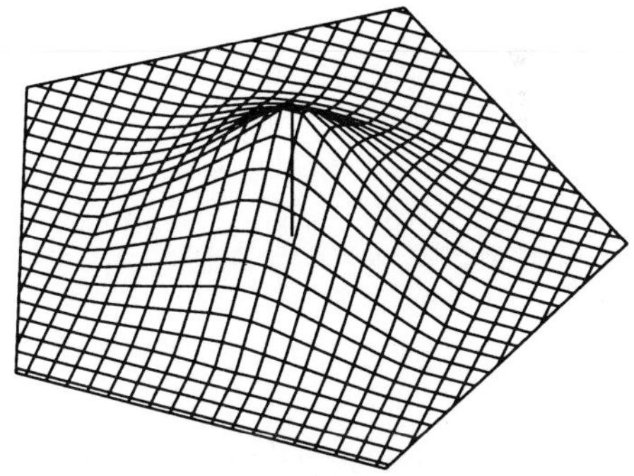

(b) A first order standard coordinate function.

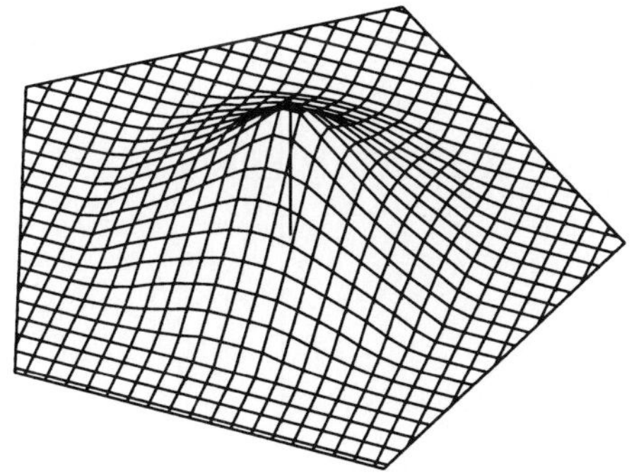

(c) A second order standard coordinate function.

Figure 6. zeroth, first and second order standard coordinate functions.

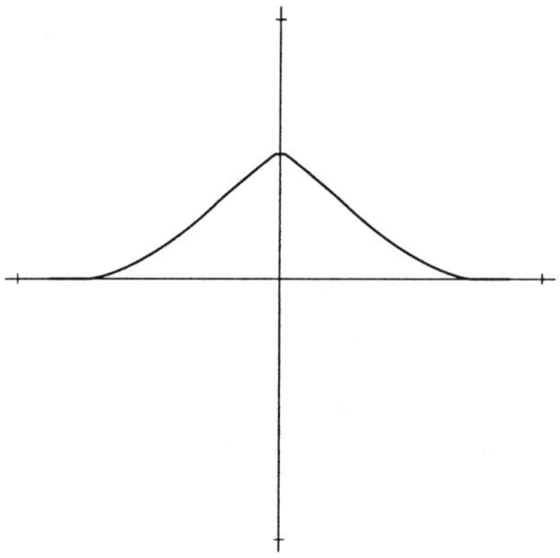

(a) A cross section of a zeroth order standard coordinate function.

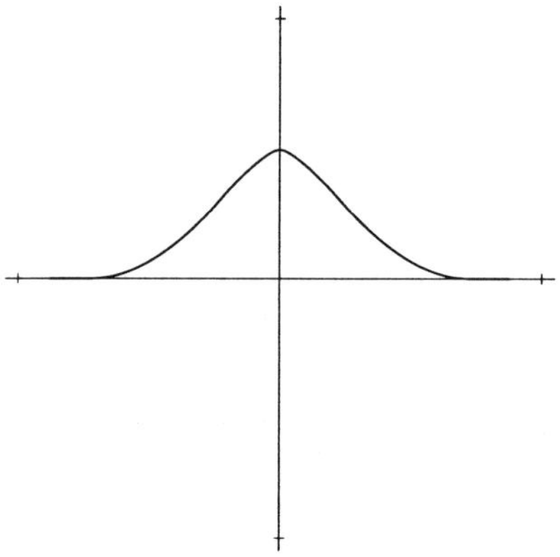

(b) A cross section of a first order standard coordinate function.

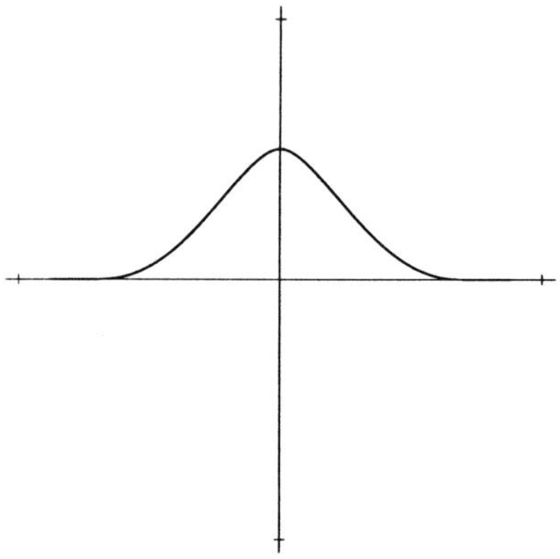

(c) A cross section of a second order standard coordinate function.

Figure 7. Cross sections of zeroth, first and second order standard coordinate functions.

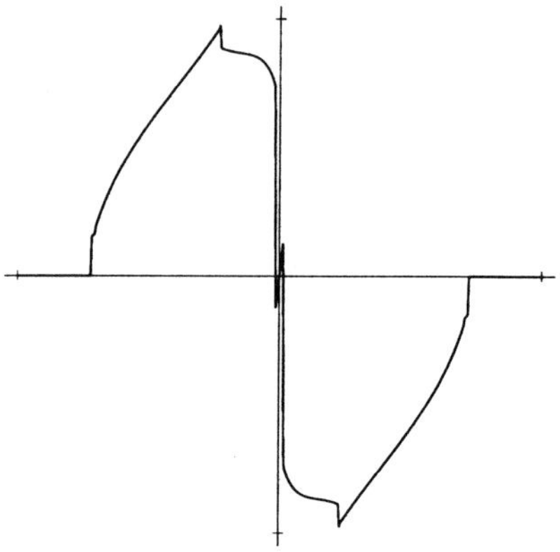

(a) Derivative function of the cross section for $\tilde{\varphi}^0$.

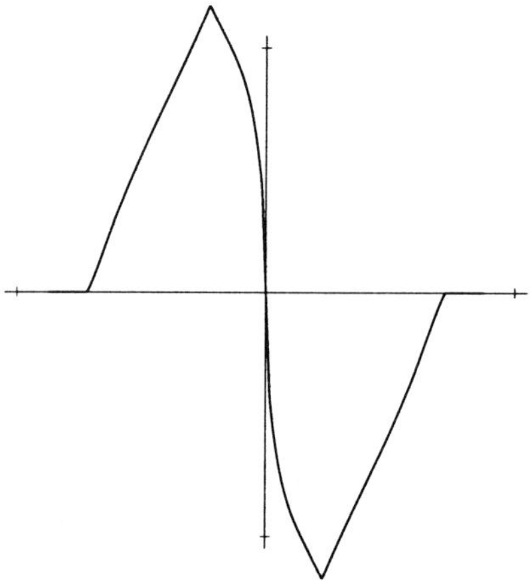

(b) Derivative function of the cross section for $\tilde{\varphi}^1$.

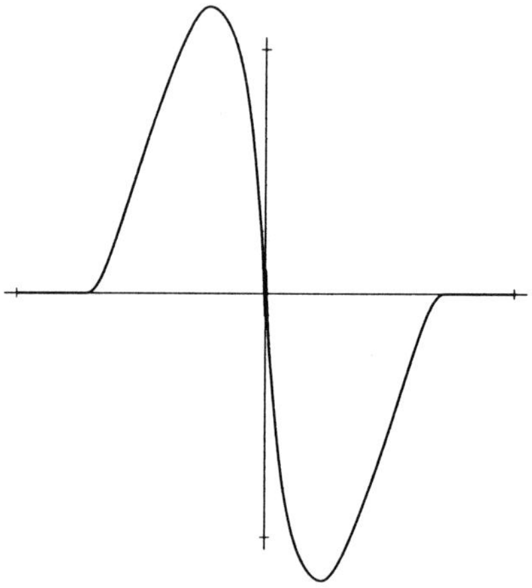

(c) Derivative function of the cross section for $\tilde{\varphi}^2$.

Figure 8. Derivative functions of the cross sections.

A lot of work must be done in future, including the following:

- To analyze the smoothness and other properties of the standard coordinates theoretically.
- To develop the applications of the Voronoi-based interpolation. Spatial surface construction is one of potential applications.

Acknowledgement

This work is supported by the Grant-in-Aid for Scientific Research of the Ministry of Education, Science, Sports and Culture of Japan.

References

[1] H. Edelsbrunner. *Algorithms in Combinatorial Geometry*. Springer-Verlag, Berlin, 1987.

[2] G. Farin. Surfaces over Dirichlet tessellations. *Computer Aided Geometric Design*, 7:281–292, 1990.

[3] H. Hiyoshi and K. Sugihara. Another interpolant using Voronoi diagrams. *IPSJ SIG Notes* 98-AL-62, 33–40, 1998, in Japanese.

[4] H. Hiyoshi and K. Sugihara. Two generalizations of an interpolant based on voronoi diagrams. *International Journal of Shape Modeling*, to print.

[5] A. Okabe, B. Boots, and K. Sugihara. *Spatial Tessellations: Concepts and Applications of Voronoi Diagrams*. John Wiley & Sons, Chichester, 1992.

[6] B. Piper. Properties of local coordinates based on Dirichlet tessellations. In G. Farin, H. Hagen, and H. Noltemeier, editors, *Geometric Modelling*, pages 227–239. Springer-Verlag, Wien, 1993.

[7] F. P. Preparata and M. I. Shamos. *Computational Geometry*. Springer-Verlag, New York, 1985.

[8] R. Sibson. A vector identity for the Dirichlet tessellation. *Mathematical Proceedings of Cambridge Philosophical Society* 87, 151–155, 1980.

[9] R. Sibson. A brief description of natural neighbour interpolation. In V. Barnett, editor, *Interpreting Multivariate Data*, pages 21–36. John Wiley & Sons, Chichester, 1981.

[10] K. Sugihara. Surface interpolation based on new local coordinates. *Computer-Aided Design*, 31:51–58, 1999.

[11] A. H. Thiessen. Precipitation averages for large areas. *Monthly Weather Report*, 39:1082–1084, 1911.

Session 5

Approximation and Interpolation Methods I

Poisson Approximation

Ron Goldman and Géraldine Morin
Department of Computer Science
Rice University
Houston, Texas 77005-1892
U.S.A.
{rng, gege}@cs.rice.edu

Abstract

Poisson approximation extends the Bernstein-Bézier scheme from polynomials to arbitrary analytic functions. A survey of some old and new results in Poisson approximation is presented here with special emphasis on algorithmic techniques for analytic functions.

1. Introduction: Probability and CAGD

Probability distributions play a central role in Computer Aided Geometric Design (CAGD). The blending functions for Bernstein-Bézier curves are taken from the binomial distribution, and the B-spline blending functions also model a simple stochastic process [5,10]. Discrete distributions $\{D_k(t)\}$ play a fundamental role in CAGD because typically curve schemes are defined by setting

$$P(t) = \sum_k D_k(t) P_k \ .$$

The defining characteristics of discrete distributions

$$\sum_k D_k(t) \equiv 1 \ \text{ and } \ D_k(t) \geq 0 \ \text{ for all } k$$

ensure that the curve $P(t)$ will be affine invariant and lie in the convex hull of its control points $\{P_k\}$.

The Poisson distribution is a limiting case of the binomial distribution as the number of Bernoulli trials approaches infinity [14]. Since the binomial distribution lies at the heart of Bézier curves and surfaces, it is natural to expect that a limiting case of this distribution would also play a prominent role in CAGD. In fact, the Poisson distribution has already appeared in Approximation Theory in the Szasz-Mirakian operator [13]. Poisson approximation has also been studied by Stone, who proved a variant of the Weierstrass Approximation Theorem for the Poisson basis functions [22].

The first mention of possible applications for the Poisson distribution in CAGD goes back to [9], where the author observes that many identities for the Bernstein basis functions extend naturally to the Poisson distribution (see too Section 2 below) and therefore that many of the salient geometric features of Bézier curves would be inherited by Poisson curves (see Section 3). He goes on to suggest that "the convergence properties of Poisson curves need to be carefully understood before they can be applied directly to problems in computer aided geometric design."

The purpose of this paper is to reexamine Poisson approximation in light of developments during the past two decades in CAGD relating to Bernstein approximation and Bézier curves. In particular, we will show how to apply the Poisson distribution to generalize the Bernstein-Bézier scheme to infinite sequences of control points and to extend polynomial techniques to arbitrary analytic functions. Our main theme is that using the Poisson representation, analytic functions can be analyzed in much the same way as polynomial functions represented in Bernstein-Bézier form. We shall provide an intuitive survey of various analytic, geometric, and algorithmic properties that Poisson approximation inherits from the Bézier scheme. For rigorous derivations of many of these results, particularly regarding issues of convergence, as well as some simple extensions to surfaces, see [17,18].

2. Identities for the Poisson Basis

The Poisson distribution $\{b_k(x)\}$ is defined by setting

$$b_k(x) = \frac{e^{-x} x^k}{k!} \qquad k = 0, 1, \ldots \ . \qquad (2.1)$$

It is well known in probability theory that the Poisson distribution is a limiting case of the binomial distribution as the number of Bernoulli trials approaches infinity. Indeed let $\{B_k^n(x)\}$ denote the binomial distribution of order n. Then

$$B_k^n(x) = \binom{n}{k} x^k (1-x)^{n-k} \quad k = 0, \ldots, n \qquad (2.2a)$$

and

$$b_k(x) = Lim_{n \to \infty} B_k^n(x/n) \qquad (2.2b)$$

(see Figure 2.1). Moreover the convergence is uniform on any bounded subinterval of $[0, \infty)$ [17].

0-7695-0562-7/00 $10.00 © 2000 IEEE

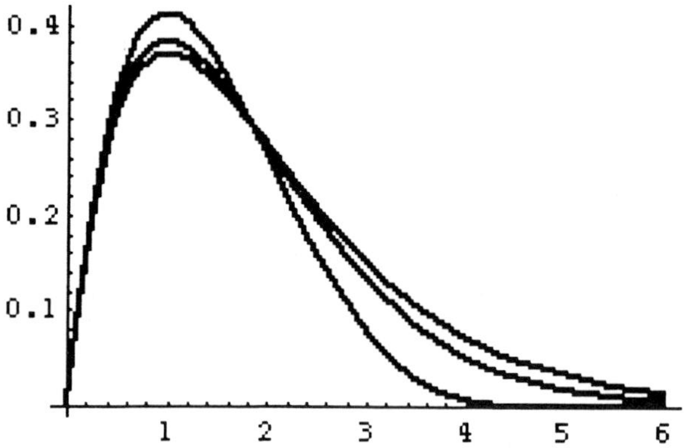

Figure 2.1. The Bernstein basis functions $B_1^5(x/5)$ (top bump) and $B_1^{15}(x/15)$ (middle bump) approach the Poisson basis function $b_1(x)$ (bottom bump).

It follows from Equations (2.2a,b) that the Poisson basis functions $\{b_k(x)\}$ inherit many of the analytic properties of the Bernstein basis functions $\{B_k^n(x)\}$. Numerous useful identities are known for the Bernstein basis functions [8,11]. Using either Equation (2.1) or Equations (2.2a,b), we can derive a comparable set of handy identities for the Poisson basis functions.

Partition of Unity

$$\textstyle\sum_k B_k^n(x) = 1 \qquad (2.3a)$$

$$\textstyle\sum_k b_k(x) = 1 \qquad (2.3b)$$

Non-Negativity

$$B_k^n(x) \geq 0 \quad 0 \leq x \leq 1 \qquad (2.4a)$$

$$b_k(x) \geq 0 \quad 0 \leq x < \infty \qquad (2.4b)$$

Interpolation

$$B_k^n(0) = \delta_{k0} \qquad (2.5a)$$

$$B_k^n(1) = \delta_{kn} \qquad (2.5b)$$

$$b_k(0) = \delta_{k0} \qquad (2.5c)$$

$$Lim_{x\to\infty} b_k(x) = 0 \qquad (2.5d)$$

Unimodality

$$Max_{0\leq x\leq 1}\{B_k^n(x)\} \text{ occurs at } x = k/n \qquad (2.6a)$$

$$Max_{0\leq x<\infty}\{b_k(x)\} \text{ occurs at } x = k \qquad (2.6b)$$

Linear Independence

$$\textstyle\sum_k c_k B_k^n(x) = 0 \Leftrightarrow c_k = 0 \text{ for all } k \qquad (2.7a)$$

$$\textstyle\sum_k c_k b_k(x) = 0 \Leftrightarrow c_k = 0 \text{ for all } k \qquad (2.7b)$$

Descartes' Law of Signs

$$zeros_{(0,1)}\{\textstyle\sum_k c_k B_k^n(x)\} \leq sign\ \ alternations(c_0,\ldots,c_n) \qquad (2.8a)$$

$$zeros_{(0,\infty)}\{\textstyle\sum_k c_k b_k(x)\} \leq sign\ \ alternations(c_0,c_1,\ldots) \qquad (2.8b)$$

Differentiation

$$\frac{dB_k^n(x)}{dx} = n\left(B_{k-1}^{n-1}(x) - B_k^{n-1}(x)\right) \qquad (2.9a)$$

$$\frac{db_k(x)}{dx} = b_{k-1}(x) - b_k(x) \qquad (2.9b)$$

Integration

$$\int_0^1 B_k^n(x)dx = \frac{1}{n+1} \qquad (2.10a)$$

$$\int_0^\infty b_k(x)dx = 1 \qquad (2.10b)$$

Reparametrization

$$B_i^n(ax) = \textstyle\sum_k B_i^k(a) B_k^n(x) \qquad (2.11a)$$

$$b_i(ax) = \textstyle\sum_k B_i^k(a) b_k(x) \qquad (2.11b)$$

Trimming

$$B_i^n(a + (1-a)x) = \textstyle\sum_{k\leq i} B_{i-k}^{n-k}(a) B_k^n(x) \qquad (2.12a)$$

$$b_i(a + x) = \textstyle\sum_{k\leq i} b_{i-k}(a) b_k(x) \qquad (2.12b)$$

Representation of Monomial Basis

$$\tbinom{n}{j} x^j = \textstyle\sum_{k\geq j} \tbinom{k}{j} B_k^n(x) \qquad (2.13a)$$

$$\frac{x^j}{j!} = \textstyle\sum_{k\geq j} \tbinom{k}{j} b_k(x) \qquad (2.13b)$$

Representation in Terms of Monomial Basis

$$B_k^n(x) = \textstyle\sum_{j\geq k} \{(-1)^{j-k} \tbinom{j}{k}\}\tbinom{n}{j} x^j \qquad (2.14a)$$

$$b_k(x) = \textstyle\sum_{j\geq k} \{(-1)^{j-k} \tbinom{j}{k}\} \frac{x^j}{j!} \qquad (2.14b)$$

Marsden Identity

$$(x-u)^n = \textstyle\sum_k (-1)^{n-k} u^{n-k} (1-u)^k B_k^n(x) \qquad (2.15a)$$

$$e^{-ux} = \textstyle\sum_k (1-u)^k b_k(x) \qquad (2.15b)$$

We shall have occasion to refer back to these identities in subsequent sections, where we will study properties of, and algorithms for, Poisson curves.

3. Properties of Poisson Curves

Bézier curves are polynomials; Poisson curves are analytic functions. A Bézier curve

$$P(t) = \textstyle\sum_k B_k^n(t) P_k \qquad 0 \leq t \leq 1 \qquad (3.1)$$

is a parametric curve defined by a finite sequence of control points $\{P_k\}$ (see Figure 3.1), but a parametric Poisson curve

$$p(t) = \textstyle\sum_k b_k(t) p_k \qquad 0 \leq t < R \qquad (3.2)$$

requires an infinite sequence of control points $\{p_k\}$ (see Figure 3.2). We say that $p(t)$ is a *Poisson curve* over the interval $[0,R)$ if $p(t)$ converges for $0 \leq t < R$. Thus unlike Bézier curves, with Poisson curves we must deal with issues of convergence. To simplify subsequent discussions, we shall generally let $R = \infty$, though analogous results typically hold for finite values of R.

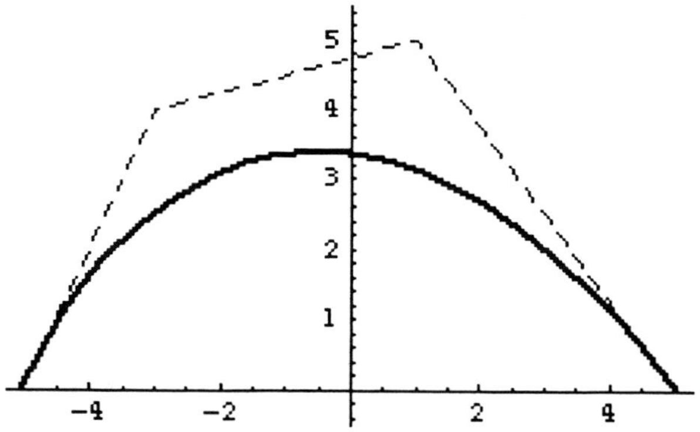

Figure 3.1: The planar cubic curve represented by the polynomials $x = -2t^3 + 6t^2 + 6t - 5$, $y = -3t^3 - 9t^2 + 12t$ parametrized over the interval [0,1] (dark), and its Bézier control polygon (dashed).

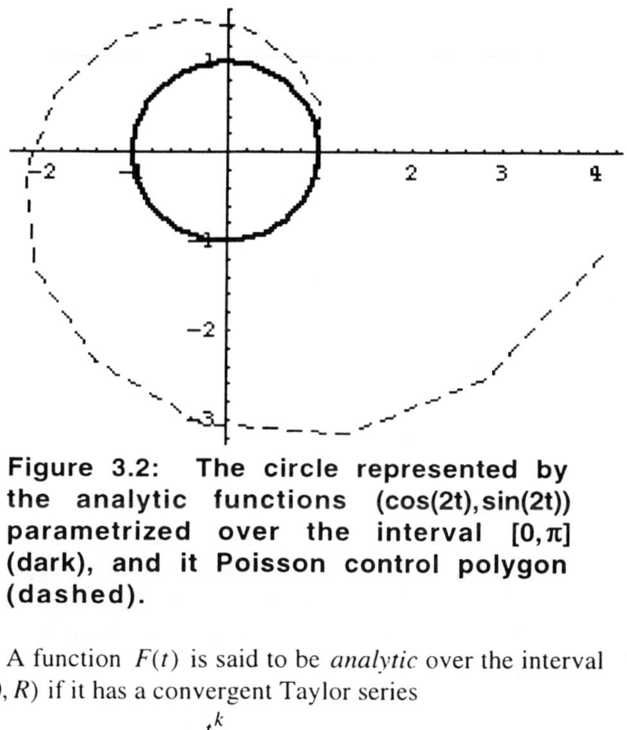

Figure 3.2: The circle represented by the analytic functions (cos(2t),sin(2t)) parametrized over the interval $[0,\pi]$ (dark), and it Poisson control polygon (dashed).

A function $F(t)$ is said to be *analytic* over the interval $[0, R)$ if it has a convergent Taylor series

$$F(t) = \sum_k F_k \frac{t^k}{k!} \qquad 0 \le t < R.$$ [1]

If $F(t)$ is analytic over the interval $[0, R)$, then so too is $e^t F(t)$. But since $e^t F(t)$ is analytic,

$$e^t F(t) = \sum_k E_k \frac{t^k}{k!}\ ; \qquad (3.3)$$

[1] Strictly speaking, it is more correct to say that $F(t)$ is *analytic* in the interval $[0, R)$ if it has a convergent power series for each $t \in [0, R)$, even if the power series are different for different values of t. To simplify our presentation, we have adopted a somewhat more elementary definition both for analytic functions and for Poisson curves. For a more rigorous treatment of this topic, see [18].

hence

$$F(t) = e^{-t}(e^t F(t)) = \sum_k E_k \frac{e^{-t} t^k}{k!} = \sum_k E_k b_k(t). \quad (3.4)$$

Thus if $F(t)$ is analytic, then $F(t)$ is a Poisson curve. Conversely, if $F(t)$ is a Poisson curve, then $e^t F(t)$ is analytic, so $F(t)$ is also analytic since $F(t) = e^{-t}(e^t F(t))$. Thus Poisson curves are equivalent to analytic functions. Moreover by Equations (3.3) and (3.4), the Poisson coefficients of an analytic function $F(t)$ are identical to the Taylor coefficients of the function $e^t F(t)$.

From Equations (2.3b)-(2.8b), it follows that Poisson curves have the following geometric properties.

Geometric Properties of Poisson Curves

- Affine Invariance
- Convex Hull Property
- Locality
- Interpolate Initial Control Point
- Non-Degenerate
- Variation Diminishing

The first two properties are self explanatory and follow directly from the fact that the Poisson blending functions form a discrete probability distribution. But not just any discrete distribution can be used as blending functions in CAGD. In Geometric Modeling we seek curves and surfaces with certain special geometric characteristics. One particularly important property is that each control point should have only a local influence over the entire curve or surface and its influence should tail off fairly rapidly as the curve or surface recedes from the control point. This condition is called *locality*.

Locality is important because it permits engineers to make local changes in the shape of a curve or surface by moving nearby control points. Thus control points should have a lot of influence on nearby segments but very little influence on far away segments of the curve or surface. This kind of locality can be guaranteed by choosing blending functions that are unimodal -- that is, functions whose graphs look like a bump with only one local maximum (see Figure 3.3). When the curve or surface parameters are near this maximum value, the corresponding control point has a large influence on the curve or surface, but far from this maximum value the affect of the corresponding control point is negligible. The binomial distribution, B-splines, and the Poisson distribution all have this simple bump-like structure; thus these distributions are natural candidates for the blending functions of curves and surfaces in CAGD.

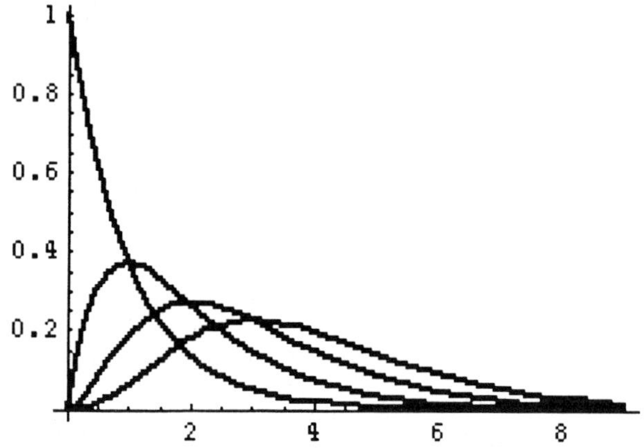

Figure 3.3. The first four Poisson basis functions. Notice that each of these functions are unimodal with only one local maximum.

Like Bézier curves, Poisson curves are non-degenerate -- that is, they cannot collapse to a single point unless all their control points are located at that same point. This property follows from the uniqueness of the Poisson expansion (Equation (2.7b)). Moreover again like Bézier curves, Poisson curves satisfy the variation diminishing property -- that is, they do not oscillate any more than the piecewise linear function (control polygon) defined by their control points. This result follows from Descartes' Law of Signs (Equation (2.8b)), which is known to be true for the Taylor polynomials [12] and hence must be true as well for the Poisson basis functions since the Poisson functions $\{b_k(t)\}$ are just the Taylor functions $\{t^k / k!\}$ multiplied by e^{-t}.

Bézier curves interpolate their first and last control points. By Equations (2.5c) and (3.2) a Poisson curve will interpolate its initial control point, but since it has an infinite number of control points, it has no final control point. Nevertheless, we shall see shortly that $Lim_{t \to \infty} b_k(t) = 0$ (Equation (2.5d)) also has important consequences for Poisson approximation.

The Weierstrass Approximation Theorem asserts that any continuous function on a compact interval can be uniformly approximated by polynomials [6]. Since the Bernstein functions $\{B_k^n(t)\}$ form a basis for the polynomials of degree n, any continuous function on a compact interval can be uniformly approximated by Bézier curves.

Analogously, since $Lim_{t \to \infty} b_k(t) = 0$, it is a consequence of the Stone-Weierstrass Approximation Theorem that any continuous function $F(t)$ defined on the interval $[0, \infty)$ for which $Lim_{t \to \infty} F(t) = 0$ can be uniformly approximated by finite sums of the form $\sum_k F_k b_k(t)$ [22]. For such functions Poisson approximation provides a very attractive alternative to the Taylor expansion (see Figure 3.4).

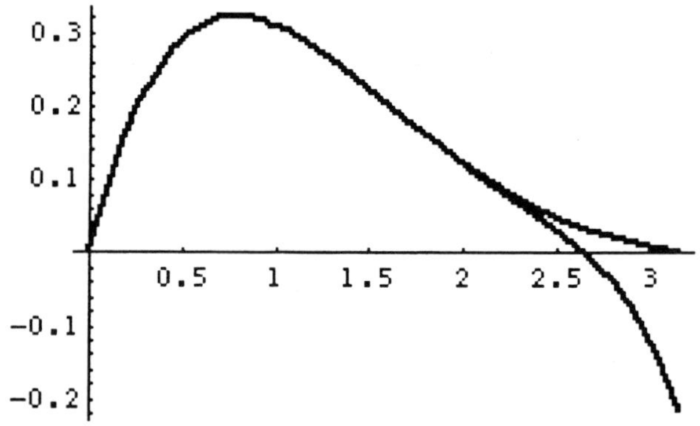

Figure 3.4. Ten term Poisson and Taylor approximations to e^{-t}sint. The Taylor expansion starts to diverges sharply from the original function around t=2.5, while the Poisson expansion continues to stay close to the original curve.

4. Algorithms for Poisson Curves

There are many standard analysis algorithms for Bézier curves, including procedures for evaluation, subdivision, differentiation, and conversion to and from Taylor form [2,4,7,15]. Here we shall see that analogous algorithms exist for Poisson curves.

The de Casteljau algorithm for a degree n Bézier curve

$$P(t) = \sum_{k=0}^{n} B_k^n(t) P_k \qquad 0 \le t \le 1$$

is both an evaluation algorithm and a subdivision procedure (see Figure 4.1). The input to this algorithm are the control points $\{P_k\}$ which we place at the base of the triangle. The edges in the diagram are labeled with the functions $1-a$ and a, and the value at each intermediate node is computed by multiplying the expressions on the arrows entering the node by the values at the nodes from which the arrows emerge and summing the results. The value at the apex of the triangle is the point $P(a)$ -- that is, the point on the curve at the parameter $t = a$. Thus this procedure generates an evaluation algorithm for Bézier curves.

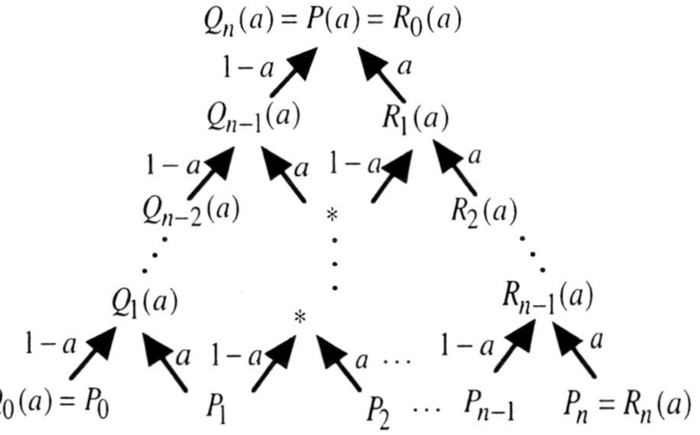

Figure 4.1: The de Casteljau evaluation algorithm for a Bézier curve of degree n. The control points are input at the base of the triangle and the point P(a) along the curve at the parameter t = a emerges at the apex of the triangle. The values that emerge along the left and right lateral edges of the diagram can be used to subdivide the curve at t = a.

The de Casteljau algorithm also generates a subdivision procedure for Bézier curves. Consider the two polynomials $P(at)$ and $P(a+(1-a)t)$ for some fixed value a where $0 < a < 1$. As the parameter t varies from zero to one, the function $P(at)$ sweeps out the left segment of the Bézier curve $P(t)$ for $0 \le t \le a$, and the function $P(a+(1-a)t)$ sweeps out the right segment of the Bézier curve $P(t)$ for $a \le t \le 1$. A subdivision algorithm is a procedure for finding the Bézier control points $\{Q_k(a)\}$ of $P(at)$ and $\{R_k(a)\}$ of $P(a+(1-a)t)$ from the control points $\{P_k\}$ of the original Bézier curve $P(t)$. From Equations (2.11a) and (2.12a) it follows easily that

$$Q_k(a) = \sum_{j=0}^{k} B_j^k(a) P_j \tag{4.1}$$

$$R_k(a) = \sum_{j=0}^{n-k} B_j^{n-k}(a) P_{j+k} \ . \tag{4.2}$$

Remarkably, these control points are already computed in the de Casteljau algorithm for the evaluation of $P(a)$; the left lateral edge of Figure 4.1 contains the control points $\{Q_k(a)\}$ for the polynomial $P(at)$ and the right lateral edge contains the control points $\{R_k(a)\}$ for the polynomial $P(a+(1-a)t)$. These results follow easily by combining the de Casteljau evaluation algorithm with Equations (2.11a) and (2.12a). Moreover it is well-known that these control polygons converge to the original Bézier curve under recursive subdivision [15]. Combining recursive subdivision with the convex hull property leads to efficient algorithms for rendering and intersecting Bézier curves.

Now consider a Poisson curve

$$p(t) = \sum_{k=0}^{\infty} b_k(t) p_k \qquad 0 \le t < \infty$$

and let

$$p_n(t) = \sum_{k=0}^{n} B_k^n(t/n) p_k \ .$$

Intuitively,

$$Lim_{n\to\infty} B_k^n(a/n) = b_k(a) \Rightarrow Lim_{n\to\infty} p_n(a) = p(a) \ . \tag{4.3}$$

(For a rigorous proof, see [17].) Hence we can use the de Casteljau algorithm for the Bézier curve $p_n(t)$ to approximate values on the Poisson curve $p(t)$ (see Figure 4.2). This approximation algorithm for Poisson curves is the analogue of the de Casteljau evaluation algorithm for Bézier curves. Notice that there is a tradeoff here between speed and accuracy; the larger the value of n, the greater the accuracy, but the slower the algorithm.

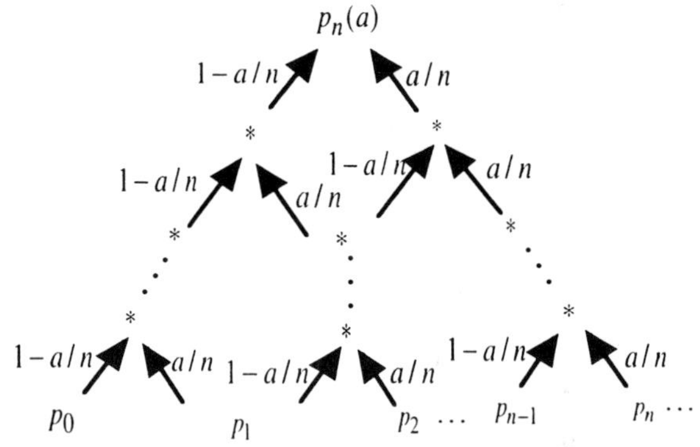

Figure 4.2. The de Casteljau approximation algorithm for a Poisson curve at t = a. The control points are input at the base of the triangle and the labels along the arrows are set to 1 − a/n and a/n. The value $p_n(a)$ that emerges at the apex of the *nth* level of the triangle approximates the value p(a) of the Poisson curve at t = a.

What about subdivision for Poisson curves? Now fix a value of a with $0 < a < 1$ and consider the Poisson curve $p(at)$. As the parameter t varies from zero to infinity, the function $p(at)$ sweeps out the same Poisson curve as $p(t)$, albeit with a different parametrization. From Equation (2.11b) it follows that the control points $\{q_k(a)\}$ for the reparametrized curve $p(at)$ are given by

$$q_k(a) = \sum_{j=0}^{k} B_j^k(a) p_j \ . \qquad (4.4)$$

Since Equation (4.4) for $q_k(a)$ is essentially identical to Equation (4.1) for $Q_k(a)$, the points $q_k(a)$ again emerge along the left lateral edge of the de Casteljau triangle (see Figure 4.3). Moreover, just like for Bézier curves, it can be shown that if we apply this algorithm recursively, then these control polygons converge to the original Poisson curve [17]. But notice that although for Bézier curves the algorithm for subdivision and the algorithm for evaluation at $t = a$ use the same de Casteljau triangle, the triangles for approximation and reparametrization at $t = a$ are different for Poisson curves. For approximation the arrows are labeled with the functions $1 - a/n$ and a/n, whereas for reparametrization the labels along the arrows are $1 - a$ and a.

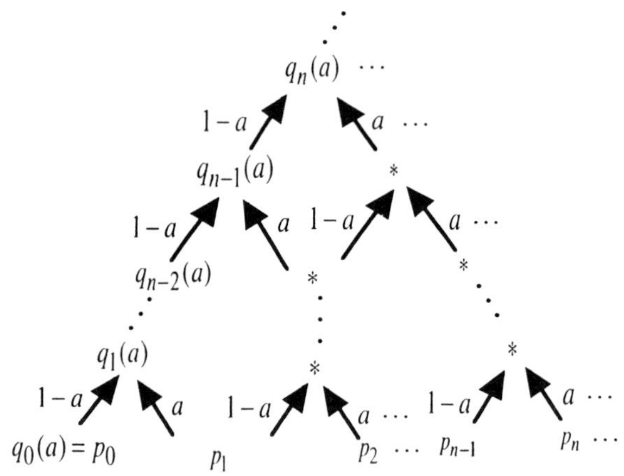

Figure 4.3. The de Casteljau reparametrization algorithm for Poisson curves.

The reparametrization algorithm for Poisson curves in Figure 4.3 is the analogue of the left subdivision algorithm for Bézier curves. But what is the analogue for Poisson curves of the right subdivision algorithm for Bézier curves? Right-sided subdivision for Bézier curves corresponds to trimming off the initial segment of the curve. The analogue for Poisson curves is to represent the curve $p(a+t)$ in terms of the Poisson basis. As t varies from 0 to ∞, the function $p(a+t)$ sweeps out the original Poisson curve from a to ∞, trimming off the initial segment of the Poisson curve $p(t)$ from $t = 0$ to $t = a$. By Equation (2.12b) the control points for this trimmed curve are given by

$$r_k(a) = \sum_{j=0}^{\infty} b_j(a) p_{j+k} \ .$$

While this formula is theoretically sound, as a practical matter we would prefer to have a finite algorithm to approximate the control points of $p(a+t)$ rather than an exact but infinite sum. Let

$$r_k^n(a) = \sum_{j=0}^{n} B_j^n(a/n) p_{j+k} \ .$$

Then, as we have already noted once before (see Equation (4.3)),

$$B_k^n(a/n) \to b_k(a) \Rightarrow r_k^n(a) \to r_k(a) \ .$$

Now we can apply the following de Casteljau like algorithm to generate the points $r_k^n(a)$ which can be used to approximate the curve $p(a+t)$:

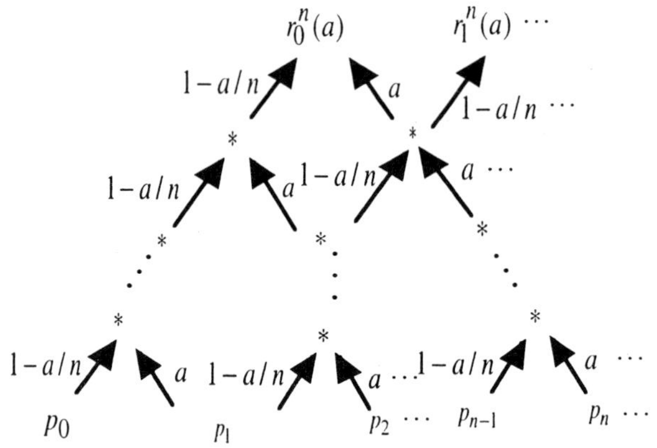

Figure 4.4. Approximation of the control points for right-sided subdivision for Poisson curves. Notice that the labels along the arrows are the same as for the approximation algorithm (Figure 4.2) and different from the labels for the left-sided subdivision algorithm (Figure 4.3) for Poisson curves.

Combining left-sided and right-sided subdivision with the convex hull property leads to efficient algorithms for rendering and intersecting Poisson curves [17,18]. Thus, using the Poisson representation, analytic functions can be rendered and analyzed in much the same way as polynomial functions represented in Bernstein-Bézier form.

In addition to evaluation and subdivision algorithms, there are well-known algorithms for converting between the Bézier and monomial bases which are similar in structure to the de Casteljau algorithm [2]. To convert from Bézier to monomial form, we employ a difference triangle (see Figure 4.5). This procedure is identical to the de Casteljau algorithm with the arrows labeled ± 1 and follows easily from Equation (2.14a). To perform the inverse transformation --- that is, to convert from monomial to Bézier form -- perform addition instead of subtraction, so that the value at each node is the sum of the nodes immediately below it to the left and right (see Figure 4.6). This procedure is identical to the de Casteljau algorithm with all the arrows labeled +1 and follows easily from Equation (2.13a). Notice that in both cases the monomial basis is given by $\{\binom{n}{k}t^k\}$ rather than by $\{t^k\}$.

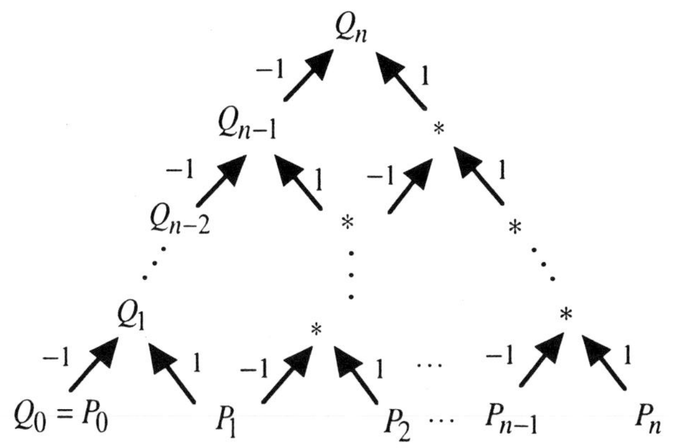

Figure 4.5. Converting from Bézier to monomial form using a difference triangle. The Bézier control points $\{P_k\}$ are placed at the base of the diagram and the monomial coefficients $\{Q_k\}$ emerge along the left lateral edge.

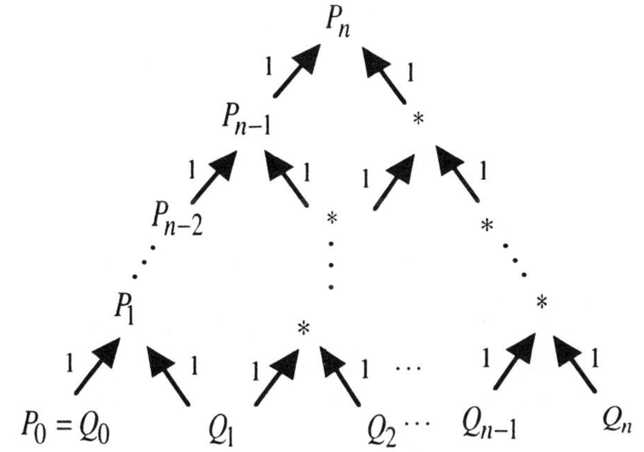

Figure 4.6. Converting from monomial to Bézier form by summing up. Here the monomial coefficients $\{Q_k\}$ are placed at the base of the diagram and the Bézier control points $\{P_k\}$ emerge along the left lateral edge.

By Equations (2.13b) and (2.14b) analogous algorithms can be applied to convert between the Taylor basis $\{t^k / k!\}$ and the Poisson basis $\{e^{-t}t^k / k!\}$. The only difference is that in this case the respective coefficients form infinite instead of finite sequences; otherwise the algorithms are exactly the same as the corresponding Bézier procedures. That is, we use a difference triangle to convert from Poisson to Taylor coefficients and we sum up to find the Poisson coefficients from the Taylor coefficients. Again the input to the algorithm is along the base of the de Casteljau triangle, and the output emerges along the left lateral edge.

Notice by the way that since the Taylor coefficients are derivatives at $t = 0$, the algorithm to convert from Poisson to Taylor form computes all the derivatives of a Poisson curve at $t = 0$. To compute all the derivative of a Poisson curve at an arbitrary parameter value $t = a$, first apply right-sided subdivision to trim the curve at $t = a$ and then convert to Taylor form.

Finally, if instead of computing all the derivatives at a single point we want to compute a single derivative at every point, notice that by Equation (2.9b) the Poisson coefficients of the derivative of $p(t)$ are given by the differences $\{p_{k+1} - p_k\}$. This observation is analogous to the well known consequence of Equation (2.9a) that the Bernstein coefficients for the derivative of a degree n Bézier curve $P(t)$ are given by $\{n(P_{k+1} - P_k)\}$.

5. Dual Functionals, Blossoming, and the Marsden Identity

Dual functionals are maps that compute the coefficients of arbitrary functions relative to a fixed basis. There are several ways to generate explicit expressions for the coefficients $\{F_k\}$ of an analytic function $F(x)$ relative to the Poisson basis. For example, by the algorithm presented in Section 4 for converting a function from the Taylor basis to the Poisson basis, it follows that

$$F(x) = \sum_{k=0}^{\infty} F_k b(t)$$
$$F_k = \sum_{j \le k} \binom{k}{j} F^{(j)}(0) \quad . \tag{5.1}$$

A similar argument based on converting from monomial to Bézier form generates an analogous formula for the dual functionals for the Bernstein bases [12]:

$$P(x) = \sum_k P_k B_k^n(x)$$
$$P_k = \sum_{j \le k} \frac{\binom{k}{j} P^{(j)}(0)}{\binom{n}{j} j!} \quad . \tag{5.2}$$

Dual functionals are important not only for generating expressions for the coefficients of arbitrary functions relative to a fixed basis, but also for generating change of basis algorithms between different bases. To achieve this goal, Equations (5.1) and (5.2) are actually too explicit.

Another way to construct the dual functionals for the Bernstein bases is to apply the Marsden identity [16] to generate the de Boor-Fix form of the dual functionals [3]. Define the bracket of two polynomials $P(u)$ and $Q(u)$ of degree n by setting

$$[P(u),Q(u)]_n = \sum_{k=0}^{n} \frac{(-1)^{n-k}}{n!} P^{(k)}(t) Q^{(n-k)}(t). \quad (5.3)$$

It is easy to check that the bracket is a bilinear operator, and that the right hand side of Equation (5.3) is independent of the choice of t because the derivative with respect to t is identically zero. Since $P(u)$ is a polynomial of degree n, by setting $t = 0$ in Equation (5.3) we obtain

$$[P(u),(x-u)^n]_n = \sum_{k=0}^{n} \frac{P^{(k)}(0)x^k}{k!} = P(x). \quad (5.4)$$

Hence by Equation (2.15a)

$$P(x) = \sum_{k=0}^{n} [P(u),(-1)^{n-k} u^{n-k} (1-u)^k]_n B_k^n(x) ,$$

so the Bézier coefficients of $P(x)$ of are given by the formula

$$P_k = [P(u),(-1)^{n-k} u^{n-k} (1-u)^k]_n . \quad (5.5)$$

Equation (5.5) is the de Boor-Fix form of the dual functionals for the Bernstein basis functions. By the way, notice that the blossom $p(x_1,...,x_n)$ of $P(x)$ can be computed from the bracket operator by setting

$$p(x_1,...,x_n) = [P(u),(x_1-u)\cdots(x_n-u)]_n, \quad (5.6)$$

since it is easy to verify that the right hand side of Equation (5.6) is symmetric, multiaffine, and, by Equation (5.4), reduces to $P(x)$ along the diagonal. Indeed it follows from Equations (5.5) and (5.6) that

$P_k = p(0,...,0,1,...,1)$, k ones and $n-k$ zeros,

which is the well-known blossoming formula for the dual functionals of the Bernstein basis [1,19,20,21]. Blossoming is a powerful tool for developing change of basis procedures. For example, subdivision algorithms, knot insertion procedures, and differentiation techniques can all easily be derived from blossoming [2].

We can take a similar approach to generate an analogue of the de Boor-Fix dual functionals for the Poisson basis. Define the bracket of two analytic functions $F(u)$ and $G(u)$ by setting

$$[F(u),G(u)] = \sum_{k=0}^{\infty} \frac{(-1)^k}{k!} F^{(k)}(0) G^{(k)}(0) . \quad (5.7)$$

It is easy to check that this bracket is a symmetric bilinear operator. Moreover since $F(x)$ is analytic,

$$[F(u),e^{-ux}] = \sum_{k=0}^{\infty} \frac{F^{(k)}(0)x^k}{k!} = F(x) . \quad (5.8)$$

Hence by Equation (2.15b)

$$F(x) = \sum_{k=0}^{\infty} [F(u),(1-u)^k] b_k(x) ,$$

so the Poisson coefficients of $F(x)$ are given by

$$F_k = [F(u),(1-u)^k] . \quad (5.9)$$

Equation (5.9) is the analogue of the de Boor-Fix form of the dual functionals for the Poisson basis. Of course, it is not difficult to show that Equation (5.9) is equivalent to Equation (5.1).

A theory of blossoming for analytic functions and the Poisson basis has yet to be developed. But we have just seen how the bracket operator can be used to define the blossom of a polynomial. All that is required to obtain Equation (5.6) for the blossom of a polynomial $P(x)$ is to replace the function $(x-u)^n$ in Equation (5.4) by its blossom $(x_1-u)\cdots(x_n-u)$. This suggests that to blossom an analytic function $F(x)$ all we need to do is to replace the function e^{-ux} in Equation (5.8) by its blossom. But what this blossom is still remains to be determined. We plan to tackle this general blossoming problem for analytic functions soon in a future paper and then to apply blossoming to develop various change of basis procedures for Poisson curves.

Acknowledgement: This work has been partially supported by NSF grant CCR-9971004.

References

1. P.J. Barry. de Boor-Fix functionals and polar forms, *Computer Aided Geometric Design*, 7:425-430, 1990.
2. P.J. Barry and R.N. Goldman. Algorithms for progressive curves: Extending B-spline and blossoming techniques to the monomial, power, and Newton dual bases, in *Knot Insertion and Deletion Algorithms for B-spline Curves and Surfaces*, Edited by R. Goldman and T. Lyche, pp.11-63, SIAM, Philadelphia, 1993.
3. C. de Boor and G. Fix. Spline approximation by quasi-interpolants, *J. Approx. Theory*, 8:19-45, 1973.
4. P. de Casteljau. *Formes a Poles*, Hermes, Paris, 1985.
5. W. Dahmen and C. Micchelli. Statistical encounters with B-splines, *AMS Series on Contemporary Mathematics*, 59:17-48, 1986.
6. P.J. Davis. *Interpolation and Approximation,* Dover, New York, 1975.
7. G. Farin. *Curves and Surfaces for Computer Aided Geometric Design: A Practical Guide (Third Edition),* Academic Press, San Diego, 1993.
8. R.T. Farouki and V.T. Rajan. Algorithms for polynomials in Bernstein form, *Computer Aided*

Geometric Design, 5:1-26, 1988.

9. R.N. Goldman. Polya's urn model and computer aided geometric design, SIAM Jour. Alg. and Disc. Meth., 6:1-28, 1985.
10. R.N. Goldman. Urn models and B-splines, *Constructive Approximation*, 4:265-288, 1988.
11. R.N. Goldman. Identities for the univariate and bivariate Bernstein basis functions, *Graphics Gems V*, Edited by Alan Paeth, pp. 149-162, Academic Press, Boston, 1995.
12. R.N. Goldman. The rational Bernstein bases and the multirational blossoms, *Computer Aided Geometric Design*, 16:701-738, 1999.
13. T. Hermann. On the Szasz-Mirakian operator, *Acta Math. Acad. Sci. Hung.*, 32:163-173, 1978.
14. N. Johnson and S. Kotz. *Urn Models and Their Applications*, John Wiley and Sons, New York, 1977.
15. J.M. Lane and R.F. Riesenfeld. A theoretical development for the computer generation and display of piecewise polynomial surfaces, *IEEE Trans. on Pattern Anal. and Mach. Intell.*, PAMI-2, pp. 35-46, 1980.
16. M.J. Marsden. An identity for spline functions with applications to variation-diminishing spline approximation, *J. Approx. Theory*, 3:7-49, 1970.
17. G. Morin and R.N. Goldman. A subdivision scheme for Poisson curves and surfaces, submitted to *Computer Aided Geometric Design*, 2000.
18. G. Morin and R.N. Goldman. Trimming for analytic functions using right sided Poisson subdivision, in preparation, 2000.
19. L. Ramshaw. *Blossoming: A Connect the Dots Approach to Splines*, Digital Systems Research Center, Report 19, Palo Alto, California, 1987.
20. L. Ramshaw. Béziers and B-splines as multiaffine maps, in: *Theoretical Foundations of Computer Graphics and CAD*, R.A. Earnshaw, ed., NATO ASI Series F, 40:757-776, Springer-Verlag, New York, 1988.
21. L. Ramshaw. Blossoms are polar forms, *Computer Aided Geometric Design*, 6:323-358, 1989.
22. M.H. Stone. A generalized Weierstrass approximation theorem, MAA Studies in Mathematics, Vol. 1, *Studies in Modern Analysis*, R.C. Buck, Editor, Prentice Hall, Inc., pp. 30-87, 1962.

Optimization of Knots for the Multi Curve B-Spline Approximation

Toni Prahasto,
Dept. of Mech. Eng.
Univ. of Diponegoro
Semarang, Indonesia
srondol@idola.net.id

Sanjeev Bedi
Dept. of Mech. Eng
Univ. of Waterloo
Waterloo, Canada
sbedi@surya.uwaterloo.ca

Abstract

This article presents a method for multi curve approximation with B-spline. The approximation is formulated as a constrained optimization problem with the least squares error as the objective function and the knot vector as the variables. The method presented in this article is designed to eliminate the well-known lethargic behaviour and to maintain the accuracy of control points. The method for fitting skeletal curves was tested on a single-section case and a multi-section case. The tests showed that the method reduces the objective function significantly. The proposed method has potential applications in the shape design of wings, turbine blades, and automotive body panels.

Nomenclature

$\mathbf{p}$	raw sample points
$\mathbf{e}$	vector of Euclidean distance between the raw sample points and the approximation points
N_i^k	the i-th B-spline basis function of order k
$\mathbf{R}$	matrix of B-spline basis function
$\mathbf{R}^+$	Moore-Penrose or pseudo inverse of matrix $\mathbf{R}$
$\mathbf{t}$	vector of parameterization of raw sample points
$\mathbf{c}$	control points of B-spline curve
$\mathbf{u}$	knot vector
$\mathbf{g}$	equations of constraints

1. Introduction

This work was motivated by a need to develop a set of skeletal curves that are ready to be skinned into a surface without needing the knot merging process. Figure 1(a) shows an example of a set of raw sample points. These points are subsequently fitted with a set of curves shown in Figure 1(b). The proposed method discussed in this article forces the curves to satisfy compatibility condition. This condition refers to a sharing of a mutual degree and a mutual knot vector. Compatibility condition eliminates the need of the process of knot-merging prior to skinning. Figure 1(c) shows the surface constructed by skinning the skeletal curves in Figure 1(b). The goal of eliminating the knot merging process prior to skinning is to reduce the number of control points of the skinned surface.

The existing approaches to this problem ignore the compatibility requirement during the approximation process. Since compatibility is realized by knot merging, the number of control points of the compatible curves may be excessive. This disadvantage will be carried over to the skinned surface that is constructed over the curves.

The method presented in this work approaches the problem with a different scheme. Compatibility is enforced by constraining the curves to share a mutual degree and a mutual knot vector. This constraint is then applied within the nonlinear least squares problem to generate a set of compatible curves. The resulting curves are compatible and thus readily available for skinning.

Golub's works show that the least-squares er-

0-7695-0562-7/00 $10.00 © 2000 IEEE

rors of an individual curve can be expressed as a nonlinear function of knots[5, 4]. For multicurve simultaneous approximation, the sum of errors for all curves is selected as the objective function, and it is a nonlinear function of mutual knots. The solution of this problem of nonlinear least squares is obtained using an optimization scheme. This scheme must consider linear constraints due to the monotonically increasing property of the knots. The goal of solving this nonlinear least squares problem is to reduce the least squares error simultaneously in all the skeleton curves.

Reduction of error will be assessed by investigating the optimization proposed in this work. This investigation served two objectives. The first objective was to measure the usefulness, in an engineering sense, of using the knots as adjustable parameters. The usefulness was measured by the decrease of least squares error. The second objective was to measure the effectiveness of incorporation of the nonlinear constraint in maintaining the accuracy and good behavior of the control points.

2. Basics of Least Squares Approximation

In this article, the disagreement between the data and the approximation curve is measured using a two-norm measure. The two-norm measure is selected for three reasons: first to allow elimination of linear parameters (the control points for B-spline curves) from the parameter set, second to simplify the expression for first derivative of the measure of disagreement with respect to its parameters, and third to provide the continuity of the first derivative of measure of disagreement. The use of two-norm yields three fundamental expressions:

$$\begin{array}{lll} \text{distance} & : & \mathbf{e} = \mathbf{p} - [\mathbf{R}]\,[\mathbf{c}] \\ \text{objective function} & : & \varepsilon = \mathbf{e}^T\mathbf{e} \\ 1^{st}\text{ derivative} & : & \dfrac{\partial \varepsilon}{\partial u} = 2\mathbf{e}^T\dfrac{\partial \mathbf{e}}{\partial u} \end{array}$$

The vector $[\mathbf{R}]\,[\mathbf{c}]$ is the B-spline curve approximating the raw sample points $\mathbf{p}$. Matrix $\mathbf{R}$ consists of the B-spline basis functions N_i^k. The number of rows of $\mathbf{R}$ is the same as the number of raw sample points. A row of $\mathbf{R}$ consists of B-spline basis functions evaluated at a parameter value that is assigned to the corresponding raw sample point. In this work, the parameters are computed using the arc-length parameterization method; this method can be found in many textbooks[9]. The vector $\mathbf{c}$ is the control points of the B-spline curve. The objective of curve approximation is to minimize the measure of disagreement, i.e. ε. The equation above clearly shows that minimization of ε can be performed by adjusting the basis function in $\mathbf{R}$ and the control points $\mathbf{c}$.

Approximation with parametric B-spline curve requires the following set of parameters: degree of basis, number of basis functions, knots, coordinates of control points, and parameterization for data. The first two parameters are discrete, and their values are almost always specified by the users instead of being computed. This work assumes that the degree and the number of B-spline basis functions are already specified. They are held constant during the proses of minimizing the ε.

Once the degree and the number of basis functions have been specified, the next step is to compute the solution, that is to solve for knots, coordinates of control points, and parameterization of raw sample points, that yields a minimum ε. Prahasto reported that adjustment of the parameterization of raw sample points contributes relatively insignificant reduction of ε[10]. Adjusting of the knots and the control points, on the other hand, is capable of reducing the ε more significantly.

Based on the result by Prahasto, this work discards the parameterization of raw sample points from the set of adjustable parameters of this least-squares problem, leaving the knots $\mathbf{u}$ and the control points $\mathbf{c}$ as the parameters of the least squares problem. The formulation of the problem thus becomes:

$$\min_{\mathbf{u},\mathbf{c}} \varepsilon = (\mathbf{p} - [\mathbf{R}]\,[\mathbf{c}])^T\,(\mathbf{p} - [\mathbf{R}]\,[\mathbf{c}]) \qquad (1)$$

where the matrix of B-spline basis function $\mathbf{R}$ is a function of the elements of the knot vector $\mathbf{u}$, i.e. $\mathbf{R} = \mathbf{R}\,(\mathbf{u})$.

Golub[4, 5] shows that this least square problem can be simplified by eliminating the linear parameter $\mathbf{c}$ from the domain of the parameters so that the actual domain of the problem is reduced to the domain of nonlinear parameters only. Therefore, adjustment of parameters is performed only on the nonlinear parameters $\mathbf{u}$.

The process of elimination of control points starts with the formulation of the explicit expression of control points. For a given matrix $\mathbf{R}$, the control points can be explicitly written as a function of the data and the pseudo inverse of matrix $\mathbf{R}$. The expression of the control point then becomes $\mathbf{c} = \mathbf{R}^{+}\mathbf{p}$ where $\mathbf{R}^{+}$ denotes the Moore-Penrose of the pseudo inverse of matrix $\mathbf{R}$. By substituting $\mathbf{c} = \mathbf{R}^{+}\mathbf{p}$ into the objective function, the problem can be stated as

$$\min_{\mathbf{u}} \varepsilon = \left(\mathbf{p} - \mathbf{R}\mathbf{R}^{+}\mathbf{p}\right)^{T}\left(\mathbf{p} - \mathbf{R}\mathbf{R}^{+}\mathbf{p}\right) \qquad (2)$$

Eqn. (2) shows that the control points have been eliminated from the expression of measure of disagreement. The proof that the minimum of the functional in Eqn. (2) is exactly identical to the minimum of the original functional can be found in Golub's articles [4, 5]; the principal requirement of this proof is that the measure of disagreement must use the two-norm.

The discussion above has indicated clearly that B-spline least squares is a nonlinear least squares problem. Solution of this problem is obtained by an optimization method. Selection of an optimization method depends on the continuity of objective function, availability of gradient, and its continuity. Golub has proven that the objective function of this nonlinear least squares problem and the gradient are continuous if matrix $\mathbf{R}$ has a constant rank within the domain of the parameters[4, 5]. The first derivative of the objective function with respect to a knot was derived by Golub and its expression is

$$\frac{\partial \varepsilon}{\partial u_i} = -2\mathbf{p}^{T}\left(\mathbf{I} - \mathbf{R}\mathbf{R}^{+}\right)\frac{\partial \mathbf{R}}{\partial u_i}\mathbf{R}^{+}\mathbf{p} \qquad (3)$$

where $\mathbf{I}$ is the identity matrix.

Therefore, the proper expressions for the objective function and its gradient have been obtained. These equations are the fundamental equations of the B-spline nonlinear least squares problem.

For the multi curve problem, the objective function is defined as the sum of the least squares errors of all approximating curves, i.e.

$$\varepsilon = \min_{\mathbf{u}} \sum_i \left(\mathbf{p}_i - \mathbf{R}_i\mathbf{R}_i^{+}\mathbf{p}_i\right)^{T}\left(\mathbf{p}_i - \mathbf{R}_i\mathbf{R}_i^{+}\mathbf{p}_i\mathbf{p}_i\right)$$

where $\mathbf{p}_i$ is the set of raw sample points representing the i-th curve, $\mathbf{R}_i$ is the matrix of B-spline basis functions defining the i-th curve, and $\mathbf{R}_i^{+}$ is the pseudo inverse of $\mathbf{R}_i$. The compatibility among the curves is obtained by constraining the $\mathbf{R}_i$'s to share a mutual knot vector $\mathbf{u}$ and a mutual degree. The first derivative of the objective function can be written as

$$\frac{\partial \varepsilon}{\partial u_i} = \sum_i -2\mathbf{p}_i^{T}\left(\mathbf{I} - \mathbf{R}_i\mathbf{R}_i^{+}\right)\frac{\partial \mathbf{R}_i}{\partial u_i}\mathbf{R}_i^{+}\mathbf{p}_i$$

The requirement that the elements of the knot vector must be non decreasing makes causes the incorporation of a set of non linear constraints over the knot vector. The definition of the constraints can be written as

$$\mathbf{g}_1 \equiv u_{i+1} - u_i \geq 0 \text{ for all } i$$

The use of finite-precision computers to solve the problem introduces a requirement of a non linear constraint that maintain the condition number of $\mathbf{R}$ to fall within a feasible range of values, i.e. its reciprocal must be significantly greater than the machine precisions of the computers. Violation of this constraint will result in a control points that have wildly large component[10]. This non linear constraint can be stated as

$$\mathbf{g}_2 \equiv \text{cond. number of } \mathbf{R_i} \leq \frac{1}{\delta} \text{ for all } i$$

where δ is a positive number significantly larger than the machine precision. In this work δ is set to 2×10^{-3} which is much greater than the machine precision of 10^{-6}.

3. Literature Survey

Jupp suggests that the optimization of knots is troubled by a lethargic phenomenon where an active constraint in $\mathbf{g}_1$ cannot be released to become inactive[6, 7, 8]. This causes poor convergence during the optimization. Jupp's suggestion is a consequence of the definition of the first derivative of the B-spline basis functions with respect to a knot. When two or more knots coincide, the gradient of the N_i^k with respect to those coincident knots is a scalar multiple of the active constraints in the $\mathbf{g}_1$. Based on this fact, Jupp concludes that the descent direction lies in the null space of the active $\mathbf{g}_1$, preventing the possibility of making these active $\mathbf{g}_1$ to become inactive. Thus, the subsequent iteration of the optimization will be trapped in the active $\mathbf{g}_1$.

In the author's opinion, however, this can only occurs if and only if the descent direction is the steepest descent direction. This direction is obtained from the equation $[\mathbf{I}][\mathbf{d}] = -\nabla\varepsilon$ where $\mathbf{d}$ is the descent direction. In this way, the descent direction $\mathbf{d}$ is the opposite of the gradient $\nabla\varepsilon$, and the lethargic phenomenon suggested by Jupp will certainly occurs.

If the identity $\mathbf{I}$ is replaced with any other matrix that is not a scalar multiple of the identity, $\mathbf{d}$ will not be a scalar multiple of $\nabla\varepsilon$. Therefore, $\mathbf{d}$ will have a component orthogonal to the $\nabla\varepsilon$. This component, in the author's opinion, allows the deactivation of the active constraints in $\mathbf{g}_1$[10]. Therefore, adjusting the knots to minimize the ε no longer concerns about the lethargic phenomenon that is suggested by Jupp.

Another work that adjusted the knots for this problem is done by Gengoux and Mekhilef[2]. They investigated nonlinear least squares with NURBS curves, in particular the optimization of knots and weights. They uses a penalty method to handle the constraints. The penalty terms represent linear constraints on the weights and knots and nonlinear constraint on the condition number of $\mathbf{R}$. The linear constraints are represented by $\mathbf{g}(\mathbf{u}, \mathbf{w}, \mathbf{t}) \leq \mathbf{0}$. The penalty term for these constraints is

$$\Phi_1 = \sum_i \alpha_i \sup(g_i, 0)^r \tag{4}$$

The power r is either 2 or 3; $r = 2$ yields once differentiable Φ whereas $r = 3$ yields twice differentiable Φ. This penalty terms are weighted by factor α_i. The penalty term representing the nonlinear constraint on the condition number of $\mathbf{R}$ is formulated indirectly as a function of control points. The term is defined as the sum of distances of two consecutive control points; the formulation of this term is

$$\Phi_2 = \beta \sum_i^{n-1} ||\mathbf{c}_{i+1} - \mathbf{c}_i||_2^2 \tag{5}$$

When the matrix $\mathbf{R}$ is ill-conditioned, the control points have wildly large components. These components will, in turn, raise the penalty term Φ_2. Therefore, Φ_2 functions as a signal and prevents the ill-conditioning of $\mathbf{R}$. However, Gengoux does not mention the analytical gradient of Φ_2 with respect to knots, weights, and parameterization of raw sample points.

In the author's opinion, availability of this gradient is crucial in the optimization with respect to the nonlinear parameters. It appears that Φ_2 does not function as it is supposed to. Gengoux's report clearly shows that whenever weights are set as the only adjustable parameters, one or more control points have very large components. This phenomena is surely caused by ill-conditioning of $\mathbf{R}$ despite the incorporation of Φ_2. It is also unclear why Gengoux concludes that the main cause of explosion of the coordinates of the control points is caused by insufficiency of number of basis.

In the author's opinion, the ineffectiveness of the method proposed by Gengoux and Mekhilef is caused by their way to prevent the ill-conditioning of $\mathbf{R}$ in an indirect way. Φ_2 appears to be ineffective in maintaining the condition number of $\mathbf{R}$ under certain feasible maximum allowable value.

4. Strategy of Optimization

The success of the optimization depends on the method to compute the descent direction and the handling of the constraints $\mathbf{g}_1$ and $\mathbf{g}_2$.

Since the objective function of least squares fitting is smooth and its analytical derivative is available, a modified BFGS descent direction is used. The modification involves a procedure to get away from the active $\mathbf{g}_1$ (coincident knots) whenever such descent direction is available. A restart procedure is also performed to handle the situation where Hessian becomes near-singular or nonpositive definite due to buildup of roundoff errors.

For the handling of constraints, a new method separating the linear and the nonlinear constraints is employed. The separation is performed to reduce the cost of computation. The new method is based on the active-set method to satisfy the linear constraints $\mathbf{g}_1$. The satisfication of the non linear constraint $\mathbf{g}_2$ is performed during the line minimization algorithm by reducing the maximum step length if a violation of $\mathbf{g}_2$ occurs. This is possible since the region where $\mathbf{g}_2$ is violated is concentrated at or near the linear constraint $\mathbf{g}_1$. The following illustration will demonstrate the concentration of the infeasible $\mathbf{g}_2$ in the feasible domain.

Matrix $\mathbf{R}$ may become ill-conditioned whenever the data supporting the basis functions are concentrated near the ends of the support of the basis functions. Figure 2 is used to explain how the concentration of data can lead to ill- conditioning of $\mathbf{R}$. The curve represents the $N_5^4(t)$ cubic B-spline basis function that is supported in the interval $[u_5, u_9)$ where $u_5 = 1$, $u_6 = 2$, $u_7 = 3$, $u_8 = 4$, and $u_9 = 5$. The parameter values whose $N_5^4(t) \neq 0$ are $t_{11} = 2.3$ and $t_{12} = 3.6$. Because these data are the only support for the $N_5^4(t)$ B-spline basis function, the 11-th and the 12-th rows of $\mathbf{R}$ are those at which the $N_5^4(t)$ B-spline basis function are nonzero; the $N_5^4(t)$ for other rows are zero. To observe the variation of the condition number of $\mathbf{R}$ with respect to the change of the knots, the u_8 and u_9 will be varied between $3 \leq u_8 \leq 5$ and $3 \leq u_9 \leq 5$ subject to the constraint of non decreasing criterion, i.e. $u_8 \leq u_9$.

Figure 3 shows the triangular area defined by $(u_8, u_9) \in [3, 5)$ subject to $u_8 \leq u_9$. The white subarea represents combinations of $(u_8, u_9) \in [3, 5)$ subject to $u_8 \leq u_9$ that yield $\text{cond}(\mathbf{R}) \leq 500$ whereas the shaded subarea represents conbination of $(u_8, u_9) \in [3, 5)$ subject to $u_8 \leq u_9$ that yield $\text{cond}(\mathbf{R}) > 500$.

This illustration demonstrates the existence of unfavorable regions within which the condition number of matrix $\mathbf{R}$ is large. In this unfavorable region, the solution of $\mathbf{Rc} = \mathbf{p}$ may suffer inaccuracy or may be nonunique. This illustration also shows that the location(s) of this unfavorable regions are close to the linear constraints, leaving the interior favorable.

This characteristic of the location of the region unfavorable with respect to $\mathbf{g}_2$ forms the justification to separate the method of satisfying the linear constraints from that of the nonlinear constraint. The separation is performed by, firstly, computing the critical step length that maintains the feasibility with respect to the linear constraint, and secondly reducing this critical step if necessary in order to maintain the feasibility with respect to the nonlinear constraint.

In Figure 4, the triangle represents the feasible domain with respect to the linear constraints. The white subarea in the triangle represents the feasible domain with respect to the nonlinear constraint, whereas the shaded subarea represents the infeasible domain with respect to the nonlinear constraint. The solid circle represent the position at the current iteration; this position satisfies both the linear and nonlinear constraints. The vectors $\mathbf{d}_1$ and $\mathbf{d}_2$ represent two possible descent directions. The figure also shows two critical steps γ_1 and γ_2 that are associated with $\mathbf{d}_1$ and $\mathbf{d}_2$ respectively.

Line minimization along the descent $\mathbf{d}_1$ can be performed at full step γ_1 since the nonlinear constraint will be satisfied at this step. In this work, the condition number of matrix $\mathbf{R}$ is computed at the full step γ.

On the other hand, line minimization along the descent $\mathbf{d}_2$ cannot be performed at full step γ_2 because violation of the nonlinear constraint will occur. To maintain the satisfaction of the linear constraint, a reduced critical step, denoted by γ_2^* in Figure 4, is used instead of the γ_2. The method of computing the reduced critical step is presented next.

Ideally, the procedure to reduce γ must be developed with a consideration to conform to the behavior of the condition number of $\mathbf{R}$ with respect to γ. However, other than just monotonically increasing with respect to increasing step length, the behavior of the condition number of $\mathbf{R}$ is difficult to be described accurately. Thus, the simplest method of reducing the maximum feasible step is by reducing it with a constant factor. The crucial step of the development of such a method is to find the factor of reduction that produces the best compromise between adequate length of maximum feasible step and fast convergence of finding the step that yields feasible condition number of $\mathbf{R}$. If the factor is too big, fast convergence is obtained at the expense of excessively short maximum step length, which in turn will reduce the rate of convergence of the overall optimization. On the other hand, an overly small factor will reduce the convergence of finding feasible maximum step length. From trial and error experiment, a reduction factor of 20% yields the best compromise.

With the scheme of constant-factor reduction and the empirical reduction factor of 20%, the mechanism to maintain the feasibility of the condition number of $\mathbf{R}$ can be setup to reduce a given γ by 20% if the γ yields infeasible condition number of $\mathbf{R}$. If the reduced γ yields a feasible condition number, it is accepted. Otherwise, it is further reduced by 20% and the process is repeated until the condition number at γ is satisfied. Work by Prahasto showed that the incorporation of maximum allowable condition number of $\mathbf{R}$ into the determination of maximum feasible step proved to be effective to prevent inaccuracy of the solution of $\mathbf{Rc} = \mathbf{p}$[10].

Armed with the BFGS descent direction, the standard LIP method, and the new method of reducing the maximum step length for line minimization in order to satisfy the $\mathbf{g}_2$, the proposed method of multi curve approximation is subjected to a pair of test cases.

5. Implementation

The main purposes of the implementation are to measure the performance of the proposed optimization method and to assess the proposed nonlinear constraint in providing accurate control points.

The proposed test case was fitted with the method described in the previous section . The termination of optimization was based on Lagrange multipliers and the standard termination criteria as suggested in[1, 11]. Arclength parameterization was used to initialize the parameterization for data. Knots were initialized using the averaging method[9]. Maximum permissible condition number for $\mathbf{R}$ was set to 500 based on a consideration that its reciprocal is still much larger than the machine's precision (10^{-6}). Tolerances of activity for the linear constraints were set to 10^{-2} for knot.

5.1. NACA 2415 Airfoil

Figure 5 shows the NACA 2415 airfoil with a chord length of 12 mm. The airfoil consist of 97 data. The degree of approximation curve is set to cubic and the number of basis is set to 12.

When the knots were optimized to fit the NACA 2415 data set with a B-spline curve, the total least squares error, i.e. $\mathbf{e}_x^T\mathbf{e}_x + \mathbf{e}_y^T\mathbf{e}_y$, decreases from 0.186 [mm^2] to 0.002 [mm^2] in 51 iterations; Figure 6 shows the decrease of error with respect to number of iteration. It is clear that the decrease of error beyond 10-th iteration is marginal (insignificant). The process of knot adjustment stops after 1225 (not shown in Figure 6 due to insignificance decrease of error) iteration due to gradient tolerance. Due to marginal decrease of error beyond 10-th iteration, the following discussion is focused on the first 10 iteration of this knot adjustment. Figure 7 shows the distributions of error along the airfoil before and after knot adjustment; solid lines represent the distribution before adjustment and circles represent the distribution after adjustment. This figure shows that, in general, the disagreement between the data and the approximation curves drops after the knot is adjusted. The maximum errors before and after knot adjustment are about 0.191 millimeters, and 0.015 millimeters respectively. Figure 8 shows the knot distributions, along the curves, before and after knot adjustment. The multiplicity of knots at the leading edge is 2, i.e. $u_8 = u_9$. Figure 9 shows the control polygons of the approximation curves before and after knot adjustment. This figure illustrates the good behavior of control points. This is a result of constraining the condition number of $\mathbf{R}$.

5.2. Compressor Blades

Figure 10 shows three airfoils defining the blade of a generic axial compressor in a four views. After the optimization, the sum of least-square error decreased from 27.5 [mm^2] to 1.53 [mm^2] in 51 iteration as shown in Figure 11. The largest distances between the data and the approximation curves decreased from 1.16 millimeters to 0.359 millimeters for hub airfoil, from 1.25 millimeters to 0.229 millimeters for mid section airfoil, and from 1.19 millimeters to 0.268 millimeters for tip airfoil as shown in Figure 12, 13, and 14. The shape of the airfoils, along with the distribution of knots, are shown in Figure 15, 16, and 17. All constraints in $\mathbf{g}_1$ are inactive.

The results of these tests clearly indicate that adjustment of knots significantly reduces the least-squares residue.This finding can be used to justify that knots should be adjusted. This is the opposite of Jupp's[6, 7, 8] and Sarkar's[13, 12] works respectively, in which adjustment of knots produced unfavorable reduction of least squares error. The favorable reduction of error in this research's knot adjustment can be attributed to the proposed method of elimination of active knot constraints as described in the previous section.

6. Conclusion

This work has presented a new method to construct a skeleton made of B-spline curves that are ready for the skinning by enforcing compatibility during the simultaneous least squares fitting of the curves. Careful optimization of knots has resulted in reduction of least squares error while maintaining the quality of control points obtained from $\mathbf{Rc} = \mathbf{p}$ and enforcing the compatibility among the curves.

Test cases have shown that a relatively good convergence is obtained by using the BFGS descent direction, the active set method to satisfy the $\mathbf{g}_1$, and the newly proposed reduced maximum step length to satisfy the $\mathbf{g}_2$. With these weaponry, a significant decrease of least squares error were obtained. The test cases also showed that the lethargic behavior can be overcome as suggested by the case of compressor blade in which all of the $\mathbf{g}_1$ are inactive.

Acknowledgement

This work was supported in part by the Islamic Development Bank of Saudi Arabia and by the University of Waterloo of Canada.

References

[1] J.E. Dennis. *Numerical Methods for Unconstrained Optimization and Nonlinear Equations.* SIAM, 1996.

[2] P.L. Gengoux and M. Mekhilef. Optimization of a NURBS representation. *Computer Aided Design*, 25(11):699–710, November 1993.

[3] P.E. Gill, W. Murray, and M.H. Wright. *Practical Optimization.* Academic Press, San Diego, California, 1981.

[4] G. Golub. Differentiation of pseudoinverses, separable nonlinear least squares problems and other tales. In M.Z. Nashed, editor, *Generalized Inverses and Applications : Proceedings of an Advanced Seminar*, pages 303–324. 1973.

[5] G.H. Golub and V. Pereyra. The differentiation of pseudo-inverse and nonlinear least square problems whose variables separate. *SIAM Journal of Numerical Analysis*, 10(2):413–432, April 1973.

[6] D.L.B. Jupp. Curve fitting by splines as an example of unconstrained optimization. In R.S. Anderssen, L.S. Jennings, and D.M. Ryan, editors, *Optimization*, pages 49–59. University of Queensland Press, 1972.

[7] D.L.B. Jupp. The lethargy theorem - a property of approximation by γ-polynomials. *Journal of Approximation Theory*, 14:204–217, 1975.

[8] D.L.B. Jupp. Approximation to data by splines with free knots. *SIAM Journal of Numerical Analysis*, 15(2):328–343, April 1978.

[9] L. Piegl and W. Tiller. *The NURBS Book.* Springer Verlag, 1997.

[10] Toni Prahasto. *Simultaneous Multi Curve Approximation with NURBS.* PhD thesis, University of Waterloo, Waterloo, Ontario, Canada, 1998.

[11] W.H Press and et. al. *Numerical Recipes in C: The Art of Scientific Computing.* Cambridge University Press, 1992.

[12] B. Sarkar and C.H. Menq. Parameter optimization in approximating curves and surfaces to measurement data. *Computer Aided Geometric Design*, 8:267–290, 1991.

[13] B. Sarkar and C.H. Menq. Smooth-surface approximation and reverse engineering. *Computer Aided Design*, 23(9):623–628, November 1991.

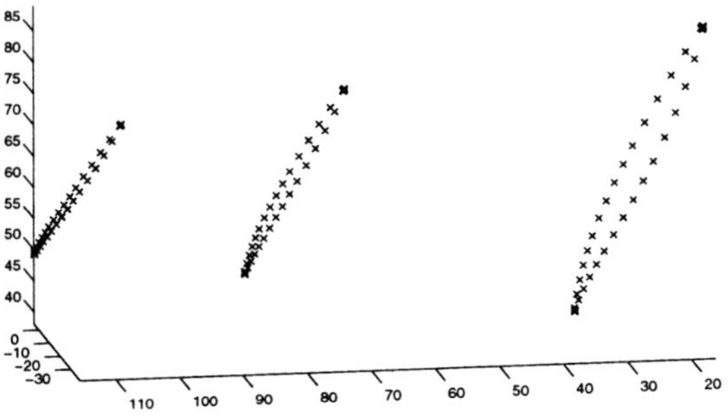

(a) data

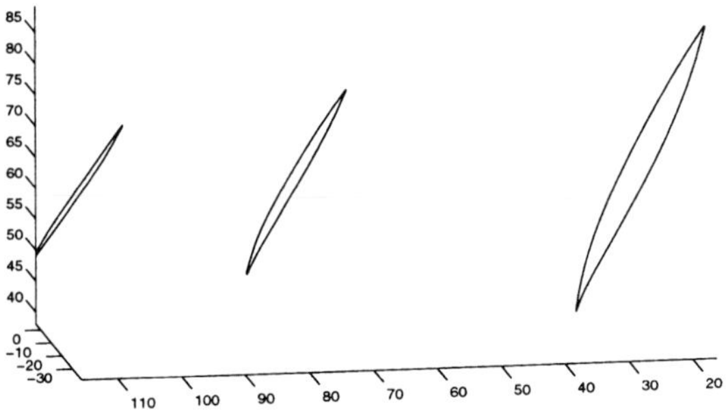

(b) fitted curves

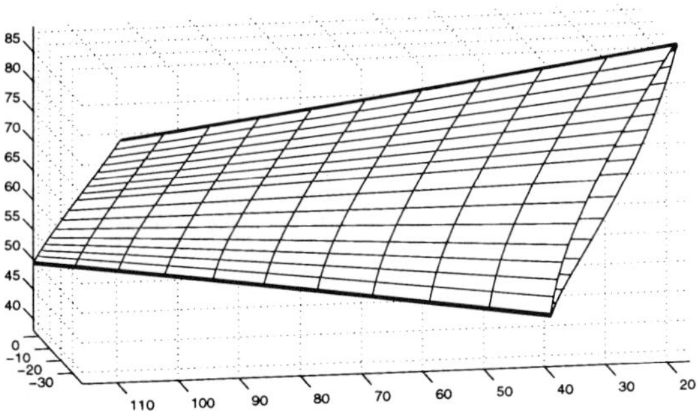

(c) skinned surface

Figure 1: Construction of Turbomachine Blades and Aircraft Wings: (a) section curves given as data, (b) approximation curve to the data, (c) skinned surface over the approximation curves

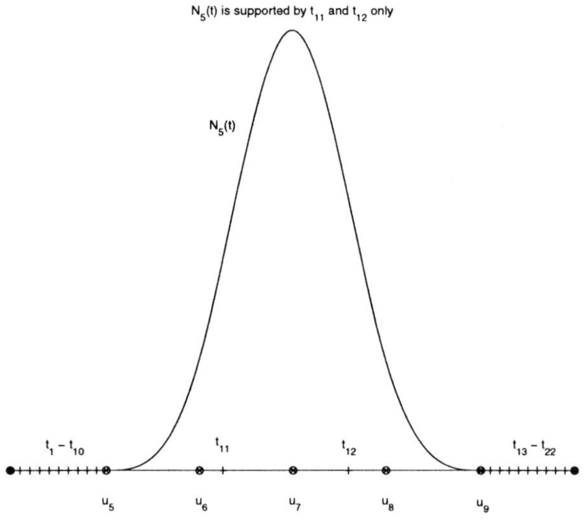

Figure 2: The B-spline Basis Function $N_5^4(t;\mathbf{u})$ and Its Supporting Data t_{11} and t_{12}

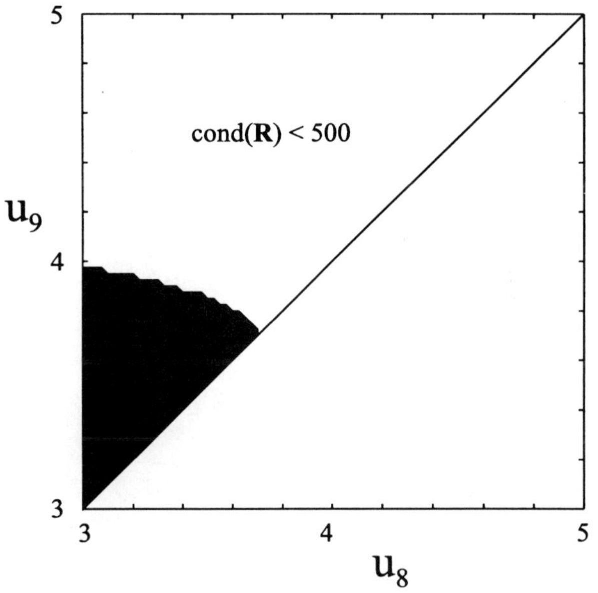

Figure 3: Triangular area defined by $(u_8, u_9) \in [3,5)$ subject to $u_8 \le u_9$. Shaded subareas represent (u_8, u_9) resulting in cond$(\mathbf{R}) > 500$. White subarea represent (t_{11}, t_{12}) resulting in cond$(\mathbf{R}) \le 500$.

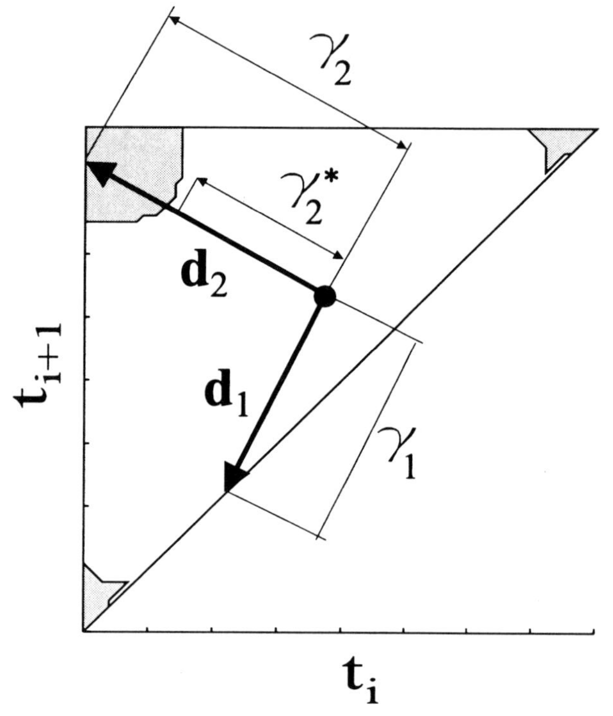

Figure 4: Reduction of Critical Step To Satisfy The Nonlinear Constraint

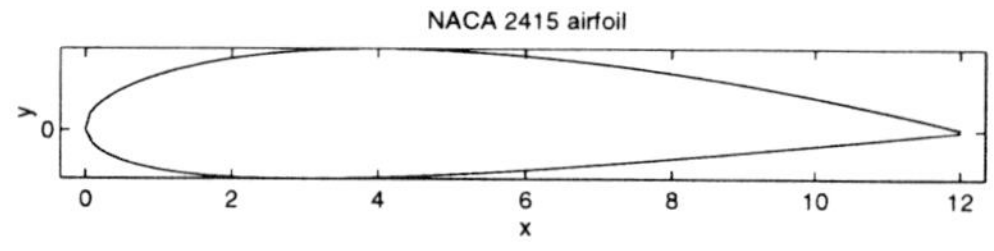

Figure 5: NACA 2415 with 12 mm chordlength

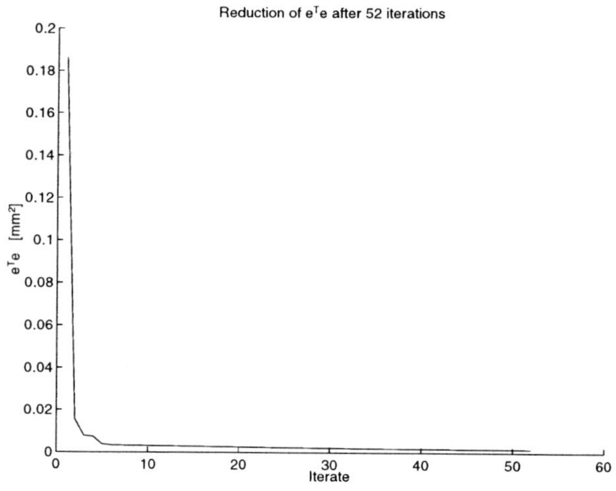

Figure 6: Reduction of least squares error of NACA airfoil vs. number of iteration obtained from optimizing the knots

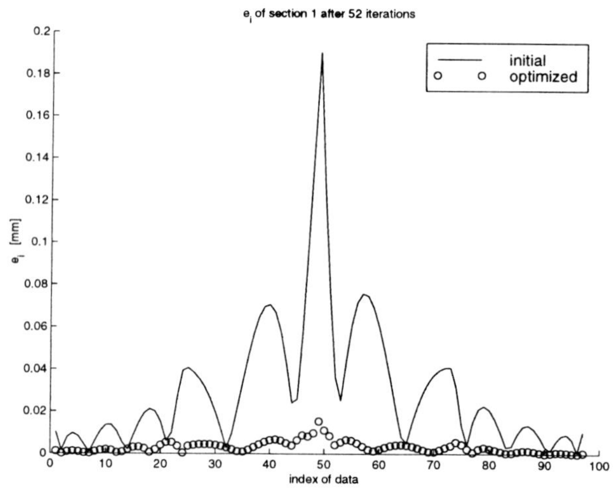

Figure 7: Error distributions of NACA airfoil before and after optimizing the knots

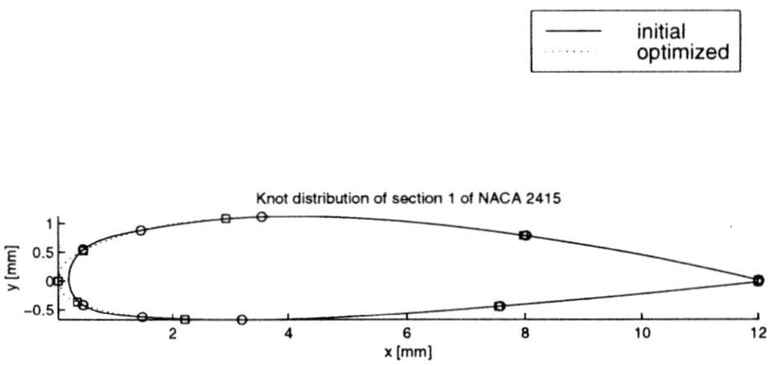

Figure 8: Distributions of Knots of NACA airfoil before and after optimizing the knots

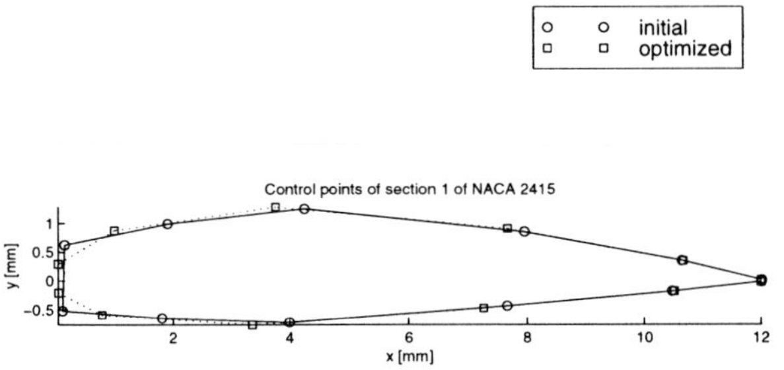

Figure 9: Control polygons of NACA airfoil before and after optimizing the knots

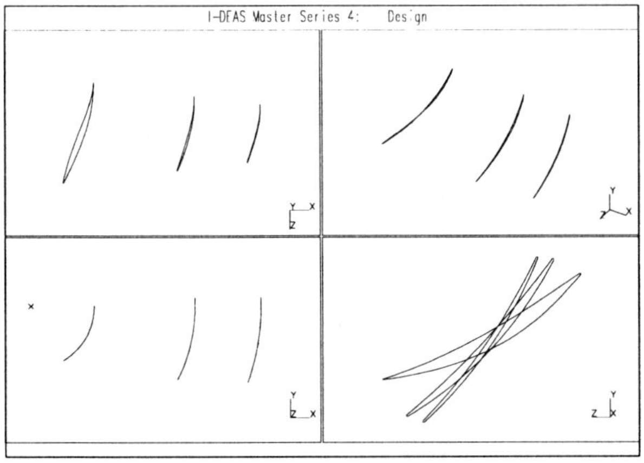

Figure 10: Data of Compressor

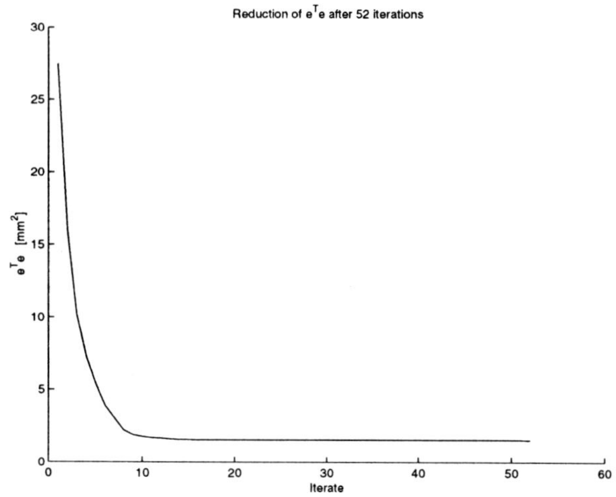

Figure 11: Reduction of error vs. number of iteration for three-curve compressor blade's airfoils obtained from optimizing the knots

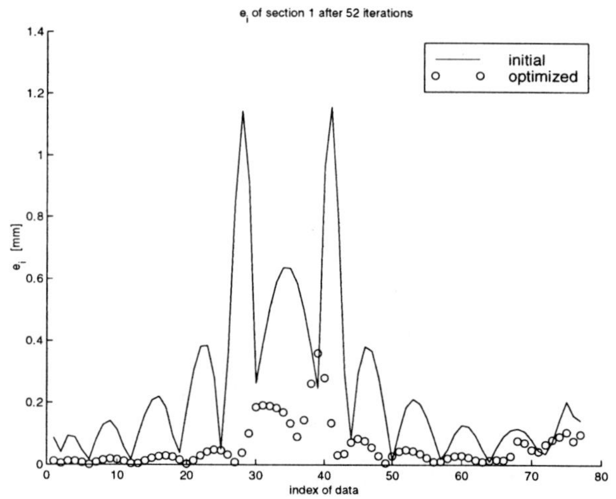

Figure 12: Error distribution of hub airfoil of three-curve compressor blade obtained from optimizing the knots.

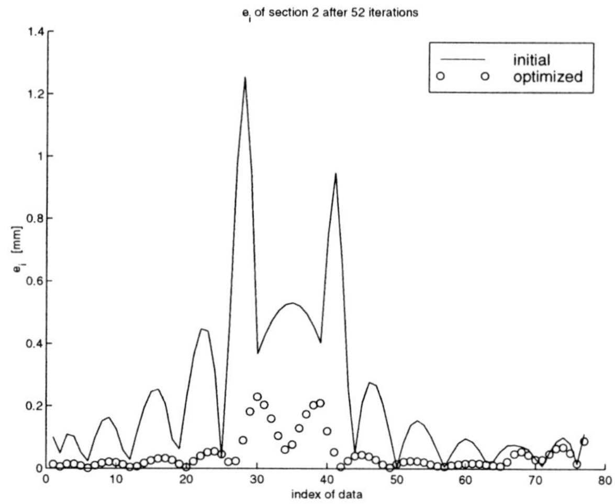

Figure 13: Error distribution of midsection airfoil of three-curve compressor blade obtained from optimizing the knots.

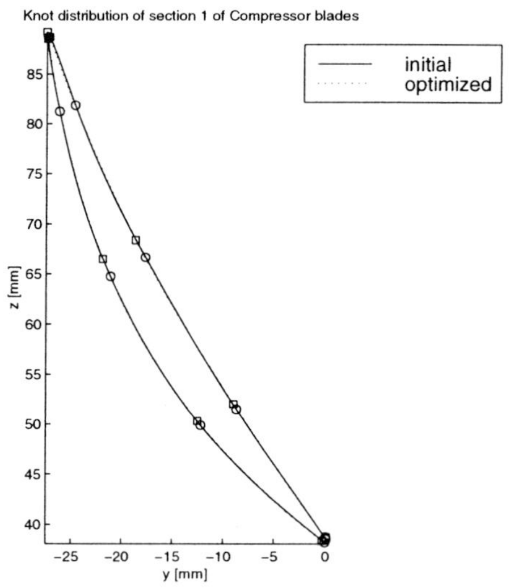

Figure 15: Fitted curves of hub airfoil of compressor blade obtained from optimizing knots.

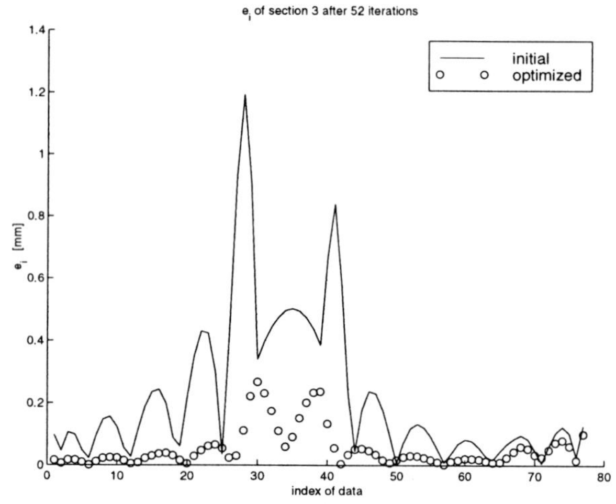

Figure 14: Error distribution of tip airfoil of three-curve compressor blade obtained from optimizing the knots.

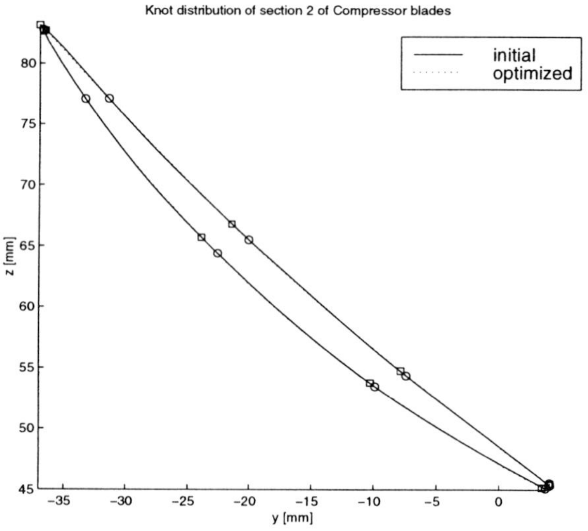

Figure 16: Fitted curves of mid section airfoil of compressor blade obtained from optimizing knots.

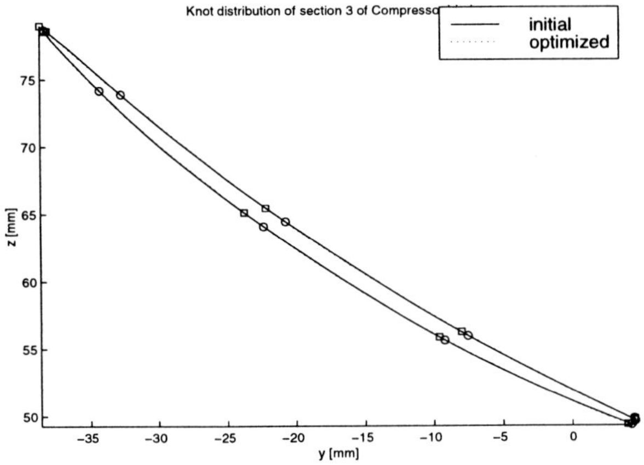

Figure 17: Fitted curves of tip airfoil of compressor blade obtained from optimizing knots.

Constrained C^1 Piecewise Bicubic Bézier Surface Interpolation

Caiming Zhang, Fuhua (Frank) Cheng and Brian A. *Barsky**
Graphics & Geometric Modeling Lab, Department of Computer Science
University of Kentucky, Lexington, KY 40506, USA
* *Department of Computer Science*
University of California, Berkeley, CA 94720, USA
E-mail: *czhang@cs.sdu.edu.en*

Abstract

The construction of a C^1 piecewise bicubic Bézier surface that interpolates a set of data points and lies on the same side of a given plane as the data points is presented. The main tasks in this construction process are the adjustment of the tangent vectors and the determination of the twist vectors at the data points. The tangent vectors are modified by decreasing their magnitudes. The twist vectors are first set by an energy minimization process and then filtered through a modification process to ensure the constructed surface lies on one side of the given plane. The technique for determining the twist vectors has an interpolation precision of cubic polynomials.

Keywords. parametric interpolation, polynomial curve/surface, energy, constraint.

1. Introduction

The problem studied in this paper is the construction of a C^1 piecewise bicubic surface $\boldsymbol{P}(u,v) = (x(u,v), y(u,v), z(u,v))$ that interpolates a given set of 3D data points $\boldsymbol{P}_{i,j}$, $i = 1, 2, \cdots, m$, $j = 1, 2, \cdots, n$, which satisfy the condition $\boldsymbol{P}_{i,j} \neq \boldsymbol{P}_{i+1,j}$ and $\boldsymbol{P}_{i,j} \neq \boldsymbol{P}_{i,j+1}$. The constructed surface is required to lie on one side of a given plane. This problem, called *constrained surface interpolation*, is a sub-area of a broader topic called *constrained shape design* in which issues such as shape ranges, convexity, and boundedness of curvature are addressed so that manufacturability issues can be incorporated early in the surface development process.

For 2D data points $\{(x_i, y_i)\}_{i=1}^{n}$, $(x_i, y_i) \neq (x_{i+1}, y_{i+1})$, the problem of constrained parametric curve interpolation has been addressed in [4][5]. In [4], a shape preserving parametric interpolant is constructed by putting together cubic curve segments and straight line segments, and requiring both a tangent direction and a curvature value to be specified at each interpolation point. In [5], the problem of constructing an interpolant to the given data points that also lies on one side of one or more given straight lines is considered. The interpolant required to lie on the same side of a line as the data points is constructed by decreasing the magnitudes of the tangent vectors at the corresponding points. This method works well for data points whose signs of convexity do not change. The ideas used in [4][5] are further developed in [6][7][8].

Recently a new approach [10] for the construction of a shape-preserving C^1 cubic spline curve that interpolates a set of data points and lies on one side of a given polygon is presented. The tangent vectors are formed by a weighted combination of two scaled chord vectors. The magnitude of each scaled chord vector is determined by requiring the constructed curve to have the desired shape and satisfy the given constraints. The curves constructed by these techniques have the shape suggested by the given data points and lie on the same side of a given polygon as the data points.

Based on the techniques of constrained C^1 curve interpolation, a method for constructing a C^1 piecewise bicubic Bézier surface that interpolates a set of data points and lies on the same side of a given plane as the data points is presented. The main tasks in this method are to modify the tangent vectors and to determine twist vector for each data point. To simplify the construction process, the modification of the tangent vectors and the determination of the twist vectors are handled separately. The tangent vectors are modified by decreasing their magnitudes. The twist vectors are first determined by an energy minimization process, and then filtered through a modification process to make the constructed surface lie on one side of a given plane. The technique for determining the twist vectors has an interpolation precision of cubic polynomials.

The remaining part of the paper is arranged as follows. A formal description of the problem is given in Section 2. The modification of the tangent vectors and the determination of the twists are discussed in Sections 3 and 4, respectively.

Modifying the twist vectors to make the constructed surface lie on one side of a given plane is discussed in Section 5. Concluding remarks are given in Section 6.

2. Problem formulation

Let $\boldsymbol{P}_{i,j} = (x_{i,j}, y_{i,j}, z_{i,j})$ with knots (u_i, v_j), $1 \le i \le m$, $1 \le j \le n$, be a set of data points satisfying the following conditions: (1) consecutive points are different, i.e., $\boldsymbol{P}_{i,j} \neq \boldsymbol{P}_{i+l,j+k}$, $l, k = 0, 1$, $l \neq k$; (2) all the data points lie on the same side of a given plane

$$L(x, y, z) = N_x x + N_y y + N_z z + N_c = 0. \tag{1}$$

Our goal is to construct a C^1 interpolant $\boldsymbol{P}(u,v) = (x(u,v), y(u,v), z(u,v))$ to the data points, with the following properties:

1. on each domain patch $[u_i,\ u_{i+1}] \times [v_j, v_{j+1}]$, $\boldsymbol{P}(u,v)$ is a bicubic Bézier patch;
2. $\boldsymbol{P}(u,v)$ lies on the same side of the given plane as the data points.

Let $\boldsymbol{P}^u_{i,j}$ and $\boldsymbol{P}^v_{i,j}$ denote the tangent vectors of $\boldsymbol{P}(u,v)$ at (u_i, v_j) along the u and v directions, respectively. The values of $\boldsymbol{P}^u_{i,j}$ and $\boldsymbol{P}^v_{i,j}$ can be computed using the three-point (Bessel) formula [1] or the technique proposed in [10]. If one uses $\boldsymbol{P}_{i,j}(u,v) = (x_{i,j}(u,v), y_{i,j}(u,v), z_{i,j}(u,v))$ to denote $\boldsymbol{P}(u,v)$ on the domain patch $[u_i, u_{i+1}] \times [v_j, v_{j+1}]$, $i = 1, 2, \cdots, m-1$, $j = 1, 2, \cdots, n-1$, then $\boldsymbol{P}_{i,j}(u,v)$ can be written as

$$\boldsymbol{P}_{i,j}(u,v) = \boldsymbol{U}\boldsymbol{B}\boldsymbol{V}^T \tag{2}$$

where

$$\begin{aligned} \boldsymbol{U} &= [B_0(s), B_1(s), B_2(s), B_3(s)], \\ \boldsymbol{V} &= [B_0(t), B_1(t), B_2(t), B_3(t)] \end{aligned} \tag{3}$$

$$\boldsymbol{B} = \begin{bmatrix} \boldsymbol{P}_{i,j} & \boldsymbol{P}_{i,j} + \frac{\delta^v_j}{3}\boldsymbol{P}^v_{i,j} & \boldsymbol{P}_{i,j+1} - \frac{\delta^v_j}{3}\boldsymbol{P}^v_{i,j+1} & \boldsymbol{P}_{i,j+1} \\ \boldsymbol{P}_{i,j} + \frac{\delta^u_i}{3}\boldsymbol{P}^u_{i,j} & \boldsymbol{P}_{i,j} + \delta^u_i\delta^v_j\boldsymbol{P}^{00}_{i,j} & \boldsymbol{P}_{i,j+1} - \delta^u_i\delta^v_j\boldsymbol{P}^{01}_{i,j+1} & \boldsymbol{P}_{i,j+1} + \frac{\delta^u_i}{3}\boldsymbol{P}^u_{i,j+1} \\ \boldsymbol{P}_{i+1,j} - \frac{\delta^u_i}{3}\boldsymbol{P}^u_{i+1,j} & \boldsymbol{P}_{i+1,j} - \delta^u_i\delta^v_j\boldsymbol{P}^{10}_{i+1,j} & \boldsymbol{P}_{i+1,j+1} + \delta^u_i\delta^v_j\boldsymbol{P}^{11}_{i+1,j+1} & \boldsymbol{P}_{i+1,j+1} - \frac{\delta^u_i}{3}\boldsymbol{P}^u_{i+1,j+1} \\ \boldsymbol{P}_{i+1,j} & \boldsymbol{P}_{i+1,j} + \frac{\delta^v_j}{3}\boldsymbol{P}^v_{i+1,j} & \boldsymbol{P}_{i+1,j+1} - \frac{\delta^v_j}{3}\boldsymbol{P}^v_{i+1,j+1} & \boldsymbol{P}_{i+1,j+1} \end{bmatrix}$$

with $B_i(s)$, $i = 0, 1, 2, 3$, being the cubic Bézier basis functions on $[0, 1]$,

$$\begin{aligned} B_0(s) &= (1-s)^3, & B_1(s) &= 3(1-s)^2 s, \\ B_2(s) &= 3(1-s)s^2, & B_3(s) &= s^3, \end{aligned}$$

and

$$\begin{aligned} s &= (u - u_i)/\delta^u_i, & \delta^u_i &= u_{i+1} - u_i, \\ t &= (v - v_i)/\delta^v_j, & \delta^v_j &= v_{j+1} - v_j \end{aligned}$$

and $\boldsymbol{P}^{00}_{i,j}$, $\boldsymbol{P}^{01}_{i,j+1}$, $\boldsymbol{P}^{10}_{i+1,j}$ and $\boldsymbol{P}^{11}_{i+1,j+1}$ are unknowns to be determined.

If one uses $\boldsymbol{P}^{uv}_{i,j}$ to denote the twist vectors of $\boldsymbol{P}(u,v)$ at (u_i, v_j), then one has the following relationships between the twist vectors and the unknowns:

$$\begin{aligned} \boldsymbol{P}^{uv}_{i,j} &= -3\boldsymbol{P}^u_{i,j}/\delta^v_j - 3\boldsymbol{P}^v_{i,j}/\delta^u_i + 9\boldsymbol{P}^{00}_{i,j}, \\ \boldsymbol{P}^{uv}_{i+1,j} &= -3\boldsymbol{P}^u_{i+1,j}/\delta^v_j + 3\boldsymbol{P}^v_{i+1,j}/\delta^u_i + 9\boldsymbol{P}^{10}_{i+1,j}, \\ \boldsymbol{P}^{uv}_{i+1,j+1} &= 3\boldsymbol{P}^u_{i+1,j+1}/\delta^v_j + 3\boldsymbol{P}^v_{i+1,j+1}/\delta^u_i + 9\boldsymbol{P}^{11}_{i+1,j+1}, \\ \boldsymbol{P}^{uv}_{i,j+1} &= 3\boldsymbol{P}^u_{i,j+1}/\delta^v_j - 3\boldsymbol{P}^v_{i,j+1}/\delta^u_i + 9\boldsymbol{P}^{01}_{i,j}. \end{aligned}$$

Thus, determining $\boldsymbol{P}^{00}_{i,j}$, $\boldsymbol{P}^{01}_{i,j+1}$, $\boldsymbol{P}^{10}_{i+1,j}$ and $\boldsymbol{P}^{11}_{i+1,j+1}$ is equivalent to determining the twist vectors $\boldsymbol{P}^{uv}_{i,j}$, $\boldsymbol{P}^{uv}_{i,j+1}$, $\boldsymbol{P}^{uv}_{i+1,j}$ and $\boldsymbol{P}^{uv}_{i+1,j+1}$. In the following, the term "twist vectors" sometimes refers to $\boldsymbol{P}^{00}_{i,j}$, $\boldsymbol{P}^{01}_{i,j+1}$, $\boldsymbol{P}^{10}_{i+1,j}$ and $\boldsymbol{P}^{11}_{i+1,j+1}$. The purpose of using eq. (2) to define a bicubic Bézier surface is to simplify the expressions in the subsequent sections.

The values of $\boldsymbol{P}^{00}_{i,j}$, $\boldsymbol{P}^{01}_{i,j+1}$, $\boldsymbol{P}^{10}_{i+1,j}$ and $\boldsymbol{P}^{11}_{i+1,j+1}$, $i = 1, 2, \cdots, m-1$, $1, 2, \cdots, n-1$, should be determined so that

1. $\boldsymbol{P}(u,v)$ is C^1 continuous, and
2. $\boldsymbol{P}(u,v)$ is below or above the given plane.

For simplicity, we only discuss how to make surface (2) below the given plane. The other case can be handle similarly. Our task will be achieved by properly modifying the tangent vectors and determining the values of the twist vectors. The modification process for the tangent vectors will be discussed in Section 3. Determining the values of the twist vectors will be done in two steps: (1) setting the initial values of $\boldsymbol{P}^{00}_{i,j}$, $\boldsymbol{P}^{01}_{i,j+1}$, $\boldsymbol{P}^{10}_{i+1,j}$ and $\boldsymbol{P}^{11}_{i+1,j+1}$ through an energy minimization process based on the requirement that $\boldsymbol{P}(u,v)$ is C^1 continuous, and (2) modifying the initial values of these vectors to ensure the resulting surface is below the given plane. The first step is performed in Section 4 and the second step is shown in Section 5. Note that if the surface $\boldsymbol{P}(u,v)$ is below the given plane $L(x,y,z) = 0$ then we must have $L(\boldsymbol{P}) < 0$.

3. Modification of tangent vectors

Straightforward computation shows that by substituting the representation of $\boldsymbol{P}_{i,j}(u,v)$ into the equation of the given plane defined in eq. (1) one gets

$$L(\boldsymbol{P}_{i,j}(u,v)) = \boldsymbol{U}\boldsymbol{L}\boldsymbol{V}^T, \tag{4}$$

where $\boldsymbol{U}$ and $\boldsymbol{V}$ are defined in (3), and

$$\boldsymbol{L} = \begin{bmatrix} L_{i,j} & L_{i,j} + \frac{\delta^v_j}{3} L^v_{i,j} & L_{i,j+1} - \frac{\delta^v_j}{3} L^v_{i,j+1} & L_{i,j+1} \\ L_{i,j} + \frac{\delta^u_i}{3} L^u_{i,j} & L_{i,j} + \delta^u_i \delta^v_j L^{00}_{i,j} & L_{i,j+1} - \delta^u_i \delta^v_j L^{01}_{i,j+1} & L_{i,j+1} + \frac{\delta^u_i}{3} L^u_{i,j+1} \\ L_{i+1,j} - \frac{\delta^u_i}{3} L^u_{i+1,j} & L_{i+1,j} - \delta^u_i \delta^v_j L^{10}_{i+1,j} & L_{i+1,j+1} + \delta^u_i \delta^v_j L^{11}_{i+1,j+1} & L_{i+1,j+1} - \frac{\delta^u_i}{3} L^u_{i+1,j+1} \\ L_{i+1,j} & L_{i+1,j} + \frac{\delta^v_j}{3} L^v_{i+1,j} & L_{i+1,j+1} - \frac{\delta^v_j}{3} L^v_{i+1,j+1} & L_{i+1,j+1} \end{bmatrix}$$

with

$$\begin{aligned} L_{i+I,j+J} &= L(\boldsymbol{P}_{i+I,j+J}), \\ L^u_{i+I,j+J} &= L(\boldsymbol{P}^u_{i+I,j+J}) - N_c, \\ L^v_{i+I,j+J} &= L(\boldsymbol{P}^v_{i+I,j+J}) - N_c, \\ L^{IJ}_{i+J,j+J} &= L(\boldsymbol{P}^{IJ}_{i+I,j+J}) - N_c, \end{aligned} \qquad I, J = 0, 1.$$

To simplify the construction process of the constrained $\boldsymbol{P}(u,v)$, the tangent vectors and the unknowns $\boldsymbol{P}^{00}_{i,j}$, $\boldsymbol{P}^{10}_{i+1,j}$, $\boldsymbol{P}^{11}_{i+1,j+1}$ and $\boldsymbol{P}^{01}_{i,j+1}$ are handled separately. This section discusses the modification of the tangent vectors.

A necessary condition for $\boldsymbol{P}_{i,j}(u,v)$ to be below $L(x,y,z) = 0$ is that its boundary curves are below $L(x,y,z) = 0$. If each of the four boundary curves, however, has one point on the given plane $L(x,y,z) = 0$, since all the data points are below the given plane, the interior of $\boldsymbol{P}_{i,j}(u,v)$ would be forced to have a flat region. To overcome this problem, the goal here is to determine the tangent vectors of $\boldsymbol{P}_{i,j}(u,v)$ so that none of its boundary curves would have points on the given plane $L(x,y,z) = 0$.

Let

$$\begin{aligned} \bar{L}^1_{i,j}(u,v) = {} & B_0(u)[B_0(v)L_{i,j}/2 + \\ & B_1(v)(L_{i,j} + \delta^v_j L^v_{i,j}/3) + \\ & B_2(v)(L_{i,j+1} - \delta^v_j L^v_{i,j+1}/3) + B_3(v)L_{i,j+1}/2], \\ \bar{L}^2_{i,j}(u,v) = {} & B_0(v)[B_0(u)L_{i,j}/2 + \\ & B_1(u)(L_{i,j} + \delta^u_i L^u_{i,j}/3) + \\ & B_2(u)(L_{i+1,j} - \delta^u_i L^u_{i+1,j}/3) + B_3(u)L_{i+1,j}/2], \\ \bar{L}^3_{i,j}(u,v) = {} & B_3(u)[B_0(v)L_{i+1,j}/2 + \\ & B_1(v)(L_{i+1,j} + \delta^v_j L^v_{i+1,j}/3) + \\ & B_2(v)(L_{i+1,j+1} - \delta^v_j L^v_{i+1,j+1}/3) + \\ & B_3(v)L_{i+1,j+1}/2], \\ \bar{L}^4_{i,j}(u,v) = {} & B_0(v)[B_0(u)L_{i,j+1}/2 + \\ & B_1(u)(L_{i,j+1} + \delta^u_i L^u_{i,j+1}/3) + \\ & B_2(u)(L_{i+1,j+1} - \delta^u_i L^u_{i+1,j+1}/3) + \\ & B_3(u)L_{i+1,j+1}/2]. \end{aligned} \tag{5}$$

The expression $\bar{L}^1_{i,j}(u,v) + \bar{L}^2_{i,j}(u,v) + \bar{L}^3_{i,j}(u,v) + \bar{L}^4_{i,j}(u,v)$ has the same values with $\boldsymbol{P}_{i,j}(u,v)$ on the four boundaries. A simple option to reach the goal is to require the following inequalities to hold:

$$\begin{aligned} \bar{L}^1_{i,j}(u,v) &\le 0, \\ \bar{L}^2_{i,j}(u,v) &\le 0, \\ \bar{L}^3_{i,j}(u,v) &\le 0, \\ \bar{L}^4_{i,j}(u,v) &\le 0. \end{aligned} \tag{6}$$

To satisfy the first inequality, for instance, one needs to choose two numbers $\alpha^0_{i,j}$ and $\alpha^1_{i,j}$ between 0 and 1 so that

$$\begin{aligned} \bar{L}^1_{i,j}(u,v) = {} & B_0(u)[B_0(v)L_{i,j}/2 + \\ & B_1(v)(L_{i,j} + \alpha^0_{i,j}\delta^v_j L^v_{i,j}/3) + \\ & B_2(v)(L_{i,j+1} - \alpha^1_{i,j}\delta^v_j L^v_{i,j+1}/3) \\ & + B_3(v)L_{i,j+1}/2] \le 0, \end{aligned} \tag{7}$$

i.e., one needs to adjust $L_{i,j} + \delta^v_j L^v_{i,j}/3 \ge 0$ and $L_{i,j+1} - \delta^v_j L^v_{i,j+1}/3 \ge 0$. To unify the process for choosing $\alpha^0_{i,j}$ and $\alpha^1_{i,j}$, these numbers will be viewed as functions of the same variable α. An intuitive choice is to set $\alpha^0_{i,j} = \alpha^1_{i,j}$. But this may result in unreasonable result, as is shown in the example at the end of this section. We define $\alpha^0_{i,j}$ and $\alpha^1_{i,j}$ in the following general forms:

$$\alpha^l_{i,j} = c^l_{i,j} + d^l_{i,j}\alpha, \; l = 0, 1. \tag{8}$$

If $L_{i,j+l} + (-1)^l \delta^v_j L^v_{i,j+l}/3 = 0$, $\alpha^l_{i,j}$ should take the value one and $d^l_{i,j}$ the value zero. Thus, $c^l_{i,j}$ and $d^l_{i,j}$ can be defined as follows:

$$\begin{aligned} c^l_{i,j} &= (-1)^{l+1} 3L_{i,j+l}/(\delta^v_j L^v_{i,j+l}), \\ d^l_{i,j} &= 1 - c^l_{i,j} \end{aligned} \tag{9}$$

and one would have

$$\begin{aligned} \bar{L}^1_{i,j}(u,v) = {} & B_0(u)[B_0(v)L_{i,j}/2 + \\ & \alpha\delta^v_j(B_1(v)d^0_{i,j}L^v_{i,j} - B_2(v)d^1_{i,j}L^v_{i,j+1})/3 \\ & + B_3(v)L_{i,j+1}/2] \le 0, \end{aligned} \tag{10}$$

where, following the formula of [5], α is the smallest positive root of the following equation

$$(\delta_j^v)^4(d_{ij}^0 d_{ij}^1 L_{i,j}^v L_{i,j+1}^v)^2\alpha^4/27+ \\ 2[L_{i,j}(d_{ij}^1 L_{i,j+1}^v)^3 - L_{i,j+1}(d_{ij}^0 L_{i,j}^v)^3](\delta_j^v)^3\alpha^3/27- \\ (L_{i,j}d_{ij}^0 L_{i,j}^v L_{i,j+1}d_{i,j}^1 L_{i,j+1}^v)(\delta_j^v)^2\alpha^2/6- \\ L_{i,j}^2 L_{i,j+1}^2/16 = 0. \tag{11}$$

If one of $L_{i,j}+\delta_j^v L_{i,j}^v/3$ and $L_{i,j+1}-\delta_j^v L_{i,j+1}^v/3$ is less than zero, for example, $L_{i,j}+\delta_j^v L_{i,j}^v/3 < 0$, then

$$\bar{L}_{i,j}^1(u,v) = B_0(u)[B_0(v)L_{i,j}/2+ \\ B_1(v)(L_{i,j}+\delta_j^v L_{i,j}^v/3) + -\alpha\delta_j^v B_2(v)L_{i,j+1}^v/3+ \\ B_3(v)L_{i,j+1}/2] \leq 0, \tag{12}$$

and α is a root of a quartic equation similar to eq. (11).

Consequently, if α_0 is a root of eq. (11), then the tangent vectors $\boldsymbol{P}_{i,j+l}^v$, $l = 0, 1$, may be redefined as follows:

$$\boldsymbol{P}_{i,j+l}^v := (c_{i,j}^l + d_{i,j}^l\alpha_0)\boldsymbol{P}_{i,j+l}^v,\ l = 0, 1, \tag{13}$$

where $c_{i,j}^l$ $d_{i,j}^l$ are defined in eq. (9). The tangent vectors $\boldsymbol{P}_{i,j}^u$, $\boldsymbol{P}_{i+1,j}^u$, $\boldsymbol{P}_{i+1,j}^v$, $\boldsymbol{P}_{i+1,j+1}^v$, $\boldsymbol{P}_{i,j+1}^u$ and $\boldsymbol{P}_{i+1,j+1}^u$ may be redefined similarly.

so that the requirement is satisfied. If α and β are set to be the same, one gets a curve tangent to the dotted line whose second and third control points, $\boldsymbol{P}_0 + \boldsymbol{P}_0'/3$ and $\boldsymbol{P}_1 - \boldsymbol{P}_1'/3$ are at the points marked by '+'. If α and β are defined by eqs. (8) and (9), one gets a more reasonable setting for the second and third control points which are at the points marked by solid squares.

Our experiments showd that the curves produced by setting $\alpha = \beta$ are visually the same as the ones produced by setting α and β using eqs. (8) and (9), but the curvature distribution of the latter one is closer to that of the original curve.

4. Construction of twist vectors

In this section, we first introduce a quartic interpolant, then discuss the process of determining the values of $\boldsymbol{P}_{i,j}^{00}$, $\boldsymbol{P}_{i,j+1}^{01}$, $\boldsymbol{P}_{i+1,j}^{10}$ and $\boldsymbol{P}_{i+1,j+1}^{11}$, based on the requirement that $\boldsymbol{P}(u,v)$ is C^1 continuous.

4.1. A quartic interpolant

Given the start point $\boldsymbol{P}_1$ and the end point $\boldsymbol{P}_2$ at $t = 0$ and $t = \Delta$, respectively, and the corresponding tangent vectors $\boldsymbol{P}_1'$ and $\boldsymbol{P}_2'$, a quartic curve $\boldsymbol{Q}(t)$ with one degree of freedom $\boldsymbol{D}$ that interpolates $\boldsymbol{P}_k$ and $\boldsymbol{P}_k'$, $k = 1, 2$, is defined by

$$\boldsymbol{Q}(t) = \varphi_0(\tfrac{t}{\Delta})\boldsymbol{P}_1 + \varphi_1(\tfrac{t}{\Delta})\boldsymbol{P}_1' + \phi(\tfrac{t}{\Delta})\boldsymbol{D} \\ + \psi_1(\tfrac{t}{\Delta})\boldsymbol{P}_2' + \psi_0(\tfrac{t}{\Delta})\boldsymbol{P}_2, \tag{14}$$

where

$$\varphi_0(t/\Delta) = (t-\Delta)^3(-3t-\Delta)/\Delta^4, \\ \varphi_1(t/\Delta) = (\Delta - t)^3 t/\Delta^3, \\ \phi(t/\Delta) = t^2(t-\Delta)^2/\Delta^3, \\ \psi_0(t/\Delta) = t^3(4\Delta - 3t)/\Delta^4, \\ \psi_1(t/\Delta) = t^3(t-\Delta)/\Delta^3$$

are basis functions. These functions satisfy the following properties:

$$\varphi_0(0) = 1,\ \varphi_0''(0) = -12/\Delta^2, \\ \varphi_0'(0) = \varphi_0(1) = \varphi_0'(1) = \varphi_0''(1) = 0, \\ \varphi_1'(0) = 1,\ \varphi_1''(0) = -6/\Delta, \\ \varphi_1(0) = \varphi_1(1) = \varphi_1'(1) = \varphi_1''(1) = 0, \\ \phi''(0) = \phi''(1) = 2/\Delta, \\ \phi(0) = \phi'(0) = \phi(1) = \phi'(1) = 0, \\ \psi_0(1) = 1,\ \psi_0''(1) = -12/\Delta^2, \\ \psi_0(0) = \psi_0'(0) = \psi_0''(0) = \psi_0'(1) = 0, \\ \psi_1'(1) = 1,\ \psi_1''(1) = 6/\Delta, \\ \psi_1(0) = \psi_1'(0) = \psi_1''(0) = \psi_1(1) = 0.$$

Thus

$$\boldsymbol{Q}''(0) = 2\boldsymbol{D}/\Delta - 12\boldsymbol{P}_1/\Delta^2 - 6\boldsymbol{P}_1'/\Delta, \\ \boldsymbol{Q}''(\Delta) = 2\boldsymbol{D}/\Delta - 12\boldsymbol{P}_2/\Delta^2 + 6\boldsymbol{P}_2'/\Delta. \tag{15}$$

The degree of freedom $\boldsymbol{D}$ in (14) can be used to optimize the shape of the quartic curve $\boldsymbol{Q}(t)$. Due to the fact that a curve with smaller energy has smoother curvature distribution and, consequently, smoother shape, we will determine the value of $\boldsymbol{D}$ by minimizing the energy of the curve. The energy form that will be used here is the integral of the squared second derivative of the quartic curve $\boldsymbol{Q}(t)$. Straightforward computation shows that

$$\int_{t_1}^{t_2} \boldsymbol{Q}''(t)^2 dt = \frac{4}{5\Delta}(2(\boldsymbol{P}_2' - \boldsymbol{P}_1')- \\ \frac{6}{\Delta}(\boldsymbol{P}_1 + \boldsymbol{P}_2) + \boldsymbol{D})\boldsymbol{D} + \boldsymbol{\alpha}, \tag{16}$$

where $\boldsymbol{\alpha}$ is an expression independent of $\boldsymbol{D}$.

Theorem 1. Eq. (14) with a degree of freedom $\boldsymbol{D}$ determined by minimizing eq. (16) reproduces cubic polynomials.

Proof: Minimizing eq. (16) gives

$$\boldsymbol{D} = 3(\boldsymbol{P}_1 + \boldsymbol{P}_2)/\Delta + \boldsymbol{P}_1' - \boldsymbol{P}_2'. \tag{17}$$

If $\boldsymbol{P}_1$, $\boldsymbol{P}_1^{'}$, $\boldsymbol{P}_2$ and $\boldsymbol{P}_2^{'}$ are taken from a cubic polynomial $\boldsymbol{P}(t)$ at $t = t_1$ and $t = t_2$, respectively, and $\boldsymbol{D}$ is defined by

$$\boldsymbol{D} = 6\boldsymbol{P}_1/\Delta + 3\boldsymbol{P}_1^{'} + \Delta\boldsymbol{P}_1^{''}/2, \tag{18}$$

or

$$\boldsymbol{D} = 6\boldsymbol{P}_2/\Delta - 3\boldsymbol{P}_2^{'} + \Delta\boldsymbol{P}_2^{''}/2, \tag{19}$$

then eq. (14) reproduces $\boldsymbol{P}(t)$ exactly.

Combining eqs. (18) and (19) gives

$$\begin{aligned}\boldsymbol{D} = {} & 3(\boldsymbol{P}_1 + \boldsymbol{P}_2)/\Delta + \boldsymbol{P}_1^{'} - \boldsymbol{P}_2^{'} \\ & + (\boldsymbol{P}_1^{'} - \boldsymbol{P}_2^{'})/2 + \Delta(\boldsymbol{P}_1^{''} + \boldsymbol{P}_2^{''})/4.\end{aligned} \tag{20}$$

If $\boldsymbol{P}(t)$ is a cubic polynomial, we have

$$(\boldsymbol{P}_1^{'} - \boldsymbol{P}_2^{'})/2 + \Delta(\boldsymbol{P}_1^{''} + \boldsymbol{P}_2^{''})/4 = 0.$$

So eq. (20) is the same as eq. (17). This completes the proof of the theorem.

4.2. $P(u, v)$ being C^1 continuous

A sufficient condition for $\boldsymbol{P}(u, v)$ to be C^1 continuous is as follows:

1. for $1 < i < m - 1$ and $1 < j < n - 1$, $\frac{\partial^2 \boldsymbol{P}_{i,j}(u,v)}{\partial u \partial v}$, $\frac{\partial^2 \boldsymbol{P}_{i-1,j}(u,v)}{\partial u \partial v}$, $\frac{\partial^2 \boldsymbol{P}_{i-1,j-1}(u,v)}{\partial u \partial v}$ and $\frac{\partial^2 \boldsymbol{P}_{i,j-1}(u,v)}{\partial u \partial v}$ have the same value at (u_i, v_j);

2. for $i = 1$, $i = m$, $2 \le j < n$, $\frac{\partial^2 \boldsymbol{P}_{i,j-1}(u,v)}{\partial u \partial v}$, and $\frac{\partial^2 \boldsymbol{P}_{i,j}(u,v)}{\partial u \partial v}$ have the same value at (u_i, v_j);

3. for $j = 1$, $j = n$, $2 \le i < m$, $\frac{\partial^2 \boldsymbol{P}_{i-1,j}(u,v)}{\partial u \partial v}$, and $\frac{\partial^2 \boldsymbol{P}_{i,j}(u,v)}{\partial u \partial v}$, have the same value at (u_i, v_j);

4. $\frac{\partial^2 \boldsymbol{P}_{1,1}(u_1,v_1)}{\partial u \partial v}$, $\frac{\partial^2 \boldsymbol{P}_{m-1,1}(u_m,v_1)}{\partial u \partial v}$, $\frac{\partial^2 \boldsymbol{P}_{m-1,n-1}(u_m,v_n)}{\partial u \partial v}$ and $\frac{\partial^2 \boldsymbol{P}_{1,n-1}(u_1,v_n)}{\partial u \partial v}$ can be any values.

The condition that $\frac{\partial^2 \boldsymbol{P}_{I,J}(u,v)}{\partial u \partial v}$, $I = i - 1, i$, $J = j - 1, j$, have the same value at (u_i, v_j) is shown below:

$$\begin{aligned}\boldsymbol{P}_{i-1,j}^{10} &= \boldsymbol{P}_{i,j}^{00} - (1/\delta_{i-1}^u + 1/\delta_i^u)\boldsymbol{P}_{i,j}^v/3, \\ \boldsymbol{P}_{i-1,j-1}^{11} &= \boldsymbol{P}_{i,j}^{00} - (1/\delta_{j-1}^v + 1/\delta_j^v)\boldsymbol{P}_{i,j}^u/3 \\ &\quad - (1/\delta_{i-1}^u + 1/\delta_i^u)\boldsymbol{P}_{i,j}^v/3, \\ \boldsymbol{P}_{i,j-1}^{01} &= \boldsymbol{P}_{i,j}^{00} - (1/\delta_{j-1}^v + 1/\delta_j^v)\boldsymbol{P}_{i,j}^u.\end{aligned} \tag{21}$$

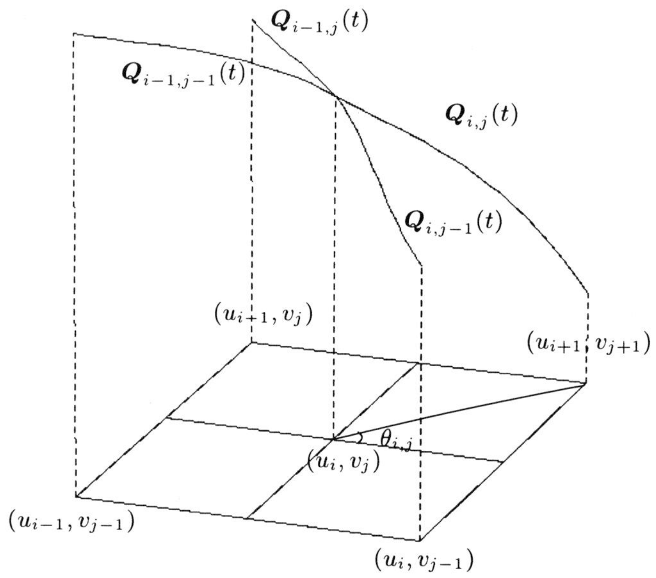

Figure 1. Four quartic curves

In the following, four quartic interpolants will be constructed. The value of $\boldsymbol{P}_{i,j}^{00}$ will be determined by minimizing the integrals of the squared second derivatives of the four quartic curves.

For $l = 0, 1$, let $l_{i-l,j-l}$ denote the direction from (u_{i-l}, v_{j-l}) to (u_{i-l+1}, v_{j-l+1}), $l_{i-l,j+l-1}$ denote the direction from (u_{i-l}, v_{j+l}) to (u_{i-l+1}, v_{j+l-1}), $\boldsymbol{Q}_{i-l,j-l}(t)$ denote the quartic interpolants to the positions and the first derivatives of $\boldsymbol{P}_{i-l,j-l}$ along $l_{i-l,j-l}$, at (u_{i-l}, v_{j-l}) and (u_{i-l+1}, v_{j-l+1}), respectively, $\boldsymbol{Q}_{i-l,j+l-1}(t)$ denote the interpolants to the positions and the first derivatives of $\boldsymbol{P}_{i-l,j+l+1}(u, v)$ along $l_{l-l,j+l-1}$, at (u_{i-l}, v_{j+l}) and (u_{i-l+1}, v_{j+l-1}), respectively, as shown in Figure 1, then from (14) we have

$$\begin{aligned}\boldsymbol{Q}_{i,j}(t) = {} & \varphi_0(\frac{t}{\Delta_{i,j}})\boldsymbol{P}_{i,j} + \varphi_1(\frac{t}{\Delta_{i,j}})\frac{\partial \boldsymbol{P}_{i,j}}{\partial l_{i,j}}(u_i, v_j) + \\ & \phi(\frac{t}{\Delta_{i,j}})\boldsymbol{D}_{i,j} + \psi_1(\frac{t}{\Delta_{i,j}})\frac{\partial \boldsymbol{P}_{i,j}}{\partial l_{i,j}}(u_{i+1}, v_{j+1}) \\ & + \psi_0(\frac{t}{\Delta_{i,j}})\boldsymbol{P}_{i+1,j+1},\end{aligned}$$

$$
\begin{aligned}
\boldsymbol{Q}_{i-1,j}(t) &= \varphi_0(\frac{t}{\Delta_{i-1,j}})\boldsymbol{P}_{i-1,j+1}+ \\
&+ \varphi_1(\frac{t}{\Delta_{i-1,j}})\frac{\partial \boldsymbol{P}_{i-1,j}}{\partial l_{i-1,j}}(u_{i-1}, v_{j+1})+ \\
&\phi(\frac{t}{\Delta_{i-1,j}})\boldsymbol{D}_{i-1,j} + \psi_1(\frac{t}{\Delta_{i-1,j}})\frac{\partial \boldsymbol{P}_{i-1,j}}{\partial l_{i-1,j}}(u_i, v_j) \\
&+ \psi_0(\frac{t}{\Delta_{i-1,j}})\boldsymbol{P}_{i,j}, \\
\boldsymbol{Q}_{i-1,j-1}(t) &= \varphi_0(\frac{t}{\Delta_{i-1,j-1}})\boldsymbol{P}_{i-1,j-1}+ \\
&\varphi_1(\frac{t}{\Delta_{i-1,j-1}})\frac{\partial \boldsymbol{P}_{i-1,j-1}}{\partial l_{i-1,j-1}}(u_{i-1}, v_{j-1})+ \\
&\phi(\frac{t}{\Delta_{i-1,j-1}})\boldsymbol{D}_{i-1,j-1} \\
&+ \psi_1(\frac{t}{\Delta_{i-1,j-1}})\frac{\partial \boldsymbol{P}_{i-1,j-1}}{\partial l_{i-1,j-1}}(u_i, v_j) \\
&+ \psi_0(\frac{t}{\Delta_{i-1,j-1}})\boldsymbol{P}_{i,j}, \\
\boldsymbol{Q}_{i,j-1}(t) &= \varphi_0(\frac{t}{\Delta_{i,j-1}})\boldsymbol{P}_{i,j}+ \\
&\varphi_1(\frac{t}{\Delta_{i,j-1}})\frac{\partial \boldsymbol{P}_{i,j-1}}{\partial l_{i,j-1}}(u_i, v_j) + \phi(\frac{t}{\Delta_{i,j-1}})\boldsymbol{D}_{i,j-1} \\
&+ \psi_1(\frac{t}{\Delta_{i,j-1}})\frac{\partial \boldsymbol{P}_{i,j-1}}{\partial l_{i,j-1}}(u_{i+1}, v_{j-1}) \\
&+ \psi_0(\frac{t}{\Delta_{i,j-1}})\boldsymbol{P}_{i+1,j-1},
\end{aligned} \tag{22}
$$

where

$$
\begin{aligned}
\frac{\partial \boldsymbol{P}_{i,j}}{\partial l_{i,j}}(u_{i+l}, v_{j+l}) &= cos\theta_{i,j}\boldsymbol{P}^u_{i+l,j+l}+ \\
&\quad sin\theta_{i,j}\boldsymbol{P}^v_{i+l,j+l}, \quad l = 0, 1, \\
\frac{\partial \boldsymbol{P}_{i-1,j}}{\partial l_{i-1,j}}(u_{i-l}, v_{j+l}) &= cos\theta_{i-1,j}\boldsymbol{P}^u_{i-l,j+l}+ \\
&\quad sin\theta_{i-1,j}\boldsymbol{P}^v_{i-l,j+l}, \quad l = 1, 0, \\
\frac{\partial \boldsymbol{P}_{i-1,j-1}}{\partial l_{i-1,j-1}}(u_{i-l}, v_{j-l}) &= cos\theta_{i-1,j-1}\boldsymbol{P}^u_{i-l,j-l}+ \\
&\quad sin\theta_{i-1,j-1}\boldsymbol{P}^v_{i-l,j-l}, \quad l = 1, 0, \\
\frac{\partial \boldsymbol{P}_{i,j-1}}{\partial l_{i,j-1}}(u_{i+l}, v_{j-l}) &= cos\theta_{i,j-1}\boldsymbol{P}^u_{i+l,j-l}+ \\
&\quad sin\theta_{i,j-1}\boldsymbol{P}^v_{i+l,j-l}, \quad l = 0, 1,
\end{aligned}
$$

with

$$
\begin{aligned}
&cos\theta_{i,j} = \delta^u_i/\Delta_{i,j}, \quad sin\theta_{i,j} = \delta^v_j/\Delta_{i,j}, \\
&\Delta_{i,j} = \sqrt{(u_{i+1} - u_i)^2 + (v_{j+1} - v_j)^2}, \\
&cos\theta_{i-1,j-1} = \delta^u_{i-1}/\Delta_{i-1,j-1}, \\
&sin\theta_{i-1,j-1} = \delta^v_{j-1}/\Delta_{i-1,j-1}, \\
&\Delta_{i-1,j} = \sqrt{(u_i - u_{i-1})^2 + (v_j - v_{j-1})^2}, \\
&cos\theta_{i-1,j} = \delta^u_{i-1}/\Delta_{i-1,j}, \\
&sin\theta_{i-1,j} = -\delta^v_j/\Delta_{i-1,j}, \\
&\Delta_{i-1,j} = \sqrt{(u_i - u_{i-1})^2 + (v_j - v_{j+1})^2}, \\
&cos\theta_{i,j-1} = \delta^u_i/\Delta_{i,j-1}, \quad sin\theta_{i,j-1} = -\delta^v_{j-1}/\Delta_{i,j-1}, \\
&\Delta_{i,j-1} = \sqrt{(u_{i+1} - u_i)^2 + (v_{j-1} - v_j)^2}.
\end{aligned}
$$

The second derivatives of $\boldsymbol{P}_{i,j}(u,v)$, $\boldsymbol{P}_{i-1,j}(u,v)$, $\boldsymbol{P}_{i-1,j-1}(u,v)$ and $\boldsymbol{P}_{i,j-1}(u,v)$, at (u_i, v_j) along the directions $l_{i,j}$, $l_{i-1,j}$, $l_{i-1,j-1}$ and $l_{i,j-1}$, respectively, are as follows:

$$
\begin{aligned}
\frac{\partial^2 \boldsymbol{P}_{i,j}}{\partial l^2_{i,j}}(u_i, v_j) &= \bar{\boldsymbol{P}}_{i,j} + 18cos\theta_{i,j}sin\theta_{i,j}\boldsymbol{P}^{00}_{i,j}, \\
\frac{\partial^2 \boldsymbol{P}_{i-1,j}}{\partial l^2_{i-1,j}}(u_i, v_j) &= \bar{\boldsymbol{P}}_{i-1,j} + 18cos\theta_{i-1,j}sin\theta_{i-1,j}\boldsymbol{P}^{00}_{i,j}, \\
\frac{\partial^2 \boldsymbol{P}_{i-1,j-1}}{\partial l^2_{i-1,j-1}}(u_i, v_j) &= \bar{\boldsymbol{P}}_{i-1,j-1} \\
&\quad + 18cos\theta_{i-1,j-1}sin\theta_{i-1,j-1}\boldsymbol{P}^{00}_{i,j}, \\
\frac{\partial^2 \boldsymbol{P}_{i,j-1}}{\partial l^2_{i,j-1}}(u_i, v_j) &= \bar{\boldsymbol{P}}_{i,j-1} + 18cos\theta_{i,j-1}sin\theta_{i,j-1}\boldsymbol{P}^{00}_{i,j},
\end{aligned} \tag{23}
$$

where

$$
\begin{aligned}
\bar{\boldsymbol{P}}_{i,j} &= cos^2\theta_{i,j}(\frac{6}{(\delta^u_i)^2}(\boldsymbol{P}_{i+1,j} - \boldsymbol{P}_{i,j})- \\
&\qquad \frac{4}{\delta^u_i}\boldsymbol{P}^u_{i,j} - \frac{2}{\delta^u_i}\boldsymbol{P}^u_{i+1,j}) \\
&- cos\theta_{i,j}sin\theta_{i,j}(\frac{6}{\delta^v_j}\boldsymbol{P}^u_{i,j} + \frac{6}{\delta^u_i}\boldsymbol{P}^v_{i,j}) \\
&+ sin^2\theta_{i,j}(\frac{6}{(\delta^v_j)^2}(\boldsymbol{P}_{i,j+1} - \boldsymbol{P}_{i,j})- \\
&\qquad \frac{4}{\delta^v_j}\boldsymbol{P}^v_{i,j} - \frac{2}{\delta^v_j}\boldsymbol{P}^v_{i,j+1}), \\
\bar{\boldsymbol{P}}_{i-1,j} &= cos\theta^2_{i-1,j}(\frac{6}{(\delta^u_{i-1})^2}(\boldsymbol{P}_{i-1,j} - \boldsymbol{P}_{i,j})+ \\
&\qquad \frac{2}{\delta^u_{i-1}}\boldsymbol{P}^u_{i-1,j} + \frac{4}{\delta^u_{i-1}}\boldsymbol{P}^u_{i,j}) \\
&- cos\theta_{i-1,j}sin\theta_{i-1,j}(\frac{6}{\delta^v_j}\boldsymbol{P}^u_{i,j} - \frac{6}{\delta^u_{i-1}}\boldsymbol{P}^v_{i,j}) \\
&+ sin\theta^2_{i-1,j}(\frac{6}{(\delta^v_j)^2}(\boldsymbol{P}_{i,j+1} - \boldsymbol{P}_{i,j})- \\
&\qquad \frac{4}{\delta^v_j}\boldsymbol{P}^v_{i,j} - \frac{2}{\delta^v_j}\boldsymbol{P}^v_{i,j+1}), \\
\bar{\boldsymbol{P}}_{i-1,j-1} &= cos\theta^2_{i-1,j-1}(\frac{6}{(\delta^u_{i-1})^2}(\boldsymbol{P}_{i-1,j} - \boldsymbol{P}_{i,j})+ \\
&\qquad \frac{2}{\delta^u_{i-1}}\boldsymbol{P}^u_{i-1,j} + \frac{4}{\delta^u_{i-1}}\boldsymbol{P}^u_{i,j}) \\
&+ cos\theta_{i-1,j-1}sin\theta_{i-1,j-1}(\frac{6}{\delta^v_{j-1}}\boldsymbol{P}^u_{i,j} + \frac{6}{\delta^u_{i-1}}\boldsymbol{P}^v_{i,j}) \\
&+ sin\theta^2_{i-1,j-1}(\frac{6}{(\delta^v_{j-1})^2}(\boldsymbol{P}_{i,j-1} - \boldsymbol{P}_{i,j})+ \\
&\qquad \frac{2}{\delta^v_{j-1}}\boldsymbol{P}^v_{i,j-1} + \frac{4}{\delta^v_{j-1}}\boldsymbol{P}^v_{i,j}),
\end{aligned}
$$

$$\bar{\boldsymbol{P}}_{i,j-1} = cos^2\theta_{i,j-1}(\frac{6}{(\delta_i^u)^2}(\boldsymbol{P}_{i+1,j} - \boldsymbol{P}_{i,j}) - \frac{4}{\delta_i^u}\boldsymbol{P}^u_{i,j} - \frac{2}{\delta_i^u}\boldsymbol{P}^u_{i+1,j}) + cos\theta_{i,j-1}sin\theta_{i,j-1}(\frac{6}{\delta_{j-1}^v}\boldsymbol{P}^u_{i,j} - \frac{6}{\delta_i^u}\boldsymbol{P}^v_{i,j}) + sin^2\theta_{i,j-1}(\frac{6}{(\delta_{j-1}^v)^2}(\boldsymbol{P}_{i,j-1} - \boldsymbol{P}_{i,j}) + \frac{2}{\delta_{j-1}^v}\boldsymbol{P}^v_{i,j-1} + \frac{4}{\delta_{j-1}^v}\boldsymbol{P}^v_{i,j}).$$

For $\boldsymbol{Q}_{I,J}(t)$, $I = i-1, i$, $J = j-1, j$, the integrals corresponding to Eq. (16) are:

$$\int_0^{\Delta_{I,J}} \boldsymbol{Q}''_{I,J}(t)^2 dt = \frac{4}{5\Delta_{I,J}}(\boldsymbol{A}_{I,J} + \boldsymbol{D}_{I,J})\boldsymbol{D}_{I,J} + \boldsymbol{\alpha}_{I,J}, \quad I = i-1, i,\ J = j-1, j, \tag{24}$$

where

$$\boldsymbol{A}_{i,j} = 2(\frac{\partial \boldsymbol{P}_{i,j}}{\partial l_{i,j}}(u_{i+1}, v_{j+1}) - \frac{\partial \boldsymbol{P}_{i,j}}{\partial l_{i,j}}(u_i, v_j)) - \frac{6}{\Delta_{i,j}}(\boldsymbol{P}_{i,j} + \boldsymbol{P}_{i+1,j+1}),$$
$$\boldsymbol{A}_{i-1,j} = 2(\frac{\partial \boldsymbol{P}_{i-1,j}}{\partial l_{i-1,j}}(u_i, v_j) - \frac{\partial \boldsymbol{P}_{i-1,j}}{\partial l_{i-1,j}}(u_{i-1}, v_{j+1})) - \frac{6}{\Delta_{i-1,j}}(\boldsymbol{P}_{i-1,j+1} + \boldsymbol{P}_{i,j}),$$
$$\boldsymbol{A}_{i-1,j-1} = 2(\frac{\partial \boldsymbol{P}_{i-1,j-1}}{\partial l_{i-1,j-1}}(u_i, v_j) - \frac{\partial \boldsymbol{P}_{i-1,j-1}}{\partial l_{i-1,j-1}}(u_{i-1}, v_{j-1})) - \frac{6}{\Delta_{i-1,j-1}}(\boldsymbol{P}_{i-1,j-1} + \boldsymbol{P}_{i,j}),$$
$$\boldsymbol{A}_{i,j-1} = 2(\frac{\partial \boldsymbol{P}_{i,j-1}}{\partial l_{i,j-1}}(u_{i+1}, v_{j-1}) - \frac{\partial \boldsymbol{P}_{i,j-1}}{\partial l_{i,j-1}}(u_i, v_j)) - \frac{6}{\Delta_{i,j-1}}(\boldsymbol{P}_{i,j} + \boldsymbol{P}_{i+1,j-1}),$$

with $\boldsymbol{\alpha}_{I,J}$ being an expression independent of $\boldsymbol{D}_{I,J}$.

If $\boldsymbol{Q}_{I,J}(t)$, $I = i-1, i$, $J = j-1, j$, in (22) are required to match the second derivatives $\frac{\partial^2 \boldsymbol{P}_{I,J}}{\partial l^2_{I,J}}(u_i, v_j)$, $I = i-1, i$, $J = j-1, j$, at (u_i, v_j), respectively, then we must have for $I = i-1, i$, $J = j-1, j$

$$\boldsymbol{Q}''_{I,J}(i-I) = \bar{\boldsymbol{P}}_{I,J} + 18cos\theta_{I,J}sin\theta_{I,J}\boldsymbol{P}^{00}_{i,j}.$$

It follows from (15) that

$$\boldsymbol{Q}''_{i,j}(0) = -12\boldsymbol{P}_{i,j}/\Delta^2_{i,j} - 6\frac{\partial \boldsymbol{P}_{i,j}}{\partial l_{i,j}}(u_i, v_j)/\Delta_{i,j} + 2\boldsymbol{D}_{i,j}/\Delta_{i,j},$$
$$\boldsymbol{Q}''_{i-1,j}(1) = -12\boldsymbol{P}_{i,j}/\Delta^2_{i-1,j} + 6\frac{\partial \boldsymbol{P}_{i-1,j}}{\partial l_{i-1,j}}(u_i, v_j)/\Delta_{i-1,j} + 2\boldsymbol{D}_{i-1,j}/\Delta_{i-1,j},$$
$$\boldsymbol{Q}''_{i-1,j-1}(1) = \frac{-12\boldsymbol{P}_{i,j}}{\Delta^2_{i-1,j-1}} + 6\frac{\partial \boldsymbol{P}_{i-1,j-1}}{\partial l_{i-1,j-1}}(u_i, v_j)/\Delta_{i-1,j-1} + 2\boldsymbol{D}_{i-1,j-1}/\Delta_{i-1,j-1},$$
$$\boldsymbol{Q}''_{i,j-1}(0) = -12\boldsymbol{P}_{i,j}/\Delta^2_{i,j-1} - 6\frac{\partial \boldsymbol{P}_{i,j-1}}{\partial l_{i,j-1}} + 2\boldsymbol{D}_{i,j-1}/\Delta_{i,j-1}.$$

Thus we have

$$\boldsymbol{D}_{I,J} = \boldsymbol{B}_{I,J} + C_{I,J}\boldsymbol{P}^{00}_{i,j}, \quad I = i-1, i, \quad J = j-1, j \tag{25}$$

where

$$\boldsymbol{B}_{i,j} = \Delta_{i,j}\bar{\boldsymbol{P}}_{i,j}/2 + 6\boldsymbol{P}_{i,j}/\Delta_{i,j} + 3\frac{\partial \boldsymbol{P}_{i,j}}{\partial l_{i,j}}(u_i, v_j),$$
$$\boldsymbol{B}_{i-1,j} = \Delta_{i-1,j}\bar{\boldsymbol{P}}_{i-1,j}/2 - 6\boldsymbol{P}_{i,j}/\Delta_{i-1,j} + 3\frac{\partial \boldsymbol{P}_{i-1,j}}{\partial l_{i-1,j}}(u_i, v_j),$$
$$\boldsymbol{B}_{i-1,j-1} = \Delta_{i-1,j-1}\bar{\boldsymbol{P}}_{i-1,j-1}/2 - 6\boldsymbol{P}_{i,j}/\Delta_{i-1,j-1} + 3\frac{\partial \boldsymbol{P}_{i-1,j-1}}{\partial l_{i-1,j-1}}(u_i, v_j),$$
$$\boldsymbol{B}_{i,j-1} = \Delta_{i,j-1}\bar{\boldsymbol{P}}_{i,j-1}/2 + 6\boldsymbol{P}_{i,j}/\Delta_{i,j-1} + 3\frac{\partial \boldsymbol{P}_{i,j-1}}{\partial l_{i,j-1}}(u_i, v_j),$$
$$C_{I,J} = 9\Delta_{I,J}\cos\theta_{I,J}\sin\theta_{I,J}.$$

Substituting eq. (25) into eq. (24), one gets

$$\int_0^{\Delta_{I,J}} \boldsymbol{Q}''_{I,J}(t)^2 dt = \frac{4}{5\Delta_{I,J}}(\boldsymbol{A}_{i,j} + \boldsymbol{B}_{I,J} + C_{I,J}\boldsymbol{P}^{00}_{i,j})(\boldsymbol{B}_{I,J} + C_{I,J}\boldsymbol{P}^{00}_{i,j}) + \boldsymbol{\alpha}_{I,J}, \tag{26}$$

where $I = i-1, i$, $J = j-1, j$. $\boldsymbol{P}^{00}_{i,j}$ is determined by minimizing the sum of the integrals of the squared second derivatives of the four quartic curves, i.e., by minimizing the following objective function

$$\sum_{I=i-1}^{i}\sum_{J=j-1}^{j}\int_0^{\Delta_{I,J}} \boldsymbol{Q}''_{I,J}(t)^2 dt,$$

which gives

$$\boldsymbol{P}^{00}_{i,j} = \frac{\sum_{I=i-1}^{i}\sum_{J=j-1}^{j} C_{I,J}(\boldsymbol{A}_{I,J} + 2\boldsymbol{B}_{I,J})/\Delta_{I,J}}{2\sum_{I=i-1}^{i}\sum_{J=j-1}^{j} C^2_{I,J}/\Delta_{I,J}}. \tag{27}$$

Hence, the values of $P^{00}_{i,j}$, $P^{01}_{i,j+1}$, $P^{10}_{i+1,j}$ and $P^{11}_{i+1,j+1}$, for $1 < i < m$ and $1 < j < n$, can be computed using eqs. (21) and (27). It is obvious from Theorem 1 that, with $P^{00}_{i,j}$, $P^{01}_{i,j+1}$, $P^{10}_{i+1,j}$ and $P^{11}_{i+1,j+1}$, $1 < i < m$, $1 < j < n$, determined by eqs. (21) and (27), $P(u,v)$ reproduces cubic polynomial surfaces.

Similarly, one can compute the values of $P^{00}_{i,j}$, $P^{01}_{i,j+1}$, $P^{10}_{i+1,j}$ and $P^{11}_{i+1,j+1}$ which correspond to boundary points. For instance, $P^{10}_{1,1}$ and $P^{00}_{2,1}$ can be obtained by minimizing the following objective function:

$$\int_0^{\Delta_{1,1}} Q''_{1,1}(t)^2 dt + \int_0^{\Delta_{2,1}} Q''_{2,1}(t)^2 dt$$

where $Q_{1,1}(t)$ is the quartic curve interpolating the positions and the first derivatives of $P_{1,1}(u,v)$ at (u_1,v_2) and (u_2,v_1), $Q_{2,1}(t)$ is the quartic curve interpolating the positions and the first derivatives of $P_{2,1}(u,v)$ at (u_2,v_1) and (u_3,v_2).

4.3. Examples

The interpolation conditions used in the examples are taken from the following two functions [2]:

$$Z_1(u,v) = 5.2exp[-81((u-0.5)^2 + (v-0.5)^2)/16]/3,$$
$$Z_2(u,v) = 5.2exp[-81((u-0.5)^2 + (v-0.5)^2)/4]/3, \tag{28}$$

with $X_i(u,v) = u$ and $Y_i(u,v) = v$, $i = 1,2$.

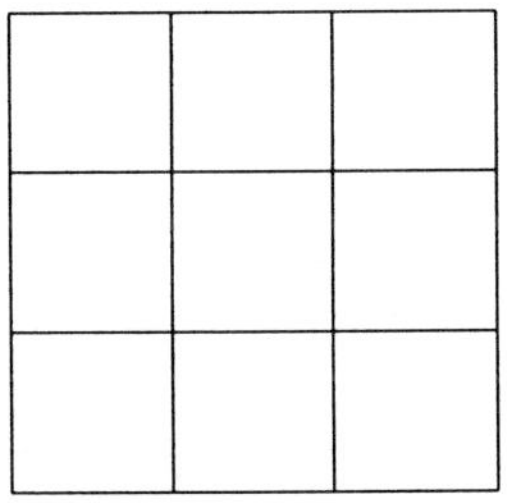

Figure 2. Network for defining interpolation conditions.

The interpolation conditions are positions and first derivatives at the vertices of the network shown in Figure 2. The network is defined on a unit square. Two interpolating surfaces constructed by the new method are compared with the original functions in the computation of the following approximate thin plate energy

$$E_i = \int\int [\frac{\partial^2 P_i(u,v)}{\partial u^2} + \frac{\partial^2 P_i(u,v)}{\partial u \partial v} + \frac{\partial^2 P_i(u,v)}{\partial v^2}]dudv,$$

where $P_i(u,v) = (X_i(u,v), Y_i(u,v), Z_i(u,v))$.

The results in part (a) and part (b) of Figures 3-4 are produced by the new method and original functions $P_i(u,v)$, $i = 1,2$, respectively. The maximum plate energies of the two original functions $P_1(u,v)$ and $P_2(u,v)$ are 19.02 and 3.65, respectively. The maximum plate energies of the two interpolants constructed by the new method are 7.44, and 3.060, respectively. Figures 3-4 also show that the surfaces constructed by the new method have smaller thin plate energies than the original functions. As the constructed interpolating surfaces are C^1 continuous only, their thin plate energies are not continuous.

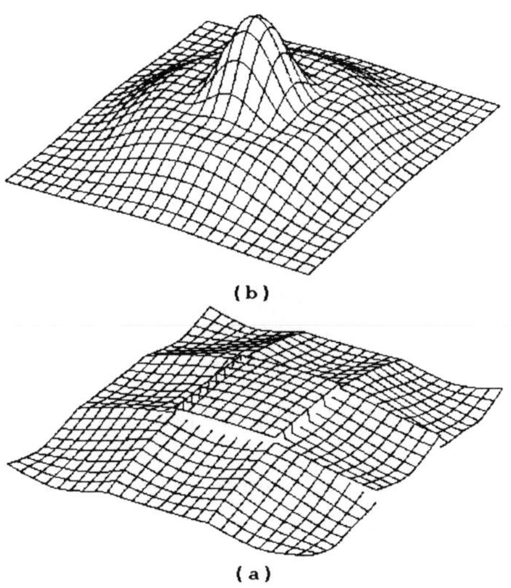

Figure 3. Thin plate energies: (a) of new method, (b) of $P_1(u,v)$

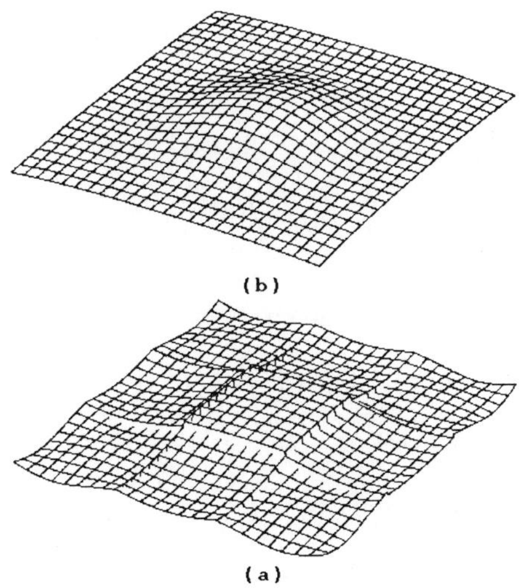

Figure 4. Thin plate energies: (a) of new method, (b) of $P_2(u,v)$

5. Modification of twist vectors

The values of $\boldsymbol{P}_{i,j}^{00}$, $\boldsymbol{P}_{i,j+1}^{01}$, $\boldsymbol{P}_{i+1,j}^{10}$ and $\boldsymbol{P}_{i+1,j+1}^{11}$ determined in the previous Section for $i = 1, 2, \cdots, m-1$ and $j = 1, 2, \cdots, n-1$ do not guarantee that $\boldsymbol{P}(u, v)$ would be below the given plane $L(x, y, z) = 0$, i.e.,

$$L(\boldsymbol{P}(u, v)) < 0.$$

Therefore, the values of these vectors have to go through a modification process to ensure that the above condition is indeed satisfied.

5.1. Modification of twist vectors

The goal is to define $\alpha_{i,j}^{00}$, $\alpha_{i,j}^{01}$, $\alpha_{i,j}^{10}$ and $\alpha_{i,j}^{11}$ so that for $(u, v) \in [u_i, u_{i+1}] \times [v_j, v_{j+1}]$, the following condition holds.

$$\begin{aligned} L(\boldsymbol{P}(u,v)) = \bar{L}_{i,j}^1(u,v) + \bar{L}_{i,j}^2(u,v) + \bar{L}_{i,j}^3(u,v) + \\ \bar{L}_{i,j}^4(u,v) + B_1(u)(B_1(v)[L_{i,j} + \alpha_{i,j}^{00}\delta_i^u\delta_j^v L_{i,j}^{00}] + \\ B_2(v)[L_{i,j+1} - \alpha_{i,j}^{01}\delta_i^u\delta_j^v L_{i,j+1}^{01}]) + \\ B_2(u)(B_1(v)[L_{i+1,j} - \alpha_{i,j}^{10}\delta_i^u\delta_j^v L_{i+1,j}^{10}] + \\ B_2(v)[L_{i+1,j+1} - \alpha_{i,j}^{11}\delta_i^u\delta_j^v L_{i+1,j+1}^{11}]) < 0. \end{aligned} \tag{29}$$

The values of $\alpha_{i,j}^{00}$, $\alpha_{i,j}^{01}$, $\alpha_{i,j}^{10}$ and $\alpha_{i,j}^{11}$ should be between 0 and 1. To determine their values, the following two cases are considered.

Case 1. $L_{i,j} + \delta_i^u\delta_j^v L_{i,j}^{00}, L_{i,j+1} - \delta_i^u\delta_j^v L_{i,j+1}^{01}, L_{i+1,j} - \delta_i^u\delta_j^v L_{i+1,j}^{10}, L_{i+1,j+1} + \delta_i^u\delta_j^v L_{i+1,j+1}^{11} \geq 0$. Getting the solutions of eq. (29) is a very complicated process. We simplify the process with the following techniques.

Similar to eqs. (8) and (9), for $(I, J) \in \{(00), (01), (10), (11)\}$, $\alpha_{i,j}^{IJ}$, is expressed in the following form:

$$\alpha_{i,j}^{IJ} = c_{i,j}^{IJ} + d_{i,j}^{IJ}\alpha, \tag{30}$$

where

$$c_{i,j}^{IJ} = (-1)^{I+J+1} L_{i+I,j+J} / (\delta_i^u\delta_j^v L_{i+I,j+J}^{IJ}),$$
$$d_{i,j}^{IJ} = 1 - c_{i,j}^{IJ}.$$

By substituting eq. (30) into eq. (29), one gets

$$\begin{aligned} L(u,v) = \bar{L}_{i,j}^1(u,v) + \bar{L}_{i,j}^2(u,v) + \bar{L}_{i,j}^3(u,v) + \\ \bar{L}_{i,j}^4(u,v) + \alpha\delta_i^u\delta_j^v\{B_1(u)B_1(v)d_{i,j}^{00}L_{i,j}^{00} - \\ B_1(u)B_2(v)d_{i,j}^{01}L_{i,j+1}^{01} \\ - B_2(u)B_1(v)d_{i,j}^{10}L_{i+1,j}^{10} + \\ B_2(u)B_2(v)d_{i,j}^{11}L_{i+1,j+1}^{11}\}. \end{aligned} \tag{31}$$

One may obtain α by solving eq. (31) using the following two algorithms.

Algorithm 1. Computing (u_k, v_k) where $L(\alpha, u, v)$ reaches its maximum value.

1. For fixed α, let $\bar{u}_0 = (u_i + u_{i+1})/2$, $\bar{v}_0 = (v_j + v_{j+1})/2$;
2. For $k = 0, 1, \cdots$, compute $(\bar{u}_{k+1}, \bar{v}_{k+1})$ using the following formula until $|\frac{\partial L(\alpha, \bar{u}_{k+1}, \bar{v}_{k+1})}{\partial u}| + |\frac{\partial L(\alpha, \bar{u}_{k+1}, \bar{v}_{k+1})}{\partial v}| < a\ small\ constant.$

$$\frac{\partial^2 L(\alpha, \bar{u}_{k+1}, \bar{v}_k)}{\partial u^2}(\bar{u}_{k+1} - \bar{u}_k) + \frac{\partial^2 L(\alpha, \bar{u}_{k+1}, \bar{v}_k)}{\partial u \partial v}(\bar{v}_{k+1} - \bar{v}_k) + \frac{\partial L(\alpha, \bar{u}_{k+1}, \bar{v}_k)}{\partial u} = 0,$$
$$\frac{\partial^2 L(\alpha, \bar{u}_{k+1}, \bar{v}_k)}{\partial u \partial v}(\bar{u}_{k+1} - \bar{u}_k) + \frac{\partial^2 L(\alpha, \bar{u}_{k+1}, \bar{v}_k)}{\partial v^2}(\bar{v}_{k+1} - \bar{v}_k) + \frac{\partial L(\alpha, \bar{u}_{k+1}, \bar{v}_k)}{\partial v} = 0.$$

Algorithm 2. Computing α where $L(\alpha, u, v) \leq 0$ using the following binary method.

1. For $\alpha_0 = 0$ and $\alpha_1 = 1$, compute the corresponding $(\bar{u}^0, \bar{v}^0)$ and $(\bar{u}^1, \bar{v}^1)$, respectively, using algorithm 1. Note that $L(\alpha_0, \bar{u}^0, \bar{v}^0) \leq 0$, $L(\alpha_1, \bar{u}^1, \bar{v}^1) \geq 0$;
2. For $\alpha_c = (\alpha_0 + \alpha_1)/2$, compute the corresponding $(\bar{u}^c, \bar{v}^c)$ using algorithm 1. If $L(\alpha_c, \bar{u}^c, \bar{v}^c) \leq 0$ then $\alpha_0 = \alpha_c$ else $\alpha_1 = \alpha_c$
3. Recursively repeat step 2 until $\alpha_1 - \alpha_0 < a\ small\ constant.$
4. Define α as follows:

$$\alpha = (\alpha_0 + \alpha_1)/2. \tag{32}$$

Now for $(I, J) \in \{(00), (01), (10), (11)\}$, $\boldsymbol{P}_{i,j}^{IJ}$ is defined by

$$\boldsymbol{P}_{i,j}^{IJ} := (c_{i,j}^{IJ} + d_{i,j}^{IJ}\alpha)\boldsymbol{P}_{i,j}^{IJ}. \tag{33}$$

where $c_{i,j}$ and $d_{i,j}$, α are defined in eqs. (30) and (32) respectively.

Case 2. some values of $L_{i,j} + \delta_i^u\delta_j^v L_{i,j}^{00}$, $L_{i,j+1} - \delta_i^u\delta_j^v L_{i,j+1}^{01}$, $L_{i+1,j} - \delta_i^u\delta_j^v L_{i+1,j}^{10}$, $L_{i+1,j+1} + \delta_i^u\delta_j^v L_{i+1,j+1}^{11}$ are less than zero. For example, $L_{i,j} + \delta_i^u\delta_j^v L_{i,j}^{00} < 0$. In this case, eq. (31) is of the following form:

$$\begin{aligned} L(u,v) = \bar{L}_{i,j}^1(u,v) + \bar{L}_{i,j}^2(u,v) + \bar{L}_{i,j}^3(u,v) + \\ \bar{L}_{i,j}^4(u,v) + B_1(u)B_1(v)(L_{i,j} + \delta_i^u\delta_j^v L_{i,j}^{00}) + \\ \alpha\delta_i^u\delta_j^v\{-B_1(u)B_2(v)d_{i,j}^{01}L_{i,j+1}^{01} \\ - B_2(u)B_1(v)d_{i,j}^{10}L_{i+1,j}^{10} \\ + B_2(u)B_2(v)d_{i,j}^{11}L_{i+1,j+1}^{11}\} \end{aligned} \tag{34}$$

Hence, one can determine $\alpha_{i,j}^{IJ}$, $I = 0, 1$, $J = 0, 1$, using algorithms 1 and 2. The other cases when more than one value of $L_{i,j} + \delta_i^u\delta_j^v L_{i,j}^{00}$, $L_{i,j+1} - \delta_i^u\delta_j^v L_{i,j+1}^{01}$, $L_{i+1,j} - \delta_i^u\delta_j^v L_{i+1,j}^{10}$, $L_{i+1,j+1} + \delta_i^u\delta_j^v L_{i+1,j+1}^{11}$ are less than zero can be handled similarly.

5.2. Example

Two interpolants to the surface $Z_2(u, v)$ defined in (28) are shown in Figure 5. The interpolation conditions are data points and first derivatives at the vertices of the network shown in Figure 2. All the data points are below a given plane. The bottom figure is the interpolant without constraint where the portion above the given plane is discarded. The top figure is the interpolant using the method presented in this paper with the given plane as constraint.

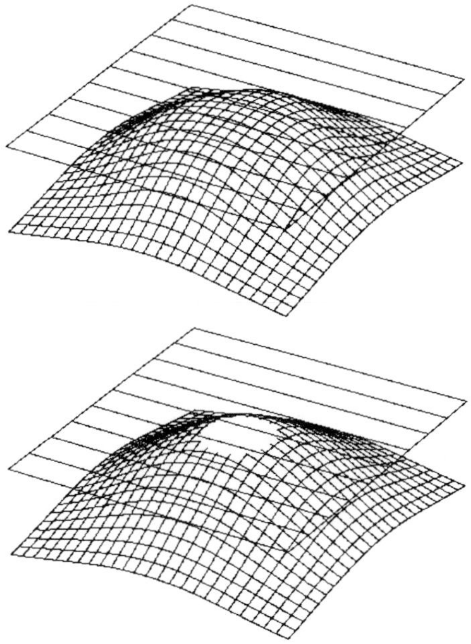

Figure 5. Bottom one: interpolant without constraint, Top one: interpolant with plane constraint.

6. Conclusion

A method for constructing a C^1 piecewise bicubic Bézier surface that interpolates a set of data points and lies on the same side of a given plane as the data points is presented. The new method has two advantages. First, the twist vector generation process has an interpolation precision of cubic polynomials. Second, a constructed surface in general has less thin plate energy and, consequently, is smoother than the original surface. Nevertheless, the new method is far from being an optimal approach. There is plenty of room to improve the presented approach in both the tangent vectors and the twist vectors modification processes.

Acknowledgments. The financial support of Ford Motor Company through a URP grant and the consulting support of Drs. Paul Stewart and Yifan Chen of the Ford Research Lab are deeply appreciated.

References

[1] de Boor, C., A practical guide to splines, Springer Verlag, New York, 1978, 318.

[2] Franke, R., A critical comparison of some methods for interpolation of scattered data. Naval Postgraduate school, Tech. Report NPS-53-79-003, 1979.

[3] Gregory, J. A. and Delbourgo, R., Piecewise rational quadratic interpolation to monotonic data, IMA J. Numer. Anal.,2 (1982), pp 123-130.

[4] Goodman, T.N.T. and Unsworth, K., Shape preserving interpolation by parametrically defined curves. SIAM J. Numer.Anal., 25(1988), pp 1453-1465.

[5] Goodman, T.N.T., Ong, B.H. and Unsworth, K., Constrained interpolation using rational cubic splines, in: Farin G., ed., NURBS for Curve and Surface Design, SIAM, 1991, 59-74.

[6] Goodman, T.N.T., Ong, B.H., Shape preserving interpolation by space curves. CAGD, 15(1997), pp 1-17.

[7] Foley, T. A., Goodman, T. N. T. and Unsworth, K., An Algorithm for shape preserving parametric interpolation curves with G^2 continuity. in: Lyche, T. and Schumaker, L. L., ed., Mathematical Methods in Computer Aided Geometric Design, 1989, pp 249-259.

[8] Ong, B.H. and Unsworth, K., On non-parametric constrained interpolation, in: Lyche, T. and Schumaker, L. L., ed., Mathematical Methods in Computer Aided Geometric Design, 1992, pp 419-430.

[9] Schmidt, J. W., and Hess, W., Positivity of cubic polynomial on intervals and positive spline interpolation. BIT, 28(1988), pp 340-352.

[10] Zhang, C., Cheng, F., Constrained shape-preserving parametric curve interpolation with minimum energy, submitted.

Session 6

Approximation and Interpolation Methods II

Approximating NURBS Curves by Arc Splines

Xunnian Yang
Department of Applied Mathematics, Zhejiang University
Yuquan, Hangzhou 310027, People's Republic of China
Email: yxn@math.zju.edu.cn

Abstract

It is desirable to approximate a smooth curve by arc spline with fewest segments within prescribed tolerance. In this paper, we present an efficient algorithm for fitting planar smooth curve by arc spline. The main idea is that we construct the optimal arc spline by optimizing the interpolating biarc curve. The scheme consists of three steps: (1)sampling the curve based on consecutive tangents deviation; (2)construct the interpolating arc spline; (3)reduce the arc number to the minimum within prescribed tolerance. The algorithm can control the approximating error efficiently and results in the fewest number of arc segments.

1. Introduction

Arc spline is a kind of geometric curve made of circular arcs and straight line segments. Arc splines are frequently used as the tool path of CNC machinery[1,2]. In programming the tool path of CNC machinery, the number of arc segments should be decreased in order to reduce the number of instructions and tool motions. Approximating general smooth curves, especially NURBS[3,4] curves by arc splines play an important role for geometric processing. Research in this area has been active in recent years[5-10].

Marciniak and Putz[11] proposed the formula for fitting a spiral by arc spline with minimum segments and the optimal arc spline can be found by solving nonlinear equations. Qiu, et al[8] used minimax approximation and numerical techniques to construct optimal circular arc interpolation for NC tool path generation. Ong et al[7] construct optimal fitting arc curve by minimizing the bounding area of B-spline curve and the arc spline curve, and nonlinear optimization method has to be used to find the solution. Meek and Walton[5,6] studied the problem of fitting quadratic NURBS curve by arc spline by employing divide and conquer strategy. Ahn et al[9] construct the optimal approximation of quadratic Bezier curves based on accurate error measurement and solving cubic equations. Chuang, et al[10] construct one-sided approximate arc spline for B-spline curve based on convex hull property of B-spline curve. Yeung and Walton[2] fit smooth arc spline to the smooth curve by fitting the dense sampled data from the continuous curves. Wallner[16] gave multiresolution analysis of arc spline based on wavelets decomposition technique.

In this paper we present a practical approach to approximate smooth curve by optimal arc spline within the prescribed tolerance. We test our algorithm by approximating NURBS curves. The method consists of following main steps:

1. Sample and interpolate the NURBS curve by arc spline with tolerance τ_1. In fact, tolerance τ_1 are always smaller than 10^{-6} and can be ignored.
2. Merging consecutive arc segments into optimal arc spline with fewest segments within tolerance τ_2. The total deviation is $\tau_1+\tau_2$.

The method owns several advantages over existing arc approximating method. Firstly, it can approximate any type of smooth parametric curves by arc spline. Secondly, constructing the interpolating arc spline by sampling the original curve are efficient and the error are controllable. At the end, the optimal approximating arc spline can be obtained in an efficient way and the segment number is much smaller compared with existing methods.

2. Preliminaries

0-7695-0562-7/00 $10.00

To fit the parametric curve by arc spline, we first construct the biarc spline interpolating the sampled points and matching the tangents at the points. A biarc curve between two sampled points consists of two smoothly connected arc segments that interpolate the two end points and the end tangents. If we sample the parametric curve with enough points and density, the fitting error from biarc curve to smooth parametric curve can be estimated from the following theorems[12]:

Theorem 1. Let C_A and C_B be the arcs that interpolate points A and B and unit tangent vector at A or at B, respectively. Any biarc that joins point A to the distinct point B and matches given tangent vectors at the two points lies between the bounding circular arcs C_A and C_B.

Theorem 2. If a given spiral of positive increasing curvature is repeatedly divided into segments such that the lengths of the segments approach zero, each segment will eventually be enclosed by the bounding circular arcs derived from that segment.

Theorem 3. As the length of a spiral approaches zero, the distance from the spiral to the approximating biarc approaches $\frac{1}{13.5}$ of the distance between the bounding circular arcs.

Proof, see ref[12].

Suppose P_A and P_B are two consecutive points associated with tangent vectors T_A and T_B on the curve. Let α and β be the angles from T_A to P_AP_B and from P_AP_B to T_B, respectively, then the maximum distance between the two circular arcs is $\frac{\|P_B - P_A\|}{2}\left|\tan\frac{\alpha}{2} - \tan\frac{\beta}{2}\right|$.

For biarc construction, we cite the formula from Su and Liu[13]. The biarc can be constructed in a local coordinate system with origin at P_A and x coordinate paralleling T_A (see Fig. 1). The conversion of points or vectors in one coordinate system to another can be achieved by direct transformation. Let O_1 and O_2 be the centers of two arc segments with radii r_1 and r_2, respectively. The contact point is denoted as P_C and the unit tangent vector at P_C is $U = (\cos\theta \quad \sin\theta)$, where θ is an angle given ahead. Then the unit normal vector N_1 at point P_A is $N_1 = (0 \quad 1)$ and the normal vector at point P_C is $V = (-\sin\theta \quad \cos\theta)$. The rest free variables for the two arcs can be computed as follows:

$$l = \|P_B - P_A\|$$

$$r_1 = \frac{l}{2\sin((\alpha+\beta)/2)} * \frac{\sin((\beta-\alpha+\theta)/2)}{\sin(\theta/2)}$$

$$r_2 = \frac{l}{2\sin((\alpha+\beta)/2)} * \frac{\sin((2\alpha-\theta)/2)}{\sin((\alpha+\beta-\theta)/2)}$$

$$P_c = P_A + r_1(N_1 - V) \tag{1}$$

$$O_1 = P_A + r_1 N_1$$

$$O_2 = P_c + r_2 V$$

If slope angles α and β have the same sign, we can then construct a C_shaped biarc panel, otherwise construct a S_shaped biarc panel. For free variable θ, we choose $\theta = \alpha$ for C_shaped biarc panel and $\theta = (3\alpha - \beta)/2$ for S_shaped biarc panel. This choice can keep the monotony of the curve curvature which play much important role for the global fairness of the curve.

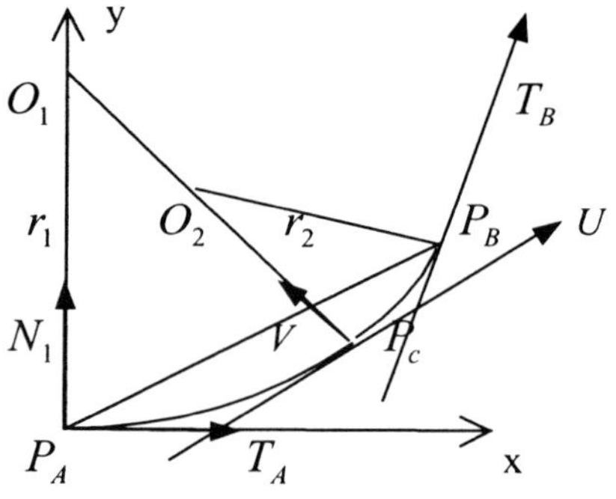

Figure 1. Biarc construction

Theorem 4. Let the curvature of a cubic spline that interpolates points P_1 and P_2 and matches the tangents at these two points be k_1 and k_2. If $k_1 < k_2$, then $k_1 < \frac{1}{r_1} < \frac{1}{r_2} < k_2$. If $k_1 > k_2$ then the inequality should be turned over.

Proof. See ref[13].

From theorem 4 we can conclude that the curvature of the interpolate biarc curve can still keep the monotone property of spirals if the segment tends to zero.

3. Fitting NURBS curves by arc spline

A NURBS curve with order k can always be represented as:

$$P(t)=\frac{\sum_{i=0}^{n} P_i w_i N_{i,k}(t)}{\sum_{i=0}^{n} w_i N_{i,k}(t)} \qquad (2)$$

where P_i, i=0, 1, ⋯, n are the control points and w_i, i=0, 1, 2, ⋯, n are the weights for the curve. The blending function $N_{i,k}(t)$ can be defined recursively,

$$N_{i,k}(t)=\frac{t-t_i}{t_{i+k}-t_i}N_{i,k-1}(t)+\frac{t_{i+k+1}-t}{t_{i+k+1}-t_{i+1}}N_{i+1,k-1}(t)$$

and $$N_{i,0}(t)=\begin{cases}1 & t_i \le t \le t_{i+1}\\ 0 & \text{otherwise}\end{cases} \qquad (3)$$

where t_i are the so-called knots forming a knot vector $\{t_0, t_1, \cdots, t_{n+k+1}\}$.

To approximate a NURBS curve, we can first divide the NURBS curve into a series of rational Bezier curve by knot insertion algorithm[14,15], and then subdivide the Bezier curve for arc spline interpolation within prescribed tolerance. A rational Bezier curve can be represented as

$$P(t)=\frac{\sum_{i=0}^{n} P_i w_i B_{i,n}(t)}{\sum_{i=0}^{n} w_i B_{i,n}(t)},$$ where

$B_{i,n}(t)=\binom{n}{i}t^i(1-t)^{n-i}$ are the Bernstein blending functions, and w_i are the weights. Let $P(t)=\frac{P_w(t)}{w(t)}$, then $P'(t)=\frac{P_w'(t)-P(t)w'(t)}{w(t)}$.

To approximate P(t) by arc spline, we sample the curve at t_i=i/N, i=0,1,⋯,N and compute the derivative at t_i simultaneously. Suppose the sampled point at t_i is P_i and the unit vector at t_i is T_i, then the distance between two bounding circular arcs interpolating P_{i-1} and P_i can be estimated based on theorem 3. If the fitting error is less than the prescribed tolerance τ_1, accept the subdivision, else sample another new point at $(t_{i-1}+t_i)/2$ and test the two subintervals respectively. This procedure can guarantee that the fitting arc spline is smooth and within prescribed tolerance. Generally, this error can be reduced to 10^{-6} to 10^{-8} and can always be ignored. Even more, since we subdivide the curve so many times that the curvature of the arc spline is monotone when the curvature of the parametric curve does.

4. Data optimization for arc spline

In this section, we'll derive explicit formula for optimizing the arc spiral consisting of arc segments with monotone curvature within the tolerance. The arc spiral is part of a chain of enclosed or enclosing circles and each arc segment except the first and last one is the part between two osculating points on the circle. We reduce the arc number by expanding the arc with biggest curvature to its limit within given tolerance for spiral with decreasing curvature, and shrinking the arc with smallest curvature for spiral with increasing curvature. When an arc is expanded or shrank, tangent continuity condition should be preserved. We implement this strategy by optimizing every three consecutive arc segments into two smoothly connected approximate arc segments within the permitted tolerance step by step. For each step, the tolerance should be recomputed for global constraints. This procedure continues until there are no two arcs can replace the current three arcs within the permitted tolerance. Then the first arc is the optimal approximate arc and we continue the optimization procedure from the second one. We obtain the optimal approximate arc spline to the arc spiral when the last three arcs are optimized into a biarc panel. The detail for optimization procedure is given below.

4.1. Arc optimization for arc spiral with decreasing curvature

Suppose O_0, O_1 and O_2 be three smooth connected arc segments with radii r_0, r_1, and r_2, respectively, and the radii satisfy $r_0 < r_1 < r_2$. The arc segments start from P_0 and end at P_3, two contact points for the curve are P_1 and P_2. To determine two new smoothly connected arc segments from the three arcs with controllable deviation, we construct the new arcs by expanding arc O_0 and shrinking arc O_2 with P_0 and P_3

fixed (see Fig2). Let $V_0 = \frac{O_0 - P_0}{\|O_0 - P_0\|}$ and $V_2 = \frac{O_2 - P_3}{\|O_2 - P_3\|}$, then the centers for two new arc segments can be obtained as: $O_a = O_0 + tV_0$ and $O_b = O_2 - uV_2$, where t and u are two variables that can be decided by continuity condition and tolerance constraint. The radii for arc O_a and O_b are $r_a = r_0+t$ and $r_b = r_2-u$, respectively.

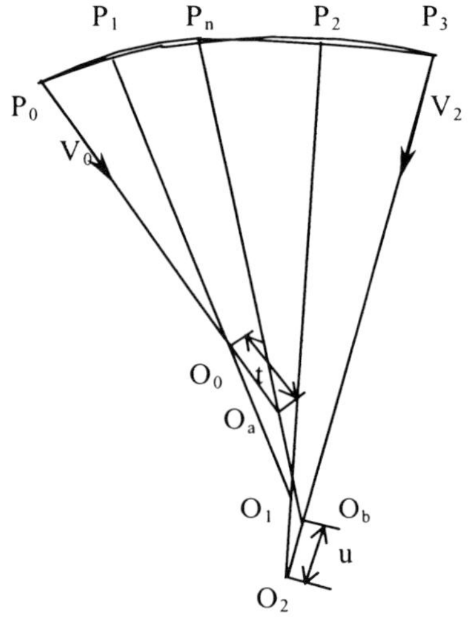

Figure 2. Optimize triarc segment by biarc

To guarantee arc segments O_a and O_b are tangent continuous at contact point, the radii for the two arcs should satisfy the following condition:

$$r_a + \|O_a - O_b\| = r_b \tag{4}$$

by expressing O_a, O_b, r_a and r_b as the function of t and u, we obtain the following formula:

$$\|O_2 - O_0 - uV_2 - tV_0\| = r_2 - r_0 - u - t \tag{5}$$

Let $D = O_2 - O_0$ and $t_0 = r_2 - r_0$, then we can derive the following formula

$$u = \frac{X_0 + X_1 t}{Y_0 + Y_1 t} \tag{6}$$

where $X_0 = t_0^2 - D^2$, $X_1 = 2(DV_0 - t_0)$, $Y_0 = 2(t_0 - DV_2)$ and $Y_1 = 2(V_0V_2 - 1)$. When t or u is chosen the two new arcs can then be obtained. Let $V_n = \frac{O_a - O_b}{\|O_a - O_b\|}$, the contacting point for arc O_a and O_b is $P_n = O_a + r_a V_n$.

To guarantee that arc O_a is within permitted tolerance ε of the initial curve, we expect that the distance from point P_1 to arc O_a is no more than ε (see Fig. 3). Let $V_b = \frac{P_1 - O_0}{\|P_1 - O_0\|}$, the nearest point on arc O_a to P_1 is P, then $P = O_0 + (r_0 + d)V_b$. Because $O_0 = O_a - tV_0$, then $P - O_a = -tV_0 + (r_0 + d)V_b$.

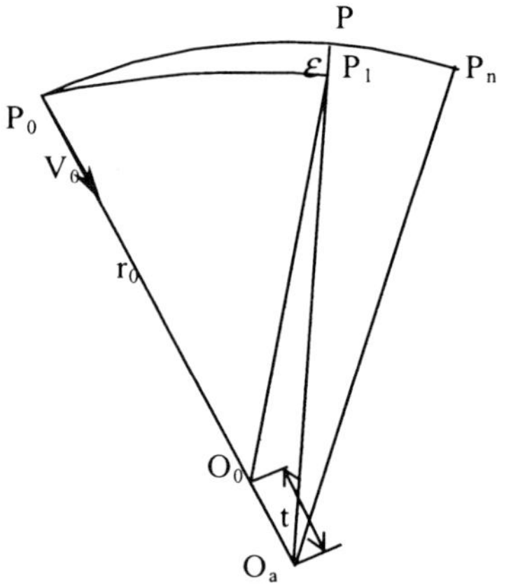

Figure 3. Tolerance constraint for arc expansion

Since the length of vector $P - O_a$ is r_a, then

$$r_a^2 = [-tV_0 + (r_0 + d)V_b]^2 \le [-tV_0 + (r_0 + \varepsilon)V_b]^2 \tag{7}$$

and because $r_a = r_0 + t$, we can conclude that

$$t \le t_b = \frac{2r_0\varepsilon + \varepsilon^2}{2[r_0 + (r_0 + \varepsilon)V_0V_b]} \tag{8}$$

To guarantee the validity of the new arc segments, when arc O_0 is expanded, another constraint is that the circle can not exceed arc O_2(see Fig. 4):

$$\|O_a - O_2\| + r_a \le r_2 \tag{9}$$

from this inequality we can conclude that

$$t \le t_c = \frac{(r_2 - r_0)^2 - (O_0 - O_2)^2}{2(r_2 - r_0) + (O_0 - O_2)V_0} \tag{10}$$

The maximum value for t can now be chosen as t=min(t_b, t_c). This also indicates that the new arc curve is still arc spiral.

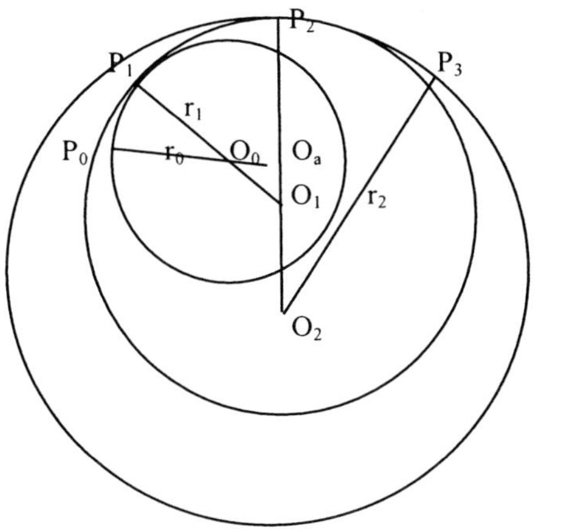

Figure 4. Expansion for arc O_0 is within arc O_2

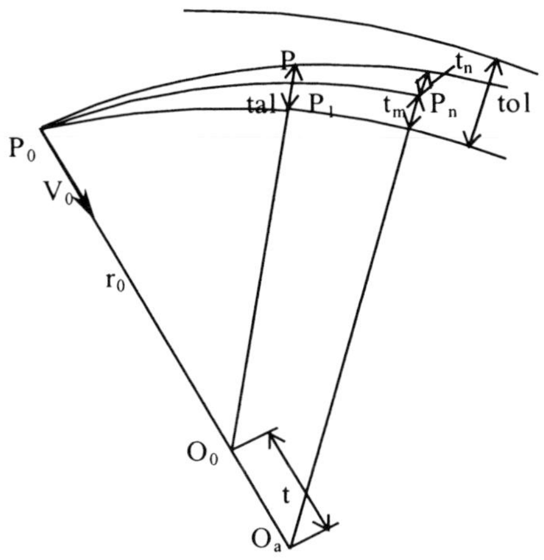

Figure 5. Compute the active tolerance

Let d_a be the distance from point P_n to arc O_1, $d_a = \|P_n - O_1\| - r_1$, and the distance from point P_2 to arc O_b is $d_b = \|P_2 - O_b\| - r_b$. Since arc O_a may be an arc by expanding a former one, then d_a should be no more than the active bounding tolerance tal, i.e., $d_a \le tal$. Because we construct arc O_b by shrinking arc O_2, then d_b should satisfy the inequality $d_b \le tbl$ to achieve the tolerance constraint. If both of these inequalities hold simultaneously, we can replace arcs O_0, O_1 and O_2 by arcs O_a and O_b, else set arc O_0 aside and choose arcs O_1, O_2 and another arc segment from the spline for further optimization.

When the optimization procedure starts, we choose the active tolerance tal and tbl both as tol, else the active tolerance should be recomputed according to the geometry of the fitting biarcs.

The distance from point P_1 to the expanded arc O_a can be computed as $t_l = r_a - \|P_1 - O_a\|$, then the permitted expansion magnitude at point P_1 is tal-t_l (see Fig. 5). Because lines O_0P_1 and O_aP_n are approximately parallel, then arc O_a should expand no more than $t_e = scl * (tal - t_l)$ at point P_n, where scl is a scaling coefficient and can be estimated as $scl = \frac{\|P_0 - P_n\|}{\|P_0 - P_1\|}$. Since point P_n lies outside of arc O_1, the distance from point P_n to arc O_1 is $t_m = \|P_n - O_1\| - r_1$, and the expansion magnitude for arc O_a at point P_n should not exceed $t_f = tol - t_m$ too. The permitted tolerance at point P_n is tal=min(t_e, t_f). It is difficult to estimate tbl accurately, we provide a well working empirical formula here. Let the arc next to arc O_2 be arc O_3 and the last end point of arc O_3 is P_4. To guarantee that point P_2 is within the current permitted tolerance tbl for latter optimization steps when $d_b > 0$, tbl for point P_3 can be estimated as $(tbl - d_b)/scl + d_a$, where $scl = \frac{\|P_2 - P_4\|}{\|P_2 - P_3\|} * 0.65$. When d_b is equal to zero, i.e., P_2 is on arc O_b, tbl can be kept unchanged.

4.2. Arc optimization for arc spiral with increasing curvature

Suppose O_0, O_1 and O_2 are three smoothly connected arc segments with increasing curvature, i.e., $r_0 > r_1 > r_2$. The derivation for approximate arcs is similar as arc spiral with decreasing curvature. We can assume that the arc spiral starts at P_0 and ends at P_3, and the two contact points are P_1 and P_2. The two approximate arcs O_a and O_b are shrank and expanded from arc O_0 and O_2, respectively. As in case 4.1, let V_0 be the unit vector pointing from P_0 to O_0 and V_2 be the unit vector pointing from P_3 to O_2. The centers for two new arcs can be expressed as $O_a = O_0 - uV_0$ and $O_b = O_2 + tV_2$. The radii for the two arcs are $r_a = r_0 - u$ and $r_b = r_2 + t$. From the smooth contacting condition we can deduce

$$r_b + \|O_b - O_a\| = r_a \quad (11)$$

Let $D = O_2 - O_0$ and $t_0 = r_0 - r_2$, from this equality we can deduce the function of t as:

$$t = \frac{X_0 + X_1 u}{Y_0 + Y_1 u} \quad (12)$$

where $X_0 = D^2 - t_0^2$, $X_1 = 2(DV_0 + t_0)$, $Y_0 = -2(t_0 + DV_2)$ and $Y_1 = 2(1 - V_0 V_2)$.

The value of u can be chosen according to tolerance constraint. Let $V_n = \frac{O_b - O_a}{\|O_b - O_a\|}$, the contacting point for two new approximate arcs is $P_n = O_a + r_a V_n$. Similar as case 3.1, the shrink magnitude for arc O_0 at point P_1 should no more than permitted tolerance ε. Let $V_b = \frac{P_1 - O_0}{\|P_1 - O_0\|}$, the tolerance constraint for variable u is

$$u \le u_b = \frac{2r_0\varepsilon + \varepsilon^2}{2[r_0(1 + V_0 V_b) + \varepsilon]} \quad (13)$$

When arc O_0 is shrank, to preserve the continuity property of new arc segments, arc O_2 should still lies inside the shrank arc, i.e. $\|O_2 - O_a\| \le r_a - r_2$. From this inequality we can deduce the following bound for u:

$$u \le u_c = \frac{t_0^2 - D^2}{2(t_0 + DV_0)} \quad (14)$$

the value of u can be chosen $u = \min(u_b, u_c)$.

Let d_a be the distance from point P_n to arc O_1, $d_a = \|P_n - O_1\| - r_1$ and the distance from point P_2 to arc O_b is $d_b = \left|\|P_2 - O_b\| - r_b\right|$. Since arc O_1 maybe the expanded arc, then d_a should satisfy the inequality $d_a \le 0$ as well as the bounding tolerance $d_a \ge (-tal)$. Similarly as case 4.1, d_b should satisfy the inequality $d_b < tbl$ for keeping point P_2 within tolerance. If all these inequalities hold simultaneously, we can replace arcs O_0, O_1 and O_2 by arcs O_a and O_b, else set arc O_0 aside and choose three arcs from the spline for further optimization.

The distance from point P_1 to the shrank arc O_a can be computed as $t_l = \|P_1 - O_a\| - r_a$, then the permitted shrink magnitude at point P_1 is tal-t_l. Let $scl = \frac{\|P_0 - P_n\|}{\|P_0 - P_1\|}$, then arc O_a should shrink no more than $t_e = scl * (tal - t_l)$ at point P_n. The distance from point P_n to arc O_1 is $t_m = \|P_n - O_1\| - r_1$ the shrink magnitude for arc O_a at point P_n should not exceed $t_f = tol - t_m$ too. The permitted maximum tolerance at point P_n is tal=min(t_e, t_f). Active tolerance tbl for point P_2 can be similarly as case 4.1.

4.3. The algorithm

For every smooth parametric curve, the approximate arc spline are always composed of a series of arc spirals with monotone curvature. Arc spirals with decreasing or increasing curvature can then be optimized by the former formulas. When three chosen arcs are replaced by two optimized arc segments within the tolerance, the radii for the spiral are still monotone. In fact, we can optimize a smooth arc curve in a uniform way. The main procedure for arc spline optimization is given below:

Algorithm Optimization-Procedure-for-Arc-Spline

input: a smooth circular arc spline and the tolerance *tol*

output: an optimal new smooth arc spline within *tol*

```
choose first two arcs from the spline as arc1 and arc2;
set tal = tol;
while (next arc != NULL)
{
    arc0 = arc1;
    arc1 = arc2;
    arc2 = next arc;
    if (arc radii are monotone)
    {
        Compute two optimized arcs arc_a and arc_b;
        if (both arc_a and arc_b are within the tolerance)
        {
            delete arc0;
            arc1 = arc_a;
```

```
        arc2 = arc_b;
        compute the tolerance tal;
}       else tal = tbl = tol;
        }       else  tal = tbl = tol;
}       /* end of while */
output the final arc spline.
```

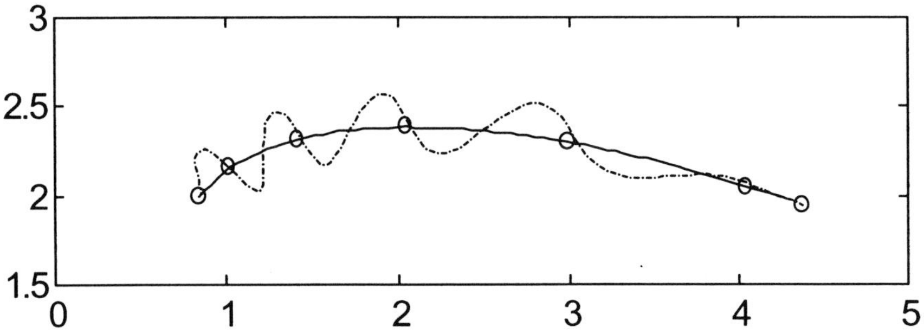

Example 1 (tol=0.001)

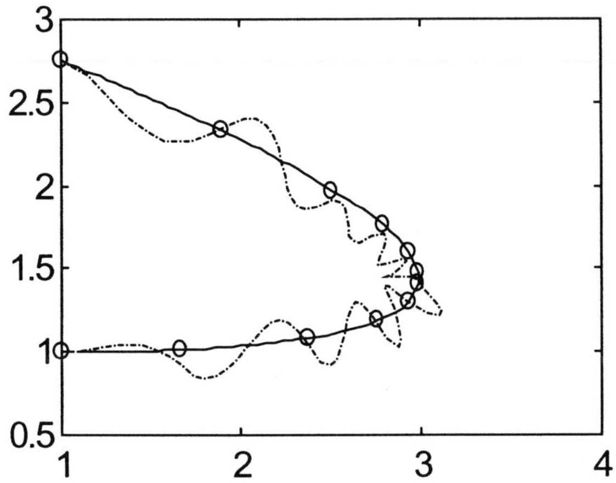

Example 2 (tol=0.001)

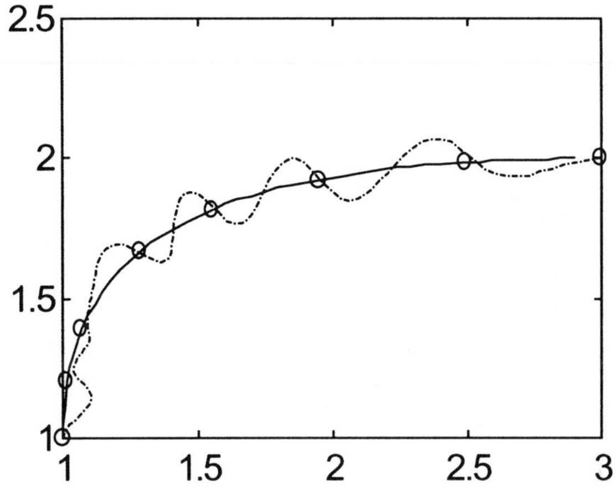

Example 3 (tol=0.0005)

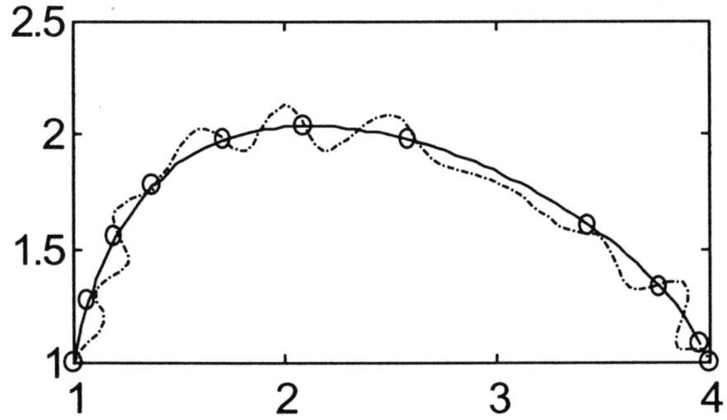

Example 4a (tol=0.0005)

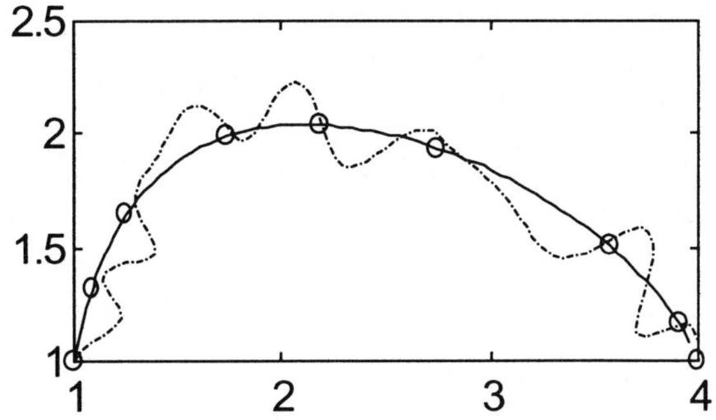

Example 4b (tol=0.001)

5. Examples

We have applied our algorithm on a number of parametric curves and obtain satisfying approximating results. We cite a few examples here to show the efficiency of our algorithm.

In Example 1 we approximate a rational quadratic Bezier curves with control points (0.85,1.99), (1.5, 2.9), (4.37,1.95) and weights 1, 0.8, 1, respectively. By sampling 51 points on the curve and interpolating the points with biarc spline consists of 100 arcs. An optimized G^1 arc spline consists of only 6 arcs is obtained with tolerance 0.001. The fitting error of this and following examples magnified 200 times is plotted in dashed curve. For Example 2, the data is the same as example 3 of Ahn et al[9], we optimize the arc spline interpolating 101 points on the curve. There are only 11 arcs are left in the final arc spline while Ahn's example consists of 15 segments. For example 3, we use the control points as (1,1), (1,2), (3,2) and weights 1, 1.2, 1. We obtain the optimal approximating arc spline consisting of 7 arc segments by first sampling the Bezier curve with 50 points and optimizing the interpolating arc spline within tolerance 0.0005. In example 4, we construct the optimal arc spline approximating a cubic rational Bezier curve with control points (1,1), (1.3,2.5), (3.5,2.2), (4,1) and weights 1, 1.2, 0.8, 1. In example 4a, we obtain optimal approximate arc spline with 10 segments within tolerance 0.0005 and in example 4b the approximate arc spline consists of just 8 arcs within tolerance 0.001.

6. Conclusions and discussion

In this paper we present a very practical algorithm for approximating smooth parametric curve by optimal G^1 arc spline within the given tolerance. The final arc spline interpolates the curve ends and matches the end tangents. We illustrate the algorithm by approximating NURBS curves with arc splines.

The crux for approximating parametric curve by arc spline is how to estimate the fitting error accurately and robustly. We employ here the theory of Meek and Walton[12] and an interpolating biarc spline within arbitrary prescribed accuracy can be obtained by sampling the parametric curve with enough density. The optimization procedure can be applied on the fitting arc spline, explicit formula have been derived to construct the optimal approximate arc spline.

If there are some arc segments with large radii or there are some straight lines in the fitting arc spline, the optimization process may be influenced by the numerical errors or more care should be taken. Approximating parametric curve by optimal one-sided arc spline can be deduced with a few modification of the current procedure.

Acknowledgements

The author owns thanks to Prof. Wang Guozhao for his support and help. This work is supported by the NNSF(19971079), Foundation of State Key Basic Research 973 Development Programming Item(G1998030600) and the Natural Science Foundation of Zhejiang Province. The author also acknowledges K. C. Wong Foundation for financial support of this project.

References

[1] K. M. Bolton, "Biarc curves", *Computer Aided Design*, 1975, 7, pp. 89-92.

[2] M. Yeung, and D. J. Walton, "Curve fitting with arc splines for NC tool path generation", *Computer Aided Design*, 1994, 26(11), pp. 845-849.

[3] G. Farin, *Curves and Surfaces for Computer Aided Geometric Design, a practical guide.2.* New York: Academic Press, 1990.

[4] L. Piegl, "On NURBS: A survey", *IEEE CG&A*, 1991, 11(1), pp. 55-71.

[5] D. S. Meek, and D. J. Walton, "Approximating quadratic NURBS curves by arc splines", *Computer Aided Design*, 1993, 25(6), pp. 371-376.

[6] D. S. Meek, and D. J. Walton, "Approximating of quadratic Bezier curves by arc splines", *Journal of Computational and Applied Mathematics*, 1994, 54, pp. 107-120.

[7] C.L. Ong, S. Wong, H.T. Loh, and X.G. Hong, "An optimization approach for biarc curve fitting of B-spline curves", *Computer Aided Design*, 1996,

28(12) pp. 951-959.

[8] H. Qiu, K. Cheng, and Y. Li, "Optimal circular arc interpolation for NC tool path generation in curve contour manufacturing", *Computer Aided Design*, 1997, 29(11), pp. 751-760.

[9] Y. J. Ahn, H. O. Kim, and K.-Y. Lee, "G^1 arc spline approximation of quadratic Bezier curves", *Computer Aided Design*, 1998, 30(8), pp. 615-620.

[10] S.-H. Chuang, and C. Z. Kao, "One-sided arc approximation of B-spline curves for interference-free offsetting", *Computer Aided Design*, 1999, 31, pp. 111-118.

[11] K. Marciniak, and B. Putz, "Approximation of spirals by piecewise curves of fewest circular arc segments", *Computer Aided Design*, 1984, 16(2), pp. 87-90.

[12] D. S. Meek, and D. J. Walton, "Approximating smooth planar curves by arc splines", *Journal of Computational and Applied Mathematics*, 1995, 59, pp. 221-231.

[13] B. Q. Su, and D. Y. Liu, *Computational Geometry*, Academic Press, USA(1989).

[14] W. Boehm, "Inserting new knots into B-spline curves", *Computer Aided Design*, 1980, 12(4), pp. 199-201.

[15] W. Boehm, "Generation of the Bezier points of B-spline curves and surfaces", *Computer Aided Design*, 1981, 13, pp. 365-366.

[16] J. Wallner, "Generalized multiresolution analysis for arc splines", in: Dahlen, M., Lyche, T. And Schumaker, L. L., eds., *Mathematical Methods for Curves and Surfaces II*, Vanderbilt University Press, pp. 537-544.

A Method of Shape Control of Curve Design

Qi Duan
Department of Mathematics and Physics
Shandong University of Technology
Jinan,250061,China

T.S.Chen
Department of Computer Science and Information Engineering
Dayeh University,Taiwan

K.Djidjeli and W.G.Price
Department of Ship Science
University of Southampton, Southampton
England, SO17 1BJ

E.H.Twizell
Department of Mathematics and Statistics
Brunel University, Uxbridge, Middlesex
England,UB8 3PH

Abstract

Constraining an interpolating curves to be bounded in a given region is an important task in curve design. In [1], the rational cubic spline with linear denominator has been used to control the interpolating curves to be bounded in the given region, but it does not work in some case. This paper deals with the weighted rational cubic spline with linear denominator for this kind of constraint, the sufficient condition for controlling the interpolating curves to be bounded in the given region are derived. An example is given in the end of this paper which shows that the constraint which can not be done by the rational spline defined in [1] could be achieved by the weighted rational spline.

Keywords and phrases: *curve design, rational interpolation, cubic spline, constrained interpolation.*

1 Introduction

Constraining and controlling the shape of curves and surfaces according to the manufacturing needs is an important task in CAD. Spline interpolation is a useful and powerful tool for curve and surface design. Many authors have studied several kinds of splines for curve and surface control [2,3,4,5,6,7,8]. In recent years, the rational spline, especially the rational cubic spline, and its application to shape control have been considered [1,9,10,11,12,13,14,15]. Since those rational splines are with parameters, it is possible to control the shape of the interpolating curves just by choosing the suitable parameters. In [1], a method is given to control the interpolating curves to be bounded in the given region by using the rational cubic spline with linear denominator, but it can not be achieved in some cases. This paper deals with the weighted rational cubic interpolation with linear denominator. Since there are more parameters in this kind of spline, it is convenient to use it to constrain the interpolating curve to lie above or below a straight or quadratic line in the subinterval.

In Section 2 of this paper, the weighted rational cubic splines with linear denominator are constructed. In Section 3, this splines are used to control the interpolating curve to lie above or below the given straight lines, the sufficient conditions for this kind of con-

0-7695-0562-7/00 $10.00 © 2000 IEEE

straint are derived, and the existence conditions show that for general cases the parameters are existed to achieve this constraint. Section 4 deals with the case that the controlling curves are quadratic curves instead of the straight lines. An example is given in Section 5.

2 Weighted Rational Cubic Interpolation

In order to construct the weighted rational cubic spline, the rational cubic spline with linear denominator[1] and the rational cubic spline based on the function values[11] will be restated briefly first.

Given a data set $\{(t_i, f_i, d_i), i = 0, 1, \ldots, n\}$, f_i and d_i are the function value and the first-order derivative value of the being interpolated function $f(t)$ defined at the knots respectively, where $t_0 < t_1 < \cdots < t_n$ is the knot spacing. Let $h_i = t_{i+1} - t_i$, $\theta = (t - t_i)/h_i$, and let α_i and β_i be positive parameters. The C^1-continuous, piecewise, rational cubic spline with linear denominator is defined in the interpolating interval $[t_0, t_n]$ as follows[1]:

$$P^*(t)\big|_{[t_i,t_{i+1}]} = \frac{p_i^*(t)}{q_i^*(t)}, \qquad i = 0, 1, \ldots, n-1, \tag{1}$$

where

$$\begin{aligned} p_i^*(t) &= (1-\theta)^3\alpha_i f_i + \theta(1-\theta)^2 V_i^* \\ &\quad + \theta^2(1-\theta)W_i^* + \theta^3\beta_i f_{i+1}, \\ q_i^*(t) &= (1-\theta)\alpha_i + \theta\beta_i, \end{aligned}$$

and

$$\begin{aligned} V_i^* &= (2\alpha_i + \beta_i)f_i + \alpha_i h_i d_i, \\ W_i^* &= (\alpha_i + 2\beta_i)f_{i+1} - \beta_i h_i d_{i+1}. \end{aligned}$$

This rational cubic spline satisfies

$$P^*(t_i) = f_i, \quad P^{*\prime}(t_i) = d_i, \quad i = 0, 1, \ldots, n.$$

It is called the rational cubic spline with linear denominator.

Denote $\Delta_i = (f_{i+1} - f_i)/h_i$, A C^1-continuous, piecewise, rational cubic function $P(t)$ based on function values[11] is defined as follows

$$P_*(t)\big|_{[t_i,t_{i+1}]} = \frac{p_{i,*}(t)}{q_{i,*}(t)}, \qquad i = 0, 1, \ldots, n-2, \tag{2}$$

where

$$\begin{aligned} p_{i,*}(t) &= (1-\theta)^3\alpha_i f_i + \theta(1-\theta)^2 V_{i,*} \\ &\quad + \theta^2(1-\theta)W_{i,*} + \theta^3\beta_i f_{i+1}, \\ q_{i,*}(t) &= (1-\theta)\alpha_i + \theta\beta_i, \end{aligned}$$

and

$$\begin{aligned} V_{i,*} &= (\alpha_i + \beta_i)f_i + \alpha_i f_{i+1}, \\ W_{i,*} &= (\alpha_i + 2\beta_i)f_{i+1} - h_i\beta_i\Delta_{i+1}, \end{aligned}$$

Obviously, the spline $P_*(t)$ defined by (2) is also the C^1-continuous, piecewise, rational cubic spline in the interpolating interval $[t_0, t_{n-1}]$ and satisfies

$$P_*(t_i) = f_i, \quad P_*'(t_i) = \Delta_i, \quad i = 0, 1, \ldots, n-1.$$

Based on the idea of adding a new parameter to the interpolation function, a weighted rational cubic spline will be constructed as follows by using the rational cubic splines defined by (1) and (2). Let

$$P(t)\big|_{[t_i,t_{i+1}]} = \lambda P^*(t) + (1-\lambda)P_*(t) = \frac{p_i(t)}{q_i(t)}, \qquad i = 0, 1, \ldots, n-2. \tag{3}$$

where

$$\begin{aligned} p_i(t) &= (1-\theta)^3\alpha_i f_i + \theta(1-\theta)^2 V_i \\ &\quad + \theta^2(1-\theta)W_i + \theta^3\beta_i f_{i+1}, \\ q_i(t) &= (1-\theta)\alpha_i + \theta\beta_i, \end{aligned}$$

and

$$\begin{aligned} V_i &= (\lambda\alpha_i + \alpha_i + \beta_i)f_i + (1-\lambda)\alpha_i f_{i+1} \\ &\quad + \lambda\alpha_i h_i d_i, \\ W_i &= (\alpha_i + 2\beta_i)f_{i+1} - (1-\lambda)h_i\beta_i\Delta_{i+1} \\ &\quad - \lambda\beta_i h_i d_{i+1}, \end{aligned}$$

with $\lambda \in R$. $P(t)$ defined by (3) is a C^1-continuous, piecewise, rational cubic spline, which is called *the weighted rational cubic spline* with linear denominator from interpolations defined by (1) and (2).

In the special case, when the knots are equally spaced, V_i and W_i become

$$\begin{aligned} V_i &= (\lambda\alpha_i + \alpha_i + \beta_i)f_i + (1-\lambda)\alpha_i f_{i+1} \\ &\quad + \lambda\alpha_i h_i d_i, \\ W_i &= (\alpha_i + 3\beta_i - \lambda\beta_i)f_{i+1} - (1-\lambda)\beta_i f_{i+2} \\ &\quad - \lambda\beta_i h_i d_{i+1}. \end{aligned}$$

Furthermore, for a real number λ, the C^2-continuous rational cubic spline defined by (3) even could be constructed by requiring

$$P''(t_i+) = P''(t_i-)$$

for $i = 1, 2, \cdots, n-2$. The conditions lead to the following tri-diagonal system of linear equations:

$$\lambda h_i \frac{\alpha_{i-1}}{\beta_{i-1}} d_{i-1} + \lambda(h_i(1+\frac{\alpha_{i-1}}{\beta_{i-1}}) + h_{i-1}(1+\frac{\beta_i}{\alpha_i}))d_i + \lambda h_{i-1}\frac{\beta_i}{\alpha_i} d_{i+1}$$
$$= h_i(1+(\lambda+1)\frac{\alpha_{i-1}}{\beta_{i-1}})\Delta_{i-1} + (h_i(\lambda-1)(1+\frac{\alpha_{i-1}}{\beta_{i-1}}) + h_{i-1}(\lambda+(\lambda+1)\frac{\beta_i}{\alpha_i}))\Delta_i + h_{i-1}(\lambda-1)\frac{\beta_i}{\alpha_i}\Delta_{i+1};$$
$$i = 1, 2, \cdots, n-2. \quad (4)$$

If the knots are equally spaced, equation (4) becomes

$$\lambda \frac{\alpha_{i-1}}{\beta_{i-1}} d_{i-1} + \lambda(2+\frac{\alpha_{i-1}}{\beta_{i-1}} + \frac{\beta_i}{\alpha_i})d_i + \lambda\frac{\beta_i}{\alpha_i} d_{i+1}$$
$$= (1+(\lambda+1)\frac{\alpha_{i-1}}{\beta_{i-1}})\Delta_{i-1} + ((\lambda-1)(1+\frac{\alpha_{i-1}}{\beta_{i-1}}) + (\lambda+(\lambda+1)\frac{\beta_i}{\alpha_i}))\Delta_i + (\lambda-1)\frac{\beta_i}{\alpha_i}\Delta_{i+1};$$
$$i = 1, 2, \cdots, n-2. \quad (5)$$

Therefore, if the successive parameters $(\alpha_{i-1}, \beta_{i-1})$ and (α_i, β_i) satisfy (4) or (5) at $i = 1, 2, \cdots, n-2$, then $P(t) \in C^2(t_0, t_{n-1})$.

3 Constraining the Interpolating Curves to be Bounded between two Given Straight Lines

Let $f(t)$ be the being interpolated function , $P(t)$ be the interpolating function defined by (3), $g(t)$ be a given straight line or piecewise straight line with joints at the partition $\Delta: t_0 < t_1 < \cdots < t_n$, and the data set $\{(t_i, f_i, d_i), i = 0, 1, \cdots, n\}$ satisfy

$$f_i \geq (\text{or } \leq) g(t_i), \qquad i = 0, 1, \cdots, n.$$

If $P(t) \geq$ (or $\leq) g(t)$ for all $t \in [t_0, t_n]$, then, $P(t)$ is called a constrained weighted rational interpolating curves above(or below) $g(t)$.

It is assumed that the knots are equally spaced for the following part of this paper. Let the weighted rational interpolation defined by (3) as the interpolating tool, there is the following theorem about the sufficient condition for the interpolating curves to lie below a given straight lines in a given interval.

THEOREM 1. Let $\{(t_i, f_i, d_i, g_i^*), i = 0, 1, \cdots, n\}$ be a given data set with $f_i \leq g_i^*$, the sufficient condition for the weighted rational cubic spline $P(t)$ defined by (3) to lie below the linear function $g^*(t)$ in $[t_i, t_{i+1}]$ is that the parameters $\alpha_i > 0, \beta_i > 0$ and weighted coefficient $\lambda \in R$ satisfy the following linear inequalities:

$$\lambda\alpha_i(f_i - f_{i+1} + h_i d_i) + \alpha_i(f_i + f_{i+1} - g_{i+1}^* - g_i^*) + \beta_i(f_i - g_i^*) \leq 0, \quad (6)$$
$$\lambda\beta_i(f_{i+2} - f_{i+1} - h_i d_{i+1}) + \beta_i(3f_{i+1} - f_{i+2} - g_{i+1}^* - g_i^*) + \alpha_i(f_{i+1} - g_{i+1}^*) \leq 0. \quad (7)$$

Proof: From (3), it is easy to know that for all $t \in [t_i, t_{i+1}]$, $q_i(t) \geq 0$. So,

$$P(t) = \frac{p_i(t)}{q_i(t)} \leq g^*(t)$$

is equivalent to

$$p_i(t) - q_i(t) g^*(t) \leq 0.$$

Let

$$U_i(t) = p_i(t) - q_i(t) g^*(t),$$

then

$$U_i(t) = (1-\theta)^3\alpha_i f_i + \theta(1-\theta)^2 V_i + \theta^2(1-\theta)W_i + \theta^3\beta_i f_{i+1} - ((1-\theta)\alpha_i + \theta\beta_i)((1-\theta)g_i^* + \theta g_{i+1}^*) \leq 0, \quad (8)$$

Since

$$((1-\theta)\alpha_i + \theta\beta_i)((1-\theta)g_i^* + g_{i+1}^*\theta)$$
$$= (1-\theta)^2\alpha_i g_i^* + \theta^2\beta_i g_{i+1}^* + \theta(1-\theta)(\alpha_i g_{i+1}^* + \beta_i g_i^*)$$
$$= (1-\theta)^3\alpha_i g_i^* + \theta(1-\theta)^2(\alpha_i g_{i+1}^* + \beta_i g_i^* + \alpha_i g_i^*) + \theta^2(1-\theta)(\alpha_i g_{i+1}^* + \beta_i g_i^* + \beta_i g_{i+1}^*) + \theta^3\beta_i g_{i+1}^*,$$

(8) becomes

$$U_i(t) = (1-\theta)^3\alpha_i(f_i - g_i^*) + \theta(1-\theta)^2 A_i + \theta^2(1-\theta)B_i + \theta^3\beta_i(f_{i+1} - g_{i+1}^*) \leq 0,$$

where

$$A_i = V_i - (\alpha_i g_{i+1}^* + \beta_i g_i^* + \alpha_i g_i^*) = \lambda\alpha_i(f_i - f_{i+1} + h_i d_i) + \alpha_i(f_i + f_{i+1} - g_{i+1}^* - g_i^*) + \beta_i(f_i - g_i^*),$$
$$B_i = W_i - (\alpha_i g_{i+1}^* + \beta_i g_i^* + \beta_i g_{i+1}^*) = \lambda\beta_i(f_{i+2} - f_{i+1} - h_i d_{i+1}) + \beta_i(3f_{i+1} - f_{i+2} - g_{i+1}^* - g_i^*) + \alpha_i(f_{i+1} - g_{i+1}^*).$$

Since

$$\alpha_i(f_i - g_i^*) \le 0,$$
$$\beta_i(f_{i+1} - g_{i+1}^*) \le 0,$$

If $A_i \le 0$, and $B_i \le 0$, then $U_i(t) \le 0$ holds for all $t \in [t_i, t_{i+1}]$.
The proof is completed.

Similar as in Theorem 1, there is the sufficient condition for the interpolating curves to lie above a given straight lines in given interval.

THEOREM 2. Let $\{(t_i, f_i, d_i, g_i), i = 0, 1, \cdots, n\}$ be a given data set with $f_i \ge g_i$, the sufficient condition for the weighted rational cubic spline $P(t)$ defined by (3) to lie above the straight line $g(t)$ in $[t_i, t_{i+1}]$ is that the parameters $\alpha_i > 0, \beta_i > 0$ and weighted coefficient $\lambda \in R$ satisfy the following linear inequalities:

$$\lambda\alpha_i(f_i - f_{i+1} + h_i d_i) + \alpha_i(f_i + f_{i+1} - g_{i+1} - g_i) + \beta_i(f_i - g_i) \ge 0, \quad (9)$$
$$\lambda\beta_i(f_{i+2} - f_{i+1} - h_i d_{i+1}) + \beta_i(3f_{i+1} - f_{i+2} - g_{i+1} - g_i) + \alpha_i(f_{i+1} - g_{i+1}) \ge 0. \quad (10)$$

Theorem 1 and Theorem 2 lead to the following theorem

THEOREM 3. Let $\{(t_i, f_i, d_i, g_i, g_i^*), i = 0, 1, \cdots, n\}$ be a given data set with $f_i \ge g_i$ and $f_i \le g_i^*$, the sufficient condition for the weighted rational cubic spline curve $P(t)$ defined by (3) to lie above the linear function $g(t)$ and below the linear function $g^*(t)$ in $[t_i, t_{i+1}]$ is that the parameters $\alpha_i > 0, \beta_i > 0$ and weighted coefficient $\lambda \in R$ satisfy the following linear system inequalities (6)(7)(9) and (10).

The following Theorem 4 is about the existence of the weighted rational cubic spline curve $P(t)$ defined by (3) to be bounded in the given region described as in Theorem 3.

THEOREM 4. Let $\{(t_i, f_i, d_i, g_i, g_i^*), i = 0, 1, \cdots, n\}$ be any given data set with $g_i < f_i < g_i^*$, there must exist parameters $\alpha_i > 0, \beta_i > 0$ and coefficient $\lambda \in R^+$ such that the weighted rational spline defined by (3) is bounded between the straight lines $g(t)$ and g_i^* in the subinterval $[t_i, t_{i+1}]$.

Before proving Theorem 4, the following Corollary 1 will be proved, it will play an important part in the proof of the theorem 4.

Corollary 1. For any real numbers $a > 0, b > 0$ and $c \in R$, the enequality

$$cx + a \ge 0,$$
$$-cx + b \ge 0.$$

must have positive solution x.

Proof: (a). If $c = 0$, it is obvious that any real number $x > 0$ could be its solution.
(b). If $c > 0$, any number x which satisfy $0 < x < \frac{b}{c}$ could be its solution.
(b). If $c < 0$, any number x which satisfy $0 < x < \frac{-a}{c}$ could be its solution.
The proof is completed.

Furthermore, denote $x^* = \max\{|b/c|, |a/c|\}$, if $c \ne 0$, any x which satisfy $0 < x < x^*$ could be its solution.

The following is the proof of the Theorem 4.
When $t \in [t_i, t_{i+1}]$, let $\zeta_i = \frac{\beta_i}{\alpha_i}$, the enequalites (6)(7)(9) and (10) can be rewritten as

$$[\lambda(f_i - f_{i+1} + h_i d_i) + (f_i + f_{i+1} - g_{i+1} - g_i)] + \zeta_i(f_i - g_i) \ge 0, \quad (11)$$
$$[\lambda(f_{i+1} - f_i - h_i d_i) + (g_{i+1}^* + g_i^* - f_i - f_{i+1})] + \zeta_i(g_i^* - f_i) \ge 0, \quad (12)$$
$$\zeta_i[\lambda(f_{i+2} - f_{i+1} - h_i d_{i+1}) + (3f_{i+1} - f_{i+2} - g_{i+1} - g_i)] + (f_{i+1} - g_{i+1}) \ge 0, \quad (13)$$
$$\zeta_i[\lambda(f_{i+1} - f_{i+2} + h_i d_{i+1}) + (-3f_{i+1} + f_{i+2} + g_{i+1}^* + g_i^*)] + (g_{i+1}^* - f_{i+1}) \ge 0. \quad (14)$$

By Corollary 1, there must exist $\lambda_i > 0$ such that for any λ which satisfy $0 < \lambda < \lambda_i$ the following two inequalities hold.

$$\lambda(f_i - f_{i+1} + h_i d_i) + (f_i + f_{i+1} - g_{i+1} - g_i) \ge 0,$$
$$\lambda(f_{i+1} - f_i - h_i d_i) + (g_{i+1}^* + g_i^* - f_i - f_{i+1}) \ge 0.$$

Thus, for any $\zeta_i > 0$ (11) and (12)hold.

For this $\lambda > 0$, using the Corollary 1 again , there must exist $\zeta_i > 0$ such that both (13) and (14) hold. Thus, there must exist $\zeta_i > 0$ such that all the inequalities (11)- (14) hold. Denote

$$\lambda^* = \min\{\lambda_0, \lambda_1, \cdots, \lambda_{n-2},\}$$

let λ which satisfy $0 < \lambda < \lambda^*$ be the weighted coefficient, then for any subinterval $[t_i, t_{i+1}]$ there must exist $\zeta_i > 0$ such that (11)- (14) hold.
The proof is completed.

4 Constraining the Interpolating Curves to be Bounded between two Given Quadratic Curves

Let $f(t)$ be the being interpolated function , $P(t)$ be the weighted rational interpolating function defined by (3), the definition of constraint is similar as in Section 3. Let $g^*(t)$ be a quadratic or piecewise quadratic function defined on $[t_0, t_n]$ with joints at the partition $\Delta : t_0 < t_1 < \cdots < t_n$. For the given data set $\{(t_i, f_i, d_i, g_i^*, g_i^{*\prime}), i = 0, 1, \cdots, n\}$ with $f_i \le g^*(t_i)$, when $t \in [t_i, t_{i+1}]$, since

$$g_i^*(t) = (1-\theta)^2 g_i^* + \theta(1-\theta)(2g_i^* + g^{*\prime}_i h_i) + \theta^2 g_{i+1}^*,$$

where

$$g_i^* = g^*(t_i),\ g_{i+1}^* = g^*(t_{i+1}),\ g^{*\prime}_i = g^{*\prime}(t_i),$$

similar as the proof for Theorem 1, it is easy to get the follows

THEOREM 5. Let $\{(t_i, f_i, d_i, g_i^*, g_i^{*\prime}), i = 0, 1, \cdots, n\}$ be a given data set, and $g^*(t)$ be a given quadratic function satisfying $f_i \le g_i^*$, then the sufficient condition for the weighted rational cubic spline $P(t)$ defined by (3) to lie below the quadratic curve $g^*(t)$ in $[t_i, t_{i+1}]$ is that the positive parameters α_i, β_i and the weighted coefficient $\lambda \in R$ satisfy the following inequality system

$$\begin{aligned} &\lambda\alpha_i(f_i - f_{i+1} + h_i d_i) + \alpha_i(f_i + f_{i+1} - 2g_i^* \\ &-g_i^{*\prime} h_i) + \beta_i(f_i - g_i^*) \le 0, \qquad (15) \\ &\lambda\beta_i(f_{i+2} - f_{i+1} - h_i d_{i+1}) + \beta_i(3f_{i+1} - f_{i+2} \\ &-2g_i^* - g_i^{*\prime} h_i) + \alpha_i(f_{i+1} - g_{i+1}^*) \le 0. \qquad (16) \end{aligned}$$

Similar as for Theorem 5, the sufficient conditions for the weighted rational interpolating curves to lie above the quadratic curve $g(t)$ in $[t_i, t_{i+1}]$ could be derived and described as the following theorem.

THEOREM 6. Let $\{(t_i, f_i, d_i, g_i, g_i'), i = 0, 1, \cdots, n\}$ be a given data set, and $g(t)$ be a given quadratic function satisfying $f_i \ge g_i$, then the sufficient condition for the weighted rational cubic spline $P(t)$ defined by (3) to lie above the quadratic curve $g(t)$ in $[t_i, t_{i+1}]$ is that the positive parameters α_i, β_i and the weighted coefficient $\lambda \in R$ satisfy the following inequality system

$$\begin{aligned} &\lambda\alpha_i(f_i - f_{i+1} + h_i d_i) + \alpha_i(f_i + f_{i+1} - 2g_i^* \\ &-g_i^{*\prime} h_i) + \beta_i(f_i - g_i^*) \ge 0, \qquad (17) \\ &\lambda\beta_i(f_{i+2} - f_{i+1} - h_i d_{i+1}) + \beta_i(3f_{i+1} - f_{i+2} \\ &-2g_i^* - g_i^{*\prime} h_i) + \alpha_i(f_{i+1} - g_{i+1}^*) \ge 0. \qquad (18) \end{aligned}$$

Combining Theorem 5 and Theorem 6, the sufficient conditions for the weighted rational interpolating curves to lie below the quadratic curve $g^*(t)$ and above the quadratic curve $g(t)$ in $[t_i, t_{i+1}]$ could be described as the following theorem.

THEOREM 7. Let $\{(t_i, f_i, d_i, g_i^*, g_i, g_i^{*\prime}, g_i'), i = 0, 1, \cdots, n\}$ be a given data set, the sufficient condition for the weighted rational cubic spline $P(t)$ defined by (3) to lie below the quadratic curve $g^*(t)$ and above the quadratic curve $g(t)$ in $[t_i, t_{i+1}]$ is that the positive parameters α_i, β_i and the weighted coefficient $\lambda \in R$ satisfy the inequality system (15)(16) (17) and (18).

About the existence of the constraint described in Theorem 7, there is the following Theorem 8 and the proof is entirely similar as Theorem 4.

THEOREM 8. Let $\{(t_i, f_i, d_i, g_i^*, g_i, g_i^{*\prime}, g_i'), i = 0, 1, \cdots, n\}$ be any given data set with $g_i < f_i < g_i^*$, there must exist parameters $\alpha_i > 0, \beta_i > 0$ and coefficient $\lambda \in R^+$ such that the weighted rational curve defined by (3) is bounded between the quadratic lines $g(t)$ and g_i^* in the subinterval $[t_i, t_{i+1}]$.

5 Numerical Examples

Example 1. Let $f(t) = sin(\pi t/2)$,$t \in [0, 4]$, and the knots are 0.0, 0.5, 1.0,1.5, 2.0,2.5, 3.0, 3.5, 4.0, i.e. $h = 0.5$. Denote the rational cubic interpolating function of $f(t)$ by (3). Let the "above" constraining curve $g^*(t)$ be

$$g^*(t) = \begin{cases} 0.08 + \frac{7}{5}t, & 0.0 \le t \le 0.5; \\ 0.48 + \frac{3}{5}t, & 0.5 \le t \le 1.0; \\ 1.68 - \frac{3}{5}t, & 1.0 \le t\ le 1.5; \\ 2.88 - \frac{7}{5}t, & 1.5 \le t \le 2.0; \\ 2.88 - \frac{7}{5}t, & 2.0 \le t \le 2.5; \\ 1.03 - \frac{33}{55}t, & 2.5 \le t \le 3.0; \\ -2.93 + \frac{33}{55}t, & 3.0 \le t\ le 3.5; \\ -5.52 + \frac{7}{5}t, & 3.5 \le t \le 4.0; \end{cases}$$

the "below" constraining curve $g(t)$ be

$$g(t) = \begin{cases} -0.08 + \frac{7}{5}t, & 0.0 \le t \le 0.5; \\ 0.29 + \frac{33}{55}t, & 0.5 \le t \le 1.0; \\ 1.61 - \frac{33}{55}t, & 1.0 \le t\ le 1.5; \\ 2.72 - \frac{7}{5}t, & 1.5 \le t \le 2.0; \\ 2.72 - \frac{7}{5}t, & 2.0 \le t \le 2.5; \\ 0.72 - \frac{3}{5}t, & 2.5 \le t \le 3.0; \\ -2.88 + \frac{3}{5}t, & 3.0 \le t\ le 3.5; \\ -5.68 + \frac{7}{5}t, & 3.5 \le t \le 4.0; \end{cases}$$

Let the weighted coefficient $\lambda = 0.6$. Denote $a_i = \alpha_i/\beta_i$, and let $a_i (i = 0, 1, \cdots, 7)$ be $1.5, 3.3, 2.4, 0.1, 1.5, 3.3, 2.4, 0.1$ respectively, then the interpolating curves are being constrained between two given piecewise linear straight lines $g^*(t)$ and $g(t)$. The following Figure 1 gives the picture of the interpolating curve which is being constrained between two given curves $g^*(t)$ and $g(t)$, and the following Table 1 gives the values of the functions $g^*(t)$, $g(t)$, $P(t)$ and $f(t)$ in the interpolating interval $[t_0, t_n]$. From them it is easy to see that the curve of the being interpolated function $f(t)$ is not bounded between the curves of $g^*(t)$ and $g(t)$, but the interpolating curve $P(t)$ of $f(t)$ could be constrained to be bounded between the two curves.

t	$g^*(t)$	$g(t)$	$P(t)$	$f(t)$
0.00	1.10	0.90	1.00000	1.00000
0.15	1.00	0.80	0.95547	0.97553
0.30	0.90	0.70	0.87521	0.90451
0.45	0.80	0.60	0.77171	0.79389
0.60	0.70	0.50	0.64650	0.65451
0.75	0.60	0.40	0.50000	0.50000
0.90	0.50	0.30	0.34913	0.34549
1.05	0.40	0.20	0.21489	0.20611
1.20	0.30	0.10	0.10486	0.09549
1.35	0.20	0.00	0.02890	0.02447
1.50	0.10	-0.10	0.00000	0.00000
1.65	0.20	0.00	0.04453	0.02447
1.80	0.30	0.10	0.12479	0.09549
1.95	0.40	0.20	0.22829	0.20611
2.10	0.50	0.30	0.35350	0.34549
2.25	0.60	0.40	0.50000	0.50000
2.40	0.70	0.50	0.65087	0.65451
2.55	0.80	0.60	0.78511	0.79389
2.70	0.90	0.70	0.89514	0.90451
2.85	1.00	0.80	0.97110	0.97553
3.00	1.10	0.90	1.00000	1.00000

Table 1. The values of $g^*(t)$,$g(t)$, $P(t)$ and $f(t)$ in [0.0,3.0].

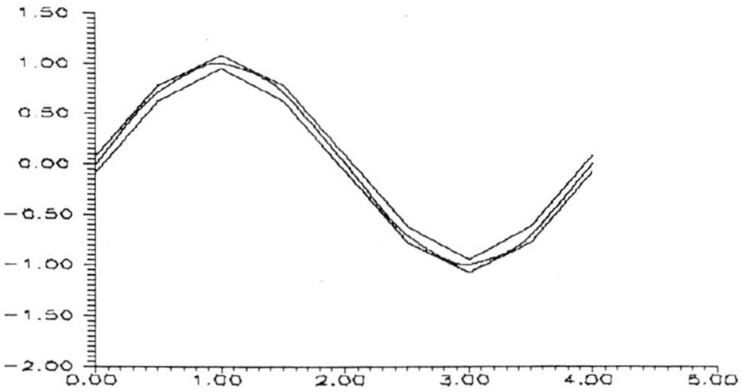

Figure 1. $P(t)$ is being constrained between two given curves.

References

[1] Qi Duan, Gongxue Xu, Aikui Liu, Xuefu Wang, and Fuhua (Frank) Cheng, "Constrained Interpolation Using Rational Cubic Spline Curve with Linear Denominators", *Korean J. of Computational and Applied Mathematics*, **6(1)**,(1999),203-215,.

[2] W. Boehm, G. Farin, and J. Kahamann, "A Survey of Curve and Surface Methods", *Computer Aided Geometric Design*, **1**, (1984) 1-60.

[3] G. M. Nielson, "CAGD's Top Ten: What to Watch", *IEEE Computer Graphics and Automation*, Jan. (1993), 35-37.

[4] L. Piegl, "On NURBS: A Survey", *IEEE Computer Graphics and Application* , **11(5)**, (1991), 55-71.

[5] B.A.Barsky, "The Beta-spline: A local Representation Based on Shape Parameters and Fundamental Geometric Measure", Ph.D thesis, University of Utah, (1981).

[6] P. Dierck, and B. Tytgat, "Generating the Bézier Points of β-spline Curve", *Comp. Aided Geom. Design*,**6**,(1989), 279-291.

[7] T.A.Foley, "Local Control of Interval Tension Using Weighted Splines", *Comp. Aided Geom. Design*, **3**,(1986),281-294.

[8] G.M.Nielson, "Rectangular ν-splines", *IEEE Comp. Graph. Appl.*, **6**,(1986), 35-40.

[9] M.Sarfraz,"Interpolatory Rational Cubic Spline with Biased Point and Interval Tension",*Comp & Graph*,**16(4)**, (1992),427-430.

[10] J.A. Gregory, M. Sarfraz, and P.K. Yuen, "Interactive Curve Design Using C^2 Rational Splines", *Comp & Graph*, **18(2)**, (1994), 153–159.

[11] Qi Duan, K. Djidjeli, W.G. Price, and E.H. Twizell, "Rational Cubic Spline Based on Function Values", *Comp & Graph*, **22(4)**,(1998), 479-486.

[12] Qi Duan, Botang Li, K.Djidjeli, W.G.Price, and E.H.Twizell, " Shape Control of Curve Design by Weighted Rational Splines", *Korean J. Comput. and Appl. Math.*, **6(3)**,(1999),537-547.

[13] Qi Duan, K. Djidjeli, W.G. Price, and E.H. Twizell, "The Approximation Properties of Some Rational Cubic Splines", *International J. of Computer Mathematics*, **72**,(1999),155-166.

[14] Qi Duan, K. Djidjeli, W.G. Price, and E.H. Twizell, "Weighted Rational Cubic Spline Interpolation and its Approximation", *J. of Computational and Applied Mathematics*, To appear.

[15] Qi Duan, Aikui Liu, and Shitian Du, "Shape Control and Energy Control of Curve Design", *J. of Computer-Aided Design & Computer Graphics*, **11(6)**,(1999),485-489.

Recursive Formulae for Hermite Polynomial Approximations to Rational Bézier Curves

Ligang Liu and Guojin Wang
State Key Laboratory of CAD&CG and Department of Mathematics
Zhejiang University, Hangzhou 310027, P.R.China
E-mail: wgj@math.zju.edu.cn

Abstract

This paper presents the algorithms for calculating the control points of Hermite polynomial approximation to rational Bézier curves and the recursive formulae between different degrees Hermite polynomial approximations based on the close relationship between hybrid polynomial approximation and Hermite polynomial approximation. The error bound for the Hermite polynomial approximation is also obtained.

1. Introduction

Bézier curves are established as a standard tool in computer aided geometric design. These polynomial curves possess of a number of very desirable properties [1]. But in mechanical engineering contours frequently contain circular arcs of more general conic sections. Contours that contain splines as well as conic sections cannot be stored in one data format if the curve description is restricted to polynomial curves. The purpose for using rational Bézier curves instead of polynomial ones is that they enjoy all the positive properties but satisfy the above points as well.

It should not be overlooked, however, that there are drawbacks. The calculation of the derivatives and integrals for the rational curves is very complicated and somewhat impossible. Moreover, it is difficult to communicate data between a CAD/CAM system based on rational Bézier curves and another one that restricts itself to polynomial curves. It is seen that the approximation of rational curves by polynomial curves has great significance in CAD/CAM applications.

In interpolation and approximation theory, Hermite polynomial function has been well studied for years[2]. It has not been used for approximating the rational Bézier curves yet. The Hermite polynomial approximation has the virtue that it interpolates the end points of the curve and their tangents and so on. But there is no satisfying method for computing the control points of higher degree Hermite polynomial approximation to rational Bézier curve so far. It is an essential task for us to get the Hermite representation of any degree in practical application.

Hybrid curves have been introduced in CAGD by Sederberg and Kakimoto[3]. They can be used in approximating rational curves by polynomial curves. In ref. [4] Wang *et al.* have investigated the relationship between hybrid polynomial approximation and the Hermite polynomial approximation. In this paper we find that the hybrid polynomial approximation relates to the Hermite polynomial approximation much more closely. We will prove that the hybrid polynomial approximation should be equivalent to the Hermite polynomial approximation under certain conditions. So we can obtain the Hermite representation of any degree based on the formulae for calculating the hybrid polynomial.

After presenting the main definitions and notations of the hybrid polynomial and the Hermite polynomial in the next seciton, Section 3 gives the formulae of computing the control points of the Hermite polynomial approxation directly. Some recursive algorithms for the control points of different degree Hermite polynomial approximation are obtained in Section 4, which are fast and more elegant than the algorithms in Section 3. Section 5 gives the error terms for the Hermite polynomial approximation. These error terms serve as the basis for the convergence conditions. Finally, in Section 6 we make the conclusion of this paper.

2. Definitions and Notations

A degree r rational Bézier curve is given by the following equation

0-7695-0562-7/00 $10.00 © 2000 IEEE

$$\mathbf{R}(t) = \frac{\mathbf{P}(t)}{w(t)} = \frac{\sum_{i=0}^{r} w_i \mathbf{R}_i B_i^r(t)}{\sum_{i=0}^{r} w_i B_i^r(t)}, \tag{1}$$

where $B_i^r(t) = \binom{r}{i}(1-t)^{r-i}t^i$ denote the i-th Bernstein basis function of degree r, $\mathbf{R}_i$ are the control points and w_i are the control point weights[1]. Throughout this paper the variable domain is $0 \le t \le 1$ unless stated otherwise.

Denote by

$$\mathbf{h}^{m,n}(t) = \sum_{i=0}^{m+n-1} \mathbf{h}_i^{m,n} B_i^{m+n-1}(t) \tag{2}$$

the degree $m+n-1$ Bézier curve that satisfies

$$\begin{aligned} \frac{d^j \mathbf{h}^{m,n}(0)}{dt^j} &= \frac{d^j \mathbf{R}(0)}{dt^j}, \quad j = 0,1,\ldots,m-1, \\ \frac{d^j \mathbf{h}^{m,n}(1)}{dt^j} &= \frac{d^j \mathbf{R}(1)}{dt^j}, \quad j = 0,1,\ldots,n-1, \end{aligned} \tag{3}$$

where $m \ge 1$ and $n \ge 1$. There are $m+n$ linear equations about the $m+n$ variables $\{\mathbf{h}_i^{m,n} \mid i = 0,1,\cdots,m+n-1\}$ in (3). So the coefficients $\{\mathbf{h}_i^{m,n} \mid i = 0,1,\cdots,m+n-1\}$ can be decided uniquely by the equations. In this case, $\mathbf{h}^{m,n}(t)$ is called *the Hermite polynomial approximation* to $\mathbf{R}(t)$ [2].

It is shown in [3] that any degree r rational Bézier curve $\mathbf{R}(t)$ can be expressed exactly as a degree $m+n$ polynomial Bézier curve

$$\begin{aligned} \widetilde{\mathbf{H}}^{m,n}(t) &\equiv \mathbf{R}(t) \\ &= \sum_{i=0, i\ne m}^{m+n} \mathbf{H}_i^{m,n} B_i^{m+n}(t) + \mathbf{V}^{m,n}(t) B_m^{m+n}(t), \end{aligned} \tag{4}$$

with the catch that one control point is chosen to serve as a moving control point — itself a rational Bézier curve with the same weights w_i as $\mathbf{R}(t)$, i.e.,

$$\mathbf{V}^{m,n}(t) = \frac{\sum_{i=0}^{r} w_i \mathbf{V}_i^{m,n} B_i^r(t)}{\sum_{i=0}^{r} w_i B_i^r(t)}, \tag{5}$$

in which $\mathbf{H}_i^{m,n} (0 \le i \le m+n, i \ne m)$ and $\mathbf{V}_i^{m,n} (0 \le i \le r)$ are control points that are uniquely determined by equation (4). Such a polynomial Bézier curve with a moving control point is called a Hybrid curve. The representation (4) leads directly to a polynomial approximation to $\mathbf{R}(t)$ by simply replacing the moving control point with a stationary control point. If we replace $\mathbf{V}^{m,n}(t)$ with a constant control point $\mathbf{H}_m^{m,n}$ in the convex hull of $\{\mathbf{V}_i^{m,n} \mid 0 \le i \le r\}$ then the Bézier curve

$$\mathbf{H}^{m,n}(t) = \sum_{i=0}^{m+n} \mathbf{H}_i^{m,n} B_i^{m+n}(t) \tag{6}$$

gives a polynomial approximation to the rational Bézier curve $\mathbf{R}(t)$. We call $\mathbf{H}^{m,n}(t)$ *the Hybrid polynomial approximation* to $\mathbf{R}(t)$. An advantage in applying hybrid polynomial approximation is that the approximation error is trivially bounded by

$$\begin{aligned} &\left\|\mathbf{R}(t) - \mathbf{H}^{m,n}(t)\right\| \\ &\le \underset{0\le t\le 1}{\mathrm{Max}}\left\|\mathbf{V}^{m,n}(t) - \mathbf{H}_m^{m,n}\right\| \cdot \underset{0\le t\le 1}{\mathrm{Max}}\, B_m^{m+n}(t). \end{aligned} \tag{7}$$

The norm used all through this paper is the Euclidean norm. If the control point weights w_i are all non-negative, then we have further

$$\begin{aligned} &\left\|\mathbf{R}(t) - \mathbf{H}^{m,n}(t)\right\| \\ &\le \underset{0\le i\le r}{\mathrm{Max}}\left\|\mathbf{V}_i^{m,n} - \mathbf{H}_m^{m,n}\right\| \cdot B_m^{m+n}\left(\frac{m}{m+n}\right). \end{aligned} \tag{8}$$

3. Calculating the Control Points of Hermite Approximation

We first give the close relationship between the ordinary Hermite polynomial approximation and the hybrid polynomial approximation expressed in the following theorem.

Theorem 1. If we replace the moving control point $\mathbf{V}^{m,n}(t)$ with a stationary control point $\mathbf{V}_0^{m,n}$ or $\mathbf{V}_r^{m,n}$, then the hybrid polynomial approximation $\mathbf{H}^{m,n}(t)$ is just equivalent to the Hermite polynomial approximation $\mathbf{h}^{m+1,n}(t)$ or $\mathbf{h}^{m,n+1}(t)$ respectively, i.e.,

$$\mathbf{H}^{m,n}(t) = \begin{cases} \mathbf{h}^{m+1,n}(t), & \text{if } \mathbf{H}_m^{m,n} = \mathbf{V}_0^{m,n}; \\ \mathbf{h}^{m,n+1}(t), & \text{if } \mathbf{H}_m^{m,n} = \mathbf{V}_r^{m,n}. \end{cases}$$

Proof. The remainder term of $\mathbf{H}^{m,n}(t)$ is

$$\begin{aligned}\Delta^{m,n}(t) &= \mathbf{H}^{m,n}(t) - \mathbf{R}(t)\\ &= \mathbf{H}^{m,n}(t) - \widetilde{\mathbf{H}}^{m,n}(t)\\ &= (\mathbf{H}_m^{m,n} - \mathbf{V}^{m,n}(t))B_m^{m+n}(t).\end{aligned} \tag{9}$$

Straightforward computation gives that

$$\frac{d^i B_m^{m+n}(0)}{dt^i} = \begin{cases} 0, \quad i = 0,1,\dots,m-1, \\ \dfrac{(m+n)!}{(m+n-i)!}(-1)^{i-m}\dbinom{i}{i-m}, \\ \qquad i = m, m+1,\dots,m+n, \end{cases} \tag{10}$$

and

$$\frac{d^j B_m^{m+n}(1)}{dt^j} = \begin{cases} 0, j = 0,1,\dots,n-1, \\ \dfrac{(m+n)!}{(m+n-j)!}(-1)^{j-n}\dbinom{j}{j-n}, \\ \qquad j = n, n+1,\dots,m+n. \end{cases} \tag{11}$$

Applying the Leibniz formula we have

$$\frac{d^i \Delta^{m,n}(0)}{dt^i} = 0, \qquad i = 0,1,\dots,m-1, \tag{12}$$

$$\frac{d^j \Delta^{m,n}(1)}{dt^j} = 0, \qquad j = 0,1,\dots,n-1. \tag{13}$$

Furthermore, if we set $\mathbf{H}_m^{m,n} = \mathbf{V}_0^{m,n}$, we have

$$\frac{d^m \Delta^{m,n}(0)}{dt^m} = \frac{(m+n)!}{n!}(\mathbf{H}_m^{m,n} - \mathbf{V}_0^{m,n}) = 0. \tag{14}$$

Noting that the degrees of $\mathbf{H}^{m,n}(t)$ and $\mathbf{h}^{m+1,n}(t)$ are both $m+n$, it means that $\mathbf{H}^{m,n}(t)$ is equivalent to $\mathbf{h}^{m+1,n}(t)$ by the uniqueness property of the Hermite polynomial approximation, i.e., $\mathbf{H}^{m,n}(t) \equiv \mathbf{h}^{m+1,n}(t)$.

In a somewhat similar way we get $\mathbf{H}^{m,n}(t) \equiv \mathbf{h}^{m,n+1}(t)$ if setting $\mathbf{H}_m^{m,n} = \mathbf{V}_r^{m,n}$. ∎

The formula for computing the control points of $\mathbf{H}^{m,n}(t)$ has been derived in [5]. By Theorem 1 the algorithm for calculating the control points of $\mathbf{h}^{m,n}(t)$ can be outlined as following.

Algorithm 1. The control points of $\mathbf{h}^{m,n}(t)$ can be calculated recursively by the formula:

$$\mathbf{h}_0^{m,n} = \mathbf{R}_0,$$

$$\mathbf{h}_i^{m,n} = \mathbf{R}_0 + \frac{1}{\binom{m+n-1}{i} w_0} \times \sum_{k=1}^{\min(r,i)} \binom{r}{k}\binom{m+n-1}{i-k} w_k(\mathbf{R}_k - \mathbf{h}_{i-k}^{m,n}),$$

$$i = 1,2,\dots,m-2,$$

$$\mathbf{h}_{m+n-1}^{m,n} = \mathbf{R}_r,$$

$$\mathbf{h}_i^{m,n} = \mathbf{R}_r + \frac{1}{\binom{m+n-1}{i} w_r} \times \sum_{k=\max(0,i+r-n-m+1)}^{r-1} \binom{r}{k}\binom{m+n-1}{i+r-k} w_k(\mathbf{R}_k - \mathbf{h}_{i+r-k}^{m,n}),$$

$$i = m+n-2,\dots,m,$$

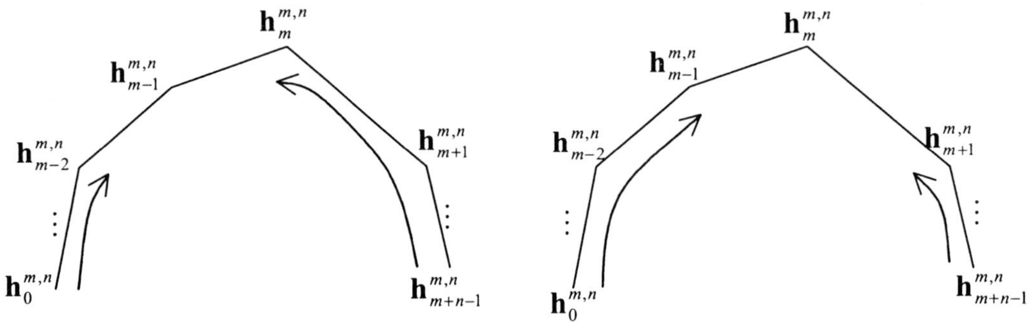

Figure 1. The sequence of caculating the control points of Hermite approximation

$$\mathbf{h}_{m-1}^{m,n} = \mathbf{R}_0 + \frac{1}{\binom{m+n-1}{m-1} w_0} \times \sum_{k=1}^{\min(r,m-1)} \binom{r}{k}\binom{m+n-1}{m-1-k} w_k (\mathbf{R}_k - \mathbf{h}_{m-1-k}^{m,n}). \blacksquare$$

Algorithm 1 allows us to solve in the following order: $\mathbf{h}_0^{m,n}, \mathbf{h}_1^{m,n}, \ldots, \mathbf{h}_{m-2}^{m,n}$; $\mathbf{h}_{m+n-1}^{m,n}, \mathbf{h}_{m+n-2}^{m,n}, \ldots, \mathbf{h}_m^{m,n}$; and $\mathbf{h}_{m-1}^{m,n}$, as described in Figure 1(a).

Algorithm 2. The control points of $\mathbf{h}^{m,n}(t)$ can be calculated recursively by the formula:

$$\mathbf{h}_0^{m,n} = \mathbf{R}_0,$$

$$\mathbf{h}_i^{m,n} = \mathbf{R}_0 + \frac{1}{\binom{m+n-1}{i} w_0} \times \sum_{k=1}^{\min(r,i)} \binom{r}{k}\binom{m+n-1}{i-k} w_k (\mathbf{R}_k - \mathbf{h}_{i-k}^{m,n}), \quad i = 1,2,\ldots,m-1,$$

$$\mathbf{h}_{m+n-1}^{m,n} = \mathbf{R}_r,$$

$$\mathbf{h}_i^{m,n} = \mathbf{R}_r + \frac{1}{\binom{m+n-1}{i} w_r} \times \sum_{k=\max(0,i+r-n-m+1)}^{r-1} \binom{r}{k}\binom{m+n-1}{i+r-k} w_k (\mathbf{R}_k - \mathbf{h}_{i+r-k}^{m,n}), \quad i = m+n-2,\ldots,m+1,$$

$$\mathbf{h}_m^{m,n} = \mathbf{R}_r + \frac{1}{\binom{m+n-1}{m} w_r} \times \sum_{k=\max(r+1-n,0)}^{r-1} \binom{r}{k}\binom{m+n-1}{m+r-k} w_k (\mathbf{R}_k - \mathbf{h}_{m+r-k}^{m,n}). \blacksquare$$

Algorithm 2 allows us to solve in the following order: $\mathbf{h}_0^{m,n}, \mathbf{h}_1^{m,n}, \ldots, \mathbf{h}_{m-1}^{m,n}$; $\mathbf{h}_{m+n-1}^{m,n}, \mathbf{h}_{m+n-2}^{m,n}, \ldots, \mathbf{h}_{m+1}^{m,n}$; and $\mathbf{h}_m^{m,n}$, as described in Figure 1(b).

4. Recursive Relationship Between Hermite Approximation

In Section 3 Algorithm 1 and Algorithm 2 give the general formulae of computing the control points of $\mathbf{h}^{m,n}(t)$ from $\mathbf{R}(t)$ directly. But it is complicated and is not convenient for practical applications. Based on the recursive relationship between hybrid approximations in [5] we can obtain the Hermite approximation representation of higher degree given a lower degree ones. The following algorithms are faster and more elegant than Algorithm 1 and Algorithm 2. For example, the new algorithms allow computation of control points in $O(m+n)$ time, whereas the above algorithms required at least $O((m+n)^2)$ time to compute the control points.

Algorithm 3 gives the formula for computing $\mathbf{h}^{m,n}(t)$ by $\mathbf{h}^{m-1,n}(t)$.

Algorithm 3. The control points of $\mathbf{h}^{m,n}(t)$ can be found as following:

$$\mathbf{h}_i^{m,n} = \frac{i}{m+n-1}\mathbf{h}_{i-1}^{m-1,n} + \left(1 - \frac{i}{m+n-1}\right)\mathbf{h}_i^{m-1,n}, \quad 0 \le i \le m+n-1, i \ne m-1, \blacksquare$$

$$\mathbf{h}_{m-1}^{m,n} = \mathbf{V}_0^{m-1,n}.$$

Algorithm 4 gives the formula for computing $\mathbf{h}^{m,n}(t)$ by $\mathbf{h}^{m,n-1}(t)$.

Algorithm 4. The control points of $\mathbf{h}^{m,n}(t)$ can be found as following:

$$\mathbf{h}_i^{m,n} = \frac{i}{m+n-1}\mathbf{h}_{i-1}^{m,n-1} + \left(1 - \frac{i}{m+n-1}\right)\mathbf{h}_i^{m,n-1}, \quad 0 \le i \le m+n-1, i \ne m, \blacksquare$$

$$\mathbf{h}_m^{m,n} = \mathbf{V}_r^{m,n-1}.$$

From the above formulae it is shown that if we degree elevate polynomial $\mathbf{h}^{m,n-1}(t)$ or $\mathbf{h}^{m-1,n}(t)$, their coefficients and those of polynomial $\mathbf{h}^{m,n}(t)$ differ only in one coefficient. Algorithm 5 gives the formula for computing $\mathbf{h}^{m,n}(t)$ by $\mathbf{h}^{m-1,n-1}(t)$.

Algorithm 5. The control points of $\mathbf{h}^{m,n}(t)$ can be found as following:

$$\mathbf{h}_i^{m,n} = \frac{i(i-1)}{(m+n-1)(m+n-2)}\mathbf{h}_{i-2}^{m-1,n-1} + \frac{2i(m+n-1-i)}{(m+n-1)(m+n-2)}\mathbf{h}_{i-1}^{m-1,n-1} + \frac{(m+n-1-i)(m+n-2-i)}{(m+n-1)(m+n-2)}\mathbf{h}_i^{m-1,n-1},$$

$$0 \le i \le m+n-1, i \ne m-1, m,$$

$$\mathbf{h}_{m-1}^{m,n} = \frac{(m-1)(m-2)}{(m+n-1)(m+n-2)}\mathbf{h}_{m-3}^{m-1,n-1} + \frac{(m-1)n}{(m+n-1)(m+n-2)}\mathbf{h}_{m-2}^{m-1,n-1} + \frac{n}{m+n-1}\mathbf{V}_0^{m-1,n-1},$$

$$\mathbf{h}_m^{m,n} = \frac{m}{m+n-1}\mathbf{V}_r^{m-1,n-1} + \frac{m(n-1)}{(m+n-1)(m+n-2)}\mathbf{h}_{m-1}^{m-1,n-1} + \frac{(n-1)(n-2)}{(m+n-1)(m+n-2)}\mathbf{h}_m^{m-1,n-1}.$$

Algorithm 3 to Algorithm 5 can be directly obtained by the recursive relationship between hybrid approximations as in [5], so the proofs are omitted here.

5. Error Bound For Hermite Approximation

To get convergence conditions for the polynomial approximation $\mathbf{h}^{m,n}(t)$, it is necessary to compute the error term of the approximation.

Theorem 2. The remainder term for $\mathbf{h}^{m,n}(t)$ is

$$\mathbf{R}(t) - \mathbf{h}^{m,n}(t) = \frac{rm}{m+n}\frac{\hat{\mathbf{V}}^{m-1,n}(t)}{w(t)}B_m^{m+n}(t) \qquad (15)$$

which is bounded by

$$\left\|\mathbf{R}(t) - \mathbf{h}^{m,n}(t)\right\| \le \frac{mr}{m+n}\frac{w_{\max}}{W_{\min}}\operatorname*{Max}_{1\le i\le r}\left\|\mathbf{V}_i^{m-1,n} - \mathbf{V}_0^{m-1,n}\right\|,$$

where $\hat{\mathbf{V}}^{m-1,n}(t)$ is a degree $r-1$ Bézier curve,

$$\hat{\mathbf{V}}^{m-1,n}(t) = \sum_{i=0}^{r-1}\frac{w_{i+1}}{i+1}(\mathbf{V}_{i+1}^{m-1,n} - \mathbf{V}_0^{m-1,n})B_i^{r-1}(t)$$

and $w_{\max} = \operatorname*{Max}_{1\le i\le r}\left|\frac{w_i}{i}\right|$, $W_{\min} = \operatorname*{Min}_{0\le t\le 1}|w(t)|$.

Proof. By Theorem 1, we have $\mathbf{H}^{m-1,n}(t) = \mathbf{h}^{m,n}(t)$ if $\mathbf{H}_{m-1}^{m-1,n} = \mathbf{V}_0^{m-1,n}$. So we have

$$\mathbf{R}(t) - \mathbf{h}^{m,n}(t) = \mathbf{R}(t) - \mathbf{H}^{m-1,n}(t)$$
$$= (\mathbf{V}^{m-1,n}(t) - \mathbf{V}_0^{m-1,n})B_{m-1}^{m+n-1}(t)$$
$$= \frac{\sum_{i=0}^{r} w_i(\mathbf{V}_i^{m-1,n} - \mathbf{V}_0^{m-1,n})B_i^r(t)}{w(t)}B_{m-1}^{m+n-1}(t)$$
$$= \frac{rt}{w(t)}\sum_{i=0}^{r-1}\left(\frac{w_{i+1}}{i+1}(\mathbf{V}_{i+1}^{m-1,n} - \mathbf{V}_0^{m-1,n})B_i^{r-1}(t)\right)B_{m-1}^{m+n-1}(t)$$
$$= \frac{mr}{m+n}\frac{\hat{\mathbf{V}}^{m-1,n}(t)}{w(t)}B_m^{m+n}(t),$$

and hence

$$\left\|\mathbf{R}(t) - \mathbf{h}^{m,n}(t)\right\|$$
$$\le \frac{mr}{m+n}\frac{\operatorname*{Max}_{1\le i\le r}\left(\frac{w_i}{i}\right)}{W_{\min}}\operatorname*{Max}_{1\le i\le r}\left\|\mathbf{V}_i^{m-1,n} - \mathbf{V}_0^{m-1,n}\right\| \blacksquare$$
$$= \frac{mr}{m+n}\frac{w_{\max}}{W_{\min}}\operatorname*{Max}_{1\le i\le r}\left\|\mathbf{V}_i^{m-1,n} - \mathbf{V}_0^{m-1,n}\right\|.$$

Similarly, we can obtain another approximation error term as in Theorem 3:

Theorem 3. The remainder term for $\mathbf{h}^{m,n}(t)$ is

$$\mathbf{R}(t) - \mathbf{h}^{m,n}(t) = \frac{rn}{m+n}\frac{\hat{\mathbf{V}}^{m,n-1}(t)}{w(t)}B_m^{m+n}(t) \qquad (16)$$

which is bounded by

$$\left\|\mathbf{R}(t) - \mathbf{h}^{m,n}(t)\right\| \le \frac{nr}{m+n}\frac{\overline{w}_{\max}}{W_{\min}}\operatorname*{Max}_{0\le i\le r-1}\left\|\mathbf{V}_i^{m,n-1} - \mathbf{V}_r^{m,n-1}\right\|,$$

where $\hat{\mathbf{V}}^{m,n-1}(t)$ is a degree $r-1$ Bézier curve,

$$\hat{\mathbf{V}}^{m,n-1}(t) = \sum_{i=0}^{r-1}\frac{w_{i+1}}{r-i}(\mathbf{V}_i^{m,n-1} - \mathbf{V}_r^{m,n-1})B_i^{r-1}(t)$$

and $\overline{w}_{\max} = \operatorname*{Max}_{0\le i\le r-1}\left|\frac{w_i}{r-i}\right|$, $W_{\min} = \operatorname*{Min}_{0\le t\le 1}|w(t)|$. ∎

From Theorem 2, we can see that the key to obtaining convergence conditions is to estimate $\left\|\mathbf{V}_i^{m,n} - \mathbf{V}_0^{m,n}\right\|$. We now derive the recursive formula for computing $\mathbf{V}_i^{m,n} - \mathbf{V}_0^{m,n}$.

Theorem 4. $\mathbf{V}_i^{m,n} - \mathbf{V}_0^{m,n}$ has the recursive formula

$$\begin{pmatrix} \mathbf{V}_1^{m+1,n+1} - \mathbf{V}_0^{m+1,n+1} \\ \vdots \\ \mathbf{V}_r^{m+1,n+1} - \mathbf{V}_0^{m+1,n+1} \end{pmatrix} = \frac{2(m+1)(n+1)}{(m+n+1)(m+n+2)} W_r^{11} \begin{pmatrix} \mathbf{V}_1^{m,n} - \mathbf{V}_0^{m,n} \\ \vdots \\ \mathbf{V}_r^{m,n} - \mathbf{V}_0^{m,n} \end{pmatrix},$$
$$m,n = 0,1,2,\cdots$$

where

$$W_1^{11} = \frac{1}{2}\left(2 - g_0 - \frac{1}{g_0}\right),$$

$$W_r^{11} = \frac{1}{2}\begin{pmatrix} 2 & g_1 & 0 & \cdots & 0 & 0 & 0 \\ \frac{1}{g_1} & 2 & g_2 & \cdots & 0 & 0 & 0 \\ 0 & \frac{1}{g_2} & 2 & \cdots & 0 & 0 & 0 \\ \vdots & \vdots & \vdots & \ddots & \vdots & \vdots & \vdots \\ 0 & 0 & 0 & \cdots & 2 & g_{r-2} & 0 \\ 0 & 0 & 0 & \cdots & \frac{1}{g_{r-2}} & 2 & g_{r-1} \\ 0 & 0 & 0 & \cdots & 0 & \frac{1}{g_{r-1}} & 2 \end{pmatrix} + \frac{1}{2}\begin{pmatrix} -g_0 & 0 & \cdots & 0 & -\frac{1}{g_0} \\ -g_0 & 0 & \cdots & 0 & -\frac{1}{g_1} \\ \vdots & \vdots & \ddots & \vdots & \vdots \\ -g_0 & 0 & \cdots & 0 & -\frac{1}{g_{r-2}} \\ -g_0 & 0 & \cdots & 0 & -\frac{1}{g_{r-1}} \end{pmatrix}, \quad (17)$$

$$g_i = \frac{(r-i)w_{i+1}}{(i+1)w_i}, \quad i = 0,1,\cdots,r-1. \quad (18)$$

Proof. By the formula for calculating $\mathbf{V}_i^{m,n}$ [5], we have

$$\mathbf{V}_i^{m+1,n+1} = \frac{\binom{m+n}{m-1}}{\binom{m+n+2}{m+1}}\mathbf{H}_{m-1}^{m,n} + \frac{\binom{m+n}{m+1}}{\binom{m+n+2}{m+1}}\mathbf{H}_{m+1}^{m,n}$$
$$+ \frac{2\binom{m+n}{m}}{\binom{m+n+2}{m+1}}[\mathbf{V}_i^{m,n} + \frac{1}{2}g_i(\mathbf{V}_{i+1}^{m,n} - \mathbf{V}_0^{m,n})$$
$$+ \frac{1}{2}g_{i-1}^{-1}(\mathbf{V}_{i-1}^{m,n} - \mathbf{V}_r^{m,n})],$$
$$i = 0,1,\cdots,r,$$

where $g_i = \frac{(r-i)w_{i+1}}{(i+1)w_i}$, $i = 0,1,\cdots,r-1$, defining $g_r = 0$ and $g_{-1}^{-1} = 0$. Let $i = 0$, we have

$$\mathbf{V}_0^{m+1,n+1} = \frac{\binom{m+n}{m-1}}{\binom{m+n+2}{m+1}}\mathbf{H}_{m-1}^{m,n} + \frac{\binom{m+n}{m+1}}{\binom{m+n+2}{m+1}}\mathbf{H}_{m+1}^{m,n}$$
$$+ \frac{2\binom{m+n}{m}}{\binom{m+n+2}{m+1}}\left[\mathbf{V}_0^{m,n} - \frac{1}{2}g_0(\mathbf{V}_1^{m,n} - \mathbf{V}_0^{m,n})\right].$$

Hence we get

$$\mathbf{V}_i^{m+1,n+1} - \mathbf{V}_0^{m+1,n+1}$$
$$= \frac{2\binom{m+n}{m}}{\binom{m+n+2}{m+1}}[-\frac{1}{2}g_0(\mathbf{V}_1^{m,n} - \mathbf{V}_0^{m,n})$$
$$+ \frac{1}{2}g_{i-1}^{-1}(\mathbf{V}_{i-1}^{m,n} - \mathbf{V}_0^{m,n}) + (\mathbf{V}_i^{m,n} - \mathbf{V}_0^{m,n})$$
$$+ \frac{1}{2}g_i(\mathbf{V}_{i+1}^{m,n} - \mathbf{V}_0^{m,n}) - \frac{1}{2}g_{i-1}^{-1}(\mathbf{V}_r^{m,n} - \mathbf{V}_0^{m,n})].$$

The above equations lead to the result. ▌

We take special interest in the case $m = n = s$. From the remainder term in Theorem 2 we know the convergence condition depends on $\left\|\mathbf{V}_i^{s,s} - \mathbf{V}_0^{s,s}\right\|$. If

$\lim_{s\to\infty}\left\|V_i^{s,s} - V_0^{s,s}\right\| = 0\ (i = 1,2,\cdots,r)$ then $\mathbf{h}^{s,s}(t)$ must converge to $\mathbf{R}(t)$ uniformly on $[0,1]$. Using recursive formula for $\mathbf{V}_i^{m,n} - \mathbf{V}_0^{m,n}$ in Theorem 4, the concrete condition for the convergence of $\mathbf{h}^{s,s}(t)$ is obtained easily.

Theroem 5. Let W_r^{11} be as defined in Theorem 4 and let $\lambda_1, \lambda_2, \cdots, \lambda_r$ be the eigenvalues of matrix W_r^{11}. If

$$|\lambda_i| < 2, \qquad i = 1,2,\cdots,r, \tag{19}$$

then $\mathbf{h}^{s,s}(t)$ converges to $\mathbf{R}(t)$ uniformly on $[0,1]$.

Proof. By Theorem 4, we get

$$\begin{pmatrix} \mathbf{V}_1^{s,s} - \mathbf{V}_0^{s,s} \\ \vdots \\ \mathbf{V}_r^{s,s} - \mathbf{V}_0^{s,s} \end{pmatrix} = \frac{s!}{(2s-1)!}\left(W_r^{11}\right)^s \begin{pmatrix} \mathbf{R}_1 - \mathbf{R}_0 \\ \vdots \\ \mathbf{R}_r - \mathbf{R}_0 \end{pmatrix}. \tag{20}$$

Let J_r^{11} be the Jordan canonical form [6] of matrix W_r^{11}, i.e.,

$$W_r^{11} = P_r^{11} J_r^{11} (P_r^{11})^{-1}, \tag{21}$$

where P_r^{11} is an $r \times r$ invertible matrix. We get from Stirling formula that

$$\frac{s!}{(2s-1)!} = \frac{2^s (s!)^2}{(2s)!} = O\left(\frac{\sqrt{s}}{2^s}\right). \tag{22}$$

If $|\lambda_i| < 2, i = 1,2,\cdots,r,$ then we have

$$\lim_{s\to\infty}\frac{\sqrt{s}}{2^s}\left(J_s^{11}\right)^s = 0, \tag{23}$$

and hence

$$\lim_{s\to\infty}\left\|\mathbf{V}_i^{s,s} - \mathbf{V}_0^{s,s}\right\| = 0, \qquad i = 1,2,\cdots,r.$$

From here the theorem follows immediately. ▌

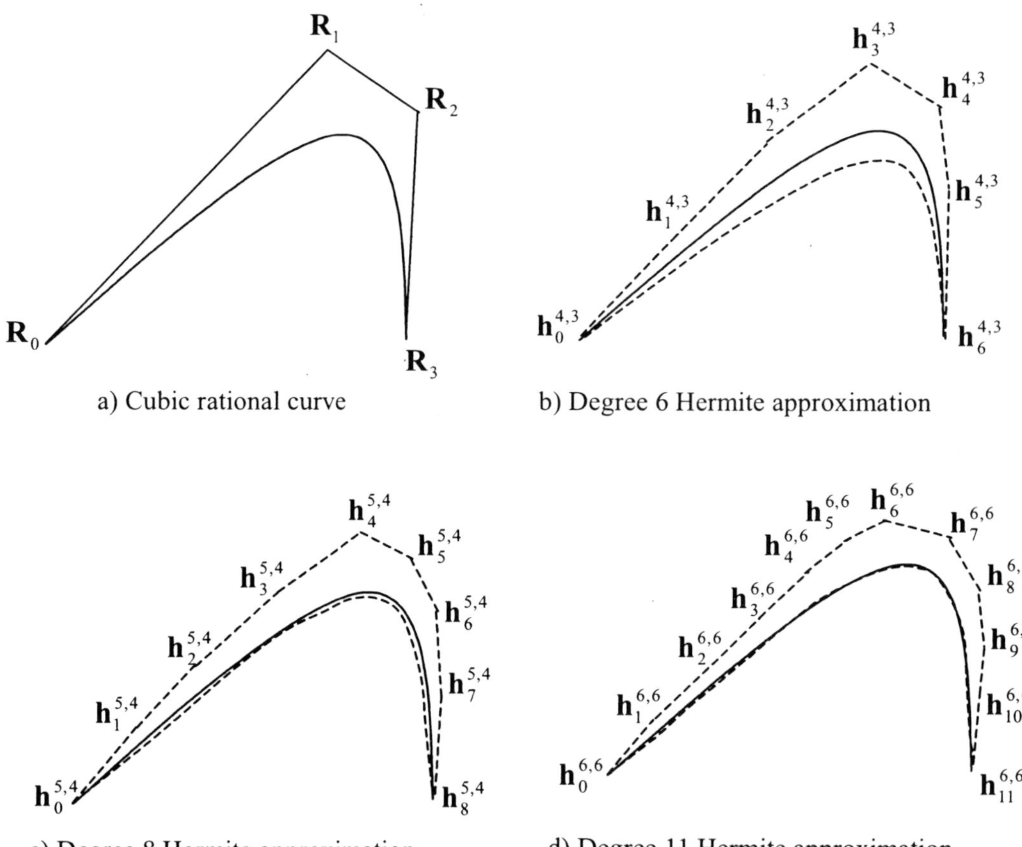

Figure 2. Convergent Hermite approximation of a cubic rational curve

6. Example and Conclusions

Let $\mathbf{R}(t)$ be a cubic rational Bézier curve, with heights $w_0 = 1$, $w_1 = 2$, $w_2 = 3$, and $w_3 = 2$, as in equation (1). By (17) and (18) we have

$$g_0 = 6, g_1 = \frac{3}{2}, g_2 = \frac{4}{9},$$

and

$$W_3^{11} = \frac{1}{2}\begin{pmatrix} 2 - g_0 & g_1 & -\frac{1}{g_0} \\ \frac{1}{g_1} - g_0 & 2 & g_2 - \frac{1}{g_1} \\ -g_0 & \frac{1}{g_2} & 2 - \frac{1}{g_2} \end{pmatrix} = \begin{pmatrix} -2 & \frac{3}{4} & -\frac{1}{12} \\ -\frac{8}{3} & 1 & -\frac{1}{9} \\ -3 & \frac{9}{8} & -\frac{1}{8} \end{pmatrix}.$$

It is easy to compute the eigenvalues of W_3^{11} as following

$$\lambda_1 = \lambda_2 = 0, \lambda_3 = -\frac{9}{8},$$

which satisfy (19). Thus the Hermite approximation to $\mathbf{R}(t)$ is convergent by Theorem 5. Figure 2 illustrates the Hermite approximation of the cubic curve $\mathbf{R}(t)$.

Hermite polynomial approximation $\mathbf{h}^{m,n}(t)$ is introduced into CAGD as a method for polynomial approximation to rational Bézier curves and surfaces. Based on the close relationship between hybrid polynomials and Hermite polynomial approximation, we have derived the formulae of calculating the control points of $\mathbf{h}^{m,n}(t)$ and their recursive algorithms. After obtaining the error bound, we get the convergence condition for the Hermite polynomial approximation.

Acknowledgments

We acknowledge the K.C. Wong Foundation for its financial support. This work was supported by the National Natural Science Foundation of China with Grant No.69673029 , Zhejiang Provincial Natural Science Foundation of China and Foundation of State Key Basic Research 973 Development Programming Item with Grant No.G1998030600.

References

[1] G. Farin, *Curves and Surfaces for Computer Aided Geometric Design*, Academic Press, Boston, 1988.

[2] P.J. Davis, *Interpolation and Approximation*, Dover, New York, 1975.

[3] T.W. Sederberg and M. Kakimoto, "Approximating Rational Curves Using Polynomial Curves", in *NURBS for Curves and Surfaces Design* (G. Farin, Ed.), SIAM, Philadelphia, 1991, pp.149—158.

[4] G.J. Wang, T.W. Sederberg and F.L. Chen, "On the Convergence of Polynomial Approximation of Rational Functions", *Journal of Approximation Theory*, Vol. 89, No. 3, 1997, pp. 267—288.

[5] G.X. Qiu, *Polynomial Approximation of Rational Bézier Surfaces*, Master's Thesis, Department of Mathematics, Zhejiang University, 1997.

[6] C.P., Marshall, *Methods of Matrix Algebra, Academic Press*, New York, 1965.

Session 7

Mesh Processing and Subdivision Surfaces I

Robust Watermarking of Polygonal Meshes

Michael G Wagner
Department of Computer Science and Engineering
Arizona State University
Tempe, AZ 85287, USA
wagner@asu.edu

Abstract

This paper presents two variations of a robust watermarking method for general polygonal meshes of arbitrary topology which can be used for copyright protection, tamper proofing or content annotation purposes. The proposed watermark is immune to translation, rotation, scaling or affine transformation of the mesh and is hard to detect unless the exact encoding parameters are disclosed. Several examples demonstrate the effectiveness of the algorithm.

1. Introduction

With state of the art information networks it is easily possible to disseminate multimedia data. As a result, it has become important to investigate possibilities for protecting intellectual property rights of multimedia content. One way to address the problem is to hide invisible data in the content that has to be protected. These so-called watermarks are widely used for traditional multimedia data types such as images, video and audio [1, 3, 4] as well as formatted text [6].

In contrast, there exists surprisingly little work on watermarking three dimensional data types. In [5] Hartung et al. present a watermarking technique for facial animation parameters. Ohbuchi et al. [8, 9, 10, 11] published a sequence of papers on watermarking algorithms for polygonal meshes. However, as detailed later, their approach is not robust enough to be used, for example, for copyright protection purposes. More recently Yeo et al. [13] introduced a fragile watermark for object verification purposes. In [2] Benedens developed the first watermark robust enough for copyright protection. Nevertheless this method requires an significant amount of data for decoding and is therefore not usable for public watermarking. Very recently, Praun et al. [12] introduced a watermarking scheme based on wavelet encoding. Although mathematically elegant, their method requires a registration procedure for decoding and is also not public.

This paper presents two variations of a robust watermarking method for general polygonal meshes of arbitrary topology. The proposed watermark is embedded in the point coordinates, does not depend on a particular point order as in [8, 9, 10, 11] and resists translation, rotation or scaling of the mesh. If the encoding procedure is combined with an affine invariant norm, the watermark is also immune to affine transformation of the mesh. The proposed watermark is public and requires in contrast to [2] and [12] knowledge of only a few parameters for proper decoding and no registration procedure. The watermark does depend on the connectivity of the mesh. Remeshing or mesh decimation attacks will degrade the watermark. However, test have shown that even in this cases, artifacts remain that can be used to detect the existence of a watermark if not the watermark itself.

2. Background

Hiding information in other information, a technique called *steganography*, serves different purposes that impose different requirements on the used techniques. Roughly speaking, there exist three main application areas.

1. *Copyright protection:* The embedded information is used to identify ownership of the data, usually by hiding a digital signature or company logo. In this case, the watermark has to be immune to a wide variety of data modifications as it might be subject to intentional attacks. Such watermarks are called *robust watermarks.*

2. *Data authentication:* If critical data is transmitted over a network it is necessary to check the authenticity of the received data. In contrast to copyright protection, data authentication requires that the watermark is removed or at least severely degraded if the data is tampered with. Watermarks of this type are called *fragile watermarks.*

0-7695-0562-7/00 $10.00 © 2000 IEEE

3. *Content annotation:* Hidden data can also be used to label specific content areas. For example, one can hide information which identifies different extremities of a human body model such as left or right arm. These labels can then be used for content based indexing and querying in multimedia database applications. Although it is unlikely that such watermarks are subject to intentional removal, they have to be robust in order to survive standard data modifications.

Watermarks are called private or public depending in whether decoding requires information not accessible to the public or not. This information can be either encoding parameters such as in [2] or an original copy of the watermarked mesh for registration as in [12].

Although this paper focuses on robust public watermarking the proposed watermarking algorithm also allows to generate fragile as well as private watermarks and is therefore suitable for all these application areas. Notice that highly robust watermarks require that they are immune to data optimization techniques. Since there exists quite sophisticated mesh optimization tools that are able to remove most of the redundancies within a mesh, one has to expect a trade-off between robustness and visual quality.

Hiding data in a polygonal mesh is a significantly different problem than hiding data in other types of multimedia objects. One reason is that there exists no natural order of the points, edges or faces of a mesh. In [8, 9, 10, 11] Ohbuchi et al. assume a predefined order of the points for their various watermarking techniques. As a result, these watermarks are destroyed as soon as the order of the points is changed.

Another disadvantage of polygonal mesh data is that meshes are usually subject to various types of transformations such a translation, rotation, scaling or more general affine transformations. Watermarks should not be destroyed by these transformations. The only possibility to achieve such robustness is to embed the data in transformation invariant mesh parameters. One example is to hide the data in the ratios between the areas of neighboring triangles or the ratios between volumes of neighboring tetrahedra as proposed in [8, 9, 10, 11]. However, apart from the fact that this requires the knowledge of a specified point order, it also works only for triangulated meshes.

The sequel of this paper introduces a watermarking technique similar to the techniques described in [2] and [12]. It is robust in terms of affine transformations but also does neither require a particular point order nor a special mesh topology for encoding. In contrast to [2] and [12] it works for manifolds as well as non-manifolds and can be used for public watermarking as well.

3. A simple watermarking algorithm

Let us consider a *mesh* $\mathcal{M} = \{\mathcal{P}, \mathcal{C}\}$ in its simplest form consisting of a set

$$\mathcal{P} = \{\mathbf{p}_i\}_{i=0,\ldots,n}, \quad \mathbf{p}_i = (x_i, y_i, z_i), \tag{1}$$

of $n+1$ points $\mathbf{p}_i$ with coordinates (x_i, y_i, z_i) and connectivity information

$$\mathcal{C} = \{\{i_k, j_k\}\}_{k=0,\ldots,m}, \quad 0 \leq i_k \leq n, 0 \leq i_k \leq n. \tag{2}$$

$\mathcal{C}$ is a set of index pairs $\{i_k, j_k\}$ each of which corresponds to one edge of the mesh. Note that we only need a wireframe model of the mesh. We also omit any associated attributes such as color, material properties, surface normals or similar as these attributes are not effected by the watermarking procedure.

Let us further define the *star* $\mathcal{S}_i$ of a point $\mathbf{p}_i$ as the set of indices of all points connected to $\mathbf{p}_i$, i.e.

$$\mathcal{S}_i = \{j | \{i, j\} \in \mathcal{C}\}, \quad i = 0, \ldots, n. \tag{3}$$

For the sake of simplicity and without loss of generality we assume that the star of a point is never empty.

One of the simplest watermarking techniques for images is to hide information in the lower bits of the pixel values, preferably in the frequency domain. It is obvious that this technique cannot be used for the point coordinates of a polygonal mesh as such a watermark would be destroyed even by translating the object. In order to overcome this problem we are going to use the vectors

$$\mathbf{n}_i = \frac{1}{|\mathcal{S}_i|} \sum_{j \in \mathcal{S}_i} (\mathbf{p}_j - \mathbf{p}_i) \tag{4}$$

averaging the vectors from the point $\mathbf{p}_i$ to all points connected by an edge for the watermarking algorithm. Here, $|\mathcal{S}_i|$ denotes the number of elements in $\mathcal{S}_i$. If the mesh does not degenerate into a plane, around $\mathbf{p}_i$ the vector $\mathbf{n}_i$ can be considered a discrete analogon to the normal vector of a smooth surface at $\mathbf{p}_i$. It is affine invariantly connected with the mesh $\mathcal{M}$.

For the same reason as above we cannot encode the hidden information in the coordinates of the vectors $\mathbf{n}_i$ directly. This watermark would be destroyed by a simple rotation of the object. As we will see, however, it is possible to hide information in the relative length of these vectors. In the following we will investigate two techniques. The first is based on the Euclidean norm of a vector and therefore not immune to affine transformations. In the second technique we generalize the algorithm by using an affine invariant vector norm which results in a more robust watermark.

The basic algorithm works as follows. First the average discrete normal vector length d is computed, i.e.

$$d = \frac{1}{n+1} \sum_{i=0}^{n} ||\mathbf{n}_i||, \tag{5}$$

where $||\mathbf{n}_i||$ denotes the Euclidean norm of $\mathbf{n}_i$. Then each individual normal vector length is converted into an integer n_i according to

$$n_i = round\left(\frac{c}{d} ||\mathbf{n}_i||\right) \tag{6}$$

where c is an arbitrary but fixed real number. In the following we will refer to c as the *main parameter* of the watermarking algorithm. Notice that the integers n_i remain unchanged if the mesh is subject to translation, rotation or scaling. This allows us to use these numbers as the target of our data hiding procedure.

The watermark itself is a function $f(\mathbf{v})$ defined over a sphere. It maps each normalized vector $\mathbf{v}$, $||\mathbf{v}|| = 1$ onto a real number w with $0 \leq w \leq 1$. For example, $f(\mathbf{v})$ could describe pixel intensity values of a company logo projected on the unit sphere or a pseudo random noise function. In the examples in section 5 we use a constant function $f(\mathbf{v}) = const$ for visualization purposes.

Similar to n_i we convert the function value of f at $\mathbf{n}_i$ to an integer, i.e.

$$w_i = round\left(2^b f\left(\frac{1}{||\mathbf{n}_i||}\mathbf{n}_i\right)\right). \tag{7}$$

The integer b denotes the number of bits used to store w_i. As for the main parameter c the parameter b is arbitrary but fixed for all $\mathbf{n}_i$.

For the encoding procedure we first choose b bits from the binary representation of n_i and replace them with the bits of the binary representation of w_i to obtain a modified value n_i' with embedded watermark. The location of these bits is again arbitrary but fixed for all n_i. Notice that it is not necessary to insert the bits into n_i in the same order as they appear in w_i. In fact they can be spread over the binary representation of n_i in a random fashion.

The robustness of the watermark strongly depends on the location of the inserted bits. If they are hidden in the lower bits of n_i or are randomly distributed the watermark will be more fragile. If they are hidden in the more significant areas in the same order as they appear in w_i, the watermark will be more robust, but in return it will have more effect on the point coordinates of the mesh.

In the final step of the encoding procedure, we change the point coordinates of the mesh such that the new discrete normal vectors $\mathbf{n}_i'$ exhibit the watermark in their normalized length. According to (6) we obtain

$$\mathbf{n}_i' = \frac{n_i\, d}{c} \frac{\mathbf{n}_i}{||\mathbf{n}_i||}, \tag{8}$$

and therefore for the new point coordinates $\mathbf{p}_i'$

$$\mathbf{n}_i' = \frac{1}{|\mathcal{S}_i|} \sum_{j \in \mathcal{S}_i} (\mathbf{p}_j' - \mathbf{p}_i'). \tag{9}$$

Equation (9) yields a system of $n+1$ linear equations in $n+1$ unknowns $\mathbf{p}_i'$. It is obvious that this system has to be singular as it is impossible to obtain the location of an object only from its discrete normal vectors. It is generally sufficient to prescribe at least one point per connected component of $\mathcal{M}$. However, test have shown that better results are achieved by leaving up to 20% of the points of $\mathcal{M}$ unchanged, i.e. by using

$$\mathbf{p}_i' = \mathbf{p}_i, \quad \mathbf{p}_i \in \mathcal{I} \tag{10}$$

instead of the corresponding equation (9) for a specified amount of points $\mathcal{I}$. This allows to encode watermarks that are more robust in exchange of introducing a reasonable amount of noise in the watermark.

For the examples in section 5 we solved the system (9) iteratively by adjusting one point at a time until a certain total minimum threshold is reached. The encoding procedure is global and due to the generally large number of points computationally expensive. However, since the encoding is done offline this does not effect the usability of the watermarking scheme. On the contrary, it adds to the robustness of the watermark.

Decoding, on the other hand, can be achieved in real time by calculating the values n_i from equation (6) and extracting the appropriate bits in which the watermark is hidden. Notice that since we changed the normal vectors $\mathbf{n}_i$ proper decoding requires that the main parameter c is adjusted to

$$c' = c\,\frac{d}{d_{new}} \tag{11}$$

where d_{new} is the average normal vector length of the modified mesh. The effects of this adjustment is shown in section 5. Further notice that if the mesh has been subject to a rotation, the decoded watermark will be rotated as well. Since the decoding requires nothing but the parameter c' and the knowledge about the exact bits in which the watermark is hidden, the method can be used for public watermarking.

4. Affine invariant watermarking

Because we used the Euclidean vector norm in the above algorithm, the watermark is degraded or even destroyed if the mesh is subject to an affine transformation such as shearing transformation. While this is sufficient for fragile watermarking, it imposes problems if the watermark has to be robust. However, this problem can be solved easily, if the used vector norm is changed to an affine invariant norm. An example of such a norm based on the inertia ellipsoid of a point set has been discussed in [7]. It is defined as follows. Consider the centroid

$$\bar{\mathbf{p}} = (\bar{x}, \bar{y}, \bar{z}) = \frac{1}{n+1} \sum_{i=0}^{n} \mathbf{p}_i \tag{12}$$

of the mesh $\mathcal{M}$. Then we define the matrix

$$V = \begin{pmatrix} x_0 - \bar{x} & y_0 - \bar{y} & z_0 - \bar{z} \\ x_1 - \bar{x} & y_1 - \bar{y} & z_1 - \bar{z} \\ \vdots & \vdots & \vdots \\ x_n - \bar{x} & y_n - \bar{y} & z_n - \bar{z} \end{pmatrix} \tag{13}$$

and further the matrix

$$A = (n+1)(V^T V)^{-1}. \tag{14}$$

It is then easy to prove (cf. [7]) that the norm defined by

$$||\mathbf{p}||_{\mathcal{P}} = (x\ y\ z) A \begin{pmatrix} x \\ y \\ z \end{pmatrix}, \quad \mathbf{p} = (x, y, z), \tag{15}$$

is invariant under affine transformations. To achieve affine invariant watermarking we simply replace the Euclidean norm $||\mathbf{p}||$ with the affine invariant norm $||\mathbf{p}||_{\mathcal{P}}$ in equations (4) to (8). Notice that this norm depends on the point set $\mathcal{P}$. However, since we generally deal with meshes with a large number of points, small changes in the coordinates or even deletion of a small number of points does not have a significant influence on the result. Tests showed that even if a watermarked mesh is cropped it will still exhibit artifacts of the watermark.

5. Examples

For the examples in this paper we have used a constant function as watermark, i.e.

$$f(\mathbf{v}) = 0.5. \tag{16}$$

In figures 1 to 6 these watermarks have been visualized by rendering the point sets

$$f(\frac{\mathbf{n}_i}{||\mathbf{n}_i||}) \frac{\mathbf{n}_i}{||\mathbf{n}_i||}, \quad \text{or} \quad f(\frac{\mathbf{n}_i}{||\mathbf{n}_i||_{\mathcal{P}}}) \frac{\mathbf{n}_i}{||\mathbf{n}_i||_{\mathcal{P}}}, \tag{17}$$

respectively. As a result the watermark appears either as a sphere with radius 0.5 or a scaled inertia ellipsoid of the mesh.

Figures 1 and 2 show a mesh with 1976 points and 3752 faces. The mesh is marked by a robust watermark using the standard Euclidean norm. The main parameter has been chosen to be 203314341, $b = 7$ and the watermark bits have been stored in bits 16 to 22 of the values n_i. After application of the watermark the maximum coordinate error is

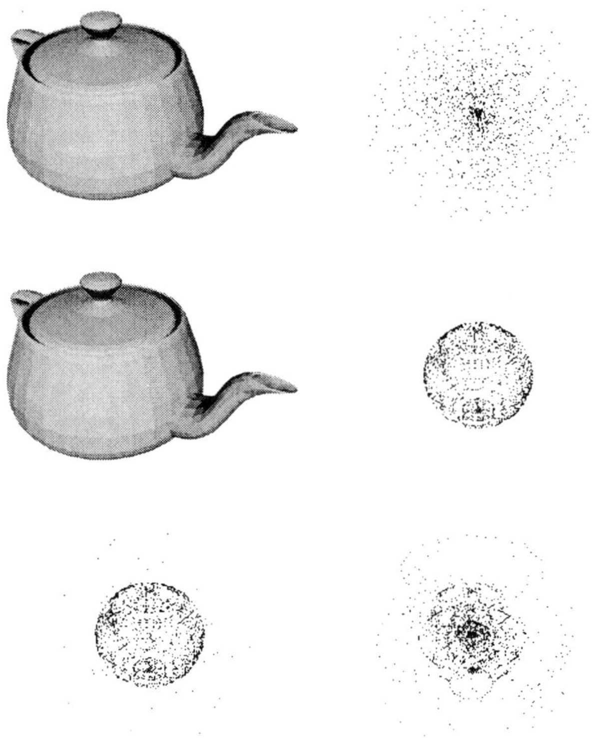

Figure 1. Mesh with 1976 points and 3752 faces. The first row shows the original mesh and the result of the watermark decoder applied to this mesh. The second row shows the mesh with encoded watermark and the original watermark. The third row shows the result of the watermark decoder if run with the adjusted and original main parameters. Since these parameters differ significantly, the original main parameter results in a severely degraded watermark signature.

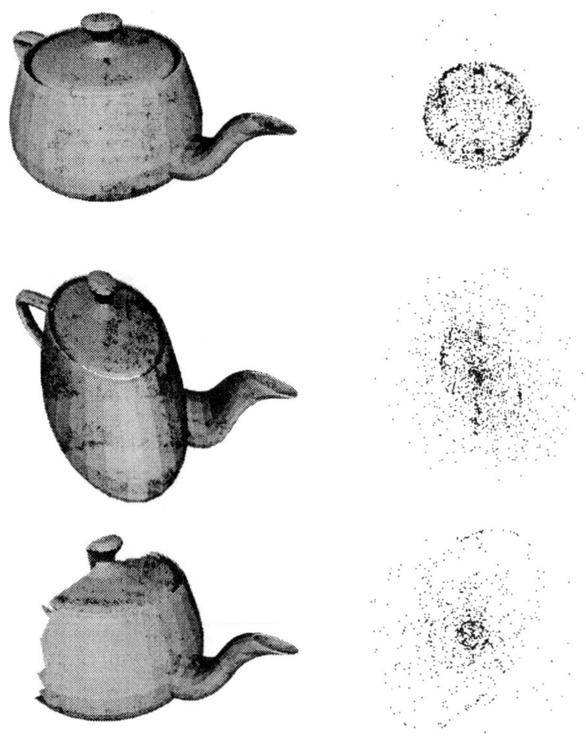

Figure 2. Effects of different mesh modifications on the embedded watermark. The first row shows the effect of reducing the coordinates to six digit precision. The second row shows the effect of an affine transformation. The third row shows the watermark after deleting half of the mesh. Although the watermark is not affine invariant and is not immune to mesh decimation, artifacts of the watermark can still be recovered.

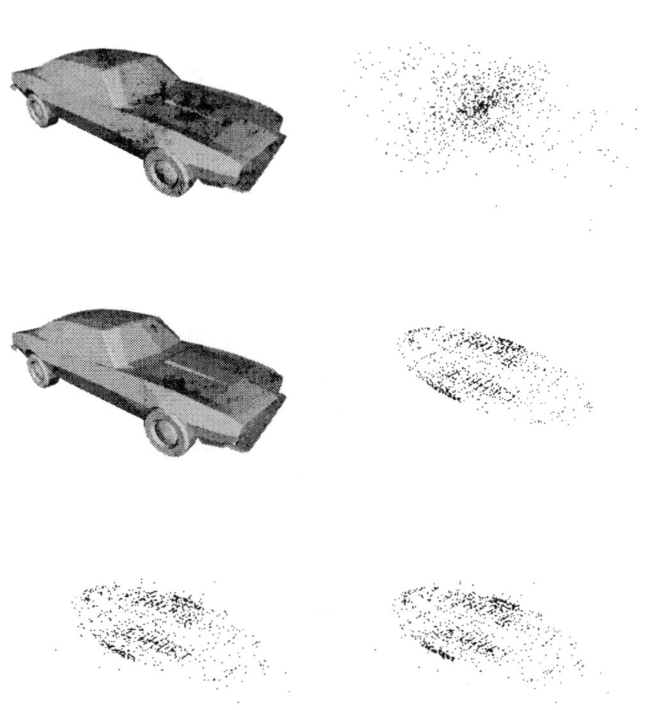

Figure 3. Mesh with 1262 points and 1215 faces. The first row shows the original mesh and the result of the watermark decoder applied to this mesh. The second row shows the mesh with encoded affine invariant watermark and the original watermark which is a scaled copy of the inertia ellipsoid of the mesh. The third row shows the result of the watermark decoder if run with the adjusted and original main parameters.

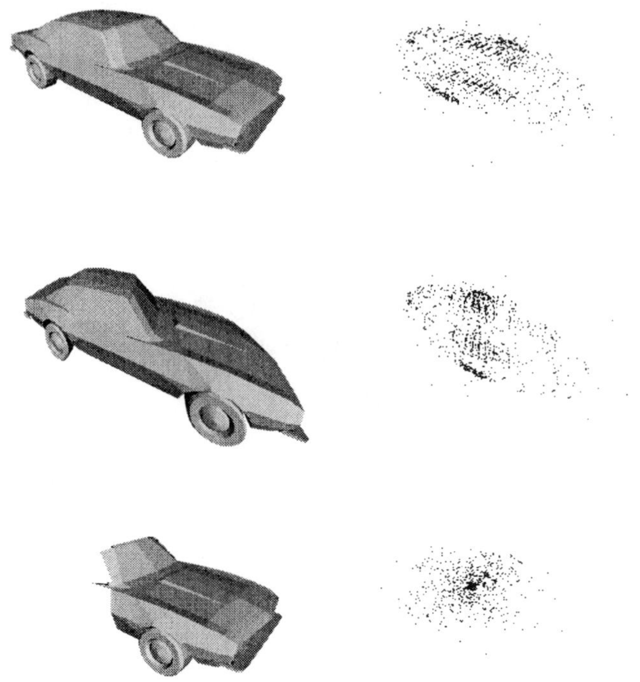

Figure 4. Effects of different mesh modifications on the embedded watermark. The first row shows the effect of reducing the coordinates to six digit precision. The second row shows the effect of an affine transformation. The third row shows the watermark after deleting half of the mesh. It is clearly visible that the watermark is immune to affine transformations of the mesh.

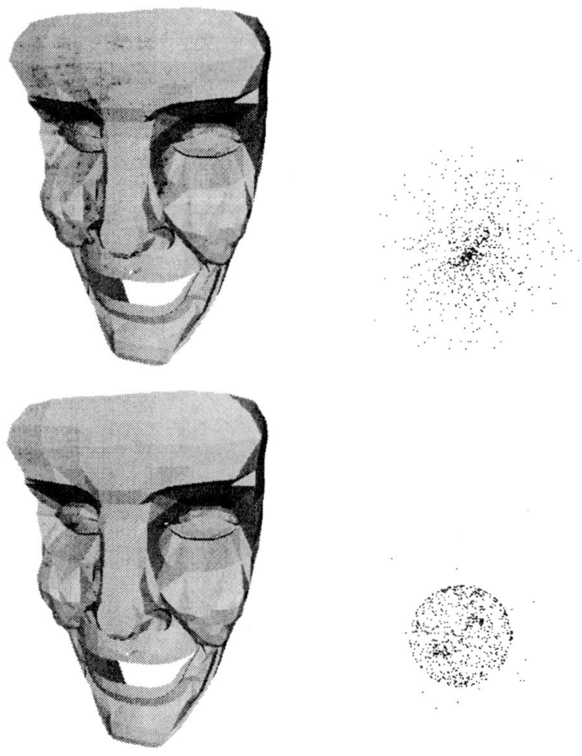

Figure 5. Mesh with 794 points and 1445 faces. The first row shows the original mesh and the result of the watermark decoder applied to this mesh. The second row shows the mesh with encoded fragile watermark and the result of the watermark decoder if run with the adjusted main parameter.

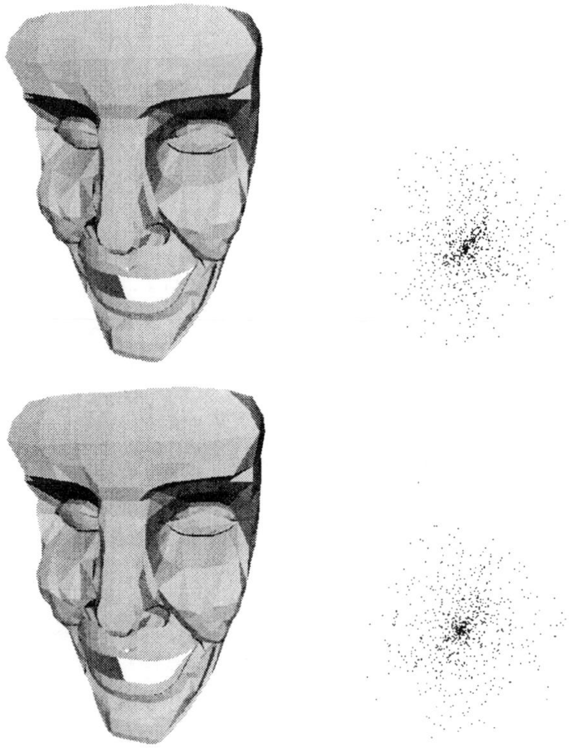

Figure 6. Effects of different mesh modifications on the embedded watermark. The first row shows the effect of reducing the coordinates to six digit precision. The second row shows the effect of a hardly noticeable affine transformation. In both cases the watermark is destroyed completely.

0.002098 at a coordinate range between -2.5 and 2.5. The adjusted main parameter is 204344101.

Similarly, figures 3 and 4 show a mesh with 1262 points and 1215 faces. The watermark is hidden using the affine invariant norm. Main parameter is 443719476, $b = 7$ and the watermark bits are again stored in bits 16 to 22. The maximum coordinate error is 0.009439 at a coordinate range between -10 and 10. The adjusted main parameter is 443766488. Notice that the mesh is transformed according to an affine transformation, the watermark is subject to the same transformation.

In both cases the watermark is severely degraded if the mesh is cropped. However, there are still artifacts visible which are sufficient to prove the existence of the watermark. The reason is that the reduction of the points influences the average normal length as well as the affine invariant norm. If the original parameters are applied, on the other hand, the decoding step will exhibit a simplified but not degraded version of the watermark. This allows to recover the watermark even in this case. The original parameters, in particular the matrix A or the average normal vector length d have to be known, however.

Finally, figures 5 and 6 show a fragile watermark. The mesh has 794 points and 1445 faces. Main and adjusted main parameter are 318513108 and the watermark has been stored in the last 6 bits of the values n_i. The maximum coordinate error is $9.2094 \cdot 10^{-6}$ at a coordinate range from -250 to 250. The watermark is removed if the mesh is subject to even the slightest affine deformation.

The examples also show the influence of a reduction of the coordinate precision to the watermark. Reducing the accuracy is a very common technique used for minimizing the size of VRML (Virtual Reality Modeling Language) files. In all examples the precision has been cut to 6 digits. It is obvious that due to the small coordinate error in the final example, the fragile watermark is deleted completely.

6. Conclusion

This paper presented a new watermarking technique for polygonal meshes. Although the proposed watermark is robust enough for most applications, there are still a lot of open questions that need to be answered. In particular it would be interesting to develop watermarks which are also immune to general projective transformations of the mesh. Furthermore, as polygonal meshes represent only one special data type, watermarking techniques for other three dimensional media need to be developed as well. This includes, for example, watermarks for general VRML files. The protection of intellectual property rights will continue to become an increasingly important issue crossing the boundaries between disciplines and research dealing with alternative data types is highly needed.

References

[1] W. Bender, D. Gruhl, N. Morimoto and A. Lu, "Techniques for data hiding", *IBM Systems Journal*, Vol. 35, Nos. 3&4, 1996, pp. 313–336.

[2] O. Benedens, Geometry-based Watermarking of 3D Models, *IEEE Computer Graphics and Applications*, Vol. 19(1), pp. 46–55.

[3] I.J. Cox, J. Kilian, T. Leighton and T. Shamoon, "Secure Spread Spectrum Watermarking for Multimedia", *IEEE Transactions on Image Processing*, Vol. 6, No. 12, 1997, pp. 1673–1687.

[4] I.J. Cox and M.L. Miller, "A review of watermarking and the importance of perceptual modeling", *Proc. SPIE Conference on Human Vision and Electronic Imaging II*, Vol. 3016, February 1997, pp. 92-99.

[5] F. Hartung, P. Eisert, and B. Girod, "Digital Watermarking of MPEG-4 Facial Animation Parameters", *Computers&Graphics*, Vol. 22, No.3, 1998.

[6] S.H. Low and N.F. Maxemchuk, "Performance Comparison of Two Text Marking Methods", *IEEE Journal on Selected Areas in Communications*, Vol. 16, No. 4, 1998.

[7] G.M. Nielson and T.A. Foley, "A Survey of Applications of an Affine Invariant Norm", in: T. Lyche and L.L. Schumaker (eds.), *Mathematical Methods in Computer Aided Geometric Design*, Academic Press, Boston, MA, 1989, pp. 445–467.

[8] R. Ohbuchi, H. Masuda and M. Aono, "Embedding Data in 3D Models", in: Steinmetz et al. (eds.), *Lecture Notes in Computer Science* No. 1309 (Proceedings of the IDMS'97, Darmstadt, Germany, September 1997), Springer, Berlin, Germany, 1997, pp. 1–11.

[9] R. Ohbuchi, H. Masuda and M. Aono, "Watermarking Three-Dimensional Polygonal Models", *Proceedings of the ACM Multimedia '97*, Seattle, WA, November 1997, Addison Wesley, Reading, WA, USA, 1997, pp. 261–272.

[10] R. Ohbuchi, H. Masuda and M. Aono, "Watermarking Three-Dimensional Polygonal Models Through Geometric and Topological Modifications", *IEEE Journal on Selected Areas in Communications*, Vol. 16, No. 4, 1998, pp. 551–560.

[11] R. Ohbuchi, H. Masuda and M. Aono, "Data Embedding Algorithms for Geometrical and Non-Geometrical Targets in Three-Dimensional Polygonal Models", *Computer Communications*, 1998.

[12] E. Praun, H. Hoppe, and A. Finkelstein, "Robust mesh watermarking", *Microsoft Technical Report* TR-99-05.

[13] B.-L. Yeo and M.M. Yeung, "Watermarking 3D Objects for Verification", *IEEE Computer Graphics and Applications*, Vol. 19(1), pp. 36–45.

On-Line Polygon Refining Using a Low Computation Subdivision Algorithm*

Yuan-Chung Lee and Chein-Wei Jen

Depart. Electronics Engineering, National Chiao-Tung University, Hsinchu, Taiwan, ROC

E-mail: {yzlee, cwjen}@twins.ee.nctu.edu.tw

Abstract

Subdivision algorithms have been extensively used in computer graphics to produce more accurate geometry model. For the purpose of integration into the context of standard graphics pipeline for real-time display, we can only make use of the available information of one triangle. In this paper we proposed a low computation cost algorithm to subdivide a polygon using vertex coordinates and vertex normal vectors of one triangle. Comparing with other suitable algorithms, our algorithm produces decent results with lowest computation complexity. We also consider the generation of interpolated normal vectors and texture mapping parameters. Detailed comparisons among other suitable algorithms are presented.

1. Introduction

With rapid progress of VLSI technology and advanced hardware architecture, the performance of 3D graphics in terms of polygon processing rate and pixel fill rate increases quickly in the new 3D graphics hardware. On the other hand to obtain better image quality, one common method is to use extra pixel fill rate for super-sampling anti-aliasing or post processing for images. Another one that can also improve image quality is to use extra polygon processing rate from rapidly enhanced geometry processors. By applying this extra polygon processing rate without much overhead, this paper proposes a method that can run-time generate more accurate 3D models with polygon subdivision.

Our target is to integrate this method into the context of Gouraud shading [1] based standard graphics libraries, such as OpenGL and Direct3D. These libraries will tessellate polygons into triangles and provide information of one triangle in their processing pipeline. Each triangle is transformed, lighted, clipped and rendered individually without knowledge of neighbor triangles. Although triangle strip and fan style primitives exist in those standard graphics libraries, they can not provide complete neighborhood information. The available information in this integration is only three vertex coordinates and three vertex normal vectors of one triangle. Our proposed method uses this available information to subdivide a triangle while keeps computation cost as low as possible. We also consider the effect of generating texture mapping parameters at subdivided vertex and present suitable subdivision methods. Besides, we propose adaptive second order interpolation for normal vector to generate appropriate normal vector at subdivided vertex.

In this paper we first review the previous works in Section 2. Then we describe our algorithm in Section 3 and show the visual result with integration into OpenGL in Section 4. Section 5 presents the comparison between various algorithms in terms of operation counts and quality. Section 6 shows the results in a recursive fashion. Finally we summarize this paper in Section 7.

2. Previous works

Surface interpolation has been developed for years. Traditional surface fitting methods, such as [2], did not utilize normal vectors and required large numbers of neighborhood information. Well-known polygon sub-division algorithms, such as Catmull-Clark subdivision surface [3], Butterfly method [4], and their extensions, also did not utilize normal vectors and required nearby information. Some recent methods [5][6] used vertex normal vectors for surface interpolation but they still needed some extra information. Several surveys [7][8] classified local surface interpolation schemes into two groups: split-domain schemes and convex combination schemes. The split-domain scheme in [9][10][11] can construct boundary curves using only vertex coordinates and their normal vectors of one triangle, but they required boundary data of adjacent patches to construct the entire triangular patch. Among the researches in convex combination schemes, Side-vertex method [12] used the

* This work was supported by the National Science Council, Taiwan, R.O.C. under Grant NSC89-2215-E009-052

0-7695-0562-7/00 $10.00 © 2000 IEEE

information of one triangle to generate both boundary curves and triangular patch. The generated tangent vectors on the boundary curves in [9][10][12] and in the normal projection method of [11] agreed in the direction but the length varied. [13][14] only constructed boundary curves, too. [13] was a boundary element method used in bio-electric and bio-magnetic source reconstruction. [14] proposed two configurations for cubic Bezier curve subdivision. [15] generated a composite surface of quartic Gregory patches. [16] used a blending of three sphere patches to produce final surface.

In summary, [9][10][11][13][14] can only be used to subdivide on the boundary, while [12][15][16] can be used to subdivide both on the boundary and in the triangular patch. Nevertheless most of them require high computation complexity that is not suitable for high performance real-time display. Besides they did not describe the detailed approaches to integrate their algorithms into standard graphics libraries. Our on-line algorithm provides a good trade-off between quality and computation cost, and has the ability to subdivide both on the boundary and in the triangular patch. We will compare these methods in Section 5.

3. On-line refining algorithm

3.1 Subdivision styles

To integrate our algorithm into graphics libraries, subdivision style must be set in advance. We adopt two subdivision styles: Ternary and Quaternary styles, as shown in Fig. 1 and 2 respectively. The arrows in the figures are the shift vectors from the triangle plane to new positions. In Ternary style, we generate one new vertex in the triangular patch and combine it with original three vertices to form three triangles. In Quaternary style, we generate three new vertices on the boundary and combine them with original three vertices to form four triangles.

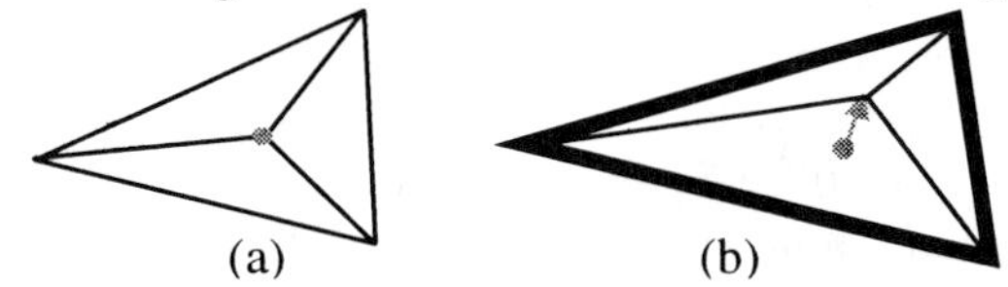

(a) (b)

Fig. 1: Ternary style. (a) planar overview, (b) real 3D illustration

(a) (b)

Fig. 2: Quaternary style. (a) planar overview, (b) real 3D illustration

After subdivision, we render these triangles with Gouraud shading planar triangle based libraries and the graphics hardware. Hence we can not elevate the continuity to G^1 degree by doing subdivision for one iteration. Ternary style subdivision will still keep the C^0 continuity. However for Quaternary style to enforce the C^0 continuity, the model must match the condition that the same normal vector at the same vertex position between adjacent triangles. It is always true in the smooth model. On some non-smooth boundaries between meshes, we should handle them specifically.

3.2 Basic techniques

Phong Shading [17] linearly interpolated the normal vector to simulate the curve surface shading for better visual quality, but it did not really affect the geometry shape. It is not implemented in standard graphics libraries, such as OpenGL and Direct3D, because of its huge computation cost for real-time animation. Our basic idea is to use the Phong-like interpolated normals with the modification to generate new vertices.

To ease the explanation of our algorithm, first we will use two vertices and their normal vectors to illustrate our techniques. Then we will extend these techniques into our subdivision styles in following subsections. The vertex normal vectors have already been normalized in the graphics library for the purpose of lighting operations. In Fig. 3, the curve $V_1M_2V_2$ is the desired curve passing through two vertices and perpendicular to the given vertex normal vectors. The available information exceeds the condition of the quadratic curve, but does not meet the requirement of the cubic curve. As a result, the curve is not unique, and there are infinite numbers of curves matching existing parameters. In the subdivision procedure, we have to find the new subdivided vertex position of M_2. So we need to make assumptions about the vertex position according to the available information.

For simplification, we calculate M_2 by approximating $\overrightarrow{M_1M_2}$ to $\overrightarrow{Q_1Q_2}$ *(*length factor*), where M_1 is the mid-point of $\overline{V_1V_2}$. Q_1 is obtained from $\overrightarrow{M_1Q_1}$, the average of $\vec{N}_1$ and $\vec{N}_2$. Q_2 is obtained from $\overrightarrow{M_1Q_2}$, the normalized vector of $\overrightarrow{M_1Q_1}$. The *length factor* is an adjusting parameter that will be described later. However, when the curve is concave as shown in Fig. 3(b), we must invert the shift vector $\overrightarrow{M_1M_2}$. We propose a simple concave detection scheme to determine whether we have to invert the shift vector by using the sign of concave detection result: $\vec{N}_1 \bullet \overrightarrow{M_1V_1} + \vec{N}_2 \bullet \overrightarrow{M_1V_2}$. If the sign of concave detection result is less than zero, we will invert the shift vector.

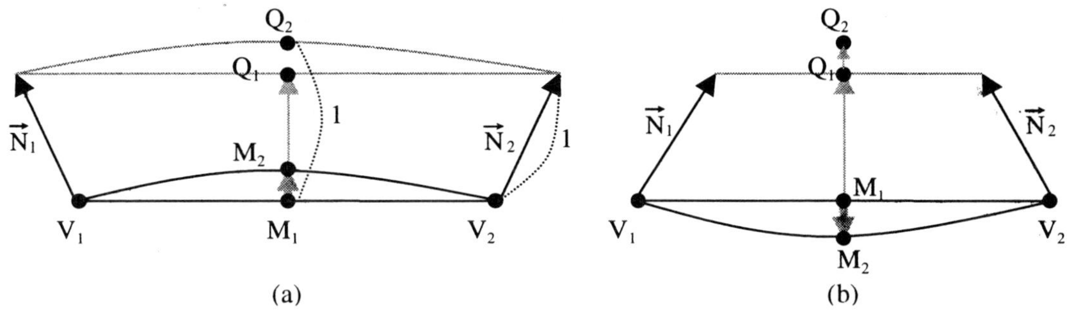

Fig. 3: (a) Convex Configuration, (b) Concave Configuration

The above techniques can be applied to those situations that have different normal vectors as shown in Fig. 4. The curve is monotonous in Fig. 4(a) and (b). Fig. 4(b) shows the extreme case that the angle between two normal vectors is approaching 180°, and the shift vector length is approximated to length factor (half of the edge length), which is like the radius of the circular arc. In Fig. 4(c) and (d), the curve is non-monotonous, and the shift vector matches the displacement condition in the middle of the arc.

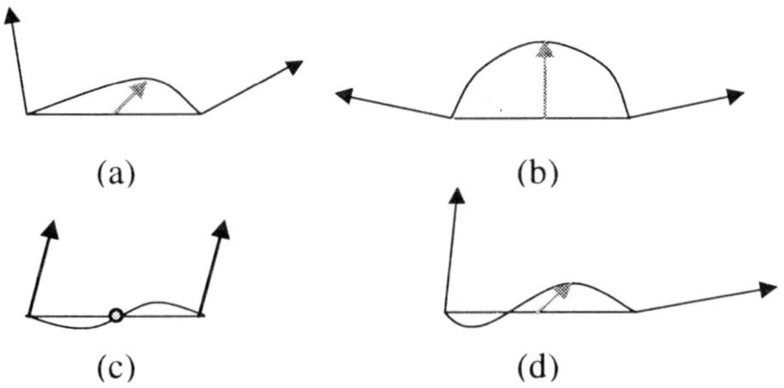

Fig. 4: Some situations illustrate shift vector

3.3 Ternary style

With the consideration of generating texture mapping coordinates for the new vertex, we propose two methods for each subdivision style to handle the shift vector.

Method 1:

Let V_1, V_2, and V_3 denote three vertices, and $\vec{N_1}, \vec{N_2}$, and $\vec{N_3}$ denote three vertex normal vectors of the given triangle. Then the center of gravity is $M_1 = (V_1 + V_2 + V_3)/3$, and its normal is assumed to be $\vec{N} = (\vec{N_1} + \vec{N_2} + \vec{N_3})/3$.

So, if we define:

Concave detection result = $\vec{N_1} \bullet \overrightarrow{M_1V_1} + \vec{N_2} \bullet \overrightarrow{M_1V_2} + \vec{N_3} \bullet \overrightarrow{M_1V_3}$,

Length factor = $\dfrac{\sqrt{\sqrt{\|\overrightarrow{V_1V_2} \times \overrightarrow{V_1V_3}\|}}}{2}$, *where* $\times$ *denotes cross product,*

and *shift vector = (Normalized of* $\vec{N}$ - $\vec{N}$ *) * length factor,* then we can calculate new vertex coordinate as:

if (concave detection result ≥ 0 *)*

new vertex coordinate $V = M_1 +$ *shift vector*

else

new vertex coordinate $V = M_1 -$ *shift vector*

We define the length factor as a parameter proportional to the square root of the triangle area. The square root of the triangle area is the quantity of length. The triangle area is calculated by using the square root of cross product of two triangle edge vectors. In addition, we can also adjust the length factor by multiplying a constant value.

In the generation of texture mapping coordinates, Method 1 simply uses average of three vertex texture coordinates, i.e., $S = (S_1 + S_2 + S_3)/3$, and $T = (T_1 + T_2 + T_3)/3$. This is not an ideal approach, but its operations are the simplest. A better way is to calculate the texture coordinate of the projected point *P* in the triangle plane shown in Fig. 5. However, we do not adopt this because it needs calculating an inverse 3x3 matrix that is not computation-efficient. The compromise way, shown below as Method 2, is to modify the shift vector to the position appropriate for the interpolated texture coordinates.

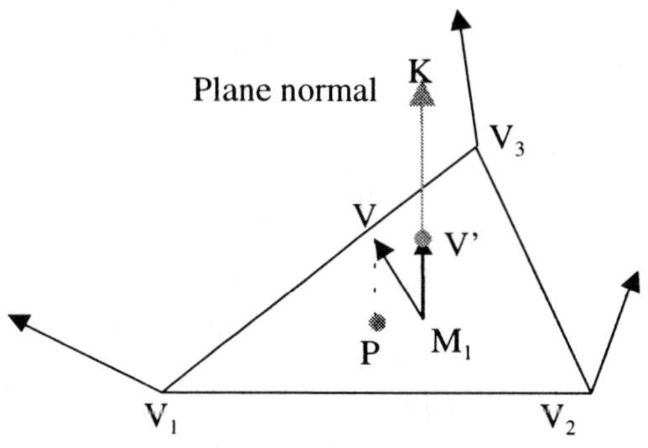

Fig. 5: Consideration of Texture Coordinate

Method 2:

The proposed Method 2 projects new vertex V onto the plane normal vector and uses the new projected point V' as the new vertex. Then it is suitable to use the average texture coordinate in the new vertex. The plane normal vector calculation is saved since we have computed it when determining length factor.

Let $\overrightarrow{M_1V}$ *be shift vector of Method 1, and* $\overrightarrow{M_1K}$ *be normalized plane normal*

New shift vector $\overrightarrow{M_1V'} = (\overrightarrow{M_1V} \bullet \overrightarrow{M_1K}) * \overrightarrow{M_1K}$

However, Ternary style subdivision has a disadvantage of producing elongated triangle shape. It might sometimes have some defect for shading. In those situations, we will adopt the Quaternary style subdivision.

3.4 Quaternary style

Method 1:

The Quaternary style subdivision is a per-edge operation. We consider two vertices with their normal vectors of the edge in a time. It is like the situation in Section 2.2. Here we define the *length factor* of Quaternary style as half of the edge length. When the angle between two normal vectors is approaching 180°, the shift vector length is approximated to length factor, which is like the radius of the circular arc as shown in Fig. 4(b). We can also multiply an extra constant value to the length factor. If the constant value is greater than one, the length of the shift vector will be larger than the original one. Besides, the concave detection result can be simplified as $(\vec{N}_1 - \vec{N}_2) \bullet \overrightarrow{M_1V_1}$

Method 2:

The extension to Method 2 of Quaternary style is more difficult because the numbers of normal vectors perpendicular to the edge are infinite. We define edge normal as the one in the plane containing the edge and the average vector of two vertex normal vectors.

New shift vector is $\vec{S} - \left(\frac{\vec{S} \bullet \overrightarrow{M_1V_1}}{\left\|\overrightarrow{M_1V_1}\right\|^2}\right)\overrightarrow{M_1V_1}$, *where* $\vec{S}$ *is the shift vector in Method 1.*

Sometimes a vertex might be assigned for two or more different normal vectors. This usually occurs on boundaries between two or more meshes. However, Quaternary style subdivision will result in cracks in those boundaries and can not keep the C^0 continuity. One solution is to blend those normal vectors at the vertex into one vector. The other is to prevent the edge from subdivision.

3.5 Second order interpolation for normal vector

The new vertex normal vector, based on linear interpolation, sometimes causes artifacts for non-monotonous curvature, as shown in Fig. 6. We adopt the quadratic interpolation proposed in [18] to solve the problem. After our simplification, the calculation in Quaternary style is:

New quadratic normal = $\vec{S} - \frac{3}{2}\left(\frac{\vec{S} \bullet \overrightarrow{M_1V_1}}{\left\|\overrightarrow{M_1V_1}\right\|^2}\right)\overrightarrow{M_1V_1}$, *where* $\vec{S}$ *is the shift vector of Method 1.*

Note that this formula is only a little modification of the formula in Quaternary style Method 2, so the calculation can be shared.

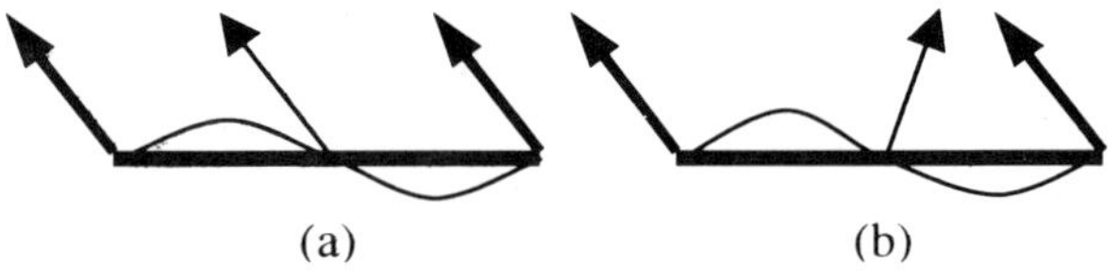

Fig. 6: (a) linear, (b) quadratic interpolation, for non-monotonous curvature

We can extend the technique to Ternary style with some change. In this case, new quadratic normal =

$\vec{S} - \frac{3}{2}(\overrightarrow{S\text{-}MV'}) = V' - \frac{1}{2}(\overrightarrow{S\text{-}MV'})$, where V' is the projected point shown in Fig. 5.

3.6 Adaptive second order interpolation for normal vector

In some monotonous curvature, the quadratic normal is no more appropriate than the linear normal. In Fig. 7, the quadratic normal will exceed the limits of two endpoint normal vectors, but it is not a realistic situation.

Fig. 7: (a) linear, (b) quadratic interpolation, for some monotonous curvature

In this paper we propose an adaptive second order interpolation scheme. Because concave detection operations will slightly determine the curvature of the surface, we use their operations for adaptive selection.

In Quaternary style: (for each edge)

If $(((\vec{N}_1 \bullet \overrightarrow{M_1V_1} \geq 0) \,\&\&\, (\vec{N}_2 \bullet \overrightarrow{M_1V_2} \geq 0))$

$\quad || \; ((\vec{N}_1 \bullet \overrightarrow{M_1V_1} \leq 0) \,\&\&\, (\vec{N}_2 \bullet \overrightarrow{M_1V_2} \leq 0)))$

use linear interpolation for normal vector

else

use quadratic interpolation for normal

In Ternary style:

if $(((\vec{N}_1 \bullet \overrightarrow{M_1V_1} \geq 0) \,\&\&\, (\vec{N}_2 \bullet \overrightarrow{M_1V_2} \geq 0) \,\&\&\, (\vec{N}_3 \bullet \overrightarrow{M_1V_3} \geq 0))$

$|| ((\vec{N}_1 \bullet \overrightarrow{M_1V_1} \leq 0) \,\&\&\, (\vec{N}_2 \bullet \overrightarrow{M_1V_2} \leq 0) \,\&\&\, (\vec{N}_3 \bullet \overrightarrow{M_1V_3} \leq 0)))$

use linear interpolation for normal vector

else

use quadratic interpolation for normal

4. Integration into OpenGL

For the simulation, we integrate our refining algorithm into Mesa [19] (an OpenGL-like 3D graphics library). The refining stage is arranged as shown in Fig. 8. The subdivision operations are all performed in eye coordinates. To minimize the overhead for refining, we put the refining stage after back-face culling stage as long as the eye coordinates of current polygon, lighting information and projection matrix are still available. By our arrangement, only the front-facing polygons in the viewing volume will be refined, which has the advantage of view-dependent refining. Besides, there are two additional switches that can be used to determine whether we should apply subdivision for the triangle. One is setting a threshold value for concave detection result, which will have the result that low curvature triangles do not need refining and thus save the computation cost. The other is setting a threshold value for the distance from the triangle to the camera (depth value) to realize level-of-detail control.

Fig. 9 shows the effect of our algorithm with Gouraud shading. Fig. 9(a) is a sphere made by 20 triangles (an icosahedron), and Fig. 9(b) is a pawn model made by 147 triangles. As shown in the figure, the lighting areas become more realistic, and the silhouettes become smoother. Fig. 9(c) shows the effect combined with texture mapping.

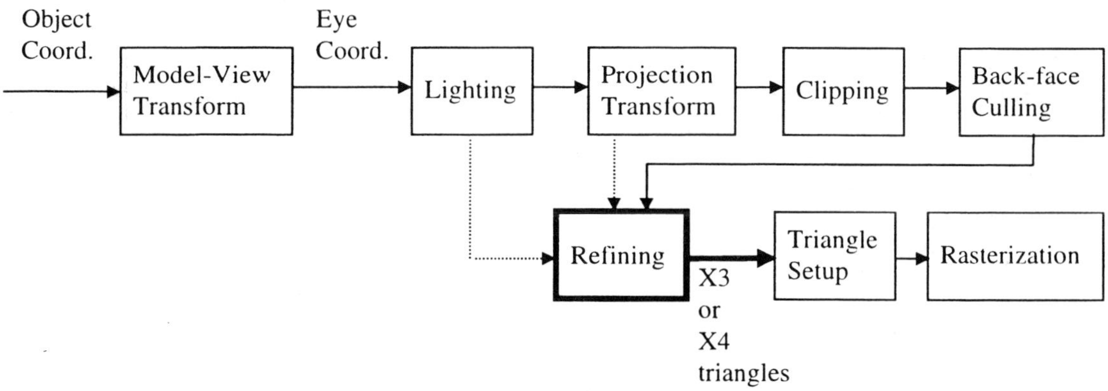

Fig. 8: Graphics pipeline with refining stage

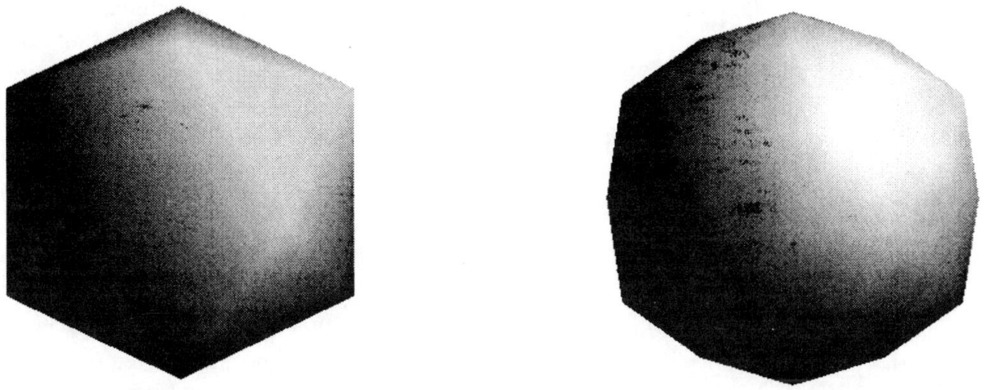

Fig. 9: (a) Left: sphere model without refining; Right: sphere model with refining

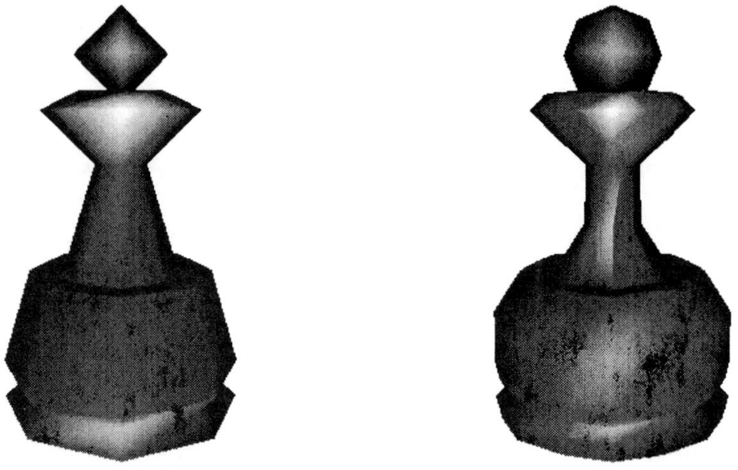

Fig. 9: (b) Left: pawn model without refining; Right: pawn model with refining

Fig. 9: (c) Sphere model with texture mapping. Left: without refining; Right: with refining

5. Algorithm comparison

5.1. Calculation analysis

To illustrate the low complexity advantage of our algorithm, we apply all suitable algorithms, which use only three vertex coordinates and three vertex normal vectors of one triangle, to our subdivision styles for comparison. Table 1 shows the total operation counts of various algorithms to refine a triangle. For Ternary style, the operations include generating one new vertex, normalizing its normal vector, and generating its texture coordinates. For Quaternary style, the operations include generating three new vertices for three edges, normalizing their normal vectors, and generating their texture coordinates. Among those algorithms, the basic techniques for Quaternary style in [9][10] and normal projection method of [11] are the same as Side-vertex method [12], so their operation counts are almost identical. In addition, [9][10][11] can not do Ternary style subdivision with only one triangle's information, hence we only list Side-vertex method [12].

In Table 1, "$1/\sqrt{\ }$" means inverse square root, "/2" means divide by 2 or simply shift right one bit, and the term "inv" means the operation to change the sign of a number after checking the sign of concave detection result. From Table 1, we find that only Quaternary style method 1 of SPHERIGON [16] has smaller computation complexity than ours. However, if we enable second order interpolation for normal vector or adaptive second order interpolation, our computation complexity will be smaller than theirs because of calculation sharing.

Table 1: Operation counts analysis

Style \ Method	Ternary style Method 1	Ternary style Method 2	Quaternary style Method 1	Quaternary style Method 2
Side-Vertex Method	*: 423 +: 243 $1/\sqrt{}$: 18	*: 441 +: 256 $1/\sqrt{}$: 19	*: 216 +: 126 $1/\sqrt{}$: 9	*: 234 +: 141 $1/\sqrt{}$: 9 /: 3
Quartic Gregory Triangle	*: 467 +: 390 $1/\sqrt{}$: 10 $\sqrt{}$: 3 /: 6	*: 485 +: 397 $1/\sqrt{}$: 11 $\sqrt{}$: 3 /: 6	*: 177 +: 120 $1/\sqrt{}$: 3 $\sqrt{}$: 3 /: 6	*: 195 +: 135 $1/\sqrt{}$: 3 $\sqrt{}$: 3 /: 9
Plane Curve method	Not used	Not used	*: 147 +: 102 $1/\sqrt{}$: 9 $\sqrt{}$: 3	*: 165 +: 117 $1/\sqrt{}$: 9 $\sqrt{}$: 3 /: 3
Virtual Triangle Refinement	Not used	Not used	Not used	*: 106 +: 99 $1/\sqrt{}$: 3 /: 6 /2: 24
Maximally Stretched	Not used	Not used	*: 189 +: 150 $1/\sqrt{}$: 3 /: 3	*: 225 +: 180 $1/\sqrt{}$: 3 /: 6
Minimal Curvature	Not used	Not used	*: 141 +: 150 $1/\sqrt{}$: 3 /: 3	*: 177 +: 180 $1/\sqrt{}$: 3 /: 6
SPHERIGON	*: 54 +: 62 $1/\sqrt{}$: 1 /: 3	*: 72 +: 75 $1/\sqrt{}$: 2 /: 3	*: 45 +: 72 $1/\sqrt{}$: 3 /: 3 /2: 27	*: 72 +: 93 $1/\sqrt{}$: 3 /: 6 /2: 27
Our On-Line Polygon Refining	*: 35 +: 52 $1/\sqrt{}$: 1 $\sqrt{}$: 2 inv: 3	*: 44 +: 54 $1/\sqrt{}$: 1 /: 1 $\sqrt{}$: 2 inv: 3	*: 45 +: 78 $1/\sqrt{}$: 3 $\sqrt{}$: 3 /2: 24 inv: 9	*: 63 +: 93 $1/\sqrt{}$: 3 /: 3 $\sqrt{}$: 3 /2: 24 inv: 9

5.2. Quality simulation

In our experiments, we find that the visual quality of various algorithms to refine triangle is similar to each other. To identify the detailed difference between the various algorithms, we use the quantitative approach to measure the variance of shift vectors. One possible simulation is to initially simplify a complicated mesh, and then refine it with various algorithms to see the error between original mesh and refined mesh. However, the method to simplify the mesh will severely affect the result. So we do not choose this simulation method.

Instead of analyzing which algorithms perform closer to the ideal mesh, we will show that the difference between these algorithms is small. We use vast amounts

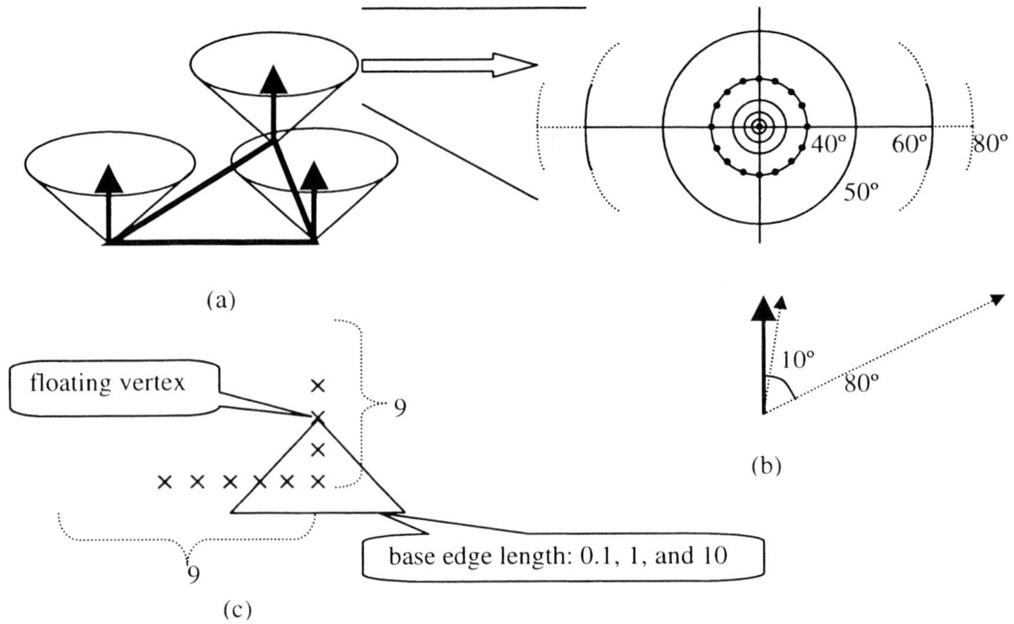

Fig. 10: Different triangles for test pattern. (a) alteration of three normal vectors, (b) illustration of cross section view for alteration of normal vector, (c) illustration of alteration of triangle shapes and sizes

of different triangles as test patterns, as shown in Fig. 10. We alter the triangle in its normal vectors, shapes, and sizes. Each normal vector is shifted by 10° shown as circles in Fig. 10(b). We take 16 samples on each circle. So there are 129 different types for each normal vector. We alter triangle shapes by moving a floating vertex to 81 different positions, and alter the length of base edge to 3 different size. The total amount of samples are about $5.2*10^8$. Furthermore, because the angle between vertex normal vector and triangle plane normal vector is seldom very large, we utilize the angle to take Gaussian weighting in the error statistics.

To show that difference between various algorithms is little, we choose one algorithm as reference and measure the shift difference between reference and other tested algorithms. For Quaternary style, the most general boundary construction algorithm is the one used in [9][10][12] and normal projection method of [11]. Their algorithms all generate tangent vectors in the same direction. [9] and [11] scaled the length of tangent vector equal to the length of the corresponded edge, and it is also suggested in [7]. So we use this algorithm as the reference. Besides, it is worthy to note that the Side-vertex method [12] provides two free parameters for user to adjust the length of tangent, and the author suggested taking both of them to be unity. However, the result of setting both of them to be unity is not good because it is not scaling invariant. The survey in [7] also pointed out this, and we will show this error in the following simulation.

To set the reference for Ternary style, we use the well-known Side-vertex method [12] with scaling tangent length to the length of the corresponded edge instead of unity. We will still show the error to set to unity in the following simulation.

Table 2 shows the experiment results, which the numbers illustrate the average ratio of shift difference to basic length.

$$\text{difference ratio} = \frac{\|(\text{reference shift vector}) - (\text{shift vector of tested algorithm})\|}{\text{basic length}} * 100\%$$

For Quaternary style, the basic length is set as half edge length. For Ternary style, the basic length is set as

$$\frac{\sqrt{\sqrt{\left\|\overrightarrow{V_1V_2} \times \overrightarrow{V_1V_3}\right\|}}}{2}.$$

Note that the numbers in Table 2 are not the measurement of deviation from the original polygon mesh. The numbers are the difference between reference algorithm and tested algorithm. The smaller number means the closer result to the reference algorithm. We do not show the deviation from the original polygon mesh because the larger deviation does not mean closer to the ideal mesh. Excessive deviation sometimes produces obvious artifacts.

As observed from Table 2, the worst algorithm is Side-vertex method with setting tangent length as unity. The algorithm is not scaling invariant, and the survey in [7] also mentioned that it produces loops in the curve if the data points are close together, as shown in Table 2(b). Other algorithms show almost stable shift difference with different base edge length in Table 2(a), (b) and (c). In Ternary style our algorithm produces less error with reference algorithm than all other algorithms while keeping the lowest computation complexity. In Quaternary style our algorithm consumes low computation cost to produce the decent result.

Table 2: Shift difference simulation.

(a) base edge length = 1.0

Style \ Method	Ternary style Method 1	Ternary style Method 2	Quaternary style Method 1	Quaternary style Method 2
Side-Vertex Method (tangent length = 1)	3.248281 %	2.580430 %	3.681000 %	3.505595 %
Quartic Gregory Triangle	10. 55633 %	8. 933927 %	7.140532 %	4.191995 %
Plane Curve method	NONE	NONE	4.375574 %	4.333332 %
Virtual Triangle Refinement	NONE	NONE	NONE	5.270199 %
Maximally Stretched	NONE	NONE	6.568363 %	5.024671 %
Minimal Curvature	NONE	NONE	7.140532 %	4.191995 %
SPHERIGON	9.943740 %	8.246331 %	4.443961 %	4.400603 %
Our On-Line Polygon Refining	9.228697 %	7.384061 %	8.571308 %	6.270763 %

(b) base edge length = 0.1

Style \ Method	Ternary style Method 1	Ternary style Method 2	Quaternary style Method 1	Quaternary style Method 2
Side-Vertex Method (tangent length = 1)	134.6371 %	90.21462 %	114.7883 %	109.3141 %
Quartic Gregory Triangle	10.55688 %	8.934078 %	7.140532 %	4.191995 %
Plane Curve method	NONE	NONE	4.375574 %	4.333332 %
Virtual Triangle Refinement	NONE	NONE	NONE	5.270199 %
Maximally Stretched	NONE	NONE	6.568363 %	5.024671 %
Minimal Curvature	NONE	NONE	7.140532 %	4.191995 %
SPHERIGON	9.943740 %	8.246331 %	4.443961 %	4.400603 %
Our On-Line Polygon Refining	9.228697 %	7.384061 %	8.571308 %	6.270763 %

(c) base edge length = 10.0

Style \ Method	Ternary style Method 1	Ternary style Method 2	Quaternary style Method 1	Quaternary style Method 2
Side-Vertex Method (tangent length = 1)	10.10471 %	8.470137 %	12.15857 %	11.62661 %
Quartic Gregory Triangle	10.55579 %	8.933839 %	7.140532 %	4.191995 %
Plane Curve method	NONE	NONE	4.375574 %	4.333332 %
Virtual Triangle Refinement	NONE	NONE	NONE	5.270199 %
Maximally Stretched	NONE	NONE	6.568363 %	5.024671 %
Minimal Curvature	NONE	NONE	7.140532 %	4.191995 %
SPHERIGON	9.943740 %	8.246331 %	4.443961 %	4.400603 %
Our On-Line Polygon Refining	9.228697 %	7.384061 %	8.571308 %	6.270763 %

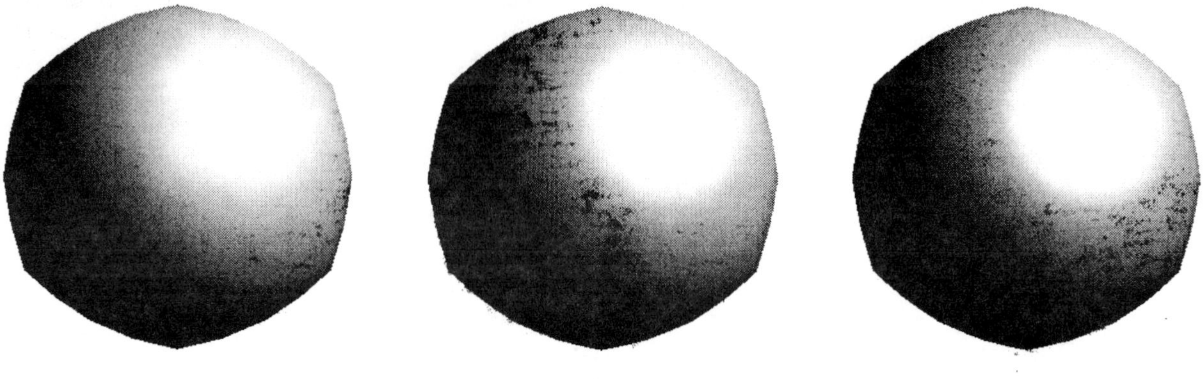

Fig. 11: (a) Sphere model with refining. Left: 2 iterations; Center: 3 iteration; Right: 4 iterations

Fig. 11: (b) Pawn model with refining. Left: 2 iterations; Center: 3 iteration; Right: 4 iterations

6. Recursive subdivision

We can apply our refining algorithm for more iterations. To examine that the shift vectors in our algorithm will not diverge, we consider following two cases. For the monotonous curvature case, i.e. not zigzag case, the angles between normal vectors in recursive subdivision are getting smaller, and the length factor will also be smaller. So the length of shift vectors will converge. For zigzag case, we will apply second order interpolation for the normal vectors. Thus we always get non-zigzag results after the first iteration. After several iterations, the shift vectors will also converge.

Fig.11 shows the results with refining for 2, 3, and 4 iterations. We get smoother visual quality with more iterations, but the improvement is getting less significant.

7. Conclusion

We propose a low computation algorithm which can run-time generate more accurate 3D meshes with polygon subdivision. It only utilizes the information of three vertex coordinates and their normal vectors in one triangle, thus it can be easily integrated into the standard 3D graphics pipeline. Besides, we also consider the generation of interpolated normal vectors and texture mapping parameters. In our analysis and simulation, our algorithm produces decent result with lowest computation complexity. Our algorithm can be extended to any multi-sided polygon with small modification. In addition, we can also use the refining recursively to get better quality with higher computation cost. Although, with Gouraud shading planar triangle based libraries and graphics hardware, we can not elevate the continuity to G^1 degree by doing subdivision for one iteration. We still get better quality than original mesh. This method can be further combined with progressive 3D mesh transmission to get better image earlier in the receiver.

References

[1] H. Gouraud, "Continuous Shading of Curved Surfaces", *IEEE Transactions on Computer*, vol. C-20, no. 6, June 1971, pp. 623-629

[2] C. Loop and T. DeRose, "Generalized B-Spline Surfaces of Arbitrary Topology", *Computer Graphics (SIGGRAPH proceedings 1990)*, 1990, pp. 347-356

[3] E. Catmull and J. Clark, "Recursively Generated B-Spline Surfaces on Arbitrary Topological Meshes", *Computer Aided Design*, Nov. 1978, pp. 350-355

[4] N. Dyn, D.Levin and J.A. Gregory, "A Butterfly Subdivision Scheme for Surface Interpolation with Tension Control", *ACM Transactions on Graphics 9(2)*, 1990, pp. 160-169

[5] C.L. Baja and I. Ihm, "Smoothing Polyhedra using Implicit Algebraic Splines", *Computer Graphics (SIGGRAPH proceedings 1992)*, 1992, pp. 79-88

[6] J. Peters, "Local Cubic and Bicubic C1 Surface Interpolation with Linearly Varying Boundary Normal", *Computer Aided Geometric Design 7*, 1990, pp. 499-516

[7] S. Mann et al., "A Survey of Parametric Scattered Data Fitting Using Triangular Interpolants", in *Curve and Surface Design*, H. Hagen, ed., SIAM, Philadelphia, 1992, pp. 145-172

[8] M. Lounsbery, S. Mann and T. DeRose, "Parametric Surface Interpolation", *IEEE Computer Graphics and Applications*, Sep. 1992, pp. 45-52

[9] L.A. Shirman and C.H. Sequin, "Local Surface Interpolation with Bezier Patches", *Computer Aided Geometric Design 4*, 1987, pp. 279-295

[10] B. Piper, "Visually Smooth Interpolation with Triangular Bezier Patches", in *Geometric Modeling: Algorithms and New Trends*, G. Farin, ed., SIAM, Philadelphia, 1987, pp. 221-233

[11] A. Kolb, H. Pottmann and H.P. Seidel, "Fair Surface Reconstruction Using Quadratic Functions", *Eurographics'95*, vol. 14, no. 3, 1995, pp. 469-479

[12] G.M. Nielson, "A Transfinite, Visually Continuous, Triangular Interpolant", in *Geometric Modeling: Algorithms and New Trends*, G. Farin, ed., SIAM, Philadelphia, 1992, pp. 235-246

[13] M. Fuchs et al., "An Improved Boundary Element Method for Realistic Volume-Conductor Modeling", *IEEE Transactions on Biomedical Engineering*, vol. 45, no. 8, Aug. 1998, pp. 980-997

[14] C.W.A.M van Overveld and B. Wyvill, "An Algorithm for Polygon Subdivision Based on Vertex Normals", *IEEE Computer Graphics International 1997*

[15] D.J. Walton and D.S. Meek, "A Triangular G1 Patch from Boundary Curves", *Computer-Aided Design*, vol. 28, no. 2, 1996, pp. 113-123

[16] P. Volino and N.M. Thalmann, "The SPHERIGON: A Simple Polygon Patch for Smoothing Quickly your Polygon Meshes", *IEEE Proceeding of Computer Animation 1998*, pp. 72-78

[17] B.T. Phong, "Illumination for Computer Generated Pictures", *Communications of the ACM 18(6)*, June 1975, pp. 311-317

[18] C.W.A.M van Overveld and B. Wyvill, "Phong Normal Interpolation Revisited", *ACM Transactions on Graphics*, vol. 16, no. 4, Oct. 1997, pp. 397-419

[19] B. Paul et al., The Mesa 3D Graphics Library, http://www.mesa3d.org

Using Most Isometric Parametrizations for Remeshing Polygonal Surfaces

U. Labsik K. Hormann G. Greiner

Computer Graphics Group, University of Erlangen-Nuremberg,
Am Weichselgarten 9, 91058 Tennenlohe, Germany
Tel: +49-9131-8529919, Fax: +49-9131-8529931
Email: labsik@informatik.uni-erlangen.de

Abstract

The importance of triangle meshes with a special kind of connectivity, the so-called subdivision connectivity *is still growing. Therefore it is important to develop efficient algorithms for converting a given mesh with arbitrary connectivity into one with subdivision connectivity. In this paper we have focused on 2-manifold triangle meshes with boundary and no holes. We discuss the importance of a parametrization with minimal distortion for the process of remeshing. Based on the concept of most isometric parametrizations we have developed a remeshing algorithm for the given class of triangle meshes. A series of examples shows the advantages of our approach.*

1. Introduction

In computer graphics, triangle meshes are a standard representation for surfaces. The use of 3D acquisition techniques like laser range scanning results in dense meshes with a large number of triangles. Because of their complexity it is necessary to use efficient algorithms for storing, transmitting and editing those meshes. Many of the algorithms in the context of multiresolution modeling need a special structure of the mesh, the so called *subdivision connectivity*. This special kind of connectivity is generated by iteratively subdividing a coarse base mesh $\mathcal{S}^0$ with a uniform refinement operator.

With most meshes not having this special kind of connectivity it is necessary to have algorithms that enable the transformation of a mesh with arbitrary connectivity into a mesh with subdivision connectivity. This process is called *remeshing*. Several approaches exist to solve this problem, e. g. [3, 12, 10]. There are many advantages resulting from this conversion. The different refinement levels automatically provide levels of detail which can be used by multiresolution algorithms like level-of-detail rendering [2], wavelets [14, 18], progressive transmission [11] or multiresolution editing [21].

In this paper we present an algorithm for remeshing triangulated, topologically disk-like surfaces. Such a triangle mesh can be parametrized over a planar domain. We demonstrate the influence of the parametrization method for the quality of the remesh. For our algorithm we use most isometric parametrizations (MIPS) [7] providing minimal distortion of the triangles in the mesh. This parametrization has the main advantage that the boundary is allowed to develop naturally and is not fixed in advance.

The paper is organized as follows. In Section 2 a short overview of related work in the field of remeshing polygonal surfaces is given. The general concept of our algorithm is explained in Section 3. In Section 4 we show how to compute a most isometric parametrization of a given triangle mesh and in Section 5 the remeshing algorithm is explained in detail. In Section 6 we demonstrate the efficiency of our algorithm by some examples. We close with a summary in Section 7.

2. Previous Work

During the last years a lot of effort has been spent on the task of remeshing polygonal surfaces. In this section we will give an overview of the most important work.

In [3] Eck et al. have presented a remeshing algorithm consisting of three steps. First, a partitioning of the given mesh $\mathcal{M}$ into a number of triangular regions has to be found. This is done by distributing vertices over $\mathcal{M}$ and growing Voronoi tiles around each vertex. With the help of the constructed Voronoi diagram a triangulation of the new vertices can be found. In the next step a parametrization of the given mesh within each triangle of the base mesh is computed. This is done with harmonic maps trying to minimize the local distortion. The remesh is achieved by uniformly subdividing each base triangle and mapping the vertices into 3-space using the parametrization.

0-7695-0562-7/00 $10.00 © 2000 IEEE

A different approach is proposed by Lee et al. in [12]. Here the mesh $\mathcal{M}$ is coarsened by vertex removal leading to a base mesh for the remeshing. A parametrization is obtained by projecting the removed vertex onto the remaining mesh. This parametrization is also only locally smooth. To achieve global smoothness of the remesh the sampling of the parametrization is not done at the dyadic points. A smoothed version of the dyadic points is computed instead by applying a variant of Loop's subdivision scheme [13] and mapping these vertices into 3-space.

The basic idea of the shrink wrapping approach [10] by Kobbelt et al. is to place a triangle mesh with subdivision connectivity around an arbitrary triangle mesh and shrink it onto the surface of that mesh. In this process two forces have to be simulated, an attracting force, moving the vertices of the subdivision in the direction of the original surface and a relaxing force, distributing the vertices over the surface. With this approach it is possible to remesh topologically sphere-like (genus-0) meshes.

3. General concept

The task of remeshing can be understood as the problem of finding an approximation operator $\mathfrak{A} : \mathbb{M} \to \mathbb{S}$ that maps from a given set $\mathbb{M}$ of meshes to the set $\mathbb{S}$ of all meshes with subdivision connectivity such that the resulting mesh is a good approximation of the original mesh, i. e., the distance of the two meshes is below a user-specified threshold. The resulting subdivision connectivity mesh is called a *remesh* of the original mesh.

In this paper we concentrate on the set of 2-manifold triangle meshes that have a boundary and no holes. In such a *spatial* mesh the intersection of two triangles may be either an edge, a vertex or empty. Additionally one edge of the mesh belongs to either two or only one triangle. In this case it is a boundary edge. The mesh $\mathcal{M}_\Psi = (P, T)$ is represented by a set of *vertices* $P = \{P_i\}$ in the *space* $\Psi \subset \mathbb{R}^3$, representing the geometry, and a set $T = \{T_j\}$ of triangles $T_j = \Delta(P_{j_0}, P_{j_1}, P_{j_2})$, defining the connectivity of the mesh.

A subdivision connectivity mesh (SCM) $\mathcal{S}_\Psi$ of level m has the further property of being a refinement of a coarse base mesh $\mathcal{S}^0$, i. e., it is the last element of a sequence of meshes $\mathcal{S}^0, \ldots, \mathcal{S}^m = \mathcal{S}_\Psi$ where each $\mathcal{S}^{l+1}$ emerges from $\mathcal{S}^l$ by uniformly subdividing each triangle of $\mathcal{S}^l$ into 4 subtriangles (cf. Fig. 1). If the vertices P^l of $\mathcal{S}^l$ are a subset of the vertices P^{l+1}, this operation is called *interpolatory subdivision*, and *non-interpolatory subdivision* otherwise. Note that all vertices of $\mathcal{S}_\Psi$ have valence 6, except for the boundary vertices and those who correspond to the vertices P^0 of the base mesh.

Two different aspects relate to the *quality* of a SCM. First, the number of triangles of the base mesh $\mathcal{S}^0$ shall be as small as possible, which results in a large number of hierarchy levels m of the remesh $\mathcal{S}_\Psi$. This assures a maximal utilization of the multiresolution techniques that can be applied to $\mathcal{S}_\Psi$. Another criterion for the quality of a SCM is the visual appearance, i. e., all triangles T_j of $\mathcal{S}_\Psi$ should have uniform size and aspect ratio.

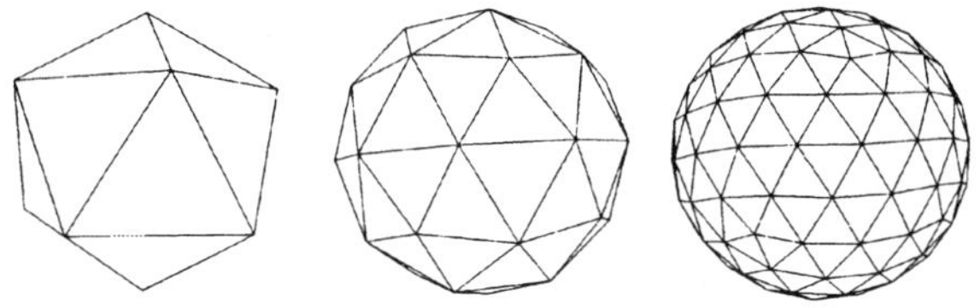

Figure 1. Starting with a coarse base mesh $\mathcal{S}_0$, a sequence of SCMs can be generated by iteratively applying the uniform subdivision operator that performs a 1-to-4 split on every triangle.

Since we consider only meshes with boundary and no holes, these meshes are topologically disk-like and can therefore be parametrized over a simply connected *planar domain* $\Omega \subset \mathbb{R}^2$ by assigning each vertex P_i a *parameter value* $p_i \in \Omega$. By triangulating these parameter values in the same way as the original data points, i. e., by defining a set of triangles $t = \{t_j\}$, $t_j = \Delta(p_{j_0}, p_{j_1}, p_{j_2})$ we obtain a *planar* mesh $\mathcal{M}_\Omega = (p, t)$ that corresponds to $\mathcal{M}_\Psi$ in a natural way. We call the function $f : \Psi \to \Omega$ that linearly maps each spatial triangle to the corresponding planar triangle (i. e., $f(T_j) = t_j$, $f(P_i) = p_i$ and $f(\mathcal{M}_\Psi) = \mathcal{M}_\Omega$) the *projection* of $\mathcal{M}_\Psi$, whereas the inverse function $F = f^{-1}$ is called the *parametrization*.

The general concept of our remeshing algorithm is to use the projection and the parametrization in order to shift the approximation problem from the space Ψ to the planar domain Ω and consider the operator $\mathfrak{A}_\Omega$ instead of $\mathfrak{A}_\Psi$ (cf. Fig. 2). After determining a projection f with minimal distortion we approximate the planar mesh $\mathcal{M}_\Omega$ with a high quality SCM $\mathcal{S}_\Omega$ that is mapped back into space by the parametrization F to yield a remesh $\mathcal{S}_\Psi = F(\mathcal{S}_\Omega)$ of the original mesh $\mathcal{M}_\Psi$. Note that this method guarantees that the vertices of the remesh lie on the surface of the given mesh.

4. Parametrization

The parametrization problem is vital for many applications in computer graphics such as surface fitting, texture mapping, and remeshing. The most important type of this problem is the parametrization of 3D data points which has been addressed in many papers before [1, 3, 4, 5, 15, 16, 17].

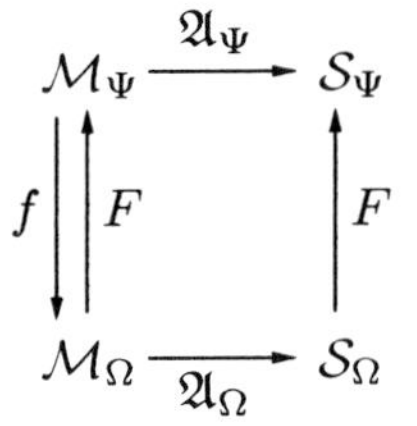

Figure 2. The general concept of our method: $\mathfrak{A}_\Psi \approx F \circ \mathfrak{A}_\Omega \circ f$.

In general, a set of points $P_i \in \mathbb{R}^3$, a certain kind of neighborhood information that defines the topology of the point set, e. g. a triangulation, and a domain Ω that is of the same topological type are given. The task is now to assign every vertex P_i a parameter value $p_i \in \Omega$ such that the topology is preserved. Typical domains are the disk $D_2 = \{x \in \mathbb{R}^2 : ||x|| \leq 1\}$ and the sphere $S_2 = \{x \in \mathbb{R}^3 : ||x|| = 1\}$, but other objects like the torus are possible as well. In this section we will concentrate on the special case of parametrizing triangulated point sets that are topologically disk-like and a simply connected planar domain Ω, but the general idea of the presented parametrization technique is also applicable for other domain types.

Using the notation of the previous section, we are looking for a projection $f : \Psi \to \Omega$. Since the given mesh $\mathcal{M}_\Psi$ is normally geometrically complex, this function will inevitably cause some deformation to the shape of the triangles T_j. A projection without distortion is called *isometric* and can only be found for developable surfaces, e. g. planes, cylinders and conical surfaces. Such a projection would be optimal for our purposes, because the parametrization $F = f^{-1}$, that is used to create the final remesh $\mathcal{S}_\Psi$ from the planar remesh $\mathcal{S}_\Omega$, would be isometric, too. Hence, $\mathcal{S}_\Psi$ would inherit the quality from $\mathcal{S}_\Omega$, since the shape of the triangles would not change. Therefore, we need to find a projection f that is "as isometric as possible".

The concept of MIPS [7] can be used to find such projections. This approach decomposes the piecewise linear projection f into *atomic linear maps* f_j that map the spatial triangle T_j to the corresponding planar triangle t_j. By introducing a local coordinate system at T_j with the third axis chosen to be perpendicular to T_j, these linear maps can be written as $f_j : \mathbb{R}^2 \to \mathbb{R}^2, x \mapsto A_j x + b_j$ and the deformation of the triangle T_j can be measured by the *2-norm condition* of the matrix A_j,

$$\kappa_2(A_j) = ||A_j||_2 \, ||A_j^{-1}||_2 = \frac{\sigma_1}{\sigma_2},$$

where $\sigma_1 \geq \sigma_2 \geq 0$ are the *singular values* of A_j. However, for practical reasons it is easier to use the *Frobenius norm condition* κ_F, which is closely related to κ_2 and can be computed by the following formula (see [17] and [7] for details)

$$\begin{aligned}\kappa_F(A_j) &= \kappa_2(A_j) + \frac{1}{\kappa_2(A_j)} \\ &= \frac{\cot\alpha_{j_0}||\ell_{j_0}||^2 + \cot\alpha_{j_1}||\ell_{j_1}||^2 + \cot\alpha_{j_2}||\ell_{j_2}||^2}{2\,\text{area}(t_j)}\end{aligned} \tag{1}$$

with the notations of Fig. 3, i. e.,

$$\ell_{j_0} = p_{j_2} - p_{j_1}, \quad \ell_{j_1} = p_{j_2} - p_{j_0}, \quad \ell_{j_2} = p_{j_1} - p_{j_0},$$

and

$$2\,\text{area}(t_j) = ||\ell_{j_1} \times \ell_{j_2}|| = \ell^x_{j_1}\ell^y_{j_2} - \ell^y_{j_1}\ell^x_{j_2}.$$

Minimization of $\kappa = \sum_j \kappa_F(A_j)$ will then lead to an optimal projection f in the sense of minimal deformation.

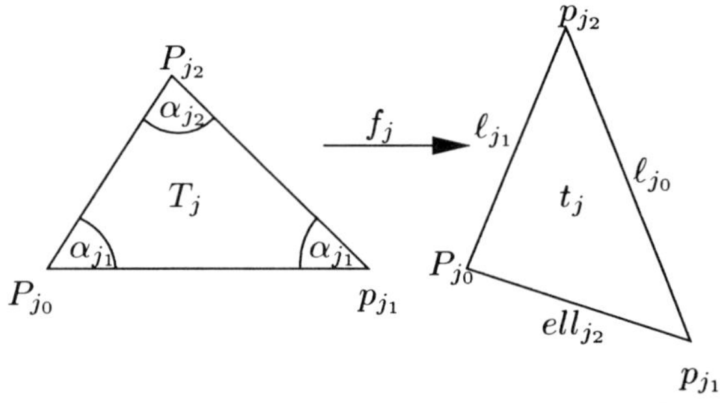

Figure 3. The atomic linear map f_j maps the spatial triangle T_j to the corresponding planar triangle t_j.

This is an optimization problem in the unknown parameter values p_i and according to Eq. (1) highly nonlinear. Instead of solving this problem globally, we use a Gauss-Seidel approach to minimize κ with a series of local optimization steps. Starting with an initial set of parameter values that can be obtained by one of the methods described in [3, 4, 5], we perform the following algorithm, illustrated in Fig. 4(a).

repeat

  choose a vertex p_i by random

  let $t_{i_1}, \ldots, t_{i_n}$ be the triangles that surround p_i

  fix the positions of all parameter values except for p_i

  minimize the local deformation energy $\kappa_i = \sum_{k=1}^{n} \kappa_F(A_{i_k})$ in order to get the optimal position $\tilde{p}_i$ of p_i

until numerical convergence

As we can see in Fig. 4(b), the local functional κ_i is convex and grows to infinity along the edges that surround the parameter value p_i. Usually the optimal $\tilde{p}_i$ can be found with two of three Newton steps. It can be proven that these local optimization steps converge to the global optimum and are stable in the sense of validity: if the initial configuration is valid (i. e., neighboring triangles do not overlap) this property is preserved by each optimization step.

A major advantage of MIPS is that the optimization step can also be performed on the parameter values at the boundary of the triangulation and does not need the boundary to be fixed in advance. Of course this may lead to overlaps of different boundary regions (cf. Fig. 5) although we never experienced it in our examples. However, if you think of the different overlapping parts being assigned to different depth layers, it is still possible to use this configuration in order to "lift" the planar SCM mesh $\mathcal{S}_\Omega$ up to the space Ψ as long as you add some depth layer information to the geometric information of the vertices p_i of $\mathcal{S}_\Omega$, e. g. by storing the number of the triangle of $\mathcal{M}_\Omega$ in which each p_i is located.

We close this section by briefly mentioning that the optimization process of MIPS can be significantly accelerated by a multiresolution approach as explained in [8]. The idea

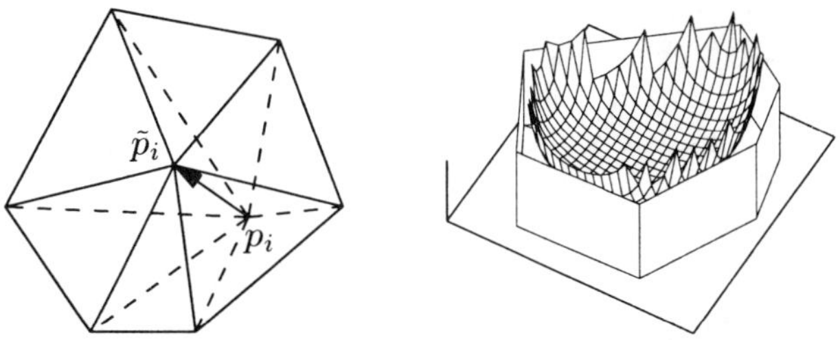

(a) (b)

Figure 4. Each optimization step finds the optimal position $\tilde{p}_i$ of the parameter value p_i (a) by minimizing the convex local functional κ_i (b).

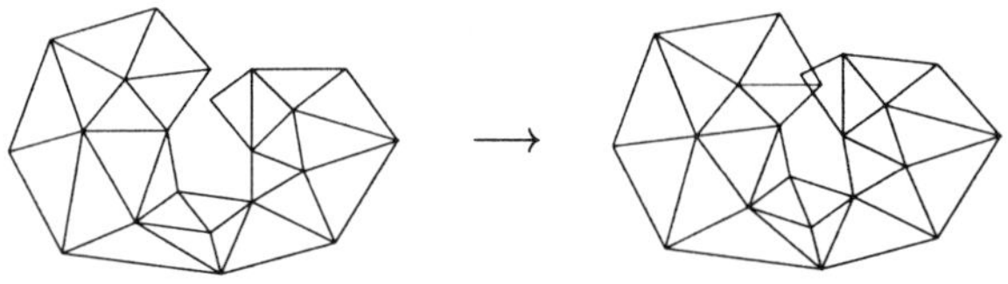

Figure 5. MIPS may generate overlapping parts in the parameter domain.

is to use the concept of Progressive Meshes [6] to build a hierarchy $\mathcal{M}_\Psi = \mathcal{M}^0, \ldots, \mathcal{M}^k$ of meshes and create the optimal projection for the coarsest mesh $\mathcal{M}^k$ first. Then the remaining vertices of $\mathcal{M}_\Psi$ are successively inserted and optimized until the final planar mesh $\mathcal{M}_\Omega$ is generated.

5. Remeshing

As described in Section 3 the task of generating a remesh $\mathcal{S}_\Psi$ with subdivision connectivity of a given mesh $\mathcal{M}_\Psi$ consists of three steps. We have already explained how to compute the parametrization of $\mathcal{M}_\Psi$. The second step is to find an appropriate remesh $\mathcal{S}_\Omega$ for the mesh $\mathcal{M}_\Omega = f(\mathcal{M}_\Psi)$ in the parameter domain Ω.

In order to find the remesh $\mathcal{S}_\Omega$ in the parameter domain we have to solve several problems which are discussed in this section. First, we have to determine an appropriate base mesh $\mathcal{S}_\Omega^0$, then we have to subdivide this base mesh until a prescribed error bound is reached. In this subdivision process we have to reconstruct the boundary polygon of $\mathcal{M}_\Omega$ with the remesh $\mathcal{S}_\Omega^l$ and the triangles $t_j \in \mathcal{S}_\Omega^l$ have to be scaled so that the corresponding triangles $T_j = F(t_j)$ are of equal size.

5.1. Constructing the base mesh

For remeshing meshes with boundary it would be possible to use just a single triangle as a base mesh. But in most cases that would not lead to a satisfying remesh of the given surface because of a high distortion of the triangles. So it is important to find a good starting mesh $\mathcal{S}_\Omega^0$ to avoid this problem. This base mesh should meet all quality criterions mentioned in Section 3, i. e., a few number of triangles being as equilateral as possible. Note that the latter criterion refers to the triangles mapped into 3-space. Furthermore the base mesh should approximate the boundary of $\mathcal{M}_\Omega$.

One important aspect of the remeshing algorithm is the reconstruction of boundary of $\mathcal{M}_\Psi$. So we start the construction of the base mesh by specifying extraordinary vertices on the boundary polygon of the given mesh $\mathcal{M}_\Psi$ which consists of a set of vertices $B = \{B_0, B_1, ..., B_n = B_0\}$. To achieve this we compute the angle $\measuredangle(\overrightarrow{B_iB_{i-1}}, \overrightarrow{B_iB_{i+1}})$

of the boundary polygon in every vertex. If the value exceeds a given threshold α the corresponding vertex $b_i = f(B_i)$ of the mesh $\mathcal{M}_\Omega$ will be a vertex $\tilde{b}_j$ of the base mesh $\mathcal{S}_\Omega^0$.

These vertices $\tilde{b} = \{\tilde{b}_i\}$ may be distributed irregularly over the boundary polygon. To avoid triangles of different size in the base mesh we now specify vertices on the boundary polygon so that the distance between all boundary vertices in the base mesh is approximately equal. The distance of two boundary vertices $\tilde{b}_i$ and $\tilde{b}_j$ of the base mesh $\mathcal{S}_\Omega^0$ is computed as follows. Using the parametrization we obtain $\tilde{B}_i = F(\tilde{b}_i)$ and $\tilde{B}_j = F(\tilde{b}_j)$ on the boundary polygon of $\mathcal{M}_\Psi$. We now compute the distance between these points by following the boundary line from one point to the other (cf. Fig. 6).

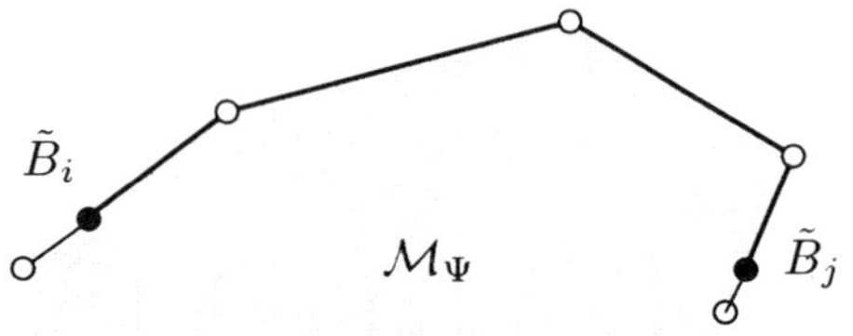

Figure 6. Computing the distance between two boundary points.

The set of vertices we have chosen up to now defines the boundary of the base mesh and has to be triangulated. Notice that the vertices also lie on the boundary polygon of $\mathcal{M}_\Omega$. Triangulating only the boundary vertices would not lead to a regular structure with almost equilateral triangles. Therefore the triangulation algorithm is allowed to insert new interior vertices, called *Steiner points*, to the base mesh. There are two constraints for the position of these interior vertices. The angles of the triangles should be close to 60° and the area representing the triangles in the original mesh $\mathcal{M}_\Psi$ should be about the same.

This algorithm enables us to compute a base mesh $\mathcal{S}_\Omega^0$ which is the starting point of the remeshing process. In practice these base meshes will have only a small number of triangles.

5.2. Border adaption

When parametrizing the mesh $\mathcal{M}_\Psi$ using the MIPS scheme, normally the boundary is allowed to develop naturally. Hence, in general case the parametrization will not be a rectangular area, not even convex. All boundary vertices of $\mathcal{S}_\Omega^l$ also belong to the boundary polygon of $\mathcal{M}_\Omega$.

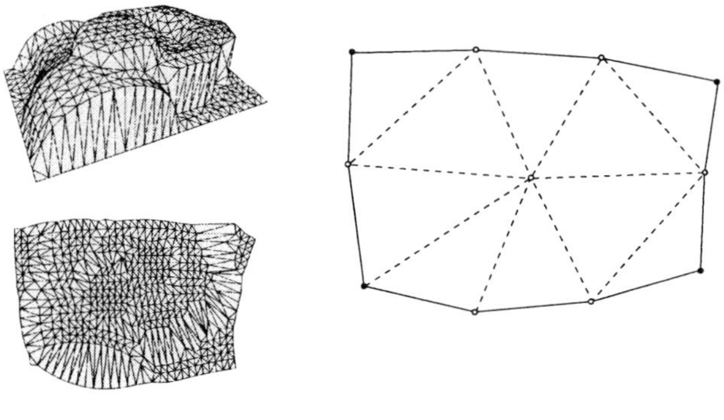

Figure 7. Constructing a base mesh. The filled vertices were found by searching for extraordinary vertices of the original mesh, the outlined boundary vertices were inserted to adapt the distances on the boundary polygon and the interior vertex was inserted to improve the mesh structure.

The subdivision scheme we use just computes the new vertices as the midpoints of the edges on the coarser level $\mathcal{S}_\Omega^l$ and retriangulates the mesh leading to $\mathcal{S}_\Omega^{l+1}$. The additional boundary vertices of $\mathcal{S}_\Omega^{l+1}$ will not be on the boundary of $\mathcal{M}_\Omega$ but somewhere inside or outside the mesh (cf. Fig. 8).

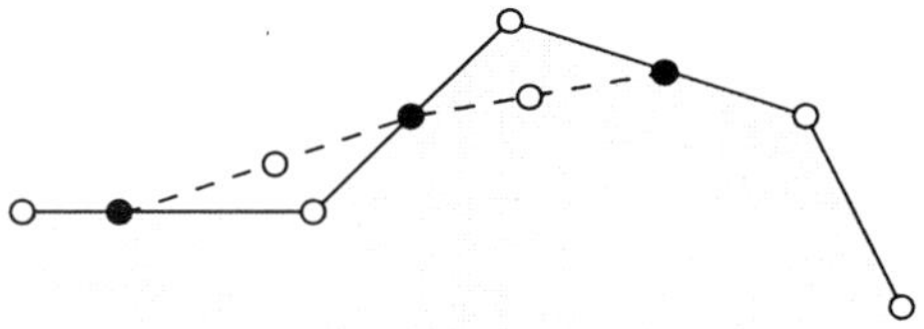

Figure 8. New boundary vertices of the remesh (dashed) have to be moved onto the boundary polygon of the parametrization.

For these vertices a new position on the boundary polygon of the parametrization has to be found. The algorithm we use works as follows. We have to find a position for the vertex $\tilde{q}$, which has been inserted between the boundary vertices $\tilde{b}_i$ and $\tilde{b}_{i+1}$. Exploiting the parametrization we obtain $\tilde{B}_i = F(\tilde{b}_i)$ and $\tilde{B}_{i+1} = F(\tilde{b}_{i+1})$ which are points on the boundary of $\mathcal{M}_\Psi$. We now compute the distance between these points over the boundary line of the mesh as described in the last section and search for the position $\tilde{Q}$ midway between these points. The new position of the vertex $\tilde{q}$ in the parameter domain is then computed by $\tilde{q} = f(\tilde{Q})$. If the distances of the boundary vertices in the mesh $\mathcal{S}_\Omega^l$ are equal then this strategy is optimal and will lead to equidistant boundary vertices in $\mathcal{S}_\Omega^{l+1}$.

5.3. Density weighted smoothing

The deformation functional κ that we use to obtain the parametrization is invariant to scalings (see Eq. (1)). This may lead to scalings of triangles between the meshes $\mathcal{M}_\Psi$ and $\mathcal{M}_\Omega$. Hence, to minimize the distortion and to get approximately the same size for each triangle T_j of $\mathcal{S}_\Psi^m$ it is essential for the remeshing algorithm to use a relaxing operator to optimize the interior vertices of the remesh. For this relaxing we use the umbrella operator $\mathfrak{U}$ [9]. The umbrella operator minimizes the membrane energy of a mesh, i. e., the surface area. The update rule for a vertex $p_i \in S_\Omega^l$ is

$$\mathfrak{U} : p_i \mapsto \tilde{p}_i = (1-\omega)p_i + \frac{\omega}{\sum_j d_{i_j}} \sum_{j=1}^{n} d_{i_j} p_{i_j} \tag{2}$$

with n being the valence of vertex p_i and $p_{i_1}, ..., p_{i_n}$ its adjacent neighbors in S_Ω^l. With the help of the density coefficients d_i the distribution of the vertices can be controlled.

In our case we want a vertex p to be moved so that neighboring triangles transformed into 3-space have approximately the same size. Therefore we use the sum of the area of the adjacent triangles T_j as the density coefficient d_i for the vertex p_i. We apply the density weighted umbrella only to the interior vertices of the mesh. The boundary vertices are not affected by this relaxing operation.

The density weighted umbrella operator is applied iteratively. With the vertices moving on the parameter domain the size of the triangles has to be recomputed after every iteration of the operator $\mathfrak{U}$. The iterations stop when the change of the vertex positions falls below a certain threshold. Then the mesh $\mathcal{S}_\Omega^l$ can be switched to the next finer level $\mathcal{S}_\Omega^{l+1}$ and the process of border adaption and relaxing iterations starts again. When reaching a certain level m the subdivision process is stopped and the mesh is transformed into 3-space by evaluating the parametrization F. Therefore we have to find the triangle $\Delta(A_\Omega, B_\Omega, C_\Omega) \in \mathcal{M}_\Omega$ which contains a vertex $p_i \in \mathcal{S}_\Omega^m$. This vertex can be expressed in barycentric coordinates

$$p_i = \alpha A_\Omega + \beta B_\Omega + \gamma C_\Omega, \quad \alpha + \beta + \gamma = 1.$$

The position of p_i in 3-space is determined by applying the same barycentric combination to the corresponding vertices A_Ψ, B_Ψ and C_Ψ of the original mesh $\mathcal{M}_\Psi$, i. e.,

$$P_i = \alpha A_\Psi + \beta B_\Psi + \gamma C_\Psi.$$

By using the MIPS parametrization which generates minimal distortion it is not necessary to use the smoothing and projecting technique in 3-space as proposed in [10]. So the risk of failing projections is excluded. Examples which are remeshed with our algorithm are shown in Section 6.

5.4. Adaptive remeshing

In order to reach the user specified error threshold for the distance of the two meshes $\mathcal{M}$ and $\mathcal{S}$ in regions of small local features the mesh $\mathcal{S}$ has to be subdivided up to a very fine level. When using uniform subdivision the number of triangles will grow exponentially. As in [12] we use a straight forward adaptive subdivision approach to keep the number of triangles small and avoid the overhead of first subdividing the mesh to a fine level and afterwards decimating triangles by a wavelet threshold like in [3, 2].

Adaptive subdivision means to split only those triangles which do not satisfy some prescribed criterion. Other triangles can remain coarse. There are a few restrictions in this approach. In order to keep the number of special configurations small we will only allow *balanced* meshes, i. e., the refinement level of two neighboring triangles may only differ by one. In order to avoid cracks in the mesh where two triangles from different levels meet, we have to use a special technique, the so-called *red-green triangulation* [19, 20]. A normal 1-to-4 split is called *green split*. To fix cracks in the mesh, triangle bisection is used, which is called *red split*. A red split is only temporary, i. e., if a red split triangle is to be further subdivided in a subsequent refinement step then the red split is undone first and a green split is applied to the original triangle.

As a subdivision criterion we compute the distance function

$$D(t_j) = \max_i \ \text{dist}(F(p_i), F(t_j)) \tag{3}$$

for every triangle of the actual mesh $\mathcal{S}_\Omega^l$. In this equation p_i are vertices of $\mathcal{M}_\Omega$ which reside in the triangle t_j of $\mathcal{S}_\Omega^l$. The computation of the distance function is performed in 3-space Ψ with the help of the parametrization F. If the maximum distance between such a vertex $F(p_i)$ and the plane defined by the triangle $T_j = F(t_j)$ is greater than a given threshold ε then the triangle will be further subdivided. With this method only a minimal number of triangles has to be refined.

For adaptive remeshing we also have to adapt the density weighted umbrella. The density of a vertex can only be computed on the lowest level of all adjacent triangles. When computing the density weighted umbrella function for a vertex, we have to check the level of the densities of all adjacent vertices and can perform the umbrella on the lowest level of all these densities. Another problem occurs on vertices where a red and a green split meet. For these vertices we can not perform the density weighted umbrella function. So we first apply the smoothing operation to all possible vertices, for the rest we compute the new position as the midpoint of the edge the vertex is dividing.

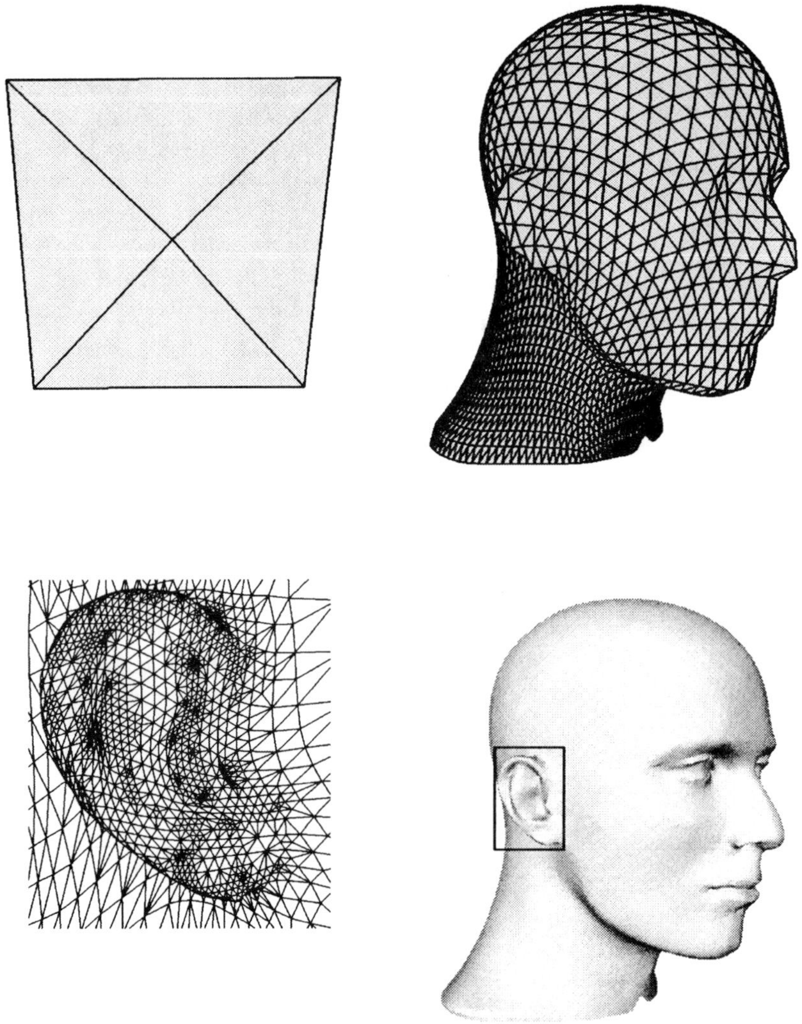

Figure 9. Remeshing the head data set. The base mesh consists of four triangles (top left). The top right image shows the 6th refinement level with uniform subdivision, below the adaptively refined mesh is presented.

6. Examples

In this section we show some of the results we have produced with our remeshing algorithm. We also illustrate the quality of the used parametrization method by showing examples with different parametrizations.

Fig. 9 shows a remesh of a head data set consisting of 21680 triangles. We used a simple mesh of four triangles as a base mesh. By uniform subdivision and smoothing with the density umbrella operator we get triangles of almost equal size in the remesh. But local detail like the ear must be sampled with a higher density. Therefore adaptive subdivision can be used. Using an error bound of $\varepsilon = 0.1\%$ leads to an adaptively refined remesh with 30530 triangles.

In Fig. 10 the remeshing process of a triangulated data set of 7938 triangles is illustrated. The base mesh consists of 10 triangles. The remesh is shown in the 5th refinement level. Using a uniform parametrization for the remeshing process leads to a mesh with highly distorted triangles. The structure of the triangles is much better when the discrete harmonic parametrization is used and can further be improved at the boundary of the mesh by using MIPS.

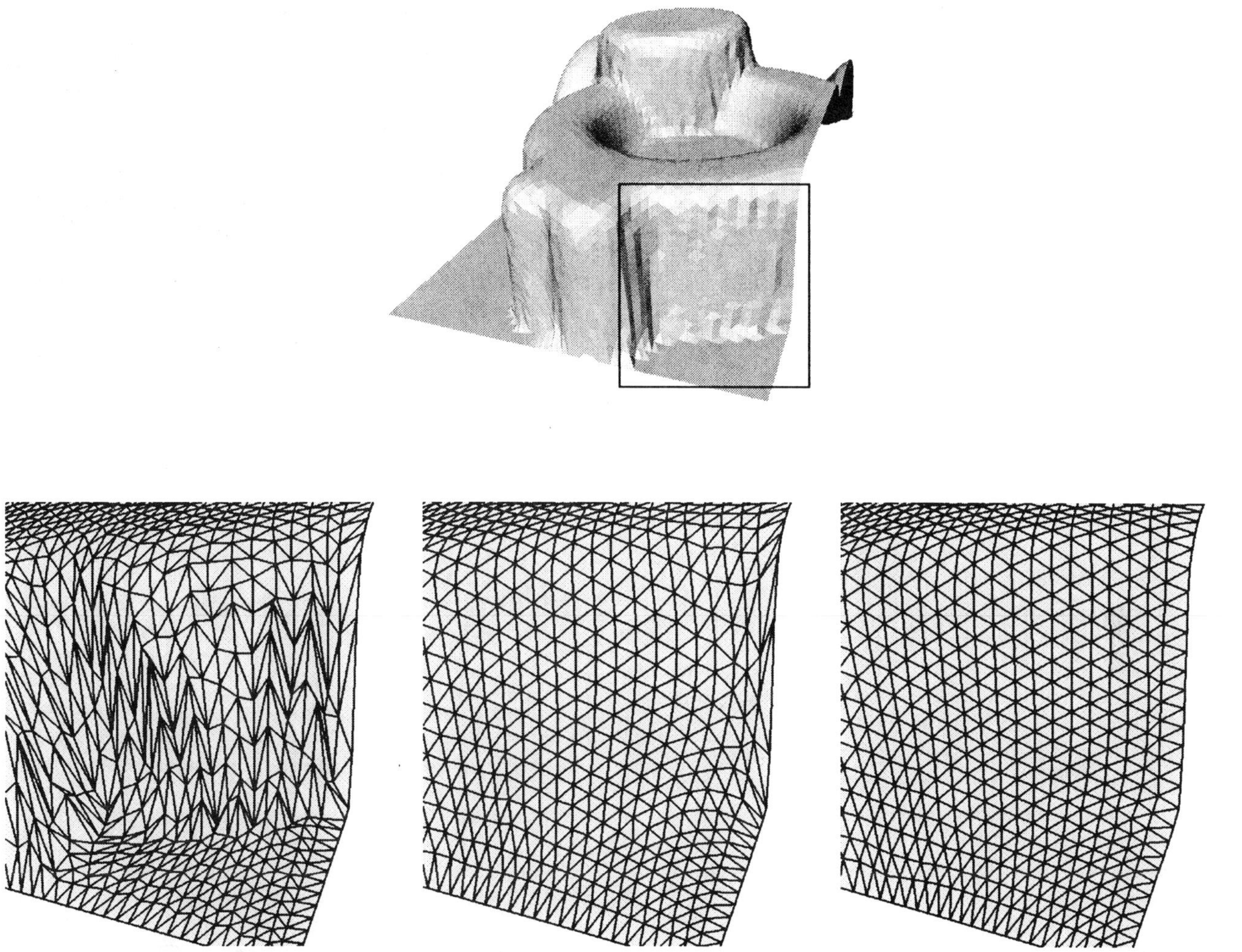

Figure 10. The quality of the remesh with different parametrizations. From left to right: uniform, discrete harmonic and most isometric.

7. Conclusion

In this paper we have presented an efficient algorithm for converting triangle meshes with arbitrary connectivity into meshes with subdivision connectivity. The algorithm works on triangle meshes with boundary and no holes. For this class of meshes we can compute a planar parametrization with minimal distortion. We now compute a remesh for this parametrization which can be mapped into 3-space again to achieve a remesh of the original mesh. By using an iterative relaxation process on every level of the remesh we adapt the remesh to the local scalings of the parametrization.

With this method all triangles of a uniformly refined remesh will have almost equal size. In highly detailed regions it is often necessary to have a higher sampling density of the original mesh. Therefore we use adaptive subdivision to compute remeshes with guaranteed error bounds.

By using the most isometric parametrization on a sphere the results for remeshing polygonal surfaces of genus 0 with the shrink-wrapping approach [10] can be improved, too. There, a parametrization produced with a density weighted umbrella was used which is not able to guarantee minimal distortion of the triangles in the parameter domain.

Future work will aim at remeshing not only disk-like objects without holes. To be able to handle meshes of arbitrary topology an appropriate segmentation of the whole object has to be found. Then the segments can be remeshed separately with the presented algorithm and merged afterwards. Therefore it is important that adjacent segments share the vertices on their common boundary.

References

[1] C. Bennis, J.-M. Vézien, and G. Iglésias. Piecewise surface flattening for non-distorted texture mapping. In *ACM Comp. Graph. (SIGGRAPH '91 Proc.)*, pages 237–246, 1991.

[2] A. Certain, J. Popović, T. DeRose, T. Duchamp, D. Salesin, and W. Stuetzle. Interactive multiresolution surface viewing. In *ACM Comp. Graph. (SIGGRAPH '96 Proc.)*, pages 91–98, 1996.

[3] M. Eck, T. DeRose, T. Duchamp, H. Hoppe, M. Lounsbery, and W. Stuetzle. Multiresolution analysis of arbitrary meshes. In *ACM Comp. Graph. (SIGGRAPH '95 Proc.)*, pages 173–182, 1995.

[4] M. S. Floater. Parameterization and smooth approximation of surface triangulations. *CAGD*, 14:231–250, 1997.

[5] G. Greiner and K. Hormann. Interpolating and approximating scattered 3D data with hierarchical tensor product B-splines. In A. Méhauté, C. Rabut, and L. L. Schumaker, editors, *Surface Fitting and Multiresolution Methods*, pages 163–172. Vanderbilt University Press, 1997.

[6] H. Hoppe. Progressive meshes. In *ACM Comp. Graph. (SIGGRAPH '96 Proc.)*, pages 99–108, 1996.

[7] K. Hormann and G. Greiner. MIPS: an efficient global parametrization method. In *Saint-Malo Proceedings*. Vanderbilt University Press, 2000. to appear.

[8] K. Hormann, G. Greiner, and S. Campagna. Hierarchical parametrization of triangulated surfaces. In B. Girod, H. Niemann, and H.-P. Seidel, editors, *Vision, Modeling and Visualization '99*, pages 219–226. infix, 1999.

[9] L. Kobbelt, S. Campagna, J. Vorsatz, and H.-P. Seidel. Interactive multi-resolution modeling on arbitrary meshes. In *ACM Comp. Graph. (SIGGRAPH '98 Proc.)*, pages 105–114, 1998.

[10] L. Kobbelt, J. Vorsatz, U. Labsik, and H.-P. Seidel. A shrink wrapping approach to remeshing polygonal surfaces. In *Comp. Graph. Forum (EUROGRAPHICS '99 Proc.)*, 1999.

[11] U. Labsik, L. Kobbelt, R. Schneider, and H.-P. Seidel. Progressive transmission of subdivision surfaces. 1999. to appear.

[12] A. Lee, W. Sweldens, P. Schröder, L. Coswar, and D. Dobkin. Multirelosution adaptive parametrization of surfaces. In *ACM Comp. Graph. (SIGGRAPH '98 Proc.)*, pages 95–104, 1998.

[13] C. Loop. Smooth sudivsion surfaces based on triangles. Master's thesis, Utah University, 1987.

[14] M. Lounsbery, T. DeRose, and J. Warren. Multiresolution analysis for surfaces of arbitrary topological type. Technical report 93-10-05, University of Washington, Department of Computer Science and Engineering, 1993.

[15] W. Ma and J. P. Kruth. Parameterization of randomly measured points for least squares fitting of B-spline curves and surfaces. *CAD*, 27:663–675, 1995.

[16] J. Maillot, H. Yahia, and A. Verroust. Interactive texture mapping. In *ACM Comp. Graph. (SIGGRAPH '93 Proc.)*, pages 27–34, 1993.

[17] U. Pinkall and K. Polthier. Computing discrete minimal surfaces and their conjugates. *Exp. Math.*, 2(1):15–36, 1993.

[18] P. Schröder and W. Sweldens. Spherical wavelets: Efficiently representing functions on the sphere. In *ACM Comp. Graph. (SIGGRAPH '95 Proc.)*, pages 161–172, 1995.

[19] M. Vasilescu and D. Terzopoulus. Adaptive meshes and shells: Irregular triangulation, discontinuities, and hierarchical subdivision. In *Proceedings of Computer Vision and Pattern Recognition conference*, pages 829–832, 1992.

[20] R. Verfürth. *A review of a posteriori error estimation and adaptive mesh refinement techniques*. Wiley-Teubner, 1996.

[21] D. Zorin, P. Schröder, and W. Sweldens. Interactive multiresolution mesh editing. In *ACM Comp. Graph. (SIGGRAPH '97 Proc.)*, pages 259–268, 1997.

Polyhedral Surface Smoothing with Simultaneous Mesh Regularization

Yutaka Ohtake
The University of Aizu
Aizu-Wakamatsu City
Fukushima 965-8580 Japan
d8011101@u-aizu.ac.jp

Alexander G. Belyaev
The University of Aizu
Aizu-Wakamatsu City
Fukushima 965-8580 Japan
belyaev@u-aizu.ac.jp

Ilia A. Bogaevski
Independent University of Moscow
Bolshoi Vlasevsky Pereulok 11
Moscow, Russia
bogaevsk@msfu.msk.ru

Abstract

A computer graphics object reconstructed from real-world data often contains undesirable noise and small-scale oscillations. An important problem is how to remove the noise and oscillations while preserving desirable geometric features of the object.

This paper develops methods for polyhedral surface smoothing and denoising with simultaneous increasing mesh regularity.

We also propose an adaptive smoothing method allowing to reduce possible oversmoothing.

Roughly speaking, our smoothing schemes consist of moving every vertex in the direction defined by the Laplacian flow with speed equal to a properly chosen function of the mean curvature at the vertex.

Keywords: mesh smoothing and denoising, mesh regularity, Laplacian smoothing, mean curvature flow

1 Introduction

Recent advances in 3D data-acquisition hardware call for development of methods to rapidly remove undesirable noise and small scale oscillations from an irregularly triangulated data.

Two the most popular approaches for smoothing and denoising of polyhedral surfaces are minimizing energy functionals associated with differential-geometric surface characteristics and Laplacian smoothing. Minimizing an energy functional is usually a computationally expensive task. Moreover, it lacks a local shape control. Laplacian smoothing is simple, fast, and so far the most common technique for mesh smoothing. The Laplacian smoothing flow, in its simplest form, moves recursively each vertex of the mesh by a displacement equal to a positive scale factor times the average of the neighboring vertices. Actually, the Laplacian smoothing flow can be considered as the gradient descent flow for a simple quadratic energy functional. However, due to its simplicity, the Laplacian flow opens many ways for modifications and improvements.

Taubin in [8] proposed to alternate two scale factors of opposite signs with the negative factor of larger magnitude in a weighted Laplacian smoothing flow. Such smoothing does not produces shrinkage and suppresses high frequencies of a discrete Laplacian operator defined on the mesh, while enhancing low frequencies.

When the scale factors are equal in magnitude, the Taubin smoothing scheme turns to the bilaplacian smoothing flow [4] which can be considered as a discrete approximation of the gradient descent flow for the thin-plate energy functional.

Another non-shrinking modification of the Laplacian smoothing flow was very recently proposed in [9].

A significant development of the Laplacian smoothing method, mesh smoothing by the mean curvature flow [3, 6] (see also references therein), came from mathematics and material science. The discrete mean curvature flow moves every vertex in the normal direction with speed equal to a discrete approximation of the mean curvature at the vertex. Smoothing by the mean curvature flow and its various modifications have became extremely popular in geometric image processing (see, for instance, [6] for references).

Laplacian smoothing, Taubin smoothing, and the discrete mean curvature flow contain a number of drawbacks. The Laplacian smoothing flow increases the mesh regularity but develops unnatural deformations while being applied to a highly irregular mesh. Smoothing by the discrete mean curvature flow is relatively independent of the mesh sampling rate but increases the mesh irregularity. Both the Laplacian and mean curvature flows do not decelerate the smoothing process and may lead to oversmoothing and loosing desirable geometric features. The Taubin smoothing scheme lacks a local shape control and enhances low frequency surface features.

In this paper we propose simple and effective polyhedral

0-7695-0562-7/00 $10.00 © 2000 IEEE

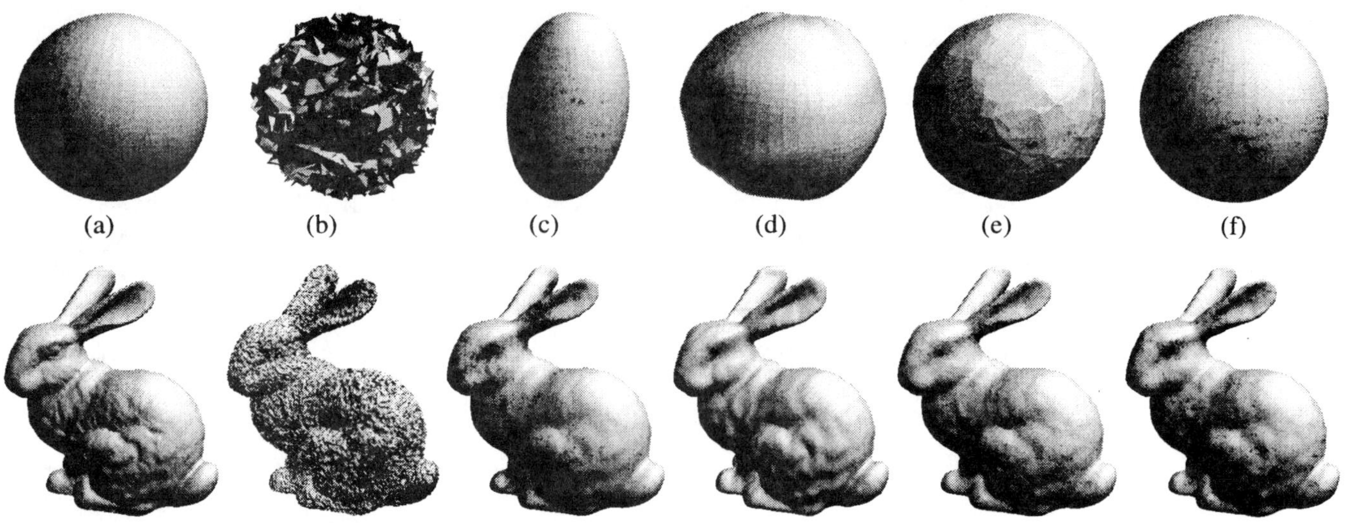

Figure 1. The top row: (a) a polyhedral sphere; (b) the sphere with a uniform noise added; (c) Laplacian smoothing develops unnatural deformations; (d) smoothing by the Taubin method converts high-frequency surface oscillations into low-frequency waves; (e) smoothing by the mean curvature flow increases the mesh irregularity; (f) smoothing according to a method proposed in this paper, see Section 3. The bottom row: (a) the Stanford bunny; (b) the bunny with a uniform noise added; (c) Laplacian smoothing with a number of iterations chosen to achieve 'a good looking' result: extra iterations will lead to oversmoothing; (d) Taubin smoothing reduces high-frequency surface oscillations but enhances low-frequency oscillations: extra iterations will lead to enhancing of surface wrinkles; (e) smoothing by the mean curvature flow with a number of iterations chosen to achieve 'a good looking' result: extra iterations will lead to oversmoothing; (f) smoothing according to a method proposed in this paper: the smoothing process slows down automatically and extra iterations will produce almost the same appearance, see Section 4.

surface smoothing schemes which combine the best properties of the Laplacian smoothing flow and discrete mean curvature flow, often outperform best existing smoothing methods, and, in addition, reduce possible oversmoothing.

Fig. 1 demonstrates some of our results.

The paper is organized as follows. In Section 2, we introduce the Laplacian flow, the Taubin smoothing method, and the discrete mean curvature flow. In Section 3, our new technique which combines the best properties of the Laplacian and discrete mean curvature flow is explained. In Section 4, we describe our technique to reduce possible oversmoothing. We discuss application of smoothing methods for stable detection of ridges and ravines on a polyhedral surfaces in Section 5. We conclude and sketch directions for future research in Section 6.

2 Laplacian Flow, Taubin Method, Bilaplacian Flow, Mean Curvature Flow

In this section we will introduce and analyze four methods for polyhedral surface smoothing: Laplacian smoothing, Taubin smoothing [8], the bilaplacian flow [4], and the mean curvature flow [3].

Laplacian Smoothing. Let us consider a triangulated surface and, for any vertex P, let us define the so-called umbrella-operator [4]

$$\mathcal{U}(P) = \frac{1}{\sum_i w_i} \sum_i w_i Q_i - P \tag{1}$$

where summation is taken over all neighbors of P, w_i are positive weights. See Fig. 2.

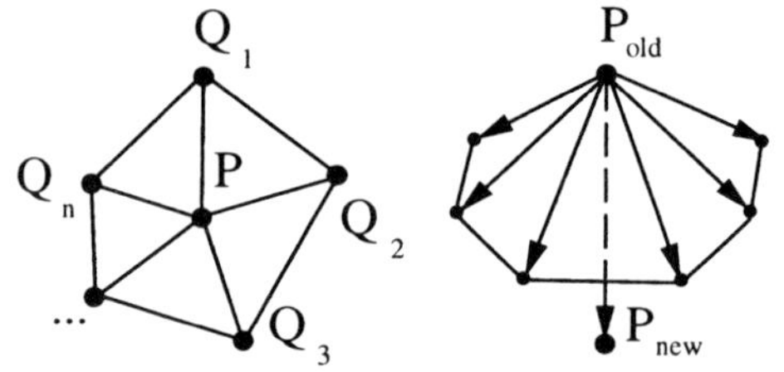

Figure 2.

The local update rule

$$P_{new} \longleftarrow P_{old} + \lambda \mathcal{U}(P_{old}) \tag{2}$$

applied typically to every inner point of the triangulated surface is called Laplacian smoothing of the surface. Here λ is a small positive number and the process (2) is executed repeatedly.

The Laplacian smoothing algorithm reduces the high frequency surface information and tends to flatten the surface.

The weights can be chosen in many different ways. The simplest choice is to set the weights equal to each other: $w_i = 1$,

$$\mathcal{U}_0(P) = \frac{1}{n}\sum_i Q_i - P, \tag{3}$$

where n is the number of neighbors. Another choice that produces good results [8] is set the weights as the inverse distances between P and its neighbors:

$$\mathcal{U}_1(P) = \frac{1}{\sum_i w_i}\sum_i w_i Q_i - P, \quad w_i = \|P - Q_i\|^{-1}. \tag{4}$$

Note that smoothing with $\mathcal{U}_0$ improves the mesh sampling rate, whereas smoothing with $\mathcal{U}_1$ worsens the rate. To demonstrate this, let us consider a plane curve $\mathbf{r}(s)$ parameterized by arclength parameter s. Consider three points on the curve

$$A = \mathbf{r}(s - \alpha), \quad O = \mathbf{r}(s), \quad B = \mathbf{r}(s + \beta)$$

with distances $a = |OA|$ and $b = |OB|$ between them. Let $d\mathbf{r}/ds = \mathbf{t}$ and $\mathbf{n} = \mathbf{t}^{\perp}$ compose the Frenet frame at O, see Fig. 3. Simple manipulations with Taylor series expansions

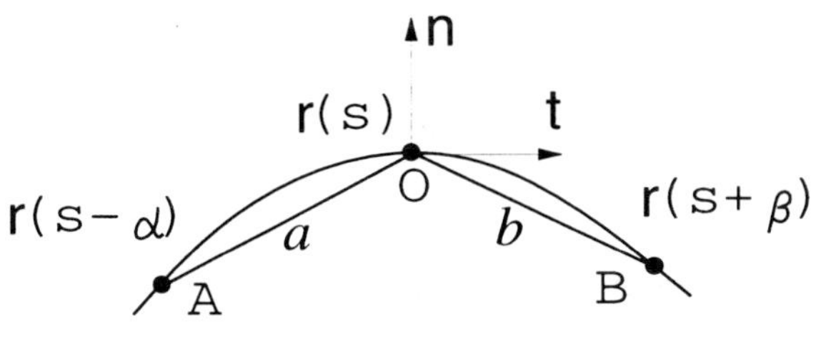

Figure 3.

and Frenet formulas show that

$$\frac{2}{a+b}\left[\frac{\overrightarrow{OA}}{a} + \frac{\overrightarrow{OB}}{b}\right] = \mathbf{n}\left(k + \frac{b-a}{3}k' + O(a,b)^2\right) + \\ + \mathbf{t}\left(\frac{a-b}{4}k^2 + O(a,b)^2\right), \tag{5}$$

$$\frac{2}{a^2+b^2}\left[\overrightarrow{OA} + \overrightarrow{OB}\right] = \\ = \mathbf{n}\left(k + \frac{b^3 - a^3}{3(a^2+b^2)}k' + O(a,b)^2\right) + \\ + \mathbf{t}\left(\frac{2(b-a)}{a^2+b^2} + \frac{a^3-b^3}{4(a^2+b^2)}k^2 + O(a,b)^2\right).$$

If, for example, point O is located closer to A than to B, $a < b$, then, due to the tangent components in the above expansions, one step of Laplacian smoothing with $\mathcal{U}_0$ shifts O closer to B and one step of the Laplacian smoothing with $\mathcal{U}_1$ shifts O closer to A.

However Laplacian smoothing with $\mathcal{U}_0$ develops unnatural deformations, see Fig. 1c (the top row).

Taubin Smoothing. Taubin in [8] proposed to alternate two scale factors of opposite signs with the negative factor of larger magnitude in the Laplacian smoothing flow. Such smoothing suppresses high frequencies of the umbrella operator (1), while preserving and enhancing its low frequencies [8]. Combining the two successive steps of the Taubin method in one local update rule we arrive at

$$P_{new} \longleftarrow (1 - \mu\mathcal{U})(1 + \lambda\mathcal{U})P_{old} = \\ = P_{old} - (\mu - \lambda)\mathcal{U}(P_{old}) - \mu\lambda\mathcal{U}^2(P_{old}), \tag{6}$$

where $\mu > \lambda > 0$, $\mathcal{U}^2$ is the squared umbrella operator

$$\mathcal{U}^2(P) = \frac{1}{\sum_i w_i}\sum_i w_i\mathcal{U}(Q_i) - \mathcal{U}(P).$$

According to [8], the best smoothing with (6) is obtained when $\mathcal{U} = \mathcal{U}_0$ or $\mathcal{U} = \mathcal{U}_1$. In our experiments, Taubin smoothing with $\mathcal{U} = \mathcal{U}_0$ works much better than with $\mathcal{U} = \mathcal{U}_1$. However, even the first Taubin smoothing scheme often produces poor results, see for example Fig. 1d (the top row) where we used $\lambda = 0.3$ and $1/\lambda - 1/\mu = 0.1$. In Fig. 1d (the bottom row) the Taubin filtering scheme demonstrates a good performance. Nevertheless, one can note enhancing of low-frequency surface wrinkles.

Bilaplacian flow. The bilaplacian flow

$$P_{new} \longleftarrow P_{old} + \lambda\mathcal{U}^2(P_{old})$$

is a discrete analog of the steepest descent flow for the the thin-plate energy functional [4]. It can be obtained from the Taubin smoothing scheme if the positive and negative scale factors are equal in magnitude. The bilaplacian flow does not enhance low-frequency surface features.

In our implementation of the bilaplacian flow we use the plain umbrella operator (3).

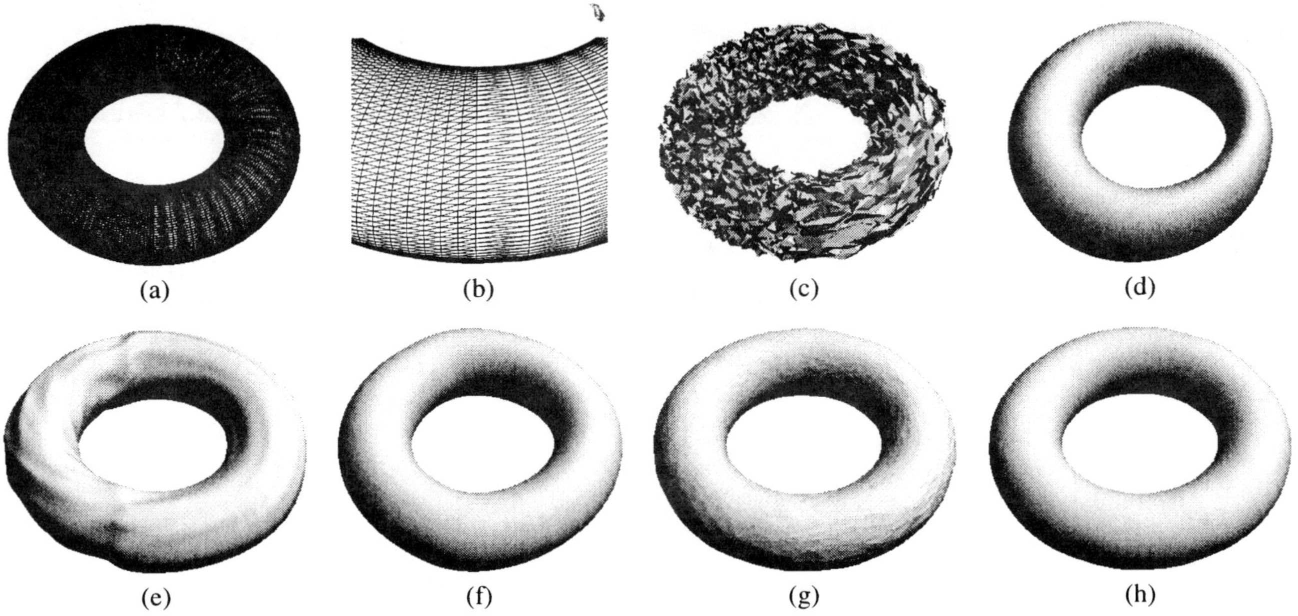

Figure 4. (a) A torus with consisting of two parts with different sampling rates; (b) a magnified view of a part of the torus; (c) the torus with a uniform noise added; (d) Laplacian smoothing deforms the initial shape; (e) smoothing by the Taubin method reduces high-frequency surface oscillations but develops low-frequency surface waves; (f) the bilaplacian flow smoothes well but slightly deforms the initial shape; (g) the mean curvature flow smoothes well but produces irregular mesh; (h) smoothing according to (11).

Mean curvature flow. Recently it was reported [3] that the smoothing procedure

$$P_{\text{new}} \longleftarrow P_{\text{old}} + \lambda H(P_{\text{old}})\,\mathbf{n}(P_{\text{old}}), \qquad (7)$$

where H is a discrete version of the mean curvature and $\mathbf{n}$ is the unit normal vector, produces better results than Laplacian smoothing (2).

According to [3], a good estimation of the mean curvature vector at a vertex P is given by

$$H\mathbf{n} = -\frac{\nabla A}{2A},$$

where $A = \sum A_i$ is the sum of the areas of the triangles surrounding P. Calculations [3] show that

$$H\mathbf{n}(P) = \frac{1}{4A}\sum_i (\cot\alpha_i + \cot\beta_i)(Q_i - P), \qquad (8)$$

where α_i and β_i are the two angles opposite to the edge Q_iP, see Fig. 5.

The two dimensional analog of this approximation is given by the left hand-side of (5), since

$$\nabla(a+b) = \nabla a + \nabla b = -\left[\frac{\overrightarrow{OA}}{a} + \frac{\overrightarrow{OB}}{b}\right]$$

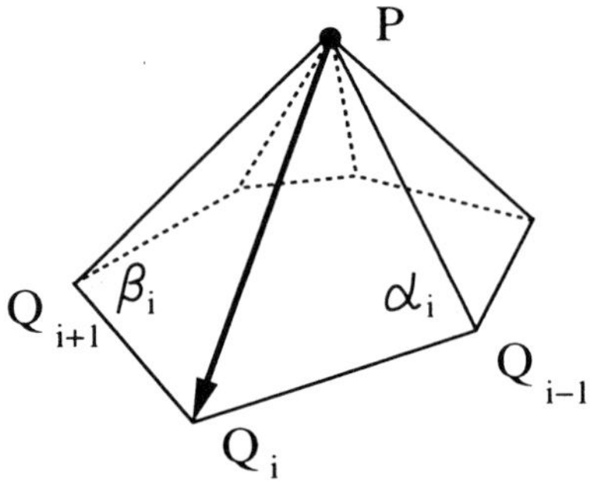

Figure 5.

Thus, similar to $\mathcal{U}_1$, smoothing by discrete curvature flow (7), (8) worsens the mesh sampling rate. See Fig. 1e (the top row) where the mean curvature flow demonstrates a good performance in smoothing a noisy sphere but produces uneven distribution of vertices.

3 Modified Mean Curvature Flow

Let us consider a family of smooth surfaces $S(u, v, t)$, where (u, v) parameterize the surface and t parameterizes the family. We suppose (u, v) to be independent of t. Let us assume that this family evolves according to the evolution

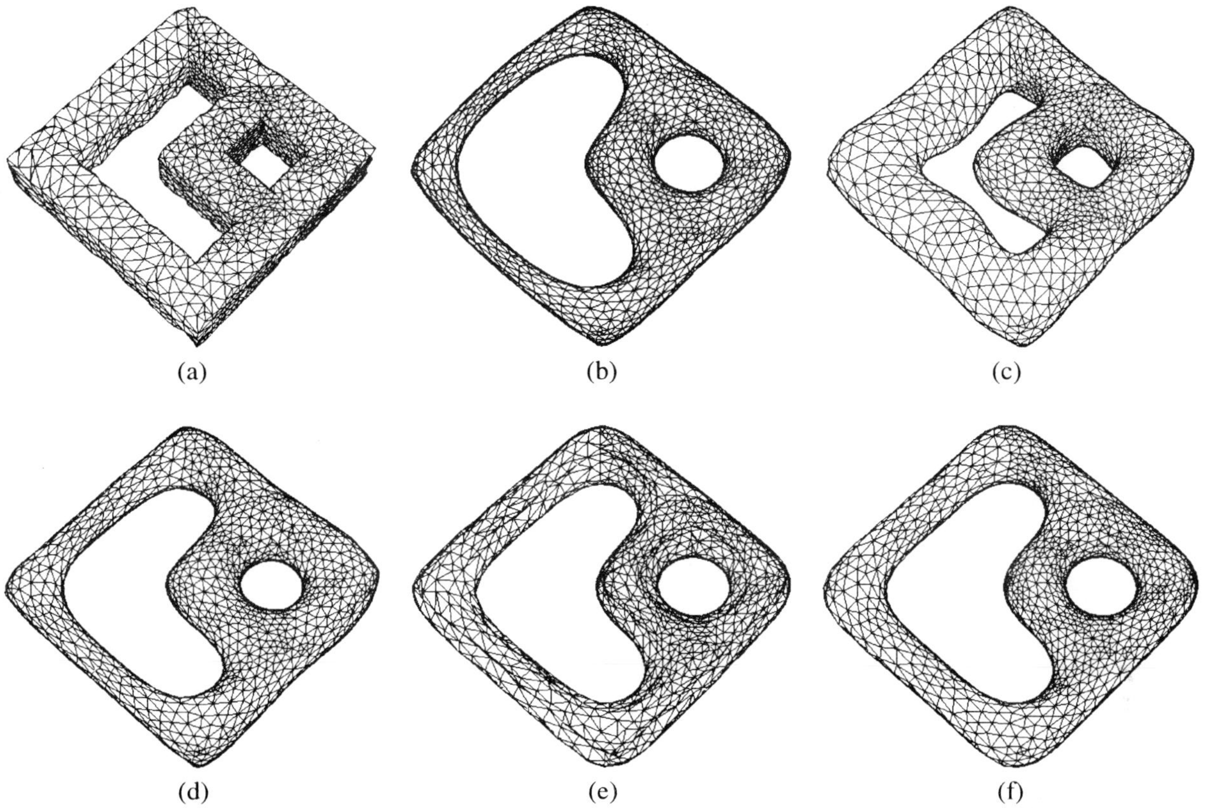

Figure 6. (a) A polyhedral two-holed torus given as a mesh having irregular connectivity and consisting of parts with different sampling rates, a small noise is also added; (b) Laplacian smoothing improves the mesh sampling rate but deforms the initial shape; (c) smoothing by the Taubin method substantially deforms the initial shape in unnatural way; (d) the bilaplacian flow also deforms the initial shape in an irregular manner. (e) the mean curvature flow produces irregular mesh; (f) smoothing according to (12), (13) produces a regular meshing surface which shape is close to the shape produced by the mean curvature flow.

equation

$$\frac{\partial S(u,v,t)}{\partial t} = F\,\mathbf{n}, \qquad S(u,v,0) = S^{(0)}(u,v), \qquad (9)$$

where $\mathbf{n}(u,v,t)$ is the unit normal vector for $S(u,v,t)$, F is a speed function, $S^{(0)}(u,v)$ is an initial surface. The family parameter t can be considered as the time duration of the evolution. Equation (9) means that the surface $S(u,v,t)$ moves along its normals with speed equal to F. Consider now the flow

$$\frac{\partial S(u,v,t)}{\partial t} = F\,\mathbf{n} + G\,\mathbf{t}, \quad S(u,v,0) = S^{(0)}(u,v), \quad (10)$$

where $\mathbf{t}$ is a vector tangent to the surface and G is a given function. Note that the tangent speed component does not affect the geometry of the evolving surface and changes only surface parameterization.

Solving (9) for a polyhedral surface by an explicit Euler scheme is more stable for polyhedral surfaces with uniform mesh sampling rates and, therefore, allows to use larger time steps to achieve faster smoothing. Note that a backward scheme (it was proposed and used for smoothing in [3]) is no better than Euler's scheme. For larger time steps where Euler's scheme is unstable the backward scheme is inaccurate [5].

Roughly speaking, *our main idea of simultaneous mesh smoothing and regularization consists of using the normal speed component* $F\,\mathbf{n}$ *for polyhedral surface smoothing and the tangent speed component to improve the mesh sampling rate.* For polygonal curve evolutions this idea was proposed and used in [1].

Let us use the discrete mean curvature flow (7), (8) for smoothing and the Laplacian flow (2), (3) for improving the mesh sampling rate. One possible implementation is to take

$$F\,\mathbf{n} = H\,\mathbf{n} \quad \text{and} \quad G\,\mathbf{t} = C\left[\mathcal{U}_0 - (\mathcal{U}_0 \cdot \mathbf{n})\,\mathcal{U}_0\right],$$

where $\mathcal{U}_0$ is the umbrella vector (3), C is a positive constant,

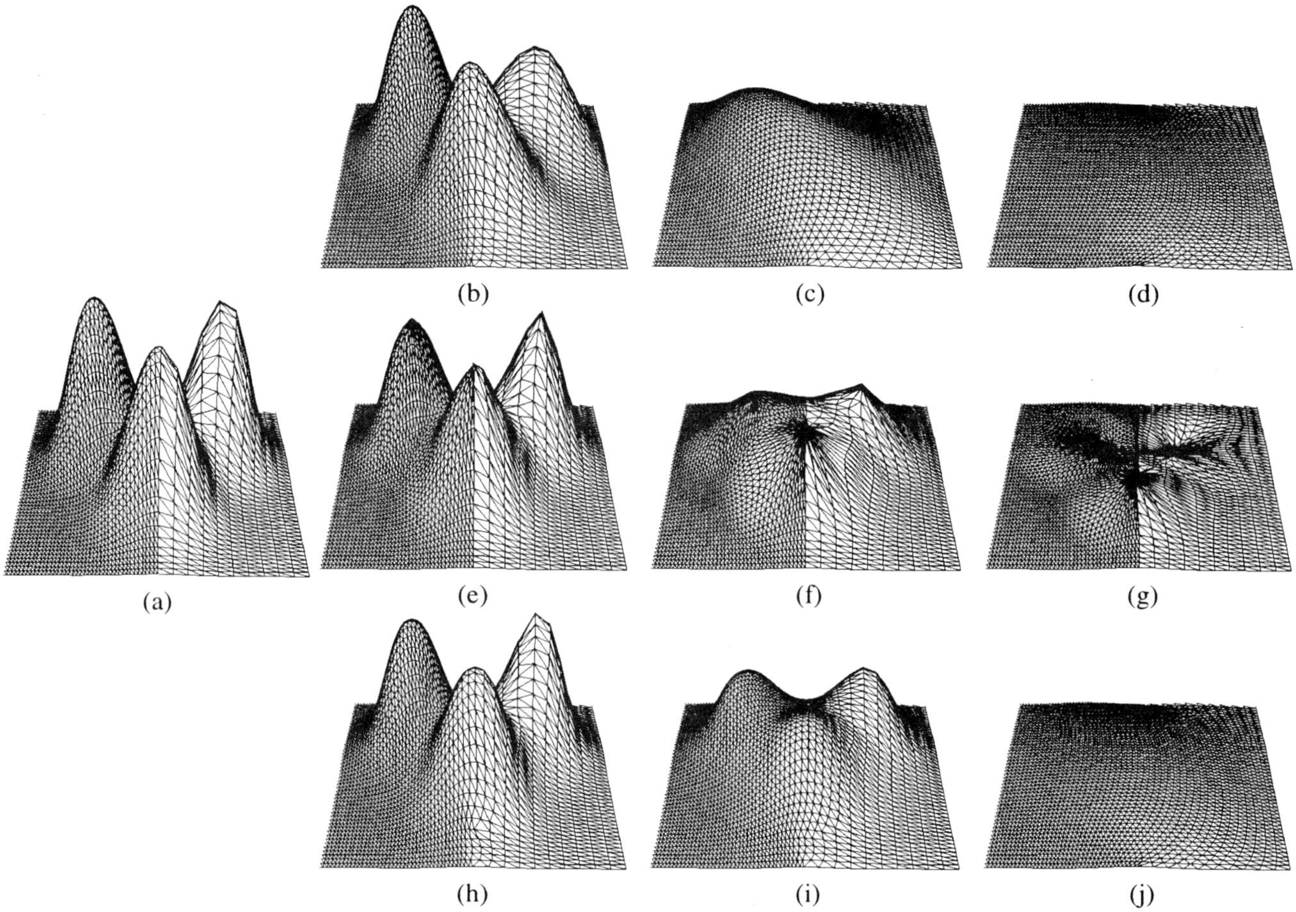

Figure 7. (a) A polyhedral surface consisting of two parts with different sampling rates. (b)–(d) Deformation of (a) into a flat patch by the Laplacian flow. (e)–(g) Deformation of (a) into a flat patch by the mean curvature flow. (h)–(j) Deformation of (a) into a flat patch by our method (12), (13).

and the dot stands for the scalar product. One can see that $\mathcal{U}_0 - (\mathcal{U}_0 \cdot \mathbf{n})\,\mathcal{U}_0$ is the projection of the umbrella vector onto the tangent plane. It leads to the local update rule

$$P_{\text{new}} \longleftarrow P_{\text{old}} + \lambda \Big\{ H(P_{\text{old}})\mathbf{n}(P_{\text{old}}) + + C\big[\mathcal{U}_0(P_{\text{old}}) - (\mathcal{U}_0(P_{\text{old}}) \cdot \mathbf{n}(P_{\text{old}}))\,\mathcal{U}_0(P_{\text{old}})\big]\Big\} \tag{11}$$

Smoothing by (11) produces shapes of the same quality as the mean curvature flow but with uniform distribution of vertices. See Fig. 4.

One can also define the parameter C as a function of surface curvatures to achieve a higher mesh sampling rate in the curved surface regions.

However, according to our experiments, a similar smoothing scheme produces better results.

Let $\mathbf{m} = \mathcal{U}_0/||\mathcal{U}_0||$ and θ is the angle between the mean curvature vector $H\mathbf{n}$ and $\mathbf{m}$: $\cos\theta = \mathbf{m} \cdot H\mathbf{n}/|H|$. Vector $\mathbf{m}$ defines a 3D analog of 2D median direction. *Our basic idea is to move the vertices in the median direction such that the normal speed component is equal to the mean curvature.* However, since for saddle vertices the median direction vector $\mathbf{m}$ and the mean curvature vector $H\,\mathbf{n}$ may have opposite normal components (i.e., $\theta > \pi/2$), see Fig. 8b, we use the following flow

$$P_{\text{new}} \longleftarrow P_{\text{old}} + \lambda\,\mathcal{F}(P_{\text{old}}), \tag{12}$$

where

$$\mathcal{F} = \begin{cases} \dfrac{|H|\mathbf{m}}{\cos\theta} & \text{if} \quad \cos\theta > \varepsilon \\ 2\,H\,\mathbf{n} - \dfrac{|H|\,\mathbf{m}}{\cos\theta} & \text{if} \quad \cos\theta < -\varepsilon \\ 0 & \text{if} \quad |\cos\theta| \le \varepsilon \end{cases} \tag{13}$$

Here ε is a small positive parameter. Geometric ideas behind (12) are explained in Fig. 8.

If the normal and median vectors are almost orthogonal to each other at a vertex ($|\cos\theta| \le \varepsilon$), we do not move the vertex at all.

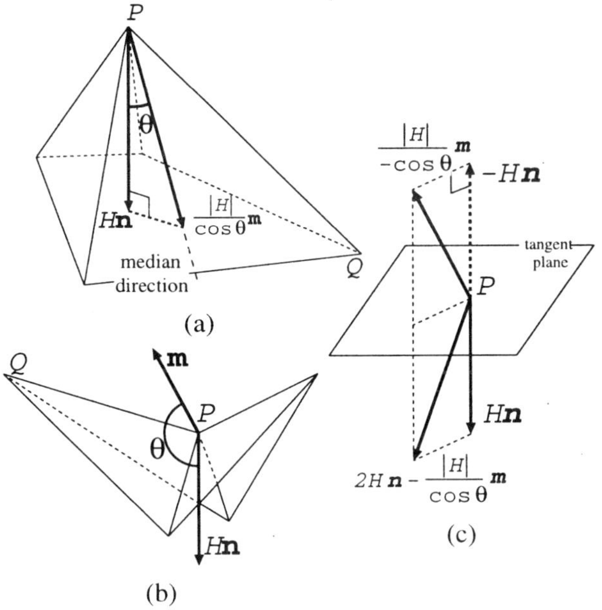

Figure 8. (a) The case when the median and normal vectors lie on the same side from the tangent plane: moving in the median direction with normal speed component equal to the mean curvature. (b) The case the median and normal vectors lie on the opposite sides from the tangent plane may happen for saddle vertices. (c) Computation of the speed vector in (b).

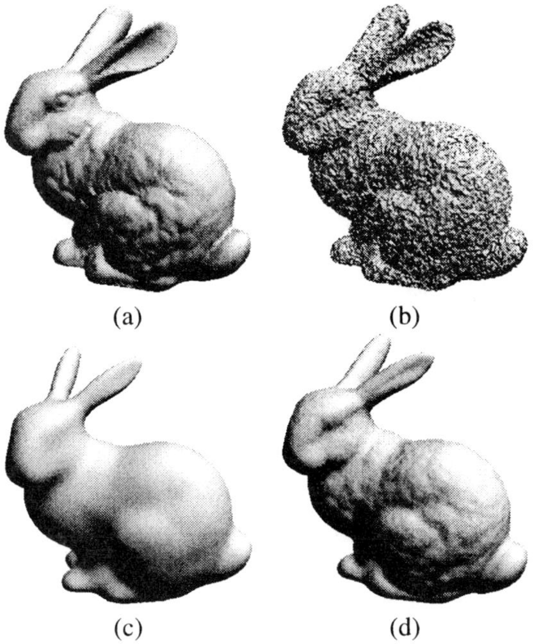

Figure 9. (a) The original Stanford bunny; (b) the bunny with a uniform noise added; (c) smoothing by (12) (Laplacian and mean curvature flows produce almost the same results); (d) smoothing by (14) with the same number of iterations as in (c).

According to our experiments, choosing $\varepsilon = 0.1$ produces good results independently of the mesh sampling rate.

Our smoothing scheme (12) demonstrates better results than the Laplacian flow, the mean curvature flow, and the Taubin smoothing scheme, see Fig. 1. See also Fig. 7 where a polyhedral surface consisting of two parts with different sampling rates is smoothed by the Laplacian flow, the mean curvature flow, and our scheme (12) and (13).

4 How to Avoid Oversmoothing

Laplacian smoothing, the mean curvature flow, and our new technique presented in the previous section smooth polyhedral surfaces by suppressing high frequency surface oscillations. Lacking a local surface control, they may lead to oversmoothing and, therefore, destroying desirable surface features. To incorporate a local control for smoothing and to reduce possible oversmoothing due to too large number of iterations, let us consider a simple but very useful modification allowing to slow down the smoothing process adaptively:

$$P_{\text{new}} \longleftarrow P_{\text{old}} + \lambda\,\widetilde{\mathcal{F}}(P_{\text{old}}) \tag{14}$$

with

$$\widetilde{\mathcal{F}}(P) = \begin{cases} (\|\mathcal{F}(P)\| - T)\,\dfrac{\mathcal{F}(P)}{\|\mathcal{F}(P)\|} & \text{if } \|\mathcal{F}(P)\| > T \\ 0 & \text{if } \|\mathcal{F}(P)\| \le T \end{cases}$$

where T is a positive threshold selected by a user. According to this modification, the smoothing is performed only on those mesh vertices P where $\|\mathcal{F}(P)\| > T$. See Fig. 9 to compare performance of smoothing schemes (12), (13) and (14) with $\mathcal{F}(P)$ defined in (13).

Of course the above simple modification can be applied to the any smoothing flow considered before.

In our experiments exposed in Fig. 9 the parameter T in (14) is the same for all vertices. It is natural to allow T be dependent on shape characteristics at the vertices: $T = T(P)$. In our experiments we define the threshold $T(P)$ at the vertex P as the arithmetic mean of the mean curvatures computed at the first ring of neighbors of P or at the first and second rings of neighbors of P. Fig. 10 demonstrates

Figure 11. (a) A Godzilla model. (b) - (f) The ridges (black) and ravines (white) detected on the Godzilla model smoothed by various smoothing schemes. (b) Smoothing by the Laplacian flow. (c) Smoothing by the Taubin method. (d) Smoothing by the bilaplacian flow. (e) Smoothing by the mean curvature flow. (f) Smoothing by our method (12), (13).

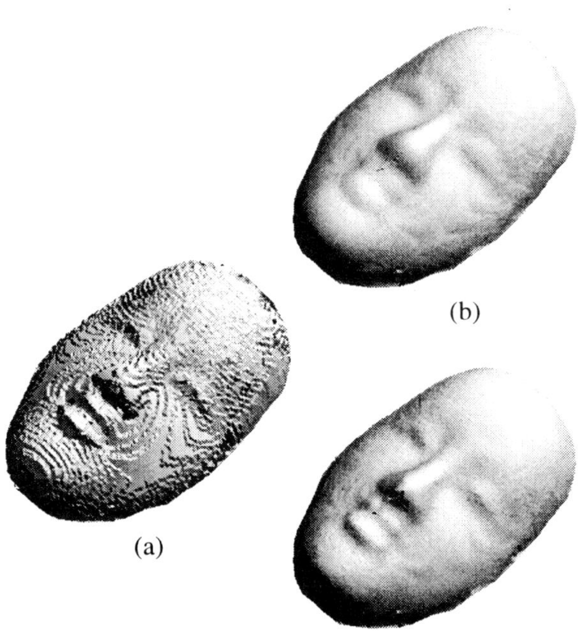

Figure 10. (a) A triangulated Noh mask model reconstructed from a scattered data generated by a laser scanner system; (b) smoothing according to (14) with constant T chosen manually; (c) smoothing according to (14) with $T(P)$ equal to the arithmetic mean of the mean curvatures of at the first and second rings of neighbors of P.

our experiments with smoothing a Noh mask triangulated model reconstructed from a cloud of points.

According to our experiments, the best smoothing strategy consists of recomputing periodically the threshold $T(P)$ for each vertex P after a fixed number of smoothing iterations. Fig. 1f (the bottom row) shows advantages of such threshold recomputing scheme.

5 Applications

We apply our smoothing technique (12), (13) to detect ridges and ravines on a smooth surface approximated by a triangular mesh.

Let us define the *ridges* as the locus of points where the maximal principal curvature attains a positive maximum along its curvature line and the *ravines* as the locus of points where the minimal principal curvature attains a negative minimum along its curvature line [2].

Practical detection of the ridges and ravines on a surface involves estimation of high-order surface derivatives and, therefore, requires a careful surface smoothing before detection. It seems natural to choose a smoothing procedure minimizing vertices drift over the surface. On the other hand, a smoothing scheme increasing the mesh regularity improves curvature estimation.

For practical detection of the ridge and ravine vertices on the mesh, we first smooth the mesh and then estimate the normal, the principal curvatures, and the principal directions at the mesh vertices. We use the method proposed in [7]. To check whether a given vertex P is a ridge vertex, we find the intersection between the polygon composed the ring of the first neighbors of P and the normal plane generated by the normal vector and maximal principal direction at P. We estimate the maximal principal curvature at the intersection points by linear interpolation. Finally we compare the the values of the maximal principal curvature at P and the intersection points.

Some results of our experiments with various smoothing schemes are shown in Fig. 11 where the ridges are colored in black and ravines are colored in white. We think that our smoothing scheme produces one of the best results.

6 Conclusion

In this paper we propose simple and effective polyhedral surface smoothing schemes which combine the best properties of the Laplacian smoothing flow and discrete mean curvature flow, are good for highly irregular meshes, outperform existing methods, and reduce possible oversmoothing.

In future we are planning to extend ideas presented in the paper to various discrete implementations of the Laplacian and mean curvature flows. Also we want to examine how the developed methods can be enhanced by bilaplacian flows and surface diffusion processes. We also hope to make our smoothing methods faster and more accurate by using the implicit Crank-Nicolson method.

Another interesting direction for future research is shape enhancement, the opposite operation to smoothing. Simple inverting of a smoothing flow is usually unstable. However, stability can be achieved by incorporating a curvature-based local control similar to that we developed in Section 4. Our preliminary results are encouraging.

Acknowledgments

We thank the anonymous reviewers for their careful reading and helpful suggestions. Ilia Bogaevski is grateful to the University of Aizu where he spent one year as a visiting researcher in 1998/99 when this research was mainly conducted. Thanks to Shin Yoshizawa for the Godzilla 3D data set.

References

[1] A. G. Belyaev, E. V. Anoshkina, S. Yoshizawa, and M. Yano. Polygonal curve evolutions for planar shape modeling and analysis. *International Journal of Shape Modeling*, to appear.

[2] A. G. Belyaev and Y. Ohtake. An image processing approach to detection of ridges and ravines on polyhedral surfaces. *Submitted for publication*, 2000.

[3] M. Desbrun, M. Meyer, P. Schröder, and A. H. Barr. Implicit fairing of irregular meshes using diffusion and curvature flow. *Computer Graphics (SIGGRAPH 99 Proceedings)*, pages 317–324, 1999.

[4] L. Kobbelt, S. Campagna, J. Vorsatz, and H.-P. Seidel. Interactive multiresolution modeling on arbitrary meshes. In *Computer Graphics (SIGGRAPH 98 Proceedings)*, pages 105–114, 1998.

[5] R. Mohtar and L. Segerlind. Accuracy-based time step criteria for solving parabolic equations. In *Modeling, Mesh Generation, and Adaptive Numerical Methods for Partial Differential Equations, IMA Volume 75*, pages 153–163. Springer, 1995.

[6] J. A. Sethian. *Level Set Methods and Fast Marching Methods*. Cambridge Univ. Press, 1999.

[7] G. Taubin. Estimating the tensor of curvature of a surface from a polyhedral approximation. In *Proc. ICCV'95*, pages 852–857, 1985.

[8] G. Taubin. A signal processing approach to fair surface design. In *Computer Graphics (SIGGRAPH 95 Proceedings)*, pages 351–358, 1985.

[9] J. Vollmer, R. Mencl, and H. Muller. Improved laplacian smoothing of noisy surface meshes. *Computer Graphics Forum (Proc. Eurographics 1999)*, 18(3):131–138, 1999.

Session 8

Mesh Processing and Subdivision Surfaces II

Approximate Shortest Path on a Polyhedral Surface Based on Selective Refinement of the Discrete Graph and Its Applications

Takashi Kanai
RIKEN
Materials Fabrication Lab.
kanai@postman.riken.go.jp

Hiromasa Suzuki
University of Tokyo
Dept. Precision Engineering
suzuki@cim.pe.u-tokyo.ac.jp

Abstract

A new algorithm is proposed for calculating the approximate shortest path on a polyhedral surface. The method mainly uses Dijkstra's algorithm and is based on selective refinement of the discrete graph of a polyhedron. Although the algorithm is an approximation, it has the significant advantages of being fast, easy to implement, high approximation accuracy, and numerically robust. The approximation accuracy and computation time are compared between this approximation algorithm and the extended Chen & Han (ECH) algorithm that can calculate the exact shortest path for non-convex polyhedra. The approximation algorithm can calculate shortest paths within 0.4% accuracy to roughly 100-1000 times faster than the ECH algorithm in our examples. Two applications are discussed of the approximation algorithm to geometric modeling.

Keywords: *geometric modeling, computational geometry, polyhedral surface, shortest path, Dijkstra's algorithm*

1. Introduction

Polyhedral surfaces, especially meshes that consist of planar triangle faces, are the fundamental geometric representation in computer graphics and related areas. Those surfaces with a large number of faces (called *dense*) can now be generated more easily due to the recent development of range scanners, GPS, and so on. For example, surface reconstruction from range images [3] and the use of such dense polyhedral surfaces have been active research topics for computer graphics applications. Geographic information systems (GIS) usually use triangular irregular networks (TINs) which are polyhedral representations of geographical features.

Finding the shortest path on such polyhedral surfaces is a fundamental problem in computational geometry, and becomes an important technique for the application of robotics, GIS, route finding, and so on. The purpose of this paper is to establish an efficient algorithm for calculating shortest paths and to use them for geometric modeling. There are many algorithms for finding the shortest path on 2D polygons and 3D surfaces, and in 3D spaces. However, the algorithms for finding the *exact* shortest path on a polyhedral surface (including the non-convex case) [17, 2] usually involve high time and space costs. It is therefore not practical to apply these algorithms to a dense polyhedral surface in most cases.

Instead, we have focused on finding the *approximate* shortest path [11, 15]. The algorithm proposed in this paper for a polyhedral surface (possibly including the non-convex case) mainly uses Dijkstra's algorithm and is based on selective refinement of the discrete graph of a polyhedron. That is, Dijkstra's algorithm is iteratively used to narrow the region in which the shortest path can exist. Our algorithm has some significant advantages: it is fast and easy to implement, it has high accuracy, and needs less space cost to get an actual path. To show some features of our algorithm, we compare the approximation accuracy and computation time between our algorithm and the Chen and Han (CH) algorithm. The CH algorithm is the fastest one for finding the exact shortest path. Unfortunately, it sometimes fails to compute for a non-convex polyhedron. In this work, we extend the CH algorithm so that it can be used for more general cases.

We also discuss applications of our algorithm to geometric modeling. It is to be noted that our algorithm can only find the approximate shortest path, so it is not suitable for applications which absolutely need the exact path (or only its length). We show two examples of our algorithm being efficiently used for modeling: One involves specification of the boundaries of a region on a polyhedral surface. It is not necessary for such boundaries to be exact paths, but they do need to be smooth and to be calculated rapidly and efficiently. The other involves the interactive length measurement of a geodesic curve on a 3D model in a case that also

0-7695-0562-7/00 $10.00 © 2000 IEEE

does not need to be exact.

2. Shortest Path Problem on a Polyhedral Surface

A survey of the shortest path problem concerning a two or higher dimensional geometric object (a surface, space, network, etc.) can be found in [16]. We mainly discuss here about finding the shortest path between two points on a polyhedral surface.

An important property of the shortest path on a polyhedron is its local optimality called *unfolding*: the path must enter and leave at the same angle to the intersecting edge. It follows that any locally optimal sub-path joining two consecutive obstacle vertices can be unfolded at each edge along its edge sequence, thus obtaining a straight segment. This property was used by Sharir and Schorr [19] who first proposed an $O(n^3 \log n)$-time algorithm (n is the number of edges) to find the exact shortest path on a convex surface. Mitchell et al. [17] have obtained an $O(n^2 \log n)$-time algorithm for general polyhedra by developing a continuous Dijkstra method for propagating the shortest path map over a surface. While Chen and Han [2] subsequently improved this to an $O(n^2)$-time algorithm, faster algorithms than these cannot presently be found.

On the other hand, some other work has been done on the approximate shortest path. This has involved less computation times than the exact approaches, but it is a tradeoff with the path accuracy. In general, the term $(1+\epsilon)$-*approx.* is usually used for evaluating the computation time for these approximate approaches, meaning that a calculated approximate path guarantees the accuracy within $(1+\epsilon)$ times longer than the exact path. In this evaluation, smaller ϵ results in better accuracy.

Algorithms theoretically exist that are fast and require less space if limited to convex polyhedra. For example, the approach proposed by Har-Peled [6] is based on the construction of a tight bounding volume covering a polyhedron. By performing $O(n)$ pre-processing, the $O((\log n)/\epsilon^{3/2} + 1/\epsilon^3)$-time computation of a $(1+\epsilon)$-approx. shortest path is possible. In the case of general polyhedra, Varadarajan and Agarwal [21] have proposed algorithms that compute a 13-approx. (ϵ =12) path in $O(n^{5/3} \log^{5/3} n)$-time or a 15-approx. (ϵ =14) path in $O(n^{8/5} \log^{8/5} n)$-time. These are based on partitioning the surface into $O(n/r)$ patches, each having at most r faces. However, no approximations with high accuracy can be obtained by these approaches.

Some practical methods for finding the approximate shortest path based on subdivided discrete graph searching have recently been proposed. The algorithm described in this paper is of this type. The common characteristic of these approaches (including ours) is to create a certain graph (called an *edge subdivision graph* or a *pathnet*) which is needed for path computation as a pre-processing step.

The $O(n \log n)$-time approach by Lanthier et al. [11] first adds some intermediate points to edges, and discrete graph G is created by using these points and the original vertices. Then Dijkstra's algorithm is applied to G. After picking up the faces under the calculated shortest path of G (called the *sleeve*), the $O(\log k)$ algorithm (k is the number of edges passed by the shortest path) proposed by Guibas and Hershberger [5] is used to refine the graph. However, [5] uses a rather special data structure called the *hourglass*, and its algorithm is difficult to implement. Mata and Mitchell [15] have proposed another approach called the pathnet algorithm which is applied with, at worst, $O(kn^3)$-time (where k is the number of rays for each vertex). One practical aspect of this algorithm is the robustness of the computation. This algorithm is sensitive to numerical errors because it requires solving a fourth-degree polynomial. An another aspect is the path accuracy, a large number of k being required to obtain an approximate shortest path close to the exact solution.

The algorithm described in this paper is similar to the algorithm of Lanthier et al. [11]. The major difference is that our algorithm is iterative: the shortest path is calculated by selective refinement of the subdivided discrete graph of a polyhedral surface.

3. Approximate Shortest Path Computation

A *polyhedral surface* is a connected union of a finite number of polygonal faces, with any two polygons intersecting along a common edge, at a common vertex, or not at all, and each edge belonging exactly to a polygon. Given polyhedral surface $\mathcal{M}$, shortest path $\mathcal{L}$ from source vertex V_S to destination vertex V_D on $\mathcal{M}$ is defined by:

$$\mathcal{L} = \{v_1, v_2, \ldots, v_n, e_1, e_2, \ldots, e_{n-1}\}, \quad (1)$$
$$V_S = v_1, \quad V_D = v_n, \quad e_i = \{v_i, v_{i+1}\},$$

where v_i and e_i denote a point and an edge on a face of $\mathcal{M}$, respectively.

Let $l = |\mathcal{L}|$ be the *length* of shortest path $\mathcal{L}$. Then:

$$l \;=\; |\mathcal{L}| = |e_1| + |e_2| + \ldots + |e_{n-1}|, \quad (2)$$

where $|e_i|$ is the Euclid distance between two connected vertices v_i and v_{i+1} of edge e_i.

3.1. Pre-processing for path calculation

Pre-processing involves creating initial discrete graph $G^0(v,e)$ from $\mathcal{M}$. $G^0(v,e)$ is composed of vertices / edges of $\mathcal{M}$ and additional vertices / edges.

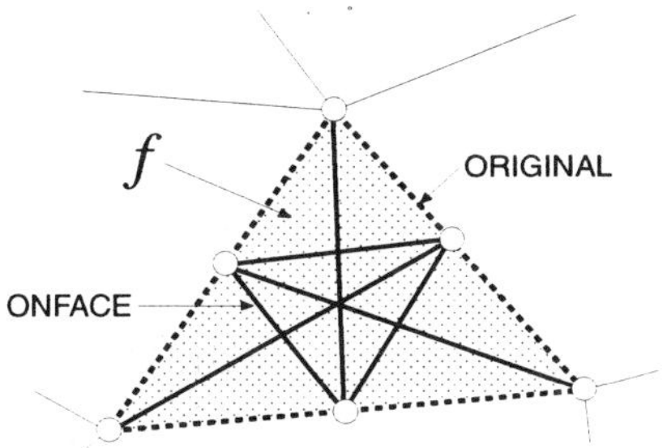

Figure 1. Edge classifications

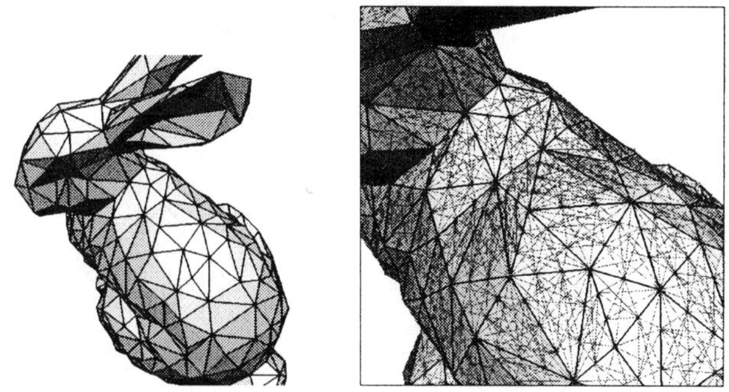

(a) (b)

Figure 2. Initial discrete graph G^0

We define here some classifications to edges on $\mathcal{M}$ to judge whether an edge should be added or not during the process of the algorithm: ORIGINAL refers to the original edges or their subdivided edges of $\mathcal{M}$, and ONFACE refers to the others, as shown in Figure 1.

G^0 is next created. First, the vertices and edges of $\mathcal{M}$ are added to G^0. Then additional vertices are sampled on each edge of $\mathcal{M}$. These are called *Steiner Points* (SPs). The number of added SPs on each edge are decided by its length and parameter γ. Let $|e|$ be the length of edge e. Then the number of SPs on e is decided as $(|e| \bmod \gamma) - 1$ so that smaller γ creates more SPs. Note that γ is a size-dependent value. To be independent of the size of $\mathcal{M}$, $\mathcal{M}$ is scaled into a unit cube in advance so that the edge length is normalized. γ is a user-defined value that decides the trade-off between the approximation accuracy and the computing time and space.

Intermediate edges are created in G^0 by using the original and additional vertices. If pair of vertices $\langle v_a, v_b \rangle \in G^0$ satisfies either 1. or 2. in **Condition 1** bellow, edge $e = \{v_a, v_b\}$ is added to G^0:

Condition 1

1. $\langle v_a, v_b \rangle$ is a pair of vertices in G that are on different edges of a face.
2. $\langle v_a, v_b \rangle$ is a pair of vertices in G that are next to each other on the same edge.

Figure 2(b) illustrates a part of initial discrete graph G^0 (2,859 vertices, 16,055 edges) of the "bunny" model (525 vertices, 999 edges) shown in Figure 2(a). In Figure 2(b), thin lines denote additional edges. We have set γ to 0.04 in this example, the average number of SPs per edge is 1.52.

3.2. Approximate Shortest Path Algorithm

By using G^0, shortest path $\mathcal{L}$ between V_S and V_D is calculated by the following algorithm. Our algorithm repeats **STEP 1 - STEP 4**, incrementing loop counter i from 0 by 1 each time.

STEP 1: Calculate shortest path on G^i

Shortest path $\mathcal{L}^i$ for discrete graph G^i and its length l^i are calculated by using Dijkstra's algorithm. The shortest path is composed of vertices and edges in G^i.

STEP 2: Generate new graph G^{i+1} from path $\mathcal{L}^i$

Shortest path $\mathcal{L}^i$ is traced and its component vertices are added to G^{i+1}. Other connected vertices of edges including these path vertices are also added if the corresponding edges are ORIGINAL. For pair of vertices $\langle v_a, v_b \rangle \in G^{i+1}$, edge $e = \{v_a, v_b\} \in G^i$ is added to G^{i+1}.

STEP 3: Add SPs and edges to G^{i+1}

m SPs are defined for edges $e \in G^{i+1}$ at even intervals in G^{i+1} if e is ORIGINAL. If pair of vertices $\langle v_a, v_b \rangle \in G^{i+1}$ satisfies both $e = \{v_a, v_b\} \notin G^{i+1}$ and either 1. or 2. in **Condition 1**, e is added to G^{i+1}. e is ONFACE if $\langle v_a, v_b \rangle$ satisfies to 1. in **Condition 1**, else e is ORIGINAL. When e is ORIGINAL, its original edges are deleted from G^{i+1}.

STEP 4: Update the graph

$G^i \leftarrow G^{i+1}$

When **STEP 1** is finished, the length of shortest path l^i of graph G^i and of l^{i-1} of its previous graph G^{i-1} are compared. The algorithm is completed if $|l^i - l^{i-1}| < \epsilon$. The addition of SPs and edges to G^0 described in the previous subsection corresponds to **STEP 3**, but the number of SPs

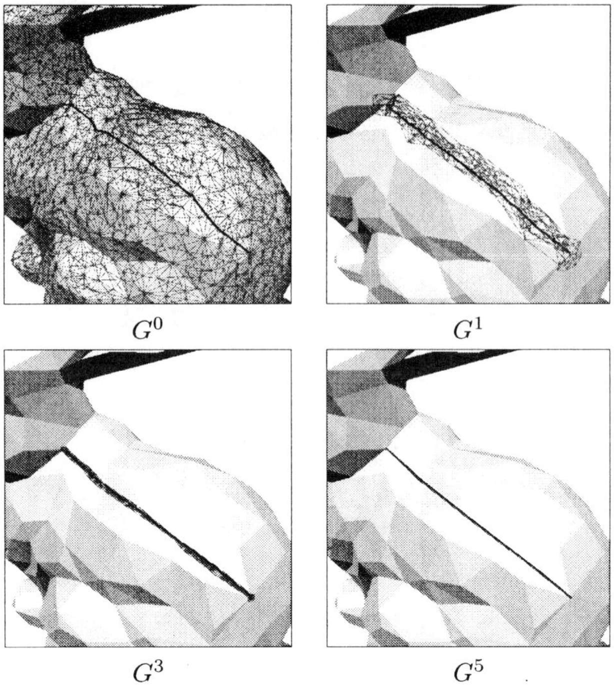

Figure 3. Iterations of the algorithm

per edge m is different. m is another variable that should be decided by the user. It is a trade-off between the decrease in the number of iterations and the increase in computing time and space. While it is also possible to set m to a different value per edge, we set m as a constant value for all edges because the length of edges in G^0 is almost uniform throughout the pre-processing stage.

The characteristic of this algorithm are presented in Figure 3 which shows the state for calculating the shortest path between two given vertices of the bunny model in Figure 2. We set $\gamma = 0.04, m = 1$ for this example, and the algorithm terminated after seven iterations. In Figure 3, the sequences of discrete graphs G^0, G^1, G^3, G^5 and the shortest path of each graph calculated in **STEP 1** are shown. The thin lines in Figure 3 represent edges of a graph calculated in **STEP 2** and **STEP 3**, while bold lines show the shortest path calculated on that graph. It can be seen that the region in which a path exists gradually becomes narrower and finally converges to a poly line.

Table 1 shows the number of elements (vertices, edges) on G^i, the length of shortest path l^i and the sum of each processing time for **STEP 1 - STEP 4**. Our algorithm searches almost a constant number of edges (from 350 to 400 in this example) on each graph G^i, except for G^0 in the first stage which has a large number of edges. It can be also seen that the total time needed for processing our algorithm largely depends on that for calculating the shortest path on G^0 in

	$\|v\|$	$\|e\|$	l^i	T (sec.)
G^0	2,859	16,055	0.52200	0.083
G^1	79	361	0.51408	0.017
G^2	75	365	0.51277	0.017
G^3	76	371	0.51252	0.017
G^4	78	393	0.51237	0.017
G^5	78	393	0.51234	0.017
G^6	70	313	0.51234	0.017

Table 1. Number of elements, length of the shortest path and processing time for each graph by our algorithm

STEP 1 by using Dijkstra's algorithm. We used a simple method for this problem based on a priority queue of partial-order trees that can be processed with $O(n \log n)$-time [1]. As this method evaluates the length of each edge, its processing time can be decided by the number of edges in G^0 and thus by γ.

To implement our algorithm, we prepared a simple graph structure for representing G, consisting of nodes for vertices and SPs, and links for (subdivided) edges of $\mathcal{M}$ and additional edges. Each node has a 3D coordinate value and a pointer to its connected links. A link has pointers to two end nodes, has its length, and has a flag for its type (ORIGINAL or ONFACE). If a node or a link comes from the original element of $\mathcal{M}$, it has a pointer to that element. If a link is ONFACE, the link has a pointer to a face of $\mathcal{M}$ under it.

4. Experimental Results and Discussion

We evaluate in this section our approximate shortest path algorithm from two viewpoints: one is the approximation accuracy and computation time with different γ and m; the other is the numerical accuracy. We also made a comparison between our algorithm and the exact computation of the shortest path on convex polyhedra as proposed by Chen and Han [2]. We discuss this algorithm in the first subsection.

4.1. Extended CH Algorithm

To evaluate to our approximate path algorithm, we implemented the Chen and Han algorithm [2] (we call it the *CH algorithm* from now on) with an extension so that it can compute the exact shortest path between any two vertices on a non-convex polyhedral surface.

The CH algorithm improves on the continuous Dijkstra method proposed by Mitchell et al. It utilizes the unfolding property (described in Section 2), whereby the shortest path and its length can be calculated by constructing a sequence

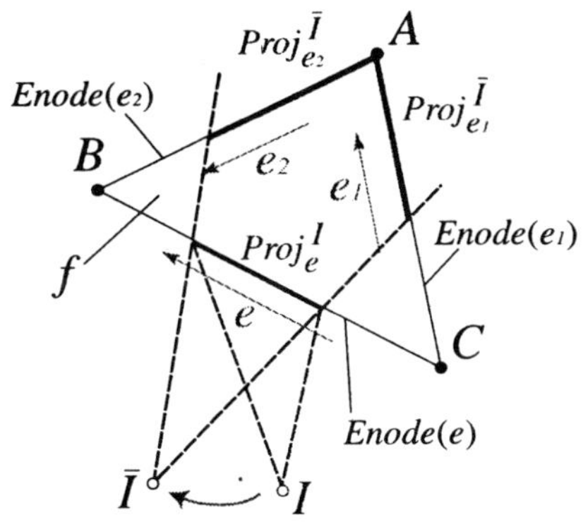

Figure 4. Relationship between $Enode(e)$ and its children in the CH algorithm

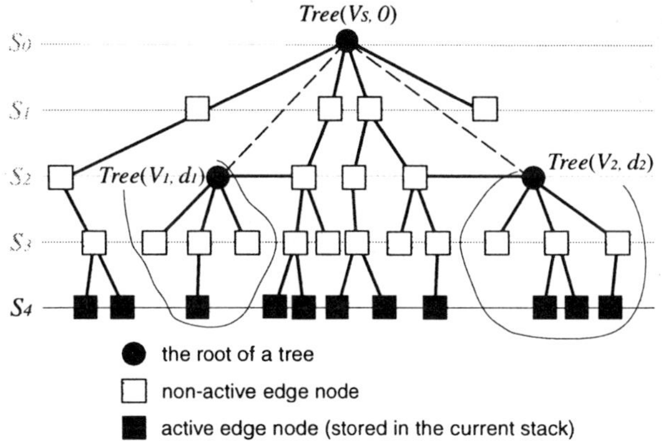

Figure 5. Construction of sequence trees in the ECH algorithm

tree structure. Its root is start vertex V_S, and leaf nodes are oriented edges. Edge node $Enode(e) = (e, I, Proj_e^I)$ is composed of oriented edge e, image I of V_S, and its projection $Proj_e^I$ to an edge. $Enode(e)$ has at most two child edge nodes. Figure 4 illustrates the relationship between $Enode(e)$ and its children $Enode(e_1), Enode(e_2)$. Each child of $Enode(e)$ can only be defined if projection $Proj_{e_1}^{\bar{I}}, Proj_{e_2}^{\bar{I}}$ of $\bar{I}$ to e_1, e_2 exists. $\bar{I}$ is calculated by rotating I around e so that $\bar{I}$ and face f are co-planar. It is said that $Enode(e)$ *covers* vertex A if both $Proj_{e_1}^{\bar{I}}$ and $Proj_{e_2}^{\bar{I}}$ exist. In addition, $Enode(e)$ *occupies* A if $|\bar{I}A|$ (the Euclid distance between A and $\bar{I}$) is the shortest. In this case, A has vertex node $Vnode(A) = (Enode(e), \bar{I}, A)$. The shortest path from V_S to V_D is calculated by tracing the edge node's parents from $Vnode(V_D)$.

The CH algorithm can always be successfully applied if a polyhedral surface is convex. For the general case, including a non-convex surface, there are some cases of V_D not having $Vnode(V_D)$ when the algorithm is terminated. The shortest path cannot be calculated in these cases. [2] points out that the shortest path of a non-convex polyhedron may pass some vertices [17] except V_S and V_D, and that it is a good choice to grow a sub-tree from a passed vertex. Unfortunately, no details for growing a sub-tree are mentioned in [2]. The problem is *how* or *when* we can begin to grow a sub-tree, because we cannot judge the vertex that the shortest path passes through during the process of the CH algorithm.

An extension for growing a sub-tree is presented next. We call it the *extended CH (ECH) algorithm.* Figure 5 illustrates sequence trees of the ECH algorithm. Root tree $Tree(V_S, 0)$ at V_S has a sequence of sub-trees. In Figure 5, such sub-trees correspond to $Tree(V_1, d_1)$ and $Tree(V_2, d_2)$. $Tree(V, d)$ has many pointers: one to the edge nodes (their number is that of the faces adjacent to V), one to the vertex nodes (their number is that of the edges adjacent to V), one to vertex V, geodesic distance d between V to V_S (this can be calculated during the process of the algorithm) and one to a parent tree. Each of the edge nodes belongs to the same tree as its parent node. A sub-tree is created if the edge node occupies a vertex V. If V with a sub-tree is occupied by another edge node, a sub-tree (including child sub-trees) and its child edge nodes are all deleted, and a new sub-tree is created.

Figure 6 shows the pseudo-code of the ECH algorithm. There are two major differences from the original: (1) the main loop is controlled by using stack S that stores edge nodes; (2) when it is judged that an edge node occupies A, A's sub-tree (including child sub-trees) are all deleted, and a new sub-tree is created. The time complexity keeps $O(n^2)$, but its efficiency is worse. The space complexity also keeps $O(n)$ if only the length of the path is needed, or $O(n^2)$ if the path itself is needed.

4.2. Approximation Accuracy and Computation Time

We next investigate how different values for user-specified variables γ and m influence the approximation accuracy and computation time. We used the simplified model of a "bunny" shown in Figure 7. To simplify the original model, we used the quadric error metric (QEM) based approach proposed by Garland and Heckbert [4]. It was difficult to use the original model because our naive implementation of the ECH algorithm needed quite a large amount of space to store the shortest path information. Figure 7(b) shows a simplified model in which the number of faces is 9,999.

```
Procedure Extended_CH_Algorithm
begin
  Create_Tree(V_S, 0, S);
  while S ≠ nil do
    Enode(e) ← top of S;
    unfold I to Ī; calculate Proj^Ī_{e1} and Proj^Ī_{e2};
    if Enode(e)'s shadow covers A then
      Enode' ← previously occupied enode;
      if Enode(e) occupies A over Enode' then
        clip off Enode's two children that is
            impossible to define the shortest sequence
            and their child enodes;
        if A has a tree then delete A's tree; end if
        insert Enode(e1) and Enode(e2) to
            Enode(e) as its children and to S';
        Create_Tree(A, d_A, S');
      else   // Enode(e) doesn't occupy A
        insert either Enode(e1) or Enode(e2) that is
            possible to define the shorest sequence;
      end if
    else   // Enode(e)'s shadow doesn't cover A
      insert either Enode(e1) or Enode(e2) that is
          not empty to Enode(e) as its child and to S';
    end if
    S ← S';
  end while
end

Procedure Create_Tree(V, d_V, S)
begin
  create Tree(V, d_V);
  forall edges e opposite to V do
    insert Enode(e) to Tree(V, d_V) as its child
        and to S;
  end for
end
```

Figure 6. Pseudo-code for the ECH algorithm

We obtained the graph data needed for the evaluation. First, 1000 pairs of vertices from the models in Figure 7(b) were randomly selected. The shortest paths and their lengths for these pairs are calculated by the ECH algorithm and by our approximation algorithm with different γ and m. All our implemention was conducted on a graphics workstation (CPU: MIPS R10000, 175 MHz).

Figure 8(a) shows a comparison of the approximation accuracy for the model in Figure 7(b). The vertical axis denotes the percentage of the average length of the shortest path of 1000 pairs to that by the ECH algorithm. Figure 8(b) shows a comparison of the computation time for the model in Figure 7(b), the vertical axis denotes the average computation time for the shortest paths. The horizontal axis in each of Figures 7(a)-(b) denotes the value corresponding to γ. Instead of using γ, which is hard to intuitively understand, we used the number of SPs per edge which was calculated from γ.

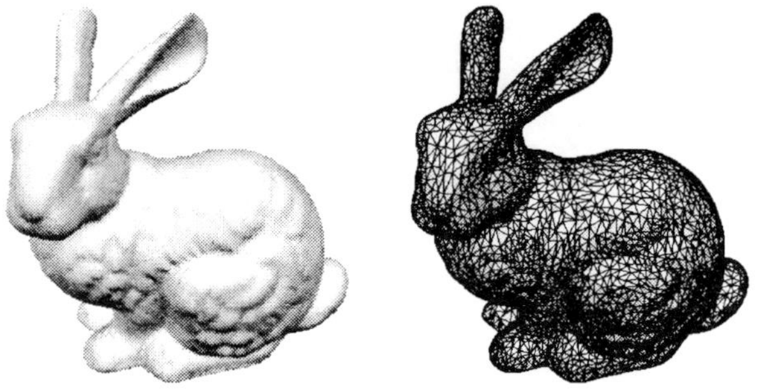

(a) (b)

Figure 7. The original "bunny" model (35,947 vertices, 69,451 faces) and its simplified model (5,047 vertices, 9,999 faces)

It can be seen from Figure 8(a) that the average lengths of the approximate shortest paths for all γ and m are within a 0.4% approximation accuracy from that of the exact shortest path. The length of a path with smaller γ approaches that of the exact path, m giving some contribution to improving the approximation accuracy. It can also be seen from Figure 8(b) that the computation time with our approximation algorithm is much less than that by the ECH algorithm (170.245 Sec.). The computation time increases with smaller γ or larger m. In particular, γ has a larger effect on the computation time than m. The computation time of a path with larger γ increases in the order of $O(n \log n)$.

Figure 8(c) shows a graph that represents the relationship between the length of the shortest path calculated by our approximation algorithm and the computation time. The horizontal axis denotes the path length, and the vertical axis denotes the computation time. To draw this graph, we classified the results into five groups according to the path length. The length interval for each group is 0.2, and we plotted the weighted average length of each group. The graph represents the characteristics of our algorithm very well. The computation time depends on the length of a path. The computation time for paths with different lengths varies in the order of $O(k \log k)$ (k is the number of edges passed by the shortest path) because our algorithm repeats the Dijkstra's algorithm for a discrete graph generated from a path. Figure 8(d) shows a graph of the relationship between the number of iterations and m. We can see from this figure that larger m decreases the number of iterations.

It can be seen from all the information in Figure 8 that our approximation algorithm can calculate the shortest path within a 0.4% accuracy roughly 100-1000 times faster than the ECH algorithm. While both γ and m contribute to the approximation accuracy to a certain extent, it is not always

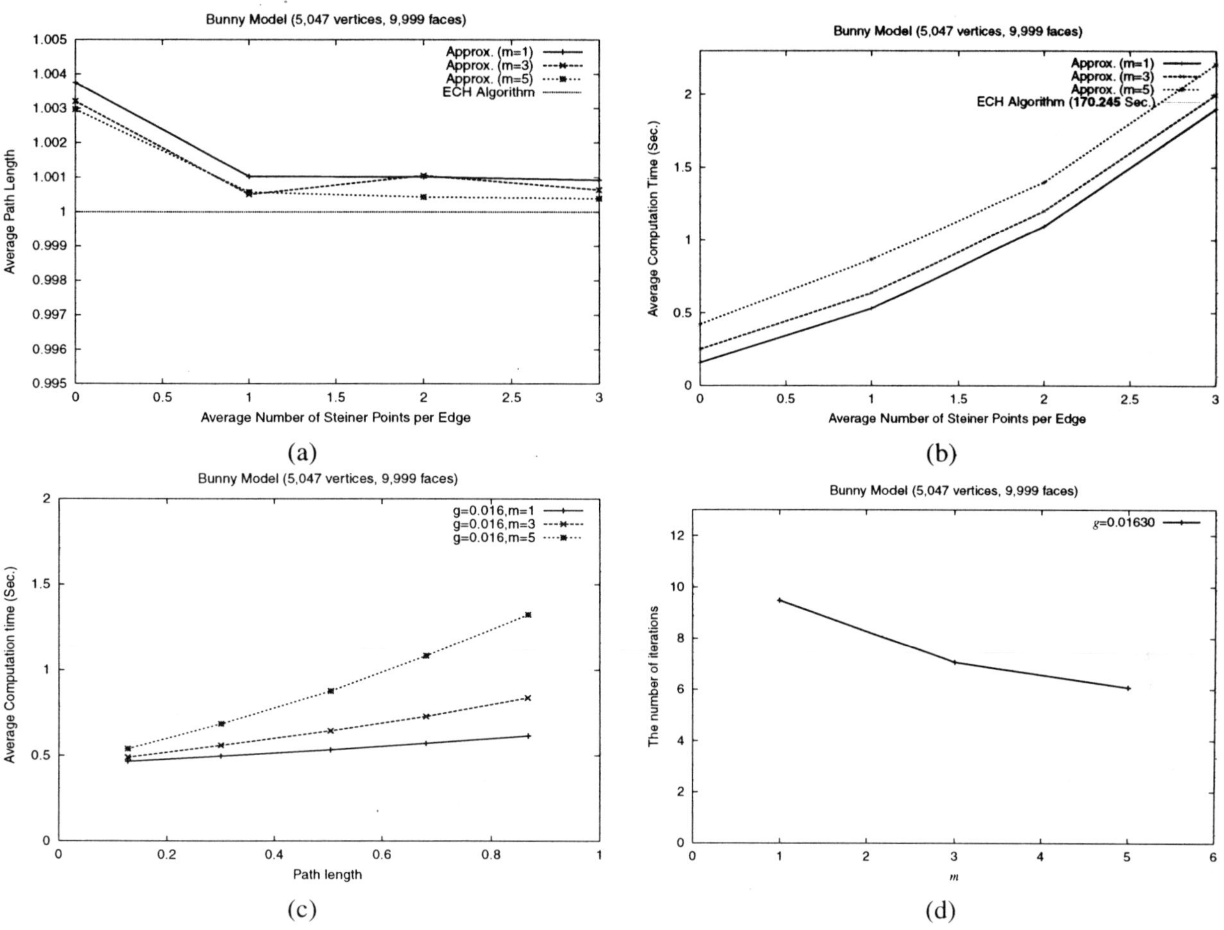

Figure 8. (a)-(b) Relationship between the path accuracy and the computation time (c) Relationship between the path distance and the computation time (d) Relationship between the number of iterations and m

the case that smaller γ or larger m gives rise to the efficiency of our algorithm because the computation time is also increased.

Furthermore, we conducted the same experiments for more simplified models in which the number of faces were about 1,000 and 5,000 respectively. Almost same results as those already described were obtained, so they are not further reported in this paper.

4.3. Numerical Accuracy

We now present experimental data about measurement of the perimeter length of regular n-gonal prisms. This data represents the superiority of our approximation algorithm over the ECH algorithm in terms of its numerical accuracy. Figure 9(a) shows a development of a regular n-gonal prism, and Figures 9(b)-(d) show regular n-gonal prisms in which $n = 10, 100$ and 1000, respectively. Perimeter length d of a regular n-gonal prism is $d = 2n \sin \frac{\pi}{n}$. In this experiment, we compare an analytical solution for d to solutions by the ECH algorithm and by our approximation algorithm.

Figure 10 shows the result of d calculated by the analytical method and by these algorithms for some sampled values of n from 0 to 15,000. You can see that the solutions by our algorithm are almost the same as those by the analytical method, while the solution by the ECH algorithm deviates greatly from it when n exceeds 3,000. This deviation originates from the method for calculating a path and its length by the ECH algorithm, which uses the 3D rotation computation and accumulates numerical errors. Our approximation algorithm merely uses the addition of the length of edges in discrete graphs. Consequently, our approximation algorithm is demonstrably stable in terms of its numerical accuracy.

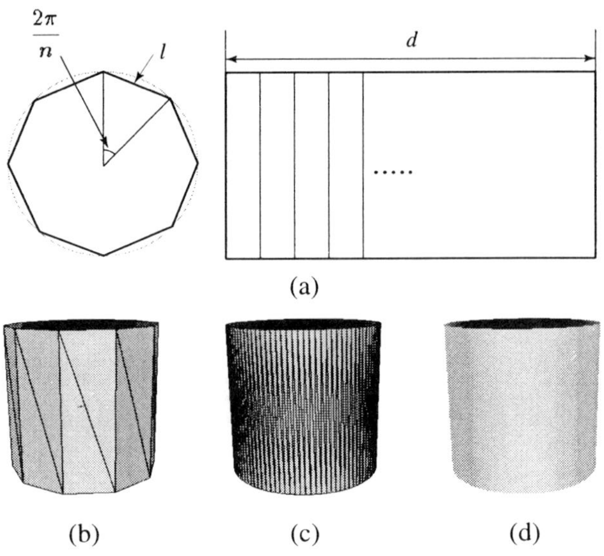

Figure 9. Regular n-gonal prisms for evaluating the numerical accuracy

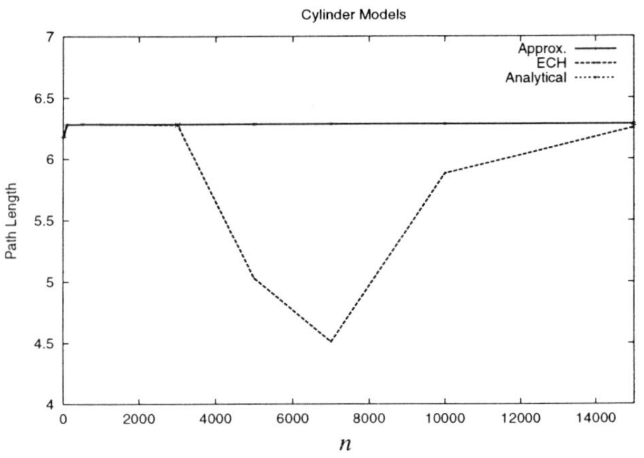

Figure 10. Comparison of numerical errors by analytical solution, our approximation algorithm and the ECH algorithm

5. Applications

We applied our approximation algorithm to two applications, one being as a tool for specifying the local region on a polyhedral surface, and the other for interactively measuring the geodesic distance on a polyhedral surface.

5.1. Region Specification on a Polyhedral Surface

Operations which specify a region of dense polyhedral surfaces, especially those generated from range images, are needed for modeling, rendering and animation. Such examples are the addition of attributes such as texture [14], conversion to a parametric surface [9], local deformation of a polyhedral surface [12, 10, 13] and 3D morphing [7, 8]. We discuss here about the use of our approximation algorithm as a tool for interactively specifying the boundaries of local regions. It is desirable for the boundaries of these regions to be smooth poly lines. If a boundary is distorted, it will negatively effect the quality of the various results in such applications. Moreover, many trial-and-error operations are often forced to users. Therefore, the algorithms used in these operations should be sufficiently fast and efficient for such interactivity.

The application of our approximate shortest path algorithm to boundary specification first involves initial discrete graph G^0 being created. Efficient calculations are possible by preserving G^0 all the time during the input and modification of a boundary. Although our approximation algorithm needs only two vertices to define a path, it is desirable that the path is composed of more than just vertices for fine and accurate specification of the boundaries. A path is generated by defining some control vertices and then by calculating shortest paths between neighboring control vertices. Figure 11 shows an example of specifying a "star" region with many boundary lines on the belly of a "horse" model (19,851 vertices, 39,698 faces). Spheres in this figure denote control vertices specified by the user. The input or modification of each control vertex leads to re-calculation of the two paths neighboring it, but this re-calculation is fast enough to make it possible to specify interactively.

5.2. Measurement of Geodesic Distance

Finding the shortest path to measure geodesic distance on a polyhedral surface is one of the most important techniques for the application of robotics, GIS, route finding, and so on. These applications often require the path or its length to be accurate. A slight deviation from the exact path might have a bad influence on the application result. On the other hand, some applications certainly exist in which approximate solutions are fully acceptable. As an example of a medical application, measurement of the magnitude of a tumor of the brain by using 3D models reconstructed from a sequence of medical images such as those by CT or MRI can be considered. Our approximation algorithm is thus also suitable for applications that need interactivity and not an exact solution.

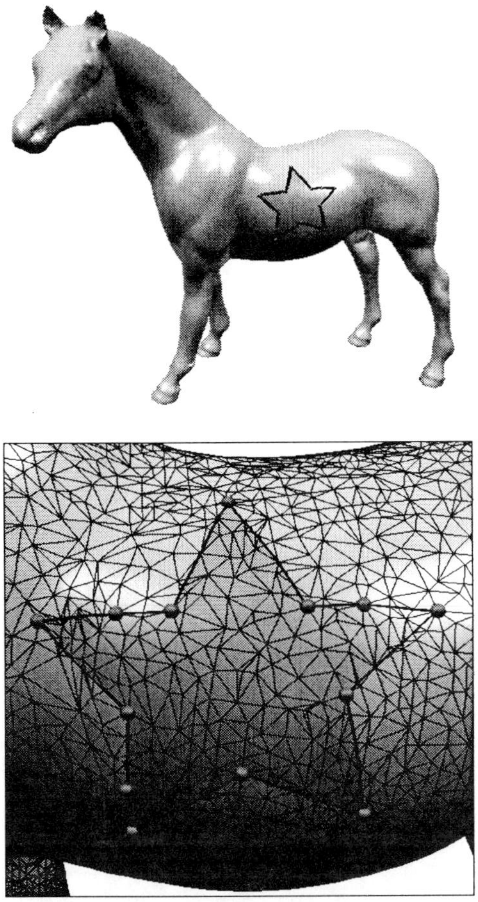

Figure 11. Generation of boundaries for specifying a "star" region

6. Conclusion and Future Work

This paper has proposed a new approach for calculating the approximate shortest path on a polyhedral surface based on selective refinement of the discrete graph. We have shown a comparison between our algorithm and the ECH algorithm in respect of the approximation accuracy, computation time and numerical accuracy. Two applications that suit the use of our approximation algorithm are discussed. The advantages of our approach are as follows:

- It is easy to implement. Only the routine of Dijkstra's algorithm and an additional graph structure are needed.
- It is fast. The computation time largely depends on a Dijkstra's algorithm. Our implementation executes in $O(n \log n)$-time. Thorup has recently proposed an $O(n)$-time method for processing Dijkstra's algorithm [20]. We need more consideration about this method.
- It provides high approximation accuracy. In our examples, an approximation accuracy within 0.4% was established.
- It is numerically robust. Our approximation algorithm merely uses the addition of the length of edges in discrete graphs.
- Although we have not described in this paper, it is also possible to calculate a weighted shortest path.

On the other hand, our algorithm has some disadvantages which we are working on to improve for the future work.

- Our approximation algorithm results in the shortest path of a *regular* polyhedral surfaces (for example, polygons arranged as a grid) being of low approximation accuracy. Such an example requires smaller γ to be set.
- The computation time depends on the length of a path, consequently, our algorithm cannot give a response with a constant time for a polyhedral surface. This may be an obstacle to some applications.
- The computation time in our algorithm largely depends on the time that searches a shortest path on a discrete graph G^0. To reduce the computation time, the use of more efficient search method alternative to Dijkstra's algorithm should be taken into consideration. For example, A* algorithm [18], one of the best-first search using the heuristic function, might be applicable.

Acknowledgement

Part of this research was supported by Elysium Co. Ltd. The bunny model was from Stanford University Computer Graphics Laboratory, and the horse model was from Cyberware Inc.

References

[1] A. V. Aho, J. E. Hopcroft, and J. D. Ullman. *Data structures and algorithms*. Addison-Wesley, 1983.

[2] J. Chen and Y. Han. Shortest paths on a polyhedron. In *Proc. 6th ACM Symp. on Computational Geometry*, pages 360–369, 1990.

[3] B. Curless and M. Levoy. A volumetric method for building complex models from range images. In *Computer Graphics (Proc. SIGGRAPH 96)*, pages 303–312, 1996.

[4] M. Garland and P. S. Heckbert. Surface simplification using quadric error metrics. In *Computer Graphics (Proc. SIGGRAPH 97)*, pages 209–216, 1997.

[5] L. J. Guibas and J. Hershberger. Optimal shortest path queries in a simple polygon. *J. Computer and System Sciences*, 39(2):126–152, 1989.

[6] S. Har-Peled. Approximate shortest paths and geodesic diameters on convex polytopes in three dimensions. In *Proc. 13th ACM Symp. on Computational Geometry*, pages 359–366, 1997.

[7] T. Kanai, H. Suzuki, and F. Kimura. Metamorphosis of arbitrary triangular meshes with user-specified correspondence. *IEEE Computer Graphics and Applications*, 2000, to appear.

[8] T. Kanai, H. Suzuki, J. Mitani, and F. Kimura. Interactive mesh fusion based on local 3D metamorphosis. In *Proc. Graphics Interface '99*, pages 148–156, 1999.

[9] V. Krishnamurthy and M. Levoy. Fitting smooth surfaces to dense polygon meshes. In *Computer Graphics (Proc. SIGGRAPH 96)*, pages 313–324, 1996.

[10] S. Kuriyama and T. Kaneko. Discrete parametrization for deforming arbitrary meshes. In *Proc. Graphics Interface '99*, pages 132–139, 1999.

[11] M. A. Lanthier, A. Maheshwari, and J.-R. Sack. Approximating weighted shortest paths on polyhedral surfaces. In *Proc. 13th ACM Symp. on Computational Geometry*, pages 274–283, 1997.

[12] A. W. F. Lee, W. Sweldens, P. Schröder, L. Cowsar, and D. Dobkin. MAPS: Multiresolution adaptive parameterization of surfaces. In *Computer Graphics (Proc. SIGGRAPH 98)*, pages 95–104, 1998.

[13] S. Lee. Interactive multiresolution editing of arbitrary meshes. *Computer Graphics Forum (Eurographics 99)*, 18(3):73–82, 1999.

[14] J. Maillot, H. Yahia, and A. Verroust. Interactive texture mapping. In *Computer Graphics (Proc. SIGGRAPH 93)*, pages 27–34, 1993.

[15] C. S. Mata and J. S. B. Mitchell. A new algorithm for computing shortest paths in weighted planar subdivisions. In *Proc. 13th ACM Symp. on Computational Geometry*, pages 265–273, 1997.

[16] J. S. B. Mitchell. Geometric shortest paths and network optimization. In J. R. Sack and J. Urrutia, editors, *The Handbook of Computational Geometry*. Elsevier Science, 1998.

[17] J. S. B. Mitchell, D. M. Mount, and C. H. Papadimitriou. The discrete geodesic problem. *SIAM J. Computing*, 16(4):647–668, 1987.

[18] E. Rich. *Artificial Intelligence*. McGraw-Hill, 1983.

[19] M. Sharir and A. Schorr. On shortest paths in polyhedral spaces. *SIAM J. Computing*, 15:193–215, 1986.

[20] M. Thorup. Undirected single source shortest paths in linear time. In *Proc. 38th Symp. on Foundations of Computer Sciences*, pages 12–21, 1997.

[21] K. R. Varadarajan and P. K. Agarwal. Approximating shortest paths on an nonconvex polyhedron. In *Proc. 38th Annual Symp. on Foundations of Computer Science*, pages 182–193, 1997.

Generating Fair Meshes with G^1 Boundary Conditions

Robert Schneider
Max-Planck Institute for Computer Sciences
Computer Graphics Group
Im Stadtwald, 66123 Saarbrücken, Germany
schneider@mpi-sb.mpg.de

Leif Kobbelt
Max-Planck Institute for Computer Sciences
Computer Graphics Group
Im Stadtwald, 66123 Saarbrücken, Germany
kobbelt@mpi-sb.mpg.de

Abstract

In this paper we present a new algorithm to create fair discrete surfaces satisfying prescribed G^1 boundary constraints. All surfaces are built by discretizing a partial differential equation based on pure geometric intrinsics. The construction scheme is designed to produce meshes that are partitioned into regular domains. Using this knowledge in advance we can develop a fast iterative algorithm resulting in surfaces of high aesthetic quality that have no local mean curvature extrema in the interior.

1. Introduction

A common method to guarantee excellent surface fairness is to minimize fairness functionals based on geometric invariants. One of the best known functionals in that category are

$$\int_A \kappa_1^2 + \kappa_2^2 \, dA,$$

punishing large curvature values and

$$\int_A \left(\frac{d\kappa_1}{d\hat{e}_1}\right)^2 + \left(\frac{d\kappa_2}{d\hat{e}_2}\right)^2 dA,$$

punishing the variation of the curvature. Here κ_1 and κ_2 are the principal curvatures and $\hat{e}_1$ and $\hat{e}_2$ are the corresponding principal curvature directions. A minimization process based on such fairing functionals - this is especially true for the latter - leads to surfaces of extraordinary quality, but due to the demanding construction process, the required computation time can be enormous [17].

A popular technique to simplify this approach is to give up the parameter independence and approximate the geometric invariants with higher order derivatives. For some important fairness functionals this results in algorithms that enable the construction of a solution by solving a linear system [9]. A representant of this category is the thin plate energy

$$\int\int f_{xx}^2 + 2f_{xy}^2 + f_{yy}^2 \, dxdy, \tag{1}$$

which can be used to create surfaces satisfying C^1 boundary conditions.

Instead of minimizing a functional, another approach first applies variational calculus to transform the minimization problem into the problem of solving a differential equation with constraints. For the functional (1) the optimal surfaces can be characterized by the partial differential equation (PDE)

$$\Delta^2 f = 0,$$

where Δ is the Laplacian operator. A solution can then be constructed with standard methods such as finite differences [2], finite elements [21] or by exploiting the knowledge that the solution can be expressed in closed form using a Green's function [1]. The PDE approach was also the starting point for operators which enable the construction of fair meshes fast enough to be used in interactive mesh modeling [10].

The parameter dependent schemes enable fast construction algorithms, but the resulting surfaces usually do not satisfy the same high aesthetic requirements as those created with geometric invariants since their quality highly depends on the underlying parameterization.

Instead of using variational calculus, the PDE approach can also be seen as a reasonable approach to the fairing problem in its own right, which is especially important for fairing based on geometric invariants. This means instead of searching for intrinsic energy functionals that lead to handy PDE's, it seems promising to search directly for simple intrinsic PDE's producing fair solutions. In our case this idea leads to the question which PDE based on geometric invariants seems suited for the creation of fair surfaces satisfying G^1 boundary constraints. We will discuss this question in section 2. Our algorithm to construct the surfaces is based on discrete data. This has been proven to be especially well suited for the construction of nonlinear splines

[13]. The idea behind the construction process is to design an algorithm that creates meshes with subdivision connectivity composed of regular patches. All patch edges besides those that lie on the boundary are geodesics and all vertices in the interior of each patch have a local 1-neighborhood with hexagonal shape. Exploiting this knowledge allows us to speed up the construction algorithm considerably. We will explain this technique in the sections 3 and 4. In section 5 we show how to calculate the regular discrete solution using an iterative scheme. Section 6 will give some implementation details and finally in section 7 we present some examples demonstrating the capability of our algorithms.

2. Problem definition

We first have to determine which PDE's seem suited to solve our fairing problem. Let us summarize what properties have to be satisfied by the PDE:

- it should be based on geometric intrinsics
- it has to be as simple as possible
- it has to allow us G^1 boundary constraints
- the resulting surface has to be fair

In many cases it is worthwhile to analyze the univariate case first, when searching for a solution of a multivariate problem. The analogous formulation of the problem in the plane is simple, because the G^1 boundary condition are especially easy to formulate. Given 2 points and 2 unit vectors, find a curve segment that interpolates the G^1-Hermite data. A differential equation that satisfies all properties in the planar case is not hard to find

$$\frac{\partial^2 \kappa}{\partial s^2} = 0, \qquad (2)$$

where κ constitutes the signed planar curvature and s the arc length. The solutions are clothoids, circles and lines. Because of the nice property that the extremal curvature values occur at the border and no local curvature extrema can appear in the interior, the curve is guaranteed to be fair [16].

At this point the question arises if this formulation can be extended to describe a solution to our original bivariate problem. An analogon to the second derivative operator is the Laplace-Beltrami operator Δ_B, which extends the planar Laplacian to a smooth surface [19]. In the planar case we know that the curvature distribution defines a curve uniquely up to a rigid motion. For surfaces Bonnet's uniqueness problem [3] suggests that the mean curvature H is a curvature measure that comes next to this property. This in mind, a candidate for such a PDE seems

$$\Delta_B H = 0. \qquad (3)$$

This PDE is simple and it is based on geometric intrinsics, but does it satisfy the two remaining properties?

For a regular surface one can always find a local conformal parameterization $\phi(u,v) \to \mathbb{R}^3$ [5], where conformal means local isometry up to a scalar factor, i. e. there is a $\lambda(u,v)$ such that $\lambda(u,v) =< \phi_u, \phi_u >=< \phi_v, \phi_v >$ and $< \phi_u, \phi_v >= 0$. With respect to such coordinates the Laplace-Beltrami operator can be expressed as

$$\Delta_B = \frac{1}{\lambda(u,v)^2}\left(\frac{\partial^2}{\partial u^2} + \frac{\partial^2}{\partial v^2}\right) = \frac{1}{\lambda(u,v)^2}\Delta, \qquad (4)$$

so beside of a nonzero factor it simplifies to the planar Laplace Operator Δ. As is shown in [5] in conformal coordinates we further get $\Delta\phi = 2\lambda(u,v)^2 H\vec{n}$, where $\vec{n}$ is the unit normal vector of the surface. This means we can interpret (3) as a nonlinear extension of

$$\Delta^2 \phi = 0.$$

As mentioned before, this equation can be used to construct surfaces with C^1 boundary constraints, so it seems promising that (3) is a candidate that allows us G^1 boundary constraints.

From the mean value property of the Laplacian follows that the extremal values of $\Delta\phi = 0$ occur on the boundary, so because of (4) this means that the extremal mean curvature values will be reached at the border and that there are no local extrema in the interior. This extends the univariate property of (2) and hence the resulting surfaces satisfy an important fairing concept [4].
This leads us to the following problem formulation:

Problem 1 *Find a surface that satisfies the G^1 boundary constraints and whose interior solves the equation*

$$\Delta_B H = 0.$$

3. Discretization background

In this section we define what is meant by a discrete solution of Problem 1. The definition leads to meshes consisting of regular patches. Following an idea presented in [20] we will exploit this structure by assigning a local parameterization to each mesh vertex which allows us to create accurate approximations of the discrete solution that can be calculated very efficiently. To simplify the notations, we will assume that the border consists of one single closed curve, but the algorithm extends to a border consisting of a number of curves (Fig. 8).

3.1. Notation and definitions

Let $p(t)$ be a parameterization of our border p, let $P = \{P_1, \ldots, P_N\}$ be the set of the N vertices characterizing the

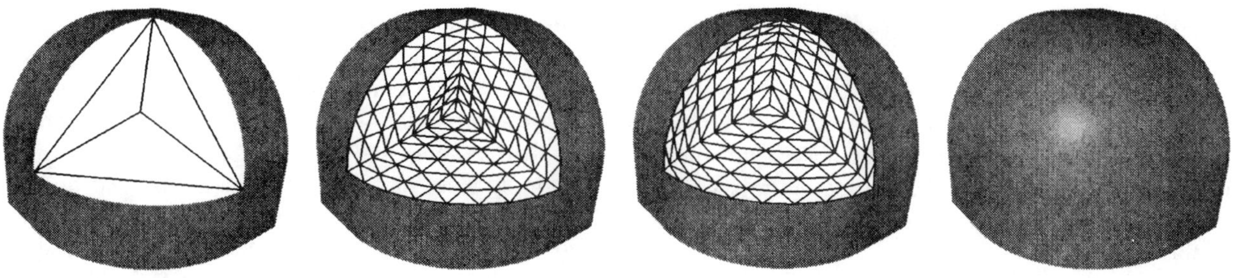

(a) (b) (c) (d)

Figure 1. Suitcase corner with 3 sides. a) A mesh $\bar{P}$ with one extraordinary vertex. b) The initial mesh M^0 is constructed by using a linear subdivision scheme on $\bar{P}$ taking into account that the border vertices Q have to be interpolated. c) Wireframe of the discrete solution after subdividing 3 times. d) Shaded solution after subdividing 5 times.

border of an N-sided hole and let $\bar{P}$ be a simple mesh that fills the hole defined by P. This $\bar{P}$ determines the mesh topology of the discrete solution (Fig. 1).

The set P divides the border curve p into N segments. Let us approximate each of the N segments by a polygon with r edges, producing a polygon Q with vertices $Q = \{Q_1, \ldots, Q_{rN}\}$ and the property $P \subset Q \subset p$. Let M be a mesh with vertices M_i whose topology is equivalent to a r-fold uniformly subdivided $\bar{P}$ interpolating the border points Q (Fig. 1 b) c)). Because of the subdivision connectivity structure we can partition the points M_i into 4 classes:

- The border vertices Q are also called interpolation vertices. Their position is fixed.
- The vertices of $\bar{P}$ that are not in P are called extraordinary vertices.
- The vertices that can be assigned to an interior edge of $\bar{P}$ are called edge vertices.
- The remaining vertices that can be assigned to the interior of a triangle face of $\bar{P}$ are called inner vertices.

For simplicity, we will denote all M_i that are no border vertices as free vertices. For vertices of valence v let $M_{i,l}, l = 1..v$ denote their adjacent neighbors. Because of the subdivision connectivity structure, edge and inner vertices M_i always have valence 6. For edge vertices we define the convention that the adjacent vertices are arranged such that $M_{i,1}$ and $M_{i,4}$ are aligned to the same edge. Let H_i resp. $H_{i,l}$ be the discrete mean curvature at M_i resp. $M_{i,l}$ and let $\vec{n}_i$ be a discrete unit normal vector at M_i. In section 4.4 and 4.8 we explain how we construct this intrinsic values.

Let us further define a mean value operator g_p for all free vertices. For edge and inner vertices this operator is

$$g_p(M_i) = \begin{cases} 1/6 \sum_{l=1}^{6} M_{i,l}, & \text{if } M_i \text{ is an inner vertex,} \\ 1/2\,(M_{i,1} + M_{i,4}), & \text{if } M_i \text{ is an edge vertex,} \end{cases}$$

for extraordinary vertices we will give its definition in section 4.7.

Finally we can define what is meant by a discrete solution of our problem

Definition 1 *A mesh M is called a regular discrete solution of problem 1, if the following conditions are satisfied:*

1. *At all interpolation points of M the G^1 boundary constraints are satisfied.*
2. *The vertices M_i of M should be regularly distributed. For all free vertices, there should be a $t_i \in \mathbb{R}$ such that $M_i = g_p(M_i) + t_i\vec{n}_i$.*
3. *At every free vertex M_i we have $\Delta_B(H_i) = 0$.*

A geodesic line has the property that its main normal vector is parallel to the surface normal vector at every point of the line [5], so we note that condition 2 implies that the edge vertices will form a geodesic net in the interior of the mesh M. The mean value operator g_p at the extraordinary vertices strongly influences the geodesic net structure. Since the inner vertices are distributed as regularly as possible, this means condition 2 guarantees that the solution is partitioned into regular patches (Fig. 1 c)), which is a property that gives us much information about the mesh M in advance. In [20] meshes consisting of regular patches were used to efficiently construct surfaces with piecewise linear mean curvature distribution, we will modify this idea to create a very efficient construction algorithm for a regular discrete solution.

3.2. Choosing the interpolation vertices

Because of condition 2 in definition 1, the discrete solution will be a collection of regular mesh patches. The interpolation vertices Q_i have to reflect that, otherwise the mesh structure will be distorted near the border and this would decrease the quality of our local parameterizations. Hence, a reasonable distribution of the interpolation vertices should satisfy

$$||Q_i - Q_{i-1}|| \approx ||Q_{i+1} - Q_i||$$

for all $Q_i \notin P$. For piecewise smooth boundary curves, uniform sampling of each of the N smooth border sides using an approximated arc length parameterization will satisfy that condition.

4. Discretization of the geometric invariants and the Laplace-Beltrami operator

For inner, edge and extraordinary vertices, our discretization technique is based on the well known idea to construct a local quadratic least square approximation of the discrete data and estimate all needed values from this approximation.

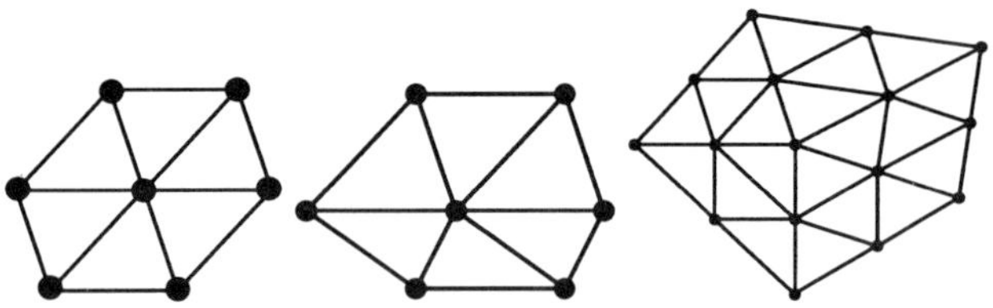

Figure 2. The parameter domains for inner, edge and extraordinary vertices.

4.1. General parameterization strategy

Instead of recalculating a local parameterization during each step in the iteration, we exploit the fact that our meshes are well structured to determine a local parameterization in advance, thus accelerating our algorithms considerably. The decision, which local parameterization should be assigned to a vertex Q_i was influenced by three mayor constraints: simplicity, regularity and uniqueness of the quadratic approximation. This constraints lead us to the following local parameterization classes. For inner vertices, the parameter domain is a regular hexagon and for edge vertices it is composed of two regular hexagon halves. Calculating the matrices needed for a least square approximation in both cases, it is easy to see, that this parameterizations always lead to a unique least square approximation. Only at the interpolation vertices it is not always sufficient to use the 1-neighborhood. This is obvious, if the valence v of the vertex is 3 or 4, but even if the valence is higher, the least square approximation can fail in the 1-neighborhood. This problem is similar to the problem that occurs when local quadratic least square approximation is used for arbitrary mesh fairing. In [25] it was proposed to determine the condition number of the matrix needed for the least square approximation and to reduce the number of the basis functions if the problem is ill conditioned. Instead of that approach, we can exploit the fact that the extraordinary vertices are located in a piecewise regular neighborhood. Using the complete 2-neighborhood consisting of regular sectors for vertices with small valence or parts of this neighborhood for vertices of high valence we can make the least square approximation unique.

4.2. Construction of a local parameterization

We now have to define how to actually calculate the local parameterizations for free vertices. In [9] Kobbelt presented an approach to approximate a local isometric parameterization by exploiting the subdivision connectivity structure of a subdivided mesh that minimizes the thin plate energy. In our case this algorithm is even better suited to construct good approximations for a local parameterization. The iteration process leads to meshes that are partitioned into regular regions making it much simpler to get good local approximations by parameter domain blending. The idea of the algorithm is to start with parameter domain seeds at those vertices, where the regular mesh structure is interrupted and to complete the local parameterizations at the remaining vertices using blending operators. In our case the parameter domain seeds have to be calculated at the extraordinary vertices and the vertices P at the border. At extraordinary vertices we can determine a local parameter domain seed by applying an exponential map on its 1-neighborhood. This operator maps a 1-neighborhood into the plane, such that the distances of the vertex to its neighbors and the ratio of the adjacent angles are preserved [25]. The 2-disc domain for extraordinary vertices as presented in Figure 2 then results by applying one linear subdivision step. At the vertices P we do not use the exponential map because the local neighborhood is open, but here we can exploit the fact that we know the tangent plane. We project the neighborhood into the tangent plane and scale the resulting vectors such that the vertex-neighbor distances are still preserved.

The remaining local parameter domains for the edge and inner vertices (as we will see in section 4.8 there is no need for a local parameterization at the remaining interpolation vertices) are then derived by linear blending the seed triangles, for a detailed description of this blending process see [9]. Instead of blending the triangle parameters described there, we got improved results when blending the edge lengths of the triangles.

4.3. Determining a least square approximation

Let us first consider the functional case. Given a local triangular parameterization domain and function values f_i at the vertices of the domain, we search for a quadratic function

$$f(x) = a_1x^2 + a_2y^2 + a_3xy + a_4x + a_5y + a_6$$

which minimizes the least square error at the domain vertices. This problem is well known and leads to a linear system $A\vec{a} = \vec{f}$, where $\vec{a}$ is the vector of the coefficients a_j and $\vec{f}$ is the vector of the function values f_i. If the domain has more than 6 vertices and the matrix A has full rank, this linear system is overdetermined and the best least square solution can be expressed as

$$\vec{a} = (A^tA)^{-1}A^tf. \tag{5}$$

In our case - we have 3-D point information instead of function values f_i - the least square approximation is constructed by applying the functional scheme to the x, y and z values of the vertices simultaneously.

4.4. Discretization of the geometric invariants

Using the coefficients of the least square approximation, we can determine the normal vector and the first and second fundamental form at the center vertex of the parameter domain and can express the mean curvature H as [5]

$$H = \frac{1}{2}\frac{eG - 2fF + gE}{EG - F^2}, \tag{6}$$

where E, F, G are the coefficients of the first fundamental form and e, f, g are those of the second fundamental form.

When we are only interested in intrinsic values, we can use the fact that an affine mapping of the parameter domain does only change the parameterization of our quadratic approximation, but no geometric invariants. This means to merely determine the normal and the mean curvature for inner vertices, it is sufficient to use an equilateral hexagon as parameter domain, thus allowing us an especially simple calculation of those values for such vertices. This is another nice advantage of the local parameterization construction.

4.5. Discretization of the Laplace-Beltrami operator

The parameterization technique presented in [9] approximates a local isometric parameterization. In our case, where we have patches with especially nice regular structure, the approximation quality of such an approach improves considerably. Moreover we do not even need a local isometric parameterization. In our case it is already sufficient to have a local conformal parameterization, a wrong scaling of the local parameter domains will not influence the solution. This means we can interpret our local parameterizations as good approximations of local conformal parameterizations.

For an conformal parameterization the equation $\Delta_B H = 0$ is equivalent to $\Delta H = 0$, so in our case we can approximate Δ_B with the planar Laplacian operator Δ. Following the discretization technique of Δ presented in [9], we can approximate Δ_B using the local quadratic approximation resulting in

$$\Delta = 2(a_1 + a_2),$$

where the constant factor is of no interest in our case. Using expression (5) this means we can find coefficients λ_i and $\lambda_{i,l}, l = 1, \ldots, v$ such that

$$\Delta_B H_i = \lambda_i H_i + \sum_{l=1}^{v} \lambda_{i,l} H_{i,l}. \tag{7}$$

For the special case when the domain is an equilateral triangle and using the fact that multiplying with a nonzero factors does not influence the results we arrive at the simple umbrella operator

$$\Delta H_i = -H_i + \sum_{l=1}^{6} \frac{1}{6} H_{i,l}. \tag{8}$$

which was used in [10] for interactive mesh modeling.

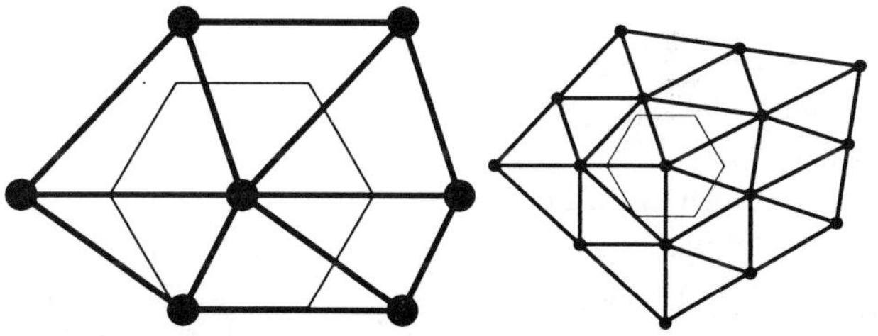

Figure 3. At edge and extraordinary vertices we can restrict the update step to the simple case where the parameter domain is an equilateral hexagon. The new parameter domain is chosen to be the maximum equilateral hexagon in the interior of the local 1-neighborhood of the center vertex.

4.6. Handling of edge and extraordinary vertices

As mentioned above, the discretization of the geometric invariants and the Laplace-Beltrami operator is especially comfortable when the parameter domain is an equilateral hexagon, but this special case has more nice properties. The equation $\Delta_B H_i = 0$ using (7) defines a sparse linear system in the unknowns H_i and such systems are usually optimally solved using iterative multigrid algorithms. Assuming the parameter domain at every vertex to be an equilateral hexagon would produce a matrix that is symmetric and positive definite and hence we would be able to use very simple iterative multigrid solver as for example based on Gauss-Seidel iteration. Gauss-Seidel is known to be a slow iterative solver when only applied at the finest level, but a fast and comfortable solver in combination with multigrid methods [10]. Unfortunately the matrix is no longer symmetric and positive definite when working with arbitrary parameter domains, hence it is no longer guaranteed to be always non singular and if the linear system is solvable it would require more involved iterative schemes like e. g. biconjugate gradient iteration.

We discovered that only the discretization at edge and extraordinary vertices affected the convergence of our multigrid solver. For such vertices we found a very nice and simple technique to avoid such problems. We first calculate the least square approximation using the mean curvature values H_i and $H_{i,l}, l = 1, \ldots, v$ of a vertex M_i and then sample this least square approximation at a regular hexagon that lies in the interior of the parameter domain (Fig. 3) resulting in the new values $\tilde{H}_{i,l}, l = 1, \ldots, 6$. An iteration step at edge and extraordinary vertices is then led back to the comfortable symmetric case (8) by using the new sampled $\tilde{H}_{i,l}$ instead of the original $H_{i,l}$. Since the $\tilde{H}_{i,l}$ lie on the quadratic least square approximation for the original parameter domain, a recalculation of the least square approximation using the new domain and the new neighbors will reproduce the same function, hence this sampling technique does not lead to false local informations. But an update step (8) with this new sampled points $\tilde{H}_{i,l}$ has again the simplicity and stability advantages of the equilateral case.
Note: It is important that the equilateral hexagon lies in the interior of the original domain, otherwise the sampling step is no longer an interpolation but an extrapolation and hence loses its stability.

4.7. Mean value operator for extraordinary vertices

The geodesic net defined by the edge vertices is determined by the position of the extraordinary vertices, so the mean value operator g_p at such vertices defines the partitioning of the discrete solution. The resampling technique presented in section 4.6 can also be applied to the local quadratic least square approximation of the 1-neighborhood of M_i resulting in new vertices $\tilde{M}_{i,l}$. We found it a reasonable strategy that the geodesic net of the discrete solution should have a shape that is similar to those of the simple mesh $\bar{P}$, so we defined the mean value operator here to be the center of the subsampled vertices $\tilde{M}_{i,l}$. Hence, for the final definition of the operator g_p we get

$$g_p(M_i) = \begin{cases} \frac{1}{6}\sum_{l=1}^{6} M_{i,l}, & M_i \text{ inner vertex,} \\ \frac{1}{2}(M_{i,1} + M_{i,4}), & M_i \text{ edge vertex,} \\ \frac{1}{6}\sum_{l=1}^{6} \tilde{M}_{i,l}, & M_i \text{ extraord. vertex.} \end{cases}$$

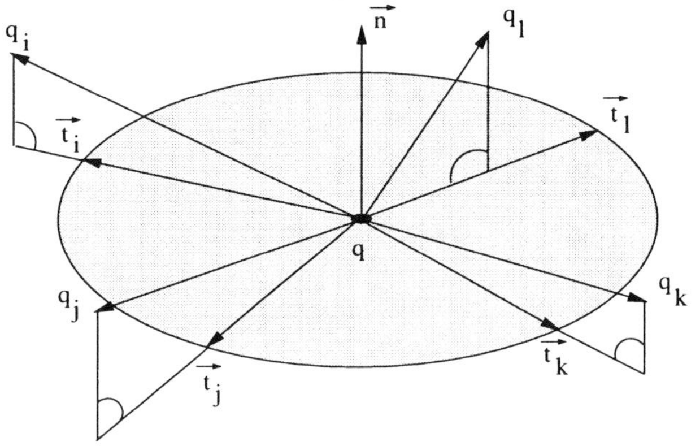

Figure 4. Projecting the neighborhood of q onto the plane defined by $\vec{n}$ and normalizing the results we get the normal curvature directions $\vec{t}_i$.

4.8. Discretization of the mean curvature at interpolation vertices

At the interpolation vertices we have a completely different situation. Here we only have one sided information of the mesh neighborhood, but therefore we know the tangent plane of the final surface. Since these vertices do not have to be updated during the iteration process, we only need the value of the discrete curvature.

Even if we would not only know the G^1 boundary conditions but the surrounding surface, we must not use outer points for the curvature estimation since the resulting surface does not have to be curvature continuous at the border. In [10] a rough approximation of the solution of $\Delta^2 f = 0$ with C^1 boundary constraints was solved by applying the umbrella operator for interior and border vertices, but this simple strategy is not suitable for high quality surfaces.

A curvature estimation algorithm that seems ideal for our needs was presented by Moreton and Sequin [17]. The idea behind this approach is to use the fact that normal curvature distribution can not be arbitrary, but is determined by the Euler-formula [5]. We describe how to discretize the mean curvature at a mesh vertex q, when the normal vector $\vec{n}$ at q is known.

Let $q_j, j = 1 \ldots m$ be the vertices adjacent to q and let $\vec{b}_x$ and $\vec{b}_y$ be an arbitrary orthonormal basis of the plane defined by the normal $\vec{n}$. To each vertex q_j we can assign a unit direction vector $\vec{t}_j$ by projecting q_j into the plane and scaling this projection to unit length (Fig. 4). For each vertex q_j we can now estimate a normal curvature $\tilde{\kappa}_j$ as the inverse of the circle radius defined by q, q_j and t_j.

Using the Euler formula, we can express the normal curvature κ_n for a direction $\vec{t}$ by the principal curvatures κ_1 and κ_2 and the principal curvature directions $\vec{e}_1$ and $\vec{e}_2$. Let t_x and t_y be the coordinates of t in the basis $\vec{b}_x, \vec{b}_y$ and let e_x and e_y be the coordinates of $\vec{e}_1$, then the normal curvature can be expressed as

$$\kappa_n = \begin{pmatrix} t_x \\ t_y \end{pmatrix}^t \cdot K \cdot \begin{pmatrix} t_x \\ t_y \end{pmatrix}$$

with

$$K = \begin{bmatrix} e_x & e_y \\ -e_y & e_x \end{bmatrix} \cdot \begin{bmatrix} \kappa_1 & 0 \\ 0 & \kappa_2 \end{bmatrix} \cdot \begin{bmatrix} e_x & e_y \\ -e_y & e_x \end{bmatrix}^{-1}.$$

The idea of Moreton and Sequin is to use the normal curvatures $\tilde{\kappa}_j$ to create a linear system and find estimates for the unknown curvature values by determining the least square solution. Let $t_{j,x}$ and $t_{j,y}$ denote the coordinates of $\vec{t}_j$, then we get

$$Ax = b$$

where

$$A = \begin{bmatrix} t_{0,x}^2 & t_{0,x}t_{0,y} & t_{0,y}^2 \\ t_{1,x}^2 & t_{1,x}t_{1,y} & t_{1,y}^2 \\ \cdot & \cdot & \cdot \\ t_{m,x}^2 & t_{m,x}t_{m,y} & t_{m,y}^2 \end{bmatrix}, \quad b = \begin{bmatrix} \tilde{\kappa}_0 \\ \tilde{\kappa}_1 \\ \cdot \\ \tilde{\kappa}_m \end{bmatrix}$$

and

$$x = \begin{pmatrix} x_0 \\ x_1 \\ x_2 \end{pmatrix} = \begin{pmatrix} e_x^2\kappa_1 + e_y^2\kappa_2 \\ 2e_xe_y(\kappa_1 - \kappa_2) \\ e_x^2\kappa_2 + e_y^2\kappa_1 \end{pmatrix}$$

Using the normal equations, the least square solution $\tilde{x}$ can be expressed as $\tilde{x} = (A^tA)^{-1}A^tb$. Since $H = \frac{1}{2}(\kappa_1 + \kappa_2) = \frac{1}{2}(x_0 + x_2)$ this means we can approximate the mean curvature as $\tilde{H} = \frac{1}{2}(\tilde{x}_0 + \tilde{x}_2)$.

5. Surface construction

We construct the surface using an iterative solver $M^k \rightarrow M^{k+1}$ starting with an initial mesh M^0. In each iteration step every free vertex M_i^k is moved to a new position M_i^{k+1}, while the interpolation vertices Q do not change their position. The equations determining the new position M_i^{k+1} are nonlinear unlike the equations usually occurring in iterative algorithms for parameter dependent functionals. The idea behind the construction of high quality surfaces based on nonlinear equations is to work with the original equations and linearize the construction process instead of simplifying the functionals itself [18, 6, 20]. This is often possible by keeping those values constant that will only have small changes during one single iteration step.

In our case we used the mean curvature linearization technique presented in [20] which is well suited when local parameterizations are available. In [20] this technique was used to enable a fast construction algorithm for meshes with subdivision connectivity having piecewise linear mean curvature distribution. As was mentioned there, this technique can be applied for two principle iteration approaches. The first approach linearizes the nonlinear equations locally, the second is based on the idea to use the available global information about the curvature distribution. We will now show how these ideas can be used to derive an iteration scheme for our fairing problem.

5.1. Direct approach

When updating the vertex M_i^k in an iteration step, the new position is determined by the two equations

$$M_i^{k+1} = g_p(M_i^k) + t_i\vec{n}_i^k \quad \text{and} \quad \Delta_B(H_i^{k+1}) = 0.$$

The first equation reduces the 3-variate to a univariate problem and the second equation is then used to determine the vertex position. Because of our discretization of the Laplace-Beltrami operator, the latter condition can be expressed as

$$\lambda_i H_i^{k+1} + \sum_{l=1}^{6} \lambda_{i,l} H_{i,l}^{k+1} = 0. \tag{9}$$

To linearize this nonlinear equation, we exploit the fact that the updated mesh vertices M_i^{k+1} will be near M_i^k in the $k+1$ iteration and hence the vertices and the mean curvature values will not change much. This in mind, we use $H_{i,l}^k$ instead of $H_{i,l}^{k+1}$ in equation (9) and simplify H_i^{k+1} by using the vertices M_i^k instead of M_i^{k+1} to calculate the coefficients E, F and G of the first fundamental form. So using (6) equation (9) leads to

$$\frac{1}{2}\frac{eG - 2fF + gE}{EG - F^2} = -\frac{1}{\lambda_i}\sum_{l=1}^{6} \lambda_{i,l} H_{i,l}^k$$

and we have to find a t_i such that this equation is satisfied. Applying (5) and replacing the normal vector $\vec{n}_i^{k+1}$ by $\vec{n}_i^k$ when calculating the coefficients of the second fundamental form e, f and g, this equation becomes linear in t_i and we can determine the new vertex M_i^{k+1} by solving this linear equation.

Since the vertices are updated along the surface normals, this approach can be interpreted as some kind of curvature flow, where the speed is determined by a local equation system. As was pointed out by Desbrun et. al. [6], flows depending only on local surface properties do not have to be numerically stable for large steps during the iteration process. To allow larger update steps, Desbrun et. al. used the backward Euler method to develop an algorithm called implicit integration for fairing of arbitrary meshes based on Laplacian and on mean curvature flow. The idea behind this approach is to use global surface information instead of considering the local neighborhood only. In [20] Schneider and Kobbelt presented an algorithm called indirect iteration, that efficiently exploits global surface properties for subdivision connectivity meshes, when fairing is based on linear curvature distribution. We show in the following that this idea can be extended to our fairing problem as well.

5.2. Indirect iteration

The central idea of the indirect iteration is to decouple the curvature information from the actual mesh geometry. In our case this means if we could iterate $\Delta_B(H_i^{k+1}) = 0$ without affecting the curvature information H_j^{k+1} along the interpolation points $Q_j^{k+1} = Q_j$, we would arrive at a solution of the following Dirichlet problem:
Find scalar values H_i^{k+1} such that $\Delta_B(H_i^{k+1}) = 0$ with given boundary values H_j^{k+1} at the interpolation vertices. So the analogon of the linear indirect approach presented in [20] can be formulated as follows:

- Determine the actual curvature values H_j^k at the interpolation vertices Q_j using the technique described in section 4.8.
- Determine mean curvature values $\tilde{H}_i^{k+1}$ for the free vertices by solving the Dirichlet problem $\Delta_B(\tilde{H}_i^{k+1}) = 0$ using the determined mean curvature values H_j^k at the border.
- Use the $\tilde{H}_i^{k+1}$ to update the free vertices M_i^k. In the update step the new position is then determined by the two equations $M_i^{k+1} = g_p(M_i^k) + t_i\vec{n}_i^k$ and $H_i^{k+1} = \tilde{H}_i^{k+1}$.

The nonlinear equation is linearized by analogous techniques as described in the direct approach, thus the new position can again be determined by solving a linear equation for t_i. As was mentioned in [20] it is important to restart at step 1 after step 3. An iteration scheme that only iterates step 3 would not converge in general, since $\tilde{H}_i^{k+1}$ are only estimates of the according values of the final solution.

In this formulation the indirect iteration is very expensive, because in every iteration step we have to solve a linear system for the next mean curvature estimates, but this can be avoided. We know that the actual discrete mean curvature values in the interior are approximations for the solution of the linear system occurring in step 2 and the quality of the approximation improves rapidly during the iteration process. Further it is not necessary to solve the linear system exactly, it is sufficient to approximate the solution. This means we can change the first and second step:

- Determine the actual curvature values H_i^k at the interpolation vertices *and* at all free vertices.
- Use the H_i^k at the free vertices as initial values and determine new mean curvature values $\tilde{H}_i^{k+1}$ by applying s iteration steps of an iterative solver for the Dirichlet problem.

If the actual mesh M^k is already a good approximation of the discrete solution, it is not necessary to choose a large s.

6. Details

Because of the subdivision connectivity of the Mesh M, all examples were created using a multigrid iteration, a technique that has proven to be excellent when hierarchic structures are available [8, 9]. Here a solution is first created on a coarse level and then serves as starting point for the iteration scheme in the next hierarchy level. Applying this strategy across several levels of the hierarchy, the convergence speed of the construction process can increase dramatically. Since the high frequency error is smoothed because of the hierachical structure, we can use the simple Gauss-Seidel algorithm when iterating step 2 of the indirect iteration scheme.

The mesh topology of the constructed discrete solution is determined by $\bar{P}$. For the N-sided hole filling problem such a triangular mesh can often be generated automatically, for holes with a more complex boundary or for problems with more than one boundary curve, such a triangular mesh has to be provided by the user.

The initial mesh M^0 is constructed using a linear subdivision scheme, this means we apply linear subdivision on the mesh $\bar{P}$ but take into account that M^0 has to interpolate the vertices Q (Fig. 1 b)). This is not only fast and convenient, but it guarantees that the vertices M_i^0 are regularly distributed. Since this approach produces peaks at the extraordinary vertices, it would not be a reasonable idea to start iterating on the finest level, but it works fine in the multigrid scheme since we only have to subdivide to the coarsest level.

As mentioned earlier the direct approach is not suited if large update steps can occur, therefore our implementation is based on the indirect iteration scheme. Since large update steps are only necessary in coarse hierarchy levels, our iteration number s is chosen small for fine hierarchy levels and large for the coarsest level. Instead of prescribing s,

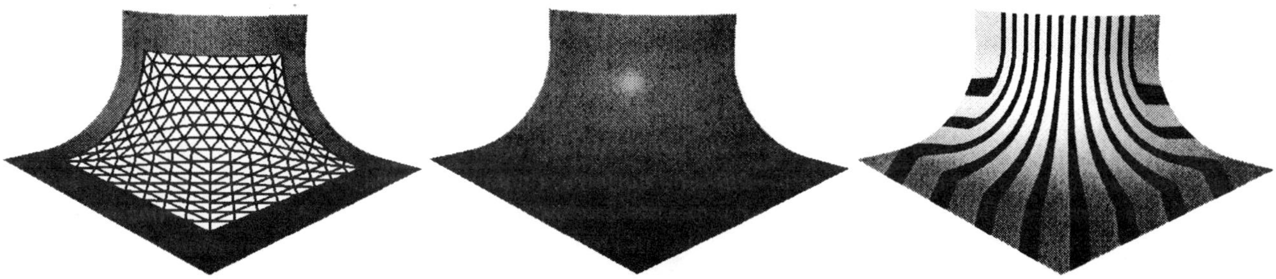

(a) (b) (c)

Figure 5. The classical house corner problem. a) A regular discrete solution of a 3 times subdivided mesh $\bar{\mathrm{P}}$ with 1 extraordinary vertex and 5 sides. b) Shaded discrete solution of a 5 times subdivided mesh. c) Lines of reflection of a 5 times subdivided mesh indicating that the solution satisfies the G^1 boundary conditions and is at least G^2 in the interior.

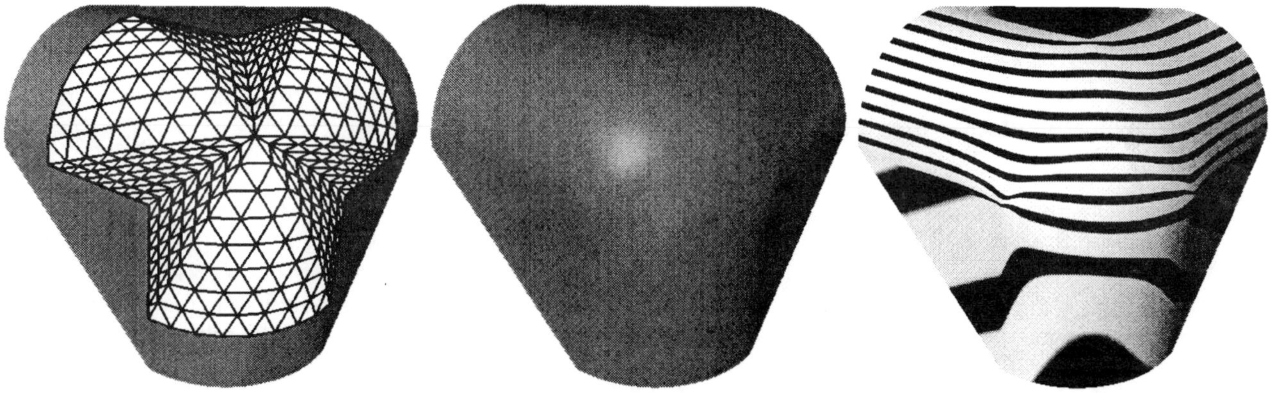

(a) (b) (c)

Figure 6. Suitcase corner with 9 sides and 3 sharp edges at the border. a) Discrete solution of a 3 times subdivided mesh. b) and c) Because of the sharp edges we subdivided 6 times. Again the lines of reflection indicate the G^1 continuity at the border and the higher continuity in the interior.

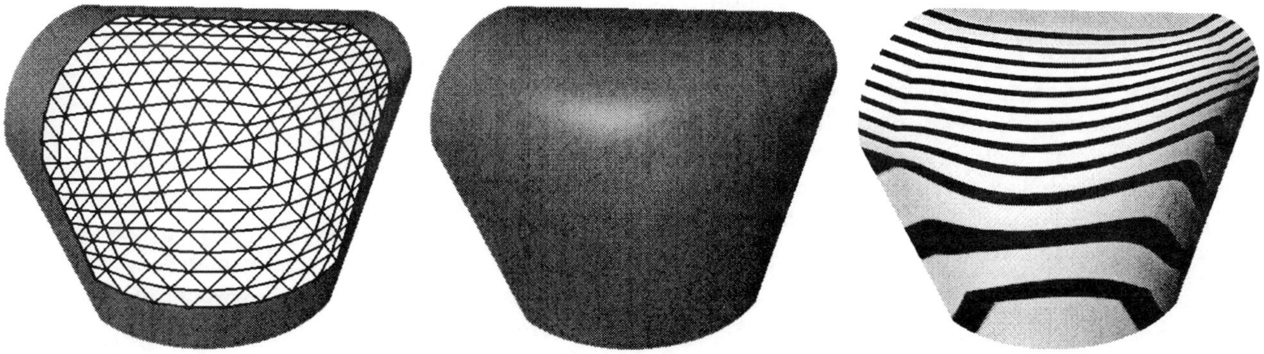

(a) (b) (c)

Figure 7. Suitcase corner with 6 sides where 3 sides are defined by cylinders with different sizes. a) Discrete solution of a 3 times subdivided mesh. b) and c) Discrete solution of a 5 times subdivided mesh.

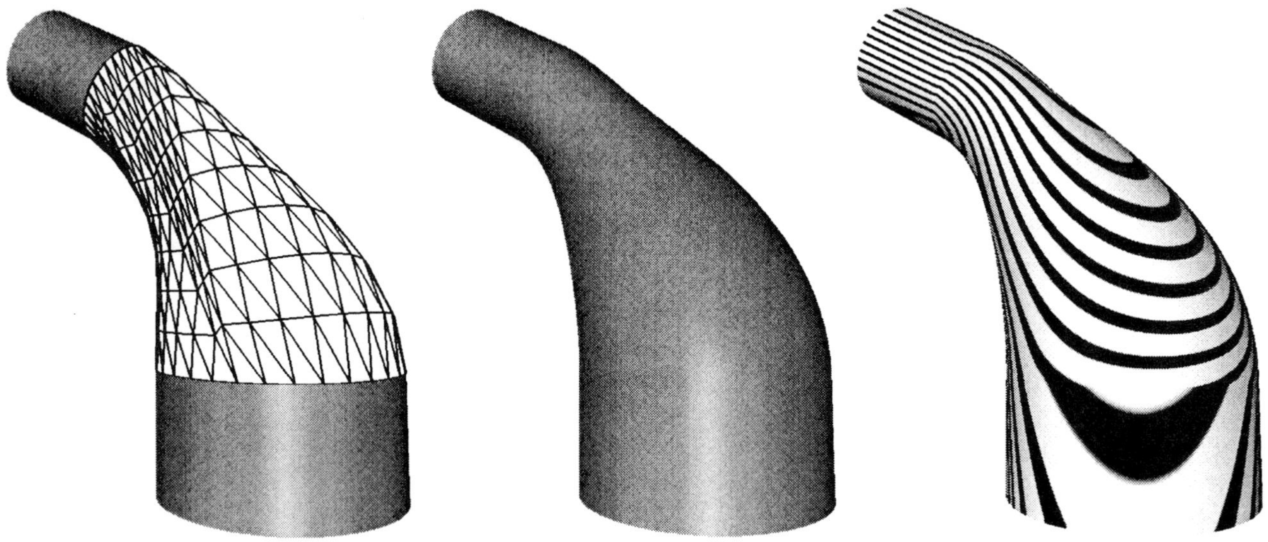

(a) (b) (c)

Figure 8. Blending two cylinders of different size. a) Discrete solution of a 3 times subdivided mesh. b) and c) Discrete solution of a 5 times subdivided mesh.

one can iterate on each level until all free vertices satisfy $\Delta_B(H_i^k) < \epsilon$ for a given tolerance ϵ.

When solving $H_i^{k+1} = \tilde{H}_i^{k+1}$ in the update step of the indirect iteration, we apply the resampling technique presented in section 4.6 also to the neighbor vertices of $M_{i,l}^k$ resulting in new vertices $\tilde{M}_{i,l}^k$ and use these vertices to update M_i^k. Thus we again profit from the simplicity and stability advantages of the equilateral case.

At edge and extraordinary vertices we cached the matrices $(A^tA)^{-1}A^t$ in (5) thus allowing us a fast least square approximation. For inner vertices this is not necessary, since the resulting formulas here are simple enough to be calculated on the fly.

When starting on the coarsest level, the initial local parameterization is constructed using the vertices of $\bar{P}$ as seeds. Since the resulting parameter domains are only rough approximations, we recalculate the parameter domains once when the mesh approximates the final solution at a coarse level. In the examples this was done when the solution of a 3 times subdivided mesh had been constructed.

7. Examples

In CAGD there are many fields of application where G^1 boundary constraints naturally appear. One of the most important is the N-sided hole filling problem, which is without doubt one of the most studied problems in geometric modeling [7, 14, 23, 12]. Solutions for some classical N-sided hole problems obtained with the proposed techniques can be seen in the figures. The lines of reflection indicate the G^2 continuity in the interior (smooth lines) and G^1 continuity at the border (continuous lines) in all our examples. As can be seen in Figure 8 our algorithm is not restricted to a border that consists of one single closed curve only. In this example we calculated the solution for a blending problem.

The construction algorithm is implemented in Java 1.2 for Windows running on a PII with 400MHz. For our presented examples the average iteration time for a 3 times subdivided mesh was ≈ 1.0 seconds and for 5 times subdivided meshes ≈ 4.0 seconds. An optimized C implementation would improve these timings without doubt.

8. Conclusion

Despite the fact that the construction idea is based on pure geometric intrinsics and the resulting high quality fairness, we presented a discrete fairing algorithm that is fast enough to become interesting in interactive design. The resulting meshes have an especially interesting structure, they are composed of regular patches where the interior patch edges form geodesics on the discrete surface.

This property allowed us to construct an efficient iteration algorithm, which should also be very useful when one does not need a discrete solution, but a surface composed of spline patches. In such a case the piecewise regular structure may be exploited when approximating or interpolating the mesh using polynomial splines. The presented algorithm allowed us G^1 boundary constraints, by increasing the degree of the intrinsic PDE it should be possible to allow higher geometric continuity constraints at the border using analogous techniques.

References

[1] Bloor, M. I. G., and M. J. Wilson, Interactive Design Using Partial Differential Equations, in *Designing Fair Curves and Surfaces*, ed. N. S. Sapidis, SIAM, Philadelphia, 1994.

[2] Bloor, M. I. G., and M. J. Wilson, Using partial differential equations to generate free-form surfaces, Computer Aided Design, 22 (1990), 202–212.

[3] Bonnet, O., *Mémoire sur la théorie des surfaces applicables*, J. Éc. Polyt 42, 1867.

[4] Burchard, H. G., J. A. Ayers, W. H. Frey, and N. S. Sapidis, Approximation with Aesthetic Constraints, in *Designing Fair Curves and Surfaces*, ed. N. S. Sapidis, SIAM, Philadelphia, 1994.

[5] do Carmo, M. P., *Differential Geometry of Curves and Surfaces*, Prentice-Hall, Inc Englewood Cliffs, New Jersey, 1993.

[6] Desbrun, M., M. Meyer, P. Schröder, and A. H. Barr, Implicit fairing of irregular meshes using diffusion and curvature flow, SIGGRAPH 99 Conference Proceedings, 317-324.

[7] Gregory, J. A., V. K. H. Lau, and J. Zhou, Smooth Parametric Surfaces and n-Sided Patches, in *Computation of Curves and Surfaces*, ed. W. Dahmen, M. Gasca and C. A. Micchelli, Kluwer Academic, 1990, 457–498.

[8] Hackbusch, W., *Iterative Lösung grosser schwachbesetzter Gleichungssysteme*, Teubner Verlag, Stuttgart, 1993.

[9] Kobbelt, L., Discrete fairing, Proceedings of the Seventh IMA Conference on the Mathematics of Surfaces, 101–131, 1996.

[10] Kobbelt, L., S. Campagna, J. Vorsatz, and H-P. Seidel, Interactive Multi-Resolution Modeling on Arbitrary Meshes, Proceedings of SIGGRAPH '98, 105–114.

[11] Kobbelt, L., A variational approach to subdivision, *Computer Aided Geometric Design* 13 (1996), 743–761.

[12] Levin, A., Filling N-sided holes using combined subdivision schemes, to appear in Proceedings of the Saint-Malo Conference '99.

[13] Malcolm, M. A., On the computation of nonlinear spline functions, SIAM J. Numer. Anal. 15, 254–282, 1977.

[14] Malraison, P., A Bibliography for N-sided surfaces, in *Mathematics of Surfaces VIII*, Information Geometers Limited, Winchester, 419–430, 1998.

[15] Mehlum, E., Nonlinear splines, *Computer Aided Geometric Design*, Academic Press, London, 173–207, 1974.

[16] Meek, D. S., and D. J. Walton, Planar G^1 Hermite interpolation with spirals, *Computer Aided Geometric Design* 15 (1998), 787–801.

[17] Moreton, H. P., C. H. Séquin, Functional optimization for fair surface design, SIGGRAPH 92 Conference Proceedings, 167–176.

[18] Pinkall, U., K. Polthier, Computing discrete minimal surfaces and their conjugates, Experimental Mathematics 2, 15–36, 1993.

[19] Rauch, J., *Partial Differential Equations*, Springer-Verlag, New York, 1991.

[20] Schneider, R., and L. Kobbelt, Discrete Fairing of Curves and Surfaces based on linear Curvature Distribution, to appear in Proceedings of the Saint-Malo Conference '99.

[21] Schwarz, H. R., *Methode der finiten Elemente*, Teubner Verlag, Stuttgart, 1991.

[22] Taubin, G., A signal processing approach to fair surface design, SIGGRAPH 95 Conference Proceedings, 351–358.

[23] Varady, T., and A. Rockwood, Geometric construction for setback vertex blending, Computer Aided Design 29 (1997), 413–425.

[24] Weimer, H., J. Warren, Subdivision Schemes for Thin Plate Splines, Computer Graphics Forum 17 (1998), 303–314.

[25] Welch, W., and A. Witkin, Free-Form shape design using triangulated surfaces, SIGGRAPH 94 Conference Proceedings, 247–256.

A Polygonal Approach for Interpolating Meshes of Curves by Subdivision Surfaces

A. H. Nasri *
Department of Mathematics
American University of Beirut
POBox 11-236, Beirut, Lebanon
Email: anasri@aub.edu.lb

Abstract

Given a polyhedral network P_0 defining a subdivision surface S and an arbitrary mesh of tagged control polygons $(cp_i)_{1\leq i\leq n}$ on P_0, this paper describes an approach to force the limit surface S to interpolate the B-spline curves of (cp_i). For each control polygon cp_i, we construct a polygonal complex whose mid-polygon is cp_i or its first subdivided one. A polygonal complex is a sequence of panels such that every two adjacent panels share exactly one edge and the mid-points of these edges make the mid-polygon of the complex. Since the complexes themselves are embodied in the original polyhedron or its first subdivided, the limit surface will interpolate the limit curves of these complexes.

Key Words: *subdivision, interpolation, B-spline, mesh of curves, polygonal complexes.*

1. Introduction

Recursive subdivision techniques are gradually becoming important tools in Computer-Aided Geometric Design and computer graphics. They have the advantage of defining free-form surfaces (referred to as subdivision surfaces) by arbitrary topological networks, thus overcoming the rigid rectangular requirement of the B-spline based schemes. Furthermore, the surfaces generated, though globally smooth, can be locally controlled by adding various features such as creases, sharp corners, interpolating points and normals [16, 17, 10]. Another important reason for the popularity of subdivision schemes is its application in multiresolution and wavelets [15], and more recently in fluid flow [31]. For modeling complex shapes, they naturally provide scalability with level-of-detail rendering and adaptive approximation. Finally, they are numerically stable and simple to implement [32].

*Supported by a URB grant #3048

A recursive division surface S is basically defined by a pair (P_0, R) where P_0 is an initial configuration and R is a refinement procedure. The configuration consists of a set of vertices, edges, and faces in the 3D space. This is often referred to as a *polyhedron*. The refinement procedure is a set of rules applied to a configuration to generate another with more vertices, edges and smaller faces than the initial one. At each level i of the refinement process, the polyhedron P_{i-1} is taken as input to the refinement R, which produces another polyhedron P_i, which may itself be taken as input to the next refinement step and so on. If R satisfies certain conditions [27, 3, 26] then at the limit the sequence of polyhedra P_i converges to a smooth surface.

There are many alternative sets of rules R. The best known examples are the Doo-Sabin [3] extension of the quadratic B-spline and the Catmull-Clark extension of the cubic B-spline [2]. There are other schemes that are extension of the box-splines such as the Loop [14] and the four-direction quadratic which was examined (though not identified) by Doo [4] and found to be rather unsatisfactory. It was later considered in more detail by Sabin [28] and by Peters and Reif [25]. All of these schemes are described as approximating the initial configuration describing the surface. Other constructions that are interpolatory in nature were considered by Dyn, Gregory and Levin [7, 6, 5], and Kobbelt [12], they are designed to have the property that every vertex lies on the limit surface. Non-uniform subdivision schemes have also emerged as in [8], [9] and more recently as in [30].

The analysis of the subdivision schemes at the extraordinary points was carried out by Doo and Sabin [3], Ball and Storry [1], Reif [27], and the C^k case was considered by Prautzsch [26].

Let us illustrate the example of the Doo-Sabin. Given a polyhedron P_i, where i is the level of subdivision, the Doo-Sabin algorithm generates a new polyhedron P_{i+1} by constructing on each old face f_i of P_i having n vertices

0-7695-0562-7/00 $10.00 © 2000 IEEE

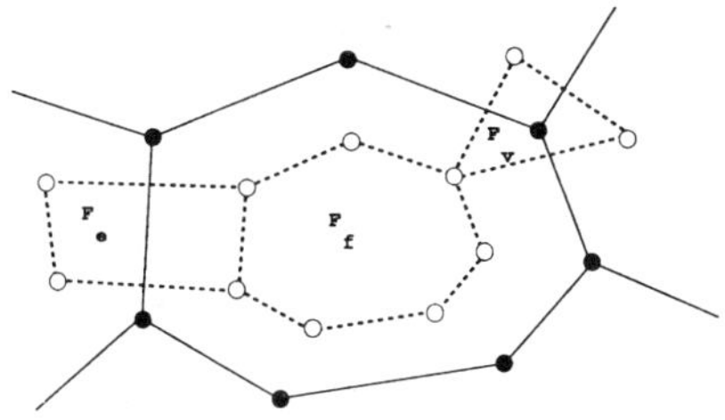

Figure 1. The three types of faces (dashed) in Doo-Sabin approach. F-face (F_f), E-face (F_e), V-face (F_v)

$(v_i)_{1\leq\ i\leq\ n}$ a new set of vertices $(w_i)_{1\leq\ i\leq\ n}$ called *images* using the following equations:

$$w_i = \sum_{j=1}^{n} v_j \alpha_{ij} \tag{1}$$

where the α_{ij} are given by:

$$\alpha_{ij} = \begin{cases} \frac{n+5}{4n} & i = j \\ \frac{3+2cos(2\pi(i-j)/n}{4n} & i \neq j \end{cases} \tag{2}$$

The topology of the new polyhedron consists of connecting the image vertices (w_i) and constructing from each old face, edge, and vertex a new face, called F-face, E-face, and V-face, respectively. The three types of faces are shown in Fig. 1.

Before subdivision schemes became practical, a number of issues had to be solved such as the interpolation of points, normals or even predefined curves. The interpolated curves can be C^0 across, in which case they are called *creases*, or C^1 across. Generating creases was addressed in [16, 24, 21] for the Doo-Sabin scheme, and in [11] for the Catmull-Clark scheme. Interpolating curves with smoothness across was discussed in [19]. Generating smooth surfaces through meshes of curves was also considered by Nasri [18, 22] for the Doo-Sabin scheme, and by Schweitzer [29] and Levin [13] for the Catmull-Clark scheme. For each region between the interpolated curves, Schweitzer used fairing techniques to generate a polyhedral network that defines a piece of surface with the same boundary curve as the given region. Levin proposed a combined subdivision scheme that repositions the points near the interpolated curves at each level of subdivision. Nasri suggested the *complex* approach [22] where a polygonal strip complex, for each curve to be interpolated, is built within the polyhedral network defining a surface. Repeated subdivision will guarantee that the polygonal complexes converge to the given curves and will be interpolated by the limit surface. After building these complexes, a process that is completed during the first subdivision, the interpolation task is achieved without modification of the subdivision rules. In fact, after the construction process is completed, the obtained network is a normal input to the subdivision procedure which does not see the interpolation issue at all.

In this paper, we describe an alternative approach called the *polygonal approach* to the curve interpolation problem. Instead of defining surfaces in terms of polyhedral networks that initially include polygonal complexes, which may not be easy to construct, we provide algorithms for constructing polygonal complexes from tagged control polygons on the network of a subdivision surface. The control polygons can define intersecting meshes of arbitrary curves whether at the interior or at the boundary of the surface. The paper is structured as follows. Section 2 introduces polygonal complexes and their mid-polygons. The convergence of polygonal complexes and their limit curves is discussed in section 3. We also highlight one possible application of the proposed method in curve interpolation using the *complex approach*. In section 4, we show how to interpolate meshes of curves using the *polygonal approach* and describe how to construct polygonal complexes from tagged control polygons. Implementation and other issues are discussed in section 5. Finally, section 6 draws conclusions and further work.

2. Polygonal complexes and their limit curves

A polygonal complex can be simply defined as a sequence of panels $(q_i)_{1\leq i\leq n}$ with the property that every two panels q_j and q_{j+1} have exactly one edge in common. If the two panels q_1 and q_n share an edge the complex is called *annular*, otherwise it is called a *strip* complex. Fig. 2 shows an example of such a complex.

Recursive subdivision of a polygonal complex consists of applying the subdivision rules to the panels and the shared edges of the complex but not to the vertices. Consequently, there are no V-faces generated in the subdivision of a polygonal complex which at the limit converges to a curve that interpolates the centroid of each panel with C^1 continuity except possibly at the n-sided ($n \neq 4$) panels where only C^0 is achieved. Since the curve is regressively defined, its control polygon is not known in general, making it difficult to control its shape. This issue was considered in [22] and it was shown that if the vertices of the panels of a polygonal complex Q_i are *symmetric* about their corresponding Chebychev points then the control polygon of the limit curve of Q_i is well defined. The following section gives the details of these panels.

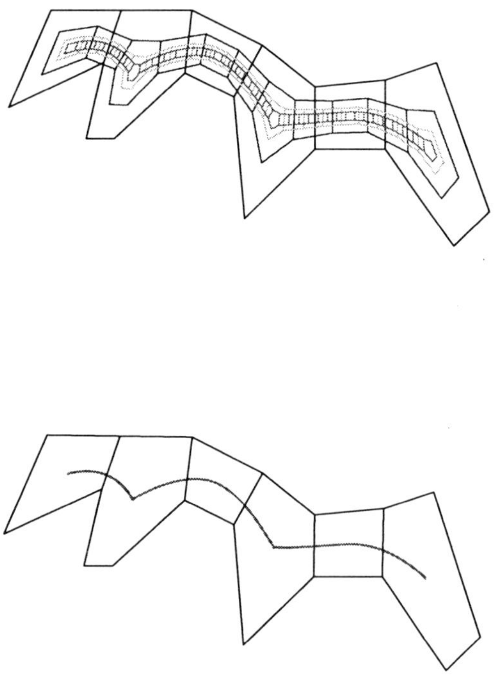

Figure 2. Convergence of a polygonal complex. Three successive subdivisions (top) and the limit curve (bottom) are shown. Note that G^1 is not achieved at the n-sided $(n \neq 4)$ face.

2.1. Symmetric panels

Let us begin by defining what is meant by Chebychev points. Given a segment $[AB]$, the set $Cb^k(A, B) = \{c_i : i = 1, 2, \ldots k\}$ of Chebychev points on $[AB]$ is given by:

$$c_i = \frac{(1 + \beta_i)A + (1 - \beta_i)B}{2} \tag{3}$$

where the values β_i are given by

$$\beta_i = \frac{\cos \frac{(2i-1)\pi}{2k}}{\cos \frac{\pi}{2k}} \tag{4}$$

Note that $A = c_1$ and $B = c_k$.

Generally speaking there are two types of symmetric panels: *singly reflected* and *doubly reflected* panels.

A panel f with n $(n = 2m)$ vertices $(v_i)_{1 \leq i \leq n}$ is called singly reflected about a segment $\hat{e} = w_1 w_m$, where w_1 and w_m are the midpoints of the edges $v_1 v_n$ and $v_m v_{m+1}$, respectively, if its vertices are symmetrically distributed about the Chebychev points defined on the edge $\hat{e}$. The edges $v_1 v_n$ and $v_m v_{m+1}$ are called *contact* edges and can be shared with other panels of the same or another polygonal complex. Accordingly, every couple of vertices v_i and v_j such that $i + j = n + 1$ are symmetric about the corresponding Chebychev point $c_i \in Cb^m(w_1, w_m)$ (see Fig. 3). These vertices are called *opposite vertices*.

A panel f of n $(n = 4m)$ vertices is called doubly reflected about two segments $\hat{e}_1 = w_1 w_3$ and $\hat{e}_2 = w_2 w_4$, where $(w_i)_{1 \leq i \leq 4}$ are the midpoints of the edges $v_n v_1$, $v_{2m} v_{2m+1}$, $v_m v_{m+1}$ and $v_{3m} v_{3m+1}$, respectively, if it is singly reflected about each of the segments $\hat{e}_1$ and $\hat{e}_2$.

The edges $v_n v_1$, $v_m v_{m+1}$, $v_{2m} v_{2m+1}$, and $v_{3m} v_{3m+1}$ are also called *contact* edges, and up to four other panels can be stitched to a doubly reflected panel. Furthermore, two contact edges are called *opposite* if their midpoints are collinear with the centroid G of a panel, otherwise they are called *adjacent*. Note that in a singly reflected panel, the contact edges are always opposite. Fig. 3 shows an example of such a face where the contact edges $v_1 v_{12}$ and $v_6 v_7$ are opposite edges, and $v_1 v_{12}$ and $v_3 v_4$ are adjacent. Furthermore, the two segments $w_1 w_2$ and $w_3 w_4$ are called the mid-segments of f and they intersect at G.

We further define the *mid-polygon* of a polygonal complex as the piecewise control polygon whose vertices are the midpoints of the contact edges. If the two contact edges of a doubly reflected panel are not opposite then its centroid is a common end-point of two pieces of the mid-polygon. In the case of a polygonal strip complex, the centroids of the end-panels are the end-points of the mid-polygon.

Fig. 4 gives examples of various types of symmetric complexes and their corresponding mid-polygons.

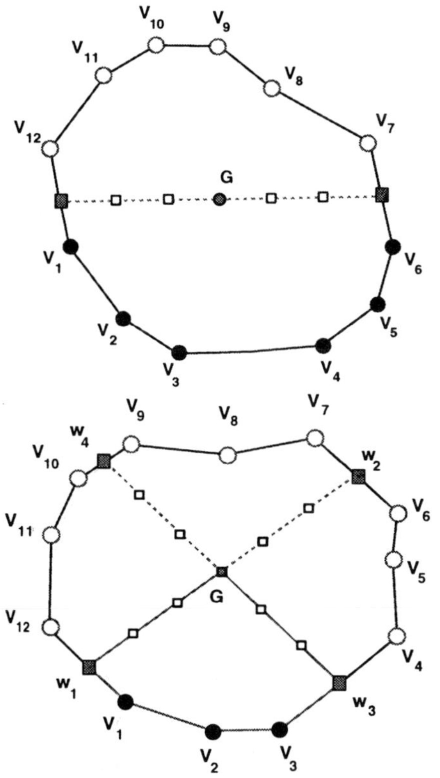

Figure 3. Panels in a symmetric polygonal complex: Singly panel (top) and doubly reflected (bottom). Original vertices are shown in solid/hollow circles, Chebychev points in hollow squares, and the centroids in shaded squares. A solid circle and its corresponding hollow circle are opposite vertices.

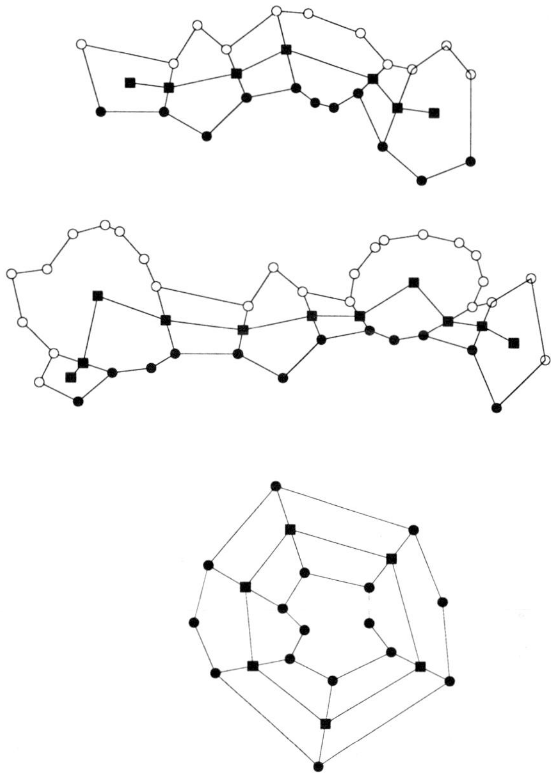

Figure 4. Various polygonal complexes and their mid-polygons: Starting from top: a strip whose panels are all singly reflected, a general form of a strip complex and an annular complex with singly reflected panels are shown with their corresponding mid-polygons (vertices in solid squares).

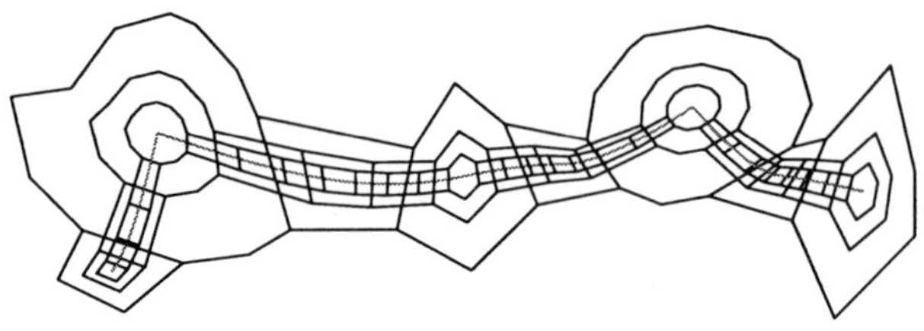

Figure 5. Convergence of symmetric polygonal complexes. A complex is shown with two successive subdivisions and its limit curve.

3. Convergence of symmetric polygonal complexes

In [22], it was shown that a symmetric polygonal complex Q_0 with a mid-polygon M_0 converges to the piecewise quadratic B-spline curve c of the control polygon M_0 such that c interpolates the centroid of each panel of Q_0 with C^1 continuity, if its two contact edges are opposite, and with C^0 only if they are adjacent. Fig. 5 shows the limit curves of a symmetric polygonal complex.

Based on this result, up to m ($m \leq 4$) symmetric polygonal complexes $(Q_i)_{1 \leq i \leq m}$ can share a doubly reflected panel p_0 in which case they converge to m corresponding curves $(c_i)_{1 \leq i \leq m}$ meeting at the centroid of p_0 such that every couple (c_i, c_j) meets with C^1 continuity if the contact edges of p_0 in Q_i and in Q_j are opposite and with C^0 continuity only if these contact edges are adjacent.

Fig. 6 shows various configurations where m polygonal complexes can share a doubly reflected panel. In the case of two polygonal complexes the limit curves can be C^1 or C^0 across the centroid of the shared panel. For three polygonal complexes, two curves will join with C^1 and the other with C^0. Finally, with four polygonal complexes, the opposite curves join with C^1 and the adjacent ones with C^0.

3.1. Curve interpolation by polygonal complexes

One more interesting application of the above results is in curve interpolation by subdivision surfaces. The idea is based on the following results obtained in [22]. Assume that a symmetric polygonal complex Q_i, whether strip or annular, is embodied in a polyhedral network P_0. Let (P_j) and (Q_i^j) be the jth subdivision of P_0 and Q_i, respectively. Let

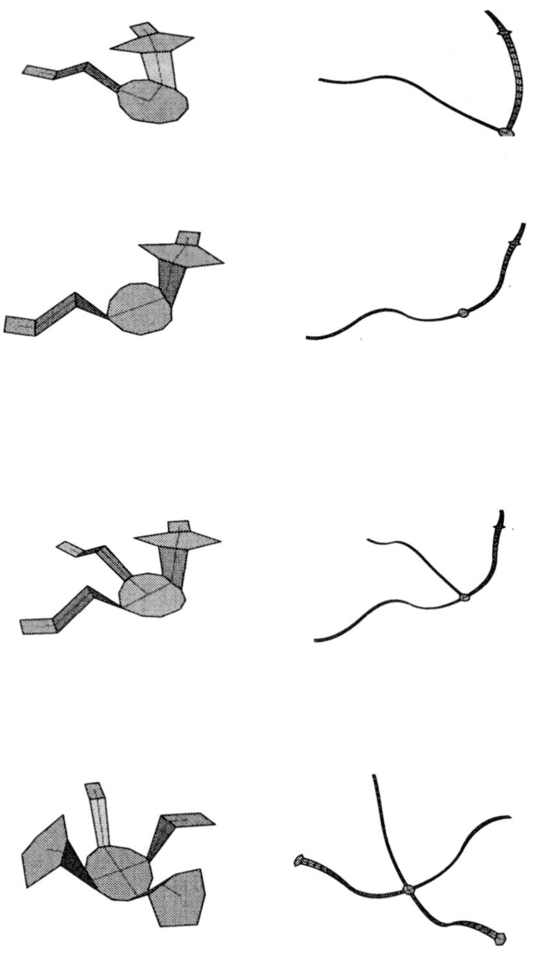

Figure 6. Convergence of polygonal complexes sharing one doubly reflected panel. From top: two polygonal complexes giving two limit curves with C^0 joint, two complexes giving two limit curves with C^1 joint, three complexes giving three limit curves, and four complexes giving four limit curves. Complexes and mid-polygons are shown on left, their third subdivisions and limit curves are shown on right.

S and C be the limit of subdivision of P_j and Q_i^j as $j \to \infty$. Then the surface S interpolates the curve C corresponding to Q_i.

Based on the above results, an approach called *the complex approach* was devised to generate subdivision surfaces through meshes of given curves. The idea consists of designing polyhedral networks that initially contain a number of polygonal complexes, corresponding to some curves to be interpolated by the limit surface. Thus, if the network contains initially, two polygonal complexes sharing a doubly reflected panel, two curves meeting at an interior point can be interpolated by the limit surface. With three polygonal complexes, an interior curve can intersect a boundary one and with four polygonal complexes, four interior-intersecting curves can be interpolated. Fig. 7 shows an example of interpolating various intersecting curves using this approach.

Another approach to interpolate curves by subdivision surfaces is the *polygonal approach* which is the subject of the following section.

4. Polygonal approach to curve interpolation

In this approach, the input is a polyhedral network P and a set of m tagged control polygon $(cp_i)_{1 \leq i \leq m}$ on P. We want to generate a subdivision surface that passes through the curves (c_i) of (cp_i). This approach, which is an extension of the methods in [19, 20], involves constructing (1) m symmetric polygonal complex, one for each cp_i and (2) a doubly reflected panel at each intersecting point where two or more cp_i meet. This panel will be shared by the polygonal complexes corresponding to these control polygons.

To illustrate, let us assume that m control polygons are to define m interpolated curves intersecting at a point G. These control polygons must then obey the following condition:

Condition The end-legs (e_i) of the control polygons must be coplanar and should meet at a common point G such that every two opposite legs are of equal length and collinear.

With this condition satisfied, for each tagged control polygon cp_i, a polygonal complex will be constructed whose limit curve will be the curve c_i of cp_i. Two construction algorithms will be discussed depending on whether the tagged control polygons correspond to curves intersecting at the interior or at the boundary of the surface.

4.1. Construction algorithm for interior intersecting curves

For each tagged control polygon cp_i, the panels of the constructed complex are modified versions of the E- and V-

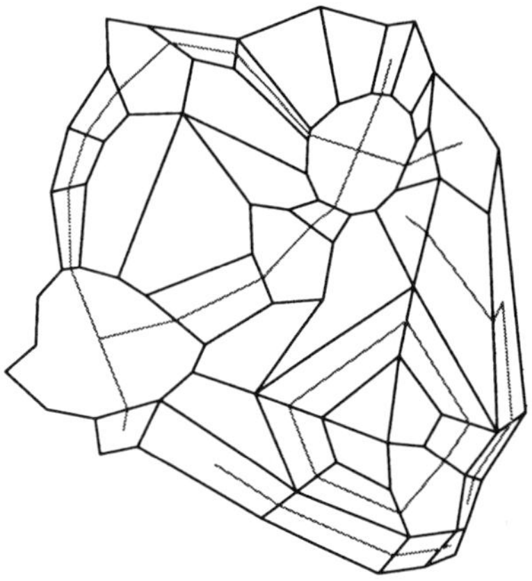

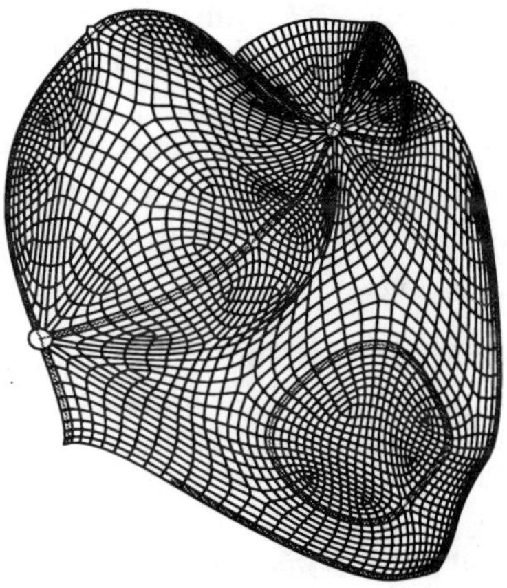

Figure 7. The complex approach: A polyhedral network with various polygonal complexes, including one annular complex, and their mid-polygons (top), and the corresponding surfaces with the interpolated curves (bottom) are shown. Note the mid-polygons are not part of the network defining the surface.

faces of the legs and vertices of cp_i. At the intersection of control polygons, the algorithm constructs a shared doubly reflected panel as follows (refer to Fig. 8 where $m = 4$):

1. Apply one step of division to the polyhedral network.

2. For each tagged control polygon, modify the corresponding E- and V-faces of its edges and vertices (other then the end-legs (e_i) and their vertices) to become singly reflected as suggested in [22, 20]. The modified panels are then panels of the complexes.

3. If the F-faces $(Ff_i)_{1 \leq i \leq m}$ of the panels $(f_i)_{1 \leq i \leq m}$ sharing the end-legs $(e_i)_{1 \leq i \leq m}$ are not coplanar then we project their vertices into the plane determined by the (e_i) and use the projected faces instead of (f_i) in the remaining steps of the algorithm. The projection is done along the direction of the normal of the plane determined by the (e_i).

4. If the F-faces (Ff_i) do not have equal number of vertices[1] then modify their topology by constructing new faces from the extra vertices as follows:

   (a) Let m be the minimum of $(n_i)_{1 \leq i \leq m}$ where n_i is the number of vertices of Ff_i. In Fig. 8, $m = 6$.

   (b) Split each face Ff_i with n_i vertices such that $n_i > m$ into two faces: one m-sided face $\hat{Ff}_i$ and another k-sided where $k = n_i - m + 2$ as shown in Fig. 8. The images of the vertices of (e_i), must not be included in the vertices of the constructed faces $\hat{Ff}_i$. This face will replace Ff_i in the rest of the algorithm.

5. Remove all edges of the F-faces (Ff_i) incident to the images of G and remove these images too.

6. Shift the leg $v_1^0 v_1^1$, which is a contact edge, generated from the two images of v_1 by a vector $p_1 c_1$, where p_1 is the midpoint of that leg and c_1 is a Chaikin point[2] on the edge $v_1 v_3$.

7. Apply step 6 to each of the legs $(v_i^0 v_i^1)$ generated from the other 3 vertices v_i and shift them accordingly.

8. Merge the remaining vertices of the (Ff_i) into one face and re-position the vertices symmetrically about their corresponding Chebychev points of the E-edges of $v_1 v_3$ and $v_2 v_4$. An E-edge of an edge connects its two Chaikin points. This results in a doubly reflected panel to be shared by all four complexes. The Chaikin points of $v_1 v_3$ and $v_2 v_4$ determine the mid-segments of this panel.

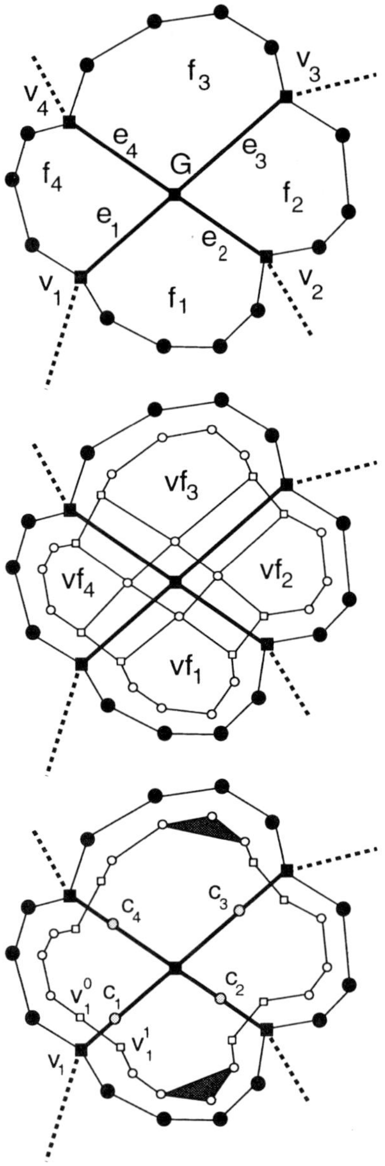

Figure 8. Construction of a shared interior doubly reflected panel: two 3-sided faces were constructed. Original configuration (top), subdivided (middle) and prior to re-positioning (bottom) are shown. Vertices are depicted in solid (original), solid squares (control polygons), hollow circles (subdivided vertices), and hollow squares (images of v_i).

[1] If they do then omit this step.

[2] A Chaikin point on a segment $v_i v_j$ is one of the two points obtained by applying one step of Chaikin's algorithm on that edge. They are given by: $\frac{3v_i + v_j}{4}$ or $\frac{v_i + 3v_j}{4}$

4.2. Construction algorithm for a boundary intersecting curves

This is the case of three control polygons intersecting at a boundary vertex G, where two of them are made of boundary edges and vertices, and the other control polygon is made of interior vertices, except G. Note that the case of two complexes can be handled analogously.

The algorithm constructs two boundary polygonal complexes, one for each of the two boundary control polygons, and one interior polygonal complex for the third control polygon. However, unlike the interior case, the panels of the boundary complexes are taken from different configurations, i.e. boundary complexes are built from modified boundary panels prior to subdivision whereas the interior complex is built from panels of the first subdivided network. This is possible because boundary vertices can be easily added along the boundary and they can be even 2-valent— such vertices cannot be used in the interior case where all vertices must be n-valent ($n \geq 3$). The algorithm begins by constructing the shared panel followed by the two boundary complexes and then the interior one (refer to Fig. 9 in the steps below.)

1. If the two boundary faces f_1 and f_2 sharing the internal edge e_3 are not coplanar then we project their vertices into the plane determined by the $(e_i)_{1 \leq i \leq 3}$ along the direction of its normal and use the projected faces instead of f_1 and f_2 in the remaining steps of the algorithm.

2. If the faces f_1 and f_2 do not have the same number of vertices then modify the topology of these faces by constructing a new one from the extra vertices as follows:

   (a) Let n_i be the number of vertices of f_i and assume, for the sake of simplicity, that $n_1 < n_2$. In Fig. 9 we have $n_1 = 6$ and $n_2 = 7$.

   (b) Split f_2 into two faces: an n_1-sided face $\hat{f}_2$ and another k-sided face, where $k = n_2 - n_1 + 2$ as shown in Fig. 9. The vertices of the new constructed face must all be interior. The face $\hat{f}_2$ is used in the rest of the algorithm instead of f_2.

3. Re-position the vertices of f_1 and f_2 symmetrically about the Chebychev points of $\hat{e}_3$, the symmetric extension of e_3 about G.

4. Reflect the $2(n_1 - 3)$ interior vertices of f_1 and f_2 about the Chebychev points of the segment $v_1 v_2$ obtained from e_1 and e_2.

5. Remove the end-legs $(e_i)_{1 \leq i \leq 3}$ and their vertices, except v_3, from the topology of the polyhedral network. The remaining vertices of f_1 and f_2 and their reflected vertices make a doubly reflected panel sp as depicted in Fig. 9. Note that v_3 will not be considered as a vertex of sp and its V-face will be handled in a special way as can be seen later.

6. For each boundary edge l_j (other than the edges e_1 and e_2) which belongs to a face q_j, construct a singly reflected panel by reflecting the interior vertices of q_j about the Chebychev points of l_j. The reflected panels including the shared doubly reflected panel sp make the two boundary complexes[3]. Note that, if the face q_j has two boundary edges it will result in a doubly reflected panel instead of a singly one. Furthermore, the 2-sided vertex common to the two boundary edges will be a C^0 point on the boundary curve.

7. Apply one step of division to the modified network taking the following points into consideration:

   (a) Since the edges $(e_i)_{1 \leq i \leq 3}$ are removed from the topology of the network, no E-faces corresponding to them are generated.

   (b) The generation of the F-face $\hat{sp}$ corresponding to the shared panel sp does not take into account the vertices v_1, v_2, v_3, nor the opposite vertex of v_3. Note that, the face $\hat{sp}$ is a shared panel of the constructed polygonal complexes.

   (c) There will be no V-face for the interior end-point v_3 of the edge e_3 (see Fig. 9).

   (d) There will be no E-faces for the edges Av_3, and v_3C, but there will an E-face for the edge[4] AC. This E-face is made of the two images of v_3 on the faces f_3 and f_4, and the two images of A and C on the shared panel sp. The constructed face is also a panel of the complex.

   (e) The V-face of A should include the images of A on the faces sharing it (there are three faces in our example, namely f_4, f_5 and sp), and the image of v_3 on f_4. Similarly, we construct a V-face that corresponds to the vertex C.

8. For the interior edges and vertices other than the end-legs e_3 and its vertices, modify the corresponding E- and V-panels as suggested in [19, 20]. This will complete the construction of the interior polygonal complex.

[3] Note that there is at least one more shared panel similar to sp since the interior curve will normally intersect the boundary twice.

[4] Note that A, v_3, and C are collinear by construction.

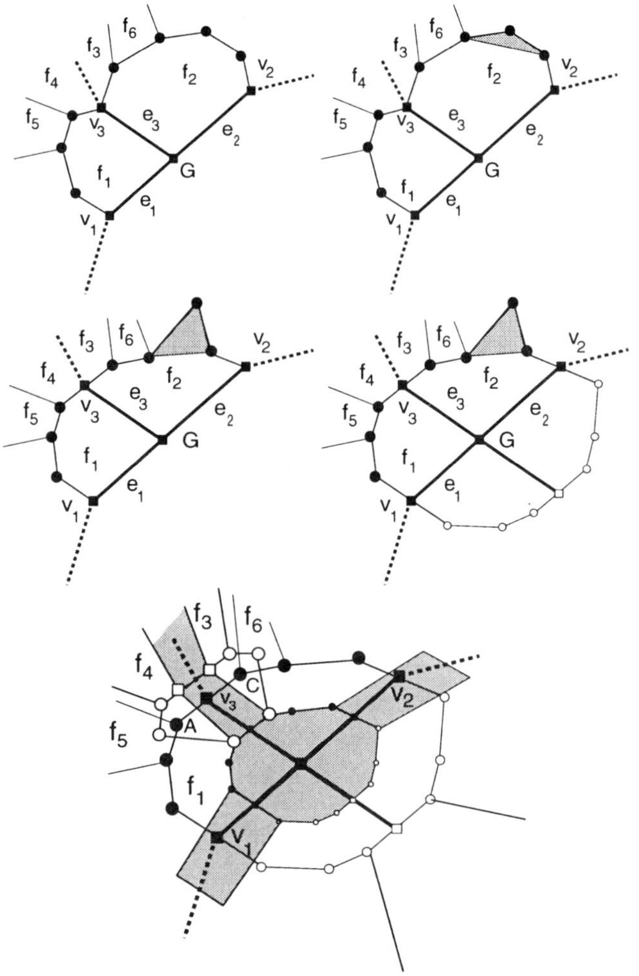

Figure 9. Construction of a boundary shared doubly reflected panel. Original configuration (top left), splitting a 3-sided face (top right), re-positioning vertices (middle left), reflecting vertices (middle right), subdivision and construction of panels adjacent to the subdivided shared panel (bottom). Vertices are depicted in solid (original), solid squares (control polygons)and reflected vertices hollow.

5. Applications

There are two approaches to use polygonal complexes in generating smooth surfaces interpolating meshes of curves: the *complex approach* and the *polygonal approach*. In the complex approach [22], polygonal complexes are originally included in the network defining a subdivision surface. In other words, it is the responsibility of the user to design a surface in terms of a polyhedral network that includes some polygonal complexes. The subdivision procedure treats these complexes as ordinary faces, vertices and edges, and the limit surface will automatically interpolate the corresponding curves, as illustrated in Fig. 7.

One difficulty that should to be mentioned here is how to design polygonal complexes, especially if control points need to be interactively modified later. Although this may be solved by using an intelligent user interface [23], the polygonal approach suggested in this paper avoids this problem by constructing polygonal complexes automatically. Therefore, instead of designing polygonal complexes, the user simply provide tagged control polygons on the given network, and the system will construct polygonal complexes having these control polygons or their subdivided one as mid-polygons. The obtained results guarantee that the limit surface will interpolate the corresponding curves.

The proposed construction algorithms were implemented on an SGI machine and the included pictures were produced. Fig. 10 shows a network with 5 tagged control polygons: four of them intersect at an interior point and the fifth one is a boundary control polygon that intersects each of the other four at their endpoints. The five constructed polygonal complexes after the first subdivided network and the limit surface interpolating the corresponding curve is also shown. Note that the construction of the shared doubly reflected panel has created an additional triangular face in the first subdivision. Fig. 11 shows a network with four tagged control polygons, their corresponding complexes, and the limit surface. Finally, Fig. 12 illustrates the surface of a phone handset interpolating a network of curves with different n-sided regions between those curves.

6. Conclusions and further work

The generation of subdivision surfaces through meshes of arbitrary curves provides another capability that makes such surfaces more attractive and practical in many applications, including surface modeling. The polygonal approach suggested in this paper provides an alternative way of using polygonal complexes to interpolate curves by subdivision surfaces. Instead of asking the user to design such complexes within the polyhedral network defining a surface, the proposed scheme constructs corresponding polygonal com-

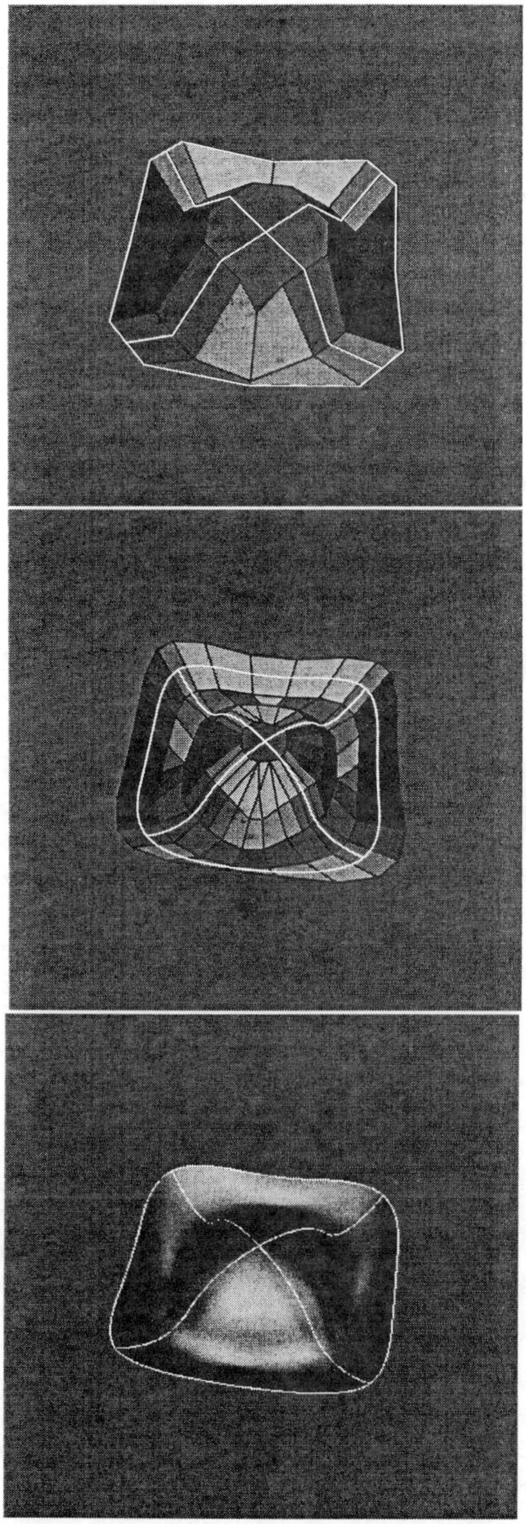

Figure 10. A polyhedral network with five tagged control-polygons (top), the network after constructing polygonal complexes (middle), and the limit surface interpolating the required curves (bottom).

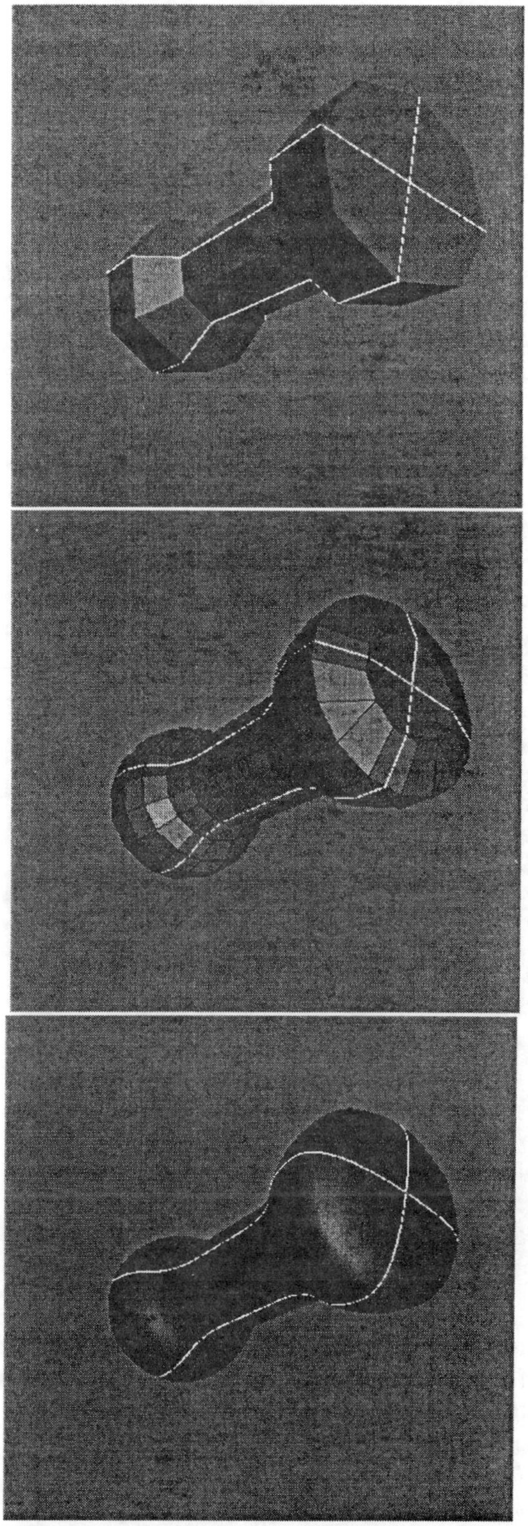

Figure 11. A polyhedral network with four tagged control-polygons (top), the network after constructing polygonal complexes (middle), and the limit surface interpolating the required curves (bottom).

plexes from tagged control polygons on the given network. This should make the surface design process much easier to handle and allow for a better control of the interpolated curves. Obviously, modifying the vertices of a control polygon is much easier than modifying the panels of its corresponding complex. However, the use of polygonal complexes in curve interpolation will generally induce a certain degree of freedom as how to position the vertices of the panels of the complexes. For example the vertices of a panel, including shared panels, can be moved around as long as opposite vertices remain symmetric about their corresponding Chebychev points. This property can be used to control the shape of the surface across the interpolated curves; a further work under consideration. Moreover, since the proposed scheme can only handle up to four interpolated curves meeting at an interior point of the surface, the extension to n $(n > 4)$ curves can be investigated. Finally, although the quadratic Doo-Sabin algorithm was used in the proposed scheme, its applications to higher order surfaces can be inspired.

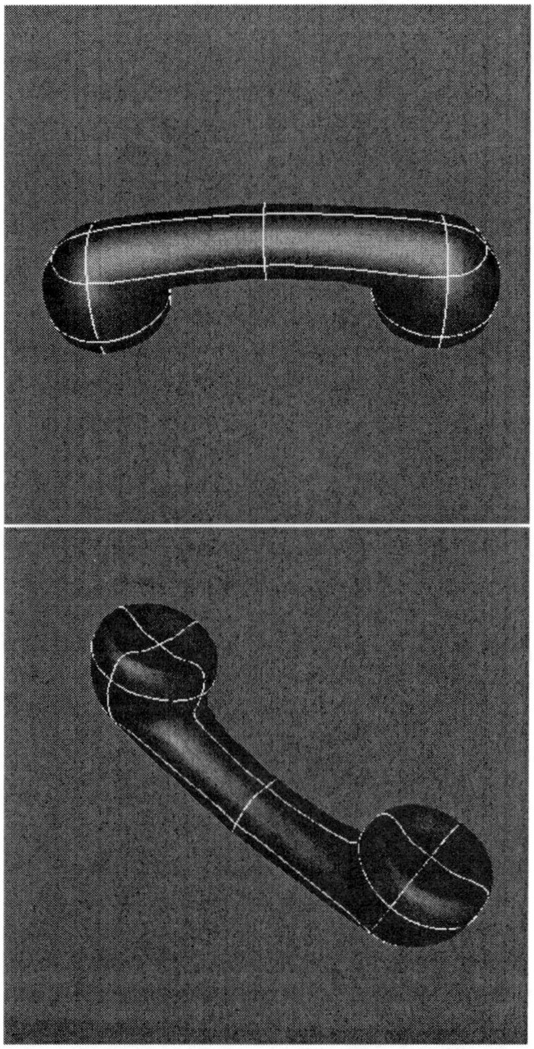

Figure 12. A phone handle interpolating an arbitrary intersecting mesh of curves with various n-sided regions beteween the interpolated curves.

7. Acknowledgments

This work was supported by a research grant from the American University of Beirut. The author would like to thank the anonymous referees for their comments. Thanks are also due to Mr. A. Charara for his help in producing some of the included pictures.

References

[1] A. Ball and D. Storry. Recursively generated b-spline surfaces. In *Proceedings Of Computer-Aided Design 84*, pages 112–119, 1984.

[2] E. Catmull and J. Clark. Recursively generated b-spline surfaces on arbitrary tmeshes. In *Seminal Graphics, Ed. Rosalee Wolfe*, pages 183–188. Acm Press, 1998.

[3] D. Doo and M. Sabin. Behaviour of recursive division surfaces near extraordinary points. In *Seminal Graphic, Ed. Rosalee Wolfe*, pages 177–181. Acm Press, 1998.

[4] D. W. Doo. A method of smoothing highly irregular polyhedrons. In *Proceedings Of Computer-Aided Geometric Design*. Bologna, 1978.

[5] D. Dyn, N. Levin. Interpolating subdivision schemes for the generation of curves and surfaces. In *Multivariate Approximation And Interpolation, Ed. W. Haubβ Mann And K. Jetter*. Birkhauser Verlag, Basel, 1990.

[6] J. Dyn, N. Gregory and D. Levin. A butterfly subdivision scheme for surface interpolation with tension control. *ACM Transaction On Graphics*, 9(2):160–169, 1987.

[7] J. Dyn, N. Gregory and D. Levin. A four-point interpolatory subdivision scheme for curve design. *Computer-Aided Geometric Design*, 4:257–268, 1987.

[8] J. Gregory and R. QU. Non-uniform corner cutting. *Computer-Aided Geometric Design*, 1996.

[9] A. Habib and J. Warren. Edge and vertex insertion for a class of c^1 subdivision surfaces. *Computer Aided Geometric Design*, 16:223–247, 1999.

[10] M. Halstead, M. Kass and T. De Rose. Efficient, fair interpolation using catmull-clark surfaces. In *Computer Graphics Proceedings, ACM SIGGRAPH 93*, pages 35–44, 1993.

[11] H. Hoppe, T. Derose, T. Duchamp, M. Halstead, H. Jin, J. Mcdonald, J. Schweitzer, and W. Stuetzle. Piecewise smooth surface reconstruction. In *Computer Graphics Proceedings, ACM SIGGRAPH 94*, pages 295–302, August 1994.

[12] L. Kobbelt. Interpolatory subdivision on open quadrilateral nets with arbitrary topology. *Computer Graphics Fourm*, 15:409–420, 1996.

[13] A. Levin. Interpolating nets of curves by somooth subdivision surfaces. In *Computer Graphics Proceedings, SIGGRAPH 99*, pages 57–64, 1999.

[14] C. Loop. Smooth spline surfaces over irregular meshes. In *Computer Graphics Proceedings, ACM SIGGRAPH 94*, pages 303–310, 1994.

[15] M. Lounsbery, T. Derose, and J. Warren. Multiresolution analysis for surfaces of arbitrary topological type. *ACM Transactions On Graphics*, 16(1):34–73, 1997.

[16] A. Nasri. Polyhedral subdivision methods for free form surfaces. *ACM Transactions On Graphics*, 8(1):89–96, 1987.

[17] A. Nasri. Interpolation on irregular networks with normal conditions. *Computer Aided Geometric Design*, 8:89–96, 1991.

[18] A. Nasri. A recursive subdivision surface through an irregular mesh of curves. Presented At The $4Th$ Siam Conferene On Geometric Design, Nashville, 1995.

[19] A. Nasri. Curve interpolation in recursively generated b-spline surfaces over arbitrary topology. *Computer Aided Geometric Design*, 1:13–30, 1997.

[20] A. Nasri. Interpolation of open curves by recursive subdivision surfaces. *The Mathematics Of Surfaces VII*, pages 173–188, 1997.

[21] A. Nasri. Interpolating meshes of boundary intersecting curves by subdivision surfaces. *The Visual Computer*, to Appear, 2000.

[22] A. Nasri. Recursive subdivision of polygonal complexes and its applications in CAGD. *Computer Aided Geometric Design*, to appear. Presented at The $5th$ Siam Conferene On Geometric Design, Nashville, 1997.

[23] A. Nasri and A. Abbas. Designing polygonal complexes for curve interpolation by subdivision surfaces. In Preparation, 1999.

[24] A. Nasri and M. Sabin. Interpolation conditions on recursive subdivision curves and surfaces: A taxonomy. Technical Report Tr-97/2, American University Of Beirut, 1999. Accepted in the 6th SIAM Conferene On Geometric Design, Alberqueque, 1999.

[25] J. Peters and U. Reif. The simplest subdivision scheme for smoothing polyhedra. *Acm Transactions On Graphics*, 16(4), 1997.

[26] H. Prautzsch. Analysis of c^k-subdivision surfaces at extraordinary points. Preprint, Ibds, Universitat Karlsruhe, 1995.

[27] U. Reif. A unified approach to subdivision algorithms near extraordinary vertices. *Computer Aided Geometric Design*, 12:153–174, 1995.

[28] M. Sabin. Recursive subdivision. In *The Mathematics Of Surfaces, Ed. J. A. Gregory*, pages 269–282. Clarendon Press, Oxford, 1986.

[29] J. Schweitzer. *Analysis And Applications Of Subdivision Surfaces*. PhD thesis, University Of Washington Seattle, 1996.

[30] T. Sederberg, J. Zheng, D. Sewell, and M. Sabin. Non-uniform recursive subdivision surfaces. In *Computer Graphics Proceedings, ACM SIGGRAPH 98*, pages 387–394, July 1998.

[31] H. Weimer and J. Warren. Subdivision schemes for fluid flow. In *Computer Graphics Proceedings, ACM SIGGRAPH 99*, 1999.

[32] D. Zorin and P. Schröder. Subdivision for modeling and animation, SIGGRAPH'99 course notes, 1999.

Catmull-Clark Surface Fitting for Reverse Engineering Applications

Weiyin Ma and Nailiang Zhao
Department of Manufacturing Engineering and Engineering Management
City University of Hong Kong, 83 Tat Chee Avenue, Kowloon, Hong Kong
E-mail: {mewma, menlzhao}@cityu.edu.hk

Abstract

Reverse engineering is an approach for reconstructing a computer model from physical object through dimensional measurement and surface modelling. Various mathematical models have be discussed for the representation of freeform surfaces in the context of reverse engineering applications. Most of the existing algorithms are, however, mainly developed for fitting isolated surfaces and one must smoothly connect these surfaces afterwards. This paper presents a procedure for simultaneously fitting smoothly connected multiple surfaces from point clouds with arbitrary topology. The final fitted surfaces are represented as Catmull-Clark surfaces, a network of smoothly connected bicubic B-spline surfaces with a finite number of B-spline subdivision surface patches next to extraordinary corner points. The final fitted surfaces are perfect G2 continuous across all surface boundaries except at a finite number of extraordinary points where G1 continuity is obtained. The algorithm is purely a linear least squares fitting procedure without any constraints.

1. Introduction

In manufacturing engineering, free-form geometry plays an important role in shape representation, product design and assembly simulation. Reverse engineering provides an alternative way for geometric shape design from physical mockup or prototypes. One could find many applications in the field of physical model-based free-form shape design. Typical examples include household utensils, toy making, motorcycle and automobile industries for initial aesthetic design. Physical mockups are also used in aircraft and ship-building industries and mold and die making industries in general for engineering tests. One often needs to correct a CAD model based on changes made to physical mockups.

As shown in Figure 1, the process of reverse engineering can usually be subdivided into three major stages, i.e. 1) data capturing, 2) data segmentation and topological modelling, and 3) surface modelling. A physical model is first measured by a co-ordinate measuring machine (CMM), a laser scanner or other digitizing equipment. The measured data are further segmented into topological areas for surface modelling. A complete CAD model is finally created through surface fitting and other geometric processing algorithms. For the examples used in this paper, the raw measured data come from a medical CT-scanner as a set of 2-D cross sectional gray-scale images, often called CT-images. The CT-images are further processed to obtain a set of 3-D surface contours of the object, known as CT-contours. Each of the CT-contours is composed of a series of surface points in 3-D space. A collection of the CT-contour points form a cloud of 3-D surface points of the measured object. The present paper is mainly focused on the last stage for smooth multiple surfaces modelling from point clouds, the central issue of reverse engineering [13].

In literature, there have been various attempts for handling objects with arbitrary topology and complex free-form shape using ordinary B-spline surfaces in the context of reverse engineering applications. Among others, Milroy et al proposed a practical method for simultaneously fitting multiple B-spline surfaces by introducing geometric continuity constraints using checkpoints along the boundaries of individual surfaces [9]. It however results in an expensive least squares fitting problem with nonlinear continuity constraints and the final fitted surfaces are approximate G^1 continuous across surface boundaries. Ma et al proposed an efficient linearly constrained least squares fitting method using general periodical knots and local check points near an extraordinary corner [8]. The approach achieves global perfect high order geometric continuity except along boundary segments near a finite number of extraordinary corners where approximate G^1 continuity is obtained. Loop and DeRose [6] also presented an approach for constructing generalized B-spline surfaces from control meshes using n-sided S-patches. This approach has not yet be applied to reverse engineering applications. Eck and Hoppe [3] also developed a constructive procedure for automatically generating B-spline surface models with arbitrary topology through mesh optimization. Eck and

0-7695-0562-7/00 $10.00 © 2000 IEEE

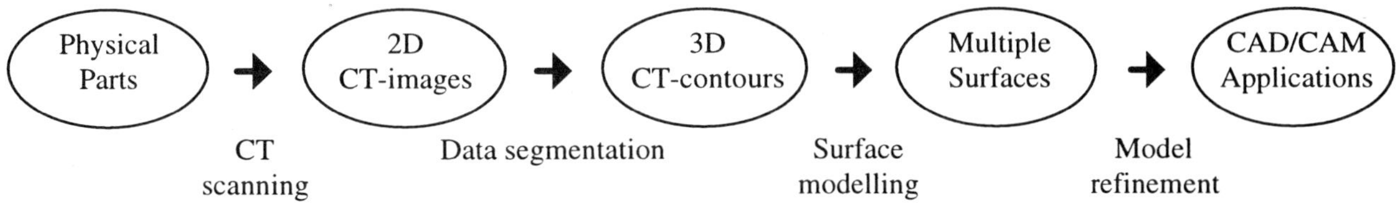

Figure 1. CT-based reverse engineering for product design and shape modelling

Hoppe's approach achieves G^1 and C^1 continuity conditions between individual surfaces.

Catmull-Clark surfaces, on the other hand, offer a promising alternative and exhibit a number of advantages over ordinary B-spline surfaces [1, 14]:

- With Catmull-Clark surfaces, it is quite easy to handle free-form objects with arbitrary topology, a challenging issue for classical splines.
- The control mesh is identical with the topological structure of the object and there is no need for handling extra constraints among individual surface patches.
- Catmull-Clark surfaces are curvature continuous between all neighboring surfaces except at a finite number of isolated extraordinary corner points where G^1 continuity is achieved.
- It possesses all elegant properties of subdivision surfaces, such as simplicity, efficiency, level-of-detail (LOD) representation and multi-resolution modelling.

Among the latest developments, Halstead et al [5] proposed an efficient scheme for constructing Catmull-Clark subdivision surfaces by interpolating a set of mesh vertices along with a subset of normal vectors. Qin et al [10] developed a physics-based model to dynamically approximate objects with arbitrary topology using Catmull-Clark surfaces through mesh refinement. One could also find some results in [12] on direct and exact evaluation of Catmull-Clark surfaces and in [11] on non-uniform subdivision of Catmull-Clark surfaces. These results are however mainly focused on shape modelling rather than surface reconstruction from range data.

This paper presents a least squares fitting approach for reconstructing a digital representation using Catmull-Clark surfaces from physical parts with arbitrary topology. Compared with approaches using ordinary B-spline surfaces, our approach is purely a linear least squares fitting problem without any geometric constraints. The resulting surfaces are a network of smoothly connected B-spline or Catmull-Clark surfaces with perfect G^2 continuity, except at a finite number of extraordinary corner points at which G^1 continuity is achieved. To our knowledge, this effective method for fitting Catmull-Clark subdivision surfaces has not been discussed previously [2,12-14]. Since the exact coordinates of a surface point and its derivatives of the resulting surfaces can be efficiently evaluated with standard B-spline basis functions at any position, the final fitted Catmull-Clark surface model is suitable for down stream applications, such as shape design, surface local updating and further NC processing.

2. General strategies

Catmull-Clark surfaces are generalization of uniform bicubic B-splines [1]. They are defined as a set of quadrilateral bicubic surfaces with special subdivision treatment for patches next to extraordinary corners. Catmull-Clark surfaces are quite suitable for modelling geometric shapes of arbitrary topology.

Subdivision schemes have gained wide applications in computer aided geometric design (CAGD) and computer graphics [2]. In connection to Catmull-Clark surfaces, there have been two major progresses in this area. One of the achievements is the understanding of continuity behaviors near extraordinary corners of subdivision surfaces [14]. The other one is the efficient evaluation of Catmull-Clark subdivision patches at arbitrary parameters [12]. These developments make subdivision surfaces an even more attractive tool for free-form surface modeling. The exact evaluation of Catmull-Clark surfaces also makes it possible to develop efficient algorithms for linear least squares fitting of Catmull-Clark surfaces from point clouds addressed in this paper. There are several major stages for reconstructing Catmull-Clark subdivision surfaces from point clouds, i.e.

- forming the mesh structure of all control points of the Catmull-Clark surfaces to be identified through topological modelling,

- setting up the observation system through data partition and parametrization using a Base Surface approach [7], and
- identifying the control points of Catmull-Clark surfaces through linear least squares fitting.

In this paper, we don't directly construct the control mesh, but only define several quadrilateral domains as global topological structure of the object. The control mesh of the Catmull-Clark surfaces will be automatically generated. The solution of all control points of Catmull-Clark subdivision surfaces is obtained through linear least squares fitting without any constraints. The whole procedure for reconstructing Catmull-Clark subdivision surfaces from range data are further explained as follows.

2.1. Topological structure modelling

At present, we define the topological mesh through an interactive interface for topological subdivision of point clouds [8]. The quadrilateral topological structure of the object is first manually defined as a network of boundary curves from the measured points $\{\overline{\mathbf{P}}_i \in R^3\}_{i=1}^{m}$. As illustrated in Figure 5 and Figure 7a, the four boundaries of each quadrilateral domain are represented as B-spline curves and are defined by fitting through a limited number of interactively selected sample points. Each boundary curve may be shared with at most two adjacent topological domains and any number of boundaries may meet at a common corner. Internal corners with number of meeting boundary curves other than four are called extraordinary corners. The number of meeting boundaries is called the valence of the extraordinary corner. An external corner with similar topology as an internal extraordinary corner could also be an extraordinary corner. These conventions are very convenient for defining complex object with arbitrary topology.

2.2. Base surface creation

After topological structure modelling, a set of approximate B-spline surfaces $\{\overline{\mathbf{S}}_j(u,v)\}_{j=1}^{s}$ are created as base surfaces for the partition and parametrization of the measured points [7]. The base surfaces are actually the initial approximation of the final fitted Catmull-Clark surfaces. These base surfaces are easily created from the four boundary curves of each quadrilateral topological domain with or without additional uni-directional interior curves for accurate approximation. To obtain a consistent topological structure, we require all surface normal vectors pointing towards the same side of the surfaces. With base surface parametrization discussed in the following sub-section, the topological structure of the final fitted surfaces will be the same as the base surfaces. The base surfaces of a computer mouse are illustrated in Figure 7c. Further details on base surface definition can be found in [7].

2.3. Data partition and parametrization

Data partition and parametrization are realized through a projection process using the base surface approach reported in [7]. After projecting all measured points $\{\overline{\mathbf{P}}_i \in R^3\}_{i=1}^{m}$ onto the multiple base surfaces $\{\overline{\mathbf{S}}_j(u,v)\}_{j=1}^{s}$, each point $\overline{\mathbf{P}}_i$ in the point clouds will be distributed to an individual topological domain and associated with an index j of a base surface with $\{j \mid 1 \le j \le s\}$. Two location parameters (u_i, v_i) of $\overline{\mathbf{P}}_i$ relative to the corresponding base surface are also simultaneously obtained. After all the measured data are processed, one obtains a set of location parameters $\left\{\{(u_{ij}, v_{ij})\}_{i=1}^{m_j}\right\}_{j=1}^{s}$ corresponding to the partitioned data point $\left\{\{\overline{\mathbf{P}}_{ij}\}_{i=1}^{m_j}\right\}_{j=1}^{s}$, where m_j is the number of measured points which belong to base surface $\overline{\mathbf{S}}_j$ with $\sum_{j=1}^{s} m_j = m$.

2.4. LSQ observation system setup

An integrated observation system for identifying the unique control points of Catmull-Clark surfaces is then set up with special treatment in handling boundary control points and patches next to extraordinary corners. This is the key issue of this paper and will be further discussed in the following section. The general procedure includes the following steps: a) setting up a sub-observation system with ordinary B-spline surfaces and measured points falling on regular patches, b) eliminating repeating control points between adjacent surfaces for maintaining G^2 continuity, c) setting up a sub-observation system with Catmull-Clark subdivision surfaces and measured points falling on extraordinary corner patches, and d) formulating an integrated observation system for the entire Catmull-Clark surfaces. It should be noted that the process is carried out automatically and in an efficient manner compared with other approaches for smooth multiple surface fitting [3,8-9].

2.5. Solution to the LSQ system

The observation system set up in the previous step for Catmull-Clark surface fitting is completely linear without any constraints. The independent control points are then solved in least-squares sense with numerical approaches. There are very stable and efficient algorithms for solving non-constrained linear least squares fitting problems. The solution for Catmull-Clark surface fitting is indeed a network of smoothly connected B-spline surfaces with perfect G^2 continuity except at a finite number of extraordinary corner points where G^1 continuity is achieved. The final fitted Catmull-Clark surfaces of a computer mouse are illustrated in Figure 7e.

3. Catmull-Clark surface fitting

Catmull-Clark surfaces are defined by B-spline subdivision surface patches next to an extraordinary corner and ordinary B-Spline surface patches elsewhere. A surface patch next to an extraordinary corner is called an extraordinary patch and all remaining surface patches are called regular surface patches. The entire network of surfaces are defined by the control point mesh of Catmull-Clark surfaces. In order to obtain the control point mesh of Catmull-Clark surfaces, we first define a set of individual B-spline surfaces $\left\{\mathbf{S}_j(u,v)\right\}_{j=1}^{s}$ over the topological domains. An integrated observation system is further set up for a least squares fitting solution of the entire control mesh. For each surface $\mathbf{S}_j$, for $1 \le j \le s$, we first process the observations on regular patches. Measured data on extraordinary patches of surface $\mathbf{S}_j$ will be further handled with subdivision surfaces.

3.1. Regular surface parches

All regular patches of an individual surface are defined by the following ordinary B-spline surface:

$$\mathbf{S}(u,v) = \sum_{i=1}^{n_u} \sum_{j=1}^{n_v} B_{ui}(u) B_{vj}(v) \cdot \mathbf{V}_{ij} \tag{1}$$

where u and v are parameters that uniquely locate a point on the surface, n_u and n_v are the numbers of control points along the u and v-directions, respectively, $\mathbf{V}_{ij}$ for $i = 1,2,\ldots,n_u$ and $j = 1,2,\ldots,n_v$ are the control points of the surface, $B_{ui}(u)$ for $i = 1,2,\ldots,n_u$ and $B_{vj}(v)$ for $j = 1,2,\ldots,n_v$, are normalized B-splines, called basis functions that can be evaluated by the recursive Cox-de Boor algorithm [4]. The two sets of basis functions are uniquely defined by knots $\xi := \left\{\xi_i \mid i = 1,2,\ldots,n_u + k_u\right\}$ and $\zeta := \left\{\zeta_j \mid j = 1,2,\ldots,n_v + k_v\right\}$, respectively, where k_u and k_v are the orders of the B-spline surface. For a single B-spline surface, there are in total $(n_u - k_u + 1) \times (n_v - k_v + 1)$ surface patches.

To agree with Catmull-Clark surfaces, we use general uniform knots with no multiplicity, i.e. periodical knots, and with $k_u = k_v = 4$ for all B-spline surfaces over quadrilateral domains. The number of control points of two neighboring surfaces along the common boundary should be the same and the two surfaces also share three rows of boundary control points as shown in Figure 2. With the above-mentioned conventions, one obtains G^2 geometric continuity along the common boundary and at a regular corner point. This approach is also commonly used for defining closed surfaces in geometric modelling.

Let $\left\{\mathbf{S}_j(u,v)\right\}_{j=1}^{s}$ denote the multiple surfaces to be fitted as defined by Equation (1) and $\left\{\left\{\overline{\mathbf{P}}_{ij}\right\}_{i=1}^{m_j}\right\}_{j=1}^{s}$ be the partitioned data points with corresponding location

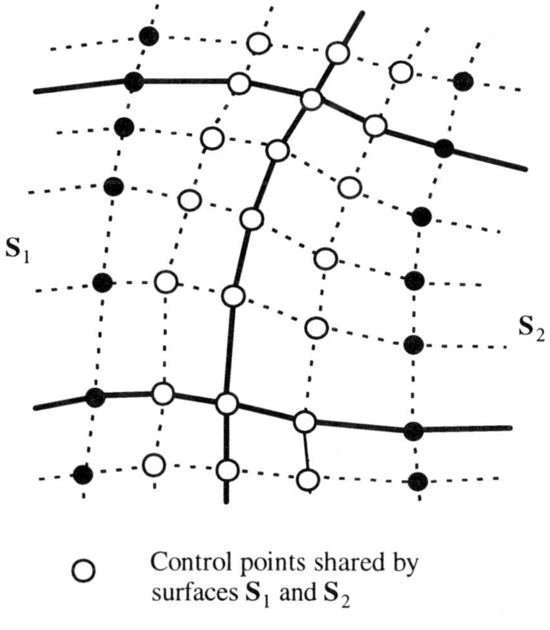

Figure 2. Control point repetition along a common boundary

parameters $\left\{\left\{(u_{ij},v_{ij})\right\}_{i=1}^{m_j}\right\}_{j=1}^{s}$. To simplify the notations, let us assume that for each set of measured points $\left\{\overline{\mathbf{P}}_{ij}\right\}_{i=1}^{m_j}$, for $j=1,2,\ldots,s$, the first $m_j^{'}$ measured points, for $m_j^{'} \le m_j$, are located on regular patches. By introducing the data points falling on regular surface patches and the corresponding location parameters into Equation (1) and rearranging the equations in matrix form, one obtains the following observation system

$$\begin{cases} \tilde{\mathbf{B}}_j \cdot \tilde{\mathbf{X}}_j = \overline{\mathbf{X}}_j \\ \tilde{\mathbf{B}}_j \cdot \tilde{\mathbf{Y}}_j = \overline{\mathbf{Y}}_j \\ \tilde{\mathbf{B}}_j \cdot \tilde{\mathbf{Z}}_j = \overline{\mathbf{Z}}_j \end{cases} \quad \text{for} \quad 1 \le j \le s \tag{2}$$

In Equation (2) $\tilde{\mathbf{X}}_j$, $\tilde{\mathbf{Y}}_j$ and $\tilde{\mathbf{Z}}_j$ represent the collection of x-, y- and z-coordinates of the control points of the j-th surface, $\overline{\mathbf{X}}_j$, $\overline{\mathbf{Y}}_j$ and $\overline{\mathbf{Z}}_j$ represent the collection of x-, y- and z-coordinates of the measured points of the j-th surface falling on regular patches, and $\tilde{\mathbf{B}}_j$ is the j-th observation matrix defined by

$$\tilde{\mathbf{B}}_j = \begin{bmatrix} B_{1j}(\cdot_{1j}) & B_{2j}(\cdot_{1j}) & \cdots & B_{nj}(\cdot_{1j}) \\ B_{1j}(\cdot_{2j}) & B_{2j}(\cdot_{2j}) & \cdots & B_{nj}(\cdot_{2j}) \\ \vdots & \vdots & & \vdots \\ B_{1j}(\cdot_{m_j^{'}j}) & B_{2j}(\cdot_{m_j^{'}j}) & \cdots & B_{nj}(\cdot_{m_j^{'}j}) \end{bmatrix} \tag{3}$$

where $(\cdot_{ij}) = (u_{ij}, v_{ij})$ for $i = 1,2,\ldots,m_j^{'}$, $j = 1,2,\ldots,s$. The s observation systems of Equation (2) can be collectively represented by the following general observation equations

$$\begin{cases} \tilde{\mathbf{B}} \cdot \tilde{\mathbf{X}} = \overline{\mathbf{X}}' \\ \tilde{\mathbf{B}} \cdot \tilde{\mathbf{Y}} = \overline{\mathbf{Y}}' \\ \tilde{\mathbf{B}} \cdot \tilde{\mathbf{Z}} = \overline{\mathbf{Z}}' \end{cases} \tag{4}$$

where $\tilde{\mathbf{X}}$, $\tilde{\mathbf{Y}}$ and $\tilde{\mathbf{Z}}$ represent the simple collection of x-, y- and z-coordinates of all related control points $\tilde{\mathbf{X}}_j$, $\tilde{\mathbf{Y}}_j$ and $\tilde{\mathbf{Z}}_j$ of all surfaces, $\overline{\mathbf{X}}'$, $\overline{\mathbf{Y}}'$ and $\overline{\mathbf{Z}}'$ represent the collection of x-, y- and z-coordinates of all the measured points falling on regular surface patches, and $\tilde{\mathbf{B}} = diag\{\tilde{\mathbf{B}}_1, \tilde{\mathbf{B}}_2, \cdots, \tilde{\mathbf{B}}_s\}$ is a general observation matrix.

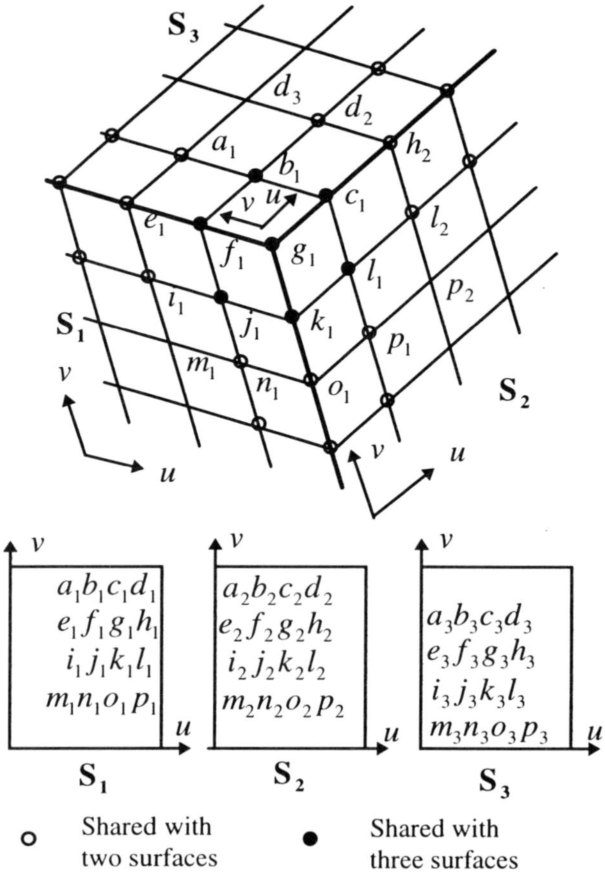

Figure 3. Control points repetition near an extraordinary corner

Table 1. The relationship between control points shared by three surfaces near an extraordinary corner showing in Figure 3.

S_1		S_2		S_3
b_1	=	c_2	=	g_3
$c_1 = h_1$	=	g_2	=	k_3
f_1	=	$b_2 = e_2$	=	f_3
g_1	=	f_2	=	j_3
j_1	=	i_2	=	e_3
k_1	=	j_2	=	$i_3 = n_3$
l_1	=	k_2	=	o_3

As discussed in the previous paragraphs, we use general uniform knots with repeating control points to obtain G^2 continuity between neighboring surfaces. This leads to the use of boundary control points being shared by two or more surfaces as shown in Figure 2 and Figure 3. It should be noticed that there are three extraordinary patches near the extraordinary corner shown in Figure 3. This leads to control points near the extraordinary corner shared by three surfaces. The relationships among relevant control points shared by three surfaces shown in Figure 3 are summarized in Table 1. As a result, the unknowns in the individual sub-observation systems are not independent from each other. After eliminating all the repeating control points in Equation (4), we obtain a revised system of linear equations as follows

$$\begin{cases} \hat{\mathbf{B}} \cdot \hat{\mathbf{X}} = \overline{\mathbf{X}}' \\ \hat{\mathbf{B}} \cdot \hat{\mathbf{Y}} = \overline{\mathbf{Y}}' \\ \hat{\mathbf{B}} \cdot \hat{\mathbf{Z}} = \overline{\mathbf{Z}}' \end{cases} \tag{5}$$

where $\hat{\mathbf{X}}$, $\hat{\mathbf{Y}}$ and $\hat{\mathbf{Z}}$ represent the collection of x-, y- and z-coordinates of all relevant and independent control points and $\hat{\mathbf{B}}$ is the revised observation matrix after the elimination of repeated control points. In addition, all control points that are not referenced by regular patches near extraordinary corners are also eliminated while constructing Equation (5). In the case of the extraordinary corner shown in Figure 3, the three corner control points d_1 of surface $\mathbf{S}_1$, a_2 of surface $\mathbf{S}_2$ and m_3 of surface $\mathbf{S}_3$ are not needed for evaluating regular patches and these corner control points do not appear in Equation (5).

3.2. Extraordinary corner patches

In case of extraordinary surface patches, we use Catmull-Clark subdivision surfaces, which ensures smooth G^2 continuity between two neighboring patches except at the extraordinary corner where G^1 continuity condition is obtained. Let N be the valence of an extraordinary corner, one can easily evaluate surface points on an extraordinary patch at any parameter locations in terms of a set of $K=2N+8$ eigenbasis functions and the K corresponding mesh control points near the extraordinary corner with the parametrization of Catmull-Clark subdivision surfaces by Stam [12]. We briefly summarize the procedure for evaluating Catmull-Clark surfaces through the example illustrated in Figure 4.

In Figure 4, there are five surfaces meeting at an extraordinary corner with valence $N=5$. After control points elimination, the control mesh of the Catmull-Clark subdivision surfaces is obtained. There are five extraordinary patches around the central extraordinary corner. We select one of these extraordinary patches, the shaded domain on surface $\mathbf{S}_1$ for an illustration. The global parameters of this extraordinary patch are defined as the same as the original B-spline surface $\mathbf{S}_1$. These parameters should be first transformed into the local parameter domain $\Omega = [0,1] \times [0,1]$, such that the extraordinary patch $\mathbf{s}(u,v)$ for $(u,v) \in \Omega$ satisfies the following conditions:

1) The patch corner $\mathbf{s}(0,0)$ must correspond to the extraordinary corner, i.e. vertex #1 in Figure 4.
2) The surface normal $\mathbf{s}_u \times \mathbf{s}_v$ of all the extraordinary patches should point to the same side as other ordinary B-spline surfaces.

The $K=2N+8$ vertices $\{\mathbf{v}_i\}_{i=1}^{K}$ are then picked out from the control mesh of the Catmull-Clark subdivision surfaces and are reordered to form the control mesh of the corresponding extraordinary patch as shown in Figure 4. In the case of $N=3$, the vertices #2 and #8 become the same vertex. If any extraordinary corner of valence $N>4$ is open, additional control vertices should be added in order to form the complete control mesh. With the above assumptions, an extraordinary patch $\mathbf{s}(u,v)$ for $(u,v) \in \Omega$ can be expressed in terms of eigenbasis functions $\{\varphi_i(u,v)\}_{i=1}^{K}$ and the control mesh $\{\mathbf{v}_i\}_{i=1}^{K}$ by the following equation

$$\mathbf{s}(u,v) = \sum_{i=1}^{K} \varphi_i(u,v) \cdot \mathbf{v}_i \tag{6}$$

The eigenbasis functions $\{\varphi_i(u,v)\}_{i=1}^{K}$ can be pre-calculated from the eigenvalues and eigenvectors of the subdivision matrix for any valence. The calculation is needed only once and the data can be stored in a file for future use. One can find further details for constructing the eigenbasis functions in [12].

In order to construct the observation sub-systems corresponding to extraordinary patches, one can introduce each of the measured points that falls on extraordinary patches into Equation (6). A collection of all observations falling on an extraordinary patch form a new sub-observation system for the corresponding patch. Combining all the extraordinary sub-observation systems with Equation (5), one obtains the global observation system for LSQ fitting of Catmull-Clark surfaces:

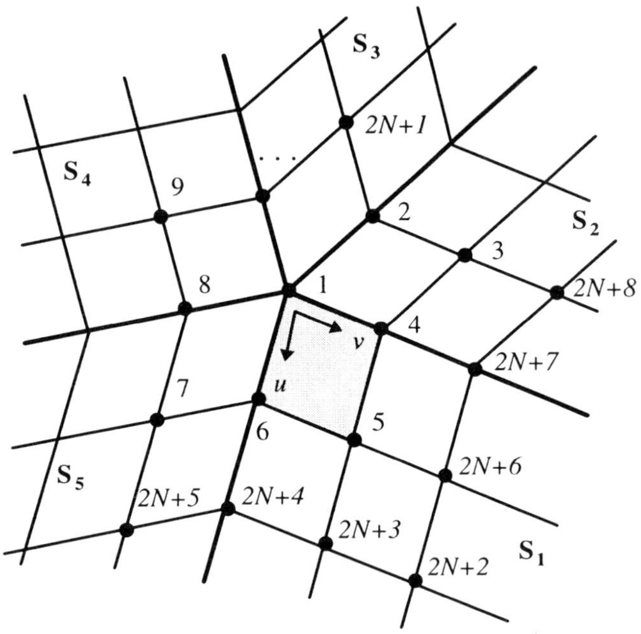

Figure 4. Construction of extraordinary surface patches

$$\begin{cases} \mathbf{B} \cdot \mathbf{X} = \overline{\mathbf{X}} \\ \mathbf{B} \cdot \mathbf{Y} = \overline{\mathbf{Y}} \\ \mathbf{B} \cdot \mathbf{Z} = \overline{\mathbf{Z}} \end{cases} \quad (7)$$

where $\mathbf{X}$, $\mathbf{Y}$ and $\mathbf{Z}$ represent the collection of x-, y- and z-coordinates of all control points of the Catmull-Clark surfaces, $\overline{\mathbf{X}}$, $\overline{\mathbf{Y}}$ and $\overline{\mathbf{Z}}$ represent the collection of x-, y- and z-coordinates of all the measured points, and $\mathbf{B}$ is the general observation matrix. Since the observation system defined by Equation (7) is purely linear without any constraints, one could solve the system with any linear least squares fitting algorithms similar to ordinary B-spline surface fitting.

4. Examples and discussions

The algorithms discussed in this paper are implemented in an own-developed modelling environment on Windows NT platform. The final fitted surfaces are represented as a B-rep surface/solid model. The B-rep model is defined as Catmull-Clark subdivision surfaces, a network of G^2 smoothly connected B-spline surfaces except at a finite number of extraordinary corner points where G^1 continuity is achieved. A mathematical library NAG from the Numerical Algorithms Groups Ltd., U.K., is used to solve the integrated observation equations.

Figure 5 illustrates two examples of fitted Catmull-Clark subdivision surfaces. In these examples, there is only one extraordinary corner located at the center of the two multiple surfaces. For the example shown in Figures 5a, 5b and 5c, the surfaces are defined with $n_u = n_v = 15$. There are in total $12 \times 12 = 144$ patches for each of the surfaces. There is only one extraordinary corner patch with valences *N=3* as shown in Figure 5c. Figure 5b shows the same fitted surfaces as those in Figure 5a in wire frame showing quadrilateral boundaries. Figure 5c illustrates all patches of the surfaces and the three extraordinary patches rendered in different colors/shadings. The example shown in Figure 5d, 5e and 5f is similar to the previous one with $n_u = n_v = 10$ and valence *N=5*.

A complete example of a computer mouse is also presented in Figure 7. The original measured data are acquired from a plastic body of a computer mouse through a medical CT-scanner. The dimension of the physical object is about $115mm \times 70mm \times 35mm$. A total number of 117 slices of contour images are obtained at an interval of $1mm$. After preprocessing, 3D CT-contours are obtained as shown in Figure 7a along with the topological structure of the mouse. For clarity, only one fifth of contours are displayed in Figure 7a. The topological structure shown in Figure 7a contains eleven inter-connected and two independent quadrilateral domains. The individual measured points used for surface fitting are illustrated in Figure 7b. All base surfaces of the quadrilateral meshes are generated without considering the continuity between the surfaces and are shown in Figure 7c. The final fitted mouse model is shown with shaded image in Figure 7d and Figure 7e. The topological boundaries are also illustrated in Figure 7d. For this example, the mouse end surface is defined with $n_u = n_v = 9$ and all other surfaces are defined with either $n_u \times n_v = 5 \times 9$ or $n_u \times n_v = 9 \times 5$. There are in total four extraordinary corners of valence *N=3*. The fitting errors are shown in Figure 7f with an amplifying scale of two. The maximum fitting error of all measured points is $1.4131mm$ and the average fitting error is $0.2458mm$. Figure 6 shows the control mesh of a part of the Catmull-Clark subdivision surfaces of the mouse. Figure 8 illustrates another example for fitting a piece of human femur bone. In this case, there are in total 15 inter-connected topological surface domains with eight extraordinary corners of valences *N=3* and *N=5*.

To speed up computation, several techniques can be adopted. As a result of the local control property of Catmull-Clark subdivision surfaces, the integrated observation system can be solved through a local fitting process without sacrificing the global optimal fitting

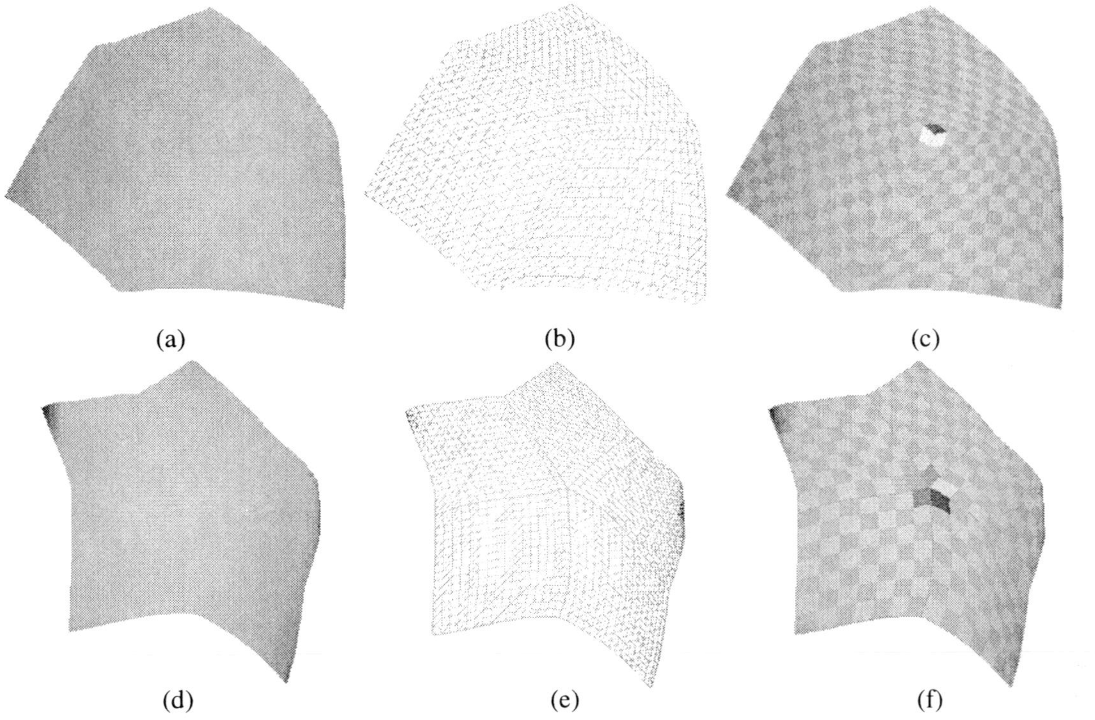

Figure 5. Catmull-Clark subdivision surfaces with valence *N=3* and *N=5*:
(a) final fitted surfaces in shaded image; (b) wire frame model showing topological boundaries; (c) extraordinary patches rendered in different colors/shadings; (a)-(c) for *N=3*; (d)-(f) for *N=5*.

conditions of the multiple surfaces. Data filtering is also particularly important for modelling objects with complex topology. In addition, the distribution of the measured points for smooth multiple surface fitting can also be filtered and optimized. For the modelling of the computer mouse, the original CT-contours contain 94,146 points. After initial filtering, there are still 50,362 measured points. The actual total number of points used for fitting the computer mouse as shown in Figure 7b is actually 288*12+576=4,032 after further filtering. In this case, we randomly selected 3*3=9 observations for each internal patches. The number of observations on a boundary patch is twice as many as an internal patch. The number of observations on a corner patch is twice as many as a general boundary patch.

The developed algorithms are quite efficient for fitting smoothly connected multiple surfaces. It takes about 1.5 minutes to carry out the whole process for fitting the mouse surfaces on a Pentium-II 400 PC with 255M RAM.

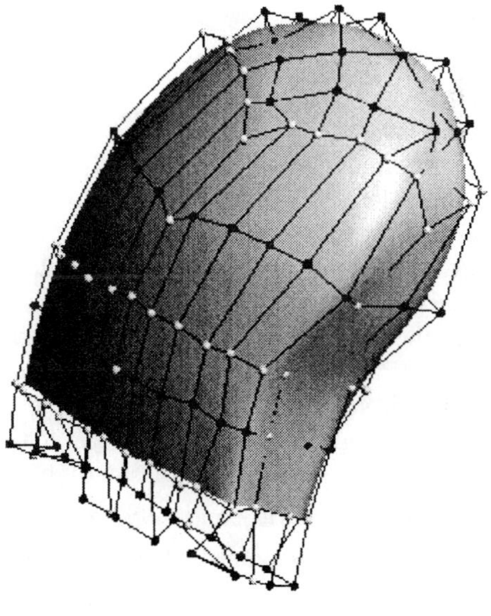

Figure 6. Catmull-Clark control mesh of a part of computer mouse

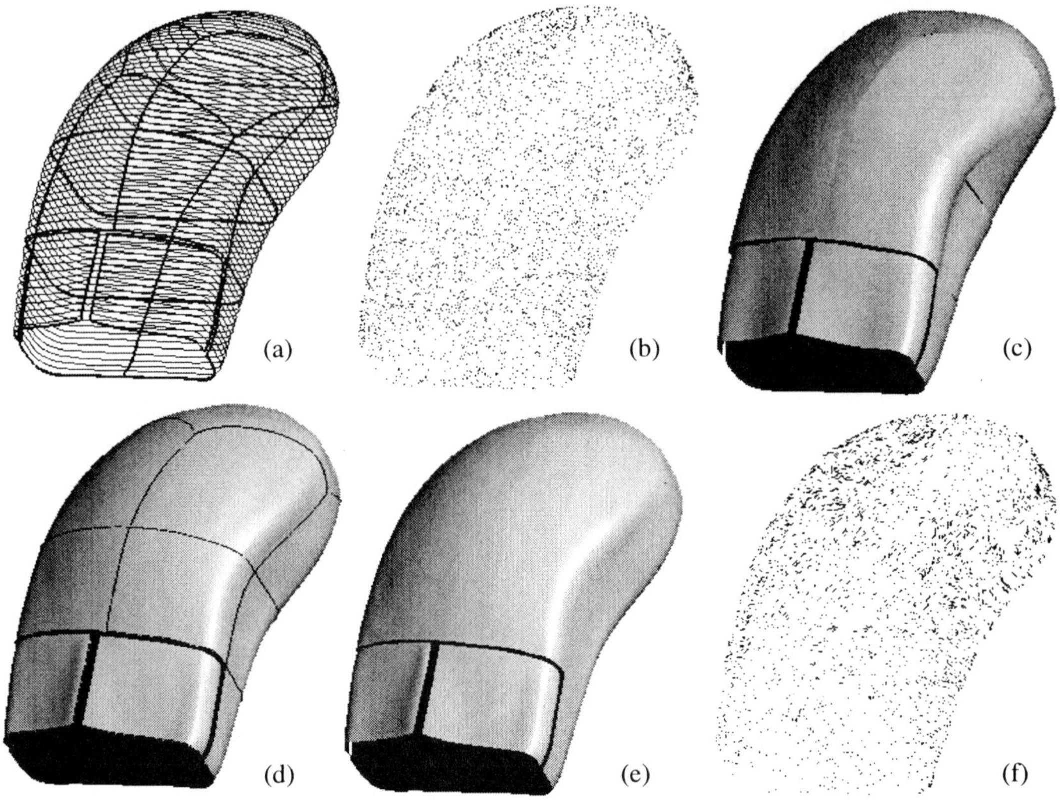

Figure 7. Construction of Catmull-Clark surfaces of a computer mouse

5. Conclusions

This paper presents an approach for reconstructing smoothly connected Catmull-Clark subdivision surfaces from measured data of physical objects. A quadrilateral topological structure is first defined through an interactive interface. A set of base surfaces are then created from the boundary curves of the topology model. The measured points are further partitioned into individual surface areas and are parametrized for surface fitting. The control mesh of Catmull-Clark subdivision surfaces is then identified automatically through unconstrained linear least squares fitting. The final fitted surfaces are smoothly connected with perfect G^2 geometric continuity except at a finite number of extraordinary points where G^1 continuity is obtained.

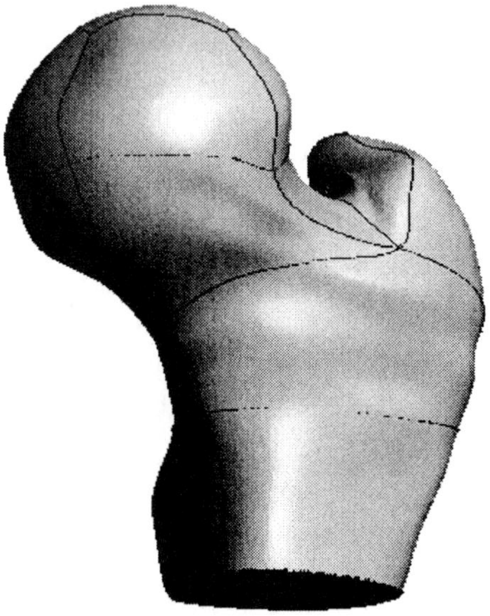

Figure 8. Fitted Catmull-Clark surfaces of a piece of human femur bone

Acknowledgments

The authors wish to thank the Research Grants Council of Hong Kong and the City University of Hong Kong for funding through research grants 9040225 and 7000861 respectively.

References

[1] Catmull, E., and J. Clark, "Recursively Generated B-spline Surfaces on Arbitrary Topological Meshes", *Computer-Aided Design*, Vol. 10, No. 6, 1978, pp. 350-355.

[2] Choi, B. K., *Surface Modeling for CAD/CAM*, Elsevier, Amsterdam, 1991.

[3] Eck, M., and H. Hoppe, "Automatic Reconstruction of B-Spline Surfaces of Arbitrary Topology Type", *Computer Graphics*, Vol. 30, 1996, pp. 325-334.

[4] Farin, G., *Curves and Surfaces for Computer Aided Geometric Design - A Practical Guide*, Academic Press, San Diego, 1992.

[5] Halstead, M., M. Kass, and T. DeRose, "Efficient, Fair Interpolation Using Catmull-Clark Surfaces", In *Proceedings of the ACM SIGGRAPH Conference on Computer Graphics*, 1993, pp. 35-44.

[6] Loop, C., and T. DeRose, "Generalized B-spline Surfaces of Arbitrary Topology", *Computer Graphics*, Vol. 24, No. 4, 1990, pp. 347-356.

[7] Ma, W., and J. P. Kruth, "Parametrization of Randomly Measured Points for the Least Squares Fitting of B-spline Curves and Surfaces", *Computer-Aided Design*, Vol. 27, No. 9, 1995, pp. 663-675.

[8] Ma, W., P. C. Leung, E. H. M. Cheung, S. Y. T. Lang, and H. Cui, "Smooth Multiple Surface Fitting for Reverse Engineering", In *Proceedings of the 32nd CIRP International Seminar on Manufacturing Systems*, Leuven, Belgium, May 24-26, 1999, pp. 53-62.

[9] Milroy, M. J., C. Bradley, G. W. Vickers, and D. J. Weir, "G^1 Continuity of B-Spline Surface Patches in Reverse Engineering", *Computer-Aided Design*, Vol. 27, No. 6, 1995, pp. 471-478.

[10] Qin, H., C. Mandal, and B. C. Vemuri, "Dynamic Catmull-Clark Subdivision Surfaces", *IEEE Transactions on Visualization and Computer Graphics*, Vol. 4, No. 3, 1998, pp. 215-229.

[11] Sederberg, T. W., J. Zheng, D. Sewell, and M. Sabin, "Non-Uniform Recursive Subdivision Surfaces", in *Proceedings of the ACM SIGGRAPH Conference on Computer Graphics*, 1998, pp. 387-393.

[12] Stam, J., "Exact Evaluation of Catmull-Clark Subdivision Surfaces at Arbitrary Parameter Values", in *Proceedings of the ACM SIGGRAPH Conference on Computer Graphics*, 1998, pp. 395-404.

[13] Várady, R. Martin, and J. Cox, "Reverse Engineering of Geometric Models - an Introduction", *Computer-Aided Design*, Vol. 29, No. 4, 1997, pp. 255-268.

[14] Zorin, D. and P. Schröder, Eds., *Subdivision for Modeling and Animation*, SIGGRAPH'99 Course Notes, Los Angeles, August 1999.

Session 9

Minkowski Roots, Parametrization, and Implicitization

Minkowski Roots of Complex Sets

Rida T. Farouki,[1] Weiqing Gu,[2] and Hwan Pyo Moon[1]

[1]*Department of Mechanical and Aeronautical Engineering, University of California, Davis, CA 95616*

[2]*Department of Mathematics, Harvey Mudd College, Claremont, CA 91711*

E-mail: `farouki@ucdavis.edu`, `gu@math.hmc.edu`, `hpmoon@ucdavis.edu`

Abstract

An n–th Minkowski root $\bigotimes^{1/n}\mathcal{A}$ of a given complex set $\mathcal{A}$ is defined by the property $\{\, \mathbf{z}_1\mathbf{z}_2\cdots\mathbf{z}_n \mid \mathbf{z}_i \in \bigotimes^{1/n}\mathcal{A} \,\} \equiv \mathcal{A}$, i.e., the set of all products of n independently chosen values from $\bigotimes^{1/n}\mathcal{A}$ is identical to $\mathcal{A}$. Hence, the n–th Minkowski power of $\bigotimes^{1/n}\mathcal{A}$ yields the original set $\mathcal{A}$. Minkowski root extractions are fundamental operations in the Minkowski geometric algebra of complex sets: depending on the nature of $\mathcal{A}$, subtle issues concerning the existence, uniqueness, and minimality or maximality of $\bigotimes^{1/n}\mathcal{A}$ may arise. For a domain $\mathcal{A}$ with a smooth boundary that is strictly logarithmically convex, we show that each connected component of the "ordinary" root $\mathcal{A}^{1/n} = \{\, \mathbf{z} \mid \mathbf{z}^n \in \mathcal{A} \,\}$ is a Minkowski n–th root. $\bigotimes^{1/n}\mathcal{A}$ has a more intricate structure, however, when $\partial\mathcal{A}$ has logarithmic inflections. For example, if $\mathcal{A}$ is a circular disk, $\bigotimes^{1/n}\mathcal{A}$ is (a single loop of) the "n–th order" ovals of Cassini or lemniscate of Bernoulli when $0 \notin \mathcal{A}$ or $0 \in \partial\mathcal{A}$, respectively. But when 0 is in the interior of $\mathcal{A}$, a composite curve (portions of the Cassini ovals and a higher–order curve) is required to describe $\bigotimes^{1/n}\mathcal{A}$.

Keywords: Minkowski geometric algebra, ordinary roots, Minkowski roots, minimal and maximal roots, logarithmic convexity, ovals of Cassini, lemniscate of Bernoulli.

1 Introduction

Minkowski geometric algebra [7, 8] is concerned with the complex sets

$$\begin{aligned} \mathcal{A} \oplus \mathcal{B} &= \{\, \mathbf{a}+\mathbf{b} \mid \mathbf{a} \in \mathcal{A} \text{ and } \mathbf{b} \in \mathcal{B} \,\}, \\ \mathcal{A} \otimes \mathcal{B} &= \{\, \mathbf{a}\times\mathbf{b} \mid \mathbf{a} \in \mathcal{A} \text{ and } \mathbf{b} \in \mathcal{B} \,\}, \end{aligned} \qquad (1)$$

populated by the sums and products of pairs of complex numbers selected from complex–set operands,[1] $\mathcal{A}$ and $\mathcal{B}$. While the Minkowski sum $\oplus$ (with "+" interpreted as vector addition) has been extensively studied [9, 11, 12, 13, 16, 22] in both its theoretical and computational aspects, interest in the theory, algorithms, and applications/interpretations of the Minkowski product $\otimes$ (with "$\times$" representing complex–number multiplication) has emerged only recently [7, 8]. A basic interpretation of the geometric algebra (1) is that it defines a natural generalization of established methods for *interval arithmetic* [17, 18] to sets of complex numbers.

However, the two–dimensional nature of the Minkowski geometric algebra endows it with much richer geometrical content than interval arithmetic. For example, it provides an intuitive language for shape description, generation, and analysis of planar geometries, and for wavefront reflection or refraction problems in geometric optics [6, 7]; it is also of interest in the stability analysis of parameter–dependent dynamical systems. As a generalization of Minkowski sums and products, we have introduced [7, 8] "implicitly–defined sets" of the form

$$\mathcal{A} \circledf \mathcal{B} = \{\, \mathbf{f}(\mathbf{a},\mathbf{b}) \mid \mathbf{a} \in \mathcal{A},\, \mathbf{b} \in \mathcal{B} \,\} \qquad (2)$$

— i.e., the set of all complex values that are generated by a bivariate function $\mathbf{f}(\cdot,\cdot)$ with arguments $\mathbf{a}$ and $\mathbf{b}$ drawn from given complex sets $\mathcal{A}$ and $\mathcal{B}$. Whereas the Minkowski sum and product can be interpreted as unions of translated and scaled/rotated instances of one set, expression (2) offers a much more versatile concept — it can be regarded as the union of a family of *conformal mappings* of one set.

Equation (2) clearly subsumes (1) as special cases, and under suitable conditions on $\mathbf{f}$ the Minkowski sum/product

[1]Our usual custom [7, 8] is to denote real variables by italic characters, complex variables by bold characters, and sets of complex numbers by upper–case calligraphic characters.

0-7695-0562-7/00 $10.00 © 2000 IEEE

algorithms can be extended to provide boundary evaluation [8] for sets of the form (2).

In the Minkowski geometric algebra, problems that are of an algebraic nature may be of as much interest as those phrased in geometrical terms. Thus, for example, we may ask whether a given set may be factorized into products of certain sets, drawn from a prescribed family of simple "prime" sets. Now the Minkowski sum and product are commutative and associative operations, and if $\mathcal{A}_1, \ldots, \mathcal{A}_n$ are given complex sets, we denote the Minkowski sum and product of all n sets by

$$\bigoplus_{k=1}^{n} \mathcal{A}_k \quad \text{and} \quad \bigotimes_{k=1}^{n} \mathcal{A}_k .$$

In particular, if $\mathcal{A}_1 = \cdots = \mathcal{A}_n$ ($= \mathcal{A}$, say) we invoke the short–hand notation $\bigoplus^n \mathcal{A}$ and $\bigotimes^n \mathcal{A}$ for the above sums and products.

We call $\bigotimes^n \mathcal{A}$ the n–th *Minkowski power* of $\mathcal{A}$. Given a Minkowski product algorithm [8], computing $\bigotimes^n \mathcal{A}$ from $\mathcal{A}$ poses no special difficulty. However, as a special type of factorization, we may invert this problem and ask: what sets exhibit the property that their n–th Minkowski powers correspond to a given set $\mathcal{A}$? Clearly, we could call such a set an n–th *Minkowski root* of $\mathcal{A}$, and thus denote it by $\bigotimes^{1/n} \mathcal{A}$ (we refer to $\bigotimes^{1/n} \mathcal{A}$ indefinitely here, since its existence and uniqueness need to be established).

The computation and properties of Minkowski roots of complex sets is our main focus in this paper. Although they are conceptually quite simple, we shall find that unexpected subtleties and interesting geometrical issues arise in formulating the Minkowski roots of even the "simplest" sets. In the case of a circular disk, for example, we can determine closed–form expressions for the Minkowski n–th roots as (subsets of) generalizations of certain classical curves, the *ovals of Cassini* and *lemniscate of Bernoulli*, provided that the origin does not lie within the interior of the disk. In the latter case, the situation becomes much more complicated, and we find that the Minkowski root must be formulated as a *composite* curve.

Our plan for this paper is as follows. In §2 we present basic definitions and properties for Minkowski roots, and we discuss their minimality/maximality and their relation to the simpler concept of "ordinary" roots of complex sets. We are mostly interested in the Minkowski roots of a planar domain $\mathcal{A}$ — the domain boundary $\partial\mathcal{A}$ is then generated by products of n points from the boundary $\partial(\bigotimes^{1/n} \mathcal{A})$ of the Minkowski root. Since the theory of envelopes of families of plane curves plays an important role in this context, we review key aspects of that theory in §3 — specifically, we derive the envelope condition for families of curves that are specified in polar form.

The important concept of *logarithmic convexity* is briefly reviewed in §4, and it is shown that the Minkowski roots of logarithmically–convex domains are particularly easy to compute. A comprehensive treatment of Minkowski square roots of circular disks is given in §5, for the three distinct cases in which the origin lies outside, on the boundary of, or inside the disk. These results are extended in §6 to the Minkowski n–th roots of circular disks.

Our present results are preliminary in nature, and many interesting practical and theoretical problems concerning Minkowski roots remain to be addressed. We conclude by briefly outlining some of these outstanding issues in §7.

2 Minkowski roots of complex sets

Since the Minkowski product operation is commutative and associative, we can uniquely define the n–th *Minkowski power* $\bigotimes^n \mathcal{A}$ of any given complex set $\mathcal{A}$ by

$$\begin{aligned} \bigotimes^n \mathcal{A} &= \underbrace{\mathcal{A} \otimes \mathcal{A} \otimes \cdots \otimes \mathcal{A}}_{n \text{ times}} \\ &= \{\, \mathbf{z}_1\mathbf{z}_2\cdots\mathbf{z}_n \mid \mathbf{z}_i \in \mathcal{A} \text{ for } i = 1,\ldots,n \,\} . \end{aligned} \tag{3}$$

Note here that each complex value $\mathbf{z} \in \bigotimes^n \mathcal{A}$ arises from a product of n complex values $\mathbf{z}_1, \mathbf{z}_2, \ldots, \mathbf{z}_n$ *independently chosen* from $\mathcal{A}$.

Correspondingly, for a complex set $\mathcal{A}$, we define an n–th *Minkowski root* $\bigotimes^{1/n} \mathcal{A}$ by the property that

$$\bigotimes^n (\bigotimes^{1/n} \mathcal{A}) = \mathcal{A} , \tag{4}$$

— i.e., it is a set of complex values whose n–th Minkowski power coincides with the given set $\mathcal{A}$. Equivalently, we may write

$$\{\, \mathbf{z}_1\mathbf{z}_2\cdots\mathbf{z}_n \mid \mathbf{z}_i \in \bigotimes^{1/n} \mathcal{A} \text{ for } i = 1,\ldots,n \} = \mathcal{A} . \tag{5}$$

We emphasize again that the complex values $\mathbf{z}_1, \mathbf{z}_2, \ldots, \mathbf{z}_n$ in this definition are *independently chosen* from $\bigotimes^{1/n} \mathcal{A}$. Also, we are careful to refer here to "a" — rather than "the" — Minkowski root. The definitions (4) and (5) are rather indirect, and it is not immediately clear whether $\bigotimes^{1/n} \mathcal{A}$ actually exists for any complex set $\mathcal{A}$ and integer n (and, if it exists, whether it is unique).

2.1 Existence and uniqueness of Minkowski roots

We can immediately conjure up some simple counter–examples to a supposition of the existence and uniqueness of n–th Minkowski roots $\bigotimes^{1/n} \mathcal{A}$ for every subset $\mathcal{A}$ of $\mathbb{C}$ and for each integer n. Consider, for example, the set $\mathcal{A} = \{1, \mathrm{i}\,\}$. One can easily verify that a Minkowski square root $\bigotimes^{1/2} \mathcal{A}$ does not exist. On the other hand, the set $\mathcal{B} = \{1, \mathrm{i}\,, -1, -\mathrm{i}\,\}$ does possess a Minkowski square root — one can easily see that $\{1, \mathrm{i}, -1\}$, $\{\mathrm{i}, -1, -\mathrm{i}\}$, $\{-1, -\mathrm{i}, 1\}$, $\{-\mathrm{i}, 1, \mathrm{i}\}$ are all Minkowski roots of this set, and hence $\bigotimes^{1/2} \mathcal{B}$ is not unique.

An even more elementary example of non–uniqueness of Minkowski roots is the trivial case of a singleton set $\mathcal{A} = \{\mathbf{z}\}$, for any complex number $\mathbf{z} = r\mathrm{e}^{\mathrm{i}\theta}$ with $r \neq 0$. Such a set clearly has n different Minkowski n–th roots — namely, the singletons $\bigotimes^{1/n}\mathcal{A} = \{r^{1/n}\mathrm{e}^{\mathrm{i}(\theta+2\pi k)/n}\}$ for $k = 0, \ldots, n-1$. The set $\mathcal{O} = \{0\}$, on the other hand, has the unique Minkowski n–th root $\bigotimes^{1/n} \mathcal{O} = \mathcal{O}$.

In the most general context — in which the complex set $\mathcal{A}$ may comprise (a combination of) discrete points, loci, or regions in the complex plane — the cardinality of the set of all Minkowski n–th roots $\bigotimes^{1/n}\mathcal{A}$ is a subtle issue, beyond our present scope. In the present paper, we are primarily concerned with the case where $\mathcal{A}$ is a connected domain in the complex plane.

2.2 Minimal and maximal Minkowski roots

Consider again the set $\mathcal{B} = \{1, \mathrm{i}, -1, -\mathrm{i}\}$. Besides the four Minkowski square roots of this set identified in §2.1, we immediately observe that $\mathcal{B} \otimes \mathcal{B} = \mathcal{B}$, and hence $\bigotimes^{1/2} \mathcal{B} = \mathcal{B}$ is another Minkowski square root of this set. Clearly, when there are several distinct Minkowski roots of a given set, they need not all be of the same cardinality, and this leads us to introduce the concepts of minimal/maximal Minkowski roots of complex sets.

Definition 1 $\mathcal{R}_{\min}$ is a *minimal* Minkowski n–th root of $\mathcal{A}$ if $\bigotimes^n \mathcal{R}_{\min} = \mathcal{A}$, but $\bigotimes^n \mathcal{R} \neq \mathcal{A}$ for each proper subset $\mathcal{R} \subset \mathcal{R}_{\min}$. $\mathcal{R}_{\max}$ is a *maximal* n–th root if $\bigotimes^n \mathcal{R}_{\max} = \mathcal{A}$, but $\bigotimes^n \mathcal{R} \neq \mathcal{A}$ for each proper superset $\mathcal{R} \supset \mathcal{R}_{\max}$.

For an arbitrary set $\mathcal{A}$, however, it is not trivial to prove that minimal and maximal Minkowski roots exist. Moreover, when they do exist, the minimal and maximal roots are not necessarily unique. Consider $\mathcal{B} = \{1, \mathrm{i}, -1, -\mathrm{i}\}$, for example. Clearly, $\bigotimes^{1/2} \mathcal{B} = \{1, \mathrm{i}, -1\}$, $\{\mathrm{i}, -1, -\mathrm{i}\}$, $\{-1, -\mathrm{i}, 1\}$, $\{-\mathrm{i}, 1, \mathrm{i}\}$ are all minimal Minkowski square roots, since we cannot delete any member from these sets while maintaining the relation $\bigotimes^2(\bigotimes^{1/2} \mathcal{B}) = \mathcal{B}$.

Similarly, $\bigotimes^{1/2} \mathcal{B} = \{1, \mathrm{i}, -1, -\mathrm{i}\}$ is a maximal Minkowski square root, since we cannot introduce any new number $\mathbf{z}$ to this set and maintain $\bigotimes^2(\bigotimes^{1/2} \mathcal{B}) = \mathcal{B}$.

In general, the n–th Minkowski roots of a complex set $\mathcal{A}$ form a *partially ordered set* with respect to the set inclusion relation $\subset$. Thus, if $\mathcal{R}_1$ and $\mathcal{R}_2$ are distinct n–th Minkowski roots of $\mathcal{A}$, we may write $\mathcal{R}_1 \prec \mathcal{R}_2$ or $\mathcal{R}_2 \prec \mathcal{R}_1$ when $\mathcal{R}_1$ is a subset of $\mathcal{R}_2$ or vice–versa. However, when neither is a subset of the other, no precedence relationship exists between them.

2.3 Minkowski roots and ordinary roots

The n–th Minkowski power (3) of the set $\mathcal{A}$ is defined to be the set of all complex values that are products of n independently–chosen complex values from $\mathcal{A}$. The set

$$\mathcal{A}^n = \{\mathbf{z}^n \mid \mathbf{z} \in \mathcal{A}\}$$

defines a simpler kind of power, which we shall call the "ordinary" n–th power of $\mathcal{A}$. Clearly, $\mathcal{A}^n$ is just the image of $\mathcal{A}$ under the conformal map $\mathbf{z} \to \mathbf{z}^n$, and one can readily see that $\mathcal{A}^n \subseteq \bigotimes^n \mathcal{A}$ for any set $\mathcal{A}$.

Similarly, we define the "ordinary" n–th root of $\mathcal{A}$ to be the set

$$\mathcal{A}^{1/n} = \{\mathbf{z} \mid \mathbf{z}^n \in \mathcal{A}\}$$

whose n–th ordinary power coincides with $\mathcal{A}$. To clarify the relation between Minkowski roots and ordinary roots, we note that if $\mathbf{z}_1, \ldots, \mathbf{z}_n$ are n values satisfying $\mathbf{z}_i^n \in \mathcal{A}$ for $i = 1, \ldots, n$, we have $\mathbf{z}_i \in \mathcal{A}^{1/n}$ for every i. However, such values do *not*, in general, also satisfy $\mathbf{z}_1\mathbf{z}_2 \cdots \mathbf{z}_n \in \mathcal{A}$, and hence they do not all belong to $\bigotimes^{1/n}\mathcal{A}$. Thus, we may deduce that

$$\bigotimes^{1/n}\mathcal{A} \subseteq \mathcal{A}^{1/n}, \tag{6}$$

where the equality holds only in special cases. For the set $\mathcal{B} = \{1, \mathrm{i}, -1, -\mathrm{i}\}$ discussed above, for example, we have

$$\mathcal{B}^{1/2} = \left\{ \pm 1, \pm \mathrm{i}, \pm\frac{1+\mathrm{i}}{\sqrt{2}}, \pm\frac{1-\mathrm{i}}{\sqrt{2}} \right\},$$

and all the Minkowski square roots of $\mathcal{B}$ given above satisfy $\bigotimes^{1/2} \mathcal{B} \subset \mathcal{B}^{1/2}$.

For any set $\mathcal{A}$ comprising a finite number of complex values, the relation (6) provides a simple, though laborious, means of identifying all Minkowski roots. Namely, we form the ordinary root $\mathcal{A}^{1/n}$ and enumerate its *power set* — i.e., the set of all subsets of $\mathcal{A}^{1/n}$. We can then test whether or not each subset satisfies the definition (5) of a Minkowski n–th root, $\bigotimes^{1/n}\mathcal{A}$.

The enumeration and testing of subsets of $\mathcal{A}^{1/n}$ is clearly a rather brutish means of finding a Minkowski root $\bigotimes^{1/n}\mathcal{A}$, although some basic principles can be invoked to help speed it up. Some of these are enumerated below (we omit the proofs, which are elementary):

Proposition 1 *For a given complex set $\mathcal{A}$, let $\mathcal{A}^{1/n}$ be the ordinary n–th root, and let $\bigotimes^{1/n}\mathcal{A}$ be a Minkowski n–th root. One can easily verify that*

1. $0 \in \mathcal{A} \Rightarrow 0 \in \bigotimes^{1/n}\mathcal{A}$;

2. *if* $\mathcal{R}_0 \subset \mathcal{R}_1 \subseteq \mathcal{A}^{1/n}$ *and* $\bigotimes^n \mathcal{R}_0 \not\subseteq \mathcal{A}$, *then* $\bigotimes^n \mathcal{R}_1 \not\subseteq \mathcal{A}$;

3. *if $\mathcal{R}_0$ is a Minkowski n–th root of $\mathcal{A}$, then the sets $\mathcal{R}_k = \{e^{2k\pi i/n}\} \otimes \mathcal{R}_0$ for $k = 1, \ldots, n-1$ are clearly also Minkowski n–th roots of $\mathcal{A}$ — we say that $\mathcal{R}_0, \ldots, \mathcal{R}_{n-1}$ are "equivalent" Minkowski roots;*

4. *if $\mathbf{w} \neq 0$, we have $\bigotimes^{1/n}(\{\mathbf{w}\} \otimes \mathcal{A}) = \{\sqrt[n]{\mathbf{w}}\} \otimes (\bigotimes^{1/n}\mathcal{A})$ for any complex root $\sqrt[n]{\mathbf{w}}$ of $\mathbf{w}$ — hence, we can "normalize" any set $\mathcal{A} \neq \{0\}$ by mapping any distinguished (non–zero) point $\mathbf{z} \in \mathcal{A}$ to 1 before we take the Minkowski n–th root.*

3 Envelopes of families of curves

The theory of envelopes of families of plane curves [1, 2, 3] plays a key role in the analysis [7, 8] of Minkowski sums and products of planar domains.

In the present context, it is particularly convenient to develop a characterization for the envelope of a family of curves described in polar coordinates.

3.1 Envelope of a family in polar coordinates

Consider the Minkowski product of two complex sets,

$$\mathcal{C} = \mathcal{A} \otimes \mathcal{B} = \{\mathbf{a}\mathbf{b} \mid \mathbf{a} \in \mathcal{A}, \mathbf{b} \in \mathcal{B}\}.$$

One can easily see [7, 8] that only points $\mathbf{a} \in \partial\mathcal{A}$ and $\mathbf{b} \in \partial\mathcal{B}$ can contribute to $\partial\mathcal{C}$. Now suppose the boundaries $\partial\mathcal{A}$ and $\partial\mathcal{B}$ are given in polar coordinates by $\mathbf{w}(\phi) = \rho(\phi)\, e^{i\phi}$ and $\mathbf{z}(\theta) = r(\theta)\, e^{i\theta}$. We then interpret

$$\mathbf{f}(\phi,\theta) = \mathbf{w}(\phi)\,\mathbf{z}(\theta) = \rho(\phi) r(\theta)\, e^{i(\phi+\theta)}$$

to be a one-parameter family of parametric curves, where each ϕ identifies a particular curve, and θ specifies position along it. This family is generated by a simultaneous rotation (by an angle ϕ) and scaling (by the factor $\rho(\phi)$) of $\partial\mathcal{B}$ about the origin, and it can be shown [7, 8] that the boundary of $\mathcal{C}$ is (a subset of) the envelope of this family of curves.

Proposition 2 *Let $\mathbf{w}(\phi) = \rho(\phi)\, e^{i\phi}$ and $\mathbf{z}(\theta) = r(\theta)\, e^{i\theta}$ be parametric curves, described in polar form, that do not pass through the origin. Then the condition*

$$\frac{\rho'(\phi_0)}{\rho(\phi_0)} = \frac{r'(\theta_0)}{r(\theta_0)} \tag{7}$$

must be satisfied if the pair of points $\mathbf{w}(\phi_0)$ and $\mathbf{z}(\theta_0)$ on the curves contribute to the boundary of their Minkowski product $\mathbf{w}(\phi) \otimes \mathbf{z}(\theta)$.

Recalling the condition $\mathbf{w}'(\phi_0)/\mathbf{w}(\phi_0) = \lambda\, \mathbf{z}'(\theta_0)/\mathbf{z}(\theta_0)$ of Proposition 8 in [7] we see that, for curves in polar form, this condition can be written simply in terms of the moduli $\rho(\phi) = |\mathbf{w}(\phi)|$ and $r(\theta) = |\mathbf{z}(\theta)|$ of the two curves, with the real proportionality constant λ set equal to 1.

Geometrically, this condition states that pairs of corresponding points on two curves, that may contribute to the boundary of their Minkowski product, are characterized by the property that *the angle between the tangent and position vectors* of the curves must be equal at such points.

Proof of Proposition 2: Writing

$$\mathbf{f}(\phi,\theta) = \mathbf{w}(\phi)\,\mathbf{z}(\theta) = \rho(\phi)\, r(\theta)\, e^{i(\phi+\theta)}.$$

we have

$$\frac{\partial \mathbf{f}}{\partial \phi} = r(\theta)\,[\,\rho'(\phi) + i\rho(\phi)\,]\, e^{i(\phi+\theta)},$$

$$\frac{\partial \mathbf{f}}{\partial \theta} = \rho(\phi)\,[\,r'(\theta) + i r(\theta)\,]\, e^{i(\phi+\theta)}.$$

We interpret $\mathbf{f}(\phi,\theta)$ as a family of plane curves, where each ϕ identifies a different curve, and θ specifies position along it. The boundary of the Minkowski product $\mathbf{w}(\phi) \otimes \mathbf{z}(\theta)$ is (a subset of) the envelope of this family. By Lemma 2 in [8], points that contribute to this envelope are identified by the condition that the derivatives $\partial\mathbf{f}/\partial\phi$ and $\partial\mathbf{f}/\partial\theta$ are parallel, i.e.,

$$\frac{\partial \mathbf{f}}{\partial \phi} = \lambda\, \frac{\partial \mathbf{f}}{\partial \theta},$$

for some real constant λ. Thus we have

$$r(\theta)\,[\,\rho'(\phi) + i\rho(\phi)\,]\, e^{i(\phi+\theta)} = \lambda\, \rho(\phi)\,[\,r'(\theta) + i r(\theta)\,]\, e^{i(\phi+\theta)},$$

and by cancelling common terms and equating real and imaginary parts on the left and right, we may deduce that $\lambda = 1$ since $\rho(\phi) \neq 0$ and $r(\theta) \neq 0$ by supposition. Hence we obtain the stated condition,

$$\frac{\rho'(\phi)}{\rho(\phi)} = \frac{r'(\theta)}{r(\theta)}. \quad \blacksquare$$

4 Logarithmically–convex domains

The Minkowski roots of *logarithmically–convex* domains are especially easy to compute — they are essentially the same as (subsets of) the ordinary roots. In order to define logarithmic convexity, we need to first review some basic concepts describing the behavior of complex–valued curves (see [8] for a detailed discussion).

The *logarithmic curvature* $\kappa_{\log}(t)$ of a smooth complex–valued curve $\gamma(t)$ is defined to be the "ordinary" curvature of its image curve, $\log \gamma(t)$, under the map $\mathbf{z} \mapsto \log \mathbf{z}$. A point on $\gamma(t)$ is called a *logarithmic inflection* when the logarithmic curvature changes sign at that point.

The logarithmic curvature is a *local* shape property of curves. To characterize *global* behavior, we use two indices. The *rotation number* R of $\gamma(t)$ for $t \in [0,1]$ is defined by

$$R = \frac{1}{2\pi}\int_0^1 \kappa(t)\,|\gamma'(t)|\,\mathrm{d}t\,,$$

where $\kappa(t)$ is the curvature of $\gamma(t)$. Hence, R indicates the fraction of a complete revolution that the tangent executes, upon traversing the curve. The second index, the *winding number* $W(\mathbf{z}_0)$ of $\gamma(t)$ with respect to a fixed point $\mathbf{z}_0$, is defined by

$$W(\mathbf{z}_0) = \frac{1}{2\pi}\int_\gamma \frac{\mathrm{d}\mathbf{z}}{\mathbf{z}-\mathbf{z}_0} = \frac{1}{2\pi}\int_0^1 \frac{\gamma'(t)}{\gamma(t)-\mathbf{z}_0}\,\mathrm{d}t\,.$$

By a logarithmically–convex domain, we mean a domain whose boundary is a logarithmically convex curve — i.e., the image of the boundary under the logarithmic map $\mathbf{z} \mapsto \log \mathbf{z}$ is convex in the usual sense[2] (note that the image curve is periodic; the portion in $-\pi \le \mathrm{Im}\,(\mathbf{z}) \le +\pi$ is repeated *ad infinitum* in strips of width 2π parallel to the real axis). A geometrical interpretation of logarithmic convexity may be phrased as follows [8]:

Definition 2 A smooth curve $\gamma(t)$ in $\mathbb{C} \setminus \{0\}$ is *logarithmically convex* if the logarithmic spiral connecting any pair of points $\gamma(a)$ and $\gamma(b)$, that has the same winding number about the origin as $\gamma([a,b])$, does not cross[3] $\gamma(t)$.

At each point of a logarithmically convex curve $\gamma(t)$, the logarithmic curvature $\kappa_{\log}$ is non–negative. Furthermore, when $\kappa_{\log}$ is positive at each point (except, possibly, for finitely–many points where $\kappa_{\log}$ vanishes), we say that $\gamma(t)$ is *strictly* logarithmically convex — no portion of the curve may coincide with the logarithmic spiral connecting two points $\gamma(a)$ and $\gamma(b)$.

Unlike “ordinary” convexity, logarithmic convexity of a domain does not necessarily imply that the domain is simply connected. For instance, a circular annulus with center at the origin is logarithmically convex, even though it is not simply connected. However, *strict* logarithmic convexity of a domain boundary does ensure simply–connectedness.

The origin plays a key role in the theory of logarithmic convexity of curves in the complex plane, since it is a branch point of the map $\mathbf{z} \mapsto \log \mathbf{z}$. The following fact will be important in proving the main result of this section:

[2]A smooth curve is convex if its curvature κ, under an appropriate sign convention, is non–negative at each point — it is *strictly* convex if κ is positive at each point (except, possibly, for finitely many isolated points where κ vanishes).

[3]This leaves open the possibility that a portion of $\gamma(t)$ coincides with the logarithmic spiral, provided $\gamma(t)$ lies on the same side of the spiral before and after this portion.

Lemma 1 *If a simply–connected domain $\mathcal{A}$ has a smooth and logarithmically convex boundary, it cannot contain the origin as an interior point (unless it is a circular disk with the origin as center).*

Proof : Let $\gamma(t)$ be a parameterization of the boundary $\partial\mathcal{A}$ with anti–clockwise orientation. For a smooth curve $\gamma(t)$, the winding number $W(0)$ and the rotation number R are related [8] through the logarithmic curvature $\kappa_{\log}(t)$ by the equation

$$R - W(0) = \frac{1}{2\pi}\int_0^1 \kappa_{\log}(t)\left|\frac{\gamma'(t)}{\gamma(t)}\right|\,\mathrm{d}t\,.$$

Since $\gamma(t)$ is a simple closed curve, it has rotation number $R = 1$. Also, the logarithmic convexity of $\gamma(t)$ implies that $\kappa_{\log}(t)$ is a non–negative function, and hence the right–hand side of the above equation is non–negative.

Now if we assume that the origin is an interior point of $\mathcal{A}$, the winding number must be $W(0) = 1$. By the above equation, we must then have $\kappa_{\log}(t) \equiv 0$. But this leads to a contradiction, because the only curve whose logarithmic curvature vanishes identically is a logarithmic spiral, and a logarithmic spiral can be a simple closed curve only in the degenerate case of a circle centered on the origin. Thus, the origin cannot lie in the interior of $\mathcal{A}$, unless $\mathcal{A}$ is a circular disk with center at the origin. ∎

This lemma indicates that, when we are dealing with a logarithmically–convex domain, we need not worry about the branch point of the complex logarithm. Hence, we can transform our problem into the logarithmic plane, and treat Minkowski products as Minkowski sums using the relation

$$\log(\mathcal{A} \otimes \mathcal{B}) = \log(\mathcal{A}) \oplus \log(\mathcal{B})\,,$$

where $\mathcal{A}$ and $\mathcal{B}$ are two logarithmically–convex domains. In particular, the n–th ordinary root $\mathcal{A}^{1/n}$ is transformed into

$$\log(\mathcal{A}^{1/n}) = \frac{1}{n}\log(\mathcal{A})\,.$$

However, we still need to take into account the fact that the complex logarithm is a multi–valued function. Since a logarithmically–convex domain $\mathcal{A}$ does not contain the origin, its logarithmic image $\log(\mathcal{A})$ comprises an infinite set of translated copies of a single connected domain $\mathcal{L}$ — i.e., we may write

$$\log(\mathcal{A}) = \bigcup_{k\in\mathbb{N}} (\mathcal{L} + 2k\pi\mathrm{i})\,.$$

Here, $\mathcal{L} + 2k\pi\mathrm{i}$ denotes a copy of $\mathcal{L}$ translated by $2k\pi\mathrm{i}$, which can also be written as $\mathcal{L} \oplus \{2k\pi\mathrm{i}\}$. Thus we have

$$\log(\mathcal{A}^{1/n}) = \bigcup_{k\in\mathbb{N}} \left(\frac{1}{n}\mathcal{L} + \frac{2k\pi\mathrm{i}}{n}\right),$$

which implies, on taking the exponential map, that

$$\begin{aligned} \mathcal{A}^{1/n} &= \bigcup_{k\in\mathbb{N}} \exp\left(\frac{1}{n}\mathcal{L}\right) \exp\left(\frac{2k\pi\mathrm{i}}{n}\right) \\ &= \bigcup_{k=0}^{n-1} \xi_k \exp\left(\frac{1}{n}\mathcal{L}\right), \end{aligned}$$

$\xi_k = \exp(2k\pi\mathrm{i}/n)$, $k = 0, \cdots, n-1$, being the n–th roots of unity. Thus, the n–th ordinary root $\mathcal{A}^{1/n}$ is the union of the n sets obtained by rotating $\exp(\mathcal{L}/n)$ by angles $2k\pi/n$ about the origin. The fact that $\mathcal{A}$ does not contain the origin guarantees that these rotated copies are mutually disjoint.

We will now establish the relation between the ordinary n–th roots $\mathcal{A}^{1/n}$ and the Minkowski n–th roots $\bigotimes^{1/n}\mathcal{A}$ for a logarithmically–convex domain $\mathcal{A}$.

Theorem 1 *Let $\mathcal{A}$ be a domain with a smooth and strictly logarithmically–convex boundary. Then each connected component $\mathcal{R}$ of its ordinary n–th root $\mathcal{A}^{1/n}$ is a Minkowski n–th root $\bigotimes^{1/n}\mathcal{A}$. Furthermore, the boundaries $\partial\mathcal{R}$ of these components are minimal Minkowski n–th roots.*

Proof : We choose a connected component $\mathcal{R}$ from the n mutually disjoint components of $\mathcal{A}^{1/n}$, and a connected component $\mathcal{S}$ from the infinitely–many components of $\log(\mathcal{R})$. Now since $\log(\mathcal{R})$ is obtained by scaling $\log(\mathcal{A})$ by $1/n$, there is a connected component $\mathcal{L}$ of $\log(\mathcal{A})$ such that $\mathcal{L} = n\mathcal{S}$. Furthermore, since $\mathcal{A}$ is strictly logarithmically convex, $\mathcal{L}$ is a strictly convex domain in the usual sense, and hence $\mathcal{S}$ is also strictly convex.

We first show that $\mathcal{R}$ is a Minkowski n–th root of $\mathcal{A}$. Since it is obvious that $\mathcal{A} = \mathcal{R}^n \subset \bigotimes^n \mathcal{R}$, it will suffice to show that $\bigotimes^n \mathcal{R} \subset \mathcal{A}$. Suppose we choose n points $\mathbf{z}_1, \ldots, \mathbf{z}_n \in \mathcal{R}$. Then each $\mathbf{z}_i$ has a unique logarithmic image $\mathbf{w}_i \in \mathcal{S}$ and, by the convexity of $\mathcal{S}$, the centroid

$$\mathbf{w}_0 = \frac{1}{n}\sum_{i=1}^{n} \mathbf{w}_i$$

of $\mathbf{w}_1, \ldots, \mathbf{w}_n$ is also contained in $\mathcal{S}$. Thus, $\sum_{i=0}^{n} \mathbf{w}_i$ is contained in $\mathcal{L}$, and by taking the exponential map we have $\prod_{i=1}^{n} \mathbf{z}_i \in \mathcal{A}$. Since the points $\mathbf{z}_1, \ldots, \mathbf{z}_n$ are freely chosen from $\mathcal{R}$, we deduce that $\bigotimes^n \mathcal{R} \subset \mathcal{A}$.

We now show that the boundary $\partial\mathcal{R}$ is also a Minkowski n–th root of $\mathcal{A}$. Since $\mathcal{R}$ is a Minkowski n–th root of $\mathcal{A}$, it is obvious that $\bigotimes^n(\partial\mathcal{R}) \subset \mathcal{A}$. So it suffices to show that $\mathcal{A} \subset \bigotimes^n(\partial\mathcal{R})$ — in other words, for each $\mathbf{a} \in \mathcal{A}$, there exist n boundary points $\mathbf{z}_1, \ldots, \mathbf{z}_n \in \partial\mathcal{R}$ such that $\prod_{i=1}^{n} \mathbf{z}_i = \mathbf{a}$. Since $(\partial\mathcal{R})^n = \partial\mathcal{A}$, we need only establish this property for *interior* points $\mathbf{a} \in \mathcal{A}$.

Now let $\mathbf{b}$ denote the logarithmic image of $\mathbf{a}$ that lies within $\mathcal{L}$. Then the problem is transformed into the question of whether $\mathbf{b}$ can be expressed as a *sum* of n points $\mathbf{w}_1, \ldots, \mathbf{w}_n$ chosen from the boundary $\partial\mathcal{S}$. Since $\mathbf{c} = \mathbf{b}/n$ is an interior point of $\mathcal{S}$, we must show that $\mathbf{c}$ is expressible as the centroid of $\mathbf{w}_1, \ldots, \mathbf{w}_n$ under the assumption that $\mathcal{S}$ is a convex domain. To establish this property, we employ induction on n.

When $n = 2$, we must show that any interior point $\mathbf{c}$ of the convex domain $\mathcal{S}$ is the mid–point of a chord. Now let $r(\theta)$ be the polar equation of $\partial\mathcal{S}$ with respect to $\mathbf{c}$, and consider the function $f(\theta) = r(\pi-\theta) - r(\theta)$ for $0 \leq \theta \leq \pi$. Clearly, $f(\theta)$ is continuous and satisfies $f(0) = -f(\pi)$. Thus, $f(\theta)$ must have a zero θ_0, and this angle identifies the chord whose mid–point is $\mathbf{c}$.

For the general n case, we first choose a boundary point $\mathbf{w}_0 \in \partial\mathcal{S}$ such that $\mathbf{d} = \mathbf{c} + (\mathbf{c} - \mathbf{w}_0)/n$ remains in the interior of $\mathcal{S}$. Such a point $\mathbf{w}_0$ always exists — for instance, a boundary point at which the distance of $\mathbf{c}$ from $\partial\mathcal{S}$ is realized always has this property. By the induction argument, $\mathbf{d}$ can be expressed as the centroid of n points $\mathbf{w}_1, \ldots, \mathbf{w}_n$ on $\partial\mathcal{S}$, i.e., $\mathbf{d} = \frac{1}{n}\sum_{i=1}^{n} \mathbf{w}_i$, and we then have

$$\mathbf{c} = \frac{\mathbf{w}_0 + n\mathbf{d}}{n+1} = \frac{1}{n+1}\sum_{i=0}^{n} \mathbf{w}_i .$$

Thus, $\mathbf{c}$ is also expressible as the centroid of $n+1$ boundary points on $\partial\mathcal{S}$.

Finally, we show that $\partial\mathcal{R}$ is a *minimal* n–th Minkowski root. Since $\mathcal{S}$ is strictly convex, the only way to express a boundary point $\mathbf{w}_0 \in \partial\mathcal{S}$ as the centroid of n points $\mathbf{w}_1, \ldots, \mathbf{w}_n$ on $\partial\mathcal{S}$ is the case where $\mathbf{w}_1 = \cdots = \mathbf{w}_n = \mathbf{w}_0$. Therefore, if we delete a single point $\mathbf{z}_0$ from $\partial\mathcal{R}$, the Minkowski power $\bigotimes^n(\partial\mathcal{R} \setminus \{\mathbf{z}_0\})$ cannot include the boundary point $\mathbf{z}_0^n$ of $\mathcal{A}$. Hence, we deduce that $\partial\mathcal{R}$ is a minimal n–th Minkowski root of $\mathcal{A}$. ■

5 Minkowski square roots of a disk

In order to gain deeper insight into Minkowski roots, it is useful to examine a simple case in some detail. Thus, we shall consider here the nature of the Minkowski square roots of a circular disk $\mathcal{D}$ in the complex plane, specified by its center $\mathbf{c}$ and radius R. Invoking the normalization process of Proposition 1 in [7] we may, without loss of generality, consider just two cases: $\mathbf{c} = 0$ and $\mathbf{c} = 1$.

The former case accommodates disks with center at the origin. When $\mathbf{c} \neq 0$, we consider the "normalized" disk $\widetilde{\mathcal{D}} = \{\mathbf{c}^{-1}\} \otimes \mathcal{D}$ with center 1 and radius $|\mathbf{c}|^{-1}R$. Then, if the latter has a Minkowski root $\bigotimes^{1/2}\widetilde{\mathcal{D}}$, the set $\bigotimes^{1/2}\mathcal{D} = \{\sqrt{\mathbf{c}}\} \otimes (\bigotimes^{1/2}\widetilde{\mathcal{D}})$ will be a Minkowski root of $\mathcal{D}$ (for both $\sqrt{\mathbf{c}}$ values). This generalizes, in a straightforward manner, to Minkowski n–th roots of $\mathcal{D}$.

Definition 3 The curve whose polar equation $r(\theta)$ satisfies

$$r^4(\theta) - 2r^2(\theta)\cos 2\theta + 1 = R^2 \tag{8}$$

is called the *ovals of Cassini*.[4] In the particular case $R = 1$, this reduces to

$$r^2(\theta) \;=\; 2\cos 2\theta\,, \tag{9}$$

which defines the *lemniscate of Bernoulli* [14, 15].

Remark 1 The ovals of Cassini (8) may have one of three morphological forms (see Figure 1 below):

(a) when $R < 1$, they comprise two disjoint loops that individually contain the points $(-1, 0)$ and $(+1, 0)$;

(b) when $R = 1$, they degenerate to a "figure–of–eight" curve (lemniscate) containing $(-1, 0)$ and $(+1, 0)$ with the origin as a double point;

(c) when $R > 1$, they comprise a single "peanut–shaped" loop containing the points $(-1, 0)$ and $(+1, 0)$.

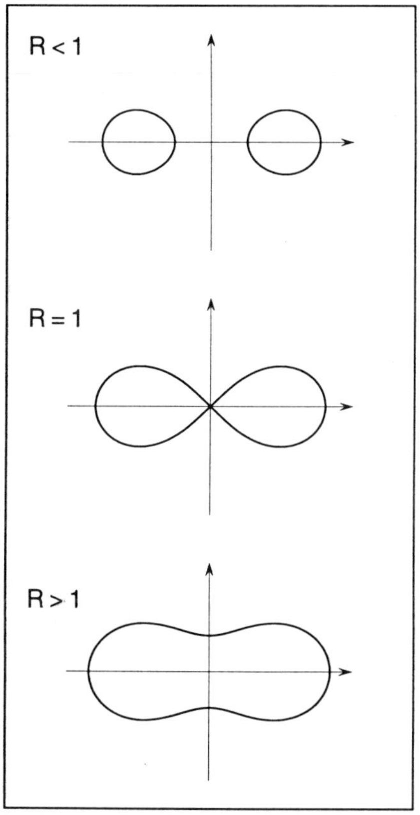

Figure 1. The three morphologies of the ovals of Cassini, for values $R < 1$, $R = 1$, and $R > 1$.

[4]The ovals of Cassini are often defined in bipolar coordinates by the equation $r_1 r_2 = k^2$, where r_1 and r_2 represent the distances of a curve point from two fixed poles. Equation (8) corresponds to the case of poles at $(-1, 0)$ and $(+1, 0)$ with $k^2 = R$.

5.1 Disk with center $(1, 0)$ and radius $R \le 1$

Let $\mathcal{D}$ be the disk with center $(1, 0)$ and radius R. In polar form, we have

$$\mathcal{D} \;=\; \{\, 1 + \rho\, \mathrm{e}^{\mathrm{i}\phi} \mid 0 \le \rho \le R,\, 0 \le \phi < 2\pi \,\}\,. \tag{10}$$

We shall first identify the ordinary square root of $\mathcal{D}$.

Lemma 2 *The ordinary square root of the disk* (10) *is the region bounded by the ovals of Cassini* (8).

Proof : Recall that $\mathcal{D}^{1/2} = \{\, \mathbf{z} \mid \mathbf{z}^2 \in \mathcal{D} \,\}$. With $\mathbf{z} = r\mathrm{e}^{\mathrm{i}\theta}$, we write

$$\begin{aligned} \mathbf{z}^2 \;=\; r^2\mathrm{e}^{\mathrm{i}\,2\theta} \;&=\; r^2(\cos 2\theta + \mathrm{i}\,\sin 2\theta) \\ &=\; 1 + \rho(\cos\phi + \mathrm{i}\,\sin\phi) \end{aligned}$$

for some $\rho \in [\,0, R\,]$ and $\phi \in [\,0, 2\pi\,)$. This gives

$$r^2 \cos 2\theta - 1 \;=\; \rho\cos\phi \quad \text{and} \quad r^2 \sin 2\theta \;=\; \rho\sin\phi\,.$$

Squaring both sides of the above equations and adding, we obtain

$$r^4 - 2r^2\cos 2\theta + 1 \;=\; \rho^2\,.$$

For $0 \le \rho \le R$, this describes a one–parameter family of Cassini ovals that completely occupy the region bounded by (8). Note that boundary points of $\mathcal{D}^{1/2}$ correspond to boundary points of $\mathcal{D}$. In the case $R = 1$, we obtain the lemniscate of Bernoulli (9) as a limiting form of the Cassini ovals (8). ∎

There are other ways to derive (8) as the equation that defines $\partial(\mathcal{D}^{1/2})$. Using polar coordinates, for example, the boundary of (10) satisfies

$$r^2 - 2r\cos\theta + 1 \;=\; R^2\,.$$

Noting that $\mathcal{D}^{1/2}$ is transformed into $\mathcal{D}$ by the map $\mathbf{z} \to \mathbf{z}^2$, we easily obtain (8) from the above equation by substituting $r \to r^2$ and $\theta \to 2\theta$.

Proposition 3 *For the disk $\mathcal{D}$ defined by* (10) *with radius $R \le 1$, a single loop of the ovals of Cassini* (8) *constitutes a Minkowski square root $\bigotimes^{1/2} \mathcal{D}$.*

Proof : This is a direct consequence of Theorem 1 — as an independent verification, we use condition (7) to show that the circle $\partial\mathcal{D}$ is the envelope of the family of curves defined by multiplying a single loop of (8) by each point on it. Let $\mathbf{f}(\phi, \theta) = \mathbf{w}(\phi)\mathbf{z}(\theta)$, where $\mathbf{w}(\phi) = \rho(\phi)\,\mathrm{e}^{\mathrm{i}\,\phi}$ and $\mathbf{z}(\theta) = r(\theta)\,\mathrm{e}^{\mathrm{i}\,\theta}$ are parameterizations of (8) satisfying

$$\begin{aligned} \rho^4(\phi) - 2\rho^2(\phi)\cos 2\phi + 1 \;&=\; R^2\,, \\ r^4(\theta) - 2r^2(\theta)\cos 2\theta + 1 \;&=\; R^2\,. \end{aligned} \tag{11}$$

We choose the loop in the right half of the complex plane, defined by positive ρ, r values with $-\frac{1}{2}\sin^{-1} R \le \phi, \theta \le +\frac{1}{2}\sin^{-1} R$ (similar arguments hold for the other loop).

Suppose first $R < 1$, so that $\rho(\phi) \ne 0$ and $r(\theta) \ne 0$. By Proposition 2, corresponding points of $\mathbf{w}(\phi)$ and $\mathbf{z}(\theta)$, whose products contribute to the envelope of the family of curves $\mathbf{f}(\phi, \theta)$, satisfy the relation

$$\frac{\rho'(\phi)}{\rho(\phi)} = \frac{r'(\theta)}{r(\theta)}. \tag{12}$$

Differentiating (11), we find that

$$\frac{r'(\theta)}{r(\theta)} = \frac{\sin 2\theta}{\cos 2\theta - r^2(\theta)},$$

and by solving (11) for $r^2(\theta)$ and substituting into the above, we obtain

$$\frac{r'(\theta)}{r(\theta)} = \pm \frac{\sin 2\theta}{\sqrt{R^2 - \sin^2 2\theta}}. \tag{13}$$

Exactly the same derivation holds for the parameterization $\mathbf{w}(\phi)$, and hence the relation (12) between ϕ and θ becomes

$$\pm \frac{\sin 2\theta}{\sqrt{R^2 - \sin^2 2\theta}} = \pm \frac{\sin 2\phi}{\sqrt{R^2 - \sin^2 2\phi}}.$$

Thus, squaring both sides and simplifying, we finally obtain

$$\sin^2 2\theta = \sin^2 2\phi,$$

For $-\frac{1}{2}\sin^{-1} R \le \phi, \theta \le +\frac{1}{2}\sin^{-1} R$, the only solution is $\theta = \phi$, and hence the envelope point is

$$\mathbf{e}(\phi) = \mathbf{f}(\phi, \phi) = \rho^2(\phi)\,(\cos 2\phi + \mathrm{i}\sin 2\phi).$$

One may easily see that $\mathbf{e}(\phi)$ describes the boundary $\partial\mathcal{D}$ of the disk (10) by setting $x(\phi) = \mathrm{Re}\,(\mathbf{e}(\phi)) = \rho^2(\phi)\cos 2\phi$ and $y(\phi) = \mathrm{Im}\,(\mathbf{e}(\phi)) = \rho^2(\phi)\sin 2\phi$. We then have

$$\begin{aligned}&[\,x(\phi) - 1\,]^2 + [\,y(\phi)\,]^2 - R^2 \\ &\quad = \rho^4(\phi) - 2\rho^2(\phi)\cos 2\phi + 1 - R^2 = 0.\end{aligned}$$

Thus, the envelope of $\mathbf{f}(\phi, \theta)$ coincides with the boundary of (10) when $R < 1$.

Finally, in the case $R = 1$ we have $\rho(\phi) = \sqrt{2\cos 2\phi}$ and $r(\theta) = \sqrt{2\cos 2\theta}$, where $-\pi/4 \le \phi, \theta \le +\pi/4$ for a single loop, and hence

$$\begin{aligned}\mathbf{f}(\phi, \theta) &= \mathbf{w}(\phi)\mathbf{z}(\theta) \\ &= \rho(\phi)r(\theta)\,\mathrm{e}^{\mathrm{i}\,(\phi+\theta)} \\ &= 2\sqrt{\cos 2\phi \cos 2\theta}\,\mathrm{e}^{\mathrm{i}\,(\phi+\theta)}.\end{aligned}$$

Solving the relation (12), we again find that $\theta = \phi$. By substituting into $\mathbf{f}(\phi, \theta)$ and proceeding along the same lines of argument, one can easily verify that the envelope of this curve family is the disk (10) with $R = 1$. ■

Figure 2 illustrates disks of radius $R < 1$ and $R = 1$, centered on $(0, 1)$, generated as the Minkowski squares of a single loop of the ovals of Cassini (8) or the lemniscate of Bernoulli (9) — i.e., as the one–parameter families of scalings/rotations of each curve defined by every point on it. Note that, for $R \le 1$, we may take *either* the left *or* the right loop of the ovals of Cassini as the Minkowski square root of the disk $|\mathbf{z} - 1| \le R$, but not *both* loops (the Minkowski square of the entire ovals of Cassini is the union of the two disks $|\mathbf{z} - 1| \le R$ and $|\mathbf{z} + 1| \le R$).

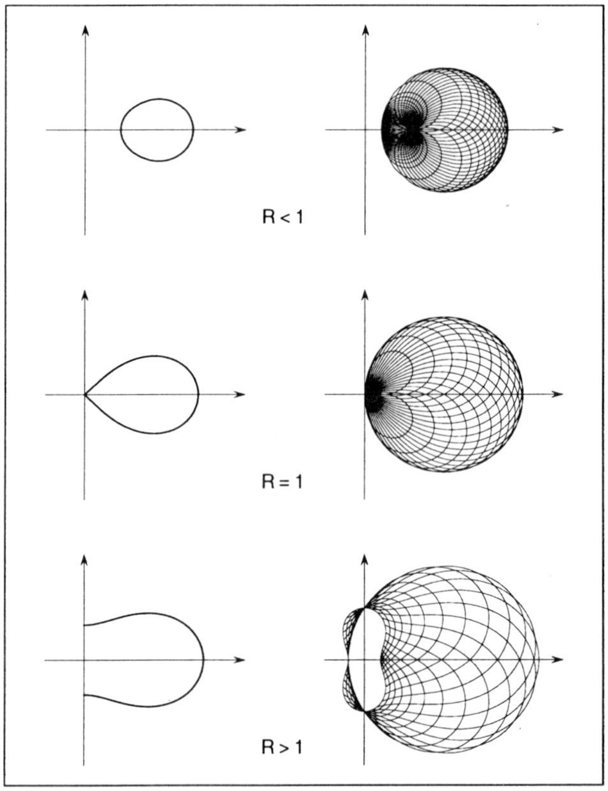

Figure 2. The right loop of the ovals of Cassini in the cases $R < 1$ and $R = 1$, with the disk $|\mathbf{z} - 1| \le R$ obtained as the Minkowski squares of these curves. Also shown is the right half of the ovals of Cassini in the case $R > 1$ — note that the Minkowski square of this locus does *not* yield the disk $|\mathbf{z} - 1| \le R$.

5.2 Disk with center $(1,0)$ and radius $R > 1$

Based on the results for $R < 1$ and $R = 1$, one might naively expect that, for $R > 1$, the Minkowski square root of the disk $|\mathbf{z} - 1| \leq R$ is simply the portion of the ovals of Cassini (8) lying in the left or right half of the complex plane. From the Minkowski square of this locus shown in Figure 2, however, we immediately see that this is incorrect. Although the circle $|\mathbf{z}-1| = R$ is (a subset of) the envelope of the family of scalings/rotations of this locus by each point on it, the Minkowski square lies on the *wrong side* of this circle in the left half of the complex plane. Moreover, the Minkowski square does not fully occupy the interior of the disk $|\mathbf{z} - 1| \leq R$. These defects arise from the logarithmic non–convexity of the disk when it contains the origin.

To understand the difference between the Minkowski square roots of disks with different radii, we investigate the nature of the logarithmic images of circles. Now when $R \leq 1$, the circles with center $(1,0)$ and radius R are all logarithmically convex. Thus, their logarithmic images are convex curves. Figure 3 shows the logarithmic images of circles with $R < 1$ and $R = 1$ (note that, since $\log \mathbf{z}$ is a multi–valued function, we choose the logarithmic branch with $-\pi < \text{Im}\,(\log \mathbf{z}) \leq \pi$ here).

Figure 3 also shows these logarithmic images scaled by a factor of $1/2$. Since taking the Minkowski square root corresponds to scaling by $1/2$ in the logarithmic plane, these scaled curves must be the logarithmic images of the Minkowski roots of the circles (i.e., of a single loop of the ovals of Cassini for $R < 1$ and of the lemniscate of Bernoulli when $R = 1$). The Minkowski sums of these scaled curves with themselves generate the convex regions bounded by the logarithmic images of the circles[5] — the families of curves that generate these Minkowski sums are shown as light curves in Figure 3.

We now apply this method to a disk with $R > 1$. Since the boundary circle contains the origin, its logarithmic image is a periodic curve whose imaginary part assumes all real values; the interior of the disk maps into the region on the left of this curve. Scaling the image curve by $1/2$ gives the logarithmic image of the right half of the ovals of Cassini. Figure 4 shows the logarithmic images of the circle and ovals of Cassini, together with the family of curves generating the Minkowski sum of the scaled curve with itself. Since the image curve is not convex, it fails to generate the desired region, even though it has the desired envelope: one can see that the family of curves lies partly on the wrong side of the envelope. Taking the exponential image, we obtain exactly the same family of curves shown at the bottom of Figure 2.

[5] This property holds for a general convex domain — i.e., if $\mathcal{A}$ is a convex domain, then by taking the Minkowski sum $(\frac{1}{2}\partial\mathcal{A}) \oplus (\frac{1}{2}\partial\mathcal{A})$, we recover the domain $\mathcal{A}$.

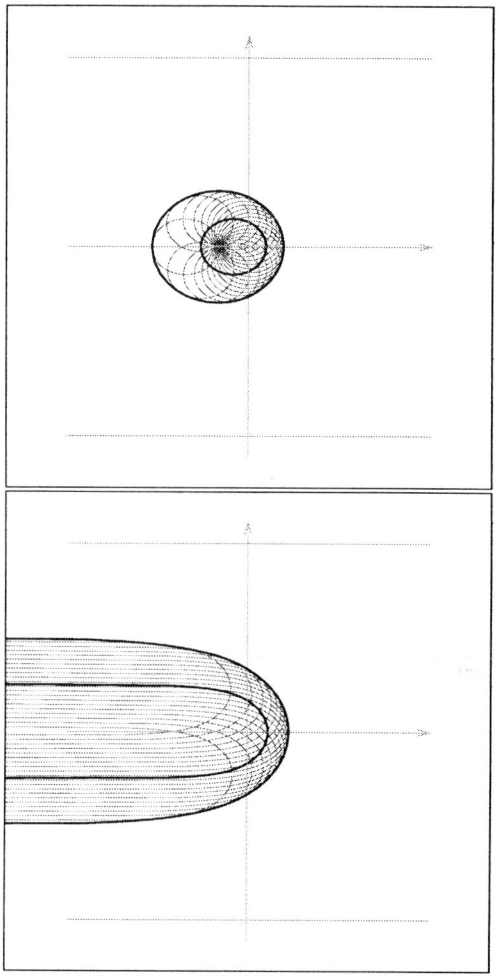

Figure 3. Logarithmic images of circles with radius $R < 1$ (upper) and $R = 1$ (lower). Also shown are the images scaled by 1/2, and the curve families generating the Minkowski sums of the scaled images with themselves.

To remedy this problem, we introduce a new curve that is a composite of parts of the ovals of Cassini and of a higher–order curve. Our goal is to construct a curve in the logarithmic plane whose Minkowski sum with itself fully occupies the region to the left of the logarithmic image of the circle.

We first show that the logarithmic image of the circle has nice symmetry properties. Since the circle centered at $(1,0)$ has logarithmic inflections at the points where it intersects the imaginary axis [8], the logarithmic image of the circle has ordinary inflections where its imaginary part is $\pm\pi/2$. For $R > 1$, the inflections are at $(\log\sqrt{R^2-1}, \pm\pi/2)$. The portion of the logarithmic image of the circle with imaginary part between $\pi/2$ and π can be obtained from the portion with imaginary part between 0 and $\pi/2$ by

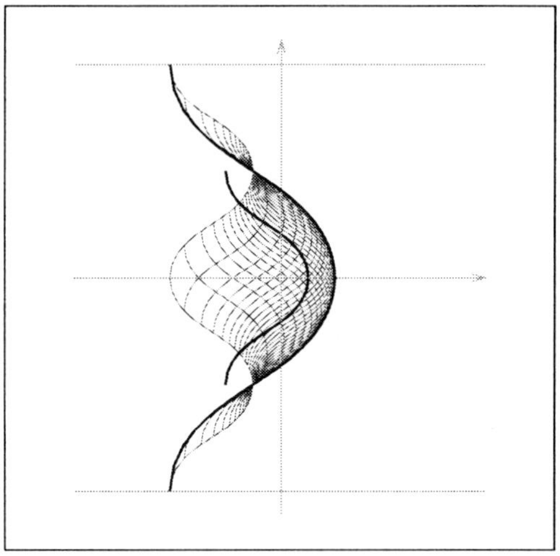

Figure 4. Logarithmic image of a circle $\mathcal{C}$ with center $(1,0)$ and radius $R>1$. Also shown is the image scaled by 1/2, and the curve family generating the Minkowski sum of the scaled image with itself. Note that $(\frac{1}{2}\log\mathcal{C})\oplus(\frac{1}{2}\log\mathcal{C})$ lies on the *wrong side* of $\log\mathcal{C}$ above and below its two inflection points.

a reflection of the latter through the inflection point[6] at $(\log\sqrt{R^2-1},\pi/2)$. An analogous symmetry holds for the portion below the real axis.

This symmetry arises from an elementary property of all circles — namely, if $\mathbf{p}$ is a fixed point inside a circle, any chord through $\mathbf{p}$ intersects the circle in two points, $\mathbf{p}_1$ and $\mathbf{p}_2$, and if r_1 and r_2 are the distances of these points from $\mathbf{p}$, the product $r_1 r_2$ is independent of the chosen chord.

Now let $\mathcal{C}$ be the circle with center $(1,0)$ and radius $R>1$. Then $\log\mathcal{C}$ is the logarithmic image of $\mathcal{C}$, and $\frac{1}{2}\log\mathcal{C}$ is the logarithmic image scaled by $1/2$. In Figure 4 we see that the convex part of $\frac{1}{2}\log\mathcal{C}$ (with imaginary part from $-\pi/4$ to $\pi/4$) generates a region to the *left* of $\log\mathcal{C}$ when we take its Minkowski sum with itself.

When we include the concave parts of $\frac{1}{2}\log\mathcal{C}$, however, the Minkowski sum of this locus with itself yields regions to the *right* of $\log\mathcal{C}$ when the imaginary part is above $\pi/2$ or below $-\pi/2$. Thus, to construct the logarithmic image of $\bigotimes^{1/2}\mathcal{C}$, we can use the convex part of $\frac{1}{2}\log\mathcal{C}$, but we must replace the concave part by portions of a new curve such that the Minkowski sum of this composite curve with itself generates *only* the region to the left of $\log\mathcal{C}$.

We construct the new curve portions such that their Minkowski sums with the convex part of $\frac{1}{2}\log\mathcal{C}$ lie on the left of $\log\mathcal{C}$. This construction employs two steps. First, we take the convex portion of $\log\mathcal{C}$, with imaginary part from 0 to $\pi/2$, and scale it by $3/2$ so that the resulting locus lies between 0 and $3\pi/4$. We then reflect this scaled curve through the inflection point $\mathbf{q}=(\log\sqrt{R^2-1},\pi/2)$. The new curve segment produced by these steps can be written[7] as $2\mathbf{q}-\frac{3}{2}\log\mathcal{C}$, and lies between $\pi/4$ and π — we also construct a new curve segment between $-\pi/4$ and $-\pi$ in an analogous manner.

Combining these new curve segments with the convex part of $\frac{1}{2}\log\mathcal{C}$, between $-\pi/4$ and $\pi/4$, yields a composite curve between $-\pi$ to π, as shown in Figure 5. The family of curves generating the Minkowski sum of this composite curve with itself is also shown in Figure 5 — note that each member of this family lies entirely to the left of $\log\mathcal{C}$.

By the symmetry of the logarithmic image of a circle, one can easily see that the Minkowski sum

$$(\tfrac{1}{2}\log\mathcal{C})\oplus(2\mathbf{q}-\tfrac{3}{2}\log\mathcal{C})$$

of the convex part of $\frac{1}{2}\log\mathcal{C}$, between 0 and $\pi/4$, and the constructed curve segment, between $\pi/4$ and π, yields the concave part of $\log\mathcal{C}$, between $\pi/2$ and π, as its boundary.

Any point $\mathbf{z}$ on $\log\mathcal{C}$ with $\mathrm{Im}(\mathbf{z})<\pi/2$ has $\mathbf{w}=2\mathbf{q}-\mathbf{z}$ with $\mathrm{Im}(\mathbf{w})>\pi/2$ as its symmetric point, where $\mathbf{q}=\log\sqrt{R^2-1}+\mathrm{i}\pi/2$ is the inflection of $\log\mathcal{C}$. For any point $\frac{1}{2}\mathbf{z}$ on $\frac{1}{2}\log\mathcal{C}$ we identify $2\mathbf{q}-\frac{3}{2}\mathbf{z}$ as the corresponding point on the new curve $2\mathbf{q}-\frac{3}{2}\log\mathcal{C}$. The Gauss maps of the curves are identical at such corresponding points, and hence $\frac{1}{2}\mathbf{z}+(2\mathbf{q}-\frac{3}{2}\mathbf{z})=2\mathbf{q}-\mathbf{z}$ lies on the boundary of their Minkowski sum. But, as noted above, the point $\mathbf{w}=2\mathbf{q}-\mathbf{z}$ lies on the concave part of $\log\mathcal{C}$, between $\pi/2$ and π.

To obtain a Minkowski root of the disk with center $(1,0)$ and radius $R>1$, we must transform the composite curve by the exponential map. In the lower part of Figure 5, we show the circle $\mathcal{C}$, the exponential image of the composite curve, and the family of curves generating its Minkowski square. The Minkowski square of the composite curve lies inside the disk, and yields the entire circle as its envelope. However, the family of curves fails to fill the *entire* disk: since the Minkowski sum of the composite curve with itself does not fully occupy the region to the left of $\log\mathcal{C}$, the corresponding Minkowski square has an "empty" region about the origin. To completely fill the disk, we need an additional element in the Minkowski root; a line segment connecting the origin to any point on the composite curve will suffice.

We will now derive explicit equations for the Minkowski square root, defined by transforming our composite curve by the exponential map. For values of the argument between $-\pi/4$ and $\pi/4$, the Minkowski square root coincides with the ovals of Cassini (8). To obtain the polar equation for the remaining portions, we take (r,θ) as polar coordinates for

[6] A "reflection through a point" $\mathbf{q}$ is equivalent to a rotation by $\pm\pi$ about $\mathbf{q}$ (i.e., $\mathbf{z}$ has the image $\mathbf{w}=2\mathbf{q}-\mathbf{z}$ under a reflection through $\mathbf{q}$).

[7] Note that, in this expression, $\log\mathcal{C}$ refers only to the convex portion from 0 to $\pi/2$.

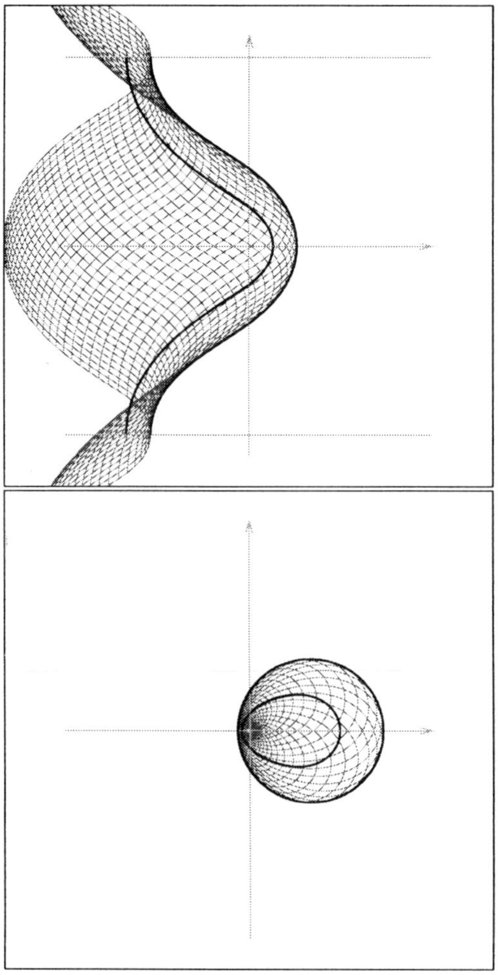

Figure 5. The upper plot shows the composite curve, whose Minkowski sum with itself generates the logarithmic image of the circle, while the lower figure shows the exponential images of this composite curve and the circle.

the Minkowski root and (ρ, ϕ) as polar coordinates on the circle with center $(1,0)$ and radius $R > 1$, so that

$$\rho^2 - 2\rho\cos\phi + 1 = R^2 . \tag{14}$$

The image of this circle has Cartesian coordinates $(\log\rho, \phi)$ in the logarithmic plane. Each point is transformed into $(\frac{3}{2}\log\rho, \frac{3}{2}\phi)$ upon scaling by $3/2$, and reflection through the inflection point $(\log\sqrt{R^2-1}, \pi/2)$ transforms it into $(\log(R^2-1) - \frac{3}{2}\log\rho, \pi - \frac{3}{2}\phi)$. Thus, (ρ, ϕ) and (r, θ) are related by

$$\log r = \log(R^2-1) - \frac{3}{2}\log\rho \quad \text{and} \quad \theta = \pi - \frac{3}{2}\phi ,$$

which yields

$$\rho = \left(\frac{R^2-1}{r}\right)^{2/3} \quad \text{and} \quad \phi = \frac{2}{3}(\pi - \theta) .$$

Finally, substituting this into (14) gives the polar equation

$$(R^2-1)r^4 + 2\left[\cos 2\theta + 3R^2\cos\tfrac{2}{3}(\pi-\theta)\right] r^2 - (R^2-1)^2 = 0 .$$

The subset of this curve for $\pi/4 \le \theta \le 7\pi/4$ defines the new curve segment that augments the appropriate part of the Cassini ovals to form the composite Minkowski root of the disk. The curve defined by (15) is shown in its entirety in Figure 6: note that, to trace the full curve, we must employ the angular range $0 \le \theta \le 6\pi$ in (15). One can easily verify that (15) defines an algebraic curve of degree 6 (it appears similar in form to an epitrochoid,[8] but we have been unable to find suitable parameters to give an exact match).

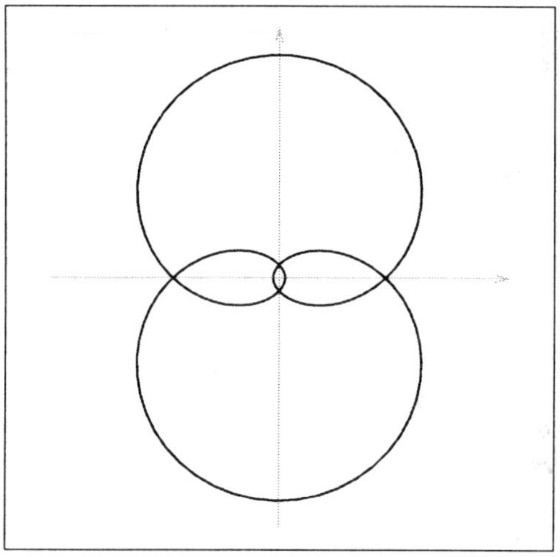

Figure 6. The curve (15) for $0 \le \theta \le 6\pi$, of which the portion $\pi/4 \le \theta \le 7\pi/4$ must be combined with the portion $-\pi/4 \le \theta \le \pi/4$ of the ovals of Cassini (8) to define a Minkowski square root for the disk (10) with $R > 1$.

5.3 Disk with center $(0,0)$ and arbitrary radius

As noted at the beginning of this section, we can obtain Minkowski square roots for *arbitrary* disks by considering just two cases — disks with center at $(1,0)$, and disks with the origin as center. Thus, for the sake of completeness, we

[8] An epitrochoid is the locus traced by a fixed point of a moving plane, carried by a circle that rolls without slipping on the exterior of a fixed base circle — see [14, 15].

now address the latter case. In fact, it suffices to consider the *unit* disk

$$\mathcal{U} = \{ \rho e^{i\phi} \mid 0 \le \rho \le 1, 0 \le \phi < 2\pi \} \tag{15}$$

since, once we know $\bigotimes^{1/2} \mathcal{U}$, a disk $\mathcal{D}$ with center $(0,0)$ and radius $R \neq 1$ clearly has the Minkowski square roots $\bigotimes^{1/2} \mathcal{D} = \{\pm\sqrt{R}\} \otimes (\bigotimes^{1/2} \mathcal{U})$.

Theorem 1 does not apply to $\mathcal{U}$, since the logarithmic curvature of $\partial\mathcal{U}$ is identically zero. Now if $\mathbf{z}_1 = \rho_1 e^{i\phi_1}$ and $\mathbf{z}_2 = \rho_2 e^{i\phi_2}$ are points of $\bigotimes^{1/2} \mathcal{U}$, the product $\rho_1\rho_2$ must assume all values between 0 and 1, and the sum $\phi_1 + \phi_2$ must independently assume (modulo 2π) all values from 0 and 2π. A trivial way to achieve this is, for $\mathbf{z}_1$, to allow ρ_1 to vary between 0 and 1 while taking a fixed value between 0 and 2π for ϕ_1 and conversely, for $\mathbf{z}_2$, to fix a value between 0 and 1 for ρ_2 while allowing ϕ_2 to vary between 0 and 2π. In other words, we can construct $\bigotimes^{1/2} \mathcal{U}$ as the union of a radial line from the origin to a point on $\partial\mathcal{U}$ and a circle of radius $0 < r < 1$ with the origin as center.

6 Minkowski n-th root of a disk

We have seen above that the Minkowski square root of a disk with center 1 and radius R is either loop of the ovals of Cassini (8), when $R < 1$, or the lemniscate of Bernoulli (9), when $R = 1$. When $R > 1$, however, the situation is more complicated — the Minkowski square root must be formulated as a composite curve. These results may be readily extended to Minkowski n–th roots, on introducing the following generalizations of (8) and (9).

Definition 4 The curve whose polar equation $r(\theta)$ satisfies

$$r^{2n}(\theta) - 2r^n(\theta)\cos n\theta + 1 = R^2 \tag{16}$$

defines the *n–th order ovals of Cassini.* For the special case $R = 1$, we obtain $r(\theta) = \sqrt[n]{2\cos n\theta}$, which defines the *$n$–th order lemniscate of Bernoulli.*

The "ordinary" ovals of Cassini (8) and lemniscate of Bernoulli (9) clearly correspond to the instance $n = 2$ of the above definition. The generalization of (8) to (16) is not found in standard "catalogs" [10, 14, 15, 21, 23] of plane curves, and the designation of this curve as the "n–th order ovals of Cassini" is our own choice. To justify (16) as a very natural and elegant generalization of (8), we shall now investigate the geometry of this curve.

We recall from §5 that the ovals of Cassini (8) can also be represented by the simple bipolar equation

$$r_1 r_2 = R, \tag{17}$$

where r_1 and r_2 are the distances of a curve point from two fixed "poles" at $(-1,0)$ and $(+1,0)$. Now, setting $\mathbf{z} = re^{i\theta}$, we may substitute $r^n = |\mathbf{z}|^n$ and $r^n\cos n\theta = \frac{1}{2}(\mathbf{z}^n + \bar{\mathbf{z}}^n)$ in equation (16), and thereby obtain

$$\mathbf{z}^n\bar{\mathbf{z}}^n - \mathbf{z}^n - \bar{\mathbf{z}}^n + 1 = (\mathbf{z}^n - 1)(\bar{\mathbf{z}}^n - 1) = |\mathbf{z}^n - 1|^2 \tag{18}$$

for the left–hand side. Hence, equation (16) describes the locus of a point $\mathbf{z}$ that satisfies $|\mathbf{z}^n - 1| = R$. Moreover, we can uniquely factorize $\mathbf{z}^n - 1$ as

$$\mathbf{z}^n - 1 = \prod_{k=1}^{n} \mathbf{z} - \boldsymbol{\xi}_k ,$$

where $\boldsymbol{\xi}_k = \exp(i\,2\pi(k-1)/n)$ for $k = 1,\ldots,n$ are the n–th roots of unity. Thus, equation (16) can be written as a natural generalization of (17), using "n–polar coordinates," in the form

$$\prod_{k=1}^{n} r_k = R, \tag{19}$$

$r_k = |\mathbf{z} - \boldsymbol{\xi}_k|$ for $k = 1,\ldots,n$ being the distances of a point on the curve (16) from the n–th roots of unity (which are simply $+1$ and -1 if $n = 2$).

Figure 7. Examples of the n–th order ovals of Cassini, defined by equation (16), for the nine cases corresponding to $n = 3, 6, 9$ and $R = 0.99, 1.00, 1.01$.

The curve (16) is geometrically, as well as algebraically, an intuitive generalization of (8). We recall that the latter comprises two distinct loops (containing $+1$ and -1) when $R < 1$, a "figure–of–eight" curve when $R = 1$, and a single loop with two "lobes" when $R > 1$. Correspondingly, the n–th order ovals of Cassini comprise n distinct loops (each containing an n–th root of unity) if $R < 1$, a flower–shaped curve with n "petals" (and an n–fold point at the origin) if

$R = 1$, and a single loop with n "lobes" if $R > 1$. Figure 7 illustrates these different forms. Note also that, for $n = 1$, equation (16) simply describes the circle of radius R with center at $+1$.

Curves defined by "n–polar equations" of the form (19), where $r_1, \ldots, r_n$ denote distances from n fixed poles in the complex plane, play a key role in analyzing convergence of polynomial approximations to given functions, constructed by interpolation of discrete sampled values. Davis [4] has called the curve defined by (19) a "lemniscate" with foci $\xi_1, \ldots, \xi_n$ and radius $\sqrt[n]{R}$ — by varying R, we obtain a family of *confocal* lemniscates.

For the case of foci located at the n–th roots of unity, we shall adhere to the nomenclature given in Definition 4 — i.e., we reserve the term "lemniscate" for the specific case $R = 1$ that exhibits an n–fold point at the origin.

Proposition 4 *If* $R \leq 1$, *a single loop of the* n*–th order ovals of Cassini* (16) *consitutes an* n*–th Minkowski root of the disk* (10).

Proof : By Theorem 1, it suffices to show that that n–th order ovals of Cassini (16) define the ordinary n–th root of the circle $|\mathbf{z} - 1| = R$ (≤ 1). In other words we must show that the relation (16) between r and θ in $\mathbf{z} = re^{\mathrm{i}\theta}$ is equivalent to $|\mathbf{z}^n - 1| = R$. This is immediately obvious from the alternate form for the left–hand side of (16) given by expression (18). ∎

We have not yet discussed higher–order Minkowski roots $\bigotimes^{1/n}\mathcal{D}$, with $n > 2$, of the disk (10) when $R > 1$. One might expect that a composite curve, as described in §5.2 for Minkowski square roots, will furnish a Minkowski n–th root. However, the complexity of such a composite curve may grow rapidly with n. For Minkowski square roots, the logarithmically convex part of the circle was generated by the Minkowski square of a Cassini oval segment, but the logarithmically concave part arose as a Minkowski product of a Cassini oval segment and a special curve segment.

For Minkowski n–th roots, we can again expect the n–th Minkowski power of a suitable Cassini oval segment to generate the logarithmically convex part of the circle. The crucial issue here is the logarithmically concave part — will a *single* specially–constructed curve segment, together with the Cassini oval, suffice for its generation, or will we need several (as many as $n - 1$) new curve segments?

The answer depends on the combinatorics of how n points are chosen from the Cassini oval and the new curve, to yield points on the circle as their products. Since we have already taxed the reader's endurance, we defer a systematic treatment of this issue to a future study.

7 Closure

Despite its fundamental nature and diverse applications, the Minkowski geometric algebra of complex sets is absent from standard texts [5, 19, 20] on "geometry of complex numbers." In earlier studies [7, 8] we have outlined basic properties of Minkowski sums, products, and implicitly–defined sets — closed–form solutions were presented for combinations with "simple" operands, and algorithms were developed for the case of general operands.

In the present paper we addressed a problem motivated by an algebraic perspective — namely, the extraction of Minkowski roots of complex sets. For a simply–connected complex domain $\mathcal{A}$ with boundary $\partial\mathcal{A}$, we showed that each connected component of the "ordinary" root $\mathcal{A}^{1/n}$ defines a Minkowski root $\bigotimes^{1/n}\mathcal{A}$ if $\mathcal{A}$ is logarithmically convex. Moreover, the *boundary* of each connected component of $\mathcal{A}^{1/n}$ defines a *minimal* Minkowski root.

When $\mathcal{A}$ is not logarithmically convex (e.g., if the origin lies in the interior of $\mathcal{A}$), however, the determination of a Minkowski root $\bigotimes^{1/n}\mathcal{A}$ is much more challenging. Even the "simple" case of a circular disk containing the origin requires (see §5.2) a careful composition of parts of the Cassini ovals, a rather enigmatic curve of degree 6, and a radial line to produce a valid Minkowski square root. For more general complex domains that are not logarithmically convex, the issues of existence, uniqueness, minimality, and algorithmic construction of Minkowski square roots remain open. Moreover, as remarked in §6, the solutions to these problems may not admit elementary generalizations to the context of Minkowski n–th roots for $n > 2$.

As the inverse to taking the power of an element, root extraction is a basic operation in any algebraic system that has a commutative and associative product. Apart from this algebraic motivation, the Minkowski n–th roots of complex sets give rise to fascinating geometrical entities, e.g., the generalization (16) of the ovals of Cassini, as Minkowski n–th roots of logarithmically–convex circular disks. Although, with hindsight, this generalization may seem natural, it is conspicuously absent from "standard" treatises on higher–order curves. We expect that elaboration of the theory of Minkowski n–th roots of complex sets will entail further interesting geometries and fruitful applications.

Acknowledgment

This work was supported in part by the National Science Foundation under grant CCR–9902669.

References

[1] V. G. Boltyanskii (1964), *Envelopes*, Macmillan, New York.

[2] J. W. Bruce and P. J. Giblin (1981), What is an envelope?, *Math. Gazette* **65**, 186–192.

[3] J. W. Bruce and P. J. Giblin (1984), *Curves and Singularities*, Cambridge Univ. Press.

[4] P. J. Davis (1963), *Interpolation and Approximation*, Blaisdell Publishing Co., New York.

[5] R. Deaux (1956), *Introduction to the Geometry of Complex Numbers* (translated from the French by H. Eves), F. Ungar, New York.

[6] R. T. Farouki and J–C. A. Chastang (1992), Curves and surfaces in geometrical optics, *Mathematical Methods in Computer Aided Geometric Design II*, (T. Lyche & L. L. Schumaker, eds.), Academic Press, pp. 239–260.

[7] R. T. Farouki, H. P. Moon, and B. Ravani (2000), Minkowski geometric algebra of complex sets, *Geometriae Dedicata*, to appear.

[8] R. T. Farouki, H. P. Moon, and B. Ravani (1999), Algorithms for Minkowski products and implicitly–defined complex sets, preprint.

[9] P. K. Ghosh (1988), A mathematical model for shape description using Minkowski operators, *Comput. Vision, Graphics, Image Process.* **44**, 239–269.

[10] F. Gomes Teixeira (1971), *Traité des Courbes Spéciales Remarquables Planes et Gauches*, Tome I, Chelsea (reprint), New York.

[11] H. Hadwiger (1957), *Vorlesungen über Inhalt, Oberfläche, und Isoperimetrie*, Springer, Berlin.

[12] A. Kaul (1993), Computing Minkowski sums, PhD Thesis, Columbia University.

[13] A. Kaul and R. T. Farouki (1995), Computing Minkowski sums of plane curves, *Int. J. Comput. Geom. Applic.* **5**, 413–432.

[14] J. D. Lawrence (1972), *A Catalog of Special Plane Curves*, Dover, New York.

[15] E. H. Lockwood (1967), *A Book of Curves*, Cambridge Univ. Press.

[16] H. Minkowski (1903), Volumen und Oberfläche, *Math. Ann.* **57**, 447–495.

[17] R. E. Moore (1966), *Interval Analysis*, Prentice–Hall, Englewood Cliffs, NJ.

[18] R. E. Moore (1979), *Methods and Applications of Interval Analysis*, SIAM, Philadelphia.

[19] T. Needham (1997), *Visual Complex Analysis*, Oxford Univ. Press.

[20] H. Schwerdtfeger (1979), *Geometry of Complex Numbers*, Dover, New York.

[21] D. H. Von Seggern (1993), *CRC Standard Curves and Surfaces*, CRC Press, Boca Raton.

[22] J. Serra (1982), *Image Analysis and Mathematical Morphology*, Academic Press, London.

[23] C. Zwikker (1963), *The Advanced Geometry of Plane Curves and Their Applications*, Dover, New York.

Rational Parametrization of Canal Surface by 4 Dimensional Minkowski Pythagorean Hodograph Curves *

Hyeong In Choi
Department of Mathematics
Seoul National University
hichoi@math.snu.ac.kr

Doo Seok Lee
Department of Mathematics
Seoul National University
dslee@math.snu.ac.kr

Abstract

Lorentzian geometry with Minkowski Pythagorean Hodograph (MPH) formalism in $\mathbb{R}^{3,1}$ *gives us a new and insightful method of rational parametrization of canal surfaces. Our previous works about MPH curves in* $\mathbb{R}^{3,1}$ *[11] shows that a curve* $\gamma(t) = (x(t), y(t), z(t), r(t))$ *in* $\mathbb{R}^{3,1}$ *can be represented by the PH representation map in* $\mathcal{C}\ell(3,1)$ *which avoids the complex root finding algorithm. Our parametrization method gives us the flexibility to represent the canal surfaces within their fiber ambiguities. This paper constitutes the first step of our ongoing work which deals with the issues for canal surfaces in a truly new and intriguing manner such as finding rotation minimizing frames. We believe this is just the tip of the iceberg and the further work will yield many valuable applications in the area of canal surfaces.*

1 Introduction

In this paper, we present a new method to study canal surfaces represented by rational or polynomial data using the concepts and tools of Lorentzian geometry. Our method uses the technique of Pythagorean hodograph curve in the 4-dimensional Minkowski space on which we based our previous paper [11], the results of which are summarized through Chapter 2, 3. This technique depends on the proper representation of $\mathcal{C}\ell(3,1)$ which is generated by the basis of 4-dimensional Minkowski space $\mathbb{R}^{3,1}$. This representation extends the well-know spin representation map to an algebra representation of $\mathcal{C}\ell^+(3,1)$. We call this map the PH representation map. By using this tool, we shed new light on the problem of the rational or polynomial parametrization of canal surfaces via Lorentzian geometry, which is the main contribution of this paper which we present in Chapter 4. As we refer later in this paper, this method constitutes the first step of our ongoing work which we hope will successfully deal with the problem of the rotation minimizing or almost minimizing rational parametrization of the rational canal surface [10].

There are lots of works related to the rational parametrization of a canal surface. Among them, Lü and Pottmann [24] proved that a pipe surface, an envelope surface by a sphere with fixed radius, with rational spine curve always admits a rational parametrization. Maekawa, Patrikalakis, Sakkalis, and Yu [25] developed an algorithm method for rational parametrization of pipe surfaces. Then Peternell and Pottmann [28] further developed another generalization that a canal surface with a rational spine curve and a radius function can be rationally parametrized. Much of our current work is based on theirs which we will discuss further in Chapter 2.

PH Formalisms

Our work is motivated by the 4-dimensional Minkowski Pythagorean hodograph(PH) curve which is one of the PH formalisms. Pythagorean hodograph curves were first introduced by Farouki and Sakkalis [19]. The PH curves have many useful properties; their arc length can be computed as a closed form, and their offset curves can be expressed as rational curves.

For a two dimensional case, Farouki and Sakkalis introduced a class of polynomial curves called the Pythagorean hodograph curves. Let $\gamma(t) = (x(t), y(t))$ be a polynomial curve in the plane. It is called a *Pythagorean hodograph (PH) curve* if

$$x'(t)^2 + y'(t)^2 = \sigma(t)^2 \tag{1}$$

for some polynomial $\sigma(t)$. The necessary and sufficient conditions for (1) to hold are that there exist polynomials $u(t), v(t)$ and $w(t)$ which satisfy

$$\begin{aligned} x'(t) &= w(t)\{u(t)^2 - v(t)^2\}, \\ y'(t) &= w(t)\{2u(t)v(t)\}, \\ \sigma(t) &= w(t)\{u(t)^2 + v(t)^2\}. \end{aligned} \tag{2}$$

0-7695-0562-7/00 $10.00 © 2000 IEEE

If these conditions are satisfied, its unit normal vector field is a rational function of the parameter t, therefore its offset curve becomes rational in t.

The three dimensional generalization of the Pythagorean hodograph curve was achieved by Farouki and Sakkalis [20] and Dietz, Hoschek and Jüttler [15]. They independently obtained more or less similar conditions for a space curve to have polynomial speed, and called it a space PH curve. In fact, it turned out that Farouki and Sakkalis' condition can be viewed as a special case of the one developed by Dietz, Hoschek and Jüttler. This condition is similar in spirit to that of the plane PH curve obtained by Farouki and Sakkalis, except that the space PH curve case is now more complicated and the algebra gets more involved. This space PH curve is useful in various contexts such as the study of pipe surfaces, rational frames, and the curve and surface parametrization on S^2. In the following paragraph, we summarize the result of the space PH curve.

A polynomial space curve $\gamma(t) = (x(t), y(t), z(t))$ is called a *space Pythagorean hodograph (PH) curve* if its hodograph $\gamma'(t)$ satisfies the Pythagorean condition

$$x'(t)^2 + y'(t)^2 + z'(t)^2 = \sigma^2(t)$$

for some real polynomial $\sigma(t)$. According to the result of Dietz, Hoschek, and Jüttler [15], a polynomial space curve $\gamma(t) = (x(t), y(t), z(t))$ has a *primitive* Pythagorean hodograph, i.e.,

$$\begin{aligned} x'(t)^2 + y'(t)^2 + z'(t)^2 &= \sigma^2(t) \\ \text{and} \quad \gcd(x'(t), y'(t), z'(t)) &= 1 \end{aligned} \tag{3}$$

for some polynomial $\sigma(t)$ if and only if there exist four polynomials $u(t)$, $\rho(t)$, $v(t)$, $w(t)$ and the hodograph $\gamma'(t)$ and the speed $\sigma(t)$ satisfies

$$\begin{aligned} \sigma(t) &= \pm(u^2(t) + v^2(t) + w^2(t) + \rho^2(t)), \\ x'(t) &= u^2(t) - v^2(t) - w^2(t) + \rho^2(t), \\ y'(t) &= 2u(t)v(t) + 2\rho(t)w(t), \\ z'(t) &= 2u(t)w(t) - 2\rho(t)v(t). \end{aligned} \tag{4}$$

Another PH formalism related to the medial axis transform is the Minkowski Pythagorean hodograph curve in $\mathbb{R}^{2,1}$. A spacelike curve $\gamma(t) = (x(t), y(t), r(t))$ in $\mathbb{R}^{2,1}$ is called a *Minkowski Pythagorean hodograph (MPH) curve* if there exists a polynomial $\sigma(t)$ such that

$$x'(t)^2 + y'(t)^2 - r'(t)^2 = \sigma(t)^2. \tag{5}$$

H.P. Moon [26] proved the necessary and sufficient conditions for a spacelike curve to be MPH, that is, a polynomial curve $\gamma(t) = (x(t), y(t), r(t))$ in $\mathbb{R}^{2,1}$ is an MPH curve with $x'(t)^2 + y'(t)^2 - r'(t)^2 = \sigma^2(t)$ if and only if there exist four polynomials $u(t), \rho(t), v(t), w(t)$ satisfying the relations

$$\begin{aligned} \sigma(t) &= \pm(u^2(t) - v^2(t) - w^2(t) + \rho^2(t)), \\ x'(t) &= -2u(t)\rho(t) - 2v(t)w(t), \\ y'(t) &= u^2(t) + v^2(t) - w^2(t) - \rho^2(t), \\ r'(t) &= 2u(t)v(t) + 2\rho(t)w(t). \end{aligned} \tag{6}$$

We will discuss more about this MPH in $\mathbb{R}^{2,1}$ in Section 3.1.

Unifying approach via Clifford Algebra

During the study of MPH curve, one can easily notice the similarity between the space PH condition and the MPH condition. This leads one to search for a unifying algebraic framework. It turns out that the Clifford algebra furnishes a *correct* and versatile framework to unify all the known incarnations together in a coherent manner. Here we will use without explanation the standard notations of the Clifford algebra. For a reference on Clifford algebra, refer to any standard algebra books. For instance, one may consult [2, 14].

The PH representation map extends the twisted adjoint representation of the spin group to a certain subset of the even Clifford algebra $\mathcal{C}\ell^+(V,Q)$, which we call $\Lambda(V,Q)$. For a explanation of $\Lambda(V,Q)$ and other subsequent notations, refer to [12, 23]. Suppose $\mathbf{x}$ is an element of $\text{Spin}(V,Q)$, then $\mathbf{x}$ can be expressed as products of unit vectors in V, i.e., $\mathbf{x} = v_i \cdots v_{2k}$ for $v_i \in V$, $Q(v_i) = \pm 1$. If we restrict the twisted adjoint representation χ to $\text{Spin}(V,Q)$, χ can be expressed in a simpler form. Namely, $\chi(\mathbf{x})(w) = \alpha(\mathbf{x})w\mathbf{x}^{-1} = \mathbf{x}w\overline{\mathbf{x}}$ for $w \in V$.

We define a new map $T : \mathcal{C}\ell^+(V,Q) \to \text{End}(\mathcal{C}\ell(V,Q))$ by

$$T(\mathbf{x})(\mathbf{y}) = \mathbf{x}\mathbf{y}\overline{\mathbf{x}},$$

for $\mathbf{x} \in \mathcal{C}\ell^+(V,Q)$ and $\mathbf{y} \in \mathcal{C}\ell(V,Q)$. When we fix $\mathbf{x}$, we can define an action $R_{\mathbf{x}}$ from V to V by

$$R_{\mathbf{x}}(\mathbf{v}) = T_a(\mathbf{x}) = \mathbf{x}\mathbf{v}\overline{\mathbf{x}}$$

for $v \in V$.

If we restrict T to $\text{Spin}(V,Q)$, T is the twisted adjoint representation χ. But in general, it is not guaranteed that $T(\mathbf{x})(V) \subset V$ for all $\mathbf{x} \in \mathcal{C}\ell(V,Q)$. So we need to restrict T to those x such that $T(x)(V) \subset V$. For that purpose, let $\Lambda(V,Q)$ be the subset of $\mathcal{C}\ell^+(V,Q)$ such that

$$\Lambda(V,Q) = \{\mathbf{x} \in \mathcal{C}\ell^+(V,Q) \ : \ T(\mathbf{x})(V) \subset V\}.$$

Clearly, $\text{Spin}(V,Q)$ is a subset of $\Lambda(V,Q)$, and $\Lambda(V,Q)$ plays the role of the Clifford group $\Gamma(V,Q)$ for the PH representation map, even though $\Lambda(V,Q)$ is not a (multiplicative) group. We adopt the notation convention to denote $\Gamma(V,Q) = \Gamma(n)$ if Q is positive definite on $\mathbb{R}^n$, and

$\Gamma(V, Q) = \Gamma(n, 1)$ if Q is the Lorentzian inner product of signature $+ + \cdots + -$ on the Minkowski space $\mathbb{R}^{n,1}$.

We now define the PH representation map.

Definition 1.1 *For a fixed unit vector a in V, the PH representation map $T_a : \Lambda(V, Q) \to V$ is defined as*

$$T_a(\mathbf{x}) = T(\mathbf{x})(a) = \mathbf{x}a\overline{\mathbf{x}}$$

for any $\mathbf{x} \in \Lambda(V, Q)$. We call a the base (vector) of the PH representation map T_a.

In the above definition, we can choose a to be any unit vector, although we usually choose a as one of the orthonormal basis vectors of V. The algebraic representation generated by T_a may be somewhat different according to different choices of $a \in V$. But since it does not change the algebraic structure itself, we will choose a convenient vector for a in subsequent presentations.

Then all the PH curves defined in the previous section can be represented by this PH representation map [11].

2 Canal Surface: Definition and Previous Works

Before we present our method, we review the works of the rational parametrization of canal surface done by Peternell and Pottmann [28]. We are indebted to their pioneering work that any canal surface with a polynomial spine curve and a polynomial radius function has a rational parametrization if it is a real canal surface, i.e., $x'(t)^2 + y'(t)^2 + z'(t)^2 - r'(t)^2 \geq 0$.

We use the following notation convention $\langle \cdot, \cdot \rangle_{\mathbb{R}^3}$ and $| \cdot |_{\mathbb{R}^3}$ to denote the Euclidean inner product and the norm in $\mathbb{R}^3$, and $\langle \cdot, \cdot \rangle_{\mathbb{R}^{3,1}}$ and $| \cdot |_{\mathbb{R}^{3,1}}$ to denote the Minkowski inner product and the norm in $\mathbb{R}^{3,1}$ respectively.

We begin with the definition of canal surface.

Definition 2.1 *A* canal surface *Φ is the envelope of one parameter family of spheres,*

$$\Sigma(t) : |(x, y, z) - (s_1(t), s_2(t), s_3(t))|_{\mathbb{R}^3} = r(t)$$

where $s(t) = (s_1(t), s_2(t), s_3(t))$ is a curve in $\mathbb{R}^3$ and $r(t)$ is a positive function which denotes the radius of spheres. Here, the curve $s(t)$ is called the spine *curve of Φ. A canal surface Φ is called a* pipe surface *if the radius function is a constant.*

The defining equation for Φ are given by the intersection of the sphere with the plane for each t:

$$\begin{aligned} \Sigma(t) &: |(x, y, z) - (s_1(t), s_2(t), s_3(t))|_{\mathbb{R}^3} = r(t), \\ \Sigma'(t) &: \langle (x, y, z) - s(t), s'(t) \rangle_{\mathbb{R}^3} = -r(t)r'(t). \quad (7) \end{aligned}$$

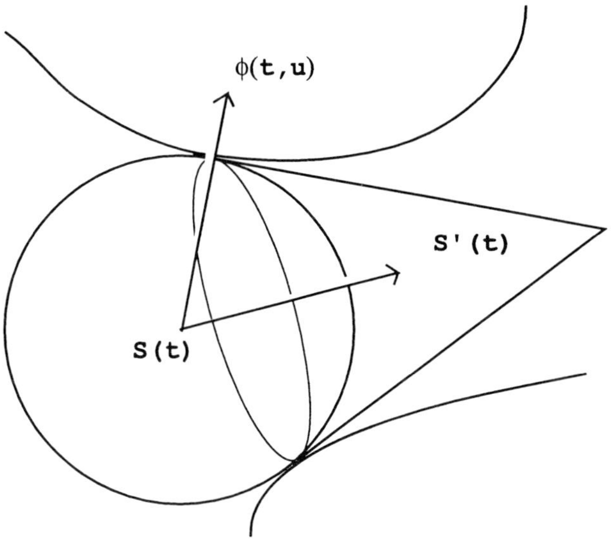

Figure 1. Geometric description of a canal surface

A real canal surface exists only if $s'(t)^2 - r'(t)^2 \geq 0$. We call the one parameter set of circles $k(t) = \Sigma(t) \cap \Sigma'(t)$ the *characteristic circles*. If we denote the unit vector in the direction $(x, y, z) - s(t)$ by $\phi(t, u)$ where the parameter u in $\phi(t, u)$ is used for the parametrization of the characteristic circle, then we get $(x, y, z) - s(t) = r(t)\phi(t)$ as in Figure 1. And the equation (7) can be rewritten by $\langle \phi(t), s'(t) \rangle_{\mathbb{R}^3} = -r'(t)$. Then the canal surface in $\mathbb{R}^{3,1}$ has the following rational parametrization:

$$\Phi(t, u) = (x(t), y(t), z(t)) + r(t)\phi(t, u).$$

Peternell and Pottmann [28] mainly use the stereographic projection in their work with canal surface. Let us summarize their method briefly. From $(0, 0, 1) \in \mathbb{R}^3$, the characteristic circle is projected to the (x, y)-plane by the stereographic projection, which is a conformal map preserving rationality. Then the coordinate of the center of the circle in the plane is given by $(s_1'(t), s_2'(t))$ and the coordinate of a point in the circle is given by $(f_1(t), f_2(t))$. Then other points in the circle are given by the reflection of $(f_1(t), f_2(t))$ by the line $U(u) = (u, 1)$ in the plane. The resulting rational parametrization $\phi(t, u)$ of the characteristic circle is given by the inverse stereographic projection of this circle in the plane to the unit sphere.

They rely on the fact that any nonnegative polynomial can be decomposed into the sum of two squared polynomials. Thus, if we have $x'(t)^2 + y'(t)^2 + z'(t)^2 - r'(t)^2 = f(t) \geq 0$ for all t, then there exist two polynomials $f_1(t)$ and $f_2(t)$ such that $f(t) = f_1(t)^2 + f_2(t)^2$. They provide an algorithm to construct the rational parametrization of the canal surface with $f_1(t)$ and $f_2(t)$. The major difficulty in applying their algorithm to a given canal surface is the step to decompose $f(t)$ into $f_1(t)^2 + f_2(t)^2$. In order to do this, one has to find all complex zeros of $f(t)$. It requires complex numerical computation and algebraic algorithm which we do not need in our method. They also mention that the rational parameterization of a given canal surface is not so amenable: the parametric grid curves generally make lots of rotation around the spine curve, which is another problem that we hope to solve.

3 4-Dimensional Minkowski PH Curves

We now summarize the result of our investigation of the 4-dimensional Minkowski PH(MPH) curve [11], which is exactly the condition in which a real canal surface exists. By using Clifford algebra $\mathcal{C}\ell^{3,1}$ and PH representation map, we define a special class of MPH curves. And we show that this special class can represent all the MPH curves, which we call the 4-dimensional MPH representation theorem.

Suppose $\gamma(t) = (x(t), y(t), z(t), r(t))$ be a polynomial curve in $\mathbb{R}^{3,1}$. As we regard the *medial axis transform* in the three dimensional Minkowski space $\mathbb{R}^{2,1}$, we can regard a point (x, y, z, r) in $\mathbb{R}^{3,1}$ as a sphere in $\mathbb{R}^3$ centered at (x, y, z) of radius r. Thus, the curve $\gamma(t)$ expresses a one parameter family of spheres in $\mathbb{R}^3$, whose envelope is a canal surface. The following is the definition of the MPH curve in $\mathbb{R}^{3,1}$.

Definition 3.1 *A polynomial curve* $\gamma(t) = (x(t), y(t), z(t), r(t))$ *is called a* Minkowski Pythagorean hodograph*(MPH) curve in* $\mathbb{R}^{3,1}$ *if there exist real polynomials* $f_1(t)$ *and* $f_2(t)$ *such that*

$$x'(t)^2 + y'(t)^2 + z'(t)^2 - r'(t)^2 = f_1(t)^2 + f_2(t)^2. \qquad (8)$$

Why should the righthand side of (8) be the sum of the squares of two polynomials? This is the new feature that cannot be seen in 2 or 3 dimensional cases where the speed of a curve is just a single square of a polynomial but is observable in 4-dimensional space. As we investigate the PH representation map in Section 3.2, the norm $N(\mathbf{x})$ for $\mathbf{x} \in \mathcal{C}\ell^{3,1}$ is no longer a scalar but the sum of a scalar with a pseudo-scalar. This means that the canal surface contains two foliations. In view of algebra, we can show inductively that a spacelike polynomial curve $\gamma(t)$ is a sum of two polynomials, as we saw in the previous chapter. (Ref. [25], [28]) That is, if we have $x'(t)^2 + y'(t)^2 + z'(t)^2 - r'(t)^2 = f(t) \geq 0$ for all t, then there exist two polynomials $f_1(t)$ and $f_2(t)$ such that $f(t) = f_1(t)^2 + f_2(t)^2$.

3.1 Clifford Algebra Formalism for $\mathbb{R}^{2,1}$

We will now briefly summarize the results of the structure of Minkowski Pythagorean hodographs in $\mathbb{R}^{2,1}$ with the corresponding Clifford algebra $\mathcal{C}\ell(2,1)$. (Ref. [11]) As we saw in the introduction, a polynomial curve $\gamma(t) = (x(t), y(t), z(t))$ is called a *Minkowski Pythagorean hodograph* (MPH) curve if there exists a polynomial $\sigma(t)$ such that $x'(t)^2 + y'(t)^2 - z'(t)^2 = \sigma(t)^2$. The necessary and sufficient condition for a polynomial curve $\gamma(t)$ to be an MPH curve is the following:

$$\begin{aligned} x'(t) &= -2p_0(t)p_3(t) - 2p_1(t)p_2(t), \\ y'(t) &= p_0(t)^2 + p_1(t)^2 - p_2(t)^2 - p_3(t)^2, \\ z'(t) &= 2p_0(t)p_1(t) + 2p_2(t)p_3(t), \\ \sigma(t) &= p_0(t)^2 - p_1(t)^2 - p_2(t)^2 + p_3(t)^2, \end{aligned} \qquad (9)$$

for some polynomials $p_0(t), p_1(t), p_2(t), p_3(t)$.

This version of the Pythagorean hodograph arises from the study of the medial axis transform of a planar region in which $(x(t), y(t))$ represents the medial axis (the spine) curve, and $z(t)$ is the radius of the maximal inscribed circle centered at $(x(t), y(t))$. The paper [5] deals with fundamental mathematical aspects of the medial axis transform. In that paper, it is proved that the medial axis transform, under rather mild assumptions on the boundary of the domain, is a finite geometric graph embedded in $\mathbb{R}^3$. An algorithm to compute the medial axis transform is presented in [6]. The medial axis transform contains all the necessary and sufficient information of the domain in the sense that the domain itself can be exactly reconstructed out of the medial axis transform and vice versa. The way to reconstruct the domain is to use the envelope formula. Namely, let $\gamma(t) = (x(t), y(t), r(t))$ be a curve portion of the medial axis transform. Interpreting $\gamma(t)$ as a one parameter family of circles with center $(x(t), y(t))$ of radius $r(t)$, its envelope curve $(\widetilde{x}(t), \widetilde{y}(t))$ can be computed by the following formula

$$\begin{aligned} x(t) &= x(t) + r(t)\frac{-r'(t)x'(t) \mp \sigma(t)y'(t)}{x'(t)^2 + y'(t)^2}, \\ y(t) &= y(t) + r(t)\frac{-r'(t)y'(t) \pm \sigma(t)y'(t)}{x'(t)^2 + y'(t)^2}, \end{aligned} \qquad (10)$$

where $\sigma(t) = \sqrt{x'(t)^2 + y'(t)^2 - r'(t)^2}$.

Using the above envelope formula, its envelope curve is rational if $\gamma(t)$ is an MPH curve, i.e.

$$x'(t)^2 + y'(t)^2 - r'(t)^2 = \sigma(t)^2 \qquad (11)$$

for some polynomial $\sigma(t)$.

This kind of approach is also well suited to the study of offset curves. Namely, once a domain is given by a one parameter family of circles, say $\gamma(t) = (x(t), y(t), r(t))$, its $\pm\delta$ offset curve is then the envelope curve constructed from the curve $\gamma_{\pm\delta}(t) = (x(t), y(t), r(t) \pm \delta)$ in $\mathbb{R}^3$. Obviously the rationality does not change even if the radius is added or subtracted by a constant.

This algebraic structure can be easily captured using the PH representation map in an appropriate Clifford algebra setting. Since the underlying vector space is the Minkowski space $\mathbb{R}^{2,1}$, the natural Clifford algebra should be $\mathcal{C}\ell(2,1)$.

We now construct the Minkowski Pythagorean hodographs from the even Clifford algebra. Recall that the map $T : \mathcal{C}\ell^+(2,1) \to \mathrm{End}(\mathcal{C}\ell(2,1))$ is defined by

$$T(\mathbf{x})(\mathbf{y}) = \mathbf{x}\mathbf{y}\overline{\mathbf{x}}$$

for $\mathbf{x} \in \mathcal{C}\ell^+(2,1)$ and $\mathbf{y} \in \mathcal{C}\ell(2,1)$. One can easily show that for any $\mathbf{x} \in \mathcal{C}\ell^+(2,1)$, $T(\mathbf{x})$ acts on $\mathbb{R}^{2,1}$ as an endomorphism, which means $\Lambda(2,1) = \mathcal{C}\ell^+(2,1)$. Let us now choose the base vector a in the definition of the PH representation map to be $\mathbf{e}_2$. Thus we work with $T_{\mathbf{e}_2} : \mathcal{C}\ell^+(2,1) \to \mathbb{R}^{2,1}$ given by $T_{\mathbf{e}_2}(\mathbf{x}) = T(\mathbf{x})(\mathbf{e}_2)$. Suppose $\mathbf{p}(t) = p_0(t) + p_1(t)\mathbf{e}_{23} + p_2(t)\mathbf{e}_{31} + p_3(t)\mathbf{e}_{12}$ is a polynomial in $\mathcal{C}\ell^+(2,1)$. Then $T_{\mathbf{e}_2}$ maps $\mathbf{p}(t)$ to a curve in $\mathbb{R}^{2,1}$ as follows:

$$\begin{aligned}
& T_{\mathbf{e}_2}(\mathbf{p}(t)) \\
= \;& (p_0(t) + p_1(t)\mathbf{e}_{23} + p_2(t)\mathbf{e}_{31} + p_3(t)\mathbf{e}_{12})\, \mathbf{e}_2 \\
& (p_0(t) - p_1(t)\mathbf{e}_{23} - p_2(t)\mathbf{e}_{31} - p_3(t)\mathbf{e}_{12}) \\
= \;& -(2p_0(t)p_3(t) + 2p_1(t)p_2(t))\mathbf{e}_1 \qquad (12) \\
& +(p_0(t)^2 + p_1(t)^2 - p_2(t)^2 - p_3(t)^2)\mathbf{e}_2 \\
& +(2p_0(t)p_1(t) + 2p_2(t)p_3(t))\mathbf{e}_3.
\end{aligned}$$

And the norm $N(\mathbf{p}(t))$ of $\mathbf{p}(t)$ is

$$\begin{aligned}
& N(\mathbf{p}(t)) \\
= \;& (p_0(t) + p_1(t)\mathbf{e}_{23} + p_2(t)\mathbf{e}_{31} + p_3(t)\mathbf{e}_{12}) \\
& (p_0(t) - p_1(t)\mathbf{e}_{23} - p_2(t)\mathbf{e}_{31} - p_3(t)\mathbf{e}_{12}) \\
= \;& p_0(t)^2 - p_1(t)^2 - p_2(t)^2 + p_3(t)^2.
\end{aligned}$$

Therefore, the image of $T_{\mathbf{e}_2}$ satisfies the Minkowski Pythagorean condition (9), and the magnitude of $T_{\mathbf{e}_2}(\mathbf{p}(t))$ is given by the norm $N(\mathbf{p}(t))$.

The next theorem shows that the converse is also true.

Theorem 3.2 *Let $\gamma(t) = (x_1(t), x_2(t), x_3(t))$ be an MPH curve in $\mathbb{R}^{2,1}$ such that $\sigma(t) > 0$ for all t. Then there exist four polynomials satisfying (12) above.*

3.2 Clifford Algebra Formalism for $\mathbb{R}^{3,1}$

We now review the results of our work with the PH formalism of $\mathbb{R}^{3,1}$ by the unifying map of PH representation map [11]. In order to generalize MPH curves to $\mathbb{R}^{3,1}$, we first discuss the structure of the Clifford algebra $\mathcal{C}\ell(3,1)$. The Clifford algebra $\mathcal{C}\ell(3,1)$ is of degree 16 and has basis $\{1, \mathbf{e}_1, \mathbf{e}_2, \mathbf{e}_3, \mathbf{e}_4, \mathbf{e}_{12}, \mathbf{e}_{13}, \mathbf{e}_{14}, \mathbf{e}_{23}, \mathbf{e}_{24}, \mathbf{e}_{34}, \mathbf{e}_{123}, \mathbf{e}_{124}, \mathbf{e}_{134}, \mathbf{e}_{234}, \mathbf{e}_{1234}\}$. The elements of this basis are distinguished according to its degree: $\mathbf{e}_1, \mathbf{e}_2, \mathbf{e}_3, \mathbf{e}_4$ are called *vectors*; $\mathbf{e}_{12}, \mathbf{e}_{13}, \mathbf{e}_{14}, \mathbf{e}_{23}, \mathbf{e}_{24}, \mathbf{e}_{34}$ are *bi-vectors*; $\mathbf{e}_{123}, \mathbf{e}_{124}, \mathbf{e}_{134}, \mathbf{e}_{234}$ are *tri-vectors*; and $\mathbf{e}_{1234}$ is called a *pseudo-scalar*. And an element $\mathbf{x}$ in the even Clifford algebra $\mathcal{C}\ell^+(3,1)$ is of the form $x_0 + x_1\mathbf{e}_{12} + x_2\mathbf{e}_{13} + x_3\mathbf{e}_{14} + x_4\mathbf{e}_{23} + x_5\mathbf{e}_{24} + x_6\mathbf{e}_{34} + x_7\mathbf{e}_{1234}$.

According to the classification results on the Clifford algebra, $\mathcal{C}\ell(3,1)$ is isomorphic to the 2×2 quaternion matrix algebra $\mathbb{H}(2)$. This is obtained from the well known isomorphism $\mathcal{C}\ell(r+1, s+1) \cong \mathcal{C}\ell(r,s) \otimes \mathcal{C}\ell(1,1)$. Thus, $\mathcal{C}\ell(3,1) \cong \mathcal{C}\ell(2) \otimes \mathcal{C}\ell(1,1)$. We already mentioned that $\mathcal{C}\ell(2) \cong \mathbb{H}$. And one can easily show that $\mathcal{C}\ell(1,1) \cong \mathrm{M}(2,\mathbb{R})$. Moreover, $\mathcal{C}\ell^+(3,1)$ is isomorphic to $\mathcal{C}\ell(2,1) \cong \mathrm{M}(2,\mathbb{C})$. This is a general fact for any n, i.e., $\mathcal{C}\ell^+(n+1,m) \cong \mathcal{C}\ell(n,m)$.

Here we are going to use the Minkowski version of the biquaternion notation. The even Clifford algebra $\mathcal{C}\ell^+(3,1)$ has the basis

$$\{1, \mathbf{e}_{23}, \mathbf{e}_{31}, \mathbf{e}_{12}, \mathbf{e}_{14}, \mathbf{e}_{24}, \mathbf{e}_{34}, \mathbf{e}_{1234}\}.$$

We write $\mathbf{e}_{23} = \mathbf{i}$, $\mathbf{e}_{31} = \mathbf{j}$, $\mathbf{e}_{12} = \mathbf{k}$ and $\mathbf{e}_{1234} = \omega$. The elements $1, \mathbf{i}, \mathbf{j}, \mathbf{k}$ are basis of the quaternion algebra $\mathbb{H}$. Thus, an element $\mathbf{x}$ in $\mathcal{C}\ell^+(3,1)$ can be written as

$$\begin{aligned}
\mathbf{x} &= \mathbf{p} + \omega\mathbf{q} \\
&= p_0 + p_1\mathbf{i} + p_2\mathbf{j} + p_3\mathbf{k} + \omega(q_0 + q_1\mathbf{i} + q_2\mathbf{j} + q_3\mathbf{k}).
\end{aligned}$$

Or by complex notation, we have $\mathbf{x} = p + wq = z_1 + z_2\mathbf{j} + \omega(z_3 + z_4\mathbf{j})$, where $z_1 = p_0 + p_1\, i$, $z_2 = p_2 + p_3\, i$, $z_3 = q_0 + q_1\, i$, and $z_4 = q_2 + q_3\, i$.

We now consider the map

$T : \mathcal{C}\ell^+(3,1) \to \mathrm{End}(\mathcal{C}\ell(3,1))$ defined in the same way as before,

$$T(\mathbf{x})(\mathbf{y}) = \mathbf{x}\mathbf{y}\overline{\mathbf{x}}$$

for $\mathbf{x} \in \mathcal{C}\ell^+(3,1)$ and $\mathbf{y} \in \mathcal{C}\ell(3,1)$.

The image of $\mathbf{x}$ under $T_{\mathbf{e}_1}$ is given by

$$\begin{aligned}
& T_{\mathbf{e}_1}(\mathbf{x}) \\
= \;& \mathbf{x}\mathbf{e}_1\overline{\mathbf{x}} \\
= \;& (p + \omega q)\mathbf{e}_1(\overline{p} + \omega\overline{q}) \qquad (13) \\
= \;& (p_0^2 + p_1^2 - p_2^2 - p_3^2 + q_0^2 + q_1^2 - q_2^2 - q_3^2)\mathbf{e}_1 \\
& +2(p_1p_2 + p_0p_3 + q_1q_2 + q_0q_3)\mathbf{e}_2 \\
& +2(p_1p_3 - p_0p_2 + q_1q_3 - q_0q_2)\mathbf{e}_3 \\
& +2(p_1q_0 - p_0q_1 + p_2q_3 - p_3q_2)\mathbf{e}_4.
\end{aligned}$$

And by using the complex notation, we have a more compact form of $T_{\mathbf{e}_1}(\mathbf{x})$ as follows:

$$\begin{aligned}
& T_{\mathbf{e}_1}(\mathbf{x}) \\
= \;& \mathbf{x}\mathbf{e}_1\overline{\mathbf{x}}
\end{aligned}$$

$$
\begin{aligned}
&= (p+\omega q)\mathbf{e}_1(\overline{p}+\omega\overline{q}) \\
&= (|z_1|^2-|z_2|^2+|z_3|^2-|z_4|^2)\mathbf{e}_1 \\
&\quad +2\mathrm{Im}(z_1z_2+z_3z_4)\mathbf{e}_2 \\
&\quad -2\mathrm{Re}(z_1z_2+z_3z_4)\mathbf{e}_3 \\
&\quad +2\mathrm{Im}(z_1\overline{z_3}-z_2\overline{z_4})\mathbf{e}_4.
\end{aligned}
$$

The pseudo-scalar ω has the following properties which makes the computations easy. Since $\omega\mathbf{e}_k = -\mathbf{e}_k\omega$ for $k = 1,\cdots,4$, it follows that ω commutes with every element of the even Clifford algebra. And $\omega^2 = -1$.

From now on, we summarize some useful propositions and properties of T-map in $\mathbb{R}^{3,1}$ from [11].

Proposition 3.3 *For any* $\mathbf{x} \in \mathcal{C}\ell^+(3,1)$ *and* $v \in \mathbb{R}^{3,1}$, *we get* $R_\mathbf{x}(v) \in \mathbb{R}^{3,1}$. *So* $\Lambda(3,1) = \mathcal{C}\ell^+(3,1)$.

We now consider the norm of $\mathbf{x}$. The norm $N(\mathbf{x})$ is given by

$$
\begin{aligned}
N(\mathbf{x}) &= (p+\omega q)(\overline{p}+\omega\overline{q}) \\
&= (|p|^2-|q|^2)+\omega(p\overline{q}+q\overline{p}).
\end{aligned}
$$

Thus, the norm of $\mathbf{x}$ has not only a scalar term but also a pseudo-scalar term. Let $f_1(\mathbf{x})$ be the scalar part of $N(\mathbf{x})$ and let $f_2(\mathbf{x})$ be the coefficient of the pseudo-scalar part.

The following proposition shows that T-map preserves orthogonalities in $\mathbb{R}^{3,1}$.

Proposition 3.4 *For any* $\mathbf{x} \in \mathcal{C}\ell^+(3,1)$, $R_\mathbf{x}$ *act on* $\mathbb{R}^{3,1}$ *as a linear transform in* $\mathbb{R}^+ \times \mathrm{SO}(3,1)$. *That is,* $R_\mathbf{x}$ *consists of a special orthogonal transform and a magnification. And the magnification factor is* $\sqrt{f_1(\mathbf{x})^2+f_2(\mathbf{x})^2}$. *More specifically, we have*

$$
\langle R_\mathbf{x}(a), R_\mathbf{x}(b)\rangle_{\mathbb{R}^{3,1}} = (f_1^2+f_2^2)\langle a,b\rangle_{\mathbb{R}^3}
$$

for vectors a, b *in* $\mathbb{R}^{3,1}$.

Proof. We can prove this by direct calculations:

$$
\begin{aligned}
\langle R_\mathbf{x}(a), R_\mathbf{x}(b)\rangle_{\mathbb{R}^{3,1}} &= -R_\mathbf{x}(a)R_\mathbf{x}(b) \\
&= -\mathbf{x}a\overline{\mathbf{x}}\mathbf{x}b\overline{\mathbf{x}} \\
&= -\mathbf{x}a(f_1+\omega f_2)b\overline{\mathbf{x}} \\
&= -\mathbf{x}(f_1-\omega f_2)ab\overline{\mathbf{x}} \\
&= \mathbf{x}(f_1-\omega f_2)\overline{\mathbf{x}}\langle a,b\rangle_{\mathbb{R}^3} \\
&= (f_1-\omega f_2)\mathbf{x}\overline{\mathbf{x}}\langle a,b\rangle_{\mathbb{R}^3} \\
&= (f_1-\omega f_2)(f_1+\omega f_2)\langle a,b\rangle_{\mathbb{R}^3} \\
&= (f_1^2+f_2^2)\langle a,b\rangle_{\mathbb{R}^3}.
\end{aligned}
$$

□

In order to understand the meaning of the pseudo-scalar, we need the following lemma.

Lemma 3.5 *For some* $\mathbf{x} \in \mathcal{C}\ell^+(3,1)$, *if* $R_\mathbf{x}$ *is the identity in* $\mathrm{SO}_+(3,1)$ *and* $N(\mathbf{x})$ *is a nonzero scalar, then* $\mathbf{x}$ *equals* ± 1 *or* $\pm\omega$.

Since we defined T in a different way from the spin representation map, the corresponding group Λ has a different structure from the Clifford group Γ. The Clifford group $\Gamma(3,1)$ is isomorphic to $\mathbb{R}^+ \times \mathrm{Spin}(3,1)$. And Proposition (3.4) shows that $\Lambda(3,1)$ equals $\mathcal{C}\ell^+(3,1)$. Here, we discuss the relation between the Clifford group $\Gamma(3,1)$ and the norm.

Proposition 3.6 *For any* $\mathbf{x} \in \Lambda(3,1)$, *assume* $N(\mathbf{x}) \neq 0$. *Then* $\mathbf{x} \in \Gamma(3,1)$ *if and only if* $N(\mathbf{x})$ *is a scalar.*

By the PH representation map, we get a class of MPH curves in $\mathbb{R}^{3,1}$ in the sense of Definition 3.1. This result is a straightforward consequence of Proposition 3.4.

Proposition 3.7 *If* $\gamma(t) = (x(t), y(t), z(t), r(t))$ *is an MPH curve defined by polynomials* $p_k(t), q_k(t)$ *for* $k = 0,\cdots,3$, *then the squared speed of* $\gamma(t)$ *measured under the Minkowski metric is expressed as*

$$
\begin{aligned}
|\gamma'(t)|^2 &= x'(t)^2+y'(t)^2+z'(t)^2-r'(t)^2 \\
&= f_1(t)^2+f_2(t)^2,
\end{aligned}
$$

where

$$
\begin{aligned}
f_1(t) &= p_0(t)^2+p_1(t)^2+p_2(t)^2+p_3(t)^2 \\
&\quad -(q_0(t)^2+q_1(t)^2+q_2(t)^2+q_3(t)^2), \\
f_2(t) &= 2(p_0(t)q_0(t)+p_1(t)q_1(t) \\
&\quad +p_2(t)q_2(t)+p_3(t)q_3(t)).
\end{aligned}
$$

Therefore, if the spine curve $s(t)$ together with the radius function $r(t)$ are defined by the PH representation map $T_{\mathbf{e}_1}(\mathbf{x}(t))$, then the canal surface generated by $s(t)$ and $r(t)$ has a rational parametrization if we directly apply the parametrization method proposed in [28] without the decomposition step.

In fact, the converse of Proposition 3.7 is true that all canal surfaces can be generated by the PH representation map shown in the next section.

3.3 4-Dimensional MPH Representation Theorem

We will state here the *4-dimensional MPH representation theorem*, Theorem 3.10, which is the converse of Proposition 3.7. As a corollary to this result, we also present a simple version which is useful if $\gamma(t)$ is a rational polynomial.

The following lemma describes the isotropy subgroup I of $T_{\mathbf{e}_1}(\mathbf{x})$. This isotropy group gives us the flexibility to handle the canal surface.

Lemma 3.8 *Isotropy subgroup* I *of* $T_{\mathbf{e}_1}(\mathbf{x})$ *is given by the following:*

$$\begin{aligned} \mathrm{I} &= \{\mathbf{x} = p + \omega q \in \mathcal{C}\ell^+(3,1) : p = \tau w_1 - \mu w_2 \mathbf{e}_{31}, \\ & q = \mu w_1 + \tau w_2 \mathbf{e}_{31}, |w_1|^2 - |w_2|^2 = \frac{1}{\tau^2 + \mu^2}, \\ & w_1, w_2 \in \mathbb{C},\ \tau, \mu \in \mathbb{R}, (\tau, \mu) \neq (0,0)\}. \end{aligned}$$

Note that the algebraic dimension of I is not 5 but 4. In fact, we can derive a more simplified formula for elements in I by dropping μ or τ when $\mu \neq 0$ or $\tau \neq 0$. For example, when $\mu \neq 0$, by the transformation $\tau \to \frac{\tau}{\mu}$, $w_1 \to \mu w_1$ and $w_2 \to \mu w_2$, we get the simpler form

$$\mathbf{x} = \tau w_1 - w_2 \mathbf{e}_{31} + \omega(w_1 + \tau w_2 \mathbf{e}_{31}), \tag{14}$$

where $|w_1|^2 - |w_2|^2 = \frac{1}{\tau^2+1}$, and $w_1, w_2 \in \mathbb{C}$. And if $\mu = 0$ or $\tau = 0$, Proposition 3.6 implies the following corollary.

Corollary 3.9 *If* $\mu = 0$ *or* $\tau = 0$, *that the corresponding elements in* I *are contained in* Spin(3, 1).

The next theorem, 4-dimensional MPH representation theorem, shows that all the MPH curves in $\mathbb{R}^{3,1}$ can be represented by the PH representation map.

Theorem 3.10 [11] *Let* $\gamma(t) = (x(t), y(t), z(t), r(t))$ *be a spacelike polynomial curve in the four dimensional Minkowski space* $\mathbb{R}^{3,1}$. *Then it is given by a PH representation map in (13), that is, there exist polynomials* $p_k(t), q_k(t)$ *for* $k = 0, \cdots, 3$, *satisfying (13).*

We also have a rational version of Theorem 3.10, which is much easier and more convenient when we begin with a rational curve $\gamma(t) = (x(t), y(t), z(t), r(t))$.

Corollary 3.11 [11] *Let* $\gamma(t) = (x(t), y(t), z(t), r(t))$ *be a spacelike rational polynomial curve in the four dimensional Minkowski space* $\mathbb{R}^{3,1}$. *Then it is given by a PH representation map in 13, that is, there exist rational polynomials* $p_k(t), q_k(t)$ *for* $k = 0, \cdots, 3$, *satisfying (13).*

4 Rational Parametrization of Canal Surfaces by Lorentzian Geometry via Clifford Formalism

Now we present the new method of rational parametrization of a canal surface using Lorentzian geometry. Using the result of the previous chapter, we already know that every canal surface with a rational spine curve and a rational radius function can be represented by the 4-dimensional PH representation map. Though by invoking the result of Peternell and Pottmann [28], we can prove that every canal surface with rational spine curve and rational radius function can be rationally parameterized, we propose another method which in the end you will see has many attractive merits.

The main tool in our method is the R-map and the result of Proposition 3.4 and the 4-dimensional MPH representation theorem. To use the Minkowski metric in $\mathbb{R}^{3,1}$, let us denote that $m'(t) = (s'(t), r'(t))$ and $c(t) = (\phi(t), -1)$. Then a real canal surface exists only if $m'(t)$ is a spacelike curve. By the 4-dimensional MPH representation theorem, we can start with $m'(t) = T_{\mathbf{e}_1}(\mathbf{x}) = R_{\mathbf{x}}(\mathbf{e}_1)$ for some polynomial $\mathbf{x} = p + \omega q$. It uses $\mathbf{e}_1$ as a base vector. Then we can ask what the role of $R_{\mathbf{x}}(\mathbf{e}_2)$, $R_{\mathbf{x}}(\mathbf{e}_3)$, and $R_{\mathbf{x}}(\mathbf{e}_4)$ are. The answer is that $R_{\mathbf{x}}(\mathbf{e}_2)$, $R_{\mathbf{x}}(\mathbf{e}_3)$, and $R_{\mathbf{x}}(\mathbf{e}_4)$ gives us a parametrization of $\phi(t, u)$. Let us explain this observation in more detail.

We must find the rational parametrization of $c(t)$ which satisfy the following property: the cone $c(t)$ is a light-like curve in $\mathbb{R}^{3,1}$, i.e. $|c(t)|_{\mathbb{R}^{3,1}} = 0$ so that a real canal surface can exist. And the equation (7) can be rewritten by the Minkowski metric in $\mathbb{R}^{3,1}$:

$$\Sigma'(t) \quad : \quad \langle m'(t), c(t) \rangle_{\mathbb{R}^{3,1}} = 0, \tag{15}$$

i.e. $c(t)$ is orthogonal to $R_{\mathbf{x}(t)}(\mathbf{e}_1)$ in $\mathbb{R}^{3,1}$.

Because the $R_{\mathbf{x}}$-map preserves norms and orthogonalities, we can think that $c(t)$ is the image of $a\mathbf{e}_1 + b\mathbf{e}_2 + c\mathbf{e}_3 + d\mathbf{e}_4$ which satisfies the following equations:

$$\begin{aligned} \langle e_1, a\mathbf{e}_1 + b\mathbf{e}_2 + c\mathbf{e}_3 + d\mathbf{e}_4 \rangle_{\mathbb{R}^{3,1}} &= 0 \\ |a\mathbf{e}_1 + b\mathbf{e}_2 + c\mathbf{e}_3 + d\mathbf{e}_4|_{\mathbb{R}^{3,1}} &= 0. \end{aligned}$$

So we get $a = 0$, and $b^2 + c^2 - d^2 = 0$.

Then by applying $R_{\mathbf{x}}$-map, the above equations are transformed to the following equations:

$$\begin{aligned} \langle R_{\mathbf{x}}(e_1), bR_{\mathbf{x}}(\mathbf{e}_2) + cR_{\mathbf{x}}(\mathbf{e}_3) + dR_{\mathbf{x}}(\mathbf{e}_4) \rangle_{\mathbb{R}^{3,1}} &= 0 \\ |bR_{\mathbf{x}}(\mathbf{e}_2) + cR_{\mathbf{x}}(\mathbf{e}_3) + dR_{\mathbf{x}}(\mathbf{e}_4)|_{\mathbb{R}^{3,1}} &= 0, \end{aligned}$$

where $b^2 + c^2 - d^2 = 0$. The relation $b^2 + c^2 - d^2 = 0$ gives a one parameter family of curves and that is a natural parametrization of the characteristic circles — a very important and useful observation. For example, we can choose a triangular parametrization where

$$(b, c, d) = (\frac{1-u^2}{1+u^2}, \frac{2u}{1+u^2}, 1).$$

If we denote the component of $b(u)R_{\mathbf{x}(t)}(\mathbf{e}_2) + c(u)R_{\mathbf{x}(t)}(\mathbf{e}_3) + d(u)R_{\mathbf{x}(t)}(\mathbf{e}_4)$ by $(U_1(t,u), U_2(t,u), U_3(t,u), U_4(t,u))$, then the parametrization of the cone $\phi(t, u)$ is given by

$$\phi(t, u) = (-\frac{U_1(t,u)}{U_4(t,u)}, -\frac{U_1(t,u)}{U_4(t,u)}, -\frac{U_1(t,u)}{U_4(t,u)}).$$

Here we note that we do not need to worry about the case that $U_4(t, u) = 0$ because if $U_4(t, u) = 0$ for some (t, u)

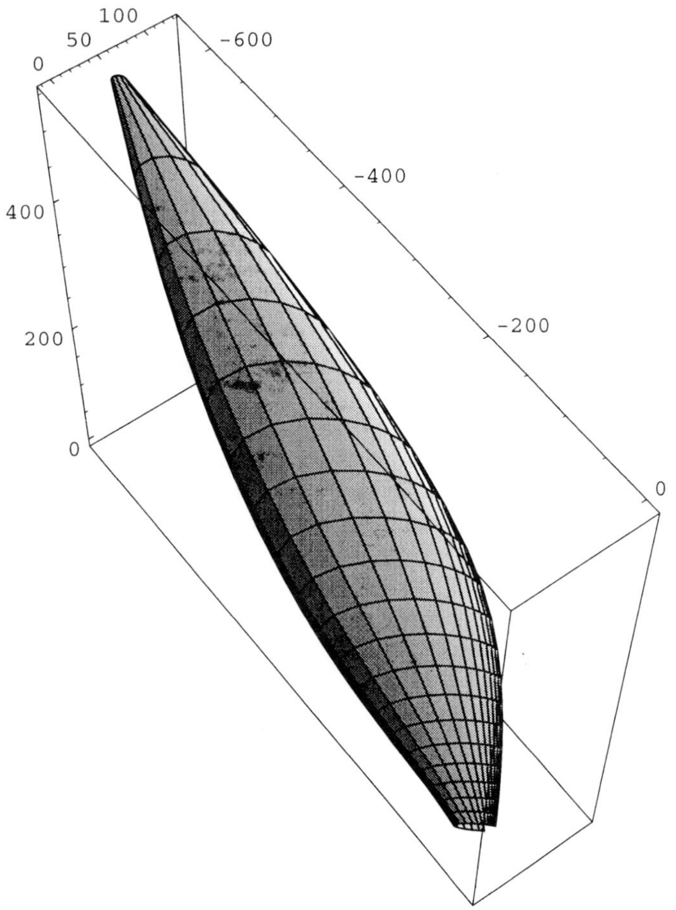

Figure 2. An example of a Canal surface.

then all $U_j(t,u) = 0$ for $j = 1,2,3$ at the same time, so these zero sets can be factored out in the above equations.

We get the rational parametrization of a canal surface by applying this $\phi(t,u)$ to the following equation:

$$\Phi(t,u) = (x(t), y(t), z(t)) + r(t)\phi(t,u).$$

We give an example of a canal surface with this parametrization method in Figure 2.

Conclusion

By now, based on the previous work on PH curves, we have proposed a new method which uses MPH representation and uses Lorentzian geometry to parameterize canal surfaces. The key point used in Chapter 4 is that R-map preserves the orthogonalities in the Minkowski space $\mathbb{R}^{3,1}$. We think it may be especially versatile when finding rotation minimizing and almost minimizing frames because we have the flexibility to represent canal surfaces within their fiber ambiguities and to choose the parametrization for $\{b, c, d\}$.

We believe this is a germ for a new approach in dealing with canal surfaces meriting further research and we are hopeful thus it will yield many new and valuable findings. We hope to be able to deal with these topics in more detail in our next paper [10].

References

[1] M. Berger. *Geometry*. Springer-Verlag, Berlin, Germany, 1987.

[2] C. Chevalley. *The Algebraic Theory of Spinors and Clifford Algebras*. Springer-Verlag, Berlin, 1997. Edited by P. Cartier and C. Chevalley.

[3] H.I. Choi, S.W. Choi, C. Y. Han, H.P. Moon, and K.H. Roh, and N.-S. Wee. Two-dimensional offsets via medial axis transform I: Mathematical Theory. preprint.

[4] H.I. Choi, S.W. Choi, C. Y. Han, H.P. Moon, K.H. Roh, and N.-S. Wee. Two-dimensional offsets via medial axis transform II: Algorithm. preprint.

[5] H.I. Choi, S.W. Choi, and H.P. Moon. Mathematical theory of medial axis transform. *Pacific Journal of Mathematics*, 181(1):57–88, 1997.

[6] H.I. Choi, S.W. Choi, H.P. Moon, and N.-S. Wee. New algorithm for medial axis transform of plane domain. *Graphical Models and Image Processing*, 59(6):463–483, November 1997.

[7] H.I. Choi, C.Y. Han, H.P. Moon, K.H. Roh, and N.-S. Wee. Medial axis transform and offset curves by Minkowski Pythagorean hodograph curves. *Computer Aided Design*, 31:59–72, 1999.

[8] H.I. Choi, C. Y. Han, H.P. Moon, K.H. Roh, and N.-S. Wee. Cubic Minkowski Pythagorean hodograph curves. preprint.

[9] H.I. Choi, C.Y. Han, H.P. Moon, K.H. Roh, and N.-S. Wee. A new algorithm for offset of plain domain via MAT. In *Differential/Topological Techniques in Geometric Modeling and Processing '98*, pages 161–185, Pohang, Korea, April 1998.

[10] H.I. Choi and D.S. Lee. in preparation.

[11] H.I. Choi, D.S. Lee, and H.P. Moon. Clifford algebra, spin representation and rational parametrization of curves and surfaces. preprint.

[12] H.P. Moon. Mikowski Pythagorean hodographs *Computer Aided Design*, 16:739–753, 1999.

[13] A. Crumeyrolle. *Orthogonal and Symplectic Clifford Algebras*. Kluwer Academic Publishers, Netherlands, 1990.

[14] R. Delanghe, F. Sommen, and V. Souček. *Clifford Algebra and Spinor-Values Functions*. Kluwer Academic Publishers, Netherlands, 1992.

[15] R. Dietz, J. Hoschek, and B. Jüttler. An algebraic approach to curves and surfaces on the sphere and on other quadrics. *Computer Aided Geometric Design*, 10:211–229, 1993.

[16] R. Dietz, J. Hoschek, and B. Jüttler. Rational patches on quadric surfaces. *Computer Aided Design*, 27(1):27–40, 1995.

[17] R.T. Farouki. The conformal map $z \to z^2$ of the hodograph plane. *Computer Aided Geometric Design*, 11:363–390, 1994.

[18] R.T. Farouki and C.A. Neff. Hermite interpolation by Pythagorean-hodograph quintics. *Mathematics of Computation*, 64:1589–1609, 1995.

[19] R.T. Farouki and T. Sakkalis. Pythagorean hodographs. *IBM Journal of Research and Development*, 34:736–752, 1990.

[20] R.T. Farouki and T. Sakkalis. Pythagorean-hodograph space curves. *Advances in Computational Mathematics*, 2:41–66, 1994.

[21] B. Jüttler. Generating rational frames of space curves via Hermite interpolation with Pythagorean hodograph cubic splines. In *Differential/Topological Techniques in Geometric Modeling and Processing '98*, pages 83–106, Pohang, Korea, April 1998.

[22] K.K. Kubota. Pythagorean triples in unique factorization domains. *American Mathematical Monthly*, 79:503–505, 1972.

[23] H. B. Lawson and M.-L. Michelsohn. *Spin Geometry*. Kluwer Princeton University Press, 1989.

[24] WM. Lü and H. Pottmann. Pipe surfaces with rational spine curves are rational. *Computer Aided Geometric Design*, 13:621–628, 1996.

[25] T. Maekawa, N.M. Patrikalakis, T. Sakkalis. Analysis and applications of pipe surfaces. *Computer Aided Geometric Design*, 15:437–458, 1998.

[26] H.P. Moon. Minkowski pythagorean hodographs. *Computer Aided Geometric Design*, 16:739–753, 1999.

[27] H.P. Moon. Computing rational offsets via medial axis transform using polynomial speed curves in $\mathbb{R}^{2,1}$. In *Differential/Topological Techniques in Geometric Modeling and Processing '98*, pages 187–203, Pohang, Korea, April 1998.

[28] M. Peternell and H. Pottmann. Computing rational parametrization of canal surfaces. *Journal of Symbolic Computation*, 23:255–266, 1997.

[29] M. Peternell and H. Pottmann. A Laguerre geometric approach to rational offsets. *Computer Aided Geometric Design*, 15:223–249, 1998.

[30] H. Pottmann. Rational curves and surfaces with rational offsets. *Computer Aided Geometric Design*, 12:175–192, 1995.

[31] H. Pottmann and M. Peternell. Applications of Laguerre geometry in CAGD. *Computer Aided Geometric Design*, 15:165–186, 1998.

[32] J. Wallner and H. Pottmann. Rational blending surfaces between quadrics. *Computer Aided Geometric Design*, 14:407–419, 1997.

Implicitization by Dixon A-Resultants

Eng-Wee Chionh
School of Computing
National University of Singapore
Singapore 117543
chionhew@comp.nus.edu.sg

Ming Zhang
Ronald N. Goldman
Department of Computer Science
Rice University
Houston, Texas 77005
mzhang@cs.rice.edu
rng@cs.rice.edu

Abstract

It is well-known that the Dixon resultant implicitizes exactly a general tensor product surface. In this paper, we show that a minor of the Dixon resultant matrix can also implicitize exactly. This occurs when the monomial support of the surface parametrization is a rectangle missing at most one sub-rectangle at each of its corners. Unlike the Sylvester dialytic method, this way of finding the implicit equation in determinant form is automatic, in bracket form, and uses much smaller matrices.

1. Introduction

Free-form surface patches in computer aided geometric design are usually given parametrically in rational form

$$X = \frac{x(s,t)}{w(s,t)}, \quad Y = \frac{y(s,t)}{w(s,t)}, \quad Z = \frac{z(s,t)}{w(s,t)}, \tag{1}$$

where

$$w(s,t), \quad x(s,t), \quad y(s,t), \quad z(s,t) \tag{2}$$

are polynomials in the parameters s and t. NURB surfaces are well-known examples. A non-degenerate rational surface can also be represented implicitly by an irreducible polynomial equation

$$p(X, Y, Z) = 0 \tag{3}$$

such that

$$p\left(\frac{x(s,t)}{w(s,t)}, \frac{y(s,t)}{w(s,t)}, \frac{z(s,t)}{w(s,t)}\right) \equiv 0. \tag{4}$$

Since the parametric form and the implicit form have their respective strengths and weaknesses, there is a need to convert the parametric form to the implicit form, a process that has become known as *implicitization* [12].

Rational parametric curves and surfaces can be implicitized quickly in determinant form using resultants. It turns out that the type of the implicitizing resultant depends a great deal on the monomials that occur in the parametric polynomials (2). To highlight this fact, we denote by $\mathcal{A}$ the set of exponent vectors (i, j) of monomials $s^i t^j$ in (2) and write

$$\begin{bmatrix} w(s,t) \\ x(s,t) \\ y(s,t) \\ z(s,t) \end{bmatrix} = \sum_{(i,j)\in\mathcal{A}} \begin{bmatrix} w_{i,j} \\ x_{i,j} \\ y_{i,j} \\ z_{i,j} \end{bmatrix} s^i t^j. \tag{5}$$

We call $\mathcal{A}$ the monomial support of (2). The convex hull of $\mathcal{A}$ in the (i, j) plane, conv($\mathcal{A}$), is known as the Newton polygon of (2). It is well known that for triangular patches using Bernstein trinomials, conv($\mathcal{A}$) is an isosceles triangle; for rectangular patches using tensor products, conv($\mathcal{A}$) is a rectangle. For example, the monomial supports for a degree 3 triangular patch and a bi-degree $(4, 3)$ rectangular patch are

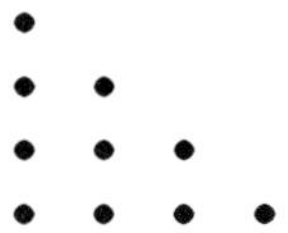

and

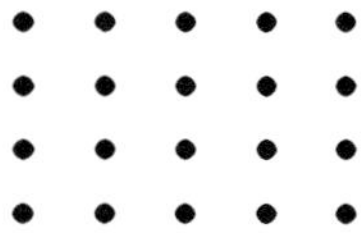

respectively in the (i, j) plane. The Macaulay quotient implicitizes triangular patches when there are no base points [4], and the Dixon resultant implicitizes rectangular patches provided there are no base points beyond the canonical ones [12, 10]. If there are additional base points, the method of Macaulay quotients or Dixon resultants fails and other non-resultant means have to be resorted to [3, 13]. Alternatively, if the additional base points are intrinsic to the monomial support $\mathcal{A}$, then the method of $\mathcal{A}$-resultants can be used. For a general discussion of $\mathcal{A}$-resultants and sparse elimination, we refer the reader to [5, 7, 8, 9, 11].

This paper describes the use of Dixon $\mathcal{A}$-resultants for implicitization. These resultants are obtained from the classical Dixon resultants by omitting an equal number of rows and columns of zero entries. The main observation is that Dixon $\mathcal{A}$-resultants arise when the monomial support $\mathcal{A}$ is obtained by removing at most one rectangular sub-support from each of the corners of a rectangular bi-degree support. For example, the following support $\mathcal{A}$

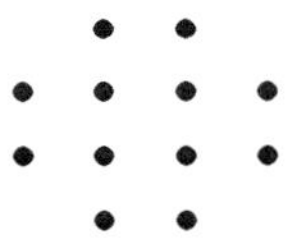

admits a Dixon $\mathcal{A}$-resultant because it is obtained by removing the four corner "rectangular" sub-supports $\{(0,0)\}$, $\{(3,0)\}$, $\{(3,3)\}$, $\{(0,3)\}$ from the rectangular support $\mathcal{A}_{3,3} = \{0,1,2,3\} \times \{0,1,2,3\}$. Note that a rectangle can degenerate to a single point or to a horizontal or vertical line segment of points.

Zube [14] discusses implicitization with $\mathcal{A}$-resultants constructed by the Sylvester dialytic or other equation augmentation method. But using Dixon $\mathcal{A}$-resultants is better. The major improvements of the Dixon method over the dialytic and the equation augmentation methods are

- there is no need to search for the dialytic monomials or the extra equations to be augmented;
- the Dixon $\mathcal{A}$-resultant is already in bracket form;
- the Dixon determinant is one-ninth the size of the Sylvester determinant.

The rest of the paper is structured as follows. Section 2 defines the classical Dixon resultant. Section 3 discusses the admissible monomial supports for the Dixon $\mathcal{A}$-resultant. Section 4 presents implicitization examples taken from [14] using the Dixon $\mathcal{A}$-resultant. The examples include Hirzebruch surfaces and cyclides. Section 5 concludes the paper with a summary of our work and an open problem for future research.

2. The Classical Dixon Resultant

In this section, we describe the construction of the classical Dixon resultant for three polynomial equations in two variables [6].

The Dixon polynomial for the three polynomials

$$\begin{aligned} f(s,t) &= \textstyle\sum_{(i,j)\in\mathcal{A}} a_{i,j}s^i t^j, \\ g(s,t) &= \textstyle\sum_{(i,j)\in\mathcal{A}} b_{i,j}s^i t^j, \\ h(s,t) &= \textstyle\sum_{(i,j)\in\mathcal{A}} c_{i,j}s^i t^j, \end{aligned} \tag{6}$$

is

$$\Delta_{\mathcal{A}}(f(s,t), g(\alpha,t), h(\alpha,\beta)) = \frac{\begin{vmatrix} f(s,t) & g(s,t) & h(s,t) \\ f(\alpha,t) & g(\alpha,t) & h(\alpha,t) \\ f(\alpha,\beta) & g(\alpha,\beta) & h(\alpha,\beta) \end{vmatrix}}{(s-\alpha)(t-\beta)}. \tag{7}$$

Note that the denominator divides the numerator because the numerator vanishes when $s = \alpha$ or $t = \beta$. The expression $\Delta_{\mathcal{A}}$ can be viewed as a polynomial in either s, t or α, β to suit the need. The monomial supports of $\Delta_{\mathcal{A}}$ in s, t and α, β are denoted as $\mathcal{R}_{\mathcal{A}}$ and $\mathcal{C}_{\mathcal{A}}$ respectively. We can also write $\Delta_{\mathcal{A}}$ in matrix form with row indices $\{s^\sigma t^\tau \mid (\sigma,\tau) \in \mathcal{R}_{\mathcal{A}}\}$ and column indices $\{\alpha^a\beta^b \mid (a,b) \in \mathcal{C}_{\mathcal{A}}\}$ as

$$\Delta_{\mathcal{A}} = \begin{bmatrix} \vdots \\ s^\sigma t^\tau \\ \vdots \end{bmatrix}^T D_{\mathcal{A}} \begin{bmatrix} \vdots \\ \alpha^a\beta^b \\ \vdots \end{bmatrix}. \tag{8}$$

We call $D_{\mathcal{A}}$ the Dixon matrix of f, g, h. Instead of calling $\mathcal{R}_{\mathcal{A}}$ and $\mathcal{C}_{\mathcal{A}}$ the monomial supports of $\Delta_{\mathcal{A}}$, we call $\mathcal{R}_{\mathcal{A}}$ the row support of $D_{\mathcal{A}}$ and $\mathcal{C}_{\mathcal{A}}$ the column support of $D_{\mathcal{A}}$.

Since the numerator in (7)

$$\begin{vmatrix} f(s,t) & g(s,t) & h(s,t) \\ f(\alpha,t) & g(\alpha,t) & h(\alpha,t) \\ f(\alpha,\beta) & g(\alpha,\beta) & h(\alpha,\beta) \end{vmatrix} = \sum_{(i,j),(k,l),(p,q)\in\mathcal{A}} |i,j;k,l;p,q| s^i t^{j+l} \alpha^{k+p} \beta^q \tag{9}$$

where

$$|i,j;k,l;p,q| = \begin{vmatrix} a_{i,j} & b_{i,j} & c_{i,j} \\ a_{k,l} & b_{k,l} & c_{k,l} \\ a_{p,q} & b_{p,q} & c_{p,q} \end{vmatrix}, \tag{10}$$

the entries of $D_{\mathcal{A}}$ are linear in the coefficients of each of f, g, and h.

For economy of space, we abbreviate $|i,j;k,l;p,q|$ to $ijklpq$ when there is no danger of confusion. For example,

$$000110 = |0,0;0,1;1,0| = \begin{vmatrix} a_{0,0} & b_{0,0} & c_{0,0} \\ a_{0,1} & b_{0,1} & c_{0,1} \\ a_{1,0} & b_{1,0} & c_{1,0} \end{vmatrix}.$$

Let $\mathcal{A}_{m,n} = \{0,\cdots,m\} \times \{0,\cdots,n\}$ be the bi-degree (m,n) monomial support. When $\mathcal{A} = \mathcal{A}_{m,n}$, we write $\Delta_{\mathcal{A}}$, $D_{\mathcal{A}}$, $\mathcal{R}_{\mathcal{A}}$, $\mathcal{C}_{\mathcal{A}}$ as $\Delta_{m,n}$, $D_{m,n}$, $\mathcal{R}_{m,n}$, $\mathcal{C}_{m,n}$ respectively. Clearly $\Delta_{m,n}$ is of degree $m-1$ in s, $2n-1$ in t, $2m-1$ in α, and $n-1$ in β. Thus

$$\begin{aligned} \mathcal{R}_{m,n} &= \{0,\cdots,m-1\} \times \{0,\cdots,2n-1\}, \\ \mathcal{C}_{m,n} &= \{0,\cdots,2m-1\} \times \{0,\cdots,n-1\}. \end{aligned} \tag{11}$$

Consequently, $D_{m,n}$ ia a square matrix of order $2mn$ — that is, $\#\mathcal{R}_{m,n} = \#\mathcal{C}_{m,n} = 2mn$. The determinant $|D_{m,n}|$ is the classical Dixon resultant of f, g, h when $A = \mathcal{A}_{m,n}$.

3. The Dixon $\mathcal{A}$-Resultant

This section presents two important results. The first result gives the row and column supports $\mathcal{R}_\mathcal{A}$ and $\mathcal{C}_\mathcal{A}$ for a monomial support $\mathcal{A}$ that is a rectangle missing some corner sub-rectangles. The second result shows that for such $\mathcal{A}$, $|D_\mathcal{A}|$ is the $\mathcal{A}$-resultant for f, g, h.

In the following discussion, we assume $\mathcal{A} \subseteq \mathcal{A}_{m,n}$.

3.1. Removal of Corner Rectangular Sub-Supports

Let $D_{m,n}(\mathcal{A})$ be $D_{m,n}$ specialized with respect to $\mathcal{A}$. That is, $D_{m,n}(\mathcal{A})$ is obtained from $D_{m,n}$ by setting $a_{i,j} = b_{i,j} = c_{i,j} = 0$ for $(i,j) \in \mathcal{A}_{m,n} - \mathcal{A}$.

When $\mathcal{A}_{m,n} - \mathcal{A}$ is a union of certain rectangular sub-supports at the corners of $\mathcal{A}_{m,n}$, the following theorem identifies $D_\mathcal{A}$ as a sub-matrix of $D_{m,n}(\mathcal{A})$ by giving explicitly its row support $\mathcal{R}_\mathcal{A}$ and column support $\mathcal{C}_\mathcal{A}$.

Theorem 1 *Let $\mathcal{A}$ be the monomial support of f, g, h; $\mathcal{R}_\mathcal{A}$ be the row support of $D_\mathcal{A}$, $\mathcal{C}_\mathcal{A}$ be the column support of $D_\mathcal{A}$; and $\mathcal{S} = \mathcal{S}_1 \cup \mathcal{S}_2 \cup \mathcal{S}_3 \cup \mathcal{S}_4$ where $\mathcal{S}_1$, $\mathcal{S}_2$, $\mathcal{S}_3$, $\mathcal{S}_4$ are respectively rectangular sub-supports at the bottom-left, bottom-right, top-right, top-left corners of $\mathcal{A}_{m,n}$ that are possibly empty and are not adjacent to each other. If $\mathcal{A} \subseteq \mathcal{A}_{m,n}$ and $\mathcal{A}_{m,n} - \mathcal{A} = \mathcal{S}$, then*

$$\mathcal{R}_\mathcal{A} = \mathcal{R}_{m,n} - \mathcal{S}_1 - \mathcal{S}_2' - \mathcal{S}_3' - \mathcal{S}_4', \qquad (12)$$
$$\mathcal{C}_\mathcal{A} = \mathcal{C}_{m,n} - \mathcal{S}_1 - \mathcal{S}_2'' - \mathcal{S}_3'' - \mathcal{S}_4'', \qquad (13)$$

where

$$\begin{aligned} \mathcal{S}_2' &= (-1,0) + \mathcal{S}_2 \\ \mathcal{S}_3' &= (-1,n-1) + \mathcal{S}_3, \\ \mathcal{S}_4' &= (0,n-1) + \mathcal{S}_4, \\ \mathcal{S}_2'' &= (m-1,0) + \mathcal{S}_2, \\ \mathcal{S}_3'' &= (m-1,-1) + \mathcal{S}_3, \\ \mathcal{S}_4'' &= (0,-1) + \mathcal{S}_4. \end{aligned} \qquad (14)$$

That is, $D_\mathcal{A}$ is a square sub-matrix of $D_{m,n}(\mathcal{A})$ and is obtained by removing $\#\mathcal{S}$ rows of zeros and $\#\mathcal{S}$ columns of zeros from $D_{m,n}(\mathcal{A})$.

A proof can be found in [2]. The theorem is illustrated pictorially in Figure 1 where the corresponding sub-supports $\mathcal{S}_i$, $\mathcal{S}_i'$, $\mathcal{S}_i''$, $i = 2,3,4$, are marked similarly.

For example, consider removing four corner monomials 1, s^2, s^2t^2, t^2 from the nine bi-degree $(2,2)$ monomials $s^it^j; i,j = 0,1,2$. Pictorially the resulting monomial, row, and column supports are

$\mathcal{A}$	$\mathcal{R}_\mathcal{A}$	$\mathcal{C}_\mathcal{A}$
4 • 3 • • • 1 • 2	4 3 • • • • 1 2	4 • • 3 1 • • 2

(15)

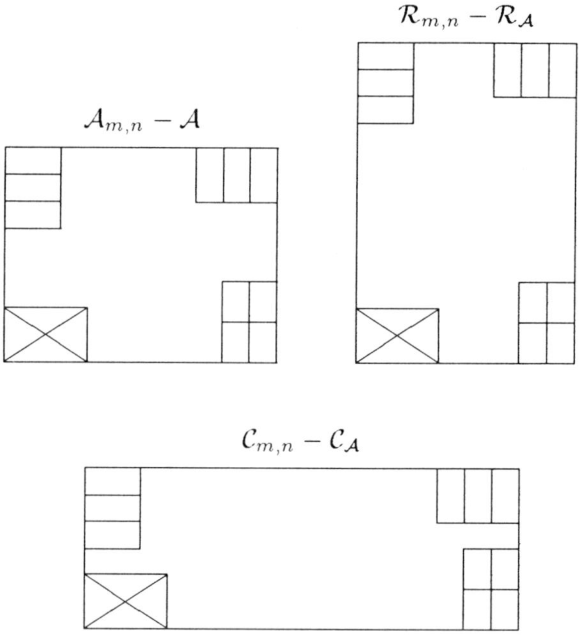

Figure 1. The corresponding sub-rectangles removed are marked similarly.

respectively. In the diagram "•" represents elements of the supports and the numbers 1, 2, 3, 4 represent elements of the corner sub-supports. Indeed, by computing the Dixon polynomial (7) directly or by using an entry formula [1], the specialized Dixon matrix $D_{2,2}(\mathcal{A})$ is found to be

$$\begin{bmatrix} 0 & 0 & 0 & 0 & 0 & 0 & 0 & 0 \\ 0 & 0 & 110110 & 101201 & 210110 & 0 & 0 & 0 \\ 0 & 0 & 120110 & 111201 & 0 & 211201 & 0 & 0 \\ 0 & 0 & 0 & 0 & 0 & 0 & 0 & 0 \\ 0 & 0 & 0 & 0 & 0 & 0 & 0 & 0 \\ 0 & 0 & 210110 & 0 & 211110 & 211210 & 0 & 0 \\ 0 & 0 & 0 & 211201 & 211210 & 211211 & 0 & 0 \\ 0 & 0 & 0 & 0 & 0 & 0 & 0 & 0 \end{bmatrix}$$

with row indices (from top to bottom)

$$1_{0,0}, t, t^2, t^3_{0,2}, s_{2,0}, st, st^2, st^3_{2,2}$$

and column indices (from left to right)

$$1_{0,0}, \beta_{0,2}, \alpha, \alpha\beta, \alpha^2, \alpha^2\beta, \alpha^3_{2,0}, \alpha^3\beta_{2,2}.$$

The indices of the zero rows and zero columns are subscripted by the responsible monomial support in $\mathcal{A}$. After removing these zero rows and zero columns, we obtain

$$D_\mathcal{A} = \begin{bmatrix} 110110 & 101201 & 210110 & 0 \\ 120110 & 111201 & 0 & 211201 \\ 210110 & 0 & 211110 & 211210 \\ 0 & 211201 & 211210 & 211211 \end{bmatrix}. \qquad (16)$$

3.2. The Determinant of $D_{\mathcal{A}}$ is the $\mathcal{A}$-Resultant

The following theorem reveals the significance of $D_{\mathcal{A}}$ of Theorem 1.

Theorem 2 *Let $\mathcal{A}$ be the monomial support of f, g, h. If $\mathcal{A}$ satisfies the conditions stated in Theorem 1, then the determinant $|D_{\mathcal{A}}|$ is the $\mathcal{A}$-resultant of f, g, h.*

A proof can be found in [2]. We call $|D_{\mathcal{A}}|$ the Dixon $\mathcal{A}$-resultant of f, g, h.

For example, if the monomial support $\mathcal{A}$ is that of (15), the Dixon $\mathcal{A}$-resultant for f, g, h is

$$|D_{\mathcal{A}}| = \begin{vmatrix} 110110 & 101201 & 210110 & 0 \\ 120110 & 111201 & 0 & 211201 \\ 210110 & 0 & 211110 & 211210 \\ 0 & 211201 & 211210 & 211211 \end{vmatrix}. \quad (17)$$

An $\mathcal{A}$-resultant for the given $\mathcal{A}$ is also presented in [14] as the determinant of

$$D'_{\mathcal{A}} = \begin{bmatrix} 1 & 0 & 0 & 0 \\ 0 & 0 & 1 & 0 \\ 0 & 0 & 0 & 1 \\ 0 & 1 & 0 & 0 \end{bmatrix} D_{\mathcal{A}} \begin{bmatrix} 1 & 0 & 0 & 0 \\ 0 & 0 & 0 & 1 \\ 0 & 1 & 0 & 0 \\ 0 & 0 & 1 & 0 \end{bmatrix}.$$

But to construct $D'_{\mathcal{A}}$, first Sylvester dialytic monomials have to be found, probably by trial and error, then that Sylvester matrix has to be transformed to $D'_{\mathcal{A}}$, requiring another non-trivial effort.

4 Implicitization by Dixon $\mathcal{A}$-Resultants

In this section we use Dixon $\mathcal{A}$-resultants to implicitize the surfaces given in [14]. They are the Hirzebruch surfaces, the diamond cyclides, the one-horned cyclides, hexagon surfaces, and pentagon surfaces. The purpose is to contrast the Dixon method against the Sylvester dialytic method and the various equation augmentation methods utilized in [14]. Note that the matrix obtained by the Dixon method is one-ninth the size of that obtained by the Sylvester method and the entries of the Dixon matrix are already in bracket form. Furthermore, unlike [14], no dialytic polygons or the augmentation equations need to be found.

4.1 The Method

To implicitize the rational surface (1), we define

$$\begin{aligned} f(s,t) &= x(s,t) - Xw(s,t), \\ g(s,t) &= y(s,t) - Yw(s,t), \\ h(s,t) &= z(s,t) - Zw(s,t). \end{aligned} \quad (18)$$

From (5), (6), and (18), we have

$$\begin{aligned} a_{i,j} &= x_{i,j} - Xw_{i,j}, \\ b_{i,j} &= y_{i,j} - Yw_{i,j}, \\ c_{i,j} &= z_{i,j} - Zw_{i,j}. \end{aligned} \quad (19)$$

In this context, the bracket (10) becomes

$$ijklpq = \begin{vmatrix} 1 & X & Y & Z \\ w_{i,j} & x_{i,j} & y_{i,j} & z_{i,j} \\ w_{k,l} & x_{k,l} & y_{k,l} & z_{k,l} \\ w_{p,q} & x_{p,q} & y_{p,q} & z_{p,q} \end{vmatrix}. \quad (20)$$

For example,

$$000110 = \begin{vmatrix} 1 & X & Y & Z \\ w_{0,0} & x_{0,0} & y_{0,0} & z_{0,0} \\ w_{0,1} & x_{0,1} & y_{0,1} & z_{0,1} \\ w_{1,0} & x_{1,0} & y_{1,0} & z_{1,0} \end{vmatrix}.$$

Theorem 3.13 of [5] (page 312) indicates that in general the $\mathcal{A}$-resultant of (18) gives the implicit equation of the rational surface.

4.2. Hirzebruch Surfaces

The monomial support of the Hirzebruch surface F_n is

$$\mathcal{A} = \{(0,0), (1,0), \cdots, (n+1,0), (0,1), (1,1)\}$$

where $n \geq 0$. Pictorially the monomial, row, and column supports of F_2 discussed in [14] are

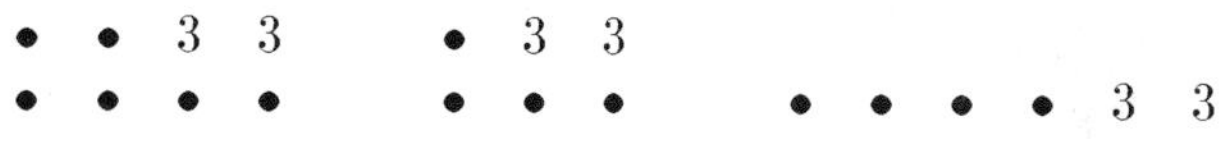

$\mathcal{A}$ $\qquad$ $\mathcal{R}_{\mathcal{A}}$ $\qquad$ $\mathcal{C}_{\mathcal{A}}$

respectively. The matrix $D_{3,1}(\mathcal{A})$ is

$$\begin{bmatrix} 100100 & 101100+200100 & 201100+300100 & 301100 & 0 & 0 \\ 110100 & 110110 & 110120 & 110130 & 0 & 0 \\ 200100 & 200110+201100+300100 & 201110+300110+301100 & 301110 & 0 & 0 \\ 0 & 0 & 0 & 0 & 0 & 0 \\ 300100 & 300110+301100 & 300120+301110 & 301120 & 0 & 0 \\ 0 & 0 & 0 & 0 & 0 & 0 \end{bmatrix}.$$

The non-zero rows are those numbered 1, 2, 3, 5 and the non-zero columns are those numbered 1, 2, 3, 4. Thus $D_{\mathcal{A}}$

is the square sub-matrix of $D_{3,1}(\mathcal{A})$ with these non-zero rows and columns and $|D_\mathcal{A}| = 0$ is the implicit equation of F_2.

For example, the rational surface (Figure 2)

$$\begin{bmatrix} w(s,t) \\ x(s,t) \\ y(s,t) \\ z(s,t) \end{bmatrix} = \sum_{(i,j)\in\mathcal{A}} \begin{bmatrix} 1 \\ i \\ j \\ i^2+j^2 \end{bmatrix} s^i t^j$$

has the implicit equation

$$\begin{vmatrix} X+Y-Z & 5X+3Y-3Z & 13X+3Y-5Z & 9X-3Y-3Z \\ X+Y-Z & X+Y-Z & X-Y-Z+2 & X-5Y-Z+6 \\ 4X+2Y-2Z & 16X+6Y-6Z-2 & 20X+6Y-6Z-8 & 8X+2Y-2Z-6 \\ 9X+3Y-3Z & 17X+5Y-5Z-6 & 13X+9Y-3Z-12 & 5X+3Y-Z-6 \end{vmatrix} = 0.$$

4.3. The Diamond Cyclides

The monomial, row, and column supports of the diamond cyclide examined in [14] are shown in (15). Its implicit equation is obtained by setting $|D_\mathcal{A}|$ of (17) to zero.

For example, the rational surface (Figure 3)

$$\begin{bmatrix} w(s,t) \\ x(s,t) \\ y(s,t) \\ z(s,t) \end{bmatrix} = \sum_{(i,j)\in\mathcal{A}} \begin{bmatrix} 1 \\ i \\ j \\ 0 \end{bmatrix} s^i t^j + \begin{bmatrix} 0 \\ 0 \\ 0 \\ 10 \end{bmatrix} st$$

has the implicit equation

$$\begin{vmatrix} -10+10X+10Y-Z & -2Z & -2Z & 0 \\ -2Z & 10+10X-10Y-Z & 0 & -2Z \\ -2Z & 0 & 10-10X+10Y-Z & -2Z \\ 0 & -2Z & -2Z & 30-10X-10Y-Z \end{vmatrix} = 0.$$

4.4. The One-Horned Cyclides

The monomial support of the one-horned cyclide presented in [14] is $\mathcal{A} = \{(0,0),(1,0),(2,0),(1,1),(1,2)\}$. Pictorially the monomial, row, and column supports are

$$\begin{array}{ccc} 4 & \bullet & 3 \\ 4 & \bullet & 3 \\ \bullet & \bullet & \bullet \end{array} \qquad \begin{array}{cc} 4 & 3 \\ 4 & 3 \\ \bullet & \bullet \\ \bullet & \bullet \end{array} \qquad \begin{array}{cccc} 4 & \bullet & \bullet & 3 \\ 4 & \bullet & \bullet & 3 \end{array}$$

$$\mathcal{A} \qquad\qquad \mathcal{R}_\mathcal{A} \qquad\qquad \mathcal{C}_\mathcal{A}$$

respectively. The matrix $D_{2,2}(\mathcal{A})$ is

$$\begin{bmatrix} 0 & 0 & 101100 & 101200 & 201100 & 201200 & 0 & 0 \\ 0 & 0 & 101200 & 111200 & 201200 & 0 & 0 & 0 \\ 0 & 0 & 0 & 0 & 0 & 0 & 0 & 0 \\ 0 & 0 & 0 & 0 & 0 & 0 & 0 & 0 \\ 0 & 0 & 201100 & 201200 & 201110 & 201210 & 0 & 0 \\ 0 & 0 & 201200 & 0 & 201210 & 201211 & 0 & 0 \\ 0 & 0 & 0 & 0 & 0 & 0 & 0 & 0 \\ 0 & 0 & 0 & 0 & 0 & 0 & 0 & 0 \end{bmatrix}.$$

The non-zero rows are those numbered 1, 2, 5, 6 and the non-zero columns are those numbered 3, 4, 5, 6. Thus $D_\mathcal{A}$ is the sub-matrix of $D_{2,2}(\mathcal{A})$ with these non-zero rows and columns and $|D_\mathcal{A}| = 0$ is the the implicit equation of the one-horned cyclide.

For example, the rational surface (Figure 4)

$$\begin{bmatrix} w(s,t) \\ x(s,t) \\ y(s,t) \\ z(s,t) \end{bmatrix} = \sum_{(i,j)\in\mathcal{A}} \begin{bmatrix} 1 \\ i \\ j \\ 0 \end{bmatrix} s^i t^j + \begin{bmatrix} 0 \\ 0 \\ 0 \\ 20 \end{bmatrix} st$$

has the implicit equation

$$\begin{vmatrix} 20Y-Z & -2Z & 40Y-2Z & -4Z \\ -2Z & 40X-20Y-Z & -4Z & 0 \\ 40Y-2Z & -4Z & 20Y-Z & -2Z \\ -4Z & 0 & -2Z & 80-40X-20Y-Z \end{vmatrix} = 0.$$

4.5. Hexagons

The monomial support of the hexagon surface described in [14] is $\mathcal{A} = \{\ (0,0),(1,0),(0,1),(1,1),(2,1),(1,2),(2,2)\ \}$. Pictorially the monomial, row, and columns supports are

$$\begin{array}{ccc} 4 & \bullet & \bullet \\ \bullet & \bullet & \bullet \\ \bullet & \bullet & 2 \end{array} \qquad \begin{array}{cc} 4 & \bullet \\ \bullet & \bullet \\ \bullet & \bullet \\ \bullet & 2 \end{array} \qquad \begin{array}{cccc} 4 & \bullet & \bullet & \bullet \\ \bullet & \bullet & \bullet & 2 \end{array}$$

$$\mathcal{A} \qquad\qquad \mathcal{R}_\mathcal{A} \qquad\qquad \mathcal{C}_\mathcal{A}$$

respectively. The matrix $D_{2,2}(\mathcal{A})$ is

$$
\left[\begin{array}{lll}
100100 & 0 & 101100 \\
110100 & 0 & 101200+110110+210100 \\
120100 & 0 & 120110+220100 \\
0 & 0 & 0 \\
0 & 0 & 0 \\
210100 & 0 & 210110+211100 \\
220100 & 0 & 220110+221100+211200 \\
0 & 0 & 221200
\end{array}\right.
$$

$$
\begin{array}{cc}
101200 & 102100 \\
101201+111200 & 102200+210110 \\
111201 & 220110 \\
0 & 0 \\
0 & 0 \\
211200 & 211110 \\
211201+221200 & 221110+211210 \\
221201 & 221210
\end{array}
$$

$$
\left.\begin{array}{ccc}
102200 & 0 & 0 \\
102201+112200+211200 & 0 & 212200 \\
112201+211201 & 0 & 212201 \\
0 & 0 & 0 \\
0 & 0 & 0 \\
211210+212200 & 0 & 212210 \\
211211+212201+221210 & 0 & 212211 \\
221211 & 0 & 221221
\end{array}\right].
$$

The non-zero rows are those numbered 1, 2, 3, 6, 7, 8 and the non-zero columns are those numbered 1, 3, 4, 5, 6, 8. Thus $D_{\mathcal{A}}$ is the square sub-matrix of $D_{2,2}(D_{\mathcal{A}})$ consisting of these non-zero rows and columns. The determinant $|D_{\mathcal{A}}|$ set to zero is the implicit equation.

For example, the rational surface (Figure 5)

$$
\begin{bmatrix} w(s,t) \\ x(s,t) \\ y(s,t) \\ z(s,t) \end{bmatrix} = \sum_{(i,j)\in\mathcal{A}} \begin{bmatrix} 1 \\ i \\ j \\ 0 \end{bmatrix} s^i t^j + \begin{bmatrix} 0 \\ 0 \\ 0 \\ 30 \end{bmatrix} st
$$

has the implicit equation

$$
\left|\begin{array}{cc}
-Z & 30Y-Z \\
30X-Z & 30X+30Y-5Z-30 \\
-Z & -4Z \\
-2Z & -30X+60Y-3Z \\
-2Z & -60X+60Y-6Z \\
0 & -2Z
\end{array}\right.
$$

$$
\begin{array}{cc}
-2Z & -Z \\
60X-30Y-3Z & -4Z \\
30X-30Y-Z+30 & -3Z \\
-3Z & -30X+30Y-Z+30 \\
-4Z & -60X+30Y-3Z+60 \\
-Z & -2Z
\end{array}
$$

$$
\left.\begin{array}{cc}
-2Z & 0 \\
60X-60Y-6Z & -2Z \\
30X-60Y-3Z+60 & -2Z \\
-4Z & -Z \\
-30X-30Y-5Z+90 & -30X-Z+60 \\
-30Y-Z+60 & -Z
\end{array}\right| = 0.
$$

4.6. Pentagons

The monomial support of the pentagon surface discussed in [14] is $\mathcal{A} = \{\,(0,0),(1,0),(2,0),(0,1),(1,1),(2,1),(0,2),(1,2),\,\}$. Pictorially the monomial, row, and column supports are

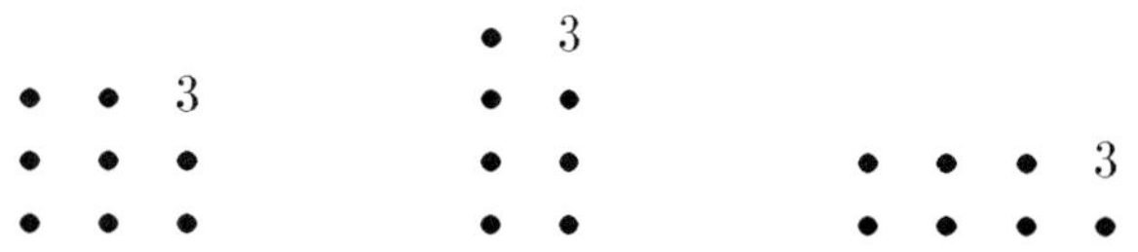

$\mathcal{A}$ $\mathcal{R}_{\mathcal{A}}$ $\mathcal{C}_{\mathcal{A}}$

respectively. The matrix $D_{2,2}(\mathcal{A})$ is

$$
\left[\begin{array}{cc}
100100 & 100200 \\
100200+110100 & 100201+110200 \\
120100+110200 & 110201+120200 \\
120200 & 120201 \\
200100 & 200200 \\
200200+210100 & 200201+210200 \\
210200 & 210201 \\
0 & 0
\end{array}\right.
$$

$$
\begin{array}{c}
101100+200100 \\
101200+110110+200200+210100 \\
120110+110210+210200 \\
120210 \\
200110+201100 \\
200210+201200+210110+211100 \\
210210+211200 \\
0
\end{array}
$$

$$
\begin{array}{c}
101200+200200 \\
101201+111200+200201+210200 \\
111201+120210+210201 \\
120211 \\
200210+201200 \\
200211+201201+210210+211200 \\
210211+211201 \\
0
\end{array}
$$

$$\begin{array}{c}
102100 + 201100 \\
110120 + 201200 + 210110 \\
120120 + 110220 + 210210 \\
120220 \\
201110 + 202100 \\
201210 + 210120 + 211110 \\
210220 + 211210 \\
0
\end{array}$$

$$\left.\begin{array}{ccc}
201200 & 202100 & 0 \\
201201 + 211200 & 210120 & 0 \\
120220 + 211201 & 210220 & 0 \\
120221 & 0 & 0 \\
201210 & 202110 & 0 \\
201211 + 211210 & 211120 & 0 \\
211211 & 211220 & 0 \\
0 & 0 & 0
\end{array}\right] .$$

Thus the implicit equation is the minor with respect to the entry $(8, 8)$ set to zero.

For example, the rational surface (Figure 6)

$$\begin{bmatrix} w(s,t) \\ x(s,t) \\ y(s,t) \\ z(s,t) \end{bmatrix} = \sum_{(i,j)\in\mathcal{A}} \begin{bmatrix} 1 \\ i \\ j \\ 0 \end{bmatrix} s^i t^j + \begin{bmatrix} 0 \\ 0 \\ 0 \\ 10 \end{bmatrix} st$$

has the implicit equation

$$\left|\begin{array}{ccc}
-Z & -2Z & 10Y - 3Z \\
10X - 3Z & 20X - 3Z & 10X + 10Y - 9Z - 10 \\
20X - 3Z & 10X - 3Z & 20X + 10Y - 7Z - 20 \\
-2Z & -Z & -2Z \\
-2Z & -4Z & 20Y - 3Z \\
-6Z & -6Z & -10X + 20Y - 9Z \\
-4Z & -2Z & -6Z
\end{array}\right.$$

$$\begin{array}{cc}
-6Z & 20Y - 3Z \\
20X - 10Y - 9Z & 10X + 20Y - 7Z - 20 \\
10X - 10Y - 5Z + 10 & 20X + 20Y - 6Z - 40 \\
-10Y - Z + 20 & -2Z \\
-6Z & 10Y - 3Z \\
-20X - 20Y - 9Z + 40 & -10X + 10Y - 5Z + 10 \\
-10X - 20Y - 3Z + 40 & -4Z
\end{array}$$

$$\left.\begin{array}{cc}
-4Z & -2Z \\
-6Z & -2Z \\
-4Z & -2Z \\
-Z & 0 \\
-2Z & -Z \\
-20X - 10Y - 3Z + 40 & -10X - Z + 20 \\
-10X - 10Y - Z + 30 & -Z
\end{array}\right| = 0.$$

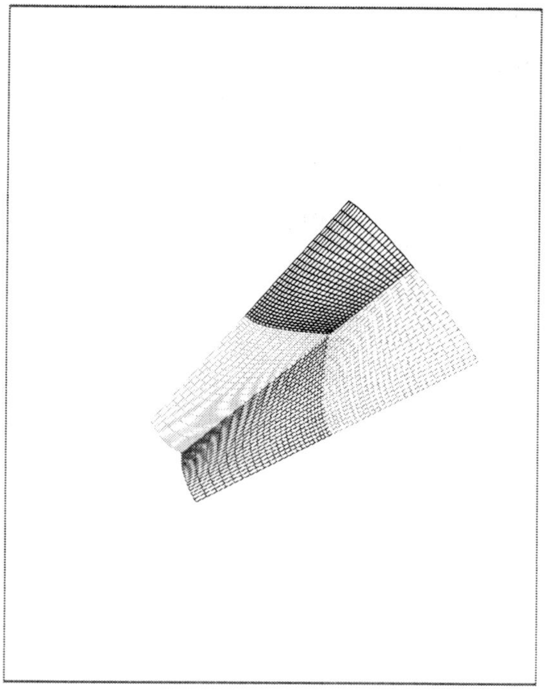

Figure 2. A Hirzebruch surface.

5. Conclusions

We have shown that a minor of the classical bi-degree Dixon matrix is the $\mathcal{A}$-resultant provided the monomial support is obtained by removing at most one rectangular sub-support at each corner of the rectangular bi-degree monomial support. By forming three auxiliary equations from the parametrization of a rational surface, the Dixon $\mathcal{A}$-resultant can be used to implicitize such general rational surfaces exactly. This way of obtaining the implicit equation in determinant form is automatic and in bracket form with smaller determinants when contrasted with the Sylvester dialytic method. These claims are illustrated with the implicitization of the rational surfaces given in [14].

Theorem 1 allows at most one rectangular sub-support at each of the corners to be removed. But the theorem should generalize to the removal of an arbitrary number of corner rectangular sub-supports at each of the corners. The Dixon matrix $D_{\mathcal{A}}$ should again be a square sub-matrix of $D_{m,n}(\mathcal{A})$, the specialized $D_{m,n}$ with respect to $\mathcal{A}$. In this case, Theorem 2 should no longer hold and the determinant $|D_{\mathcal{A}}|$ should be a multiple of the $\mathcal{A}$-resultant if it does not vanish. The properties of such a $D_{\mathcal{A}}$ also seem worth some further investigation.

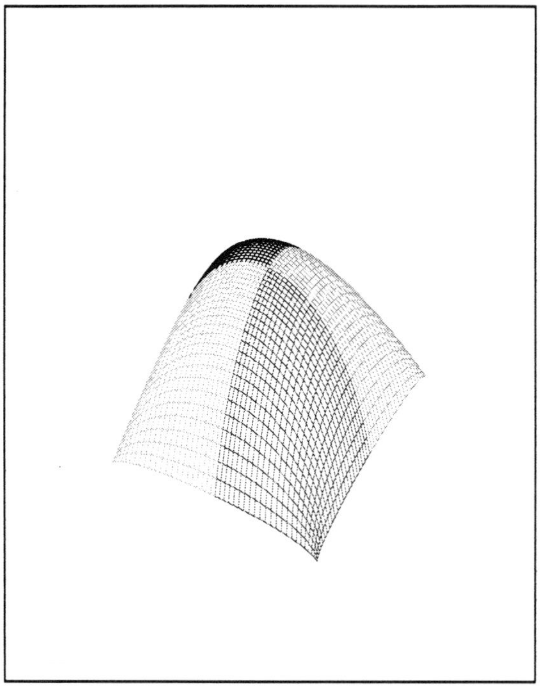

Figure 3. A diamond cyclide surface.

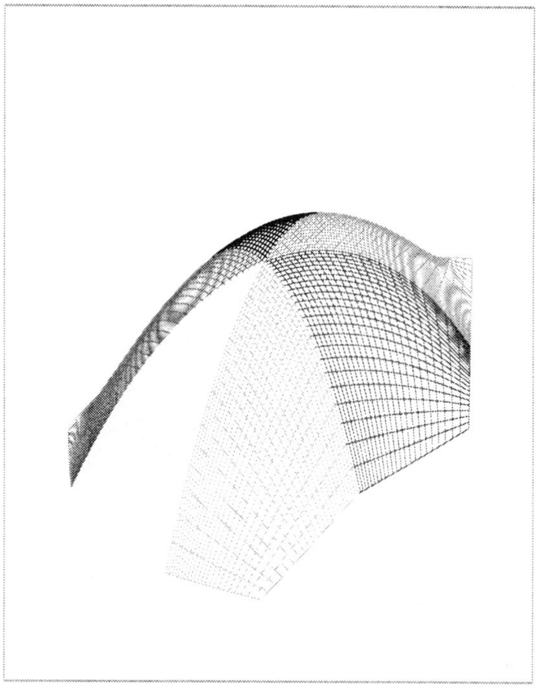

Figure 5. A hexagon surface.

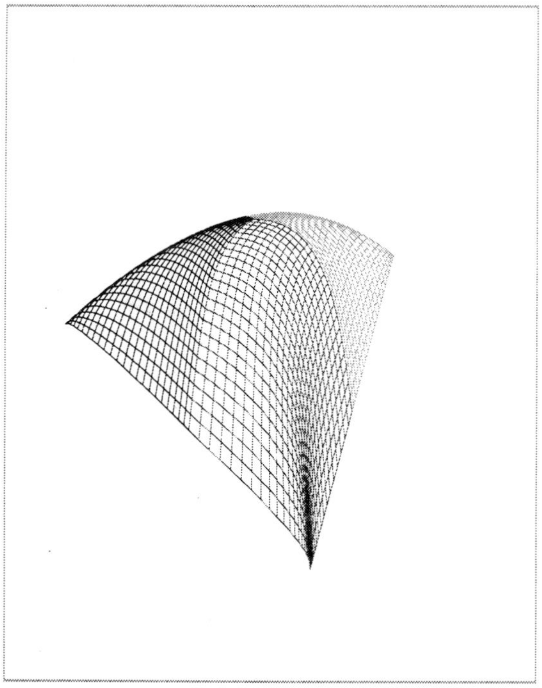

Figure 4. A one-horned cyclide surface.

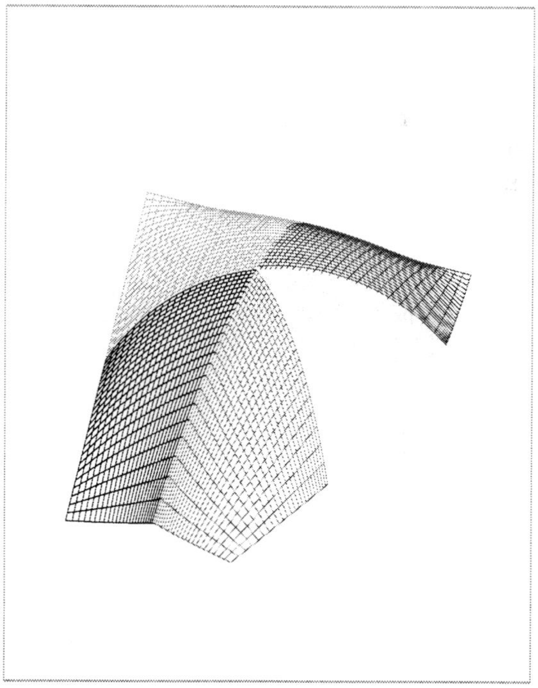

Figure 6. A pentagon surface.

References

[1] E. W. Chionh. Concise parallel Dixon determinant. *Computer Aided Geometric Design*, 14:561–570, 1997.

[2] E. W. Chionh. Rectangular corner cutting and Dixon A-resultants, 1999. Submitted for publication.

[3] E. W. Chionh and R. N. Goldman. Implicitization techniques for rational surfaces. In *Proceedings of the 5th International Conference on Engineering Computer Graphics and Descriptive Geometry*, Melbourne, Australia, August 1992.

[4] E. W. Chionh and R. N. Goldman. Using multivariate resultants to find the implicit equation of a rational surface. *The Visual Computer*, 8:171–180, 1992.

[5] D. Cox, J. Little, and D. O' Shea. *Using Algebraic Geometry*. Springer-Verlag, New York, 1998.

[6] A. L. Dixon. The eliminant of three quantics in two independent variables. *Proc. London Math. Soc.*, 6:49–69, 473–492, 1908.

[7] I. Z. Emiris and B. Mourrain. Matrices in elimination theory. *J. Symbolic Computation*, 28:3–44, 1999.

[8] I. M. Gelfand, M. M. Kapranov, and A. V. Zelevinsky. *Discriminants, Resultants and Multidimensional Determinants*. Birkhäuser, Basel, 1994.

[9] M. Kapranov, B. Sturmfels, and A. Zelevinsky. Chow polytopes and general resultants. *Duke Math. J.*, 67:189–218, 1992.

[10] D. Manocha and J. F. Canny. Algorithms for implicitizing rational parametric surfaces. *Computer Aided Geometric Design*, 9:25–50, 1992.

[11] T. Saxena. *Efficient Variable Elimination Using Resultants*. PhD thesis, State University of New York, 1997.

[12] T. W. Sederberg, D. C. Anderson, and R. N. Goldman. Implicit representation of parametric curves and surfaces. *Computer Vision, Graphics and Image Processing*, 28:72–84, 1984.

[13] T. W. Sederberg and F. L. Chen. Implicitization using moving curves and surfaces. *Computer Graphics Proceedings*, pages 301–308, 1995.

[14] S. Zube. The n-sided toric patches and A-resultants. Technical report, Johannes Gutenberg-Universität Mainz, 1999. Preprint-reihe des Fachbereichs Mathematik.

Session 10

Computation in Geometric Modeling

Interval Methods in Geometric Modeling

Adrian Bowyer[1] Jakob Berchtold[2] David Eisenthal[1] Irina Voiculescu[3]
Kevin Wise[4]

Abstract

This paper is about using interval computations in location, simplification, and root-finding for multivariate implicit functions that are used as shape primitives in a set-theoretic (that is, a CSG) geometric modeller. Three problems are discussed, and solutions to them presented: a) the location and simplification of the surfaces of semi-algebraic sets (surfaces involving some transcendental functions will be dealt with as well); b) the generalization of Newton-Raphson using intervals; and c) interval ray-tracing.
Examples are presented for both conventional three-dimensional geometric models and for CSG models in higher dimensions representing configuration-space maps for moving and colliding three-dimensional objects.
Keywords: Interval arithmetic, CSG, implicit functions, ray-tracing, multidimensional geometric modeling, svLis.

1. Introduction

Interval arithmetic is a technique for doing simultaneous calculations on whole ranges of numbers. One of its key strengths is its ability to calculate the range of values that a function will take given a range of input values. The results are always either exact or conservative—they never miss any possible answers.

This is not the place to go into detail on how interval arithmetic is done (though in a moment we will give definitions and examples), so the reader who needs further information on this is referred to other literature. The best-known work on interval arithmetic is the textbook by Moore [9]. He was mainly concerned with the use of intervals to bound errors in floating-point arithmetic: an important application of them in numerical analysis, but not one that will concern us here. For more geometrical applications of interval arithmetic—its use in this paper—the reader is referred to Snyder's book [14], and to our own [4]. Philip Milne's thesis [8] (which is available as a technical report) also has a useful chapter of interval definitions and discussion.

1.1. Interval definition and interval arithmetic operations

An interval, u, is written as a pair of numbers, $u = [\underline{u}, \overline{u}]$. Usually $\underline{u} \leq \overline{u}$ and this represents the set of all real values from $\underline{u}$ to $\overline{u}$ including $\underline{u}$ and $\overline{u}$ themselves—the interval $[\underline{u}, \overline{u}]$ is closed. Open intervals not containing their ends are written thus: $(\underline{u}, \overline{u})$; of course an interval may be closed at one end and open at the other.

It is convenient to adopt the convention (introduced by Vuillemin [15]) that the complement of an interval is represented by swapping the bounds, so an interval $[\underline{u}, \overline{u}]$ in which $\underline{u} > \overline{u}$ represents the complement of $(\overline{u}, \underline{u})$ (again, if square brackets are used the complement is closed, if round, open). The complement of an interval is all real numbers except those in the interval—the real line with a gap, if you will.

The ordinary monadic and dyadic operators of arithmetic ($+ - \times \div$ and exponentiation) are defined on intervals such that the result of an operation is the interval containing all possible answers that could be got from performing the operation on any of the numbers in the interval (for monadic operations) or from any pair of numbers, one from the first interval operand, and one from the second for dyadic operations.

Thus (if u and v are finite)

$$u + v = [\underline{u} + \underline{v}, \overline{u} + \overline{v}],$$

If one of u or v is infinite (for example as a result of the complement convention described above) the bounds can still be summed, but if the resulting interval is finite then the bounds have crossed over and the result is all real numbers, $\perp$. If both u and v are infinite the result is $\perp$.

Subtraction immediately follows from addition and the fact that

$$-u = [-\overline{u}, -\underline{u}].$$

[1]Department of Mechanical Engineering, University of Bath, BA2 7AY, U.K.
[2]Spatial Technology Ltd, Cambridge, U.K.
[3]Oxford University Computing Laboratory, Parks Road, Oxford, OX1 3QD, U.K.
[4]Department of Continuing and Distance Education, University of Bath, BA2 7AY, U.K.
e-mail: g_mod@bath.ac.uk

Multiplication can involve a certain amount of work, as four possible answers sometimes have to be generated and inspected if either of the input intervals contains both positive and negative numbers. Again, if u and v are finite

$$u.v = [min(\underline{u}\underline{v}, \overline{u}\overline{v}, \underline{u}\overline{v}, \overline{u}\underline{v}), max(\underline{u}\underline{v}, \overline{u}\overline{v}, \underline{u}\overline{v}, \overline{u}\underline{v})].$$

A look-up table (due to Milne [8]) can be used to give optimally tight bounds for multiplications that include infinite intervals:

$\times$ u / v	$[-,-]$	$[-,+]$	$[+,-]$	$[+,+]$
$[-,-]$	$[\overline{u}\overline{v}, \underline{u}\underline{v}]$	$[\overline{u}\underline{v}, \underline{u}\underline{v}]$	$[\overline{u}\overline{v}, \underline{u}\overline{v}]$	$[\overline{u}\underline{v}, \underline{u}\overline{v}]$
$[-,+]$	$[\underline{u}\overline{v}, \underline{u}\underline{v}]$	$[min(\overline{u}\underline{v}, \underline{u}\overline{v}), max(\underline{u}\underline{v}, \overline{u}\overline{v})]$	$\perp$	$[\overline{u}\underline{v}, \overline{u}\overline{v}]$
$[+,-]$	$[\overline{u}\overline{v}, \overline{u}\underline{v}]$	$\perp$	$[min(\underline{u}\underline{v}, \overline{u}\overline{v}), max(\underline{u}\overline{v}, \overline{u}\underline{v})]$	$[\underline{u}\underline{v}, \underline{u}\overline{v}]$
$[+,+]$	$[\underline{u}\overline{v}, \overline{u}\underline{v}]$	$[\underline{u}\overline{v}, \overline{u}\overline{v}]$	$[\underline{u}\underline{v}, \overline{u}\underline{v}]$	$[\underline{u}\underline{v}, \overline{u}\overline{v}]$

As with addition, it is necessary to check that the results in the table are valid: if one of u and v contains 0 and the other is infinite, then the answer is $\perp$. In the corners of the table it is necessary to check that if just one of u or v is infinite then the result is infinite; otherwise the bounds have swapped and we need to return $\perp$.

Reciprocals are simple. In all cases

$$1/u = [1/\overline{u}, 1/\underline{u}]$$

and division immediately follows from this and the multiplication table above.

It might be thought that raising to integer powers could just be done with multiplication, but, for example, unfortunately $u^2 \neq u.u$. To see this consider $u = [-1,2]$. If we multiply $[-1,2].[-1,2]$ we get $[-2,4]$. But if we square any number in $[-1,2]$ the answer must lie in $[0,4]$. Even powers have to produce non-negative results. Once again the optimal answer is due to Milne [8]:

$\uparrow$ n / u	odd	even
$[-,-]$	$[\underline{u}^n, \overline{u}^n]$	$[\overline{u}^n, \underline{u}^n]^\dagger$
$[-,+]$	$[\underline{u}^n, \overline{u}^n]$	$[0, max(-\underline{u}, \overline{u})^n]$
$[+,-]$	$[\underline{u}^n, \overline{u}^n]$	$[min(\underline{u}, -\overline{u})^n, \infty]$
$[+,+]$	$[\underline{u}^n, \overline{u}^n]$	$[\underline{u}^n, \overline{u}^n]^\dagger$

The intervals marked † are replaced with $[0,\infty]$ whenever u is infinite.

Functions like $\log(u)$ and e^u are trivial to implement as they are monotonic (assuming u is positive in the case of log). All that is needed is to apply the function to the bounds. Periodic functions like sine require more work, but are comparatively simple. Clearly if u is infinite then $\sin(u) = [-1,1]$. Otherwise the lower bound of $\sin(u)$ is $min(\sin(\underline{u}), \sin(\overline{u}))$ unless the range of u is greater than $(3\pi/2 - \underline{u})$ **mod** 2π, in which case the lower bound is -1; the upper bound is $max(\sin(\underline{u}), \sin(\overline{u}))$ unless the range of u is greater than $(\pi/2 - \underline{u})$ **mod** 2π, in which case the upper bound is 1. Other transcendental functions can be implemented similarly.

2. Using intervals to locate complicated surfaces for geometric modeling

Many methods of geometric modeling use the zeros of implicit functions to define shapes in three dimensions. One of the earliest techniques was due to Ricci [13] who used implicits to create approximations to unions and intersections of shapes and to blend them smoothly. Soft objects and metaballs [18] also generate implicits to represent geometry, and Pasko's work on *R-functions* [10][11] does likewise.

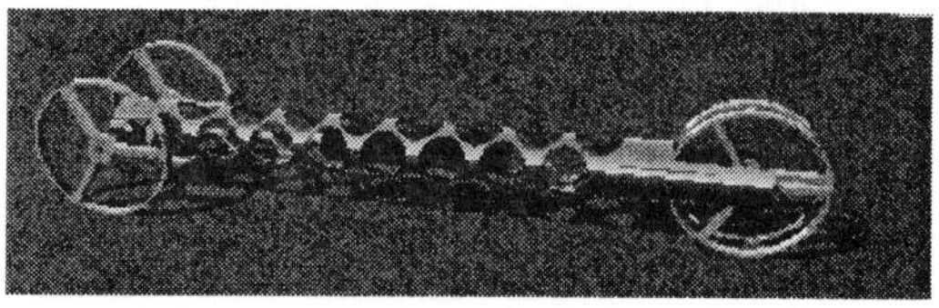

Figure 1: A sVLIS model of a spring-powered vehicle, part of an undergraduate design exercise at the University of Bath.

Our own geometric modeller, sVLIS [3] (which was used for all the examples in this paper), constructs shapes by taking Boolean combinations of primitives represented by implicit functions. Figure 1 shows a simple sVLIS model.

A problem that needs to be solved if implicits are to be used for surface definition is that of locating the implied surface in space. It is very easy to membership-test a point against an implicit function—all that is required is to evaluate the function at the point. A positive result (by convention) indicates that the point is outside the solid, and a negative inside. This simplicity is the principal reason for the use of implicits in CSG/Boolean modeling schemes; the representations and data for these are, after all, little more than a definition of the result of a membership test for any point that might be chosen in space. But this doesn't help very much if such unevaluated models are, say, to be rendered—for such problems the surface is required.

Interval arithmetic provides an efficient way to extend the idea of point membership testing to the testing of whole regions of space. Specifically, the point becomes a box defined by three intervals. These are substituted into an implicit function that defines a shape. If the resulting interval is all negative or all positive, it is guaranteed that the surface, the zeros, of the implicit do not pass through the box. If the resulting interval contains zero, the surface may lie (partly) in the box, though interval arithmetic is conservative (see below), and we may sometimes get a false positive. It seems intuitively obvious that as the box gets smaller and smaller the conservativeness problem reduces too, as the box is approaching a point for which the implicit evaluation is exact.

Figure 2: A SvLIS model of a cyclide. The red box has unit-length sides and is centred at the point (5.09008, 3.03746, 1.12317), which is on the cyclide's surface.

Figure 2 shows a svlis model of a cyclide and a unit-side-length box. Figure 3 shows what happens to the range[1] of the interval that the cyclide's implicit function produces as the three intervals representing the box are fed into it. The box is reducing in size, and is centred on a point on the cyclide's zero. As can be seen the response is roughly linear, even though the cyclide is a degree four function.

SvLIS uses such an interval technique combined with recursive division of a large box surrounding the whole region of interest. The intervals representing boxes are fed into the implicit functions representing primitives of geometric models. If the interval produced is entirely positive it represents the null set, and thus *in that box* the Boolean expression for the model can be simplified using the rules of Boolean logic. Similarly an entirely negative interval represents the universal set, and once again a Boolean simplification can be obtained.

Thus a complicated model in a big box is recursively simplified as the box is divided into smaller ones. If the interval arithmetic is sometimes conservative, this doesn't

[1]The upper minus the lower bound.

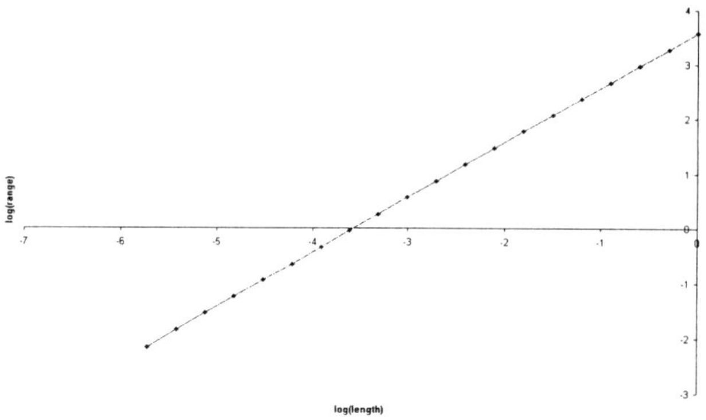

Figure 3: The variation in the range of the output interval generated by the cyclide as intervals representing smaller and smaller cubes are fed into its implicit function.

matter very much as

1. Nothing will be lost; the only problem is that a potential simplification will not be performed; and

2. As the boxes reduce in size, so does the conservativeness problem as demonstrated above, and so, even if a simplification is missed, it will often occur slightly further down the tree.

3. Identical functions are different with intervals

Clearly

$$1 + 2u - u^2 = 1 + u(2 - u).$$

The expression on the left is the power form of the polynomial, and on the right is the Horner form. But, for example, if $u = [0, 1]$ the power form gives $[0, 3]$ as the answer and the Horner form gives $[1, 3]$. If an expression is evaluated using interval arithmetic, the form of the expression can affect the result. The answer is always correct, in the sense that the interval obtained is guaranteed to contain the range of values that the function will take for all input values of (in this case) u. But rearranging the function can give tighter bounds on the result, and in general the result will be conservative, not exact.

It is sometimes said that this conservativeness effect is a consequence of repeated occurrences of interval variables in expressions, but this is not always so. We have reported elsewhere [1] on comparisons between the conservativeness of interval arithmetic for implicit functions held in the power form, the standard form[2] (when this is available), and in the Bernstein form (see Lorentz [7]).

Figure 4 shows a two-dimensional example taken from [1]. It gives the results of taking the polynomial for which the power form is:

$$5x^5y - 9xy^4 + 2x^2y^3 + 5xy^3;$$

this is:

$$\begin{gathered} 5x^5y(1-y)^3 + 15x^5y^2(1-y)^2 + \\ 4[\tfrac{5}{4}x(1-x)^4 + \\ \tfrac{11}{2}x^2(1-x)^3 + 9x^3(1-x)^2 + \\ \tfrac{13}{2}x^4(1-x) + \tfrac{11}{2}x^5]y^3(1-y) + \\ [-4x(1-x)^4 - 14x^2(1-x)^3 - \\ 18x^3(1-x)^2 - 10x^4(1-x) + 3x^5]y^4 \end{gathered}$$

in the Bernstein form.

As can be seen the latter is considerably more complicated and has many repeated terms. The figure shows what happens if the unit square is uniformly subdivided into $50 \times 50 = 2500$ smaller squares. The x and y intervals representing all these little squares were evaluated in the power and the Bernstein forms of the polynomial. The red region is the squares which evaluated negative. The blue region is positive. The green region is that for which the output interval contains zero. The matrix under the figure shows the counts of the classified boxes for both forms of the polynomial. The Bernstein form correctly classifies

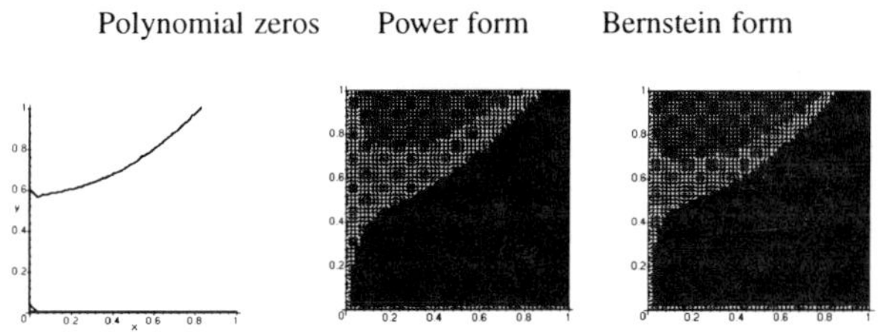

		Bernstein		
		$[-,-]$	$[-,+]$	$[+,+]$
Power	$[-,-]$	325	0	0
	$[-,+]$	93	418	66
	$[+,+]$	0	0	1598

Figure 4: Box classification for $5x^5y - 9xy^4 + 2x^2y^3 + 5xy^3$. The table shows counts of the box classifications for the power and the Bernstein form. $[-,-]$ is a negative output interval, $[-,+]$ one that contains 0, and $[+,+]$ one that is all positive. The leading diagonal is the boxes that are classified the same by both methods. Off that diagonal, the 93, for example, is the count of boxes that the Bernstein form classifies as definitely negative, but which the power form gives as maybe containing 0, and so on.

$93 + 66 = 159$ more boxes than the power form. In this example, there are no boxes which the power form classifies correctly and which the Bernstein form does not.

Given the known numerical stability of the Bernstein form, one might expect it to do well with interval arithmetic. But when one sees the complexity of the Bernstein expression above in comparison with the simplicity of the power form, and one knows that repeated terms can lead to excessive conservativeness in interval arithmetic, one might think such optimism misplaced. However, the actual results show that surprisingly the very complicated Bernstein form not only does well, it does better than the much simpler power form.

The standard form, incidentally, is always good, and this is what sVLIS uses whenever possible. However, encouraged by the above results, one of us [2] has written a version of sVLIS that uses the standard form for simple shapes like cylinders and so on, and the Bernstein form for blend surfaces and all other complicated shapes. This allows sVLIS 's interval-arithmetic-driven recursive division scheme to operate better than on power-form polynomials. It does not increase its range nor its accuracy very much, but it *does* increase its efficiency.

4. The interval version of Newton's method

One particularly useful geometric application of interval methods is an interval version of Newton's method for finding a zero of a differentiable function. Unlike the conventional version of Newton's method, the interval version is guaranteed to isolate and to find *all* roots in a given interval.

Given a differentiable function $f(x)$ and an interval u_i in which roots of f need to be found, Walster [16] gives the interval version of Newton's algorithm as follows:

1. Let $m = (\underline{u}_i + \overline{u}_i)/2$
2. $u_{i+1} = u_i \cap \left[m - \frac{f(m)}{f'(u_i)}\right]$

Looping round this generates a series of intervals that converge on all the roots in the original u_i. Note that the division in Step 2 means that, if the denominator contains zero, the resulting interval will be complemented—the whole real line with a gap (see above). When the intersection with the original interval is performed this can generate two resulting intervals. It is this process that generates the extra intervals needed to bound all the roots.

Running this algorithm using ordinary single-precision floating-point arithmetic on the polynomial obtained by multiplying out

$$(x-1)(x-2)(x-3)(x-4) = 0$$

[2]The standard form for a circle, for example, is $(x-x_c)^2 + (y-y_c)^2 - r^2 = 0$.

with an initial interval $u = [-2, 7]$ and printing all intervals found with a range less than 0.0005 gives the following:

[0.999816, 1.0002]
[1.99976, 2.00016]
[2.99999, 3.00001]
[3.99999, 4]

To get this the algorithm examined 142 intermediate intervals.

Hansen [6] and Walster [16] give the following list of remarkable properties of this algorithm:

1. If there is no zero of f in u_i this will be proved in finitely many steps.
2. Every zero of f in u_i will always be found and correctly bounded.
3. If $0 \notin f'(u_i)$ at least half of u_i is eliminated at the next step.
4. If $0 \notin f'(u_i)$ then asymptotic convergence is quadratic.

Of course, even with double precision, the algorithm is susceptible to rounding error. However, it is easily implemented using rational arithmetic which eliminates this problem completely at the expense of slower execution. A good half-way house would be to ensure that floating point rounding is always conservative in the size of the intervals generated; that is, to round $\underline{u}$ down and $\overline{u}$ up. However this is not always easy given the inaccessibility of the machine's floating-point unit controls via compilers.

5. Interval ray tracing

A CSG model, such as those built by svLis, that uses Boolean operators acting on implicit functions as primitives can always generate a potential value for any point in space for a whole model, not just for the primitives that compose it. This is a consequence of the fact, as described by Quarendon *et. al.* [12], that it is possible to replace the Boolean union operator with the minimum function, and the intersection operator with the maximum function[3]. If this is done, then any point can be substituted into a model; the implicit functions for primitives give potentials in the usual way and these are combined using *max* and *min* to give an overall potential at the point for the whole model.

[3]Assuming the convention that negative potential means solid; the functions are swapped if the opposite convention is adopted.

Note that, while this potential gives some information about the distance between the point and the surface of the model, and it also (if this information is tracked through the *max* and *min* operations) identifies the primitive that generated the actual potential at the point, it does not give actual distance to the model, nor does it identify the closest primitive to the point. However more useful information is returned than that obtained from a simple membership test.

Once again, we can do the same trick of substituting three intervals representing a box for the three coordinates of the point in the calculation. The result is an interval that gives a conservative range of potentials for the whole model in the given box, unions and intersections and all.

Among other things, this allows us to program a ray-tracer for the model that is very simple, even crude, but whose elegance is a consequence of that very crudeness.

Imagine a ray passing through the space occupied by a svLis-type model. Figure 5 shows a two-dimensional example. The model consists of the union of the red ellipse and rectangle. The ray is the blue line running from top left to bottom right.

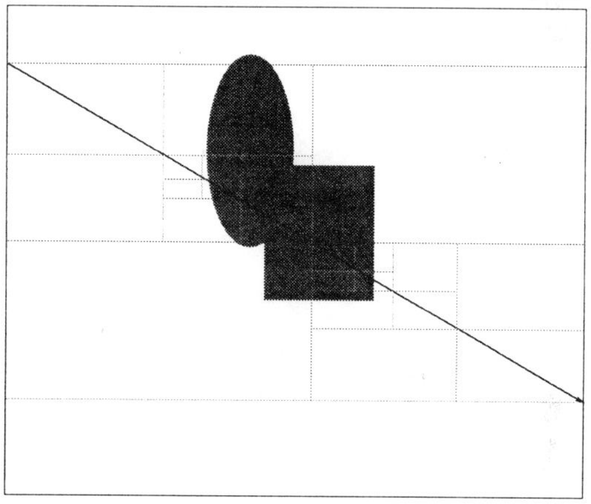

Figure 5: Interval ray tracing.

The ray tracer starts by treating the ray as a diagonal of a large box—the largest green rectangle in the figure. The box is represented by intervals, and, in the case of svLis, is obtained simply by computing the line segment of the ray's intersection with the box surrounding the whole model, which svLis models all have[4]. These intervals are substituted into the set-theoretic expression for the model and the resulting interval is examined. If it is all positive the ray misses the model and no further work is needed. If it is all negative the ray segment is en-

[4]In the absence of this any segment sufficiently large to include everything of interest will do; it doesn't have to bound the model particularly tightly.

tirely within the model and again no further work needs be done.

If the interval contains zero, the ray may strike the surface of the model. In this case the ray is bisected, and the two smaller boxes of which its two halves are diagonals are computed. These are substituted into the model's set-theoretic expression as before. This process is repeated recursively until the box being considered is sufficiently small to represent a struck surface.

Clearly, if only the first struck surface is required (that is the one with the lowest ray parameter value), then the recursion proceeds in depth-first manner examining those boxes representing the low-end of the ray first. As soon as a surface is found, the remaining recursion stack can be discarded. If all intersection points are required, the stack is retained and the recursion continues to find them all.

The process can be made more efficient by using the Boolean simplification on the set-theoretic contents of the boxes described above. In practice it is usually not worth doing this at every step in the recursive binary division of the ray; calling the Boolean simplification procedure once for about every three steps down seems to produce the fastest execution for most models.

The advantages of this ray tracer are that:

1. It is very compact to code. Given the modeller (which includes interval and box classes), our implementation is about 25 lines of C++.

2. It works independently of the form of the model's primitives—these can be any functions including sines, exponentials and so on as well as polynomials. This makes it inherently very versatile.

We have recently completed a multidimensional version of SVLIS called SVLIS-M [5]. This is the basis for a larger project investigating the use of its multidimensional CSG representation for configuration-space mapmaking, constraints analysis, symmetry analysis, feature recognition, and other geometric modeling problems.

A configuration-space map [17] is a complete representation of the kinematics of a system of moving rigid bodies. Specifically, a configuration space is a space in which each degree of freedom of each rigid body is given a separate dimension; a single point in this space completely defines the positions and orientations of all the bodies. A configuration space map, then, is a partition of the configuration space of the bodies into two types of region: *safe* regions, where the moving bodies are free and don't touch, and *prohibited* regions where two or more of the bodies interpenetrate. The boundaries between these types of region (usually, though not always, surfaces embedded in the configuration space) represent configurations where two or more bodies are just touching.

Figure 6 shows a fixed cuboid (blue), and an object consisting of the union of two tetrahedra (brown) that is free to move. SVLIS-M was used to compute the con-

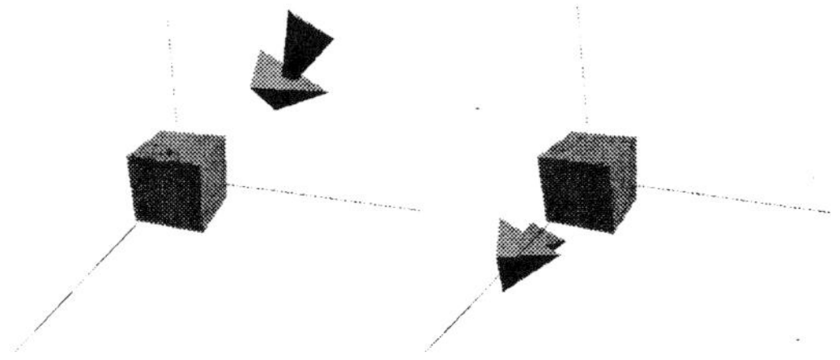

Figure 6: Initial and final configurations of the two-tetrahedra block. The red, green and blue lines are the coordinate axes.

figuration space map for these objects. As the moving brown object is free both to translate and to rotate, the

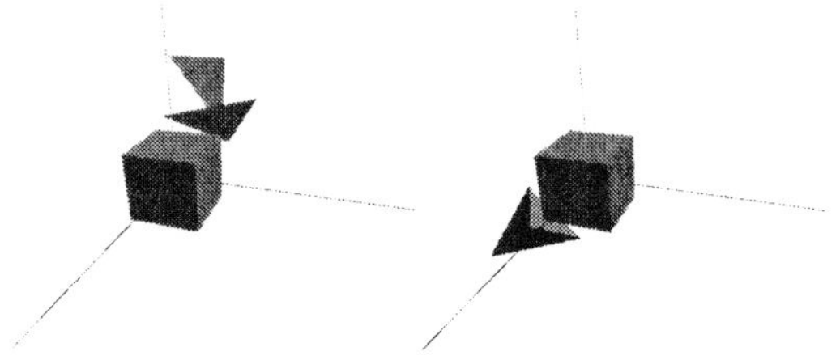

Figure 7: The interval ray tracer working in the 6-dimensional configuration-space map of the fixed blue cuboid and the moving brown two-tetrahedra block. The ray was the straight line between the two configuration points representing the positions in Figure 6. Here are the configurations where the 6-D ray just enters, and just leaves, the prohibited region of the configuration-space map.

configuration-space is six dimensional, and the map is an extremely complicated CSG expression in this 6-D space unioning and intersecting sinusoids and helicoids.

Figure 7 shows the interval ray tracer working in this six-dimensional space. The two points in the configuration space representing the positions of the brown object in the left- and right-hand pictures in Figure 6 were used as the ends of a line segment in the configuration space. Of course tracking along this line segment represents both translations and rotations of the movable object as its configuration point passes through all 6 dimensions.

The interval ray-tracer was then used to find the points (in this case two) where that ray entered the prohibited region and left it—the two configurations where the brown object just enters and just leaves the blue cuboid as it moves from its initial position and orientation to its final one.

6. Conclusions

Interval arithmetic methods have wide application in geometric computations. They can be used to examine the potential values of implicit functions being used for shape definition over whole regions of space, they can be used for multiple real root isolation in a way that guarantees to find *all* roots, and they allow CSG expressions to be localized and simplified using Boolean substitution rules and the consequent implementation of a compact and efficient simple ray tracer.

7. Acknowledgements

The authors would like to thank the University of Bath for supporting two of them (Berchtold and Voiculescu), and the UK's Engineering and Physical Sciences Research Council for the grant under which sVLIS-M was developed.

References

[1] J. Berchtold, I. Voiculescu and A. Bowyer, Interval arithmetic applied to multivariate Bernstein–form polynomials, Technical report Number 31/98, School of Mechanical Engineering, University of Bath, 1998 or http://www.bath.ac.uk/~ensab/G_mod/Bernstein/interval.html

[2] J. Berchtold, The Bernstein basis in set–theoretic geometric modeling, PhD thesis, University of Bath, to be submitted.

[3] A. Bowyer, sVLIS: set–theoretic kernel modeller, (Information Geometers Ltd, 1995) and http://www.bath.ac.uk/~ensab/G_mod/Svlis/

[4] A. Bowyer and J. Woodwark, Introduction to Computing with Geometry, (Information Geometers Ltd, 1993).

[5] D. Eisenthal, K Wise and A. Bowyer, SVLIS-M - a multidimensional CSG modeller, http://www.bath.ac.uk/~ensab/G_mod/Svm/

[6] E.R. Hansen, Global Optimization using interval analysis, Marcel Dekker Inc., New York, 1992.

[7] Lorentz, G. G., Bernstein Polynomials, Chelsea Publishing Company, New York, 1986.

[8] P. S. Milne, On the algorithms and implementation of a geometric algebra system, University of Bath Computer Science technical report 90-40, 1990 (e-mail: tech-report@maths.bath.ac.uk).

[9] R.E. Moore, Methods and applications of Interval Analysis, SIAM 1979.

[10] A. Pasko and V. Savchenko, Blending operations for functionally-based constructive geometry. Proc. CSG 94, Winchester, UK. (151–161), Information Geometers Ltd., 1994.

[11] A. Pasko, V. Adzhiev, A. Sourin and V. Savchenko, Function representation in geometric modeling: concepts, implementation and application, in: Visual computer, **11**, (428–446), 1995.

[12] P. Quarendon, S. Todd and J. Woodwark, Winning primitive in point-membership tests, IBM Technical Disclosure Bulletin 33, 3A (206-207), August 1990.

[13] A. Ricci, A constructive geometry for computer graphics, in: The Computer Journal, **16**, 2 (157–160), 1973.

[14] J. M. Snyder, Generative modeling for computer graphics and CAD (Academic Press, Inc., 1992).

[15] J.E. Vuillemin, Exact computer arithmetic with continued fractions, Proc. ACM conference on LISP and Functional Programming, Snowbird, Utah, 1988.

[16] G.W. Walster, Interval Arithmetic: the new floating-point arithmetic paradigm, available at http://www.mscs.mu.edu/~globsol/walster-papers.html

[17] K. Wise and A. Bowyer, A survey of global configuration-space mapping techniques for a single robot in a static environment, *in press* International Journal of Robotics Research.

[18] G. Wyvill, C. McPheeters and B. Wyvill, Soft Objects, in: Proceedings of computer graphics, Vol. Advanced computer graphics (Springer, Tokyo, 1986) 113–128.

Bounding Interval Rational Bézier Curves with Interval Polynomial Bézier Curves

Falai Chen
Department of Mathematics
University of Science and Technology of China
Hefei, Anhui, 230026, China
chenfl@ustc.edu.cn

Thomas W. Sederberg
Computer Department
Brigham Young University
Provo, Utah 84602, USA
tom@cs.byu.edu

Wenping Lou
Department of Mathematics
University of Science and Technology of China
Hefei, Anhui, 230026, China
wplou@263.net

Abstract

In this paper, we put forward and study the problem of bounding an interval rational Bézier curve with an interval polynomial Bézier curve. We propose three different methods—Hybrid Method, Perturbation Method and Linear Programming Method to solve this problem. Examples are illustrated to compare the three different methods. The empirical results show that the Perturbation Method and the Linear Programming Method produce much tighter bounds than the Hybrid Method, though they are computationally several times more expensive.

1 Introduction

In the communities of Approximation Theory and Computer Aided Geometric Design(CAGD), there is considerable interest in approximating functions (or curves/surfaces) with functions (or curves/surfaces) of simpler forms, for example, in approximating rational functions (or curves) with polynomial functions (or curves). Much literature has focused on these problems from different points of views, e.g., from pure approximation theory to applications in CAGD. However, as far as the authors are aware, almost all the related work is concerned with how accurate the approximation is and seldom consider the problem of how to transfer the approximation errors into subsequent applications. This problem can be important in some applications, e.g., in tolerance analysis in CAD and in numerical analysis. To solve this problem, Sederberg, et. al. [2] introduced interval forms of curves and surfaces. Based on the new representations of curves and surfaces, several authors developed robust algorithms for geometric operations such as curve/curve intersections in CAD/CAM systems ([3]–[8]).

In this paper, we are interested in the following problem, a problem that has potential applications in CAD and numerical analysis:

Given an interval rational function (or curve), bound it with an interval polynomial function (or curve) such that the bound is as tight as possible.

We will develop three different methods–Hybrid Method, Perturbation Method and Linear Programming Method to solve the problem in the following sections. Testing and comparisons are made between the three different approaches through illustration of examples. The empirical results show that the Hybrid Method generally produces much looser bounds than the other two methods, though it is computationally less expensive.

2. Interval Bézier Curves

An *interval* $[a, b]$ is the set of real numbers $\{x | a \leqslant x \leqslant b\}$. Interval arithmetic operations are defined by

$$[a, b] * [c, d] = \{x * y | x \in [a, b] \text{ and } y \in [c, d]\}, \quad (1)$$

where $*$ represents an arithmetic operation, $* \in \{+, -, \cdot, /\}$. One can verify that

$$\begin{aligned} [a, b] + [c, d] &= [a + c, b + d] \\ [a, b] - [c, d] &= [a - d, b - c] \\ [a, b] \cdot [c, d] &= [\min(ac, ad, bc, bd), \\ &\quad \max(ac, ad, bc, bd)] \\ [a, b]/[c, d] &= [a, b] \cdot [1/d, 1/c], \quad 0 \notin [c, d]. \end{aligned} \quad (2)$$

Interval arithmetic is an important tool in numerical analysis, and it has many applications in other areas. For details, the reader is referred to Moore's book *Interval Analysis* [1].

0-7695-0562-7/00 $10.00 © 2000 IEEE

An *interval polynomial* is a polynomial whose coefficients are intervals:

$$[p](t) := \sum_{k=0}^{n}[a_k, b_k]B_k^n(t), \quad 0 \leqslant t \leqslant 1, \tag{3}$$

where $B_k^n(t) = \binom{n}{k}t^k(1-t)^{n-k}$, $k = 0, 1, \ldots, n$ are Bernstein bases.

An interval polynomial can be also expressed in the form

$$[p](t) = [p_{\min}(t), p_{\max}(t)], \quad 0 \leqslant t \leqslant 1, \tag{4}$$

where

$$p_{\min}(t) = \sum_{k=0}^{n} a_k B_k^n(t)$$

and

$$p_{\max}(t) = \sum_{k=0}^{n} b_k B_k^n(t)$$

$p_{\min}(t)$ and $p_{\max}(t)$ are called *lower bound* (denoted by $lb([p](t))$) and *upper bound* (denoted by $ub([p](t))$) of $[p](t)$ respectively.

The *width* of an interval polynomial can be defined by

$$W([p](t)) = ||p_{\max}(t) - p_{\min}(t)||, \tag{5}$$

where the norm $||.||$ is the standard norm, such as $||.||_1$, $||.||_\infty$ and $||.||_2$.

An *interval rational function* is defined

$$[r](t) := \frac{\sum_{k=0}^{n}[a_k, b_k]\omega_k B_k^n(t)}{\sum_{k=0}^{n}\omega_k B_k^n(t)}, \quad 0 \leqslant t \leqslant 1, \tag{6}$$

where ω_k, $k = 0, 1, \ldots$ are weights (It is also possible to define the weights to be intervals. Since the algorithms in the following sections are similar for both cases, we assume the weights are real numbers in this paper). It is similar to define the upper bound, the lower bound and the width of an interval rational function.

An *interval polynomial Bézier curve* is a polynomial Bézier curve whose control points are vector-valued intervals (i.e., rectangular regions in a plane):

$$[\mathbf{P}](t) = \sum_{i=0}^{n}[\mathbf{P}_i]B_i^n(t), \tag{7}$$

where $[\mathbf{P}_i] = [a_i, b_i] \times [c_i, d_i] = ([a_i, b_i], [c_i, d_i])$, $i = 0, 1, \ldots, n$ are the interval control points.

Similarly, an *interval rational Bézier curve* is defined

$$[\mathbf{R}](t) = \frac{\sum_{i=0}^{n}[\mathbf{R}_i]\omega_i B_i^n(t)}{\sum_{i=0}^{n} w_i B_i^n(t)}, \tag{8}$$

where $[\mathbf{R}_i]$ and $w_i > 0, i = 0, 1, \ldots, n$ are interval control points and weights respectively.

An interval (polynomial or rational) Bézier curve defines a region (a slender tube) in the plane which consists of all the Bézier curves whose control points satisfy $\mathbf{P}_i \in [\mathbf{P}_i]$ for $i = 0, 1, \ldots, n$. Figure 1 shows a sample cubic interval polynomial Bézier curve. An interval polynomial (or rational) Bézier curve can be viewed as a vector-valued interval polynomial (or rational function). For example, the interval polynomial Bézier curve (7) can be rewritten in the form

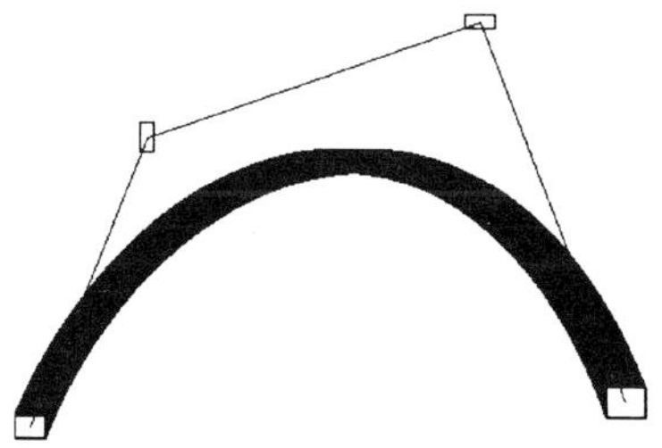

Figure 1. A sample interval Bézier curve

$$[\mathbf{P}](t) = ([x](t), [y](t)), \tag{9}$$

where $[x](t)$ and $[y](t)$ are interval polynomials:

$$[x](t) = \sum_{k=0}^{n}[a_k, b_k]B_k^n(t) \tag{10}$$

$$[y](t) = \sum_{k=0}^{n}[c_k, d_k]B_k^n(t) \tag{11}$$

With this form, the problem of bounding interval rational Bézier curves with interval polynomial Bézier curves can be conveniently converted to the problem of bounding interval rational functions with interval polynomials.

3. Bounding Interval Rational Bézier Curves with Interval Polynomial Bézier Curves

Before dealing with the problem of bounding interval rational Bézier curves with interval polynomial Bézier curves, we solve the problem of bounding an interval rational function with an interval polynomial.

Problem 1 *Given an interval rational function*

$$[r](t) := \frac{\sum_{k=0}^{n}[r_k]\omega_k B_k^n(t)}{\sum_{k=0}^{n} w_k B_k^n(t)}, \tag{12}$$

find an interval polynomial

$$[p](t) := \sum_{k=0}^{m}[p_k]B_k^n(t) \tag{13}$$

such that $[p](t)$ *bounds* $[r](t)$

$$[r](t) \subset [p](t), \quad \text{for } 0 \leqslant t \leqslant 1, \tag{14}$$

and that the width of $[p](t)$ *is as tight as possible.* $[p](t)$ is called an *interval polynomial bound* of $[r](t)$.

Since the process of finding an upper bound is similar to that of finding a lower bound, we can solve the following problem instead

Problem 2 *Given a rational function of degree n*

$$R(t) := \frac{\sum_{k=0}^{n} R_k \omega_k B_k^n(t)}{\sum_{k=0}^{n} \omega_k B_k^n(t)}, \tag{15}$$

find a polynomial of degree m

$$P(t) := \sum_{k=0}^{m} P_k B_k^n(t) \tag{16}$$

such that

$$P(t) \geqslant R(t), \qquad 0 \leqslant t \leqslant 1 \tag{17}$$

and

$$||P(t) - R(t)|| \tag{18}$$

is minimized.

In the following, we will propose three different methods–Hybrid Method, Perturbation Method and Linear Programming Method to solve the above problem.

3.1. Hybrid Method

In [2], Sederberg, et al proposed a method called the *hybrid curve method* to approximate rational functions with polynomials. Indeed, this method also provides a way to compute an upper polynomial bound of a rational function. The main idea of the Hybrid Method is as follows.

Any rational function $R(t)$ can be expressed as a polynomial, one of whose coefficients is a rational function:

$$\begin{aligned} R(t) &= \frac{\sum_{k=0}^{n} r_k \omega_k B_k^n(t)}{\sum_{k=0}^{n} \omega_k B_k^n(t)} \\ &\equiv \sum_{k=0,k\neq l}^{m} P_k B_k^m(t) + M(t) B_l^m(t), \end{aligned} \tag{19}$$

where $M(t)$ is a rational function of degree n:

$$M(t) = \frac{\sum_{k=0}^{n} M_k \omega_k B_k^n(t)}{\sum_{k=0}^{n} \omega_k B_k^n(t)}, \tag{20}$$

and $l = [m/2]$ is an integer.

The coefficients of P_i, $i = 0, 1, \ldots, m, i \neq l$ and M_i, $i = 0, 1, \ldots, n$ can be computed as follows:

For $i = 0, 1, \ldots, l-1$,

$$\begin{aligned} P_0 &= R_0, \\ P_1 &= R_0 + \frac{n\omega_1(R_1 - P_0)}{m\omega_0}, \\ &\vdots \\ P_i &= R_0 + \frac{\sum_{j=1}^{\min(n,i)} \binom{n}{j}\binom{m}{i-j}\omega_j(R_j - P_{i-j})}{\binom{m}{i}\omega_0}. \end{aligned} \tag{21}$$

For $i = m+n, \ldots, n+l+1$,

$$\begin{aligned} P_m &= R_n, \\ P_{m-1} &= R_n + \frac{n\omega_{n-1}(R_{n-1} - P_m)}{m\omega_n}, \\ &\vdots \\ P_{i-n} &= R_n + \frac{\sum_{j=\max(0,i-m)}^{n-1} \binom{n}{j}\binom{m}{i-j}\omega_j(R_j - P_{i-j})}{\binom{m}{i-n}\omega_n}, \end{aligned} \tag{22}$$

and

$$M_i = \sum_{j=\max(0,i+l-m)}^{\min(i+l,n)} \frac{\binom{n}{j}\binom{m}{i+l-j}\omega_j(R_j - P_{i+l-j})}{\binom{m}{l}\binom{n}{i}\omega_i}, \quad i = 0, 1, \ldots, n. \tag{23}$$

Theorem 1 *Let* $M_{max} = max_{\,0 \leqslant k \leqslant n} M_k$ *and* $M_{min} = min_{\,0 \leqslant k \leqslant n} M_k$. *Then polynomial*

$$P(t) = \sum_{k=0,k\neq l}^{m} P_k B_k^m(t) + M_{max} B_l^m(t) \tag{24}$$

is an upper bound of $R(t)$, *and*

$$||P(t) - R(t)||_\infty \leqslant c(M_{max} - M_{min}), \tag{25}$$

where c is a constant:

$$c = \binom{m}{l}\left(\frac{l}{m}\right)^l \left(\frac{m-l}{m}\right)^{m-l}. \tag{26}$$

Proof: Since

$$M(t) \leqslant M_{\max}, \qquad 0 \leqslant t \leqslant 1,$$

from (19), one has

$$R(t) \leqslant \sum_{k=0,k\neq l}^{m} P_k B_k^m(t) + M_{\max} B_l^m(t) = P(t),$$

i.e., P(t) is a polynomial upper bound of $R(t)$.

On the other hand,

$$||P(t) - R(t)||_\infty \leqslant (M_{\max} - M_{\min}) \max_{0 \leqslant t \leqslant 1} B_l^m(t).$$

Since $B_l^m(t)$ attains maximum value at $t = \frac{l}{m}$ in $[0, 1]$, (25) follows immediately.

The Hybrid Method generally provides a loose bound for a rational function, and it has the disadvantage that there is a restricted condition under which the polynomial converges to the rational function when $m \to \infty$. For the specific convergence condition, the reader is referred to [10].

3.2. Perturbation Method

The Hybrid Method presented in the last subsection generally produces quite a loose bound. In this subsection, we will propose an improved algorithm to solve Problem 2.

Given rational function (15), we perturb $R(t)$ with another rational function $\epsilon(t)$ such that $R(t)+\epsilon(t)$ is a polynomial

$$R(t)+\epsilon(t)=P(t):=\sum_{k=0}^{m} P_k B_k^m(t), \qquad (27)$$

where

$$\epsilon(t)=\frac{\sum_{k=0}^{m+n}\epsilon_k\omega'_k B_k^{m+n}(t)}{\sum_{k=0}^{m+n}\omega'_k B_k^{m+n}(t)} \qquad (28)$$

and

$$\omega'_k=\frac{1}{\binom{m+n}{m}}\sum_{i+j=k}\omega_i\binom{k}{i}\binom{m+n-k}{m-i}. \qquad (29)$$

From (27), we have

$$\sum_{k=0}^{n} R_k\omega_k B_k^n(t)+\sum_{k=0}^{m+n}\epsilon_k\omega'_k B_k^{m+n} = \sum_{i=0}^{m} P_i B_i^m(t)\sum_{j=0}^{n}\omega_j B_j^n(t). \qquad (30)$$

Writing both sides of the above equation in Bernstein form and comparing the coefficients, one get

$$\epsilon_k=\sum_{i=\max(k-n,0)}^{\min(k,m)}(P_i-R_{k-i})\alpha_{k,i}, \qquad (31)$$

$k=0,1,\ldots,m+n$, where for $\max(k-n,0)\leqslant i\leqslant \min(k,m)$,

$$\alpha_{k,i}=\frac{\omega_{k-i}\binom{k}{i}\binom{m+n-k}{m-i}}{\sum_{j+l=k}\omega_l\binom{k}{j}\binom{m+n-k}{m-j}}, \qquad (32)$$

otherwise, $\alpha_{k,i}=0$. Now we wish to make the perturbation $\epsilon(t)$ is as small as possible. To this end, we minimize $\sum_{k=0}^{m+n}\epsilon_k^2$. Let

$$L(P_0,\ldots,P_m):=\sum_{k=0}^{m+n}\epsilon_k^2 := \left(\sum_{i=\max(k-n,0)}^{\min(k,m)}(P_i-R_{k-i})\alpha_{k,i}\right)^2 \qquad (33)$$

From

$$\frac{\partial L}{\partial P_l}=2\sum_{k=0}^{m+n}\sum_{i=\max(k-n,0)}^{\min(k,m)}(P_i-R_{k-i})\alpha_{k,i}\alpha_{k,l} =0,\quad l=0,1,\ldots,m, \qquad (34)$$

we obtain a system of linear equations for P_i, $i=0,1,\ldots,m$:

$$\sum_{i=0}^{m}\left(\sum_{k=\max(i,l)}^{\min(i+n,l+n)}\alpha_{k,i}\alpha_{k,l}\right)P_i = \sum_{i=0}^{m}\left(\sum_{k=\max(i,l)}^{\min(i+n,l+n)}\alpha_{k,i}\alpha_{k,l}\right)R_{k-i}, \qquad (35)$$

$l=0,1,\ldots,m$. By solving the above equations and substituting P_i into (31), the values of ϵ_i are obtained.

Now a polynomial upper bound of $R(t)$ can be computed as follows.

Theorem 2 *Let $\epsilon_{max}=\max_{0\leqslant i\leqslant m+n}\epsilon_i$ and $\epsilon_{min}=\min_{0\leqslant i\leqslant m+n}\epsilon_i$. Then polynomial $P(t)-\epsilon_{min}$ is an upper bound of $R(t)$, and*

$$\|P(t)-\epsilon_{min}-R(t)\|_\infty\leqslant\epsilon_{max}-\epsilon_{min}. \qquad (36)$$

Proof: From (27), we have

$$R(t)\leqslant P(t)-\epsilon_{\min},$$

which means $P(t)-\epsilon_{\min}$ is an upper bound of $R(t)$.

To get (36), one need only notice that $P(t)-R(t)=\epsilon(t)$.

3.3. Linear Programming Method

In this approach, we take

$$\|P(t)-R(t)\|_1=\frac{1}{m+1}\sum_{k=0}^{m}P_k-c \qquad (37)$$

as the minimization target, where $c=\int_0^1 R(t)dt$ is a constant.

To satisfy condition (17), one must have

$$\sum_{k=0}^{m}P_k B_k^m(t)\sum_{k=0}^{n}\omega_k B_k^n(t)\geqslant\sum_{k=0}^{n}R_k\omega_k B_k^n(t), \qquad (38)$$

for any $t\in[0,1]$. Or equivalently,

$$\sum_{k=0}^{m+n}\sum_{i+j=k}P_i\omega_j\binom{k}{i}\binom{m+n-k}{m-i}B_k^{m+n}(t) \geqslant\sum_{k=0}^{m+n}\sum_{i+j=k}R_j\omega_j\binom{k}{i}\binom{m+n-k}{m-i}B_k^{m+n}(t).$$

Thus a sufficient condition for (17) to hold is

$$d_k := \sum_{i+j=k} (P_i - R_j)\omega_j \binom{k}{i}\binom{m+n-k}{m-i} \geqslant 0, \quad (39)$$

$k = 0, 1, \ldots, m+n$. An immediate consequence of the above discussion is the following

Theorem 3 *Let P_i, $i = 0, 1, \ldots, m$ be the solutions of the following linear programming problem*

$$\begin{cases} Min & \sum_{i=0}^{m} P_i \\ s.t. & d_k \geqslant 0, \quad k = 0, 1, \ldots, m+n. \end{cases} \quad (40)$$

Then polynomial $P(t)$ defined in (16) is an upper bound of $R(t)$, and

$$||P(t) - R(t)||_\infty \leqslant max_{\,0 \leqslant k \leqslant m+n} \frac{d_k}{\omega_k''}, \quad (41)$$

where

$$\omega_k'' = \sum_{i=max(k-n,0)}^{min(k,m)} \omega_{k-i} \binom{k}{i}\binom{m+n-k}{m-i}, \quad (42)$$

k=0,1, ... ,m+n.

Proof: We need only prove the second part of the theorem. To estimate $||P(t) - R(t)||_\infty$, we write $P(t) - R(t)$ in a rational Bernstein polynomial:

$$P(t) - R(t) = \frac{\sum_{k=0}^{m+n} \frac{1}{\binom{m+n}{m}} d_k B_k^{m+n}(t)}{\sum_{k=0}^{m+n} \omega_k' B_k^{m+n}(t)},$$

where w_k' is defined as in (29). Now (41) follows directly from the above equation.

Remarks: The Hybrid Method automatically solves the *constrained upper bound problem*, i.e., in addition to satisfying (15—18), the following constraints must be held:

$$\begin{cases} P^{(i)}(0) = R^{(i)}(0) \\ P^{(i)}(1) = P^{(i)}(1) \end{cases} \quad i = 0, 1, \ldots, \mu, \quad (43)$$

where $\mu \leqslant [\frac{m-1}{2}]$. The Perturbation Method and the Linear Programming Method can also be adapted to solve the constrained upper bound problem. For the Perturbation Method, just set $\epsilon_k = 0$, $k = 0, 1, \ldots, \mu, m-\mu, \ldots, m$, and solve for the corresponding P_k from (31). The remaining P_k can be obtained from (35) with $l = \mu + 1, \ldots, m-\mu-1$. For the Linear Programming Method, P_k, $k = 0, 1, \ldots, \mu, m-\mu, \ldots, m$ can be solved by setting the corresponding $d_k = 0$, and the remaining P_k are obtained by solving the linear programming problem (40).

3.4. Bounding Interval Rational Bézier Curves with Interval Polynomial Bézier Curves

In the previous subsections, we derived three different methods to bound an interval rational function with an interval polynomial. These three methods lead directly to algorithms of bounding an interval rational Bézier curve with an interval polynomial Bézier curve. The main results are based on the following fact.

Theorem 4 *Given an interval rational Bézier curve of degree n*

$$[\mathbf{R}](t) = ([x](t), [y](t)). \quad (44)$$

If

$$[\bar{x}](t) = \sum_{i=0}^{m} [\bar{x}_i] B_i^m(t) \quad (45)$$

and

$$[\bar{y}](t) = \sum_{i=0}^{m} [\bar{y}_i] B_i^m(t) \quad (46)$$

are degree m interval polynomial bounds of $[x](t)$ and $[y](t)$ respectively, then interval polynomial Bézier curve

$$[\mathbf{P}](t) = ([\bar{x}](t), [\bar{y}](t)) = \sum_{i=0}^{m} ([\bar{x}_i], [\bar{y}_i]) B_i^m(t) \quad (47)$$

bounds interval rational Bézier curve $[\mathbf{R}](t)$, i.e., $[\mathbf{R}](t) \subset [\mathbf{P}](t)$.

Proof: Straightforward.

To measure how tight a bound is, we define the *bounding error* as follows

$$\begin{aligned} e([x](t), [\bar{x}](t)) &= \max(||ub([x](t)) - ub([\bar{x}](t))||_\infty, \\ &\qquad ||lb([x](t)) - lb([\bar{x}](t))||_\infty) \\ e([y](t), [\bar{y}](t)) &= \max(||ub([y](t)) - ub([\bar{y}](t))||_\infty, \\ &\qquad ||lb([y](t)) - lb([\bar{y}](t))||_\infty) \\ e([\mathbf{R}](t), [\mathbf{P}](t)) &= \max(e([x](t), [\bar{x}](t)), \\ &\qquad e([y](t), [\bar{y}](t))) \end{aligned} \quad (48)$$

It is a little hard to compute the bounding errors by the above definitions. Fortunately, Theorems 1, 2 and 3 give a good estimation for the bounding error by each of the three methods.

For a given interval rational Bézier curve $[\mathbf{R}](t)$ and tolerance $\epsilon > 0$, we use one of the three methods to find an interval polynomial Bézier curve $[\mathbf{P}](t)$ to bound $[\mathbf{R}](t)$. If the bounding error is larger than ϵ, we subdivide $[\mathbf{R}](t)$ at parameter value $t = 1/2$ and then bound each segment with an interval polynomial Bézier curve respectively. This process is continued until the bounding error is less than ϵ for each segment. At last, we find a piecewise interval polynomial Bézier curve which bound the original interval rational Bézier curve $[\mathbf{R}](t)$.

4. Examples and Comparison

In this section, we will compare the bounding errors and the computational costs between the three methods–Hybrid Method (HM), Perturbation Method (PM) and Linear Programming Method (LPM) for bounding an interval rational Bézier curve with an interval polynomial Bézier curve through illustration of examples. We implemented these examples on a 266 MHZ Pentium PC.

Example 1 Let $[r](t)$ be a cubic interval rational function with the following interval coefficients and weights:

$$\begin{array}{ll}
[a_0, b_0] = [0, 1/4], & \omega_0 = 1, \\
[a_1, b_1] = [1, 6/5], & \omega_1 = 2, \\
[a_2, b_2] = [-1, -3/4], & \omega_2 = 2, \\
[a_3, b_3] = [1, 7/6], & \omega_3 = 1
\end{array}$$

We use degree $m = 4, 5$ and 6 interval polynomials to bound $[r](t)$ respectively. The bounding errors (B.E.) and the computational time (C.T.) in CUP seconds for the three methods are list in Table.1. Figure.2–Figure.10 show the corresponding figures.

Table 1. Comparison of the bounding errors and the computational costs (I)

m	HM		PM		LPM	
	B.E.	C.T.	B.E.	C.T.	B.E.	C.T.
$m=4$	3.00	0.02	0.53	0.11	1.02	0.17
$m=5$	3.68	0.02	0.33	0.11	0.13	0.22
$m=6$	2.25	0.02	0.22	0.16	0.10	0.22

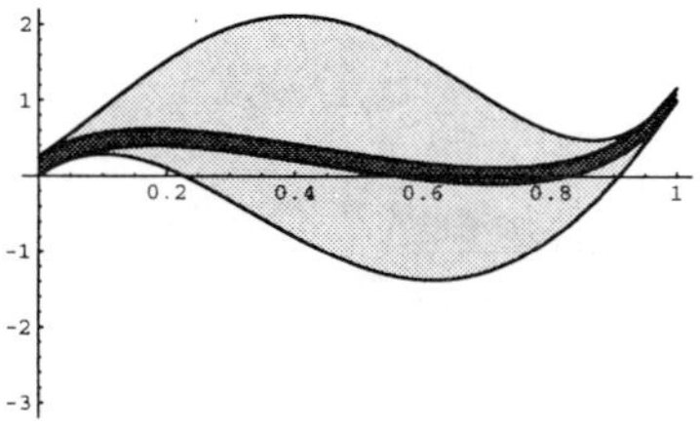

Figure 2. Hybrid Method $m = 4$

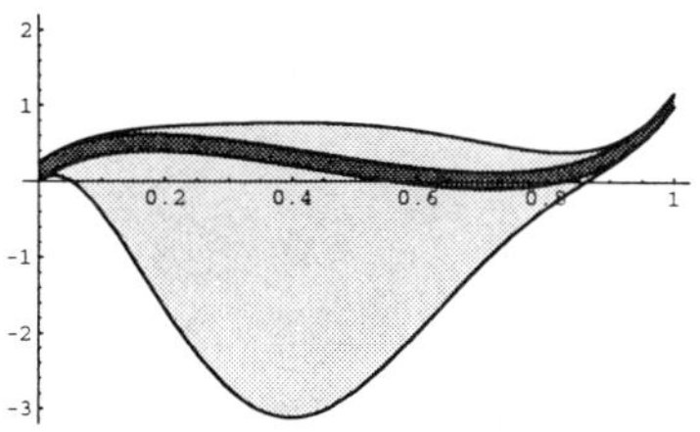

Figure 3. Hybrid Method $m = 5$

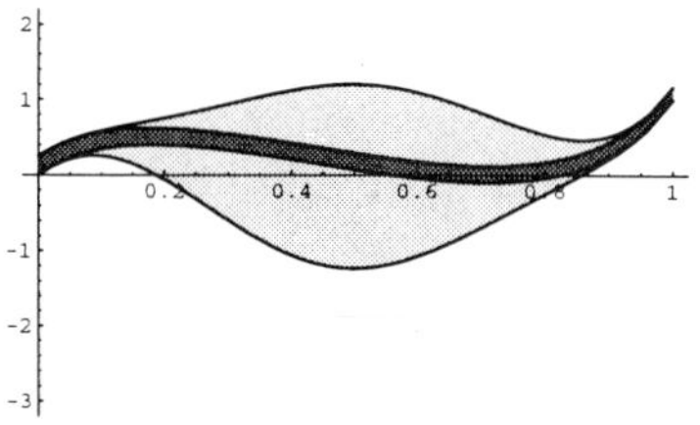

Figure 4. Hybrid Method $m = 6$

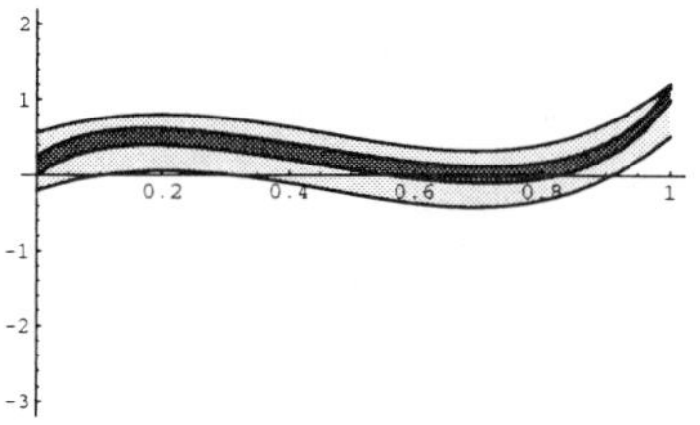

Figure 5. Perturbation Method $m = 4$

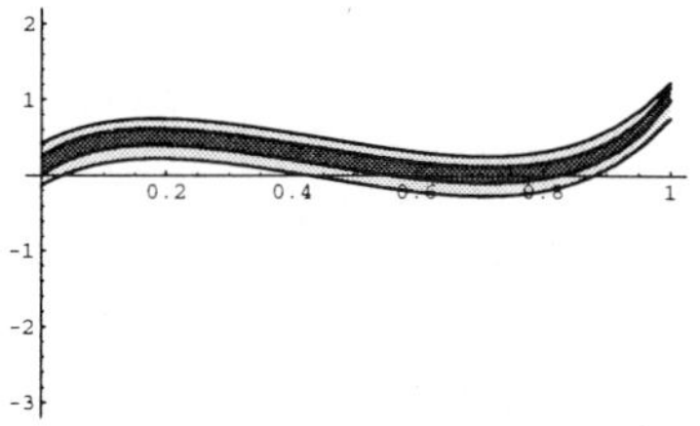

Figure 6. Perturbation Method $m = 5$

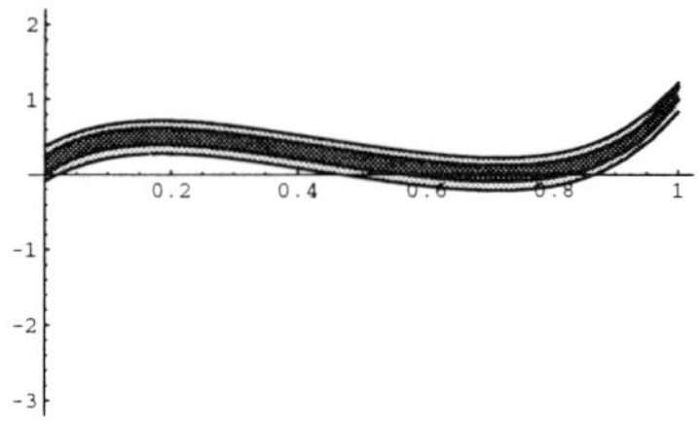

Figure 7. Perturbation Method $m = 6$

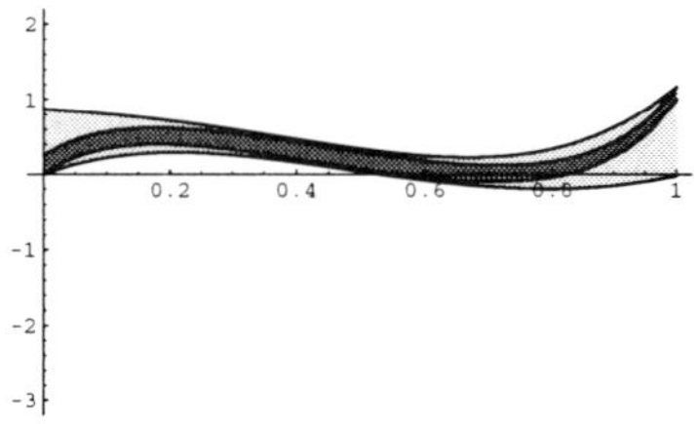

Figure 8. Linear Programming Method $m = 4$

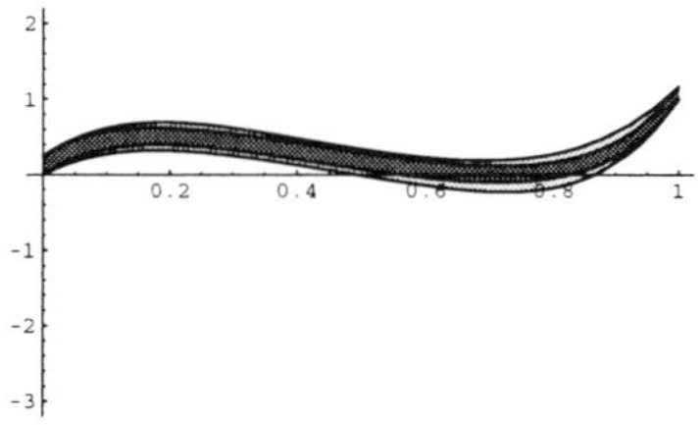

Figure 9. Linear Programming Method $m = 5$

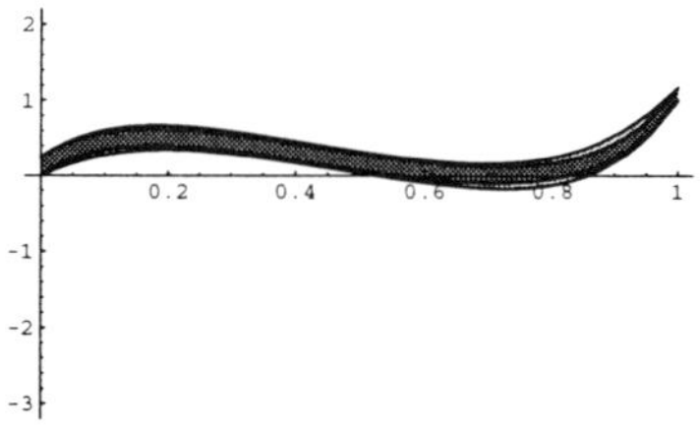

Figure 10. Linear Programming Method $m = 6$

Example 2 In this example, we consider a case where the Hybrid Method diverges while the other two methods converges. The interval rational function $[r](t)$ is defined by

$$\begin{array}{ll} [a_0, b_0] = [-1/8, 1/8], & \omega_0 = 1, \\ [a_1, b_1] = [7/8, 1], & \omega_1 = 3, \\ [a_2, b_2] = [7/4, 2], & \omega_2 = 8, \\ [a_3, b_3] = [7/8, 1], & \omega_3 = 2, \\ [a_4, b_4] = [-1/8, 1/8], & \omega_4 = 1 \end{array}$$

The degree 4, 5 and 6 interval polynomial bounds by the three methods are shown in Figure 11-Figure 19. The corresponding bounding errors and computational time are list in Table 2.

Table 2. Comparison of the bounding errors and the computational costs (II)

m	HM		PM		LM	
	errors	time	errors	time	errors	time
$m = 4$	4.32	0.02	0.68	0.10	1.18	0.24
$m = 5$	12.99	0.02	0.54	0.14	0.97	0.24
$m = 6$	5.54	0.02	0.46	0.16	0.22	0.28

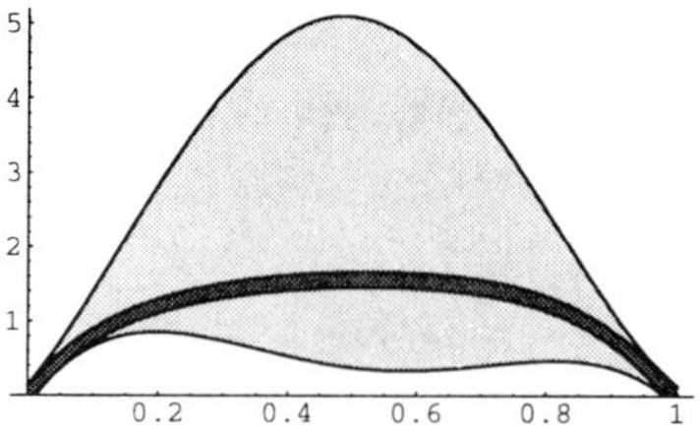

Figure 11. Hybrid Method $m = 4$

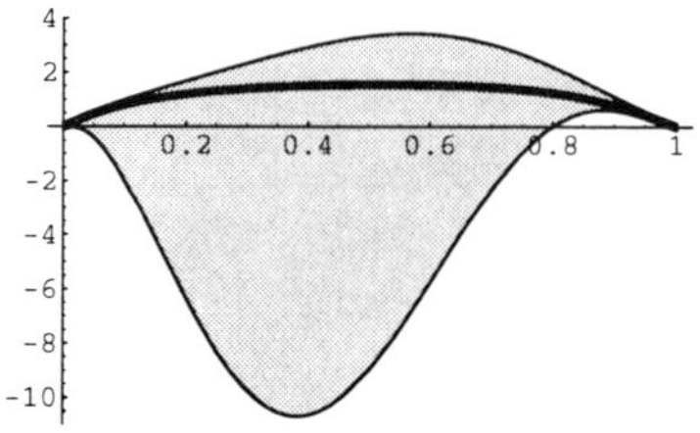

Figure 12. Hybrid Method $m = 5$

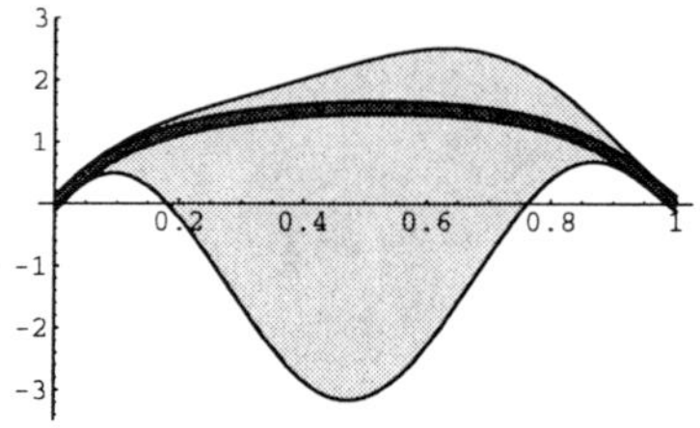

Figure 13. Hybrid Method $m = 6$

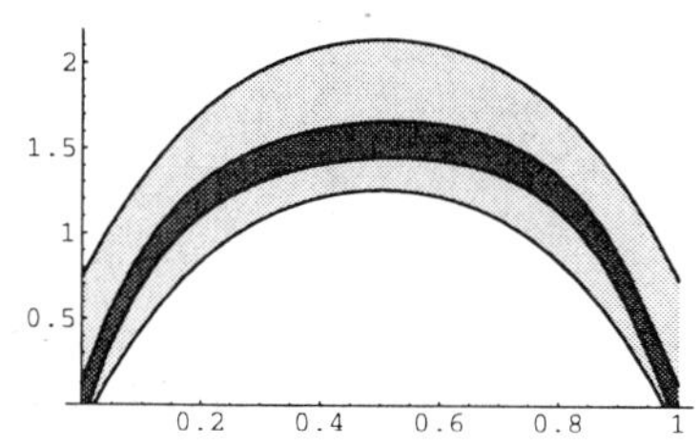

Figure 14. Perturbation Method $m = 4$

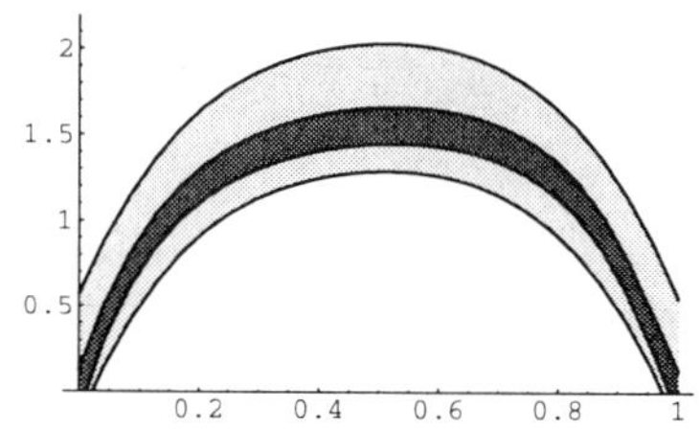

Figure 15. Perturbation Method $m = 5$

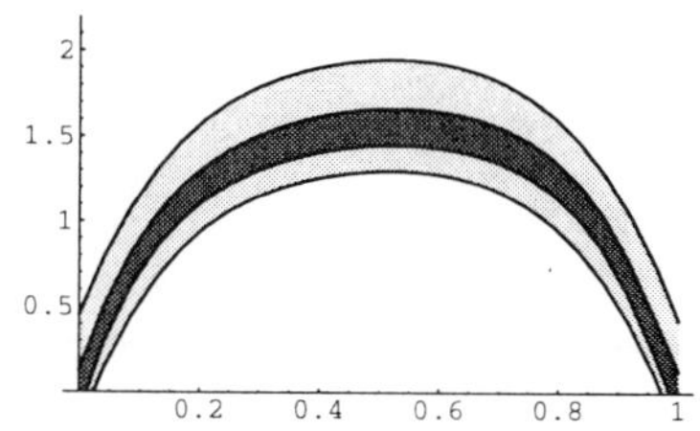

Figure 16. Perturbation Method $m = 6$

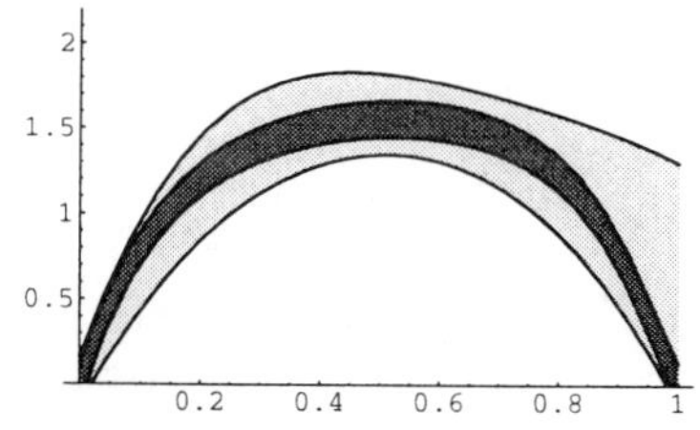

Figure 17. Linear Programming Method $m = 4$

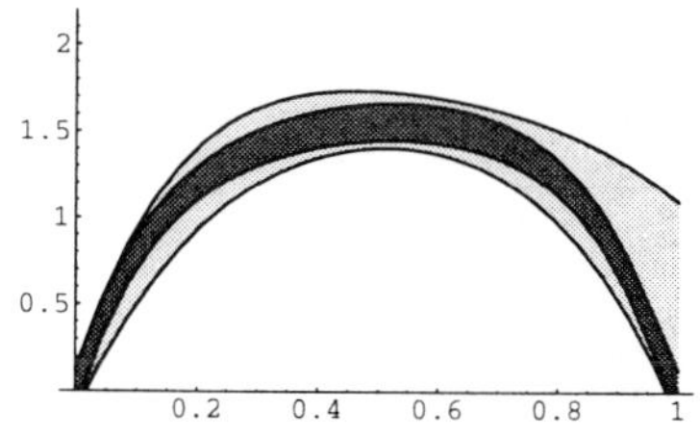

Figure 18. Linear Programming Method $m = 5$

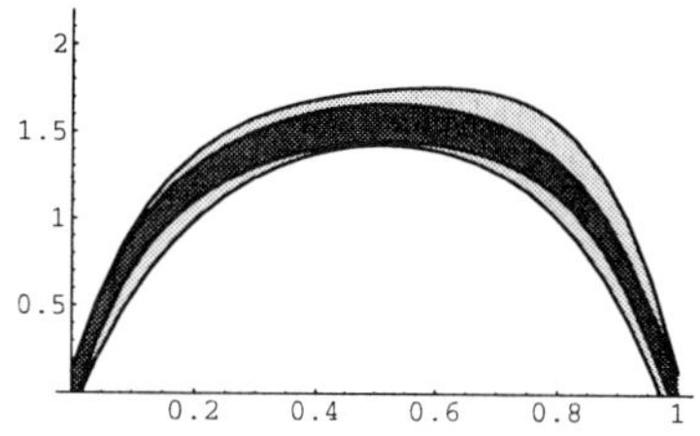

Figure 19. Linear Programming Method $m = 6$

Example 3 In the last example, we use piecewise interval polynomial curves to bound an interval rational curve. Let the control points and weights of the interval rational curve be:

$$\begin{array}{ll} [\mathbf{R}_0] = ([50,70],[340,355]), & \omega_0 = 1, \\ [\mathbf{R}_1] = ([190,220],[100,120]), & \omega_1 = 2, \\ [\mathbf{R}_2] = ([340,360],[330,340]), & \omega_2 = 2, \\ [\mathbf{R}_3] = ([430,455],[120,130]), & \omega_3 = 1 \end{array}$$

We recursively subdivide the interval rational curve at the parameter value $t = \frac{1}{2}$ and then bound each segment using an interval polynomial curve of degree $m = 4$ and 5 respectively. Table 3 and 4 list the corresponding bounding errors and the computational time for the three methods.

Table 3. Comparison of the bounding errors and computational costs (III) ($m = 4$)

number of	H M		P M		L M	
subdivisions	errors	time	errors	time	errors	time
0	423.75	0.03	74.68	0.22	123.3	0.44
1	31.82	0.05	11.06	0.33	6.89	0.71
2	3.24	0.11	1.28	0.71	0.73	1.38
3	0.23	0.27	0.093	1.32	0.051	2.69
4	0.012	0.44	0.0051	2.69	0.0026	5.44
5	0.0012	0.88	0.00049	5.38	0.00026	10.93

From the above examples, we can draw a conclusion that the bounding errors by the Perturbation Method and the Linear Programming Method are much smaller than those by the Hybrid Method, even though they are computationally several times more expensive. Consequently, if we use a piecewise interval polynomial curve to bound an interval rational curve such that the bounding error is less than some given tolerance, the Hybrid Method will generally need more segments than the other two methods. Furthermore, examples show that the Hybrid Method has a much more restrict convergence condition than the other two methods do.

Table 4. Comparison of the bounding errors and computational costs (III) ($m = 5$)

number of	H M		P M		L M	
subdivisions	errors	time	errors	time	errors	time
0	386.73	0.03	36.77	0.22	15.83	0.38
1	29.74	0.05	5.41	0.44	3.19	0.77
2	1.53	0.11	0.40	0.88	0.25	1.59
3	0.055	0.27	0.017	1.76	0.011	3.13
4	0.0018	0.49	0.00059	3.46	0.00042	6.32
5	.00012	1.05	.000041	6.97	.000029	12.58

5. Conclusions

In this paper, we put forward a new problem in CAD and interval analysis communities–bounding an interval rational function (or Bézier curve) with an interval polynomial function (or Bézier curve). We proposed three different methods to solve the problem, and comparisons of bounding errors and computational costs are made between these three methods. The experimental results show that the Perturbation Method and the Linear Programming Method produce much tighter bounds than the Hybrid Method does, although they are computationally several times more expensive.

Further research problems include: (1) for the Perturbation Method and the Linear Programming Method, explore whether the interval polynomial bounds converge to the original interval rational curve when m goes to infinity; and (2) generalizing the results in this paper to surface cases.

Acknowledgement

The first and third authors are supported by the National Natural Science Foundation of China (grants 69603009 and 19971087) and the 973 Project on Mathematical Mechanics (grant G1998030600). The authors are sincerely thankful for the financial support of K.C. Wong Foundation.

References

[1] R.E. Moore, *Interval Analysis*, Prentice Hall, Englewood Cliffs, NJ, 1966.

[2] T.W. Sederberg, and R.T. Farouki, "Approximation by interval Bézier curves", *IEEE Comput. Graph. Appl.*, 1992, 15(2), 87-95.

[3] C.Y. Hu, T. Maekawa, E.C. Sherbrooke, and N.M. Patrikalakis, "Robust interval algorithm for curve intersections", *Computer Aided Design*, 1996, 28(6/7), 495-506.

[4] C.Y. Hu, N.M. Patrikalakis, and X. Ye, "Robust interval solid modeling: Part I: representations", *Computer Aided Design*, 1996, 28(10), 807-817.

[5] C.Y. Hu, N.M. Patrikalakis, and X. Ye, "Robust interval solid modeling: Part II: boundary evaluation", *Computer Aided Design*, 1996, 28(10), 819-830.

[6] C.Y. Hu, T. Maekawa, N.M. Patrikalakis, and X. Ye, "Robust interval algorithm for surface intersections", *Computer Aided Design*, 1997, 29(9), 617-627.

[7] S.T. Tuohy, T. Maekawa, G. Shen, and N.M. Patrikalakis, "Approximation of measured data with interval B-splines", *Computer Aided Design*, 1997, 29(11), 791-799.

[8] F. Chen, and W. Lou, "Degree Reduction of Interval Bézier curves", accepted by *Computer Aided Design*.

[9] T.W. Sederberg, and M. Kakimoto, "Approximating rational curves using polynomial curves", *NURBS for Curve and Surface Design*, G. Farin, eds., SIAM, Philadelphia, 1991, 149-158.

[10] G. Wang, T.W. Sederberg, and F. Chen, "On the convergence of polynomial approximation of rational functions", *Journal of Approximation Theory*, 1997, 89, 267-288.

Fast Conversion of Dynamic B-spline Curves into a Set of Power Form Polynomial Curves

Deok-Soo Kim*, Joonghyun Ryu*, Hyunchan Lee**, Hayong Shin***,
Joonyoung Park****, Taeboom Jang*
** Department of Industrial Engineering, Hanyang University*
dskim@email.hanyang.ac.kr
*** Department of Industrial Engineering, Hongik University*
hclee@wow.hongik.ac.kr
****Technical Computing Center, Chrysler Corporation*
HS30@daimlerchrysler.com
***** Department of Industrial Engineering, Dongguk University*
jypark@cakra.dongguk.ac.kr

Abstract

Computation of the characteristic points such as inflection points or cusp on a curve is often necessary in CAGD applications. When a curve is represented in a B-spline form, such computations can be made easier once it is transformed in a set of polynomial curves in a power form. Once a curve is represented in a power form, a point evaluation can be also made faster due to Horner's rule even though some issues of stability remains. In addition, the implicitization process of a parametric curve using a resultant usually requires the geometry represented in a power form.

Usual practice of the transformation of a B-spline curve into a set of piecewise polynomial curves in a power form is done by either a knot refinement followed by basis conversions, or applying a taylor expansion on the B-spline curve for each knot span.

Presented in this paper is a new algorithm, called Direct Expansion, for the problem. The algorithm first locates the coefficients of all the linear terms that make up the basis functions in a knot span, and then the algorithm directly obtains the power form representation of basis functions by expanding the summation of products of appropriate linear terms. Then, a polynomial segment of a knot span can be easily obtained by the summation of products of the basis functions within the knot span with corresponding control points. Repeating this operation for each knot span, all of the polynomials of the B-spline curve can be transformed into a power form.

Experiment shows that the proposed algorithm outperforms the conventional approach when the degree of a given B-spline curve is relatively low. However, when one or more control points of a B-spline curve is continuously moving, the computational gain for the same problem turns out to be much more significant. Therefore, the presented algorithm can be very useful for the curve and surface implicitization as well as intersections when the curves or surfaces are dynamically changing.

1. Introduction

In the computer graphics and geometric modeling environment, it is often necessary to manipulate B-spline curves or surfaces by converting the B-spline representation into a set of piecewise polynomial curves or surfaces in a power form. It is known that the faster computation of the characteristic points on a curve, such as inflection points and cusps, can be facilitated by the conversion of a B-spline curve into a set of piecewise polynomial curves in a power form. Note that the subdivision of a parametric curve at these characteristic points facilitates the fast computation of intersection points between curves [1]. Once a curve is represented in a power form, a point evaluation can be also made faster due to Horner's rule even though some issues of stability remains [2]. In addition, this form of curve is supported by IGES as an entity type 112 [3].

Due to the relative advantages of implicit representation of curves or surfaces over parametric one in some geometric calculations such as a point inclusion problem, a parametric form needs to be converted to an implicit form [4]. The implicitization process, which employs a resultant, usually requires the geometry to be represented in a power form [5]. Since this operation is

0-7695-0562-7/00 $10.00 © 2000 IEEE

often computationally demanding, the reduction of computation should not be ignored.

Discussed in this paper is the transformation of a B-spline curve into a set of piecewise polynomials in a power form, which is known to be a tedious task [6]. Two types of curves are considered for the defined problem: a static curve and a dynamic curve. In this paper, a curve is called dynamic when one or more of the control points of the curve are changing. Otherwise a curve is called static.

There are a few existing approaches to this problem on static cases. Since a static B-spline curve can be converted into a set of piecewise Bezier curves by a knot refinement [7][8][9][10], applying a basis conversion operation to each piece will produce a set of polynomials in a power form. The same problem can be also solved by applying Taylor expansion of the B-spline at the beginning of each knot interval with appropriate number of terms [11]. Another approach is to represent NURBS curves in a matrix form [12] [13]. It is interesting to note that there are only a few, however, not many, reports dealing with this problem. We think that the problem might have been thought as either simple or not as important as other issues in CAGD or computer graphics, or both. In addition, the authors are not aware of any discussions for a dynamic case.

The main idea of the proposed algorithm, called DE (Direct Expansion) algorithm, is as follows: After locating the coefficients of all the linear terms that make up the basis functions in a knot span, the algorithm directly obtains the power form representation of basis functions by expanding the summation of products of appropriate linear terms. Then, the piecewise polynomial curves in a power form can be easily obtained by multiplying the basis functions with corresponding control points [14]. The proposed algorithm outperforms the existing approaches for both static and dynamic curves. Especially, the computational gain turns out to be significant for dynamic curves.

This paper is organized as follows: Section 2 provides the basic introduction to basis functions of B-spline curve and introduces the key concepts to develop the algorithm. In Section 3, an algorithm to find the paths to multiply the linear terms is provided. In Section 4, the concept for the calculation of the power from polynomial curve segment of a given static B-spline curve is discussed. Then, the discussions on the experiment for a static curve is provided. In Section 6, the discussions and experiments for the dynamic curve counterpart are presented. And the paper concludes.

2. Representation of basis functions

A B-spline curve of degree p with (m + 1) knots is defined by

$$\mathbf{C}(t) = \sum_{i=0}^{m-p-1} \mathbf{P}_i N_{i,p}(t) \text{ for } 0 \le t \le 1 \quad (1)$$

where $\mathbf{P}_i$ and $N_{i,p}(t)$ are control points and B-spline basis functions of degree p on a knot vector $U = \{0,...,0,t_{p+1},...,t_{m-p-1},1,...1\}$, respectively[15]. In Equation (1), $N_{i,p}(t)$ is defined as the following recurrence formula.

$$N_{i,p}(t) = \frac{t - t_i}{t_{i+p} - t_i} N_{i,p-1}(t) + \frac{t_{i+p+1} - t}{t_{i+p+1} - t_{i+1}} N_{i+1,p-1}(t) \quad (2)$$

$$N_{i,0}(t) = \begin{cases} 1 & if \quad t_i \le t < t_{i+1} \\ 0 & otherwise \end{cases}$$

where $i = 0, 1, ..., m-p-1$. Equation (2) shows that the B-spline basis function of degree p consists of a linear combination of two B-spline basis functions of degree $p-1$, where the coefficients are linear functions in t. Illustrated in Figure 1 is the graph representation of calculation for B-spline basis functions.

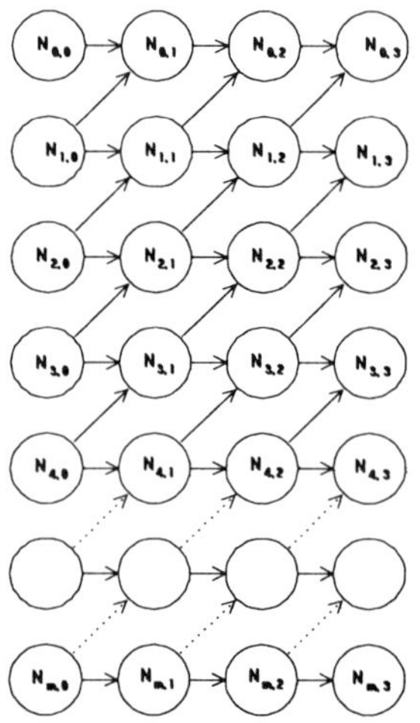

Figure 1. A graph representation of calculation for B-spline basis functions

Equation (2) can be now rewritten as

$$N_{i,p}(t) = e_h(t;i,p) N_{i,p-1}(t) + e_v(t;i,p) N_{i+1,p-1}(t) \quad (3)$$

where

$$e_h(t;i,p) = \frac{t - t_i}{t_{i+p} - t_i}, \quad e_v(t;i,p) = \frac{t_{i+p+1} - t}{t_{i+p+1} - t_{i+1}} \quad (4)$$

The subscripts h and v are short for *horizontal* and *vertical* and the meaning will be discussed later.

Definition 1. A *truncated basis polynomial, $h_m(t)$, $m=0,1,..p$,* is an active polynomial segment of $N_{j,p}$, $i-p \le j \le i$ in $[t_i, t_{i+1})$.

Note that there are $p+1$ truncated basis polynomials in each knot span of nonzero length for B-spline curve of degree p. The proposed algorithm collects all the $e_h(t;i,p)$'s and $e_v(t;i,p)$'s in a knot span, and efficiently performs the necessary multiplications between appropriate $e_h(t;i,p)$'s and $e_v(t;i,p)$'s so that the result to be the desired truncated basis polynomials arranged in a power form in the knot span. Then, the B-spline curve in the knot span can be easily transformed into a polynomial curve in a power form by the summation of the multiplications between appropriate control points and truncated basis functions. Therefore, a B-spline curve can be transformed by repeating the above-mentioned operation for each knot span of nonzero length.

Therefore, the fundamental question is how to efficiently extract all truncated basis polynomials. The approach taken is exploiting the repeated form of the recurrence formula given by Equation (2). Let us focus our interest in one knot span for the time being. Illustrated in Figure 2 is a subgraph of Figure 1, which shows the necessary computations for four truncated basis polynomials in a knot span $[t_3,t_4)$. Arrows in the figure should be interpreted as appropriate linear polynomials, which are defined by Equation (3) and eventually comprise four truncated basis polynomials in $[t_3,t_4)$. The root and leaf nodes correspond to a knot span and the truncated basis polynomials, respectively.

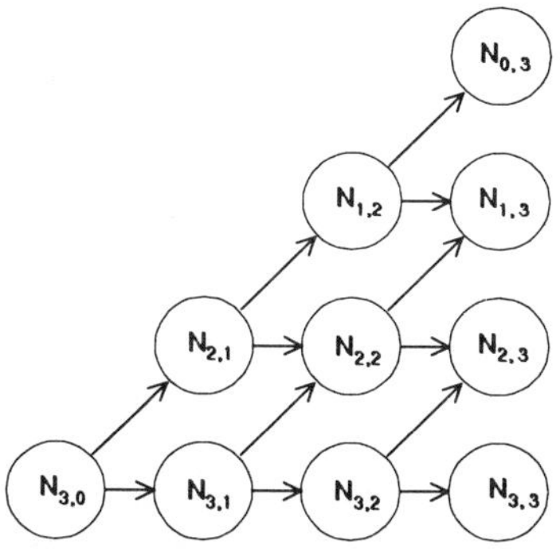

Figure 2. Triangular scheme to get the truncated basis polynomials in $[t_3,t_4)$

Definition 2. A *transition edge* is the edge of the directed graph of Figure 1. A transition edge starting from $N_{i,j}$ and ending at $N_{k,l}$ is called a *horizontal transition edge* and has an *edge value* of $e_h(t;k,l)$ if $k = i$ and $l = j + 1$. Similarly, a *vertical transition edge* has an *edge value* of $e_v(t;k,l)$ if $k = i - 1$ and $l = j + 1$.

Note that once a transition edge is determined to be horizontal or vertical, the corresponding edge value can be directly obtained from the subscripts of ending vertex. For example, an edge from $N_{3,2}$ to $N_{2,3}$ is vertical and therefore the edge value is $e_v(t;2,3) = \frac{(t_{2+3+1}-t)}{(t_{2+3+1}-t_{2+1})} = \frac{(t_6-t)}{(t_6-t_3)}$. Similarly an edge from $N_{3,2}$ to $N_{3,3}$ is horizontal and has an edge value of $e_h(t;3,3) = \frac{(t-t_3)}{(t_{3+3}-t_3)} = \frac{(t-t_3)}{(t_6-t_3)}$.

Definition 3. A *path* π from $N_{i,j}$ to $N_{k,l}$ is the set of transition edges consisting of a path from $N_{i,j}$ to $N_{k,l}$, and a path set $\Pi = \{\pi_a, a = 1,2,..,n\}$ where n is the number of all these possible paths.

Definition 4. Given $N_{i,p}$, $N_{j,0}$, and Π, a *graph primitive* is $\bigcup_i \pi_i$.

The union operation in the previous definition implies the superposition of all paths so that the result yields a topologically merged graph. It turns out that a truncated basis polynomial corresponds to a graph primitive. In $[t_3,t_4)$, for example, there exist four truncated basis polynomials. Each truncated basis polynomial, $h_m(t)$, m=0,1,2,3, corresponds to the summation of the product of all the linear polynomials in paths from $N_{3,0}$ to $N_{i,3}$ (i=0,1,2,3). For example, there exist three possible paths from $N_{3,0}$ to $N_{1,3}$: $N_{3,0} \rightarrow N_{2,1} \rightarrow N_{1,2} \rightarrow N_{1,3}$, $N_{3,0} \rightarrow N_{2,1} \rightarrow N_{2,2} \rightarrow N_{1,3}$ and $N_{3,0} \rightarrow N_{3,1} \rightarrow N_{2,2} \rightarrow N_{1,3}$. Thus, truncated basis polynomial $h_1(t)$ can be obtained as follows and can be easily verified in Figure 3. Thus, the following theorem holds.

$$\begin{aligned} h_1(t) = {} & e_v(t;2,1)\, e_v(t;1,2)\, e_h(t;1,3) \\ & + e_v(t;2,1)\, e_h(t;2,2)\, e_v(t;1,3) \\ & + e_h(t;3,1)\, e_v(t;2,2)\, e_v(t;1,3) \end{aligned} \quad (5)$$

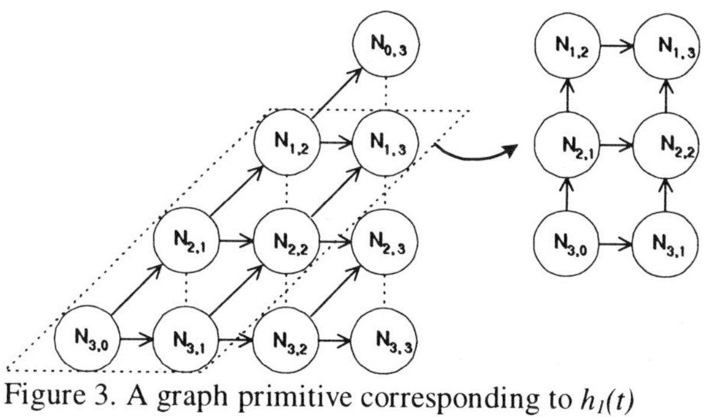

Figure 3. A graph primitive corresponding to $h_1(t)$

Theorem 1. $h_m(t)$,$m=0,\ldots,p$, can be represented as

$$h_m(t) = \sum_{k=1}^{n_p} P_k(t)$$

$$P_k(t) = \prod_{l_1=1}^{n_h} e_h^{l_1}(t; p_h, q_h) \prod_{l_2=1}^{n_v} e_v^{l_2}(t; p_v, q_v) \qquad (6)$$

where n_p is the number of all the possible paths from $N_{i,0}$ to $N_{j,p}$. n_h and n_v are the number of horizontal and vertical transition edges in each path. $n_h + n_v = p$. p_h, p_v, q_h, and q_v represents appropriate integers.

Note that $n_p = \dfrac{p!}{(i-j)!(p-i+j)!}$, and the number of graph primitives in a knot span for a curve of degree p is $(p + 1)$. In Figure 4, four graph primitives for Figure 3, a case for cubic curve, are presented. Each graph primitive corresponds to one of four truncated basis polynomials in $[t_3, t_4)$.

According to the theorem, if all the possible paths in each graph primitive of a knot span can be enumerated, then each truncated basis polynomial of the knot span can be obtained by adding the products of the linear polynomials for each path. Furthermore, once the degree of curve is fixed, these paths are identical for all the knot spans of non-zero length. Therefore, the path enumeration step of the algorithm is needed only once. The only remaining question is how all the possible paths can be enumerated.

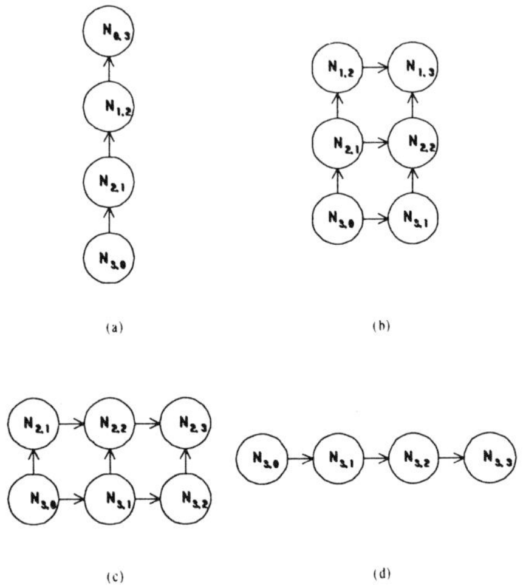

Figure 4. Four graph primitives for Figure 3

3. Path enumeration

The problem that enumerates all the possible paths in each graph primitive can be formulated as the enumeration of all the possible sequences of size p that consist of 0's and 1's. The numbers of 0's and 1's are the numbers of grids of graph primitive horizontally and vertically, respectively. Simple algorithm for this task is available [14].

Enumeration of all paths in one knot span is the union of paths in all graph primitives in the knot span, which is an enumeration of all possible 0's and 1's. For example, all paths for a cubic B-spline curve in a knot span are {1, 1, 1}, {1, 1, 0}, {1, 0, 1}, {0, 1, 1}, {1, 0, 0}, {0, 1, 0}, {0, 0, 1}, and {0, 0, 0}. The number of all paths in a knot span for a curve of degree p is

$$\sum_{j=i-p}^{i} \frac{p!}{(i-j)!(p-i+j)!} = \sum_{k=0}^{p} \frac{p!}{k!(p-k)!} = \sum_{k=0}^{p} \binom{p}{k} = 2^p$$

Therefore, a cubic case has $2^3 = 8$ paths in a knot span. Note that the path enumeration is needed only once for a given B-spline curve regardless knot vector size. Once the paths for a knot span are determined, simple change of subscripts will produce correct paths for an arbitrary knot span. This property holds even for the dynamic curves.

4. Direct expansion of a static curve

Once all of the truncated basis polynomials are obtained, the B-spline curve can be easily transformed into a set of piecewise polynomial curves by multiplying truncated basis polynomials with corresponding control points for each knot span. Note that there are a number of nodes in common between consecutive triangular schemes of neighboring knot spans.

5. Experiments for static curves

Experiments are performed to compare the performance of the proposed algorithm (DE : Direct Expansion) with KR (Knot Refinement) and TE (Taylor Expansion) approaches. Illustrated in Figure 5 is the computation vs. degrees of curves with knot vector size of 50. This figure shows that DE and TE outperform KR when the degree is lower.

However, it should be noted that DE deteriorates quite rapidly compared to TE. It is observed that DE shows a quadratic-like increase whereas KR and TE show only a linear-like increase w.r.t. the degrees of curves. This is due to the fact that there are 2^p number of paths for a B-spline curve in a knot span, when the degree of the curve is p. Thus, the computation time of truncated basis polynomials increases for each nonzero length knot span doubly.

Figure 6 shows that DE is superior to the others for degrees of three or four regardless knot vector size. X-axis is the size of knot vector and the Y-axis is the computation time.

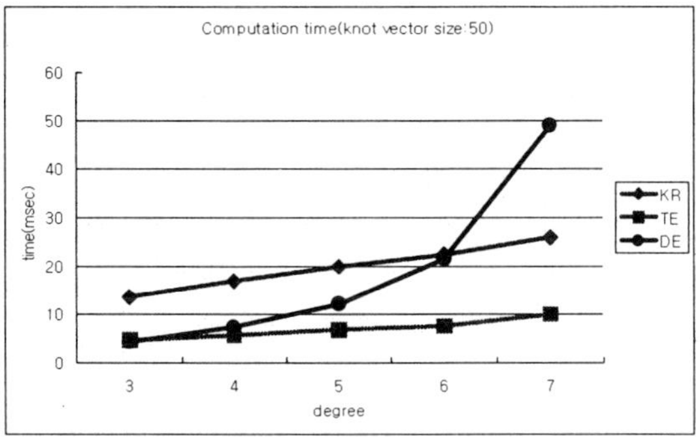

Figure 5. Time vs. degrees (knot vector size : 50)

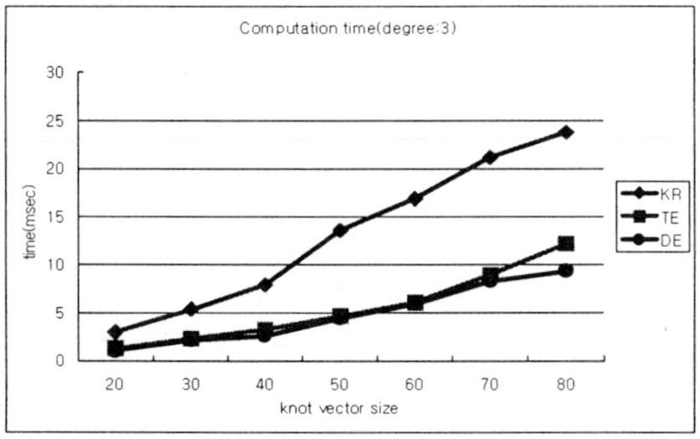

a. degree: 3

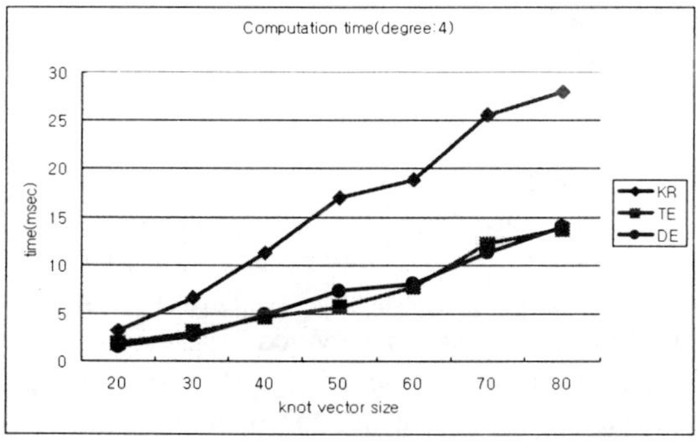

b. degree: 4

Figure 6. Time vs. different knot vector sizes

6. Direct expansion of a dynamic curve

The computational gain of DE for a dynamic curve, of which one or more of the control points vary, seems more significant. Suppose, as an example, one control point of a given cubic B-spline curve has been changed. Then, the curve shape will be changed over four contiguous knot spans.

Since $(p + 1)$ truncated basis polynomials for each knot span are fixed, the new polynomial in a power form can be obtained by i) multiplying the difference vector, starting at the old control point and ending at the new control point, with corresponding truncated basis polynomial, and ii) adding the previous result, which is a polynomial in a power form, to the old polynomial in the knot span. Repeating this operation for $(p + 1)$ knot spans will produce $(p + 1)$ pieces of new polynomials in a power form. Note that the process is both very simple to program and very efficient in its speed. There are $(p + 1)$ knot spans for the calculation and each knot span takes only $(p + 1)$ multiplications. Therefore, $O(p^2)$ operations are only needed to get all the new polynomial curves.

If KR is used, all of the four curve segments, which will change their shape, have to be recalculated. In other words, knot refinement for the four knot spans and basis conversions for the four coefficients for each knot span. Needless to say, this approach will take much more time than DE.

Similar observation can be made for TE. For each of $(p + 1)$ number of knot spans, new polynomial curves should be computed completely again. For a knot span, there are $(p + 1)$ number of coefficients to compute and each coefficient needs the calculation of factorial function as well as derivative information. Usually, a derivative for a B-spline curve is calculated by evaluating the hodograph counterpart at an appropriate parameter value, which is again an operation for $O(p)$ if the evaluation is done based on Horner's rule. Otherwise, it is also $O(p^2)$ operation. Therefore, TE approach is either $O(p^3)$ or $O(p^4)$. However, to have $O(p^3)$ time behaviour, the hodograph itself should be a polynomial in a power form, again.

The following figures illustrate another perspective of the experiments. Each figure represents for different degrees of curves: 3, 4, 5, and 6. The x-axis is the number of control points with changed coordinate values. Y-axis is the computation time taken by the re-calculation of new polynomial segments due to the changed control points. For each algorithm, we believe that the necessary minimum arithmetic is only performed, even though we do not ignore the possible variations due to the implementation details. What is shown in the figure is surprising. The time taken by the proposed DE approach is ignorable compared to the conventional KR approach. For a fixed degree, the computational gain of DE approach gets more significant as the number of changed control points get higher. As the degree of the curve gets higher, the gain gets higher, as was pointed earlier. (Note that the scales of y-axis are same except

Figure 7.d. We had different scale to make the relative behaviour of different algorithms clear.)

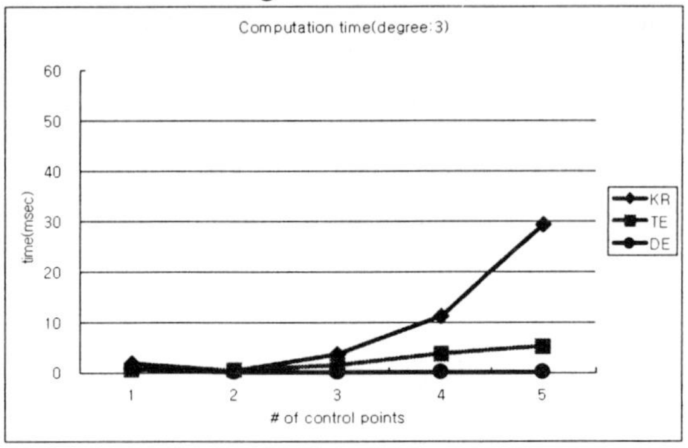

a. degree: 3

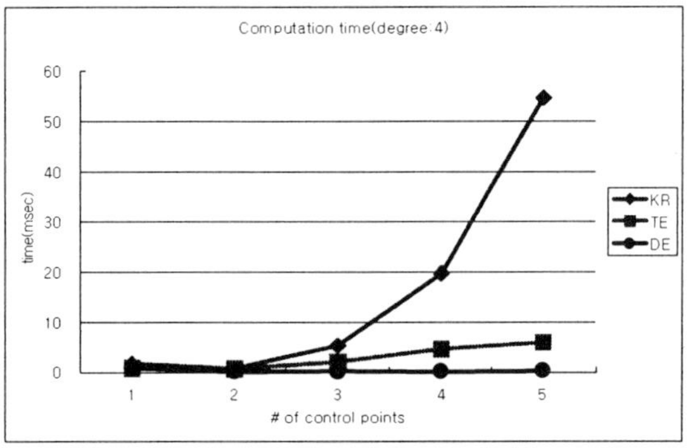

b. degree: 4

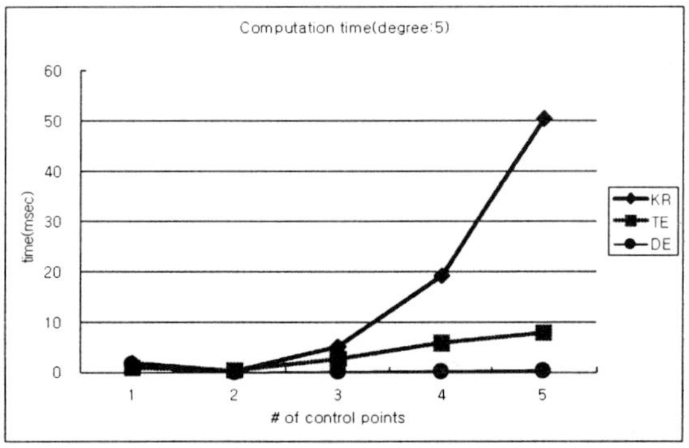

c. degree: 5

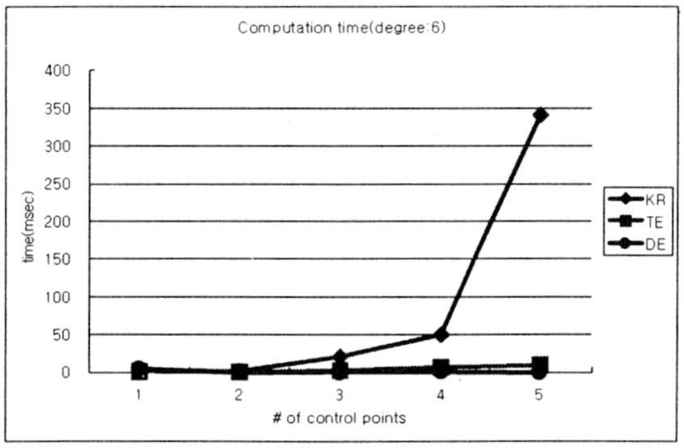

d. degree: 6

Figure 7. Computation time vs. the number of changed control points
(The number of all control points of given B-spline curve is 10.)

Similar problem can be defined for B-spline surfaces, and the approach presented in this paper can be easily extended. It is our expectation that our approach will show more significant computational properties for the problem.

7. Conclusions

In the computer graphics and CAGD, it is often necessary to manipulate B-spline curves or surfaces by converting the B-spline representation into a set of piecewise polynomial curves or surfaces in a power form. It is known that the faster computation of the characteristic points on a curve, such as inflection points and cusps, can be facilitated by the conversion of a B-spline curve into a set of piecewise polynomial curves in a power form. Once a curve is represented in a power form, a point evaluation can be also made faster due to Horner's rule even though some issues of stability remains. Also, the implicitization process, which employs a resultant, usually requires the geometry to be represented in a power form. Since this operation is often computationally demanding, the reduction of computation should not be ignored. In addition, the power form of curves is supported by IGES as an entity type 112.

In this paper, a new algorithm for converting a B-spline curve to piecewise polynomial curves in a power form is presented. By defining several new concepts such as truncated basis polynomials, transition edges, and so on, the theoretical basis is provided.

We claim that the proposed algorithm outperforms the conventional approach when the degree of the curve is relatively low. When the curve is dynamically changing its shape, the speed of computation becomes rather important. In this case, the theoretical analysis and experiments show that the advantage of DE algorithm becomes much more significant. Also, our algorithm applied to a dynamic surface can be more significant in its speed.

The fact that the performance of algorithm deteriorates rapidly as the degree of the curve gets higher guides the future research direction of this algorithm. In addition, the extensions of this algorithm to rational B-spline curves and surfaces are straightforward.

8. Acknowledgements

This research was supported by Center of Innovative Design Optimization Technology (iDOT), Korea Science and Engineering Foundation.

9. References

[1] Kim, D.-S., Lee, S.-W., and Shin, H., "A cocktail algorithm for planar Bezier curve intersections," *Computer Aided Design*, Vol. 30, No. 13, pp. 1047-1051, 1998.

[2] Farin, G., *Curves and Surfaces for Computer-Aided Geometric Design*, 3rd Ed., Academic Press, 1997.

[3] *The Initial Graphics Exchange Specification (IGES)*, Version 5.2, ANSI Y14.26M, 1993.

[4] Bloomenthal, J., *Introduction to implicit surfaces*, Morgan Kaufmann Publishers, Inc., 1997.

[5] Sederberg, T. W., *Implicit and parametric curves and surfaces for computer aided geometric design*, Ph. D. Thesis, Purdue University, 1983.

[6] Lee, K., *Principles of CAD/CAM/CAE Systems*, Addison-Wesley, 1999.

[7] Boehm, W. and Prautzsch, H., "The insertion algirthm," *Computer-Aided Design*, Vol. 12, No. 4, July, pp.58-59, 1985.

[8] Boehm, W., "On the efficiency of knot insertion algorithms," *Computer Aided Geometric Design*, Vol. 2, Nos. 1-3, July, pp.141-143, 1985.

[9] Cohen, E., Lyche, T. and Riesenfeld, R., "Discrete B-splines and subdivision techniques in computer-aided geometric design and computer graphics," *Computer Graphics and Image Processing* Vol. 14, No. 2, pp.87-111, 1980.

[10] Goldman, R.N., "Blossomming and knot insertion algorithm for B-spline curves," *Computer Aided Geometric Design*, Vol.7, pp.69-81, 1990.

[11] Lasser, D. and Hoschek, J., *Fundamentals of Computer Aided Geometric Design*, A. K. Peters, 1993.

[12] Choi, B. K., Yoo, W. S. and Lee, C. S., "Matrix representation for NURB curves and surfaces," *Computer-Aided Design*, Vol. 22, No. 4, May, pp.235-240, 1990.

[13] Yamaguchi, F., *Curves and Surfaces in Computer Aided Geometirc Design*, Springer-Verlag, 1988.

[14] Kim, D.-S., Ryu, J., Jang, T., Lee, H., and Shin, H., "Conversion of a B-spline curve into a set of piecewise polynomial curves in a power form," *Korea Israel Bi-National Conference on Geometric Modeling and Computer Graphics in the World Wide Web Era*, pp.195-201, Seoul, Korea, 1999.

[15] Piegl, L. and Tiller, W., *The NURBS Book*, Springer, 1995.

Session 11

Geometric Processing and Tolerancing

On Optimal Tolerancing in Computer-Aided Design

H. Pottmann [a], B. Odehnal [a], M. Peternell [a], J. Wallner [a], R. Ait Haddou [b]

[a] Institute of Geometry, Vienna University of Technology, Wiedner Hauptstraße 8–10, A-1040 Wien, Austria

[b] Department of Mechanical Engineering, University of Calgary, 25000 University Dr. NW, Calgary, Alberta, Canada T2N 1N4

Abstract

A geometric approach to the computation of precise or well approximated tolerance zones for CAD constructions is given. We continue a previous study of linear constructions and freeform curve and surface schemes under the assumption of convex tolerance regions for points. The computation of the boundaries of the tolerance zones for curves / surfaces is discussed. We also study congruence transformations in the presence of errors and families of circles arising in metric constructions under the assumption of tolerances in the input. The classical cyclographic mapping as well as ideas from convexity and classical differential geometry appear as central geometric tools.

1. Introduction

In the past decades, a huge amount of research has been devoted to geometric problems arising in Computer Aided Design. Almost always one assumes precise input or one is not so much concerned about the effect of input errors or tolerances on the output. However, for practical purposes, this is a fundamental question. A closer look exhibits a possible reason for a lack of research in this area: even for simple constructions, the analysis of tolerances and their interdependencies is quite complex.

Before outlining the contributions of the present paper, let us briefly review the state of the art in geometric tolerancing.

Tolerance logics are tables with the values of the dimensions (distances, radii, angles ...) for a special drawing (designed value, minimal/maximal value and mean value). Each drawing requires its own table. One disadvantage is the (semi-)manual creation of the tolerance logics. The designer has to specify the values of possibly hundreds of dimensions, a time consuming and risky process, since only a well trained designer may be able to find the critical tolerance combinations (see [16] and references therein).

An alternative is the *stochastic approach*, where one assumes a distribution for each dimension. Then, with e.g. a Monte-Carlo simulation one derives an estimation of the minimal/maximal distance of two points on the designed object. Again the designer has to find critical situations. An advantage of the method is a sensitivity analysis which one can derive from the simulation and its applicability to 3D problems [44].

Worst case approaches with no manual interaction to find critical tolerance situations give the designer an easy to use tool for the tolerance analysis. Requicha [38] introduced such a model based on *tolerance zones*. Tolerance zones are also a very intuitive visualization technique for error propagation in geometric constructions.

Robust geometric computing using tolerance zones and *interval arithmetic* has recently received much attention [1, 17, 20, 18, 40, 41, 42, 43]. Interval arithmetic possesses a lot of computationally attractive features, but it tends to overestimate the error and thus may produce too large tolerance zones.

For 2-dimensional CAD constructions, important progress on worst case tolerancing has been made by C. U. Hinze [16], who analyzed several frequently appearing elementary constructions in the plane involving straight lines and circles. His tolerance zones for points are convex and bounded by segments of straight lines and circles. An application to collision problems involving toleranced objects has been given by M. Aichinger [2].

Recently, J. Wallner et al. [46] proposed to use more general convex tolerance zones for points and showed how linear constructions in arbitrary dimensions and a variety of curve and surface design schemes can be effectively analyzed using ideas from convexity. Moreover, the dramatic change in the level of difficulty when switching to metric constructions has been illustrated by the seemingly simple problem of constructing a circle through three points (in the plane or in space). In fact, this problem is so complex that only the special case of balls for the tolerance regions of the three input points has been analyzed.

In the present paper, we would like to continue these developments with a major focus on metric constructions and the computation of boundaries of tolerance zones. After briefly reviewing the basic facts on linear constructions and tolerance zones for freeform curves and surfaces, we present a surprisingly simple way of computing curvatures of tolerance zone boundaries for curves in the plane. This allows us to apply high accuracy approximation schemes to represent such boundaries. An extension to surfaces is just outlined and shall be discussed in a separate contribution. Moreover, we study planar displacements in the presence of input errors.

We then introduce to a very useful concept from geometry: the cyclographic mapping. It is applied to the tolerance anlysis of constructions involving circles. Finally, a brief outlook on future research in the wide unexplored area of geometric tolerancing is given.

The present contribution is thought as a presentation of basic results as well as ideas and tools from geometry which simplify the study of tolerance zones in geometric constructions. Many topics we touch need further research. Moreover, in order to get the necessary understanding, we discuss certain constructions in more generality than the actual implementation might exhibit. For example, we currently admit arbitrary convex bodies as tolerance zones for points. Knowing the type and complexity of the arising constructions, we will later decide which restrictions have to be made for a practical realization of our tolerance analysis.

2. Linear combinations of points and applications to freeform curves and surfaces

2.1. Support function and linear combination of convex bodies

Since we will restrict our study to convex tolerance zones for points, we present a brief outline of necessary facts from convex geometry.

One of the main tools for studying compact convex bodies of $\mathbb{R}^d$ is the *support function*. This concept has been used for a long time and there is large amount of literature, including some monographs on convex geometry (see e.g. [15] for a detailed overview of the whole field of convex geometry).

We call a plane ε a *support plane* of a convex body K if K has a point in common with ε and K is entirely contained in one of the two closed half-spaces defined by ε. For all unit vectors n there is a unique plane $\varepsilon(n)$ orthogonal to n which is a support plane of K such that K lies in the half-space of ε which is indicated by $-n$. The oriented distance of this plane to the previously fixed origin is the value of the support function $s(n)$. Then ε has the equation $x \cdot n = s(n)$ and K lies in the half-space $x \cdot n \leq s(n)$.

The domain of a support function is the unit sphere S^{d-1} of $\mathbb{R}^d$. In the plane ($n = 2$) it is sometimes useful to make the domain of the support function the interval $[0, 2\pi]$, where an angle ϕ is identified with the appropriate point of the unit circle. We will never do this, because we sometimes evaluate the support function at vectors n and $-n$, which are opposite points of the unit sphere. In the case $n = 2$ opposite points correspond to angles $\phi, \phi + \pi$, and $-\phi$ means something different. In order to avoid confusion, let us state explicitly that the minus sign always indicates the *opposite point* of the unit sphere.

If K is a convex body, λK is the set of all λx with $x \in K$. For two convex bodies K_1, K_2, their *Minkowski sum* $K = K_1 + K_2$ is defined as the set of all $x_1 + x_2$ with $x_1 \in K_1$, $x_2 \in K_2$.

For real numbers t_1, t_2 we then can define the body $K = t_1 K_1 + t_2 K_2$. Especially we define *affine combinations* $(1-t)K_1 + tK_2$ of convex bodies, and *convex combinations*, which are affine combinations with $0 \leq t \leq 1$.

Among the basic properties of support functions are the following: If $s : S^{d-1} \to \mathbb{R}$ is the support function of K, then $K' = \lambda K$ has the support function $s'(n) = \lambda s(n)$ if $\lambda \geq 0$. The support function s' of $-K$ is given by $s'(n) = s(-n)$. If s_1, s_2 are the support functions of K_1, K_2, resp., and $0 \leq t \leq 1$, then the convex body $K = (1-t)K_1 + tK_2$ has the support function $s = (1-t)s_1 + ts_2$. From this it easily follows that the support function of the convex body $\lambda_1 K_1 + \lambda_2 K_2$ equals $\lambda_1 s_1 + \lambda_2 s_2$ if both λ_1, λ_2 are nonnegative.

As an application of this, we derive the support function s of a linear combination $K = \sum \lambda_i K_i$ of convex bodies K_i with support functions s_i,

$$K = \sum_{\lambda_i > 0} \lambda_i K_i + \sum_{\lambda_i < 0} \lambda_i K_i = \sum_{\lambda_i > 0} \lambda_i K_i + \sum_{\lambda_i < 0} (-\lambda_i)(-K_i).$$

The support function s_i' of $-K_i$ is given by $s_i'(n) = s_i(-n)$. Clearly, the support function of a *positive* linear combination of convex bodies is the same linear combination of their support functions. Thus, the convex body K possesses the support function

$$s(n) = \sum_{\lambda_i > 0} \lambda_i s_i(n) + \sum_{\lambda_i < 0} (-\lambda_i) s_i(-n). \qquad (1)$$

We see that *Minkowski sums* are the basic entity for linear combinations of convex bodies. For computational issues of Minkowski sums in the plane, we refer the reader to Kaul and Farouki [21] and Lee et al. [23]. In the latter paper, Minkowski sums are studied for arbitrary planar domains. This is much more subtle than Minkowski sums of convex bodies, which is one reason for our limitation to convex tolerance zones for points. Instead of the tolerance region of a point we will also speak of a 'fat point'. Likewise, tolerance regions to curves or surfaces will be briefly called 'fat curves/surfaces'.

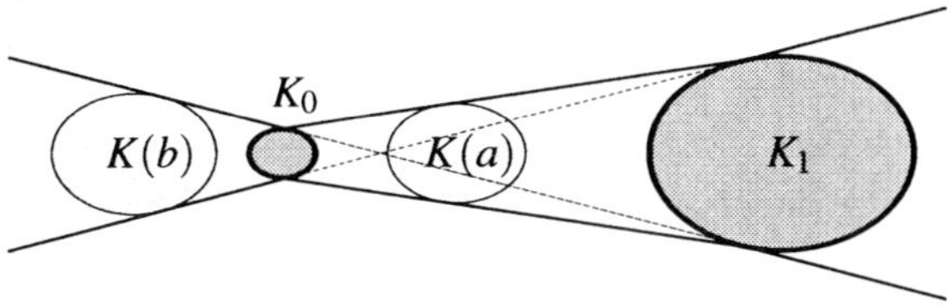

Figure 1. Boundary of a fat line in the plane connecting fat points K_0 and K_1 and tolerance regions to affinely combined points $K(a)$ and $K(b)$ ($a = 0.35$, $b = -0.5$)

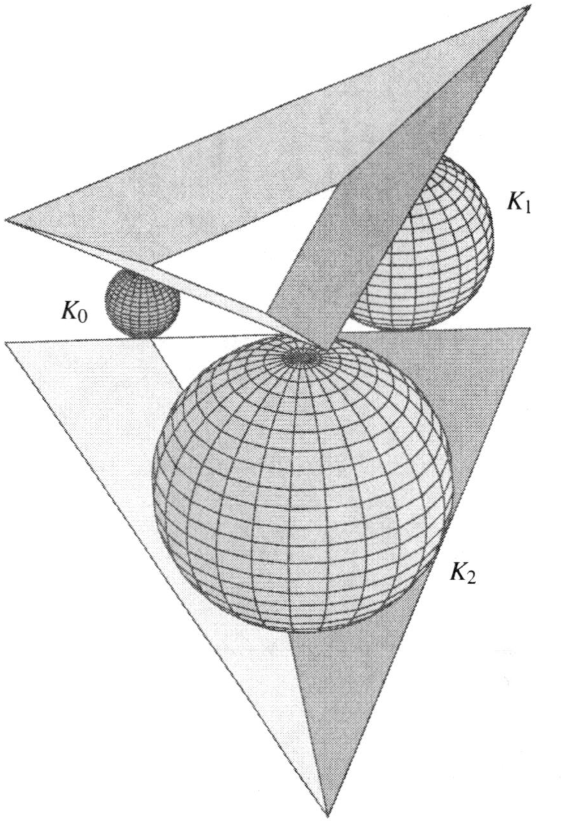

Figure 2. Boundary of the tolerance region for the spanning plane of three fat points (balls)

2.2. Curves and surfaces defined by control points

Let us mention some applications of linear combinations of convex bodies to geometric tolerancing. For more details, we refer the reader to Wallner et al. [46].

Spanning line and plane

Given the convex tolerance regions K_0, K_1 to two points in $\mathbb{R}^d$, the tolerance regions for the points on the *connecting line* are given by

$$K(t) = (1-t)K_0 + tK_1.$$

For each t, this is a convex domain, and the union of all domains $K(t)$ is the tolerance zone of the connecting line. It is bounded by (part of) the envelope of $K(t)$.

Let us assume disjoint regions K_0, K_1. Then, in the plane, the tolerance region of the connecting line of the two fat points K_0, K_1 is bounded by parts of the four common support lines (tangents for smooth K_i) (see Fig. 1).

In 3-space, the bounding surface of the tolerance region to the spanning line consists of parts of *developable surfaces* joining the two convex surfaces K_1, K_2. Their computation is briefly discussed below.

Points of the *spanning plane of three fat points K_i in $\mathbb{R}^3$* are associated to the tolerance regions

$$K(u,v) = uK_0 + vK_1 + (1-u-v)K_2.$$

Assuming disjoint regions K_i, the boundary of the tolerance zone to the plane has eight planar faces, which are parts of the common support planes of the three convex bodies K_0, K_1, K_2. This is illustrated in Fig. 2 with three balls K_0, K_1, K_2. Note that the points of tangency of the common tangent planes with the spheres are not at the boundary of the tolerance region for the spanning plane.

Developable surfaces joining two curves or surfaces

Consider two curves or surfaces c_1, c_2. Then, a developable surface joining the two objects is the envelope of all planes which touch both c_1 and c_2. Here, a tangent plane to a curve is a plane through a tangent of the curve. A developable surface contains a one-parameter family of straight lines (rulings, generators). The generators are the connections of the contact points $p_i \in c_i$ of common tangent planes of c_1 and c_2 (Fig. 3).

We may view c_i as dual objects, i.e., as sets of their tangent planes. The computation of a joining developable surface is therefore dual to the computation of an intersection curve of two surfaces. Just like the intersection curve may have different topological components, the connecting developable may also consist of different components.

Consider two non-intersecting convex bodies K_0, K_1 in $\mathbb{R}^3$ with support functions s_0, s_1. We admit piecewise smooth bodies and thus speak of support planes instead of tangent planes. The normals n of common support planes satisfy

$$s_0(n) - s_1(n) = 0, \quad (2)$$

if K_0 and K_1 lie on the same side of the support plane. Normals to common support planes, where K_0 and K_1 lie on

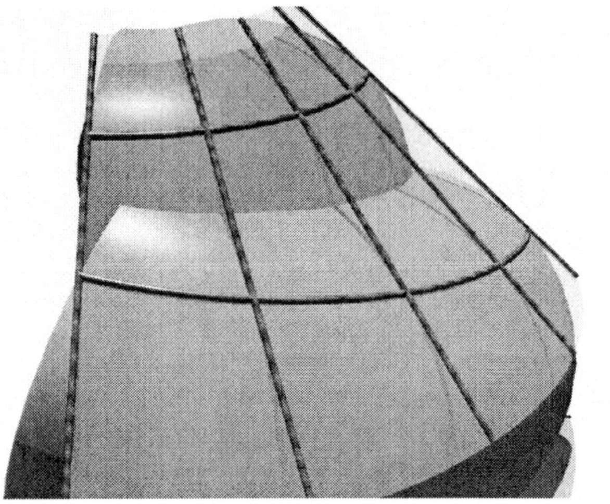

Figure 3. Connecting developable surface of two surfaces

different sides, are characterized by

$$s_0(n) + s_1(-n) = 0. \tag{3}$$

Equation (2) belongs to the first component of the joining developable. Its part between K_0 and K_1 lies in the boundary of the tolerance zone for the line segment $(1-t)K_0 + tK_1$, $t \in [0,1]$. Equation (3) is describing the normals of the developable surface, which contributes to the boundary of the tolerance zone for $t < 0$ and $t > 1$.

The normals satisfying (2) or (3) form curves on the unit sphere. These are the Gaussian images of the two components of the joining developable surface. Algorithmically, we have to compute the zero set of a function on the unit sphere. In general, parameterizations of these spherical curves, say $k_1(t)$ and $k_2(t)$, can be found only by approximation algorithms. The developable surfaces are then enveloped by planes $x \cdot k_i(t) = s_0(k_i(t))$ for $i = 1,2$. Then, algorithms for working with this dual representation may be employed. For the latter topic and approximation with rational developable surfaces, see e.g. [35, 37]. In the present application, simple boundaries such as those composed by cones of revolution are an advantage [24, 25].

K_0 and K_1 may degenerate to planar convex domains. Such a case is illustrated in Figure 4, where K_i lie in parallel planes, say $x_3 = 0$ and $x_3 = 1$. Just the component to equation (2) between K_0 and K_1 $(0 \leq x_3 \leq 1)$ is displayed. A plane $x_3 = t$, $t \in [0,1]$ cuts the developable surface in the boundary of $K(t)$. In a parallel projection onto $x_3 = 0$ we see the planar figure of a set of convex domains $K(t) = (1-t)K_0 + tK_1$.

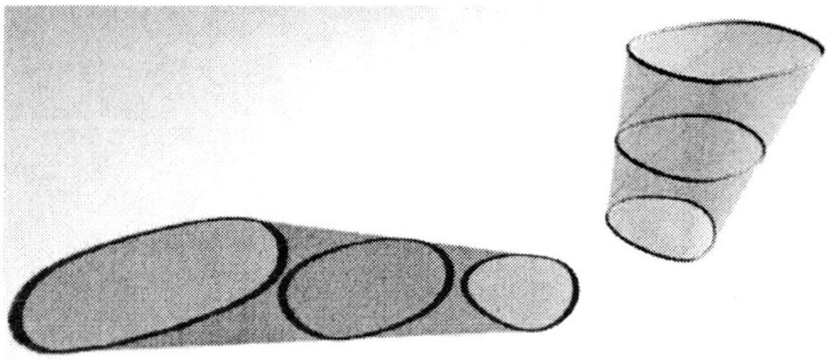

Figure 4. Developable surface and convex combination of convex domains

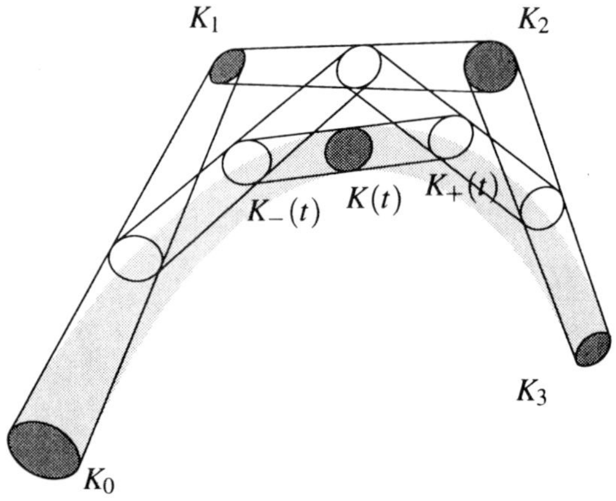

Figure 5. Tolerance zone to Bézier curve with $0 \leq t \leq 1$

Tolerance regions of Bézier curves

A large amount of curve and surface schemes in Computer Aided Geometric Design consists, in principle, of appropriate linear combinations of control points b_i. This allows us to apply the same methods as above to the tolerance analysis for many freeform curves and surfaces [46].

Let us look at the example of *Bézier curves*.

A control point b_i of a Bézier curve corresponds to a convex body K_i as its tolerance zone. It shall have support function s_i. The tolerance zone for the curve point with parameter value t is the convex body

$$K(t) = \sum_{i=0}^{n} B_i^n(t) K_i.$$

The sign of the ordinary Bernstein polynomials is given by

$$\text{sgn}(B_i^n(t)) = \text{sgn}(t)^{n-i}\,\text{sgn}(1-t)^i$$

and so the support function s of $K(t)$ is given by

$$s(n) = \sum B_i^n(t) s_i^\star(n), \tag{4}$$

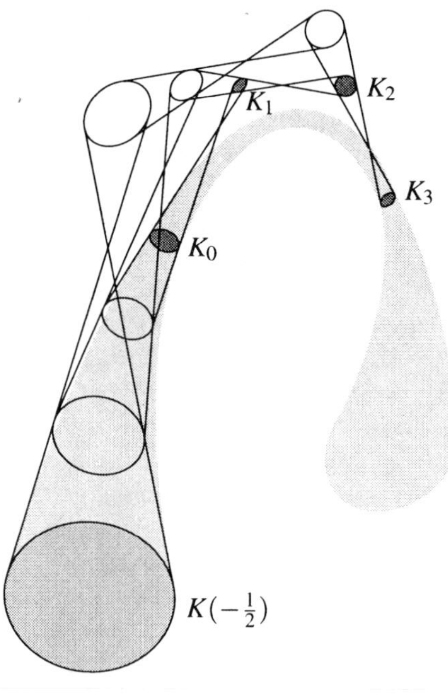

Figure 6. Tolerance zone to Bézier curve with $-1/2 \leq t \leq 3/2$

where

$$s_i^\star(n) = \begin{cases} s_i(n) \text{ for } & t<0,\ n-i \text{ even, or} \\ & t>1,\ i \text{ even, or} \\ & 0 \leq t \leq 1 \\ -s_i(-n) & t<0,\ n-i \text{ odd, or} \quad t>1,\ i \text{ odd.} \end{cases}$$

This can be symbolized by the following diagram:

○●○●○●○● | ●●●●●●●● | ●○●○●○●○
$t<0$ $0<t<1$ $1<t$

Here, the dots are symbolizing the signs of the basis functions $B_0^n, \ldots, B_n^n$ ($n = 7$ in our example), and fat points indicate a positive sign.

A figure of the union of all possible K's when t ranges in an interval, can be seen in Fig. 5 and Fig. 6. Fig. 5 also illustrates the computation of points and tangents of the boundary of the tolerance zone: In the plane, it is based on common tangents to the tolerance regions $K_-(t), K_+(t)$ which arise in the de Casteljau algorithm. In $\mathbb{R}^3$, the curve, along which the region $K(t)$ touches the envelope ('characteristic curve') lies on the joining developable surface of $K_-(t), K_+(t)$ (see connecting line of two fat points). From Fig. 6 it is clearly seen that in the case $t \notin [0,1]$, the size of $K(t)$ is rapidly increasing with the distance of t to the standard interval $[0,1]$. So the often cited 'optimality' of the Bernstein basis is again seen to be valid only in the interval $[0,1]$ (see [11]).

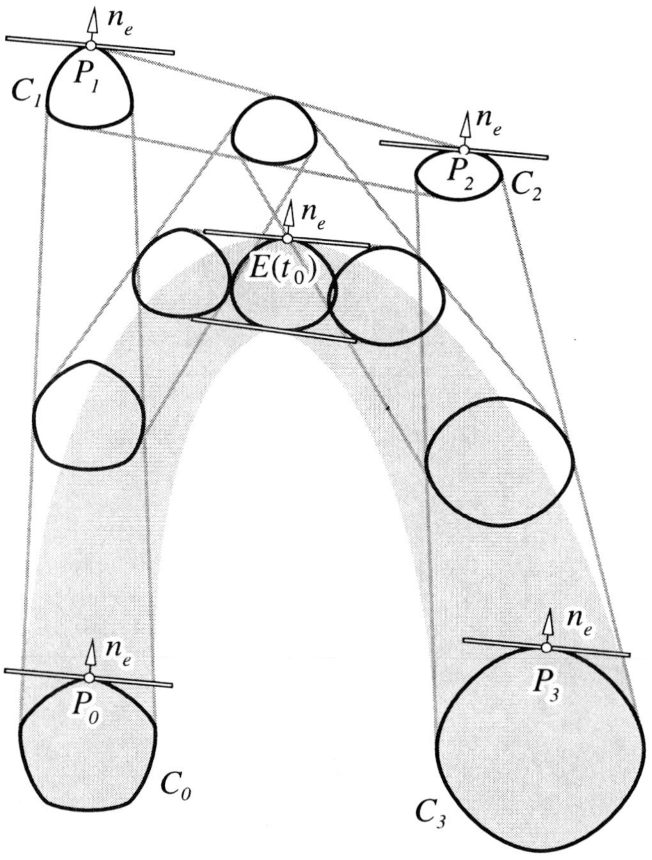

Figure 7. To the computation of the curvature of the boundary of a fat Bezier curve at a point $E(t_0)$

More details on the computation of the tolerance zone boundaries are found in [46], but there are still open questions in this area. One will be addressed in the following subsection.

Curvature of tolerance zone boundaries

Let us first focus on a fat Bézier curve in the plane and discuss the computation of its boundary. Apart from special cases discussed below, closed form expressions of the boundary will not be obtained since the problem of computing common tangents of the fat points K_-, K_+ in the de Casteljau construction will in general not have a closed form solution. Hence we are interested in a good approximation of the boundary, for which it is desirable to know the curvatures at discrete boundary points. Then, appropriate curve approximation schemes such as high accuracy geometric Hermite interpolation with polynomial cubics [8] or osculating arc splines [28] may be applied.

We consider a fat Bézier curve of degree n, whose fat

control points K_i shall have C^2 boundary curves denoted by C_i. For convenience, we limit ourselves to the parameter interval $[0,1]$; extensions follow immediately from the discussions above. The one-parameter family of fat curve points $K(t)$ has convex C^2 boundary curves $C(t)$, whose envelope E (eventually after trimming away parts in case of self-intersections of E) forms, together with parts of $C(0)$ and $C(1)$ the boundary of the fat Bézier curve (see Fig. 5).

We are interested in the curvature of E at an envelope point $E(t_0) \in C(t_0)$. Let n_e be the outward unit normal of E and $C(t_0)$ at $E(t_0)$ and consider those points P_i on the control domain boundaries C_i, which also possess the outward unit normal n_e (see Fig. 7). Points P_i, $i = 0, \ldots, n$ are considered as control points of a Bézier curve B. It follows immediately from (4), that the curve point $B(t)$ is exactly that point of the tolerance zone boundary $C(t)$ whose unit normal is n_e (for all $t \in [0,1]$). Moreover, since the curvature radius of a convex curve is the sum of the support function and its second derivative, we see that the curvature radius ρ_c of $C(t_0)$ at $E(t_0)$ is computed from the curvature radii ρ_i of C_i at P_i by

$$\rho_c = \sum_{i=0}^{n} B_i^n(t_0)\rho_i. \tag{5}$$

With the curvature radius ρ_b of the Bézier curve B at $B(t_0) = E(t_0)$, we are then able to compute the curvature radius ρ_e of the envelope at $E(t)$:

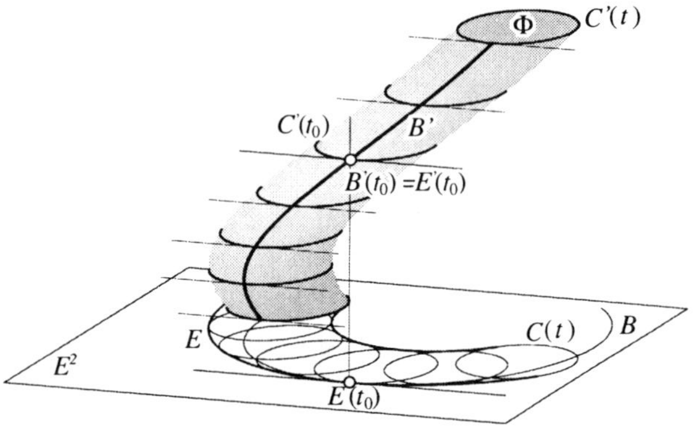

Figure 8. Spatial interpretation of curvature construction of zone boundaries

Theorem 1. *The curvature radius ρ_e of the boundary of a fat Bézier curve at a point $E(t_0)$ with normal n_e is computed as follows. Compute the curvature radii ρ_i of the control domain boundaries C_i at points P_i with normal n_e, and compute the curvature radius ρ_b of the Bézier curve with control points P_i at the curve point $E(t_0)$ to parameter t_0. Then,*

$$\rho_e = \rho_b + \sum_{i=0}^{n} B_i^n(t_0)\rho_i. \tag{6}$$

All curvature radii are the inverse curvatures, i.e., taken with sign.

Proof: Our proof is based on an interpretation of the planar figure as projection of a spatial one. We translate each curve $C(t)$ by the vector $(0,0,t)$ up to height $z = t$ in 3-space, see Figure 8. The translated domain boundaries, denoted by $C'(t)$, form the contour curves $z = t = const$ of a surface Φ in $\mathbb{R}^3$. The silhouette of Φ for orthogonal projection into $z = 0$ is the envelope E. The lifted points $B'(t) \in C'(t)$ of the auxiliary Bézier curve B form a Bézier curve B' on the surface Φ. The tangents to the level curves $C'(t)$ at its points $B'(t)$ are parallel and form a general cylinder surface which touches Φ along the curve B'. Hence, the tangents of B' are conjugate (in the sense of differential geometry) to the horizontal cylinder generators. In particular, at the lifted envelope point $E'(t_0)$ (point on the contour of Φ for z-parallel projection), the tangent to the level curve $C'(t_0)$ and the tangent to the Bézier curve B' are conjugate. By a theorem of H. Schaal on the curvature radius of the silhouette of a surface [39, 45], we conclude that the sum of the curvature radii of the projections $C(t_0), B$ of the two curves at the silhouette point $E(t_0)$ equals the curvature radius ρ_e of the silhouette. With (5), this proves the claim of the theorem.

It is obvious that an analogous theorem holds for other linear curve schemes (B-splines, interpolatory C^2 cubic splines,...) as well. The result can also be generalized to surfaces. Here, curvature radii (inverse curvatures) are replaced by inverse Weingarten maps of surfaces which characterize the curvature behaviour. The derivation of this result is beyond the scope of the present paper and shall be given elsewhere. We just mention that a special case concerns curvature computation for general offset surfaces [34].

Remark. The limitation to C^2 boundaries C_i of the fat control points is not necessary. Replacing normals by normals to support lines, we can compute points $P_i \in C_i$. If, at a point P_i curve C_i has a tangent discontinuity (and n_e is not normal to a boundary curve segment ending there), we assign curvature radius 0 to it. If n_e is normal at P_i and there is a curvature discontinuity of C_i, we compute with right and left limit of the curvature and obtain right and left limit curvature for the envelope.

'Interval' curves and surfaces

If the coordinates of control points are known to lie in certain intervals, then an equivalent formulation of this is that the control point itself may vary in some orthogonal parallelotope whose faces are parallel to the coordinate hyperplanes. In the plane, interval bounds on the coordinates of points are equivalent to working with tolerance zones of points, which are rectangles with edges parallel to the coordinate axes.

The procedure of linear combination, especially of convex combination of such parallelotopes again produces convex orthogonal parallelotopes. If the linear combination is with nonnegative factors, then the coordinates of the faces of the box containing the eventual curve or surface point can be computed by the *same* linear combination. Interval Bézier curves are studied in [40]. Interval B-splines and their applications to curves and surfaces were studied in [17, 20, 41, 42, 43]. There, the authors also address the boundary computation in the plane, which is simpler than above, since the boundary consists of special Bézier or B-spline curves and straight line segments parallel to the coordinate axes. It may be a good exercise to verify this with help of the spatial interpretation given above.

Ball curves and surfaces

Another class of convex bodies closed under the operation of linear combination are the Euclidean balls of radius r, which in dimension one coincide with the intervals, and which in dimension two are usually called disks. If the linear combination of control points which yields the eventual curve/surface point is nonnegative, then both the midpoint and the radius of the eventual balls is obtained by the *same* linear combination of the midpoints and radii of the control point balls. Planar 'disk' Bézier curves are studied in [26]. In dimensions 2 and 3, ball Bézier curves are cyclographic images of Bézier curves C' in $\mathbb{R}^3$ and $\mathbb{R}^4$, respectively (see [36] and the following section). We should also remark that the boundary of the domain to a planar disk Bézier curve can be parameterized using polynomials and square roots; it is in general not a rational curve. Only for special curves C' in $\mathbb{R}^3$, which are a counterpart to the Pythagorean-hodograph (PH) curves of Farouki and Sakkalis [14, 10], these envelopes are rational [29]. Surprisingly, canal surfaces which appear as boundaries of ball Bézier curves in $\mathbb{R}^3$ always possess a rational parameterization [32].

General offsets

A further simple special case occurs if all control domains are *translates* of a single convex domain D. Then, all fat points on the fat curve/surface are translates of D.

In case that D is a ball of radius r, the fat curve/surface is bounded by the two-sided *offset* at distance r to the center curve/surface. In the planar curve case, all auxiliary Bézier curves B are translates of the center curve. Hence, we get the well-known result that the curvature radius of an offset is the sum of the curvature radii of the progenitor curve and the offset distance. Offsets, in particular questions on their rationality, have been a very active research area in the past (see e.g. [10, 27, 33]).

Analogously, an arbitrary centrally symmetric convex domain D gives rise to *general offsets* as boundaries. The curvature formula, even for surfaces, is known in this case [34]. As for classical offsets, it can be formulated with center curve/surface and D only. Rationality of general offsets has also been discussed recently ([3, 4, 34]).

Constructions with linear schemes

So far we are able to compute tolerance regions for linear curve and surface schemes including the special cases of spanning line and plane of points. If a construction involves just these basic elements and intersection operations, the intersections of the tolerance domains have to be computed. This is, particularly in $\mathbb{R}^3$, computationally expensive. Moreover, the resulting tolerance regions to intersection points need not be convex. We then replace them by their convex hull. Limitation to special convex domains as tolerance regions for points (such as the 2D approach in [16]) may help to speed up such computations. However, note that we are interested in sharp enclosures of tolerance zones and thus need restrictions that still allow us to get good approximations. This will be a topic for future research.

2.3. Displacements in the Euclidean plane

A *toleranced translation* in the plane may be given by translation vectors v which describe some convex body V. Then, the image of a fat point K under the translation is the *Minkowski sum* $K+V$.

Let us now study a *toleranced rotation*. We merge Cartesian coordinates (x_1, x_2) of points in the plane into complex numbers $x = x_1 + ix_2$. In the complex notation, a rotation around a point c with angle ϕ maps x to x' via

$$x' = c + (x-c)e^{i\phi} = c(1-e^{i\phi}) + xe^{i\phi}. \quad (7)$$

We now assume that the angle ϕ is precise, but the center c and point x are fat points (convex domains). Writing the complex number $a = 1 - e^{i\phi}$ in the form $a = \rho e^{i\psi}$, we see that $c(1-e^{i\phi}) = ca$ describes a convex domain, obtained from c by a rotation around the origin with angle ψ and a scaling with factor ρ. Clearly, $xe^{i\phi}$ is a rotated copy of x. Thus, we find the following result: *The fat image point is a convex domain, obtained as Minkowski sum of a rotated copy of the fat preimage point and a rotated and scaled copy of the fat rotation center.*

In case that the rotation angle ϕ is also subject to tolerances, say ϕ in some interval I, we can use recent work by Farouki, Moon and Ravani [13, 12] to study the fat image points. These are in general no longer convex. In order to stay within our approach, we have to form the convex hull of the image domains.

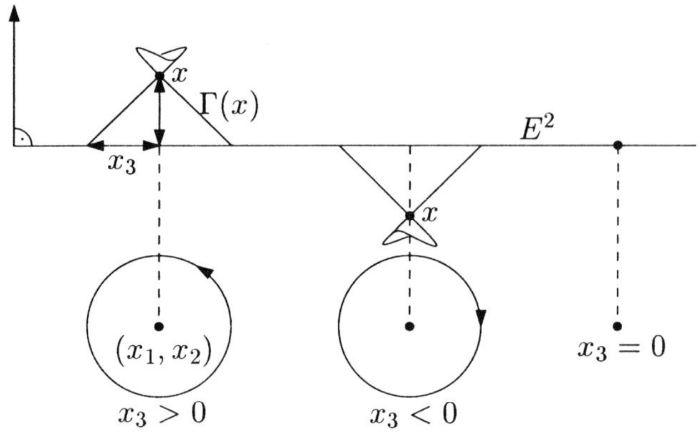

Figure 9. Cyclographic mapping, top and front view

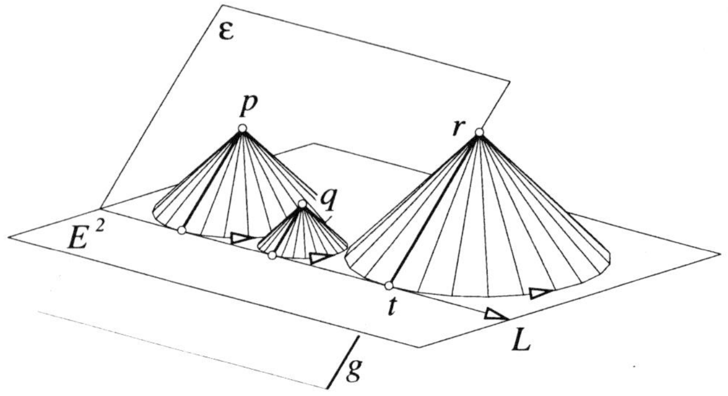

Figure 10. γ-cones tangent to a γ-plane

3. The cyclographic mapping

In a tolerance analysis of planar metric constructions, tolerance zones for circles play a central role. These are formed by k-parameter ($k = 1,2,3$) sets of circles. In this section, we will briefly introduce to a concept from classical geometry, the so-called *cyclographic mapping*, which is very helpful both for understanding circle constructions and for computing tolerance zones. For a detailed exposition of this interesting classical topic we refer to the monographs by Coolidge [6] and Müller and Krames [31]. In connection with different applications in CAGD, the cyclographic mapping has recently been applied by several authors [22, 33, 36].

The *cyclographic mapping* ζ maps a point $x = (x_1, x_2, x_3)$ in $\mathbb{R}^3$ to an oriented circle in the plane: this circle $c = \zeta(x)$ has midpoint (x_1, x_2) and radius x_3. The sign of x_3 determines the orientation of the circle (Fig. 9); $x_3 = 0$ characterizes a point in the plane. The image circle of x can also be obtained as follows: consider a cone of revolution with vertex x, whose axis is parallel to the x_3-axis and whose generators form an angle $\gamma = \pi/4$ with the axis (and the reference plane $E^2 : x_3 = 0$). In the following, we will call such a cone a *γ-cone*. Then, c is the intersection of the plane E^2 and the γ-cone $\Gamma(x)$ with vertex x. This is geometrically a type of projection, but with help of a cone instead of a single projection line.

The orientation of a circle c induces orientations of its tangents; we say that c is in oriented contact with its tangents. Given an oriented line $L \subset E^2$, we may ask for all oriented circles c touching L. Their cyclographic preimages are the points of a plane ε through L, whose inclination angle against E^2 is also $\gamma = \pi/4$ (Fig. 10). Such a plane shall be called a *γ-plane*. Hence, we may say that corresponding to oriented lines in E^2 we have γ-planes in $\mathbb{R}^3$. Oriented contact between circle and line in the plane is seen as incidence between point and γ-plane in the spatial model, which is called *cyclographic model*. All oriented circles touching an oriented line L at a given point t form a one-parameter family. The cyclographic preimage is a ruling g of a γ-cone, which we call a *γ-line*.

Consider an oriented circle c_1, which is cyclographic image of a point p_1: $c_1 = \zeta(p_1)$. Then, the points of the γ-cone $\Gamma(p_1)$ are exactly those which are mapped to oriented circles $c \subset E^2$ that touch c_1. Hence, the construction of oriented circles touching three circles $c_i, i = 1,2,3$ amounts to computing the intersection of the three corresponding γ-cones $\Gamma(p_i); \zeta(p_i) = c_i$. This leads to (algebraically counted) two solutions and also solves the Apollonian problem of constructing all circles which are tangent to three given circles. Since there are four essentially different orientation combinations of the three input circles, the Apollonian problem has (algebraically) eight solutions. This spatial approach to the solution of the Apollonian problem ([7]) has been the starting point for studying the cyclographic mapping.

We should mention why three γ-cones (surfaces of order 2), intersect in just two (finite) points, although a careless application of Bezout's theorem might yield $2 \cdot 2 \cdot 2 = 8$ solutions. This is so, since any two γ-cones share a conic Ω at infinity (the cones are translates of each other!). This ideal conic Ω plays an important role for projective and algebraic considerations.

For our application, orientations are in most cases not necessary. Then, we can either admit only nonnegative radii (confine to the positive half-space $x_3 \geq 0$ in the spatial model) or work with both orientations (associate a pair of points $(m_1, m_2, \pm r)$ to a circle in E^2).

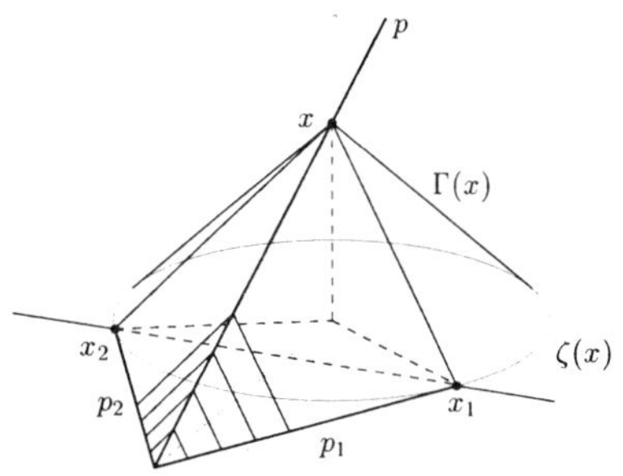

Figure 11. Cyclographic image of a straight line

3.1. One-parameter sets of circles

A one-parameter set of circles in E^2 corresponds to a *curve* p in the cyclographic model. We are interested in the domain $D(p)$ of the plane which is swept by the image circles $\zeta(x)$ for all points $x \in p$. We call the envelope of the circles $\zeta(x)$ the 'cyclographic image' of the curve p and denote it by $c(p)$.

Let us start with the simplest example, i.e., a *straight line* p. If the line p is inclined under an angle $\alpha < \pi/4$ against E^2, the γ-cones $\Gamma(x)$ for $x \in p$ possess two common tangent planes $\varepsilon_1, \varepsilon_2$ as envelope, namely the two γ-planes through p. The two planes ε_i intersect E^2 in two lines p_1, p_2, which bound the domain $D(p)$ traced out by the circles $\zeta(x)$, $x \in p$ (Fig. 11). In connection with an interpretation from special relativity, the current type of lines is called 'space-like'.

If p is a γ-line ('light line'), we have a unique γ-plane ε through p and all image circles touch $\varepsilon \cap E^2$ at $p \cap E^2$. Here and in the case of a line p with inclination angle $\alpha > \pi/4$ ('time-like' line, where we have no real envelope), the image circles $\zeta(x)$, $x \in p$, cover the entire plane E^2. If we restrict to a segment p with end points a, b on such a line, then the domain $D(p)$ is bounded by the image circles $\zeta(a), \zeta(b)$ of the end points.

For an arbitrary regular C^1 curve p, we pick a point x on it and consider its tangent t. Since the envelope computation requires first order derivatives only and t is a first order approximant to p at x, the following construction yields the contact points x_1, x_2 of $\zeta(x)$ with the envelope (Fig. 12): compute the intersection point $T = t \cap E^2$ and pass the tangents t_1, t_2 through T to the image circle $\zeta(x)$; their contact points x_1, x_2 are the envelope points, the tangents t_1, t_2 are tangent to the envelope. The lines xx_1 and xx_2 may be considered as rulings of a developable surface through the curve p whose rulings have inclination angle $\gamma = \pi/4$. Such a γ-*developable* can also be viewed as developable surface joining p and the ideal conic Ω.

The domain $D(p)$ is bounded by (parts of) of the following curves: the envelopes to segments of p with space-like tangents and the image circles of the end points of p. For piecewise C^1 curves, we also have to consider the image circles of points with tangent discontinuities. Note that trimming of these curves may be necessary.

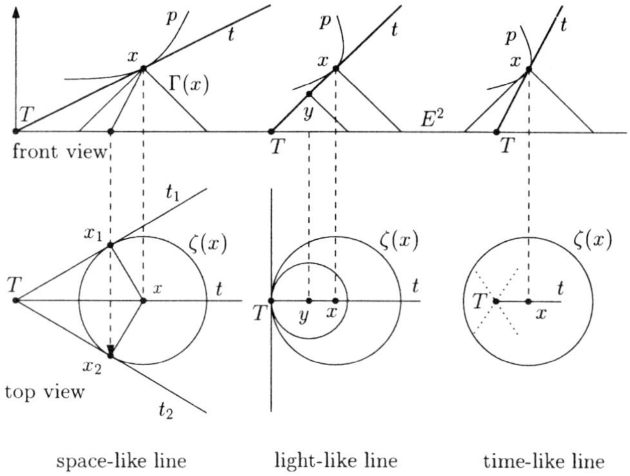

Figure 12. Cyclographic image of point plus tangent

It might be surprising that curve segments p, all whose tangents are γ-lines, are not considered for the boundary computation: the tangent surface of such a curve p is a γ-developable which intersects E^2 in the envelope curve $c(p)$. The image circles $\zeta(x)$ are osculating circles of the envelope and hence contained in each other (since we require a regular curve p). Thus, we just have to map the end points of these curve segments in order to get the corresponding domain boundary in E^2.

For more details on the cyclographic mapping of curves, both from a differential geometric and an algebraic point of view, we refer to the literature [6, 31, 36].

Example 3.1. An application of the cyclographic mapping of curves are *disc Bézier curves* in the plane. The tolerance region boundaries for the control points are circles (say with positive radius), which determine points p_i, $i = 0, \ldots, n$ in the cyclographic model. The spatial Bézier curve segment p with control points p_i and parameterized over the standard interval $[0, 1]$ has a cyclographic image curve which – eventually after trimming – forms together with parts of the end circles $\zeta(p_0)$ and $\zeta(p_n)$ the boundary of the tolerance zone $D(p)$ to the planar Bézier curve segment.

Remark. We pointed out that trimming of the envelope $c(p)$ may be necessary. This occurs in case of self-intersections of $c(p)$ or intersections of different components of $c(p)$. There are 'local' types of self-intersections arising in the neighborhood of typically two cusps of the

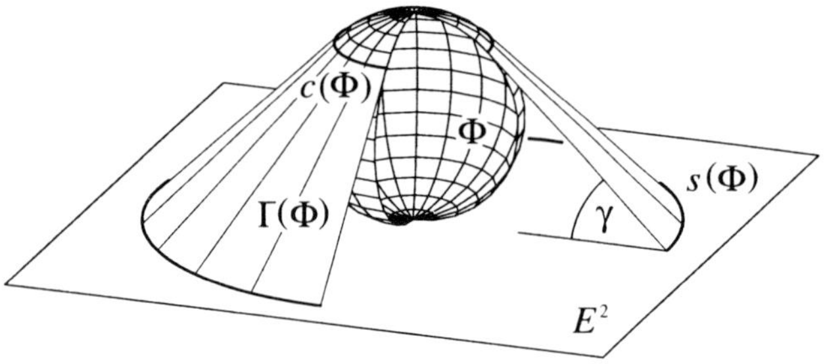

Figure 13. Cyclographic contour and silhouette of a surface (outer part)

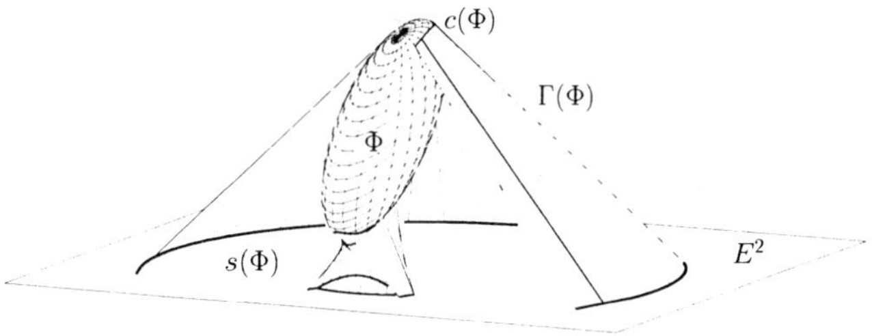

Figure 14. γ-developable tangent to an ellipsoid

envelope (swallow–tail). Those cannot occur if there are no singularities of the envelope; conditions for an envelope without singularities may be found with curvature formula (6). 'Global' types of self-intersections are harder to detect, but under certain global shape conditions it may be possible to prove results which help to speed up tests for self-intersections. Such results could be derived recently in case of offsets, e.g., cyclographic images of curves p in 'horizontal' planes $x_3 = \text{const}$ [47]. Trimming can also be caused by intersections of different parts of the envelope, e.g., those to different segments of the curve p, which are separated by segments with time-like tangents.

3.2. Two-parameter sets of circles

A two-parameter set of circles is represented by a *surface* Φ in the cyclographical model. We are interested in the planar domain $D(\Phi)$ traced out by the circles $\zeta(x)$, $x \in \Phi$, of the system.

Contributions to the boundary of D may come from the cyclographic image of the boundary of the surface Φ or from the *cyclographic silhouette* of the surface. The first issue has been discussed already, and thus we are now focusing on the cyclographic silhouette. Similar to the definition of contour and silhouette of a surface with respect to a central or parallel projection, we use the following definition. The *cyclographic contour* $c(\Phi)$ is the set of all points of the surface $\Phi \in \mathbb{R}^3$, whose tangent plane is a γ-plane. If such points exist, the generic case is the one in which these points form a curve c on the surface (Fig. 13). Since the tangent planes along c form the constant angle $\pi/4$ with the x_3-direction, the contour is also an *isophote* for illumination parallel to the x_3-axis. The tangent planes along c possess as envelope a γ-developable $\Gamma(\Phi)$, which may be viewed as connecting developable surface of Φ and the ideal conic Ω. The intersection curve of $\Gamma(\Phi)$ with the reference plane E^2 is the cyclographic silhouette $s(\Phi)$ of the surface Φ.

Looking at the cyclographic image of a plane, we see that only a γ-plane results in a 2-parameter set of circles with an envelope (the intersection of the plane with E^2). From this it follows immediately that a nontrivial envelope of the circle system (i.e., one which does not come from the boundary of Φ) can only stem from the cyclographic contour, i.e., it must be the cyclographic silhouette.

For trimming the boundary $D(\Phi)$ the following simple fact is helpful: The cyclographic image of parts of the boundary curve b of Φ can contribute to $D(\Phi)$ only if the tangent planes at its points have an inclination angle less or equal $\pi/4$ against E^2.

Example 3.2. The simplest nontrivial surfaces which may be used to approximate a smooth surface Φ are *quadrics*. Hence, it is interesting to discuss the cyclographic contour and silhouette of a quadric Φ. We have to construct the connecting developable surface $\Gamma(\Phi)$ between Φ (surface of algebraic class 2) and the conic at infinity Ω (as a set of tangent planes, also a surface of class 2; it is dual to the point set of a quadratic cone). Hence, the connecting developable is dual to the intersection curve of two surfaces of order 2 and hence it is algebraic (in general not rational) and has class four. This means that, algebraically counted, four tangent planes pass through a general point. Hence, the intersection of $\Gamma(\Phi)$ with the plane E^2, i.e., *the cyclographic silhouette is an algebraic curve of class* 4. The class may be less in case of degree reductions. We do not go into such details here, but mention that this study is dual to degree reductions of central projections of the intersection curve of two quadrics. The developable surface $\Gamma(\Phi)$ possesses Ω as a self-intersection curve. In general, there are three further self-intersections, which are conics. For a quadric with midpoint m, these conics lie in planes through m. Clearly, the silhouettes are therefore also cyclographic images of conics. For more details on the silhouettes, which are known in classical geometry as *hypercycles*, see W. Blaschke [5].

Figure 14 illustrates the case of an ellipsoid. In this example we have two topological components of the boundary of $D(\Phi)$ belonging to the two topological components of the γ-developable $\Gamma(\Phi)$. However, the (untrimmed) boundary is just one irreducible algebraic curve and the γ-developable

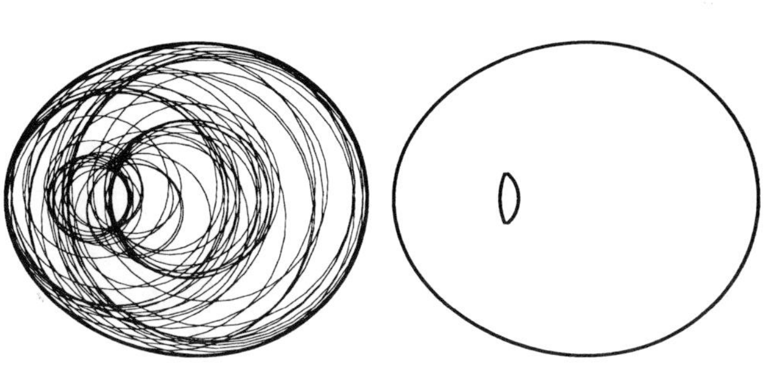

Figure 15. 2-parameter set of circles bounded by the untrimmed part of the cyclographic silhouette (left), trimmed silhouette (right)

is an irreducible algebraic surface. Figure 15 shows some circles in the corresponding 2-parameter set of circles. It illustrates that fact that the (trimmed) cyclographic silhouette appears as boundary of the domain covered by the set of circles.

Remark. Although it is not necessary for the computation of the silhouette, it is interesting to ask for the *cyclographic contour* of a quadric. The polarity κ with respect to a quadric maps each point of the quadric to its tangent plane and vice versa. Hence, κ maps the tangent planes of $\Gamma(\Phi)$ to the points of the contour $c(\Phi)$. By linearity of κ, it follows that *the cyclographic contour of a quadric is in general an algebraic curve of order four* (image of a developable surface of class 4). It may be computed as the intersection of Φ with the polar image of $\Omega \subset \Gamma(\Phi)$, which is a quadratic cone. The same argumentation applies for arbitrary isophotes and thus we see the classical result that *isophotes of quadrics are algebraic curves of order* ≤ 4.

3.3. Three-parameter sets of circles

These are the cyclographic images of *solids* in $\mathbb{R}^3$. Contributions to the boundary of domain $D(s)$ covered by the circle system can only be cyclographic images of boundary surfaces, edge curves and vertices of the solid.

Example 3.3. We discuss the tolerance region D for a circle given by a fat midpoint M and an interval $I = [r, R]$ for the radius (r shall be larger than the diameter of M). The set of all circles with a midpoint in the convex domain M and a radius $\rho \in I$ has the following solid S as cyclographic preimage: S is bounded by the two planes $x_3 = r$, $x_3 = R$ and a cylinder surface with x_3-parallel rulings and base curve through the boundary curve m of M. These three surfaces cannot have γ-planes as tangent planes and thus need not be considered for a contribution to the boundary of D. The contribution can only come from the two convex boundary curves m_r

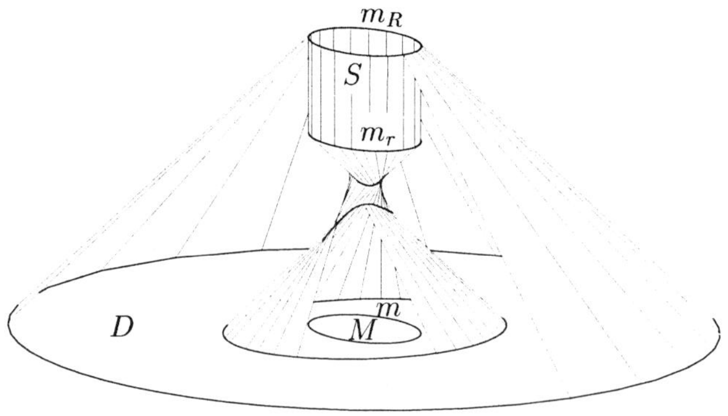

Figure 16. Cyclographic preimage of the tolerance region of a circle with fat center and radius $\rho \in [r, R]$

and m_R which are translates of m by vectors $(0,0,r)$ and $(0,0,R)$, respectively. The complete cyclographic image of m_r is the two-sided offset of m at distance r, and analogously the image of m_R is the two-sided offset at distance R. The γ-developable through m_r, which leads to the 'exterior' offset of m at distance r (defined by $m + rn$ with outside unit normals n of the curve m) cuts at points of m_r locally into the solid; hence it does not contribute to D. The same holds for the interior offset of m at distance R. The exterior offset of m at distance R is always convex. It bounds the outer boundary curve of the ring-shaped domain D. The inner boundary curve is the trimmed interior offset of m at distance r. This simple result can be obtained easily without the cyclographic model (see also [16]). We used it here just to give an example of the concept. Figure 16 shows an axonometric view.

4. Tolerance zones for circles in the plane

4.1. Circles through a fat point

For a tolerance analysis of circle constructions with help of the cyclographic model, the following problem is of fundamental importance: Given a convex domain K (fat point), find the cyclographic preimage S of all circles through the fat point, i.e., circles intersecting K.

A circle intersecting K, which contributes to the boundary of S must possess a circle in an arbitrarily small neighborhood of itself, which does not intersect K. Hence, such a circle must be tangent to K. All circles touching K have as cyclographic preimage two γ-developables Γ_1, Γ_2 through the boundary k of K. Confining to a positive radius, we may restrict to the part in $x_3 > 0$. Let Γ_1 belong to circles which are centered outside K and touch K from outside. Together with $E^2 : \ x_3 = 0$, the surface Γ_1 confines an unbounded

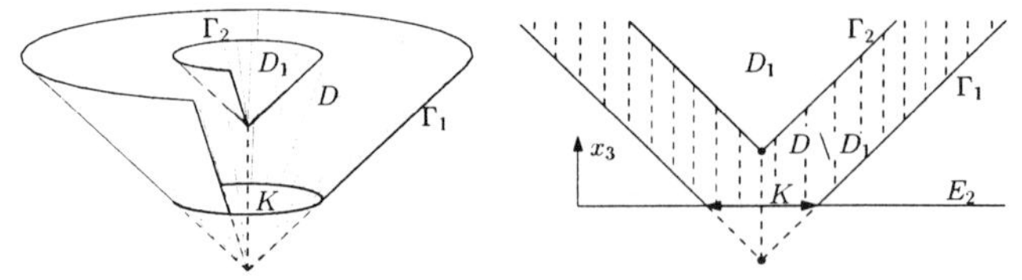

Figure 17. Cyclographic preimage of circles through a fat point (disk), axonometric and front view

convex domain D in $\mathbb{R}^3$.

Now, we have to remove that subset D_1 of D, whose cyclographic image circles cover K. Here and in the following we say that a circle 'covers' a set A, if the closed circular disk contains A. The set D_1 is bounded by the part of Γ_2, whose cyclographic image circles touch k and cover K. This part is therefore the one which is visible for parallel projection in direction $(0,0,-1)$. In general, it contains edges which are self-intersections of the complete surface Γ_2 (see Fig. 18). *D_1 is convex*: Given two points $a,b \in D_1$, their cyclographic image circles cover K. The images circles of the points x on the straight line segment with end points a,b cover the intersection of the circular disks bounded by $\zeta(a)$ and $\zeta(b)$. Hence, $\zeta(x)$ also covers K. Thus, $x \in D_1$, and we have proved the convexity of D_1.

Theorem 2. *The cyclographic preimage of all circles (with positive radius) through a fat point K is the difference set $D \setminus D_1$ of two unbounded convex domains in $\mathbb{R}^3$. D is bounded by the convex domain K in the plane E^2 and part of a γ-developable Γ_1 through the boundary k of K; its points are cyclographic preimages of circles touching k from outside. D_1 is bounded by the part of the second γ-developable Γ_2 through k, which is visible for projection parallel to the vector $(0,0,-1)$. This boundary possesses cyclographic image circles, which touch k and cover K.*

Example 4.1. As the simplest example, let us consider a circular disc K with center (m_1,m_2) and radius r. The two γ-developables through the circle k are two γ-cones Γ_1 (with vertex $(m_1,m_2,-r)$) and Γ_2 (with vertex (m_1,m_2,r)). D is bounded by K and the (smooth) part of Γ_1 which lies above E^2. D_1 is bounded by the upper half of the cone Γ_2 (Fig. 17).

Example 4.2. Another example is illustrated in Figures 18 and 19. The fat point K in E^2 is bounded by an ellipse k. The complete γ-developable through k is an algebraic developable surface of class four (connection with the ideal conic Ω). The surface has four conics as self-intersections: one is the ideal conic Ω and another one is the ellipse k. The remaining two self-intersections lie in z-parallel planes through the axes of the ellipse. In the upper halfspace $z \geq 0$,

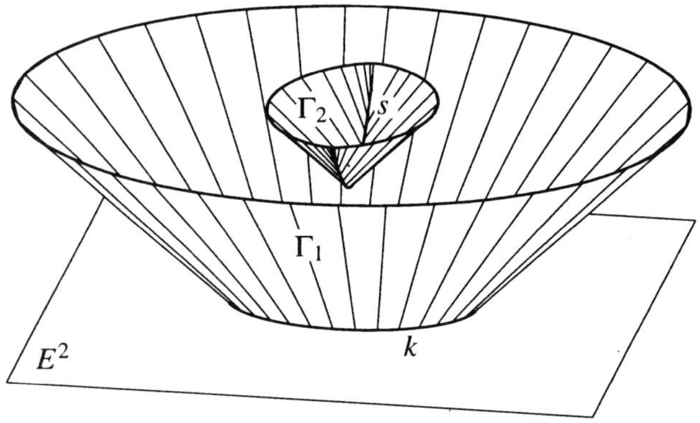

Figure 18. Boundary surfaces of cyclographic preimage of circles through a fat point (bounded by an ellipse k)

the outer part Γ_1 is free of self intersections, but the other component Γ_2 has two conic arcs as self-intersection; one of them appears as edge of the boundary of D_1 (Fig. 19).

Fat circle through three fat points

As an application of the previous considerations, let us study the region D covered by all circles that intersect three convex domains K_1,K_2,K_3. They shall be arranged such that they form a *fat triangle*. This requires that they may not be viewed as collinear, i.e., no line intersects all three domains K_i. In particular, this implies that the zones K_i are pairwise disjoint.

The cyclographic preimage of this 'fat circumcircle' is the intersection S of the domains $S(K_i)$, characterizing circles which intersect K_i according to Theorem 2. The bounding surfaces of S are γ-developables only; hence their cyclographic silhouettes are the boundary curves k_i of K_i. The edges occurring as intersections of two γ-developables also possess just parts of k_i as cyclographic images. Hence, other contributions to the boundary of D can only be parts of circles, namely the cyclographic images of the vertices of S. Not all of these vertices need to generate part of the boundary. One of the vertices, say v_{---} belongs to a circle, which covers all three K_i. Another one, v_{+++} has as cyclographic image a circle touching the K_i from outside. It is not trivial to prove that these vertices (and image circles) exist. The straight line segment $v_{---}v_{+++}$ generates circles all of which intersect K_i, $i = 1,2,3$. These circles cover a domain which contains all K_i and thus the boundaries k_i of K_i do not contribute to the boundary of D. Together with Theorem 2 on the boundary of $S(K_i)$ we can state the following result. For its formulation we further note: A circle touches a domain K_i from outside or covers it iff it intersects K_i in exactly one point.

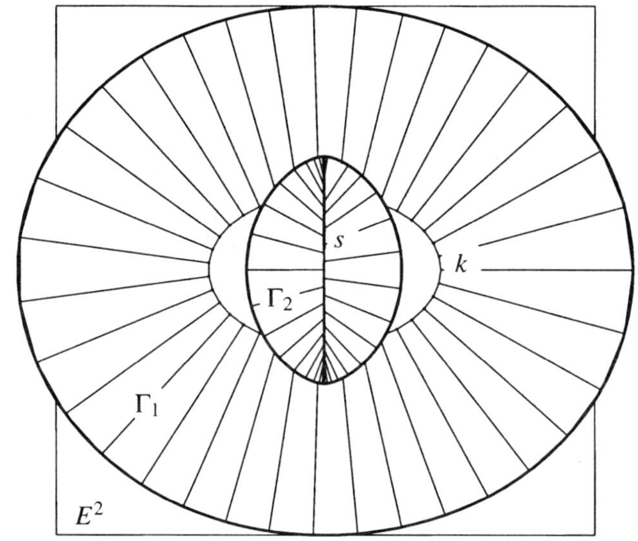

Figure 19. Top view of Γ_1, Γ_2, ellipse k and self-intersection s of Γ_2

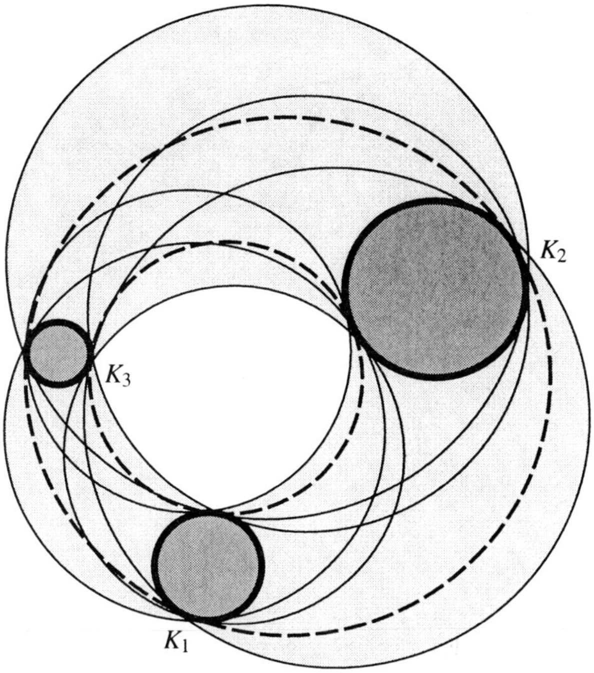

Figure 20. Fat circle through three fat points

Theorem 3. *The domain covered by all circles through three fat vertices K_1, K_2, K_3 of a triangle is bounded by parts of circles which intersect K_i in exactly one point.*

In case of three circular disks K_i, there are 8 candidate circles, but two of them, namely circles touching all three k_i from outside or covering all three (the cyclographic images of the vertices v_{---} and v_{+++}) do not contribute to the boundary [46] (see Fig.20). An analogous statement can be made about $\zeta(v_{---})$ and $\zeta(v_{+++})$ for general convex domains K_i.

For practical calculations it seems advantageous to work with piecewise linear and circular boundary curves k_i of the fat points. Then the computation of the vertices of the solid S in $\mathbb{R}^3$ can be done efficiently. It amounts only to intersecting γ-planes or γ-cones and thus needs no numerical approximation.

4.2. Circles tangent to a toleranced line

Consider a fat line spanned by two fat points K_1, K_2. The fat line consists of all lines intersecting both K_1 and K_2. This set of lines is a set of points in dual projective plane and thus we can speak of its boundary B. Dual to the boundary of the intersection of two domains (2-parameter point sets) in the plane, we see that B consist of *all lines which touch at least one of the convex domains K_i and intersects the other one*. If the boundaries of K_i are smooth, B is piecewise smooth as a set of lines: its envelope consists of four segments $c_{11}, c_{12}, c_{21}, c_{22}$ on the boundary curves of K_i (between the contact points of common tangents). Discontinuities of the contact point with the envelope appear at the common tangents (see Fig. 21).

We now look at the cyclographic preimage of the set of circles which touch the fat line from a given side. All circles touching an oriented line L have their cyclographic preimages in a γ-plane through L. Since we are confining to a positive radius, we need just the half-plane in $x_3 > 0$. In this way, we associate a γ-plane with each line in the fat line. The domain S traced out in space is bounded by parts of the γ-planes to the common tangents of K_1, K_2 and parts of the γ-developable surfaces through the curve segments c_{ij}.

To study the domain S, we may intersect it with a plane $\varepsilon : x_3 = r = const$. It belongs to circles with radius r. The orthogonal projection of such a horizontal slice $\varepsilon \cap S$ into E^2 is the tolerance region for the centers of circles with radius r which touch the fat line.

For $r = 0$, we get the boundary of the tolerance zone $S \cap E^2$ to the fat line; here, the curves c_{ij} do not contribute to the boundary. However, increasing r, we see that the mentioned developable surfaces contribute to the boundaries of the slices; these curved boundary parts are of course offsets of c_{ij} at distance r (Fig. 21).

As an application we may discuss the *fat circle touching three given fat lines*. Each fat line may be given as connection of two fat points; thus, altogether we admit 6 input points (not just a triangle). We have to intersect the three domains S_1, S_2, S_3 corresponding to the contacting circles

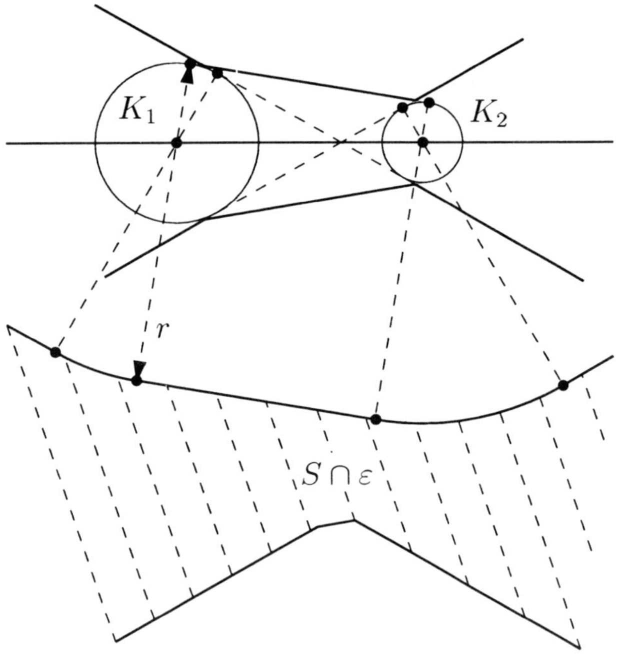

Figure 21. Tolerance domain for the centers of circles with fixed radius r which are tangent to a toleranced line

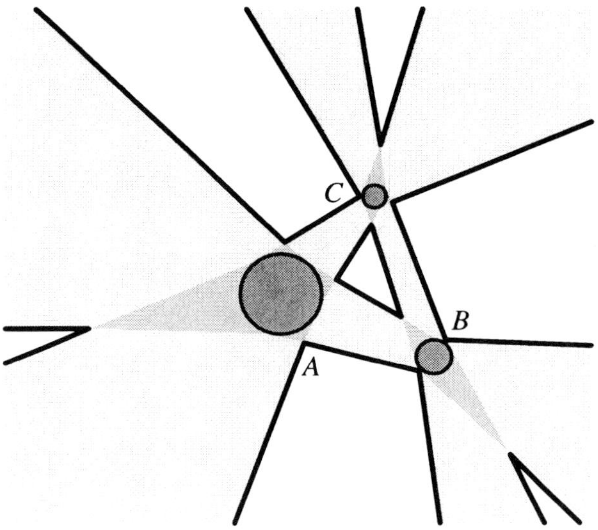

Figure 22. Fat triangle

of each of the three lines. The intersecting solid S has as its boundary surfaces only γ-planes and γ-developable surfaces. Hence, there will be no nontrivial cyclographic contours and silhouettes. This proves the following result:

Theorem 4. *Let us consider three fat lines L_A, L_B, L_C, each of them given as connection of two fat points $[A_1, A_2]$, $[B_1, B_2]$ and $[C_1, C_2]$, respectively. Then, the domain covered by all circles which are tangent to the three fat lines at a given side is bounded by parts of circles which touch triples of lines H_A, H_B, H_C of the following type: H_A touches either A_1 or A_2 (at a boundary point whose tangent intersects A_2 or A_1, resp.) or a common tangent of A_1, A_2; analogous statements hold for H_B, H_C.*

Further research will be necessary to make a sharper statement and to rule out parts of A_i, B_i, C_i on the boundary.

With an appropriate definition of the side from which the circle shall touch the three given lines, the resulting tolerance zone is the tolerance zone for the inscribed circle of a fat triangle. Its vertices are the intersections of the tolerance zones to the given lines. However, given three fat points A, B, C, defining the three fat connecting lines, and then doing the present construction does *not* give the tolerance region for the inscribed circle to the triangles with the given fat vertices A, B, C . This is so, because the tolerance regions to the edges of the fat triangle have intersections which cover the given fat points A, B, C, but they are in general larger than those (see Fig. 22).

4.3. Tangents to a fat circle

Let us consider a fat circle represented in the cyclographic model by some solid $S \in \mathbb{R}^3$. We would like to pass tangents through a given point in the plane to the circles in this 3-parameter family and ask for the *tolerance zones of the contact points*.

To start with a simple case, we look at tangents parallel to a given (at the moment) precise line $L \in E^2$ with a unit normal vector (n_1, n_2). The two lines, parallel to L, which touch a circle c with center (p_1, p_2) and radius p_3 have contact points $(p_1, p_2) \pm p_3(n_1, n_2)$. These points may be viewed as images of the cyclographic preimage $p = (p_1, p_2, p_3)$ of c under a projection parallel to vectors $l_1 = (n_1, n_2, 1)$ or $l_2 = (n_1, n_2, -1)$. The projection lines are γ-lines in planes orthogonal to L. We see that in the present situation *the tolerance zones for the contact points are the images of the cyclographic preimage S under projections parallel to vectors l_1 and l_2.*

This is extended to the case of an inprecise direction of L. Instead of l_1, we have the rulings of a part Γ_1 of a γ-cone as possible projection rays, and analogously for l_2. Let a_1, b_1 be the end rulings of the cone segment Γ_1. Now, the tolerance zone for one of the contact points contains at its boundary the following curves: *parts of the boundaries of the projections of S parallel to a_1, b_1 and part of the cyclographic image of S.* For the latter, only those cyclographic contour points on S are admitted at which a normal of S is parallel to a normal of the cone segment Γ_1, i.e., a tangent

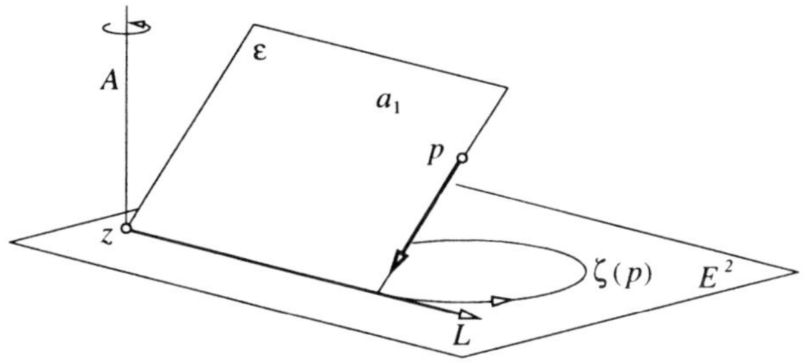

Figure 23. Construction of tangents to a fat circle

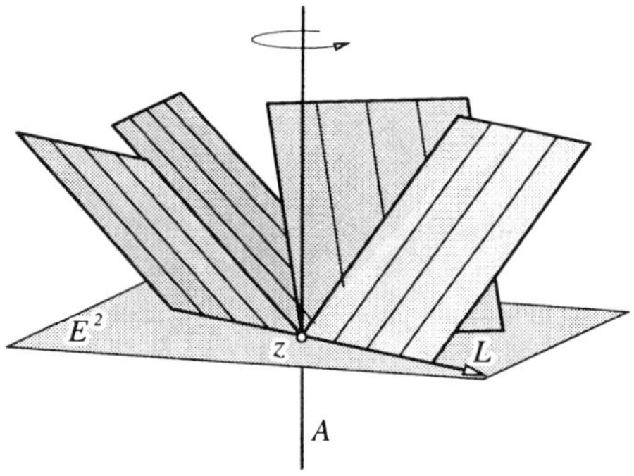

Figure 24. Rotationally symmetric γ-congruence

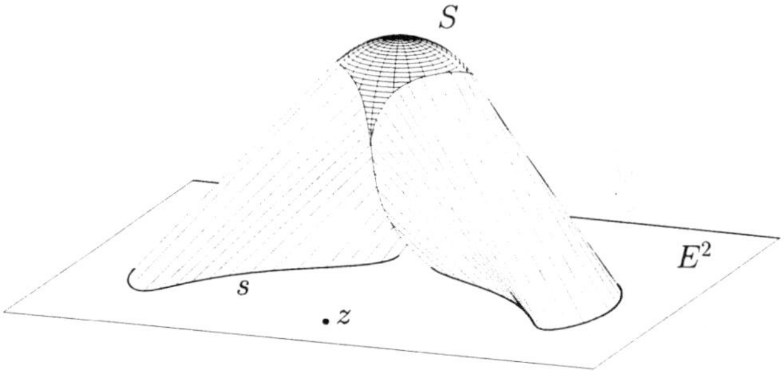

Figure 25. Cyclographic preimage S of fat circle and construction of tolerance region of contact points of tangents through z to the fat circle

of the cyclographic image is within the given range of directions of L.

Contact points of tangents through a given (precise) point $z \in E^2$ can also be successfully discussed with help of the cyclographic model. Consider an oriented line L through z. Then, the cyclographic preimage of the circles touching L are points p in a γ-plane ε through L. The corresponding contact points are obtained as projections of p with γ-lines in the plane ε (Fig. 23). Rotating the pencil of γ-lines in ε around the x_3-parallel line A through z, we obtain a two-parameter set of lines, i.e. a line congruence C. We call it *rotationally symmetric γ-congruence with axis A*, see Figure 24. In line geometric terms, C is an algebraic congruence with bundle degree two (algebraically counted, two lines of C pass through a general point) and field degree two (2 lines of K lie in a general plane).

The two lines of C through a given point $p \in \mathbb{R}^3$ intersect E^2 in the contact points of those tangents of $\zeta(p)$, which pass through p. Hence, the contact regions are again obtained via a projection, but now the projecting lines are taken from the congruence C. The boundary of the projection of the solid S, representing the fat circle, is formed by image curves of edges and silhouettes of surfaces, arising as intersections of those lines $l \in C$, which are tangent to S. We see that the nontrivial part of the boundary again amounts to the determination of the silhouette with respect to some projection. Let us summarize:

Theorem 5. *The tolerance regions of the contact points of tangents from a fixed point z to the circles of a k-parameter family are obtained as follows: Project the cyclographic preimage of the circle set via the rotationally symmetric congruence of γ-lines with axis $A \ni z$ into E^2.*

This generalized projection deserves further studies, both from an algebraic and differential geometric viewpoint. We also mention that the tangents from a fat point can be discussed in a similar way. However, at each point in space, we then have two segments of γ-cones for the projection.

Aspects of the computation of contours of surfaces with respect to line congruences have been studied by Elber and Cohen [9] in connection with accessibility in NC machining.

Example 4.3. Consider a fat circle, whose cyclographic preimage is a sphere S. In the plane E^2, the fat circle consists of a two-parameter set of circles. We construct the tolerance zones for the contact points of tangents from a fixed point z to these circles according to Theorem 5. Fig. 25 shows the two components of the ruled surface formed by those lines in the rotationally symmetric γ-congruence, which touch the sphere S. The intersection with E^2 yields the boundaries of the tolerance zones for the two contact points, see Figure 26. Whereas a parallel projection maps a convex surface onto a convex silhouette, the present generalized projection obviously does not possess this convexity preserving property.

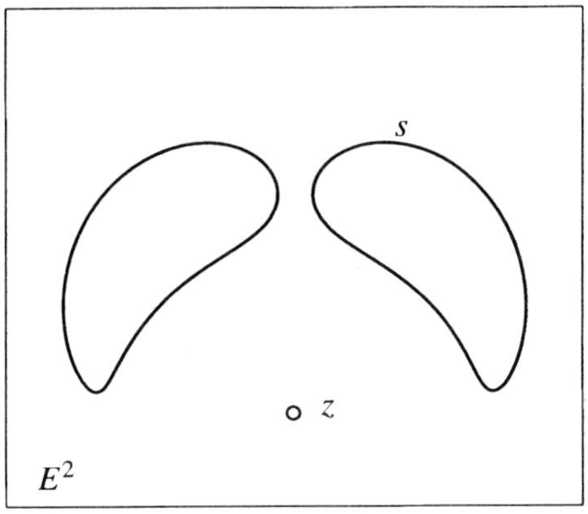

Figure 26. Tolerance region of contact points of tangents through z to a fat circle

5. Conclusion and future research

We have outlined geometric concepts for a worst case tolerance analysis of CAD constructions. From a theoretical viewpoint, linear constructions as well as linear freeform curve schemes are understood very well.

We have also shown how the cyclographic mapping helps to study planar constructions involving circles. Here, further research is needed to get a better understanding of the global behaviour of the resulting tolerance zones.

Missing planar constructions concern normals to lines and reflections in the plane, which will be subject of a separate contribution. Very much remains to be done in three dimensions, both concerning metric constructions and reliable approximations of tolerance zone boundaries of surfaces.

Another important issue which needs to be addressed in the future is an efficient implementation. For that, it will be necessary to make limitations on the shapes which are used to represent the tolerance zone boundaries.

Acknowledgements

This research has been supported in part by projects P12252-MAT and P13648-MAT of the Austrian Science Fund. R. Ait-Haddou acknowledges support by a fellowship from Vienna University of Technology and by the Geomechanics Project at the University of Calgary. We would also like to thank N. Pomaroli for his help with some figures.

References

[1] S. Abrams, W. Cho, C.-Y. Hu, T. Maekawa, N. Patrikalakis, E. Sherbrooke, and X. Ye. Efficient and reliable methods for rounded-interval arithmetic. *Computer-Aided Design*, 30:657–665, 1998.

[2] M. Aichinger. *A collision problem for worst case tolerance objects*. PhD thesis, Kepler University Linz, 1996.

[3] R. Ait-Haddou, L. Biard, and M. A. Slawinsky. Pythagorean hodograph curves in the Minkowski plane I: geometric characterization. *Submitted to Computer Aided Geometric Design*.

[4] R. Ait-Haddou, L. Biard, and M. A. Slawinsky. Pythagorean hodograph curves in the Minkowski plane II: dual Bézier characterization. *Submitted to Computer Aided Geometric Design*.

[5] W. Blaschke. Untersuchungen über die Geometrie der Speere in der Euklidischen Ebene. *Monatshefte für Mathematik und Physik*, 21:3–60, 1910.

[6] J. Coolidge. *A Treatise on the Circle and the Sphere*. Clarendon Press, Oxford, 1916.

[7] B. Cousinery. *Géométrie Perspective*. Paris, 1828.

[8] C. de Boor, K. Höllig, and M. Sabin. High accuracy geometric Hermite interpolation. *Computer Aided Geometric Design*, 4:269–278, 1987.

[9] G. Elber and E. Cohen. A unified approach to accessibility in 5-axis freeform milling environments. In B. C. G.J. Olling and R. Jerard, editors, *Machining impossible shapes*, pages 33–41, Boston, 1998. Kluwer.

[10] R. Farouki. Pythagorean-hodograph curves in practical use. In R. E. Barnhill, editor, *Geometry Processing for Design and Manufacturing*, pages 3–33, Philadelphia, 1992. SIAM.

[11] R. Farouki and T. Goodman. On the optimal stability of the Bernstein basis. *Mathematics of Computation*, 216:1553–1566, 1996.

[12] R. Farouki, H. Moon, and B. Ravani. Algorithms for Minkowski products and implicitly–defined complex sets. *Preprint*, 1999.

[13] R. Farouki, H. Moon, and B. Ravani. Minkowski geometric algebra of complex sets. *Geometriae Dedicata*, 2000.

[14] R. Farouki and T. Sakkalis. Pythagorean hodographs. *IBM J. Res. Develop.*, 34:736–752, 1990.

[15] P. Gruber and J. Wills. *Handbook of Convex Geometry*, volume 1,2. North Holland, Amsterdam, 1993.

[16] C. U. Hinze. *A contribution to optimal tolerancing in 2-dimensional computer aided design*. PhD thesis, Kepler University Linz, 1994.

[17] C. Hu, T. Maekawa, E. Sherbrooke, and N. Patrikalakis. Approximation of measured data with interval B-splines. *Computer-Aided Design*, 29:791–799, 1997.

[18] C.-Y. Hu, N. Patrikalakis, and X. Ye. Robust interval solid modeling: part I, representations. *Computer-Aided Design*, 28:807–817, 1996.

[19] C.-Y. Hu, N. Patrikalakis, and X. Ye. Robust interval solid modeling: part II, boundary evaluation. *Computer-Aided Design*, 28:819–830, 1996.

[20] C.-Y. Hu, N. Patrikalakis, and X. Ye. Robust interval algorithm for surface intersections. *Computer-Aided Design*, 29:617–627, 1997.

[21] A. Kaul and R. Farouki. Computing Minkowski sums of plane curves. *Intl. J. Computational Geometry & Applications*, 5:413–432, 1995.

[22] R. Krasauskas and C. Mäurer. Studying cyclides with Laguerre geometry. *Comput. Aided Geom. Design*, 17(2):101–126, 2000.

[23] I. Lee, M. Kim, and G. Elber. Polynomial/rational approximation of Minkowski sum boundary curves. *Graphical Models and Image Processing*, 60:136–165, 1998.

[24] S. Leopoldseder. *Cone spline surfaces and spatial arc splines*. PhD thesis, Vienna Univ. of Technology, 1998.

[25] S. Leopoldseder and H. Pottmann. Approximation of developable surfaces with cone spline surfaces. *Computer-Aided Design*, 30:571–582, 1998.

[26] Q. Lin and J. Rokne. Disk Bézier curves. *Comp. Aided Geometric Design*, 15:721–737, 1998.

[27] T. Maekawa. An overview of offset curves and surfaces. *Computer Aided Design*, 31:165–173, 1999.

[28] D. S. Meek and D. J. Walton. Planar osculating arc splines. *Comp. Aided Geometric Design*, 13:653–671, 1996.

[29] H. Moon. Minkowski Pythagorean hodographs. *Comp. Aided Geometric Design*, 16:739–753, 1999.

[30] R. Moore. *Interval Analysis*. Prentice-Hall, Englewood Cliffs, NJ, 1966.

[31] E. Müller and J. Krames. *Vorlesungen über Darstellende Geometrie II: Die Zyklographie*, volume 2. Deuticke, Wien/Leipzig, 1929.

[32] M. Peternell and H. Pottmann. Computing rational parametrizations of canal surfaces. *Journal of Symbolic Computation*, 23:255–266, 1997.

[33] H. Pottmann. Rational curves and surfaces with rational offsets. *Comp. Aided Geometric Design*, 12:175–192, 1995.

[34] H. Pottmann. General offset surfaces. *Neural, Parallel & Scientific Comp.*, 5:55–80, 1997.

[35] H. Pottmann and G. Farin. Developable rational Bézier and B-spline surfaces. *Computer Aided Geometric Design*, 12:513–531, 1995.

[36] H. Pottmann and M. Peternell. Applications of Laguerre geometry in CAGD. *Comp. Aided Geometric Design*, 15:165–186, 1998.

[37] H. Pottmann and J. Wallner. Approximation algorithms for developable surfaces. *Comp. Aided Geometric Design*, 16:539–556, 1999.

[38] A. A. G. Requicha. Towards a theory of geometric tolerancing. *Int. J. of Robotics Research*, 2:45–60, 1983.

[39] H. Schaal. Zur Konstruktion der Krümmungskreise des scheinbaren Umrisses einer Fläche bei Zentral- und Parallelprojektion. *Sitzungsberichte d. Bayrischen Akademie d. Wissenschaften, math.-naturwiss. Klasse*, 17:277–310, 1960.

[40] T. Sederberg and R. Farouki. Approximation by interval Bézier curves. *IEEE Computer Graphics*, 12:87–95, 1992.

[41] G. Shen and N. Patrikalakis. Numerical and geometric properties of interval B-splines. *Int. J. Shape Modeling*, 4:31–62, 1998.

[42] S. Tuohy, T. Maekawa, G. Shen, and N. Patrikalakis. Representation of geophysical maps with uncertainty. In N. M. Thalmann and D. Thalmann, editors, *Communicating with Virtual Worlds*, volume 4 of *Proc. of CG International '93*, Tokyo, 1993. Springer.

[43] S. T. Tuohy and N. M. Patrikalakis. Nonlinear data representation for ocean exploration and visualization. *Journal of Visualization and Computer Animation*, 7:125–139, 1996.

[44] J. U. Turner. *Tolerances in computer-aided geometric design*. PhD thesis, Rensselaer Polytechnic Institute, Troy, NY, 1987.

[45] H. Vogler. Ein einfacher Beweis eines Satzes von H. Schaal über den Krümmungsmittelpunkt des scheinbaren Umrisses einer Fläche. *Elemente d. Math.*, 17:79–81, 1962.

[46] J. Wallner, R. Krasauskas, and H. Pottmann. Error propagation in geometric constructions. *Computer-Aided Design, to appear*, 2000.

[47] J. Wallner, T. Sakkalis, T. Maekawa, H. Pottmann, and G. Yu. Self-intersections of offset curves and surfaces. *Preprint*, 1999.

A Computational Model for Nonrational Bisector Surfaces: Curve-Surface and Surface-Surface Bisectors

Gershon Elber
Department of Computer Science
Technion, Israel Institute of Technology
Haifa 32000, Israel
E-mail: gershon@cs.technion.ac.il

Myung-Soo Kim
Dept. of Computer Engineering
Seoul National University
Seoul 151-742, Korea
E-mail: mskim@comp.snu.ac.kr

Abstract

The bisector of two rational surfaces in $\boldsymbol{R}^3$ is, in general, nonrational; and so is the bisector of a rational curve and a rational surface. Thus, bisector surfaces in these two cases must be approximated numerically. Unfortunately, they are algebraic surfaces of very high degree and numerical approximation is non-trivial.

This paper suggests a new computational model for constructing curve-surface and surface-surface bisectors in $\boldsymbol{R}^3$. The curve-surface bisector problem is reformulated as the search for a trivariate zero-set; and the surface-surface bisector problem is reduced to that of finding the common zero-set of two four-variate functions.

1. Introduction

The medial axis of a 3D object consists of various bisector surfaces: point-point, point-curve, point-surface, curve-curve, curve-surface, and surface-surface bisectors. The point-point bisector is the plane of symmetry between two points. The point-curve bisector is a rational developable surface if the original curve is rational [12]. The point-surface bisector is a rational surface if the original surface is rational [6]. And the bisector of two rational space curves is also a rational surface [3]. In contrast, curve-surface and surface-surface bisectors are in general nonrational, even if the input curves and surfaces are all rational: they are algebraic surfaces of such a high degree that even numerical approximation is not trivial. In the plane, the bisector of two rational curves is nonrational. Farouki and Ramamurthy [8, 13] present a robust numerical tracing algorithm for these bisectors; however, it is not easy to extend this approach to similar problems in higher dimensions.

We suggest an alternative way of representing nonrational bisector surfaces in terms of contraints represented in uvt or $uvst$-space, where u, v, s, t are the parameters of the input curves and surfaces. An important advantage of this alternative representation is that the degrees of constraint equations in the uvt or $uvst$-space are considerably lower than those of bisector surfaces in the xyz-space. This paper is an extension of the authors' approach to the two-dimensional case [4].

Given two planar rational curves $C_1(u)$ and $C_2(v)$, of degrees d_1 and d_2, respectively, Elber and Kim [4] reformulate the curve-curve bisector problem as a search for the zero-set of a polynomial of degree $2(d_1 + d_2) - 2$ in the uv-plane. So, for example, given two cubic curves in the plane, the problem reduces to the solution of a polynomial equation of degree 10. This is a considerably lower degree than that of the bisector curve in the xy-plane which will now be discussed.

No analysis of the degree of curve-curve bisectors of arbitrary degree has yet been given as a formal proof. However, using Mathematica [15], we observed that the bisector curve in the xy-plane has an algebraic degree of $7d_1d_2 - 3(d_1 + d_2) + 1$, for $1 \le d_1 \le 3$ and $1 \le d_2 \le 6$. (Because of the lack of memory space, we have not been successful in checking the degree analysis for higher values of d_1 and d_2.) This means that the bisector of two cubic polynomial curves has degree 46, which is too high to be useful in practice. Thus, this exploratory degree analysis supports the approach of Elber and Kim [4].

Curve-surface and surface-surface bisectors will be of even higher degree in xyz-space. Therefore, previous work on bisector surfaces has focused on restricted types and configurations [2, 5, 6, 12]. In this paper, we take a first step towards computing the curve-surface and surface-surface bisectors for arbitrary rational curves and surfaces.

We represent the geometric constraints on the curve-surface bisector using a polynomial equation, given in terms of the parameters of the initial curve and surface. These constraints on the bisector are represented as the zero-set of a trivariate polynomial function, which is an algebraic surface in uvt-space. The curve-surface bisector is then constructed by transforming the zero-set into xyz-space using a rational map. The surface-surface bisector is represented by the common zero-set of two four-variate polynomial equations, constructed in terms of the parameters of two input surfaces. These equations represent two algebraic hypersurfaces in $uvst$-space and their intersection is the common zero-set. There is a rational map from the $uvst$-space to the bisector surface; thus, the construction of bisector surfaces is straightforward once the common zero-set has been found.

The zero-set in the parameter space has geometric properties different from those of the bisector surface in the xyz-space, and the error in computing the zero-set may be greatly magnified in the construction of the bisector surface. Nevertheless, using the rational map from the zero-set to the bisector surface, we can predict the error propagation and employ higher precision when necessary. Thus the accuracy of bisector construction mainly depends on the error in approximating the zero-set. The algebraic degree of constraint equation(s) for a zero-set can be very high; however, the Bernstein representation handles high-degree equations in a stable way. Computing time is still the main bottleneck in the overall procedure. The reasonable performance of our algorithm for the examples demonstrated in this paper (from several seconds to a few minutes) shows that the basic approach is promising, while there are various ways of improving the performance in future work.

The rest of this paper is organized as follows. Section 2 reformulates the curve-surface and surface-surface bisector problems as zero-set finding problems. In Section 3, we describe numerical methods for computing the zero-sets. In Section 4, we present some experimental results that demonstrate the effectiveness of our approach. We conclude this paper with some final remarks in Section 5.

2 Computational Model

We will now propose a computational model for the curve-surface and surface-surface bisectors in $\boldsymbol{R}^3$. The bisector surface of a rational space curve and a rational surface will be considered in Section 2.1, and that of two rational surfaces in Section 2.2.

2.1 Curve–Surface Bisector

Let $C(t)$ be a regular rational curve and $S(u,v)$ be a regular rational surface. We assume that they are sufficiently differentiable, as necessary. Now consider a bisector point $\mathcal{B}$ between $S(u,v)$ and $C(t)$:

$$0 = \left\langle \mathcal{B} - S(u,v), \frac{\partial S(u,v)}{\partial u} \right\rangle, \quad (1)$$

$$0 = \left\langle \mathcal{B} - S(u,v), \frac{\partial S(u,v)}{\partial v} \right\rangle, \quad (2)$$

$$0 = \left\langle \mathcal{B} - C(t), \frac{dC(t)}{dt} \right\rangle, \quad (3)$$

$$0 = \langle \mathcal{B} - C(t), \mathcal{B} - C(t) \rangle \quad (4)$$

$$- \langle \mathcal{B} - S(u,v), \mathcal{B} - S(u,v) \rangle. \quad (5)$$

Equations (1) and (2) tell us that the bisector point $\mathcal{B}$ is located on the normal line of $S(u,v)$, and Equation (3) implies that $\mathcal{B}$ is contained in the normal plane of $C(t)$. Moreover, Equation (5) constrains the bisector point $\mathcal{B}$ to the symmetry plane of $S(u,v)$ and $C(t)$:

$$2\langle \mathcal{B}, S(u,v) - C(t) \rangle + ||C(t)||^2 - ||S(u,v)||^2 = 0, \quad (6)$$

which is obtained by cancelling the term $||\mathcal{B}||^2$ out of Equation (5). This symmetry plane contains all points equidistant from the two points $S(u,v)$ and $C(t)$.

Equations (1)–(5) are all linear in $\mathcal{B}$. These equations would produce a rational representation for the bisector $\mathcal{B}$ if the problem were given in $\boldsymbol{R}^n$, for $n \geq 4$. Unfortunately, we are restricted to $\boldsymbol{R}^3$, which does not provide sufficient degrees of freedom for a rational solution.

Given a curve and a surface in general position, consider the intersection between the normal line of the surface $S(u,v)$ and the normal plane of the curve $C(t)$. This point is given as the solution of Equations (1)–(3). The point of intersection $\mathcal{B}(u,v,t)$ has a rational representation as a trivariate volume in $\boldsymbol{R}^3$. When we substitute this trivariate representation into Equation (6), we get the following constraint equation:

$$0 = 2\langle \mathcal{B}(u,v,t), S(u,v) - C(t) \rangle + ||C(t)||^2 - ||S(u,v)||^2. \quad (7)$$

When $S(u,v)$ and $C(t)$ are both rational, this equation represents an algebraic surface in uvt-space.

A rational representation of $\mathcal{B}(u,v,t)$ is obtained by solving Equations (1)–(3). The solution, based on Cramer's rule, involves the determinants of 3×3 matrices. We take a direct geometric approach and derive a closed-form solution for the intersection point. The intersection point must be located on the normal line of the surface $S(u,v)$:

$$\mathcal{B} = S(u,v) + \alpha n(u,v), \tag{8}$$

where $n(u,v) = \frac{\partial S(u,v)}{\partial u} \times \frac{\partial S(u,v)}{\partial v}$ is an unnormalized normal vector of $S(u,v)$, and $\alpha \in \boldsymbol{R}$. Equations (1) and (2) are now clearly satisfied. Then, by substituting $\mathcal{B}$ into Equation (3), we get

$$0 = \left\langle S(u,v) + \alpha n(u,v) - C(t), \frac{dC(t)}{dt} \right\rangle,$$

or equivalently,

$$\alpha = \alpha(u,v,t) = \frac{\left\langle C(t) - S(u,v), \frac{dC(t)}{dt} \right\rangle}{\left\langle n(u,v), \frac{dC(t)}{dt} \right\rangle}. \tag{9}$$

The point $\mathcal{B}(u,v,t) = S(u,v) + \alpha(u,v,t)n(u,v)$ is the intersection point between the normal line of the surface and the normal plane of the curve. By substituting this point of intersection into Equation (5) or Equation (6), which enforce an equal distance from $S(u,v)$ and $C(t)$, we get the following constraint equation:

$$\begin{aligned} 0 &= \langle S(u,v) - C(t) + \alpha(u,v,t)n(u,v), \\ &\qquad S(u,v) - C(t) + \alpha(u,v,t)n(u,v) \rangle \\ &\quad - \langle \alpha(u,v,t)n(u,v), \alpha(u,v,t)n(u,v) \rangle \qquad (10) \\ &= \|S(u,v) - C(t)\|^2 + \\ &\qquad 2\alpha(u,v,t) \langle n(u,v), S(u,v) - C(t) \rangle . \end{aligned}$$

The zero-set of Equation (10) satisfies all four constraints of Equations (1)–(5). The first three constraints are satisfied by the proper selection of $\alpha(u,v,t)$ in Equation (9), and the fourth constraint is satisfied by the substitution of $\alpha(u,v,t)$ into the constraint, with the result shown in Equation (10).

Now we have reduced the problem of constructing a curve-surface bisector in $\boldsymbol{R}^3$ into the search for a zero-set of a trivariate function, as shown in Equation (10). In Section 3, we consider a numerical method for computing this zero-set. When the zero-set has been found, then, for each point (u,v,t), the corresponding bisector point can be computed by the rational map

$$\mathcal{B}(u,v,t) = S(u,v) + \frac{\left\langle C(t) - S(u,v), \frac{dC(t)}{dt} \right\rangle}{\left\langle n(u,v), \frac{dC(t)}{dt} \right\rangle} n(u,v). \tag{11}$$

2.2 Surface–Surface Bisector

Let $S_1(u,v)$ and $S_2(s,t)$ be two regular rational surfaces which are sufficiently differentiable, as necessary. We consider the bisector surface $\mathcal{B}$ of $S_1(u,v)$ and $S_2(s,t)$ that satisfies the following five constraint equations:

$$\begin{aligned} 0 &= \left\langle \mathcal{B} - S_1(u,v), \frac{\partial S_1(u,v)}{\partial u} \right\rangle, && (12) \\ 0 &= \left\langle \mathcal{B} - S_1(u,v), \frac{\partial S_1(u,v)}{\partial v} \right\rangle, && (13) \\ 0 &= \left\langle \mathcal{B} - S_2(s,t), \frac{\partial S_2(s,t)}{\partial s} \right\rangle, && (14) \\ 0 &= \left\langle \mathcal{B} - S_2(s,t), \frac{\partial S_2(s,t)}{\partial t} \right\rangle, && (15) \\ 0 &= \langle \mathcal{B} - S_1(u,v), \mathcal{B} - S_1(u,v) \rangle - \\ &\qquad \langle \mathcal{B} - S_2(s,t), \mathcal{B} - S_2(s,t) \rangle . && (16) \end{aligned}$$

Equations (12) and (13) mean that the bisector point $\mathcal{B}$ is located on the normal line of $S_1(u,v)$, and Equations (14) and (15) imply that $\mathcal{B}$ is on the normal line of $S_2(s,t)$. Additionally, Equation (16) constrains the point $\mathcal{B}$ to the symmetry plane of $S_1(u,v)$ and $S_2(s,t)$.

Equations (12)–(16) are all linear in $\mathcal{B}$. If the problem were given in $\boldsymbol{R}^n$, for $n \geq 5$, then these five equations would produce a rational representation for the surface-surface bisector. However, the problem is restricted to $\boldsymbol{R}^3$ and the computation of the surface-surface bisector in $\boldsymbol{R}^3$ is not easy.

Let $\mathcal{B} = S_1(u,v) + \alpha n_1(u,v)$, $\alpha \in \boldsymbol{R}$, be a bisector point located on the normal line of $S_1(u,v)$ (see Equation (8)), where $n_1(u,v) = \frac{\partial S_1(u,v)}{\partial u} \times \frac{\partial S_1(u,v)}{\partial v}$ is an unnormalized normal vector of $S_1(u,v)$. Then, Equations (12) and (13) are automatically satisfied. By substituting $\mathcal{B}$ into Equation (16), we get the equation

$$\begin{aligned} 0 &= \langle \alpha n_1(u,v), \alpha n_1(u,v) \rangle \\ &\quad - \langle S_1(u,v) + \alpha n_1(u,v) - S_2(s,t) \\ &\qquad S_1(u,v) + \alpha n_1(u,v) - S_2(s,t) \rangle . \end{aligned}$$

or, equivalently,

$$\alpha(u,v,s,t) = \frac{\langle S_1(u,v) - S_2(s,t), S_1(u,v) - S_2(s,t) \rangle}{-2 \langle n_1(u,v), S_1(u,v) - S_2(s,t) \rangle}. \tag{17}$$

Finally, we substitute this representation of $\alpha(u,v,s,t)$ into $\mathcal{B} = S_1(u,v) + \alpha n_1(u,v)$, and update Equations (14) and (15) with this new representation of $\mathcal{B}$. Let $\Delta(u,v,s,t)$ be defined as follows:

$$\begin{aligned}&\Delta(u,v,s,t)\\&= -2(S_1(u,v) - S_2(s,t))\\&\qquad \langle n_1(u,v), S_1(u,v) - S_2(s,t)\rangle\\&\quad + n_1(u,v)\langle S_1(u,v) - S_2(s,t), S_1(u,v) - S_2(s,t)\rangle.\end{aligned}$$

Then, Equations (14) and (15) reduce to

$$0 = \left\langle \Delta(u,v,s,t), \frac{\partial S_2(s,t)}{\partial s}\right\rangle, \tag{18}$$

$$0 = \left\langle \Delta(u,v,s,t), \frac{\partial S_2(s,t)}{\partial t}\right\rangle. \tag{19}$$

The common zero-set of Equations (18) and (19) satisfies all five constraints of Equations (12)–(16). Hence, we have reduced the surface-surface bisector problem of $\boldsymbol{R}^3$ into a common zero-set finding problem for the two four-variate functions of Equations (18) and (19). In Section 3, we consider numerical methods for computing this zero-set. When the zero-set has been found, then, for each point (u,v,s,t) in the zero-set, the corresponding bisector point can be computed by the rational map

$$\begin{aligned}&\mathcal{B}(u,v,s,t)\\&= S_1(u,v)\\&\quad - \frac{\langle S_1(u,v) - S_2(s,t), S_1(u,v) - S_2(s,t)\rangle}{2\langle n_1(u,v), S_1(u,v) - S_2(s,t)\rangle} n_1(u,v).\end{aligned} \tag{20}$$

3 Constructing the Bisector Surfaces

We now consider the problem of zero-set finding for multivariate functions, using an approach similar to those presented in References [9, 11, 14]. We are interested in computing the zero-set of the trivariate function introduced in Section 2.1, and the common zero-set for the pair of four-variate functions introduced in Section 2.2. In both cases two degrees of freedom remain; thus, the bisectors are surfaces in $\boldsymbol{R}^3$. In general they are nonrational algebraic surfaces and must be approximated numerically.

We will now subdivide the multivariate function(s) until the function values are bounded within a certain resolution. Section 3.1 discusses this subdivision stage, which ends up with a set of discrete points approximating the zero-set. We then apply a numerical improvement to this set of points, which is explained in Section 3.2. Finally, we fit a bivariate surface through the points, which will approximate the curve–surface and surface–surface bisectors. This surface fitting stage is discussed in Section 3.3

3.1 Subdivision Stage for Zero-Set Finding

Consider n multivariate polynomial functions $\mathcal{F}_i(u_1, u_2, \cdots, u_{m-1}) = 0$, $(i = 1, \cdots, n)$, in $\boldsymbol{R}^m$, where $n \leq m$. And let a point in the $(m-1)$-dimensional parameter space of $\mathcal{F}_i$ be denoted by $\boldsymbol{u} = (u_1, u_2, \cdots, u_{m-1})$. Now let us consider all simultaneous solution points, $\boldsymbol{u}^s \in \boldsymbol{R}^{m-1}$, such that $\mathcal{F}_i(\boldsymbol{u}^s) = 0$, for all $i = 1, \cdots, n$.

We will assume that $\mathcal{F}_i$, $i = 1, \cdots, n$, are represented as B-spline or Bézier multivariate scalar surfaces. Then, the domain of $\mathcal{F}_i(\mathbf{u})$ contains zeros only if the control coefficients of $\mathcal{F}_i$ have different signs; this is because the value of zero must be contained in the convex hull of the control coefficients of $\mathcal{F}_i$. Using this convex hull property, a subdivision-based approach, as described in Algorithm 3.1, will generate a set of discrete points that converges to the simultaneous zero-set of $\mathcal{F}_i$, $i = 1, \cdots, n$.

Let τ be a subdivision tolerance that determines the maximum number of points to be generated in the subdivision stage. The parameter domain is a unit hypercube of dimension $m - 1$, which is sliced into $\approx \frac{1}{\tau}$ segments in each coordinate direction. The result of this subdivision will be a total of $\left(\frac{1}{\tau}\right)^{m-1}$ cells. Each such cell generates one point (i.e., the center of each cell); thus, we will end up with at most $\left(\frac{1}{\tau}\right)^{m-1}$ approximate solution points to the zero-set at the end of this subdivision stage.

3.2 Numerical Improvement Step for Zero-Set Finding

This section discusses the numerical improvement for the approximated solutions. While essentially a Newton-Raphson step is described, we take a more geometrically intuitive approach.

Let $\boldsymbol{u}^0 = (u_1^0, u_2^0, \cdots, u_{m-1}^0)$ be an approximate solution as computed in Section 3.1. That is, $\mathcal{F}_i(\boldsymbol{u}^0) \approx 0$, for $i = 1, \cdots, n$. The normal space of $\mathcal{F}_i$ is a line in $\boldsymbol{R}^m$, and at $\boldsymbol{u}^0$ this line is parallel to

$$\boldsymbol{n}_i(\boldsymbol{u}^0) = \left(\frac{\partial \mathcal{F}_i}{\partial u_1}(\boldsymbol{u}^0), \frac{\partial \mathcal{F}_i}{\partial u_2}(\boldsymbol{u}^0), , \cdots, \frac{\partial \mathcal{F}_i}{\partial u_{m-1}}(\boldsymbol{u}^0), -1\right)$$

where $\frac{\partial \mathcal{F}_i}{\partial u_j}(\boldsymbol{u}^0)$ denotes the partial derivative of each $\mathcal{F}_i$ with respect to u_j, evaluated at $\boldsymbol{u}^0$. Complementing the normal's one-dimensional space at $\boldsymbol{u}^0$ is the tangent hyperplane

Algorithm 3.1
Input:
$\mathcal{F}_i$, $i = 1, \cdots, n$, n multivariate explicit constraints;
τ, tolerance of subdivision process;
Output:
$\mathcal{S}$, a set of points in the parametric space of $\mathcal{F}_i$ such that each point $\mathbf{u}^s \in \mathcal{S}$ yields an approximated simultaneous zero over all $\mathcal{F}_i$;
ZeroSetSubdiv($\mathcal{F}_i$, $i = 1, \cdots, n$, $\left\{(u_j^{min}, u_j^{max})\right\}$, $j = 1, \cdots, (m-1)$, τ)
begin

$\Delta^{max} u \Leftarrow \max_j \left(u_j^{max} - u_j^{min}\right)$;
if ($\Delta^{max} u < \tau$) **then begin**
return $\left\{\left(\frac{u_1^{min}+u_1^{max}}{2}, \frac{u_2^{min}+u_2^{max}}{2}, \cdots, \frac{u_{m-1}^{min}+u_{m-1}^{max}}{2}\right)\right\}$;
end
else begin
$p_i^{min}, p_i^{max} \Leftarrow$ Minimal and maximal control coefficients of $\mathcal{F}_i$;
if ($\exists i$ such that $p_i^{min} p_i^{max} > 0$) **then**
return ϕ;
else begin
$\mathcal{F}_i^1, \mathcal{F}_i^2 \Leftarrow \mathcal{F}_i$ subdivided at $\frac{u_j^{min}+u_j^{max}}{2}$, along the jth parametric direction, where $\Delta^{max} u$ has been detected, $i = 1, \cdots, n$;
return ZeroSetSubdiv($\mathcal{F}_i^1$, $i = 1, \cdots, n$,
$\left\{(u_0^{min}, u_0^{max}), \cdots (u_j^{min}, \frac{u_j^{min}+u_j^{max}}{2}), \cdots (u_n^{min}, u_n^{max}), \right\}$) $\cup$
ZeroSetSubdiv($\mathcal{F}_i^2$, $i = 1, \cdots, n$,
$\left\{(u_0^{min}, u_0^{max}), \cdots (\frac{u_j^{min}+u_j^{max}}{2}, u_j^{max}), \cdots (u_n^{min}, u_n^{max}), \right\}$);
end
end

of $\mathcal{F}_i(u_1, u_2, \cdots, u_{m-1})$, a function of variety $(m-1)$ with parameters $\boldsymbol{u} = (u_1, u_2, \cdots, u_{m-1})$, that is equal to

$$\begin{aligned}\mathcal{P}_i(\boldsymbol{u}^0) \quad : \quad & \frac{\partial \mathcal{F}_i}{\partial u_1}(\boldsymbol{u}^0)u_1 + \frac{\partial \mathcal{F}_i}{\partial u_2}(\boldsymbol{u}^0)u_2 + \cdots \\ & + \frac{\partial \mathcal{F}_i}{\partial u_{m-1}}(\boldsymbol{u}^0)u_{m-1} \\ & - u_m - \left\langle \boldsymbol{n}_i(\boldsymbol{u}^0), \overline{\boldsymbol{u}^0} \right\rangle = 0,\end{aligned}$$

where $\overline{\boldsymbol{u}^0} = (u_1^0, u_2^0, \cdots, u_{m-1}^0, \mathcal{F}_i(\boldsymbol{u}^0))$.

Using a first-order approximation, each function $\mathcal{F}_i$ constrains the solution to be on the hyper-plane $\mathcal{P}_i(\boldsymbol{u}_0)$. The planes $\mathcal{P}_i(\boldsymbol{u}^0)$ also contain u_m as a free variable and we recollect that $\mathcal{F}_i = 0, \forall i$. So this requirement can be simply embedded as the additional constraint $u_m = 0$. The final result is $n+1$ linear constraints.

Consider the case that $m = n+1$ which leads to a fully constrained system of m linear equations formulated from the n tangent planes $\mathcal{P}_i(\boldsymbol{u}^0)$ of the functions $\mathcal{F}_i(\boldsymbol{u})$, and also one final constraint derived from $u_m = 0$, the zero being sought. Therefore, we have m linear equations and m unknowns. Note that m planes, in general position in $\boldsymbol{R}^m$, intersect at a single point, which corresponds to the result of one iteration of a multidimensional Newton Raphson procedure in $\boldsymbol{R}^m$.

For example, consider the case of two scalar bivariate constraints in $\boldsymbol{R}^3$, $\mathcal{F}_i(u_1, u_2) = 0$, $i = 1, 2$. Additionally, let $u_3 = 0$ and $\mathcal{P}_i(u_1, u_2, u_3) = 0$, $i = 1, 2$. Now we have three linear constraints and three unknowns. Three planes in general position in $\boldsymbol{R}^3$ intersect at a single point, which is also the result of one iteration of the Newton Raphson procedure.

Now let $m > n+1$. The linear system of equations is under-determined. Note that $(n+1)$ planes in $\boldsymbol{R}^m$ intersect in an $(m-n-1)$-dimensional variety, $\mathcal{V}$. In order to find the closest solution to the current position, $\boldsymbol{u}^0$, out of the possible solution set, $\mathcal{V}$, one must employ a linear solver that yields a minimal L_2-norm solution. SVD decomposition [1] and QR factorizations [1] are two options for solving this under-determined problem, producing a minimal L_2-norm solution.

A QR factorization procedure was used in preparing the examples of this paper. The tolerance employed in the numeric stage is several magnitudes smaller than that of the subdivision stage. Thus the vast majority of points were successfully improved to that tolerance in very few iterations, becoming more than an order of magnitude closer to the solution in each iteration. Since we use a surface fitting procedure in the next step, we don't need to improve all points from the subdivision stage. We can simply purge away those few points that fail to improve. Examples are presented in Section 4.

3.3 Fitting a Surface to the Bisector Surface

Once a set of points that satisfy the simultaneous zeros has been constructed, we map these points to the bisector, and then fit a surface to them. In contrast to the problem of regular scattered data interpolation, this problem permits a simple parameterization of data points.

Points in the simultaneous zero-set map directly to the bisector. Thus, we can employ the parameterization of the zero-set as the parameterization of the bisector surface. This approach has been taken in all the examples presented in Section 4.

4 Experimental Results

We will now demonstrate the effectiveness of our approach by presenting several examples of curve-surface and surface-surface bisectors in $\boldsymbol{R}^3$, computed using the algorithm presented in this paper. Each example shows the set of discrete points that results from the simultaneous solution of the multivariate constraint equations. An approximate bisector surface that interpolates these points is shown in gray.

Two simple examples of curve-surface bisectors appear in Figure 1: a line-plane bisector is shown as a cone in Figure 1(a), and a curve-plane bisector is shown in Figure 1(b). Figure 2 shows a more complex bisector surface equidistant from a freeform curve and a convex surface. Two surface-surface bisectors are presented in Figure 3: the bisector of a plane and a convex surface is shown in Figure 3(a), and the bisector of a plane and a concave surface is shown in Figure 3(b). Note that the bisector surface of Figure 3(b) exhibits a self-intersection. Finally, Figure 4 shows the bisector of two freeform rational surfaces.

Computing each of the above examples took from several seconds to a few minutes on a Silicon Graphics 150 MHz R5000 Indy machine. The most time-consuming step is the generation of the set of discrete points that approximates the zero-set; the number of discrete points generated for each example is also listed in Table 1.

In all the examples presented so far, there is a one-to-one correspondence between the zero-set and an input surface. Thus, the mapping from an input surface to the surface of zero-set, and to the bisector surface, is well-defined. But, this is not always the case. For example, when the two initial surfaces intersect, as shown in Figure 5, both sides of each surface must be considered in computing the bisector. In this example, each point on each initial surface (except those on the intersection curve) corresponds to two different points on the surface of zero-set; thus, a point on one of the initial surfaces generates two points on the bisector surface. Note that this bisector surface intersects itself at the intersection of the two initial surfaces. Near the intersection curve, it is difficult to fit a surface through discrete points sampled on the bisector surface because we are no longer able to employ the parameterization of either of the original surfaces. The fitted surfaces are not single sheets, but rather two sheets that intersect each other. Nonetheless, as long as the correspondence between points on the two initial surfaces is one-to-one, the simple surface-fitting approach proposed in this paper works very well even when there are self-intersections in the bisector as shown in Figure 3(b).

5 Conclusions

In this paper, we have proposed a new computational scheme for the construction of bisector surfaces in $\boldsymbol{R}^3$: between a rational curve and a rational surface, and also between two rational surfaces. Although the approach is promising, there are some drawbacks in the current procedure. In particular, we need to resolve

1. the parameterization of multi-sheeted bisector surfaces, and

2. the global error measure in the approximation of a bisector surface using surfaces fitted through the discrete sample points. Moreover, we need to relate the errors of the zero-set finding in the parameter space (e.g., Equation (10)) to the errors in the bisector surface itself.

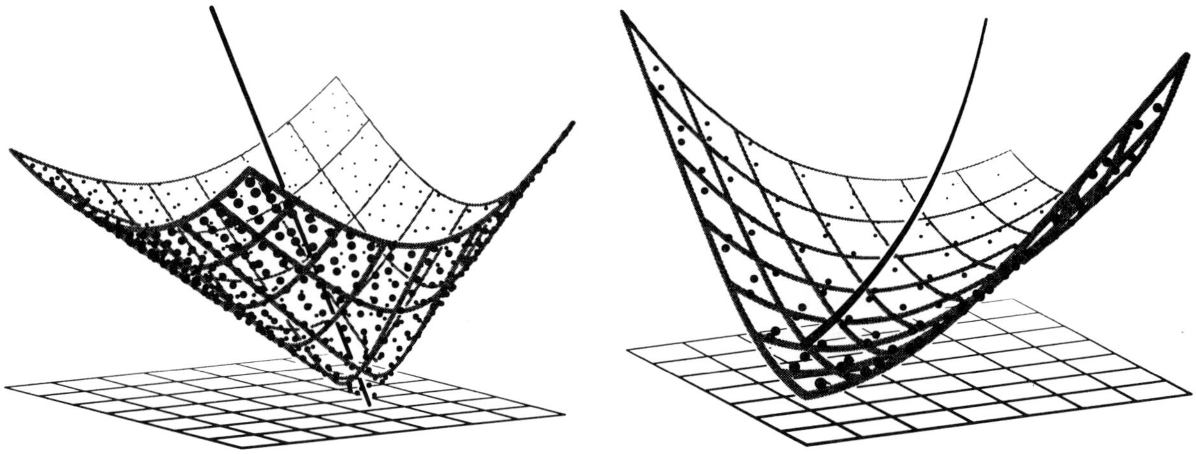

(a)

(b)

Figure 1. (a) Bisector surface (in gray) of a (tilted) line and a plane; and (b) bisector surface (in gray) of a (tilted) quadratic curve and a plane. The exact bisector of (a) is a cone.

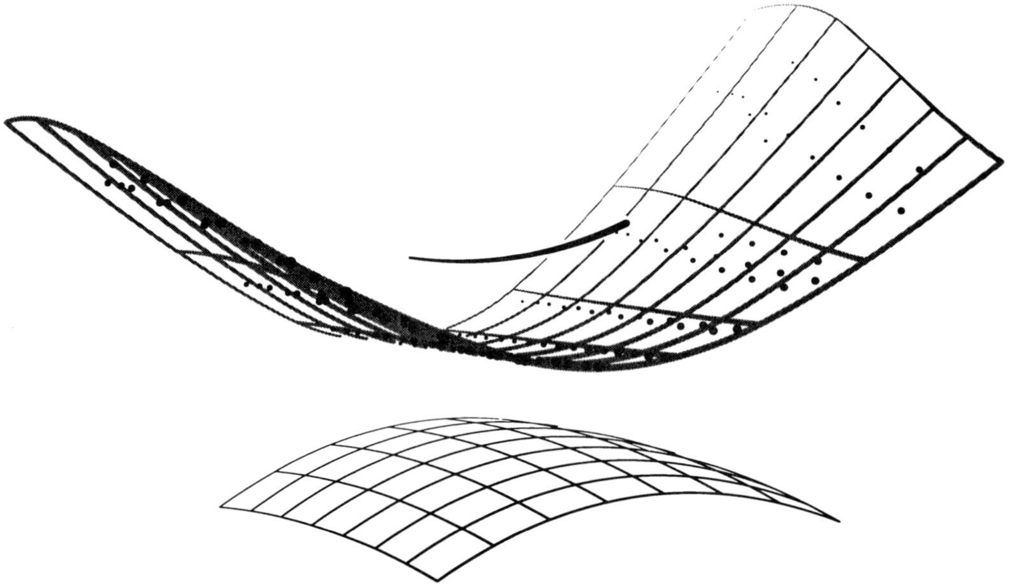

Figure 2. Bisector surface (in gray) of a quadratic curve and a convex surface.

In the current implementation, we represent the constraint equations exactly using high degree polynomials. Approximation with low degree polynomials would improve the computational efficiency significantly. Moreover, an optimal distribution of discrete points on the bisector surface is an important issue to consider. These improvements will be considered in future work.

Acknowledgments

The authors thank anonymous referees for useful comments on related previous work and some possible improvements. This research was supported in part by the Fund for Promotion of Research at the Technion, Haifa, Israel, and by the Korean Ministry of Science and Technology (MOST) under the National Research Laboratory Project.

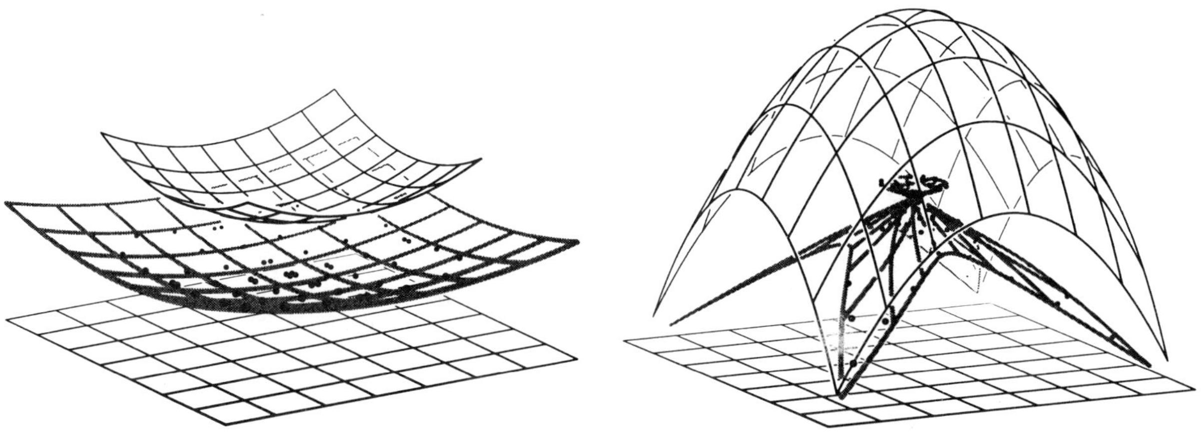

(a) (b)

Figure 3. (a) Bisector surface (in gray) of a convex surface and a plane; and (b) bisector surface (in gray) of a concave surface and a plane.

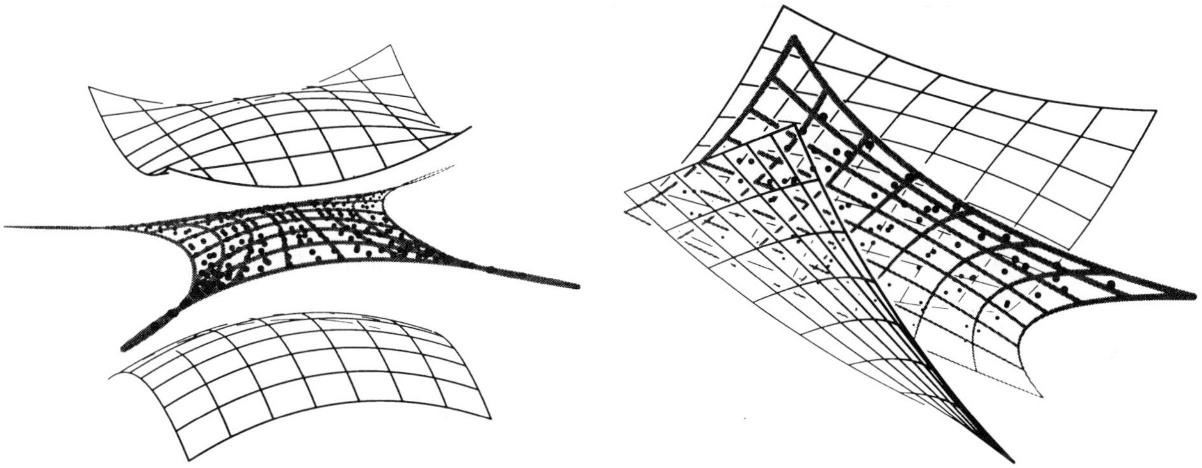

(a) (b)

Figure 4. Bisector surfaces (in gray) between two freeform surfaces.

References

[1] G. H. Golub and C. F. Van Loan. Matrix Computation. The John Hopkins University Press, Baltimore and London, 2nd Ed., 1996.

[2] D. Dutta and C. Hoffmann. On the skeleton of simple CSG objects. *ASME J. of Mechanical Design*, vol. 115, pp. 87–94, 1993.

[3] G. Elber and M.-S. Kim. The bisector surface of freeform rational space curves. *ACM Trans. on Graphics*, vol. 17, no. 1, pp. 32–49, January 1998.

[4] G. Elber and M.-S. Kim. Bisector curves of planar rational curves. *Computer-Aided Design*, vol. 30, no. 14, pp. 1089–1096, December 1998.

[5] G. Elber and M.-S. Kim. Rational bisectors of CSG primitives. *Proc. 5th ACM/IEEE Symposium on Solid Modeling and Applications*, Ann Arbor, Michigan, pp. 246–257, June 1999.

[6] G. Elber and M.-S. Kim. Computing rational bisectors. *IEEE Computer Graphics and Applications*, vol. 19, no. 6, pp. 76–81, November/December 1999.

Figure	1 (a)	1 (b)	2	3 (a)	3 (b)	4 (a)	4 (b)
# Points	558	122	132	64	77	318	80

Table 1. Number of points computed on the zero-set of each example.

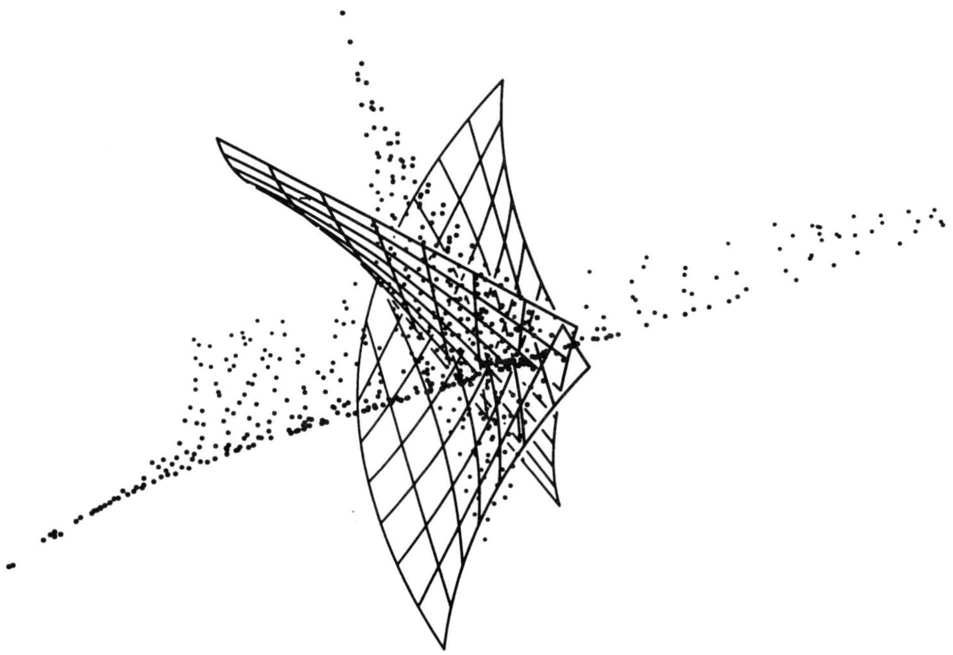

Figure 5. A case of a surface–surface bisector that produces multiple sheets of bisector.

[7] R. Farouki, and J. Johnstone. The bisector of a point and a plane parametric curve. *Computer Aided Geometric Design*, vol. 11, no. 2, pp. 117–151, April 1994.

[8] R. Farouki, and R. Ramamurthy. Specified-precision computation of curve/curve bisectors. *Int. J. of Computational Geometry & Applications*, vol. 8, nos. 5–6, pp. 599–617, October–December 1998.

[9] C.-Y. Hu, T. Maekawa, E. C. Sherbrooke, and N. M. Patrikalakis. Robust Interval Algorithm for Curve Intersections. *Computer-Aided Design*, vol. 28, nos. 6/7, pp. 495–506, June/July 1996.

[10] IRIT 7.0 User's Manual. The Technion—IIT, Haifa, Israel, 1997. Available at `http://www.cs.technion.ac.il/~irit`.

[11] J. Lane and R. Riesenfeld. Bounds on a Polynomial. BIT, vol. 21, pp. 112–117, 1981.

[12] M. Peternell. Geometric properties of bisector surfaces. To appear in *Graphical Models and Image Processing*, 2000.

[13] R. Ramamurthy. *Voronoi Diagrams and Medial Axes of Planar Domains with Curved Boundaries*. Ph.D. Thesis, Dept. of Mechanical Engineering, The Univ. of Michigan, Ann Arbor, USA, 1998.

[14] E. C. Sherbrooke and N. M. Patrikalakis. Computation of the Solutions of Nonlinear Polynomial Systems. *Computer Aided Geometric Design*, vol. 10, no. 5, pp. 379–405, October 1993.

[15] S. Wolfram. *Mathematica*, 2nd Ed., Addison-Wesley, (1991).

Surface Slicing Algorithm for Rapid Prototyping and Machining

Cha-Soo Jun, Dong-Soo Kim
Department of Industrial Engineering and Research Center for Aircraft Parts Technology, Gyeongsang National University, South Korea
csjun@nongae.gsnu.ac.kr, dskim@cecc-1.gsnu.ac.kr
JiSeon Hwang
Hyundai Motor Co., South Korea
and Tien-Chien Chang
Department of Industrial Engineering, Purdue University, U.S.A

Abstract

Presented in this paper is an algorithm to obtain the intersections of a free-form surface with a series of parallel planes. When sectioning the surface with parallel planes, the change of the topology of the intersection curves is caused by characteristic points of the surface. There are seven types of characteristic points: interior maximum, interior minimum, interior saddle, boundary maximum, boundary minimum, boundary max-saddle, and boundary min-saddle points. The starting points of the intersections are found efficiently and robustly using the characteristic points.

The characteristic points as well as the intersection contours can be used to evaluate the machining information for process planning, to generate NC tool path for surfaces, and to generate slices for rapid prototyping.

1. Introduction

Surface/surface intersection (SSI) is one of the most important and difficult problems in CAD/CAM. There exist various methods for solving this problem [Mullenheim 91, Chang 94, Abdel 96, Grandine 97, Abdel 97 and Ma 98]. The methods can be classified into four main categories, namely algebraic, lattice evaluation, marching and recursive subdivision [Barnhill 87, 89 and Pratt 86].

Several researchers have addressed the problem of plane/surface intersection (PSI), a special case of SSI. Lee and Fredericks [Lee 84] applied a recursive subdivision technique to this problem. Q. Fu [Fu 90] investigated the necessary and sufficient conditions for non-empty surface intersection. Lee and Chang [Lee 92] obtained intersection points by a lattice search method and connected the points by a contour growth method. Farouki [Farouki 86] introduced a semi-algebraic method for sectioning a parametric surface with a conic surface. Blinn [Blinn 78] described a heuristic method for determining the local extremum perpendicular distances from the surface to the section plane, and employed these to identify all closed loops of the section curve. This technique has been refined by Hoitsma and Roche [Hoitsma 84], who located the extrema by algebraic methods.

An important case for the plane/surface intersection is the intersection of a surface with a set of parallel planes. We can find several tasks including this problem:

- Surface slicing for laminated object manufacturing (LOM) and stereolithography [Feygin 92 and Dolenc 92],
- NC tool path generation [Li 98],
- Intersections of hunting planes to evaluate machining information [Lee 92a],
- Contour lines for geographical models.

The existing PSI algorithms for a single plane may solve this problem by repeating them as many times as the number of planes. We have not found any research considering this situation seriously.

If we investigate two sets of the intersections obtained by sectioning a surface with two parallel planes, we may find not only geometric changes but topological changes, such as appearance of new curves, disappearance of existing curves, joining of two curves, joining of two ends of an open curve to be a closed curve, and splitting of a curve. For example, a new closed curve is created if a section plane moves down under a local maximum point of the surface. The topological changes are caused by local extreme points of the surface, which will be called the characteristic points (CPs) of the surface in this paper. Using the properties of the CPs, we developed a new and efficient algorithm for finding all the intersection curves between a free form surface and a series of parallel planes.

The proposed algorithm consists of three main steps:

(1) To detect all the CPs of the surface and to sort them along the direction of the normal vector of the section planes.

(2) To find the starting points of the intersection curves using the properties of the CPs and the information of the section curves of the previous section plane.

(3) To trace the intersection curves from the starting points.

This paper focuses on investigating the properties of the CPs of the surface and making use of them to find the starting points. Illustrative examples and discussions are also provided. As for tracing, we adopted a method similar to the one described in [Barnhill 87], and we will not address it in this paper.

2. Properties of characteristic points of surface

For $\mathbf{r}, \mathbf{n} \in \mathbb{R}^3$, $h_i \in \mathbb{R}$, a section plane P_i and a surface S are defined as

$$P_i = \{ \mathbf{r} \mid \mathbf{r} \cdot \mathbf{n} = h_i \},\ i = 1, 2, \ldots, Np \qquad (1)$$

$$S = \{ \mathbf{r} \mid \mathbf{r} = (x(u,v), y(u,v), z(u,v)) \}, \qquad (2)$$

where $\mathbf{n}$ is the normal vector of the section planes, and Np is the number of planes. The section planes are ordered such as $h_1 > h_2 > \ldots > h_{Np}$. The intersection for P_i is defined in u,v-domain as

$$C_i = \{(u,v) \mid S \cap P_i \} = \{(u,v) \mid \mathbf{r}(u,v) \cdot \mathbf{n} = h_i\}. \qquad (3)$$

C_i may be a collection of several disjoint curves. The half space H_i constructed by a section plane P_i is defined by

$$H_i = \{ \mathbf{r} \mid \mathbf{r} \cdot \mathbf{n} \geq h_i \}. \qquad (4)$$

The intersection regions for P_i is defined in the domain as

$$D_i = \{(u,v) \mid S \cap H_i \}. \qquad (5)$$

The complementary set of D_i, called non-intersection region, is defined as

$$Q_i = D_i^{\,c} = U - D_i, \qquad (6)$$

where U is the entire domain of the surface.

D_i and Q_i may be a collection of several disjoint areas respectively. An intersection loop denotes a boundary of a single intersection region, and a non-intersection loop denotes a boundary of a single non-intersection region. A loop consists of a single closed curve, or it consists of one or more open curve(s) and some portions of the domain boundary. A particular portion of the domain boundary belongs to either an intersection loop or an non-intersection loop, while an intersection curve is shared by both an intersection loop and a non-intersection loop.

The characteristic points (CPs) in this paper denote the points on the surface which cause topological changes when a section plane moves in -**n** direction. There are seven types of CPs: interior maximum point (IXP), interior minimum point (INP), interior saddle point (ISP), boundary maximum point (BXP), boundary minimum point (BNP), boundary max-saddle point (BXSP), boundary min-saddle point (BNSP). The properties of CPs are described in Table 1 and Figure 1. The properties in Table 1 are individual ones, and the CPs may be interrelated. Some of the general properties of CPs are described as follows:

Property 1. C_i and C_j have the same topology if there is no CP on the surface between P_i and P_j.

Property 2. The number of CPs on the boundary is always even. If the boundaries are connected to a loop, the CPs are alternating between max-points (BXP or BXSP) and min-points (BNP or BNSP) on the loop.

Property 3. The end points of an open curve lie on the boundary of the domain.

Property 4. If there exist two or more IXPs in a closed intersection curve, there must be one or more ISP(s) in the curve.

Property 5. The Intersection region for the lower plane P_i includes the intersection region for the higher plane P_{i-1}, i.e., $D_i \supset D_{i-1}$ *iff* $h_i < h_{i-1}$.

3. Detection of boundary characteristic points

At an extreme point u_0 on the boundary curve $\mathbf{r}(u)$,

$$\mathbf{r}_u(u_0) \cdot \mathbf{n} = 0, \qquad (7)$$

where $\mathbf{r}_u = d\mathbf{r}/du$, if $\mathbf{r}(u)$ is differentiable at u_0.

As for tangent discontinuous points such as vertices, we can identify CPs by investigating both tangent vectors at those points. To find extrema on the boundary curve is a process of univariate optimization. First, we subdivide a boundary curve recursively into unimodal segments such that each segment does not have more than one extremum. Next, we classify the segments into convex, concave, and monotonic segments by considering the positions and tangents of the end of the segments, then discard monotonic segments. For a non-monotonic segment $\mathbf{r}(u)$, $u_0 \leq u \leq u_1$, we apply the following procedure to find a CP:

(i) Choose a tolerance $\varepsilon > 0$ to be used for terminating the search.

(ii) Let $\xi_0 = \mathbf{r}_u(u_0) \cdot \mathbf{n}$; $\xi_1 = \mathbf{r}_u(u_1) \cdot \mathbf{n}$

(iii) Approximate $\mathbf{r}(u)$ to a cubic Hermite curve $\mathbf{h}(u)$, using the position and tangent vectors of the two ends of $\mathbf{r}(u)$. Then, find an extreme point $u^* \in [u_0, u_1]$ such that the quadratic equation $\mathbf{h}_u(u^*) \cdot \mathbf{n} = 0$.

(iv) Let $\xi = \mathbf{r}_u(u^*) \cdot \mathbf{n}$. If $|\xi| < \varepsilon$ then go to step(v). If $(\xi_0 - \xi_1)\,\xi > 0$, let $u_0 = u^*$ and go to step(ii). Otherwise, let $u_1 = u^*$ and go to step (ii).

(v) Determine the type (BXP, BNP, BXSP, BNSP) of the CP by considering the position of the CP, the positions of the two ends of the segment, and the cross-boundary tangent vector of the surface at the CP (see the Figure 1). Note that horizontal inflection points are ignored.

Table 1. Properties of CPs

Properties / Types of CPs	When the section plane passing through the CP	When the section plane moving down the CP	
		Intersection Curve	Intersection region or loop
Interior maximum point (IXP)	An isolated point in the interior	Appearance of a closed curve	Appearance of an intersection region
Interior minimum point (INP)	An isolated point in the interior	Disappearance of a closed curve	Disappearance of a hole in the intersection region
Interior saddle point (ISP)	A self-intersection of the intersection curves	Switching of intersection curves	Joining of an intersection region(s) in the interior
Boundary maximum point (BXP)	An isolated point on the boundary	Appearance of an open curve	Appearance of an intersection region
Boundary minimum point (BNP)	An isolated point on the boundary	Disappearance of an open curve	Replacing the open curve with the portion of the border
Boundary max-saddle point (BXSP)	A tangent point of the border with the intersection curve	Splitting of a curve	Inserting a portion of the border into the existing loop
Boundary min-saddle point (BXNP)	A tangent point of the border with the intersection curve	Joining of a curve(s)	Joining of an intersection region(s) at boundary

4. Detection of interior characteristic points

To find the interior CPs of the surface is a process of bivariate optimization. Here we use a cyclic coordinate method [Bazaraa 79]. First, we subdivide a surface recursively into subpatches such that they do not have more than one extremum. Then, considering the position and normal vectors of the four corners of each subpatch, we may discard the subpatch that has no possibility to possess an extremum. For each remaining subpatch $\mathbf{r}(u,v)$, $u_0 \leq u \leq u_1$, $v_0 \leq v \leq v_1$:

(i) Choose a tolerance $\varepsilon > 0$ *to be used for terminating the search. Choose an initial point* $u^* = (u_0 + u_1)/2$, $v^* = (v_0 + v_1)/2$.

(ii) Find an extreme point (u^*,v^*) *such that* $\mathbf{r}_u(u^*,v^*) \cdot \mathbf{n} < \varepsilon$ *along the u-directional isoparametric curve* $\mathbf{r}(u,v^*)$ *using the boundary CP search method described above. If* $\mathbf{r}_v(u^*,v^*) \cdot \mathbf{n} < \varepsilon$, *go to step (iv). Otherwise, go to step (iii).*

(iii) Find an extreme point (u^*,v^*) *such that* $\mathbf{r}_v(u^*,v^*) \cdot \mathbf{n} < \varepsilon$ *along the v-directional isoparametric curve* $\mathbf{r}(u^*,v)$. *If* $\mathbf{r}_v(u^*,v^*) \cdot \mathbf{n} < \varepsilon$, *go to step (iv). Otherwise, repeat step (ii).*

(iv) Determine the type (IXP, INP, ISP) of the CP by considering the position of the CP and u- and v-directional isoparametric curves at the CP (see Figure 1).

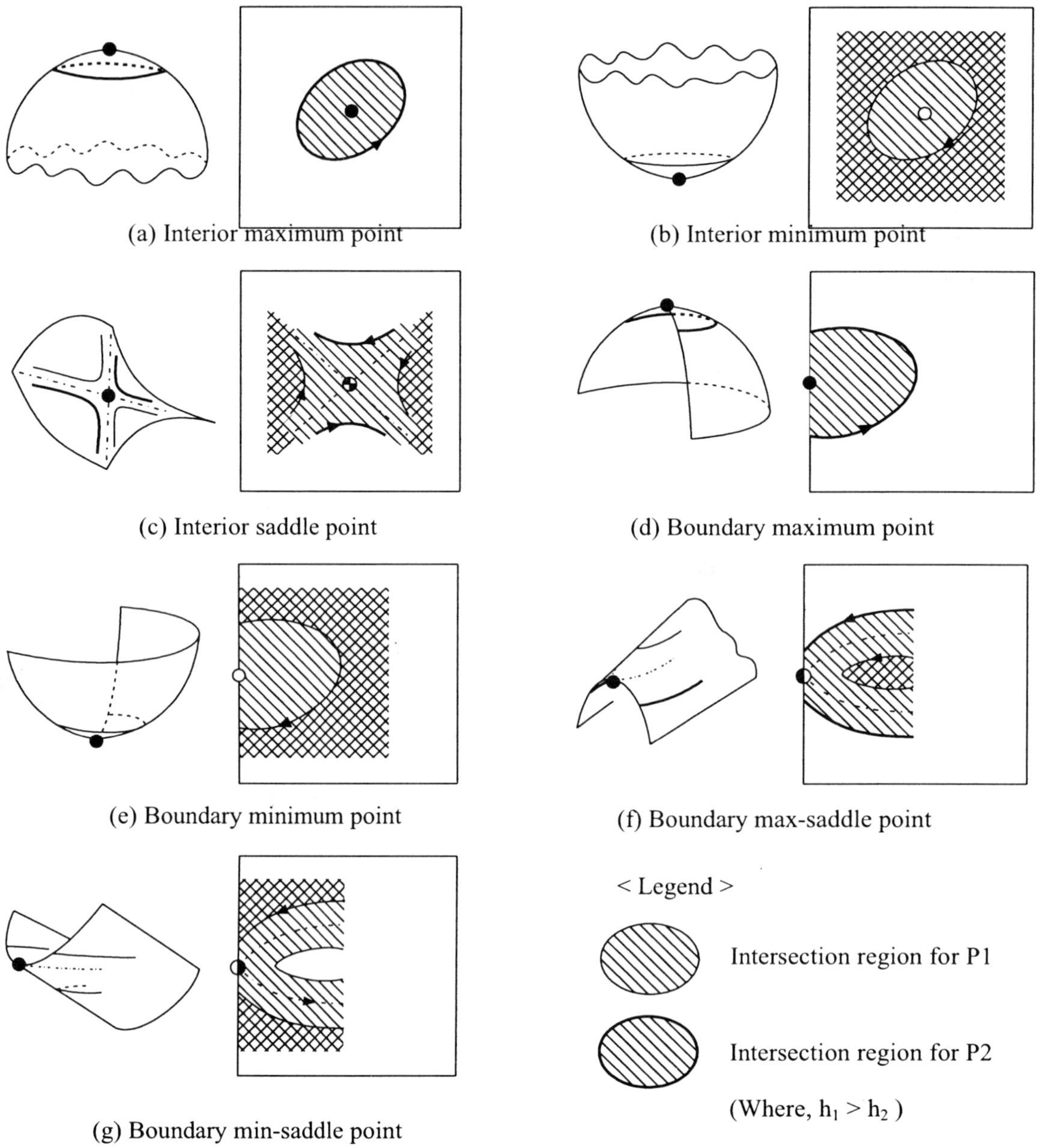

Figure 1. Characteristic points

5. Finding starting points of intersection curves

The entire domain is partitioned into several intersection and non-intersection regions. Each region is bounded by one or more loops. Figure 2 shows intersection loops and loop structure, which is adopted from the trimmed surface loop structure [Choi 91]. As a convention, intersection loops have an anti-clockwise direction and non-intersection loops a clockwise direction. Each loop has at least one CP. In Figure 2, a rectangled number denotes a CP number, which is sorted along the normal of the section planes. The CPs are used to find the intersection of the surface for a section plane P_i, after obtaining the intersection for P_{i-1}.

1) Case of no CP between two section planes: simple expansion of the existing loops

Because there is no topological change in this case, the intersection loops of a new plane P_i are obtained simply by expanding the loops of P_{i-1}

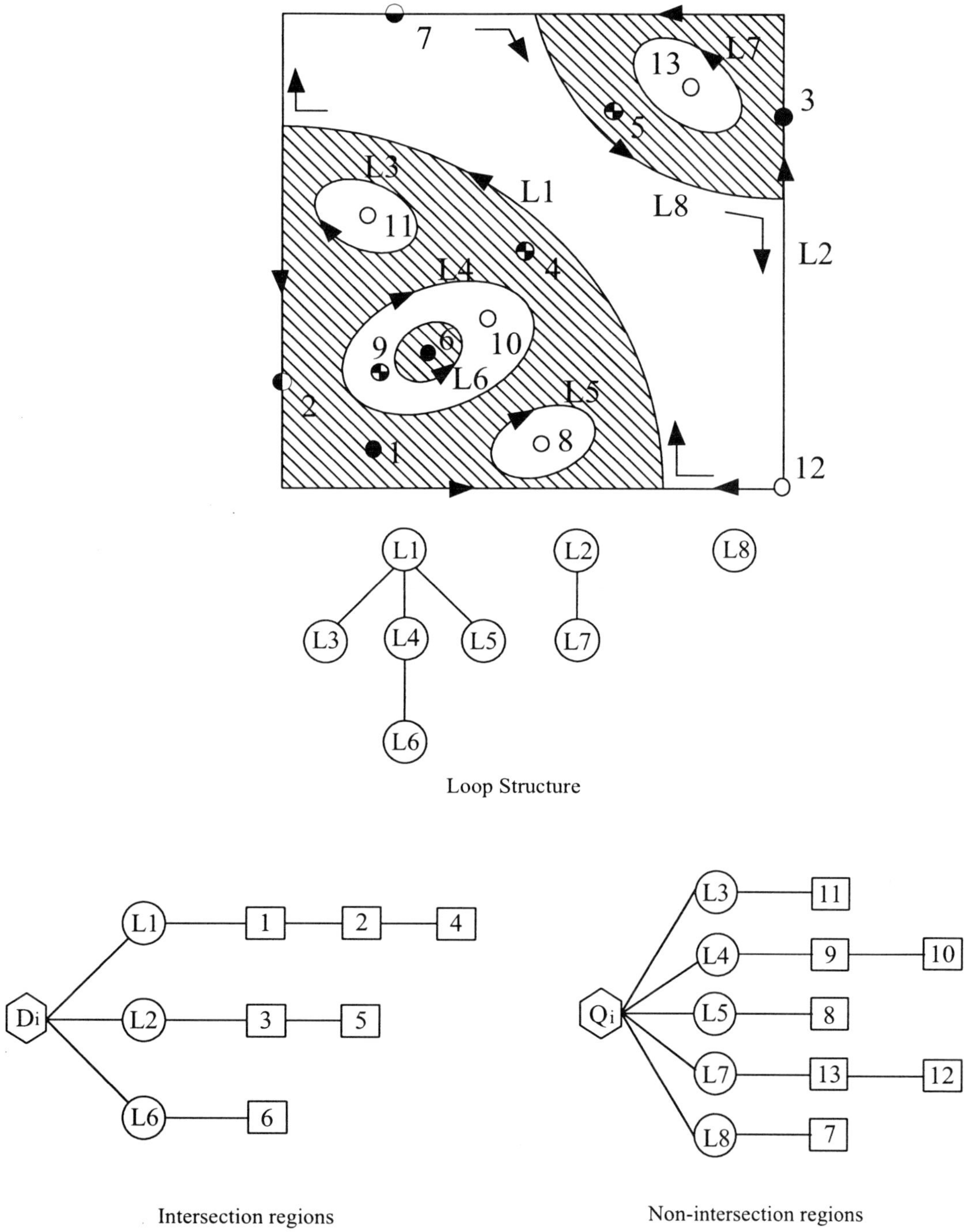

Figure 2. Intersection regions and loop structure

(i) For an open-intersection loop (consisting of open intersection curves and some portions of the domain boundary), a starting point of the new loop is found by searching along the domain boundary from an end of an old intersection curve (see Figure 3(a)). In Figure 3, a small rectangle denotes a starting point, and a cross mark denotes an inclusion test.

(ii) For a closed-intersection loop, a starting point is found by searching in the outer normal direction of the loop at any point on it (Figure 3(a)).

(iii) For a hole in a loop (a closed non-intersection loop), a starting point is found by searching in the direction of an interior minimum point which is located in the hole (Figure 3(b)).

2) Case of one CP between two section planes

(1) Interior maximum point (IXP): appearance of a new closed curve

A starting point of the new loop is obtained by searching in an arbitrary direction from the IXP (Figure 3(c)). The existing loops are expanded simply as described above.

(2) Interior minimum point (INP): disappearance of a closed curve

The hole loop which has the INP vanishes. The existing loops are expanded simply (Figure 3(d)).

(3) Interior saddle point (ISP): joining of loops

An ISP combines the existing loops. It may make a hole in a loop by joining two different portions of a loop (Figure 3(e)), or it may join two or more loops into a new loop (Figure 3(f)). An n-fold saddle point has n valleys and n hills. Thus, it joins n portions of loops (or a loop). A procedure for new intersection loops is as follows:

(i) For each valley of a saddle point:

Search for a starting point while going down the valley.

If this point is on the loops that have already been obtained from this ISP, discard it.

Otherwise, trace a new loop from it.

(ii) For each old loop:

If it is included in the new loop obtained in step i), remove it.

Otherwise, expand it simply.

(iii) For each CP below the present plane, rearrange its non-intersection loop.

(4) Boundary maximum point (BXP): appearance of a new open curve

A starting point is found by searching along the boundary as shown in Figure 3(g).

(5) Boundary minimum point (BNP): disappearance of an open curve

An open curve including the BNP is removed.

(6) Boundary max-saddle point (BXSP): splitting of a curve

A starting point of the new loop is found by searching along the domain boundary from one end of an old intersection curve (Figure 3(i)). Then, we construct a loop from the starting point and remove old loops which are included in the new loop. Figure 3(i) shows a 1-fold BXSP, Figure 3(j) a 2-fold BXSP.

(7) Boundary min-saddle point (BXSP): joining of a curve

In this case, two loops (or two portions of a loop) on both side of the BXSP along the boundary are joined. As shown in Figure 3(k), if two different loops join into one, we find a starting point along the boundary. If two portions of a loop join, a hole is made. Thus, we find a starting point by searching in the direction of an INP (interior minimum point) in the hole.

3) Case of more than one CP between two section planes

The effects of CPs are overlapped in this case. Especially, saddle points (ISP, BXSP, BNSP) increase the complexity because they combine loops. Therefore we consider them before others.

(i) For each ISP, BXSP, and BNSP between two planes in the reverse order of height, from the lowest to the highest one:

- Construct a new loop including it. (refer to (3), (6), (7) of case 2).

- Rearrange the CPs located in the non-intersection regions.

(ii) For each old loop, if it is included in the new loops, remove it.

(iii) For each IXP and BXP between two planes, construct a new closed loop including it.

(iv) For each INP and BNP between two planes, assign its loop.

6. An extension of the proposed algorithm

The proposed algorithm can be employed to obtain the intersection of a polyhedral surface with parallel planes. The CPs are detected simply by comparing the height of each vertex to that of its adjacent vertices. The starting point search procedure is very similar to the procedure for parametric surfaces. This extended version of the algorithm can be applied to slice STL data in a stereolithography apparatus (SLA) [Dolenc 92].

7. Examples

The proposed algorithm has been implemented in C language on a Windows NT machine and applied to some examples. Figure 4(a) shows a portion of a sphere. It has an IXP, four BXSPs, and four BNPs. The highest plane passes through the IXP exactly, and an isolated point is made. The hyperbolic paraboloid shown in Figure 4(b) has two BXPs, two BNPs, and an ISP. Two intersection curves meet at the ISP because a section plane passes through it. Note that none of the corner points are CPs in this example. Figure 4(c) shows a bicubic Ferguson surface. Figure 4(d) shows a torus, which has an IXP, an INP, and two ISPs. Originally its domain has boundaries, but we ignored the boundaries by cycling the domain value by setting 2π equal to 0. Figure 4(e) shows a polyhedral surface, and Figure 4(f) shows the intersection curves for it.

Figure 5 shows two more complicated surfaces. The first one has two ISPs, and the other has nine ISPs.

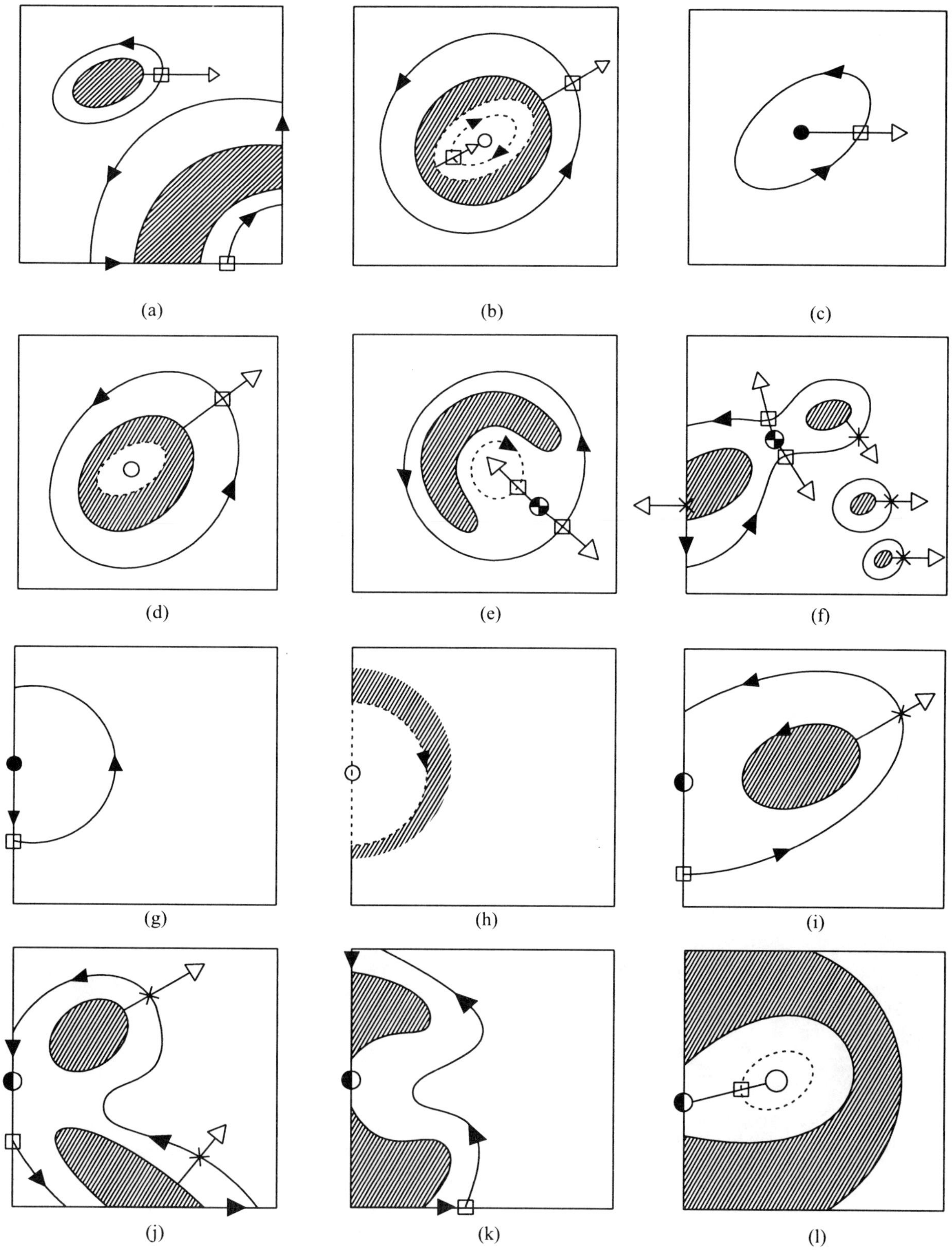

Figure 3. Search for intersections

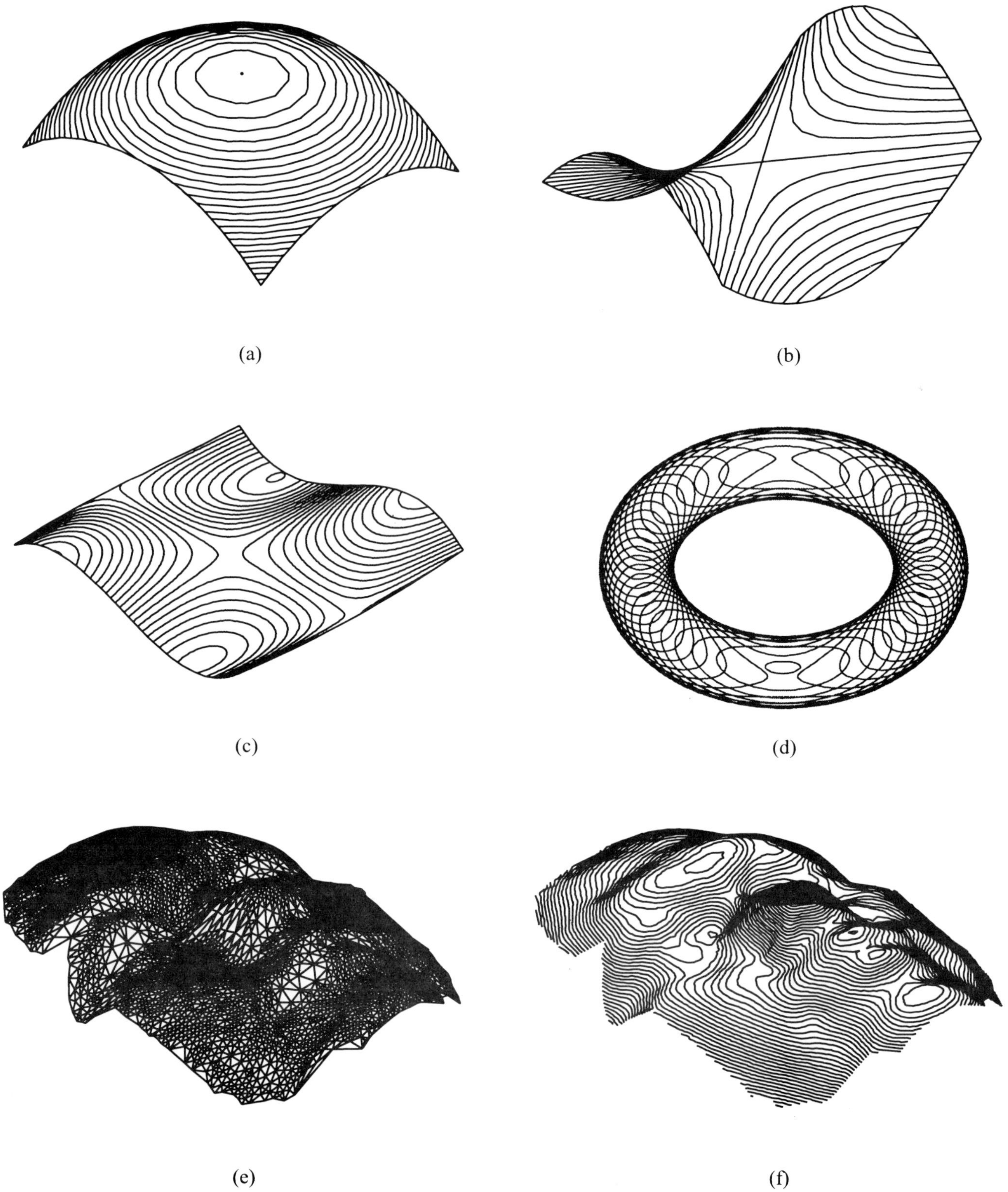

Figure 4. Examples

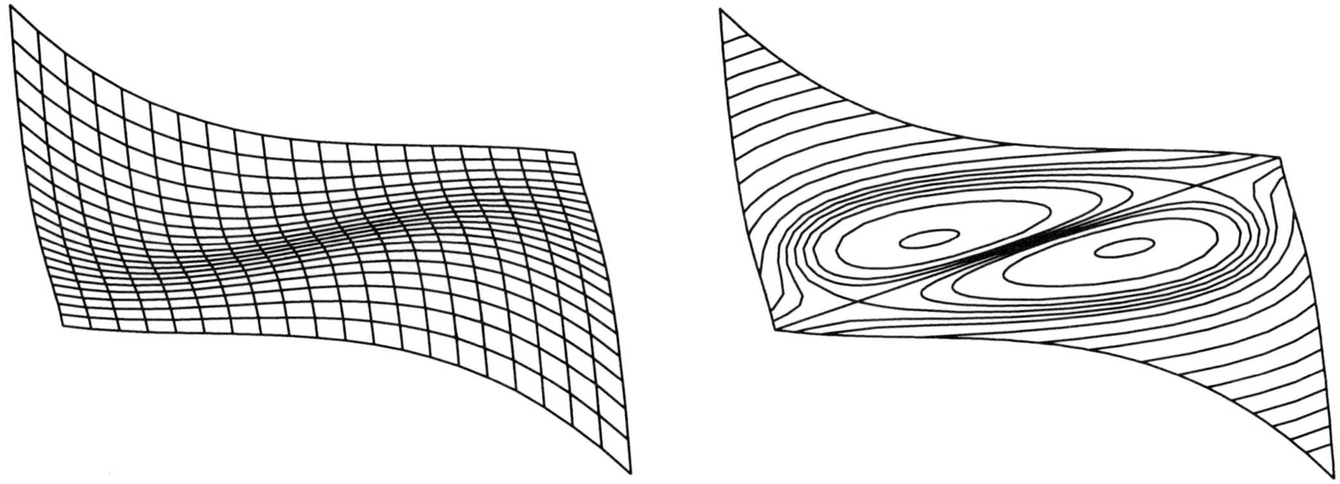

(a)

(b)

Figure 5. Examples

8. Conclusions and future researches

We have presented an algorithm for intersecting a general parametric surface with a series of parallel planes. The algorithm worked well on a variety of examples. The main feature of the algorithm is that the time-consuming exhaustive search for a surface is conducted only once regardless of the number of section planes. As a result, the algorithm is efficient. In addition, the algorithm treats robustly singular points such as isolated points, self-intersection points, and cusps, which have been considered to be vexing problems. However, there are a few unresolved issues which need further investigation. For example, the singular case where an intersection is a planar area is difficult to accommodate. We are still working on the problems.

From the algorithm, we get not only the geometric information for the intersection curves but also their topological information and characteristic points, which can be used to evaluate machining information of the surfaces.

9. References

[1] Abdel-Malek, K. and Yeh, H-J.(1997), On the determination of starting points for parametric surface intersections, *Computer Aided Design*, 29(1), 21-35.

[2] Barnhill, R.E. and Farin,G., Jordan,M. and Piper,B.R. (1987), Surface/Surface Intersection, *Computer Aided Geometric Design*, 4(1-2), 3-16.

[3] Barnhill, R.E. and Kersey, S. N.(1990), "A marching method for parametric surface/surface intersection", Computer Aided Geometric Design,7(1-4),257-280.

[4] Bazaraa, M. and Shetty, C.M.(1979), *Nonlinear Programming - Theory and Algorithm*, John Wiley & Sons.

[5] Blinn, J.(1978) *Computer Display of Curved Surface*, Ph.D. thesis, University of Utah, 1978.
[6] Chang, L.C., Bein, W.W. and Angel, E.(1994), Surface intersection using parallelism, *Computer Aided Geometric Design*, 11(1), 39-69.
[7] Choi, B.K.(1991), *Surface Modeling for CAD/CAM*, Elsevier.
[8] Dolenc,A.,Hovtun,R., and Makela,I.(1992), Better Software for Rapid Prototyping with INSTANTCAM, in *Human Aspects in Computer Integrated Manufacturing,* Olling,G.J. and Kimura, J. (ed), Elsevier Sci. Pub. 449-456.
[9] Farouki, R.(1986), The characterization of parametric surface sections, *Computer Vision, Graphics, and Image Processing,* 33, 209-236.
[10] Farouki, R.(1987), Direct surface section evaluation, in *Geometric modeling*, Farin,G.(ed), SIAM, Philadelphia.
[11] Feygin, M., Hsieh, B., and Melkanoff, M.A.(1992), Laminated object Manufacturing(LOM): A New Tool in the CIM World, in *Human Aspects in Computer Integrated Manufacturing,* Olling,G.J. and Kimura, J. (ed), Elsevier Sci. Pub. 457-464.
[12] Fu, Q.(1990), The Intersection of a bicubic Bezier patch and a plane, *Computer Aided Geometric Design*, 6(1-4), 137-153.
[13] Grandine, T. and Klein, F. W. IV (1997) A new approach the surface intersection problem, *Computer Aided Geometric Design*, 14, 111-134.
[14] Hoitsma, D.H. and Roche, M.(1984), The computation of all plane/surface intersections for CAD/CAM applications, in *Computer Aided Geometric modeling*, NASA C.P. 2272, 15-18.
[15] Houghton, E.G., Emnett, R.F., Factor, J.D. and Sabharwal,C.L.(1984), Implementation of a divide -and-conquer method for intersection of parametric surface, in *Surfaces in Computer Aided Geometric Design '84*, Barnhill,E.(ed.), North Holland, Amsterdam.
[16] Lee, R.B. and Fredericks, D.A.(1984) Intersection of parametric surfaces and a plane, *IEEE Computer Graphics and Applications*, 4(8),112-117.
[17] Lee, Y-S. and Chang T-C.(1992a) A contour method for intersections of hunting-planes to evaluate machining information, *Journal of Design and Manufacturing*, 2, 119-133.
[18] Lee, Y.S., Choi, B.K. and Chang, T.C.(1992b) Cuter distribution and cutter selection for sculptured surface cavity machining, *International Journal of Production Research*, 36(6), 1447-1470.
[19] Li, H., Dong, Z. and Vickers, G. W.(1998) "Optimal Tool Path Generation for 2 1/2 Milling of Dies and Molds", *Sculptured Surface Machining Conference*, Nov.9-11, 1998, Chrysler Technology Center.
[20] Ma, Y. and Lee, Y.S.(1998) Detection of loops and singularities of surface Intersections, *Computer-Aided Design*, 30(14), 1059-1067.
[21] Mullenheim, G.(1991), On determining start points for a surface/surface intersection algorithm, *Computer Aided Geometric Design*, 8(5), 401-408.
[22] Patrikalakis, N.M.(1993) Surface-to-surface intersections, *IEEE Computer Graphics and Applications,* pp. 89-95.
[23] Pratt, M.J. and Geisow, A.D.(1986) Surface/surface intersection problem, in *The Mathematics of Surface*, Oxford University Press, Oxford, pp.117-142.

Session 12

Geometric Operations

G^n–blending with Rolling Ball Contact Curves

Erich Hartmann
Darmstadt University of Technology
Dept. of Mathematics
Schlossgartenstrasse 7, 64289 Darmstadt, Germany
ehartmann@mathematik.tu-darmstadt.de

Abstract

A blending method for surfaces is introduced which produces pencils of parametrically defined blending surfaces which have G^n–continuous contact to the base surfaces along curves determined by a rolling ball of constant or variable radius. The performance is numerical.

1 Introduction

The classical method for blending two intersecting surfaces is to replace the intersection curve and its vicinity by a part of a canal surface which is generated by a rolling ball and which has G^1-continuous (tangent planes are continuous) contact to the surfaces. In literature there are many attempts for simulating this method by free form surfaces, for example [1], [8]. An extended survey of blending methods for parametric surfaces is contained in [9]. All these methods produce G^1–continuous surfaces. Some mention possibilities for raising the smoothness, but no examples are given. In the recent paper [5] a rather simple method is introduced which produces pencils of blending surfaces having G^n–continuous contact to the base surfaces. But an essential precondition of this method is the existence of convenient parameterizations of the surfaces to be blended.
The aim of this paper is to combine these two methods such that the generated blending surfaces have G^n–continuous contact along curves which were determined by a rolling ball with constant or variable radius. Of course this aim is achievable only by numerical evaluation of the parametric representation. An essential tool is the numerical parameterization of curves and surfaces given in [4].

For the convenience of the reader the used results of previous papers are given in Section 2. Section 3 describes the new blending method and contains several applications.

2 G^n–blending and numerical parameterization

2.1 Parametric G^n–blending surfaces

Definition:
Two regular C^n–continuous surfaces $\Phi_1 : \mathbf{x} = \mathbf{S}_1(s,t)$, $\Phi_2 : \mathbf{x} = \mathbf{S}_2(s,t)$, have C^n*–contact* at a common point $\mathbf{p} = \mathbf{S}_1(s_1,t_1) = \mathbf{S}_2(s_2,t_2)$ if all partial derivatives up to the n-th order of $\mathbf{S}_1$ and $\mathbf{S}_2$ at point $\mathbf{p}$ are the same:

$$\left.\frac{\partial^k \mathbf{S}_1}{\partial s^i \partial t^{k-i}}\right|_{(s_1,t_1)} = \left.\frac{\partial^k \mathbf{S}_2}{\partial s^i \partial t^{k-i}}\right|_{(s_2,t_2)},$$

$i = 0, 1.., k, \quad k = 0, 1, ...n.$

Definition:
Two regular C^n–continuous surfaces Φ_1, Φ_2 have G^n*–contact* at a common point $\mathbf{p}$ if they have C^n–contact at point $\mathbf{p}$ with respect to some local reparameterizations of the two surfaces. (cf. [7], [10])

[5] contains the following result.

Result:
Let be $\Phi_1 : \mathbf{x} = \mathbf{S}_1(s,t),\ s,t \in [0,1],\ \Phi_2 : \mathbf{x} = \mathbf{S}_2(s,t),\ s,t \in [0,1]$ *two regular* C^n*–continuous surface patches (base surfaces) and* $f(t),\ t \in [0,1]$ *the real function (blending function)*

$$f(t) = \frac{(1-t)^{n+1}}{(1-t)^{n+1} + t^{n+1}},\ t \in [0,1], \quad n \geq 0.$$

Then the surface patch

$$\Psi : \mathbf{x} = \mathbf{S}(s,t) := f(t)\mathbf{S}_1(s,t) + (1 - f(t))\mathbf{S}_2(s,t),$$

$s,t \in [0,1]$*, has* C^n*–contact to the surface patch* Φ_1 *along the curve* $\mathbf{S}_1(s,0),\ s \in [0,1]$ *and to* Φ_2 *along the curve* $\mathbf{S}_2(s,1)\ s \in [0,1]$.

0-7695-0562-7/00 $10.00 © 2000 IEEE

The proof is straight forwarded differentiating using the essential properties of function f: $f(0) = 1$, $f(1) = 0$, $f^{(k)}(0) = f^{(k)}(1) = 0$ for $k = 1, ..., n$.

Remark:

a) The curves $\mathbf{S}_1(s, 0)$ and $\mathbf{S}_2(s, 1)$ $s \in [s_1, s_2]$ are the *contact curves* between the base surfaces and the blend surface.

The curves $\mathbf{S}_1(s, 1)$ and $\mathbf{S}_2(s, 0)$ $s \in [s_1, s_2]$ are called *auxiliary curves.*

Contact curves and auxiliary curves are boundaries of the surface patches actually involved.

b) Curves $\mathbf{x} = \mathbf{S}(s = const, t)$ are *profile curves* which locally define the shape of the blend surface.

c) For the application of the blending method to the examples below there is i.g. need of parameter transformations to meet the precondition "$s \in [s_1, s_2]$, $t \in [0, 1]$". So the blend surface $\mathbf{S}$ will have i.g. only G^n– (instead of C^n–) contact to the base surfaces.

In order to get a design parameter we apply the following non linear parameter transformations on parameter t:

$$p_1(t) := \frac{t}{1 - \lambda + \lambda t}, \quad \text{and} \quad p_2(t) := \frac{t(1-\lambda)}{1 - \lambda t},$$

$0 \leq \lambda < 1$.

The application is as follows:

$$\mathbf{S}(s, t) = f(t)\mathbf{S}_1(s, p_1(t)) + (1 - f(t))\mathbf{S}_2(s, p_2(t)).$$

Functions p_1, p_2 have the property: $p_2(1 - t) = 1 - p_1(t)$ which is used below. Parameter λ plays the role of a *thumbweight* (cf. [9]). Raising λ brings the blend surface closer to the auxiliary curves (cf. [5]).

Example 1:

Let Φ_1 and Φ_2 be two tensor product Bézier patches with parametric representations

$\mathbf{x} = \mathbf{B}_1(u, v) = (10v - 5, 10u - 5, 6(u - u^2 + v - v^2))$,

$\mathbf{x} = \mathbf{B}_2(u, v) = (6(u - u^2 + v - v^2), 10u - 5, 10v - 5)$

respectively.

The boundary curves on Φ_1 are:

curve of contact: $v = v_{11} = 0.5$,

auxiliary curve: $v = v_{12} = 0.3$,

trim curves: $u = 0$ and u=0.75,

The boundary curves on Φ_2 are:

curve of contact: $v = v_{21} = 0.9$,

auxiliary curve: $v = v_{22}(u) = 0.8(1 - (0.5 - u)^2)$.

trim curves: $u = 0$ and $u = 0.75$.

In order to meet the preconditions of the blending method we use the following parameter transformations:

$\mathbf{S}_1(s, t) := \mathbf{B}_1(s, v_{11} + (v_{12} - v_{11})t), \quad t \in [0, 1]$,

$\mathbf{S}_2(s, t) := \mathbf{B}_2(s, v_{21} + (v_{22}(s) - v_{21})t), \quad t \in [0, 1]$

$s \in [0, 0.75]$.

Hence

$\Phi : \mathbf{x} = \mathbf{S}(s, t) := f(t)\mathbf{S}_1(s, t) + (1 - f(t))\mathbf{S}_2(s, t)$,

$s \in [0, 0.75]$, $t \in [0, 1]$,

is a G^n–blending surface.

(f is the G^n–blending function defined above.)

Figure 1a) shows the base surfaces, the curves of contact (thick) and the auxiliary curves (dashed). Figure 1b) shows a G^2–blending surface for thumbweight $\lambda = 0.65$.

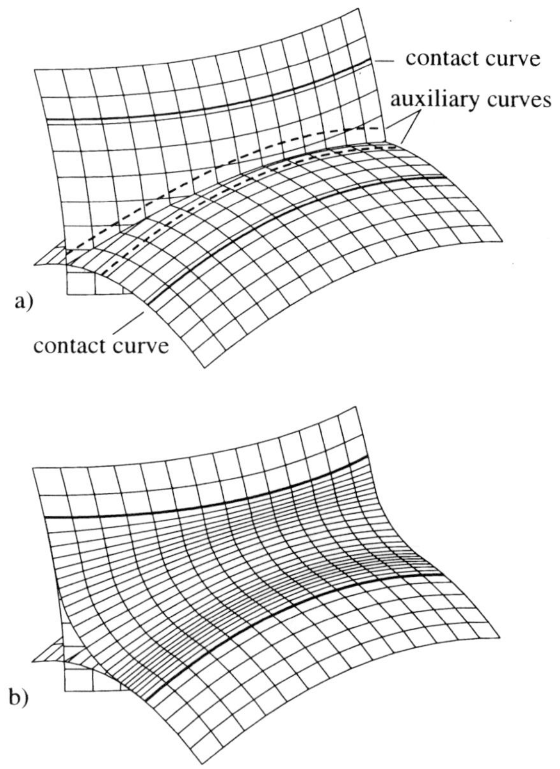

Figure 1. G^2–blending surface between two tensor product Bézier surfaces with thumbweight $\lambda = 0.65$ (b)) and the boundaries of the patches used for blending (a))

2.2 Numerical parameterization of curves and surfaces

[4] introduces C^{n-1}–continuous parameterizations of C^n–continuous, $n > 1$, curves (planar or spatial) and surfaces, respectively, which are based on the idea of normalforms. The evaluation of these parameterizations has to be done numerically.

The numerical parameterization of an arc Γ_0 of a planar or spatial C^2–continuous curve (implicit, parametric,...) is as follows

1. Choose a Bézier or B-spline curve or another suitable parametric curve $\Gamma : \mathbf{x} = \gamma(t), t \in [0,1]$ that is a rough approximation of Γ_0.

2. **Point** $\gamma_0(t)$ is the foot point of point $\gamma(t)$ on curve Γ_0, i.e. $\gamma(t) - \gamma_0(t)$ is orthogonal to Γ_0 and $||\gamma(t) - \gamma_0(t)||$ is minimal. (An appropriate algorithm for determining a foot point on a curve is contained in [3].)

The numerical parameterization of a surface patch Φ_0 of a C^2–continuous surface (implicit, parametric,...) is as follows

1. Choose a Bézier or B-spline surface or another suitable parametric surface $\Phi : \mathbf{x} = \mathbf{S}(u,v), (u,v) \in [0,1] \times [0,1]$ which is a rough approximation of the surface patch Φ_0 to be parameterized.

2. **Point** $\mathbf{S}_0(u,v)$ is the foot point of point $\mathbf{S}(u,v)$ on surface Φ_0. (For an appropriate algorithm see [3].)

3 Blending with rolling ball contact curves

The blending **method**:

1. Let be Φ_1, Φ_2 two parametric (or implicit) surfaces. We assume that the offset surfaces for distances 0..1 exist and the intersection curve of the offsets are C^2–continuous for any distance $t \in [0,1]$.

2. Any intersection curve of pairs of offsets (corresponding to the same distance t) can be numerically parametrized by a suitable common Bézier or B-spline curve $\Gamma : \gamma(s), s \in [0,1]$ (cf. Section 2.2, see Figure 2). Hence we get a (numerical) parameterization $\mathbf{V}(s,t), s,t \in [0,1]$ of the Voronoi surface (points which have equal distance to both the surfaces, see Figure 2, cf. [6]) which is C^{n-1} if Φ_1, Φ_2, Γ are C^n.

3. The parameterization $\mathbf{V}(s,t)$ is used for a (numerical) parameterization (see Section 2.2) of the given surfaces Φ_1, Φ_2 which we assign by $\mathbf{S}_1(s,t), \mathbf{S}_2(s,t),\ s,t \in [0,1]$.

4. We apply the C^n–blending method out of Section 2.1 and get the blending surface Ψ :

$$\mathbf{x} = \mathbf{S}(s,t) := f(t)\mathbf{S}_1(s,1-t) + (1-f(t))\mathbf{S}_2(s,t),$$

$s,t \in [0,1]$, with blending function $f(t)$ given in Section 2.1. The curves of contact $\mathbf{S}_1(s,1), \mathbf{S}_2(s,1), s \in [0,1]$ can be considered as curves of contact of a rolling ball of *fixed* radius $r = 1$ (see Figure 3, 4). The curve containing the midpoints of the rolling balls , the *spine curve*, is the curve $\mathbf{V}(s,1)$. The auxiliary curves coincide with the intersection curve of the given surfaces Φ_1, Φ_2.

5. If one replaces the boundary $t = 1$ of the surface patches by a suitable curve $t = t(s)$, then the curves of contact can be considered as curves of contact of a rolling ball of *variable* radius $r = t(s)$ (see Figure 5.)

The **algorithm** for determining point $\mathbf{S}(s,t)$ of blending surface Ψ:

1. Calculate $\gamma(s)$ (point of the auxiliary curve)

2. Determine foot point $\mathbf{V}(s,t)$ of $\gamma(s)$ on the intersection curve of the offset surfaces belonging to distance t.
Determine foot point $\mathbf{V}(s,1-t)$ of $\gamma(s)$ on the intersection curve of the offset surfaces belonging to distance $1-t$.

3. Determine foot point $\mathbf{S}_1(s,1-t)$ of $\mathbf{V}(s,1-t)$ on surface Φ_1.
Determine foot point $\mathbf{S}_2(s,t)$ of $\mathbf{V}(s,t)$ on surface Φ_2.

4. **Point** on blending surface:

$$\mathbf{S}(s,t) = f(t)\mathbf{S}_1(s,1-t) + (1-f(t))\mathbf{S}_2(s,t),$$

where $$f(t) = \frac{(1-t)^{n+1}}{(1-t)^{n+1} + t^{n+1}}.$$

5. For the examples shown below the parameter transformations $p_1(t), p_2(t)$ are used:

$$\begin{aligned}\mathbf{S}(s,t) &= f(t)\mathbf{S}_1(s,1-p_1(t)) + (1-f(t))\mathbf{S}_2(s,p_2(t)),\\ &= f(t)\mathbf{S}_1(s,p_2(1-t)) + (1-f(t))\mathbf{S}_2(s,p_2(t)),\end{aligned}$$

where $p_1(t) := \frac{t}{1-\lambda+\lambda t}$, and $p_2(t) := \frac{t(1-\lambda)}{1-\lambda t}$, $0 \leq \lambda < 1$.

Example 2:
Figure 2 shows the numerical parameterization of the Voronoi surface of the two Bézier patches given in Example 1 (Section 2). The auxiliary curve is the Bézier curve of degree 4 with control points: $(-3.3,-4.9,2)$, $(-2.5,-3.5,2.7)$, $(-2,-2,3.4)$, $(-2,0,3.6)$, $(-2.5,2.4,3)$.
The rolling ball has radius $r = 2$.
Figure 3 shows the (numerical) re-parameterization of the two Bézier patches by the Voronoi surface and a resulting G^2–blending surface.

Figure 4 shows two further examples. Figure 4b is an application of the method on an example involving the **implicit** surface $x^4 + y^4 + z^4 = 1$. (For the representation of the offset surfaces of an implicit surface see [2].)

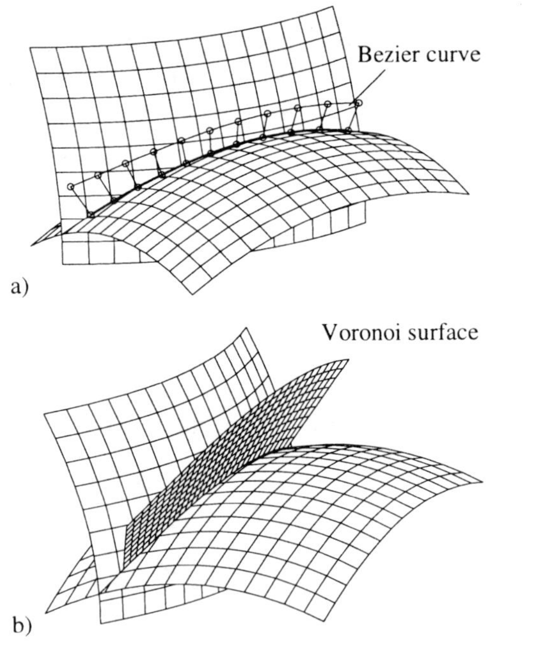

Figure 2. Parameterization of the intersection curve (a)) and the Voronoi surface (b))

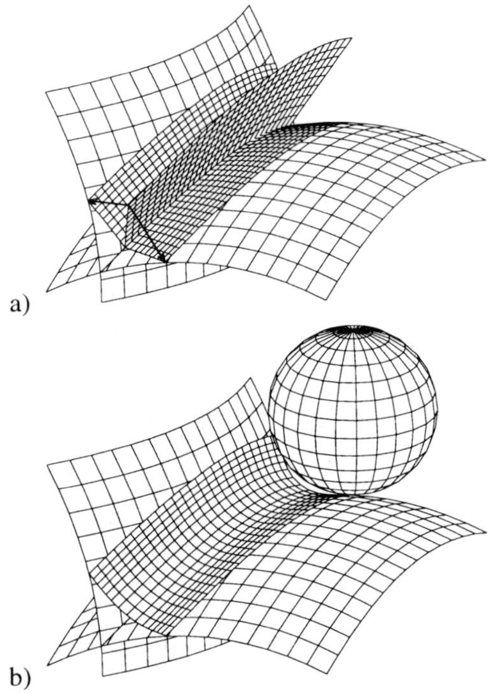

Figure 3. Re-parameterization (a)) and G^2–blending surface (b))

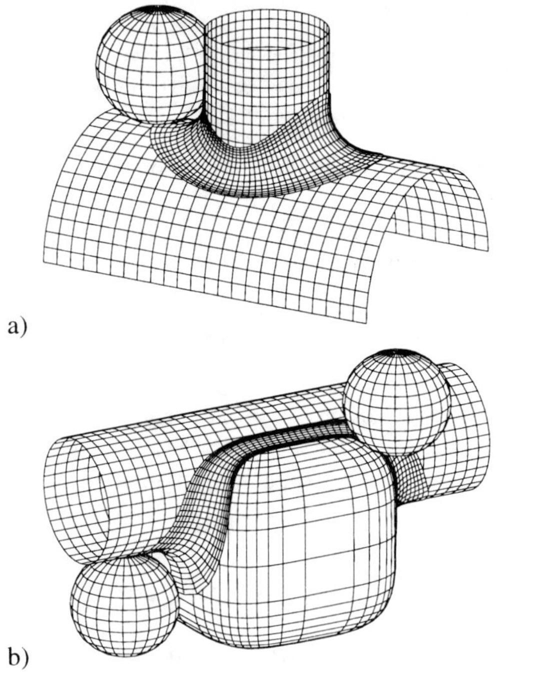

Figure 4. G^2–blending of two further examples with constant radius

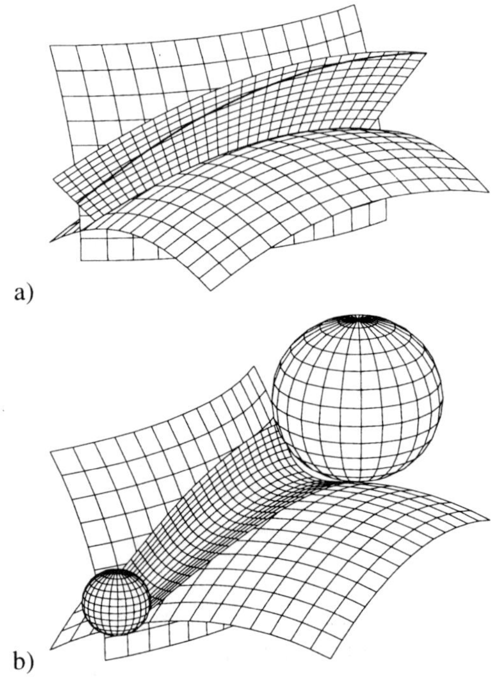

Figure 5. Variable radius: a) curve of ball centers, b) G^2–blending surface

Figure 5 shows a blending surface of the Bézier patches (Example 1) with variable radius rolling ball contact curves. Radius: $r = 2(0.4 + 0.6s)$.

Figure 6 shows further examples of G^n–blending surfaces with variable radius contact curves. The radius is oscillating.

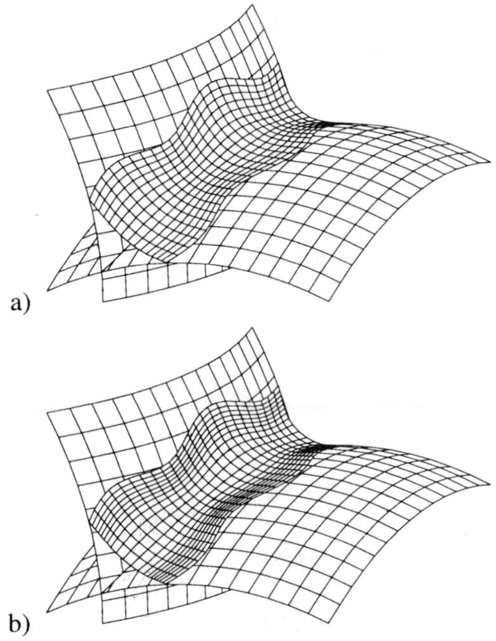

Figure 6. G^1 (a)) and G^3 (b)) blending surfaces with variable radius rolling ball contact curves

4 Conclusion

A numerical blending method is introduced which meets the advantages of the classical rolling ball method (with constant or variable radius) and the demand of G^n–continuity. Theoretical background is the idea of the normalform of a surface (cf. [3]. The performance of the method is based on suitable foot point algorithms (cf. [3]).

References

[1] Choi,B.K., Ju,S.Y.: *Constant-radius blending in surface modelling.* Computer–Aided Design 21 (1989), 213–220.

[2] Hartmann,E.: *Numerical Implicitization for Intersection and G^n-continuous Blending of surfaces.* Comp. Aided Geom. Des. 15 (1998), 377–397.

[3] Hartmann,E.: *On the curvature of curves and surfaces defined by normalforms.* Comp. Aided Geom. Des. 16 (1999), 355–376.

[4] Hartmann,E.: *Numerical Parameterization of Curves and Surfaces.* Appears in Comp. Aided Geom. Des. (2000)

[5] Hartmann,E.: *Parametric G^n–blending Curves and Surfaces.* submitted.

[6] Hermann,T., Lukács,G., Várady,T.: *Techniques for Variable Radius Rolling Ball Blends.* in Mathematical Methods for Curves and Surfaces, Daehlen,Lyche,Schumaker (eds.), Vanderbilt Univ. Press (1995), 225–236.

[7] Mazure, M.-L.: *Geometric contact curves and surfaces.* Comp. Aided Geom. Des. 11 (1994), 177–195.

[8] Pegna,J., Wilde,D.J.: *Spherical and Circular Blending of Functional Surfaces.* ASME J. of Offshore Mechanics & Arctic Engineering 112 (1990), 134–142.

[9] Vida,J., Martin,R.R., Várady,T.: *A survey of blending methods that use parametric surfaces.* Computer–Aided Design 26 (1994), 341–365.

[10] Wolter,F.-E., Tuohy,S.T.: *Curvature computations for degenerate surface patches.* Comp. Aided Geom. Des. 9 (1992), 241–270.

Blending between Right Circular Cylinders with Parabolic Cyclides

Kenji Ueda
RICOH Company, Ltd.
1-1-17, Koishikawa, Bunkyo-ku, Tokyo 112-0002, JAPAN
ueda@src.ricoh.co.jp

Abstract

There are right circular cylinders touching a parabolic cyclide along a curve, and all the cylinders have the same point of contact with the cyclide. So a pair of cylinders with a point of contact are blended by a parabolic cyclide. In this paper, a method for blending between a pair of right circular cylinders with a parabolic cyclide is presented. A blend with a parabolic cyclide is a kind of variable-radius rolling-ball blend. The contour lines of the blending surface are central conics.

1. Introduction

Dupin cyclides play an important role in CAGD by virtue of their properties [2, 7]. They have been recognized as surfaces for blending natural quadrics intersections [1, 13]. The blend with cyclides is based on the tangent cones [8] to lines of curvature on cyclides. As the lines of curvature of cyclides are the isoparametric curves of cyclide patches [5], the blending surfaces are represented by cyclide patches.

While ordinary cyclides are bounded surfaces, parabolic cyclides extend to infinity. As parabolic cyclides have straight lines of curvature, they are useful in blending with cylinders or planes [2, 11].

On the other hand, it is known that there are an infinite number of cyclides touching a cyclide along curves from the viewpoint of Laguerre geometry [3, 4, 6]. Generally, the curves of contact are not lines of curvature. There are also parabolic cyclides touching the other parabolic cyclides. The surfaces touching parabolic cyclides include right circular cylinders and cones. We can construct G^1 continuous surfaces with them.

In this paper, a method for blending between a pair of right circular cylinders with a parabolic cyclide is presented as a practical application of touching cyclides. First, parametric and implicit forms of parabolic cyclides are presented. Right circular cones touching parabolic cyclides are shown with their parametric forms in section 2. Then, section 3 introduces blending surfaces and investigates the contour lines on the surface. The shapes and triangular Bézier form of the blending surfaces are investigated in section 4 and 5. Finally, a method for selecting appropriate cyclides for blending given pairs of cylinders is presented.

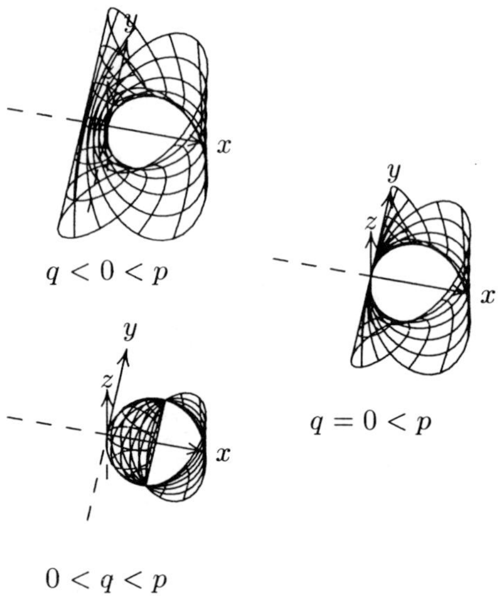

Figure 1. Parabolic cyclides and a sphere.

2. Parabolic cyclides

The parametric form $\mathbf{r}(u, v)$ of parabolic cyclides is given by [6, 9]

$$\mathbf{r}(u,v) = \begin{pmatrix} x(u,v) \\ y(u,v) \\ z(u,v) \end{pmatrix} = \frac{r_1(u)\mathbf{c}_2(v) - r_2(v)\mathbf{c}_1(u)}{r_1(u) - r_2(v)}$$

$$= \frac{1}{u^2+v^2+1} \begin{pmatrix} pu^2 + qv^2 \\ (p-(q-p)v^2)u \\ (q-(p-q)u^2)v \end{pmatrix}, \qquad (1)$$

0-7695-0562-7/00 $10.00 © 2000 IEEE

where the curves $\mathbf{c}_1(u)$ and $\mathbf{c}_2(v)$ are the focal parabolas

$$\mathbf{c}_1(u) = \frac{q}{2}\begin{pmatrix}1\\0\\0\end{pmatrix} + \frac{p-q}{2}\begin{pmatrix}u^2\\2u\\0\end{pmatrix}, \tag{2}$$

$$\mathbf{c}_2(v) = \frac{p}{2}\begin{pmatrix}1\\0\\0\end{pmatrix} + \frac{q-p}{2}\begin{pmatrix}v^2\\0\\2v\end{pmatrix}, \tag{3}$$

and the scalar functions $r_1(u)$ and $r_2(v)$ are the radius functions

$$r_1(u) = \frac{q-(p-q)u^2}{2}, \quad r_2(v) = \frac{p-(q-p)v^2}{2}. \tag{4}$$

Parabolic cyclides are recognized as canal surfaces with the spine curve $\mathbf{c}_1(u)$ and the variable-radius $r_1(u)$ or with spine curve $\mathbf{c}_2(v)$, and variable-radius $r_2(v)$. Parabolic cyclides are classified into three types. They are called ring cyclides for $pq < 0$, horn cyclides for $pq > 0$, and thorn cyclides for $p = 0$ or $q = 0$. In the case of $p = q$, Equation (1) represents spheres. As parabolic cyclides are unbounded, Figure 1 illustrates them for $(u, v) \in ([-\infty,\infty] \times [-1,1]) \cup ([-1,1] \times [-\infty,\infty])$.

The unit normal vector $\mathbf{n}(u, v)$ to the parabolic cyclide $\mathbf{r}(u, v)$ is given by [10]

$$\mathbf{n}(u,v) = \frac{\mathbf{r}_u \times \mathbf{r}_v}{|\mathbf{r}_u \times \mathbf{r}_v|} = \frac{1}{u^2+v^2+1}\begin{pmatrix}u^2+v^2-1\\2u\\2v\end{pmatrix}. \tag{5}$$

The parabolic cyclide $\mathbf{r}(u, v)$ is expressed as the linear combination of thorn cyclides $\mathbf{p}(u, v)$ and $\mathbf{q}(u, v)$.

$$\mathbf{r}(u,v) = p\,\mathbf{p}(u,v) + q\,\mathbf{q}(u,v), \tag{6}$$

where

$$\mathbf{p}(u,v) = \frac{u}{u^2+v^2+1}\begin{pmatrix}u\\1+v^2\\-uv\end{pmatrix}, \tag{7}$$

$$\mathbf{q}(u,v) = \frac{v}{u^2+v^2+1}\begin{pmatrix}v\\-vu\\1+u^2\end{pmatrix}. \tag{8}$$

The cyclide $\mathbf{r}(u, v)$ is also expressed as the offset surface of the thorn cyclides

$$\begin{aligned}\mathbf{r}(u,v) &= (p-q)\mathbf{p}(u,v) + \frac{q}{2}\left(\mathbf{n}(u,v) + \begin{pmatrix}1\\0\\0\end{pmatrix}\right)\\ &= (q-p)\mathbf{q}(u,v) + \frac{p}{2}\left(\mathbf{n}(u,v) + \begin{pmatrix}1\\0\\0\end{pmatrix}\right). \end{aligned}\tag{9}$$

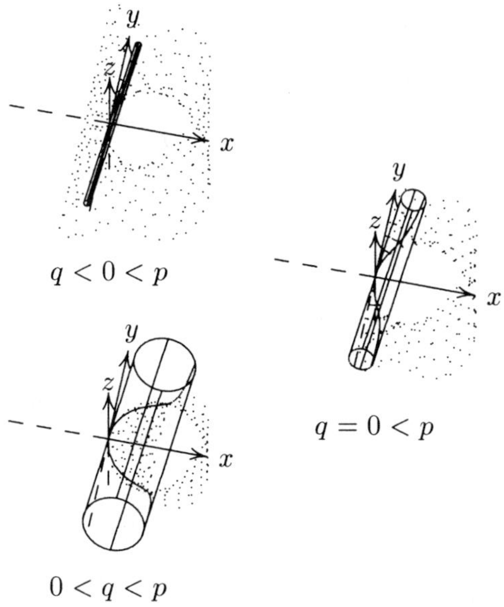

Figure 2. Cylinders touching parabolic cyclides.

By eliminating the parameters u and v of parametric form (1) of parabolic cyclides, we obtain the implicit form of parabolic cyclides.

$$(x-q)y^2 + (x-p)z^2 + x(x-p)(x-q) = 0. \tag{10}$$

Equation (10) implies that the contour lines at $x = \alpha$ are central conics [12]

$$(\alpha-q)y^2 + (\alpha-p)z^2 + \alpha(\alpha-p)(\alpha-q) = 0. \tag{11}$$

From the x-value of Equation (1), the contour lines are the image of the central conics $(p-\alpha)u^2 + (q-\alpha)v^2 - \alpha = 0$ in the parameter space.

3. Right Circular Cylinders Touching Parabolic Cyclides

Suppose $\mathbf{C}$ is a right circular cylinder $(X-R)^2 + Y^2 = R^2$ with the radius

$$R[\theta] = \frac{p\cos^2\theta + q\sin^2\theta}{2}, \qquad (-\frac{\pi}{2} \le \theta \le \frac{\pi}{2}). \tag{12}$$

The cylinder $\mathbf{C}$ has a parametric form

$$\begin{pmatrix}X(u,v)\\Y(u,v)\\Z(u,v)\end{pmatrix} = \begin{pmatrix}R[\theta]\left(\frac{u^2-\cos^2\theta}{u^2+\cos^2\theta}+1\right)\\ R[\theta]\frac{2u\cos\theta}{u^2+\cos^2\theta}\\ -(p-q)v\cos\theta\end{pmatrix}. \tag{13}$$

The rotated cylinder $\mathbf{c}[\theta](u,v)$ of the cylinder $\mathbf{C}$ about the x axis with the angle θ is expressed as

$$\mathbf{c}[\theta](u,v) = \begin{pmatrix} 1 & 0 & 0 \\ 0 & \cos\theta & -\sin\theta \\ 0 & \sin\theta & \cos\theta \end{pmatrix} \begin{pmatrix} X(u,v) \\ Y(u,v) \\ Z(u,v) \end{pmatrix}. \quad (14)$$

The image curve $\mathbf{s}(u)$ of the straight line $v = u\tan\theta$ in the parameter space is given by

$$\begin{aligned} \mathbf{s}(u) &= \mathbf{c}[\theta](u, u\tan\theta) \quad (15) \\ &= \frac{u}{u^2+\cos^2\theta} \begin{pmatrix} (p\cos^2\theta + q\sin^2\theta)u \\ (p\cos^2\theta + (p-q)u^2\sin^2\theta) \\ (q-(p-q)u^2)\sin\theta\cos\theta \end{pmatrix}. \end{aligned}$$

The curve $\mathbf{s}(u)$ is also equal to the image of the straight line $v = u\tan\theta$ in the parameter space of the parabolic cyclide $\mathbf{r}(u,v)$, namely, $\mathbf{r}(u, u\tan\theta) = \mathbf{s}(u)$. Hence the cylinder is intersecting with the parabolic cyclide $\mathbf{r}(u,v)$.

Moreover, the unit normal vector to the cylinder $\mathbf{c}[\theta]$ is equal to the unit normal vector to the cyclide $\mathbf{r}(u,v)$ along the curve $\mathbf{s}(u)$, namely,

$$\begin{aligned} \frac{\mathbf{c}_u \times \mathbf{c}_v}{|\mathbf{c}_u \times \mathbf{c}_v|} &= \frac{1}{u^2+\cos^2\theta} \begin{pmatrix} u^2-\cos^2\theta \\ 2u\cos^2\theta \\ 2u\cos\theta\sin\theta \end{pmatrix} \\ &= \mathbf{n}(u, u\tan\theta). \quad (16) \end{aligned}$$

Therefore, the cylinder and the parabolic cyclide make contact with each other along the curve $\mathbf{s}(u)$. Figure 2 illustrates circular cylinders contacting with different types of parabolic cyclides. The curves of contact are coiled around the cyliders. The circular cylinder is not only passing through the origin, but also the singular points of the parabolic cyclide.

From the derivative of the radius $R[\theta]$ of the circular cylinder,

$$\frac{d}{d\theta}R[\theta] = -\frac{p-q}{2}\sin 2\theta, \quad (17)$$

the radius has the minimum value $R[-\pi/2] = R[\pi/2] = q/2$ and the maximum value $R[0] = p/2$ for $q < p$.

Of course, the contour lines of the cylinder at $x = \alpha$ are straight lines

$$y\cos\theta + z\sin\theta = \pm\sqrt{\alpha(p\cos^2\theta + q\sin^2\theta - \alpha)}, \quad (18)$$

which are the generatrices of the cylinder and the image of the straight lines in the parameter space,

$$u = \pm\sqrt{\frac{\alpha}{p\cos^2\theta + q\sin^2\theta - \alpha}}\cos\theta. \quad (19)$$

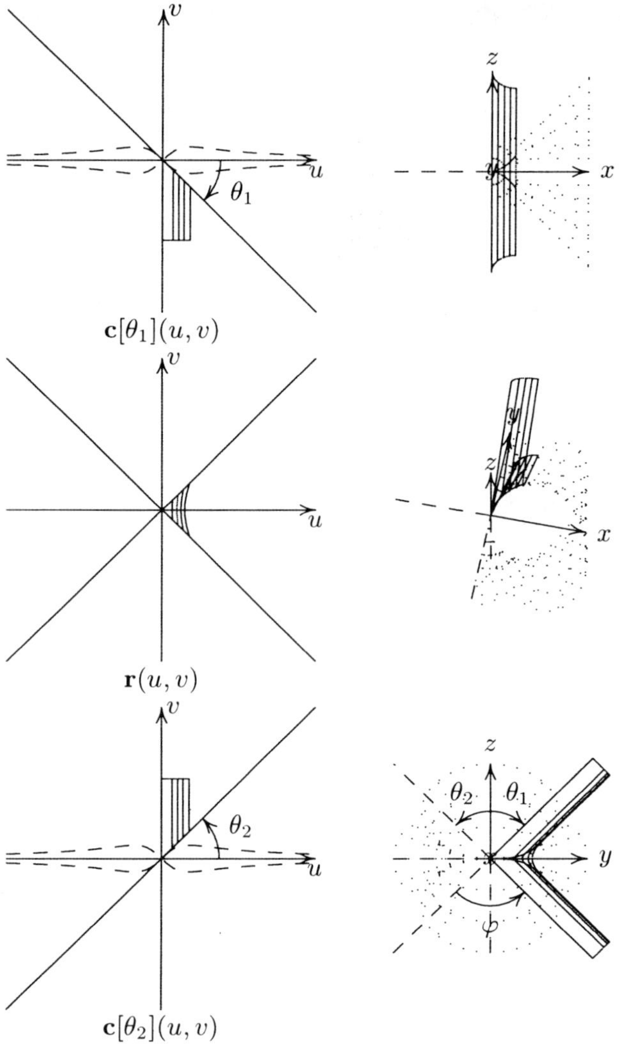

Figure 3. Blending with a thorn cyclide.

4. Cylinder-cylinder blend with parabolic cyclides

It has been shown that there are right circular cylinders touching a parabolic cyclide along a curve, This means that two of the cylinders making contact at the origin are blended geometrically with a region of the parabolic cyclide. The whole cylinders, however, can not be blended with a definite surface, because the curve of contact is not closed.

In this section, cylinder-cylinder blends with different types of cyclides of the shape parameter $0 < p$ and $q < p$ are discussed.

The contour lines of the blended surface at $x = \alpha$ consist of straight lines and central conics which are connected with G^1 continuity. As parabolic cyclides are canal surfaces, the blend is a kind of variable-radius rolling-ball blend [13].

4.1. Blending with thorn cyclides

For $0 = q < p$, the radius $R[\theta]$ is positive,

$$R[\theta] = \frac{p}{2}\cos^2\theta > 0. \tag{20}$$

Hence, all the cylinders exist in the region of positive x in 3D space. The cylinders touching a thorn cyclide are intersecting.

We can blend a pair of quarter cylinders with a thorn cyclides, as illustrated in Figure 3. The left hand side shows the parameter spaces of the cylinders $\mathbf{c}[\theta_1]$, $\mathbf{c}[\theta_2]$ and the thorn cyclide $\mathbf{r}(u, v)$. The angle between the rotation angles of the cylinders is φ. The angle φ is the angle between the generatrices of the cylinders passing through the contact point, namely, the origin. The right hand side shows the blended surface in 3D space from three viewpoints. The blended surface consists of the images of the regions in the parameter spaces.

The contour lines at $x = \alpha$ of thorn cyclides are hyperbola, because the central conics (11) are the hyperbolas for $0 < \alpha < p$,

$$\frac{y^2}{\alpha(p-\alpha)} - \frac{z^2}{\alpha^2} = 1. \tag{21}$$

The blending surface, which is a region on the thorn cyclide, is the image of the region bounded by two straight lines and a hyperbola in the parameter space in the figure.

4.2. Blending with horn cyclides

For $0 < q < p$, the radius $R[\theta]$ is also positive,

$$R[\theta] = \frac{p\cos^2\theta + q\sin^2\theta}{2} > 0. \tag{22}$$

There are pinch points $(q, \pm\sqrt{q(p-q)}, 0)$ on horn cyclides. The points are the image of the straight lines $u = \pm\sqrt{q/(p-q)}$ in the parameter space. The surface is subdivided into two parts by the pinch points. One part is horned and bounded. Another is the rest of the surface and is unbounded. The bounded part is inside any cylinder touching the horn cyclide and the unbounded part is outside the cylinder.

The central conics (11) are the ellipses

$$\frac{y^2}{\alpha(p-\alpha)} + \frac{z^2}{\alpha(q-\alpha)} = 1, \qquad (0 < \alpha < q), \tag{23}$$

on the bounded part and are the hyperbolas

$$\frac{y^2}{\alpha(p-\alpha)} - \frac{z^2}{\alpha(\alpha-q)} = 1, \qquad (q < \alpha < p), \tag{24}$$

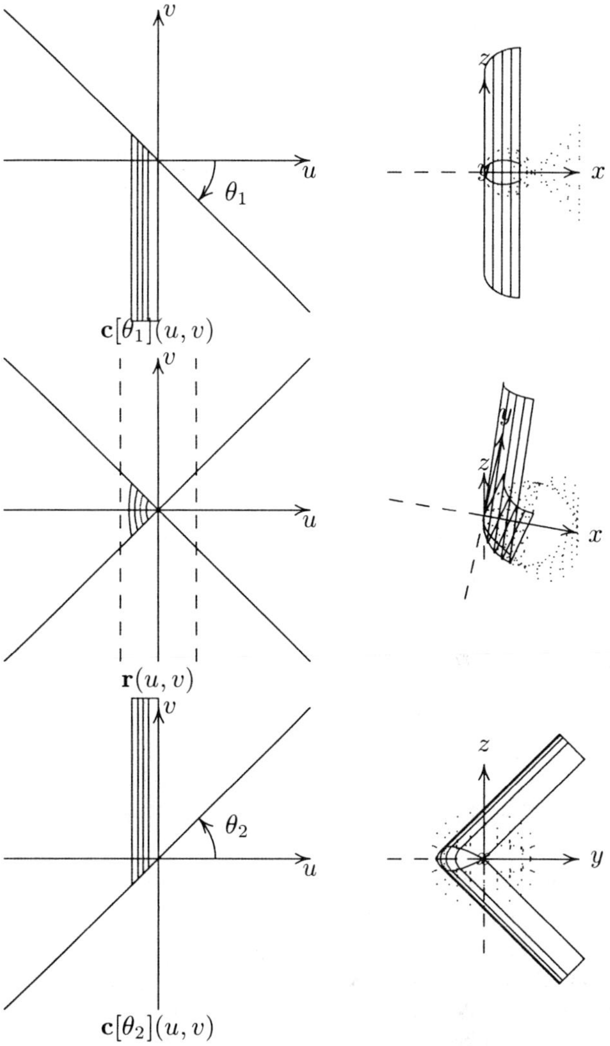

Figure 4. Blending with a horn cyclide.

on the unbounded part. Figure 4 illustrates an example of a blend of a pair of cylinders with a horn cyclide. As the blending surface is a region of the bounded part of the horn cyclide. the contour lines are ellipses. The dashed lines in the parameter space of the horn cyclide $\mathbf{r}(u, v)$ are the preimage of the pinch points.

4.3. Blending with ring cyclides

For $q < 0 < p$, the sign of radius $R[\theta]$ is not always positive,

$$R[\theta] = \begin{cases} +, & |\tan\theta| < \sqrt{\frac{p}{-q}}, \\ 0, & |\tan\theta| = \sqrt{\frac{p}{-q}}, \\ -, & |\tan\theta| > \sqrt{\frac{p}{-q}}. \end{cases} \tag{25}$$

The cylinders of radius 0 are the two intersecting straight lines at $x = 0$,

$$y = z \tan\theta = \pm\sqrt{\frac{p}{-q}}z, \tag{26}$$

and the images of $v = \pm\sqrt{p/(-q)}u$. The straight lines are not only the degenerated right circular cylinder, but also the degenerated hyperbolas on the parabolic ring cyclide.

The central conics (11) are the hyperbolas

$$\frac{z^2}{(-\alpha)(\alpha - q)} - \frac{y^2}{(-\alpha)(p-\alpha)} = 1, \quad (q < \alpha < 0), \tag{27}$$

and

$$\frac{y^2}{\alpha(p-\alpha)} - \frac{z^2}{\alpha(\alpha - q)} = 1, \qquad (0 < \alpha < p). \tag{28}$$

From a different point of view, the cylinders touching parabolic cyclides are the offset surfaces of the degenerated cylinders. Because a parabolic cyclide is the offset surface of the other parabolic cyclides.

5. Bézier triangles of blending surfaces

By combining the blending surfaces described in the previous section, we are able to blend half cylinders. Figure 5 illustrates a combination of blends with a thorn and a horn cyclide. The contour lines are consists of straight lines and central conics.

The blending surface is a region of a parabolic cyclide, as illustrated in Figures 3 and 4. From Equation (1), the x-value of a parabolic cyclide at $x = \alpha$ satisfies

$$x(u,v) = \frac{pu^2 + qv^2}{u^2 + v^2 + 1} = \alpha. \tag{29}$$

This eqation expresses central conics. Hence, the blending surface patch is the image of a region bounded by straight lines and a central conic.

The regions in the parameter space can be expressed as triangular rational Bézier surfaces of degree 2, as illustrated in Figure 6. An example of the control points $\{\mathbf{b}_{ijk}\}$ are expressed as

$$\begin{aligned} &&& \mathbf{b}_{200} = (\underline{u}, \underline{v}), \\ && \mathbf{b}_{101} = (\tfrac{\overline{v}}{\tan\theta}, \overline{v}), \quad & \mathbf{b}_{110} = (\overline{u}, \overline{v}), \\ \mathbf{b}_{002} = (0,0), \quad & \mathbf{b}_{011} = (\tfrac{\overline{v}}{\tan\theta}, 0), \quad && \mathbf{b}_{020} = (\overline{u}, 0), \end{aligned} \tag{30}$$

where

$$\underline{u} = \sqrt{\frac{\alpha}{(p-\alpha)\cos^2\theta + (q-\alpha)\sin^2\theta}}\cos\theta = \frac{\underline{v}}{\tan\theta}, \tag{31}$$

$$\underline{v} = \sqrt{\frac{\alpha}{(p-\alpha)\cos^2\theta + (q-\alpha)\sin^2\theta}}\sin\theta, \tag{32}$$

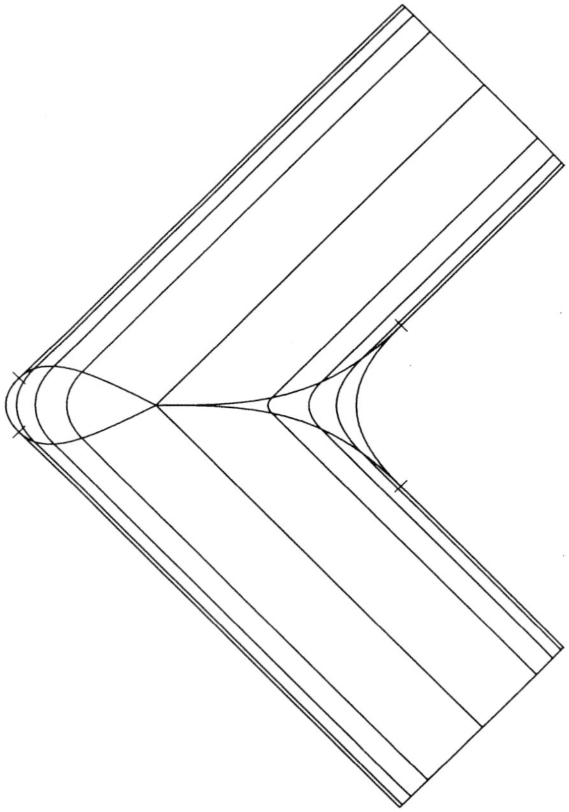

Figure 5. Blending half-cylinders with two cyclides.

$$\overline{u} = \sqrt{\frac{\alpha}{p-\alpha}}, \tag{33}$$

$$\overline{v} = \frac{\sqrt{\alpha}(\sqrt{p\cos^2\theta + q\sin^2\theta - \alpha} - \sqrt{p-\alpha}\cos\theta)}{(q-\alpha)\sin\theta}. \tag{34}$$

Their weights are given by

$$\begin{aligned} &&& w_{200} = 1, \\ && w_{101} = \overline{w}, \quad & w_{110} = \overline{w}, \\ w_{002} = 1, \quad & w_{011} = 1, \quad && w_{020} = 1, \end{aligned} \tag{35}$$

where

$$\overline{w} = -\frac{b \pm \sqrt{b^2 - ac}}{2a} \qquad (0 < \overline{w}), \tag{36}$$

$$a = (p-\alpha)\overline{u}^2 + (q-\alpha)\overline{v}^2 - \alpha, \tag{37}$$

$$b = (p-\alpha)(\overline{u} + \underline{u})\overline{u} + (q-\alpha)\underline{v}\overline{v} - 2\alpha, \tag{38}$$

$$c = (p-\alpha)(\overline{u} + \underline{u})^2 + (q-\alpha)\underline{v}^2 - 4\alpha. \tag{39}$$

Therefore, the blending surface patch are obtained as triangular rational Bézier surfaces of degree 6 on a parabolic cyclide.

Figure 7 illustrates an example of blending surface patches with its pre-image and central conics in the parameter space. As the isoparametric lines of the region for a parameter are straight lines parallel to the u-axis, their images are circular arcs on the cyclide.

6. Relationship between the pair of cylinders

In section 4, examples of blends with parabolic cyclides have been shown. It is necessary to select the appropriate cyclide for various arrangements of the pair of cylinders. In this section, the relationships between a pair of cylinders and a parabolic cyclide are discussed.

The pair of cylinders $\mathbf{c}[\theta_1]$ and $\mathbf{c}[\theta_2]$ are of radii

$$R[\theta_1] = \frac{p\cos^2\theta_1 + q\sin^2\theta_1}{2},$$
$$R[\theta_2] = \frac{p\cos^2\theta_2 + q\sin^2\theta_2}{2}. \quad (40)$$

Let φ be the angle between the two rotation angles θ_1 and θ_2 of the cylinders, namely,

$$\theta_2 = \theta_1 + \varphi, \qquad (-\frac{\pi}{2} \le \theta_1 \le 0,\ 0 < \varphi \le \frac{\pi}{2}). \quad (41)$$

The intersection curve on the cylinders $\mathbf{c}[\theta_1]$ is obtained by solving the equation

$$(x - R[\theta_2])^2 + y^2 = R[\theta_2]^2, \quad (42)$$

where

$$\begin{pmatrix} x \\ y \\ z \end{pmatrix} = \begin{pmatrix} 1 & 0 & 0 \\ 0 & \cos(-\theta_2) & -\sin(-\theta_2) \\ 0 & \sin(-\theta_2) & \cos(-\theta_2) \end{pmatrix} \mathbf{c}[\theta_1](u,v). \quad (43)$$

Hence, the intersection is the image of the curve (44) in the parameter space.

$$v = \frac{2R[\theta_1]\cos\theta_1\cos\varphi \pm \sqrt{D(u)}}{(p-q)\cos\theta_1\sin\varphi}\,\frac{u}{u^2+\cos^2\theta_1}, \quad (44)$$

where $D(u) = 2R[\theta_1](Au^2 + C)$,

$$A = -(p-q)\sin(2\theta_1+\varphi)\sin\varphi, \quad (45)$$
$$C = [(p-q)\cos(\theta_1+2\varphi)\cos\theta_1 + (p\sin^2\varphi + q\cos^2\varphi)]\cos^2\theta_1.$$

The dashed lines in the parameter spaces of the cylinders in Figure 3 are the preimage of the intersection between two cylinders.

Suppose that the ratio of the radii $R[\theta_1]$ and $R[\theta_2]$ of the two cylinders is equal to a constant r,

$$R[\theta_2] = rR[\theta_1]. \quad (46)$$

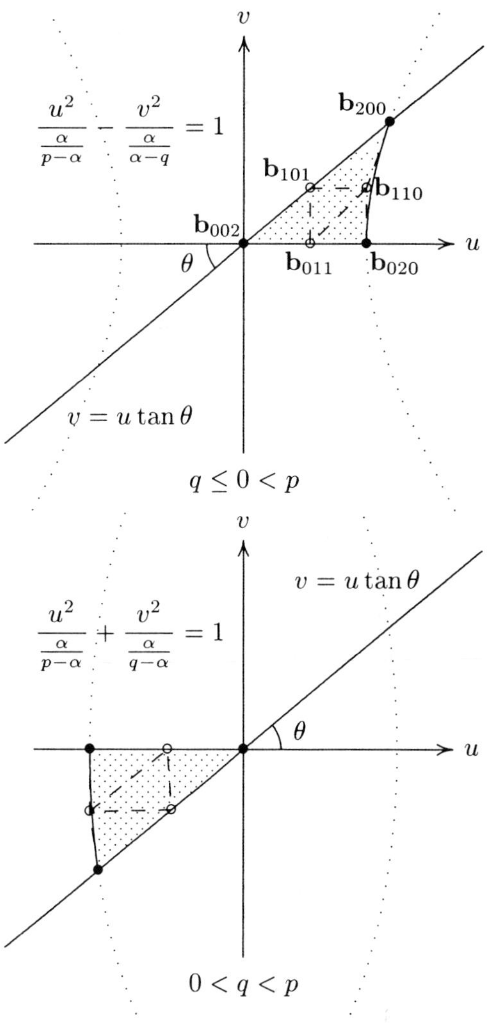

Figure 6. Pre-images of blend surface patches.

Equation (46) is transformed into

$$F \equiv A\sin 2\theta_1 + B\cos 2\theta_1 + C = 0, \quad (47)$$

where s is a new shape parameter defined by $s = q/p\,(\neq 1)$ and

$$A = \sin 2\varphi,\ B = r - \cos 2\varphi,\ C = \frac{1+s}{1-s}(r-1). \quad (48)$$

By using the half-angle formulae for trigonometric functions, Equation (47) is transformed into

$$F = \frac{(C-B)\tan^2\theta_1 + 2A\tan\theta_1 + (C+B)}{1+\tan^2\theta_1} = 0. \quad (49)$$

Hence, the angle θ_1 is obtained by solving Equation (49), if the shape parameter s of a parabolic cyclide is given.

A method for deciding the shape parameter s of a parabolic cyclide and the rotation angle θ_1 for a given arrangement of a pair of cylinders by specifying them with the angle φ and the ratio r is shown in the rest of this section.

6.1. Cylinders with the same radius ($r = 1$)

When the two cylinders intersect and have the same radius, i.e., $r = 1$, Equation (47) is transformed into

$$F = 2 \sin \varphi \sin(2\theta_1 + \varphi) = 0. \tag{50}$$

If the rotation angles satisfy

$$\theta_1 = -\frac{\varphi}{2}, \qquad \theta_2 = \theta_1 + \varphi = \frac{\varphi}{2}, \tag{51}$$

any parabolic cyclide is able to blend a pair of intersecting cylinders with the same radius and a point of contact.

In this case, Equation (44) of the intersection is simplified, because $A = 0$, that is, $D(u)$ is a constant.

6.2. Orthogonal generatrices ($\varphi = \pi/2$)

When the generatrices of two cylinders are orthogonal at the origin, Equation (47) becomes

$$(r + 1) \cos 2\theta_1 + \frac{1 + s}{1 - s}(r - 1) = 0. \tag{52}$$

If we choose $r = -1$ and $s = -1$, Equation (52) is satisfied. Hence, two cylinders with opposite radii are blended with the ring cyclide of $p + q = 0$ for any rotation angle θ_1.

In the case of $r \neq -1$, a rotation angle θ_1 is calculated by

$$\cos 2\theta_1 = \frac{1 + s}{1 - s} \frac{1 - r}{1 + r}. \tag{53}$$

For example, the rotation angle θ_1 is 0 for the parabolic cyclide with a shape parameter of $s = r$.

6.3. Other cases ($r \neq 1$ and $0 < \varphi < \pi/2$)

In the case of $C - B = 0$, the quadratic equation (47) yields

$$(\cos^2 \varphi - r)s = (\cos^2 \varphi - 1). \tag{54}$$

Hence, we can obtain the values

$$s = \frac{\cos^2 \varphi - 1}{\cos^2 \varphi - r}, \ \tan \theta_1 = -\frac{C + B}{2A} = \frac{\cos 2\varphi - r}{\sin 2\varphi}, \tag{55}$$

when $r \neq \cos^2 \varphi$.

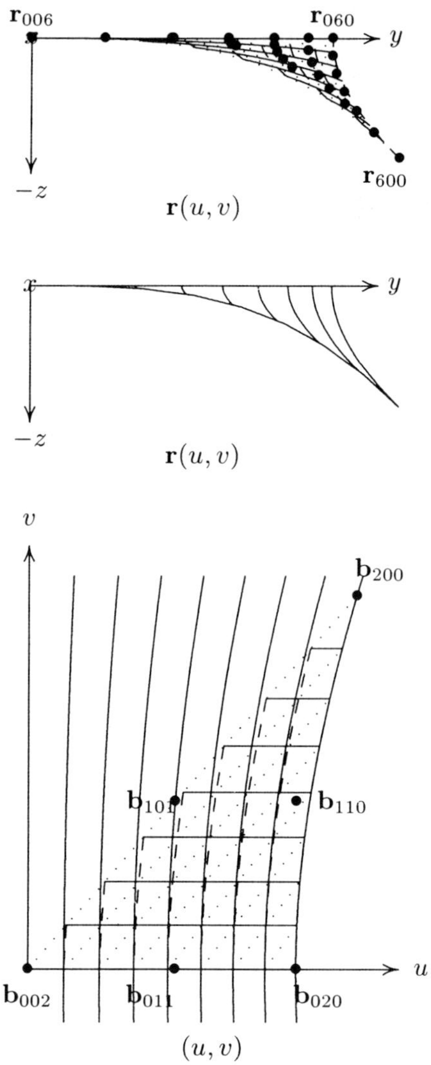

Figure 7. A triangular Bézier patch with its pre-image.

As the discriminant D of the quadratic equation in (47) is $A^2 - (C - B)(C + B)$, the equations $r = \cos^2 \varphi$ and $D = 0$ yields

$$s = \frac{1 + \cos^2 \varphi \pm \sin \varphi \sqrt{1 + 3\cos^2 \varphi}}{2 \cos^2 \varphi}, \tag{56}$$

and

$$\tan \theta_1 = -\frac{A}{C - B} = -\frac{\sin \varphi \pm \sqrt{1 + 3\cos^2 \varphi}}{2 \cos \varphi}. \tag{57}$$

6.4. Blend parameter

There may be freedom of choosing the shape of the blending parabolic cyclide for given two cylinders, as described in this section. For example, any parabolic cyclide is able to blend a pair of cylinders with the same radius as shown in Section 6.1.

From Equation (13) of the cylinder $\mathbf{C}$, the point for the parameter $(u, v) = (\cos\theta, \cos\theta\tan\theta)$ is the intersection of the curve $\mathbf{s}(u)$ of contact and the plane of $x = R[\theta]$, namely

$$\begin{pmatrix} X(\cos\theta, \sin\theta) \\ Y(\cos\theta, \sin\theta) \\ Z(\cos\theta, \sin\theta) \end{pmatrix} = \begin{pmatrix} R[\theta] \\ R[\theta] \\ -(p-q)\sin\theta\cos\theta \end{pmatrix}. \quad (58)$$

The value $Z(\cos\theta, \sin\theta)$ represents the expanse of blending surface at $x = R[\theta]$, Hence, the following ratio of the expanse to the radius can be accepted to control the shape of blending surface.

$$\frac{|Z(\cos\theta, \sin\theta)|}{|R[\theta]|} = \left| \frac{-(p-q)\sin\theta\cos\theta}{R[\theta]} \right|. \quad (59)$$

By using the shape parameter $s = q/p$ for parabolic cyclides introduced in Section 5, the ratio is rewritten as

$$\frac{|Z(\cos\theta, \sin\theta)|}{|R[\theta]|} = \left| \frac{2(1-s)\tan\theta}{1 + s\tan^2\theta} \right|. \quad (60)$$

We can calculate the shape parameter s to control the shape of the blending surface by specifying this blend parameter.

7. Conclusion

The parameterization for right circular cylinders touching a parabolic cyclide along a curve has been shown. The curve of contact is represented as the image of the same straight line passing through the origin in their parameter spaces.

A method for blending the intersections of a pair of right circular cylinders around the point of contact with a parabolic cyclide also has been shown. The method is able to blend cylinders not only with the same radius but also with different radii. The blending method is a kind of variable-radius rolling-ball blendings with a parabola as the spine curve.

The contour lines of the blending cyclide are central conics connecting the generatrices of cylinders with G^1 continuity. The blending surface bounded by a contour line is represented by a triangular Bézier patch of degree 6.

References

[1] V. Chandru, D. Dutta and C. M. Hoffmann: Variable radius blending using Dupin cyclides, *Geometric Modeling for Product Engineering*, M. J. Wozny, J. U. Turner and K. Preiss (eds.), North-Holland, 39–57, 1990.

[2] D. Dutta, R. R. Martin and M. J. Pratt: Cyclides in surface and solid modeling, *IEEE Computer Graphics and Applications*, 13(1):53–59, 1993.

[3] R. Krasauskas and C. Mäurer, Studying cyclides with Laguerre geometry, *Computer Aided Geometric Design* 17:101–126, 2000.

[4] C. Mäurer and R. Krasauskas, Joining cyclide patches along quartic boundary curves, *Mathematical Methods for Curves and Surfaces II*, M. Dælen, T. Lyche and L. L. Schumaker (eds.), Vanderbilt University Press, 359–366, 1998.

[5] R. R. Martin: Principal patches - A new class of surface patch based on differential geometry, *EUROGRAPHICS '83*, P. J. W. ten Hagen (ed.), North-Holland, 47–55, 1994.

[6] M. Peternell and H. Pottmann, Designing rational surfaces with rational offsets, *Advanced Topics in Multivariate Approximation*, F. Fontanella, K. Jetter and P. J. Laurent (eds.) World Scientific, 275–286, 1996.

[7] M. J. Pratt, Cyclides in computer aided geometric design, *Computer Aided Geometric Design* 7:221–242, 1990.

[8] M. J. Pratt, The Virtues of Cyclides in CAGD, *Mathematical Methods in Computer Aided Design II*, T. Lyche and L. L. Schumaker (eds.), Academic Press, 457–473, 1992.

[9] M. J. Pratt, Cyclides in computer aided geometric design II, *Computer Aided Geometric Design* 12:131–152, 1995.

[10] M. J. Pratt, On a class of Pythagorean-normal surfaces with planar lines of curvature, *The Mathematics of Surfaces VIII*, R. Cripps (ed.), Information Geometers, 281–295, 1998.

[11] Y. L. Srinivas and D. Dutta: Rational parametric representation of parabolic cyclide: Formulation and applications, *Computer Aided Geometric Design*, 12: 551–566, 1995.

[12] K. Ueda: Central conics on parabolic Dupin cyclides, The Fourth International Conference on Curves and Surfaces, Saint-Malo, 1999.

[13] J. Vida, R. R. Martin and T. Varady: A survey of blending methods that use parametric surfaces, *Computer-Aided Design*, 26:341–365, 1994.

Constrained Shape Scaling of Multi-Surface Objects

Pifu Zhang, Caiming Zhang and Fuhua (Frank) Cheng

Graphics & Geometric Modeling Lab, Department of Computer Science
University of Kentucky, Lexington, Kentucky 40506-0046
E-mail: *pifu@csr.uky.edu*

Abstract

A method to scale a multi-surface object while holding the shape and size of specific features (trimming curves) unchanged is presented. The new method includes an earlier version for an one-NURBS-surface object as a special case by considering more general objects and more general features. The constrained scaling process is attach-and-deform based. The new surface is constructed by attaching the original features to a scaled version of the given object. The attaching process requires several transformations and a deformation of the scaled object. The resulting object has the same features as the original object while reflecting the shape and curvature distribution of the scaled object. The presented approach maintains a NURBS representation for each component surface of the resulting object and hence, is compatible with most of the current data-exchange standards. Test results on several car body surfaces with trimming curves are included. The quality of the resulting surfaces is examined using the highlight line model.

Keywords: constrained scaling, constrained deformation, trimming curves, NURBS surfaces, strain energy

1. Introduction

Constrained shape modification refers to the process of altering the shape of an object while holding certain features unchanged. The altering process may involve scaling and/or deformation. This area has attracted the interest of industry recently in that it provides a possibility to reshape an existing model without affecting certain important features and, consequently, avoid expansive redesign process.

Constrained shape deformation has been studied for a while. Topics that have been considered include axial deformation and free-form deformation. Both of which have been thoroughly investigated [1][7][8] [9][10][13] [15][16][17] [18][20][24] [27][28][29] [32][35] and some of the techniques have already been implmented by popular CAD/CAM software packages such as ACIS and CATIA.

Constrained shape scaling, as a design tool, has not been studied so extensively yet. The only known results are two techniques for trimmed NURBS surfaces [33][34] with trimming curves being considered as features. In the first case [33], a attach-and-deform based approach is used, i.e., the new surface is formed by scaling the given NURBS surface according to the scaling requirement and then attaching the original features to the scaled surface at appropriate locations. The attaching process requires a minor deformation of the scaled surface to ensure complete attachment. The second case [34] is fix-and-stretch based, i.e., the new surface is formed by fixing selected regions of the given trimmed NURBS surface that contain the trimming curves and stretching the remaining part of the surface to reach certain boundary condition. The stretching process of the second approach and the deformation process of the first approach are both performed through an optimization process to ensure that the resulting surface reflects the shape and curvature distribution of the scaled version of the given NURBS surface. The second method is more robust (can tolerate bigger scaling factors) while the first method is more efficient. Both methods can only consider trimming curves not intersecting the boundary of the surface.

In this paper, we consider a more general problem: constrained scaling of an multi-surface object. The component surfaces of the object are trimmed NURBS surfaces and the features to be held unchanged are trimming curves of the object that could intersect the boundary curves of the object. An atttach-and-deform based approach will be presented. Hence, the new method includes [33] as a special case. The reason for using an atttach-and-deform based approach is that in industry, especialy in automibile industry, it is rare to use large scaling factors in the redesign process (imaging stretching the wheel base of a vehicle by tw-` ~ ~
is actually a more important factor.

The remaining part of the paper is arran
A formal description of the problem is giv
The basic idea of the proposed method is pr
tion 3. Techniques needed in constructing
are described in Sections 4-7. Test results o

0-7695-0562-7/00 $10.00 © 2000 IEEE

method are shown in Section 8 and concluding remarks are given in Section 9.

2. Problem formulation

The problem to be solved in this paper may be described as follows: Given an object O defined by a set of *component surfaces* $S_k(u,v)$, $k = 1,2,...,M$, and a set of *features* $O(C_l)$, $l = 1,2,...,N$, construct a new object $\bar{O}$ whose representation is a scaled version of the given object O, but carries all the original features. The component surfaces are trimmed NURBS surfaces and the features are trimming curves of O.

More specifically, each component surface S_k of O is a trimmed NURBS surface of degree p_k in u direction and degree q_k in v direction with a domain D_k defined as follows:

$$S_k(u,v) = \frac{\sum_{i=0}^{m_k}\sum_{j=0}^{n_k} w_{i,j}^k Q_{i,j}^k N_{i,p_k}(u) N_{j,q_k}(v)}{\sum_{i=0}^{m_k}\sum_{j=0}^{n_k} w_{i,j}^k N_{i,p_k}(u) N_{j,q_k}(v)}, \quad (1)$$

$$(u,v) \in D_k \subseteq [0,1] \times [0,1]$$

where $Q_{i,j}^k$ are 3D control points, $N_{i,p_k}(u)$ and $N_{j,q_k}(v)$ are B-spline basis functions of degree p_k and q_k, respectively, and $w_{i,j}^k$ are weight functions. $N_{i,p_k}(u)$ and $N_{j,q_k}(v)$ are defined with respect to the knot vectors $\tau^k = \{\tau_0^k, \tau_1^k, \ldots, \tau_{m_k+p_k+1}^k\}$, and $\sigma^k = \{\sigma_0^k, \sigma_1^k, \ldots, \sigma_{n_k+q_k+1}^k\}$, respectively, with $\tau_0^k = \ldots = \tau_{p_k}^k = \sigma_0^k = \ldots = \sigma_{q_k}^k = 0$ and $\tau_{m_k+1}^k = \ldots = \tau_{m_k+p_k+1}^k = \sigma_{n_k+1}^k = \ldots = \sigma_{n_k+q_k+1}^k = 1$. The domain D_k is a region of $[0,1] \times [0,1]$ bounded by a sequence of connected (curved) edges E_i^k, $i = 1,...,d_k$. Each edge is a NURBS curve.

The features to be held unchanged are defined as follows. For each l, $1 \le l \le N$, C_l is a set of 2D curves

$$C_l = \{C_{1,i_1}^l, C_{2,i_2}^l, \ldots, C_{h_l,i_{h_l}}^l\}$$

with C_{j,i_j}^l being a NURBS curve in the domain of the component surface S_{i_j}. These 2D curves C_{j,i_j}^l, when mapped onto the object O by their corresponding component surfaces S_{i_j}, form a closed sequence of connected 3D curves

$$O(C_l): S_{i_1}(C_{1,i_1}^l) \to S_{i_2}(C_{2,i_2}^l) \to \ldots \to S_{i_{h_l}}(C_{h_l,i_{h_l}}^l) \to S_{i_1}(C_{1,i_1}^l)$$

$O(C_l)$ is a trimming curve of O and is called a feature of O. The features of O do not self-intersect and they do not intersect each other. A feature could intersect several component surfaces. It could also have segments coinciding with the boundary curves of some component surfaces. A feature is called an *interior feature (trimming curve)* if it does not intersect the boundary curves of the object O Otherwise, it will be called a *corner feature (trimming curve)* or a *boundary feature (trimming curve)* depending on if the feature contains a corner point of the object. Note that a feature could intersect the boundary curves of some component surfaces and still be an *interior feature*. An example with all three types of features is shown in Figure 1.

If the scaling factors in x, y and z directions are s_x, s_y and s_z, respectively, then the new object $\bar{O}$ is supposed to be equal to $\hat{O} = T_s(O)$ (or, as close to $\hat{O} = T_s(O)$ as possible)

$$T_s(O) = \{T_s(S_k) \mid 1 \le k \le M\}$$

where T_s is a scaling matrix defined as follows

$$T_s = \begin{bmatrix} s_x & 0 & 0 \\ 0 & s_y & 0 \\ 0 & 0 & s_z \end{bmatrix} \quad (2)$$

and $T_s(S_k)$ represents the scaled version of S_k. The requirement that the new object carries all the original features $O(C_l)$ means that $O(C_l)$ are also trimming curves of the new object subject to some translation and rotation.

In industrial applications, a trimming curve of a free-form surface is usually represented as a linear polygon in the domain of the surface with vertices of the polygon being points of the curve. We follow the same approach in this work. The vertices of this approximating polygon for a feature are called *feature vertices*.

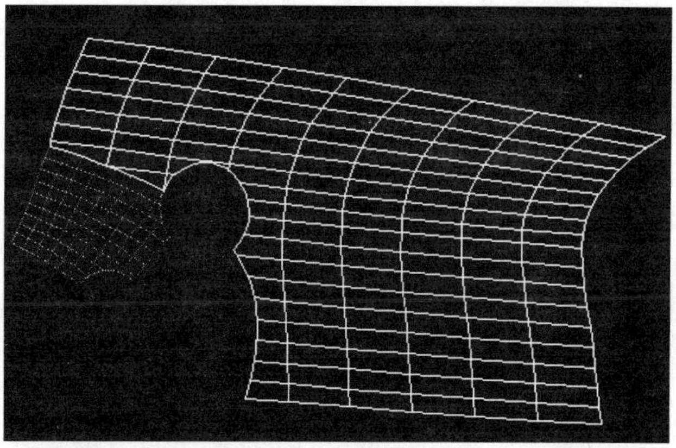

Figure 1. A multi-surface object with three features.

3. Basic idea

In general, due to change of the curvature distribution after a scaling process, it is not possible for the required new object $\bar{O}$ to have exactly the same shape and dimension as the scaled object $\hat{O}$ while carrying all the original features. An approximation method has to be used to construct $\bar{O}$. In this work, the new object will be constructed using an *attach-and-deform* [33] based approach. The main idea of this approach is to attach the original features to the scaled object at appropriate locations. The attaching process requires several geometric operations and a deformation of the scaled object. The deformation is performed with constraints to ensure the following three conditions are satisfied: (1) the features are attached to the object completely, (2) the boundary continuity conditions between the scaled object and adjacent objects are held unchanged, and (3) the internal continuity conditions between adjacent component surfaces of the object are maintained. The deformed component surfaces of the object are again NURBS surfaces. The deformation process will also ensure the resulting object reflects the shape and curvature distribution of the scaled object. An algorithm is presented below. The main steps of the algorithm are the same as the ones given in [33] but more general cases are considered in each step.

1. Subdividing component surfaces which intersect the trimming curves
2. Scaling object O with T_s
3. Relocating features $O(C_l)$
4. Setting up objective function
5. Setting up constraints
6. Performing constrained surface deformation

The third step is the most critical step because it also determines the outcome of the deformation process. We shall use $\hat{O}$ to represent the scaled object $T_s(O)$, $\bar{O}$ to represent the deformed $\hat{O}$, and similar notations for the component surfaces. Details of the above steps, except Step 2, are given in the subsequent sections.

4. Subdividing component surfaces

This step is to perform subdivision on component surfaces that are overlaid by some of the features. The purpose of the subdivision process is to facilitate the deformation process to be performed later on.

Each component surface that is overlaid by some of the features is recursively subdivided until the following two conditions are satisfied:

1. Each subpatch of the component surface intersects at most one feature.
2. The number of feature vertices contained in each subpatch of the component surface is at most $(p_k+1)(q_k+1)$ where p_k and q_k are the degrees of the component surface in u and v directions, respectively.

The first condition is to provide enough flexibility for setting up the *feature constraint* and the second condition is to avoid over-determined systems in the deformation process.

The first condition is satisfied if the dimension of each subpatch is smaller than or equal to one half of the minimum of Ω: Ω is the smallest distance between vertices of different features $O(C_l)$. The second condition has to be tested after each level of recursive subdivision once the first condition is satisfied.

It is possible to perform subdivision on boundary spans and spans that intersect the features only. This would reduce the subdivision time to certain extent. However, the curvature distribution in this case is not as good as the results of uniform subdivision on all the spans (This is reasonable because a movement of a control point in a large patch causes shape change in a larger area).

Without loss of generality, we shall use the same notations for the control points and parameter knots for the component surfaces even though both of them might have been changed after the subdivision process.

5. Relocating features

This step is to move each original feature (trimming curve) $O(C_l)$ $(l = 1, ..., N)$ to an appropriate location that is not only as close to the scaled object $\hat{O} = T_s \circ O$ as possible but also with an appropriate orientation. The first requirement is to ensure that only a small deformation is required to include the original feature as a feature of the scaled object. The second requirement is to ensure that deformation would not cause much distortion of the curvature of the scaled object.

Three types of features are to be considered. Namely, *Interior features*, *Boundary features* and *Corner features* (see Section 2 for their definitions). Different feature types require different relocation techniques. The relocation scheme presented in [33] considers interior features for a single surface only.

5.1. Interior features

Three steps, two translations and one rotation, are needed to relocate an interior feature $O(C_l)$ to its new location. Let $\{P^l_{i,k_i} \mid i = 1, 2, ..., n_l\}$ be the vertices of the feature $O(C_l)$ where P^l_{i,k_i} are points on the component surfaces S_{k_i}. First, the feature $O(C_l)$ is translated from P^l, its centroid, to $T_s P^l$, the centroid of the scaled feature $T_s \circ O(C_l)$. The mean normal vector N_l of the feature $O(C_l)$,

$$N_l = \frac{1}{n_l} \sum_{i=1}^{n_l} N^l_{i,k_i}, \tag{3}$$

where N^l_{i,k_i} are normal vectors of the feature $O(C_l)$ at its vertices P^l_{i,k_i}, is then rotated about the vector $U_l = N_l \times \bar{T}_s N_l$ until it is in the same direction as $\bar{T}_s N_l$, the mean normal vector of the scaled feature, where

$$\bar{T}_s = \begin{bmatrix} s_y s_z & 0 & 0 \\ 0 & s_x s_z & 0 \\ 0 & 0 & s_x s_y \end{bmatrix} \quad (4)$$

and s_x, s_y and s_z are given scaling factors. These two steps align an original feature with the corresponding feature on the scaled object in both centroid and direction. We will use the same notations for the translated and rotated original feature, including their vertices.

The third step is to move the translated and rotated feature $O(C_l)$ along the vector $\bar{T}_s N_l$ to be as close to the scaled object as possible. This step is performed using the least squares method.

5.2. Boundary feature

Four steps, one translation and three rotations, are needed to relocate a boundary feature $O(C_l)$ to its new location. Let A and B be the intersection points of the feature $O(C_l)$ with the boundary of the given object. and P^l be the mid-point of the boundary edge from A to B. Step one is to translate the feature $O(C_l)$ from P^l to $T_s P^l$, where T_s is defined in eq. (4). For the sake of notation, we will use A, B and P^l to refer to the new locations of these points after translation too. Note that, after the translation, A and B may not lie on the boundary of the scaled object.

Let C be a point on the boundary of the scaled object with the following properties: (1) the distance between C and P^l is the same as the distance between A and P^l, (2) there is no such point on the boundary edge between C and P^l, and (3) C is closer to A than the other points satisfying the above two conditions (see Figure 2 for the location of such a C). The second step is to rotate the line segment P^lA about the vector $P^lC \times P^lA$ until P^lA is in the same direction as the line segment P^lC. After the rotation, A would coincide with C and, hence, is a point on the boundary of the scaled object. Again, we will use A and B to represent their new locations after the rotation. The next step is to put B on the boundary of the scaled object as well.

Let D be a point on the boundary of the scaled object with the following properties: (1) the distance between D and A is the same as the distance between B and A, (2) there is no such point on the boundary edge between D and A, and (3) D is closer to B than the other points satisfying the above two conditions (see Figure 2 for the location of such a D). The third step is to rotate the line segment AB about the vector $AB \times AD$ until AB is in the same direction as the line segment AD. This step forces B to coincide with D and, hence, puts B on the boundary of the scaled object as well.

Let N_l be the normal of the object O at P^l and $\bar{T}_s N_l$ the normal of the scaled object $T_s(O)$ at $T_s(P^l)$. N_l is defined as the orientation of $O(C_l)$. During the above rotation processes, the orientation of $O(C_l)$ are changed too. However, for the sake of notation, we use N_l to denote the new orientation of the feature after the rotations as well. The last step is to align the new orientation of the feature, N_l, with $\bar{T}_s N_l$. This is done by rotating N_l about the vector $U_l = N_l \times \bar{T}_s N_l$ until it is the same as $\bar{T}_s N_l$ with pivot point P^l.

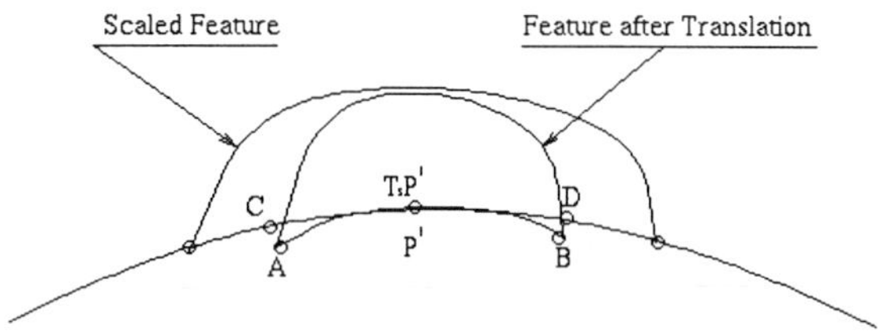

Figure 2. Translation of A Boundary Feature.

5.3. Corner feature

The relocation process for a corner feature is similar to that of a boundary feature except the first step. To ensure that the corner point of the feature remains a corner point of the scaled object, the first step is performed by translating the corner point of the feature from its location with the original object to the new location with the scaled object.

The second, the third and the fourth steps are then performed to ensure that the intersection points A and B of the feature with the boundary of the given object lie on the boundary of the scaled object, and the orientation of the feature is the same as the normal vector of the scaled object at the corner point.

We will use R_l to represent the relocating transformation costructed for the feature $O(C_l)$.

6. Setting up objective function

The relocation process does not guarantee that the reloated features would lie on the scaled object $\hat{O}$ completely. A deformation of the scaled object is necessary to ensure complete attachment without changing the shape of the features. The new object $\bar{O}$ produced by deforming the scaled object, consequently, contains all the original features (subject to some transformations).

To obtain a new object that is essentially the same as the scaled object, the deformation process should make the difference between the new object $\bar{O}$ and the scaled object $\hat{O}$ as litte as possible, while ensure the overall shape and curvature distribution of the new object are as much the same as the scaled object as possible. The deformation process should also hold all external and internal smoothness conditions unchanged.

The deformation process, consequently, should be carried out as a global optimization process with constraints so that the above requirements are satisfied. The objective function for the global optimiztion process will be developed based on a physical model where the deformation energy between the new object and the scaled object is minimized. The displacement function is the difference of the new object and the scaled object:

$$V(u,v) = (\bar{O} - \hat{O}). \tag{5}$$

Following the theory of mechanics [22][30] and our previous work [33], the deformation energy of a thin plate can be defined as the sum of spring potential energy, bending energy and stretching energy. Hence, the deformation energy of the displament function in our case is

$$\begin{aligned} E(V) &= \frac{1}{2}\int\int_D F(u,v)\, dudv \\ &= \frac{1}{2}\sum_{k=1}^{M}\int\int_{\bar{D}_k} F(u,v)\, dudv \end{aligned} \tag{6}$$

where D is the domain of the object, $\bar{D}_k$ is the domain of the component surface S_k and

$$F(u,v) = \alpha F_1(u,v) + \beta F_2(u,v) + \gamma F_3(u,v) \tag{7}$$

with $F_1(u,v)$, $F_2(u,v)$ and $F_3(u,v)$ representing the bending, stretching and spring components of the deformation process, respectively. These quantity are defined as follows:

$$\begin{aligned} F_1(u,v) &= (\frac{\partial^2 V}{\partial u^2})^2 + (\frac{\partial^2 V}{\partial v^2})^2 + (\frac{\partial^2 V}{\partial u \partial v})^2 + (\frac{\partial^2 V}{\partial u^2})(\frac{\partial^2 V}{\partial v^2}) \\ F_2(u,v) &= (\frac{\partial V}{\partial u})^2 + (\frac{\partial V}{\partial v})^2 + (\frac{\partial V}{\partial u})(\frac{\partial V}{\partial v}) \\ F_3(u,v) &= V^2 \end{aligned} \tag{8}$$

They have impact on the amount of surface displacement, variation of surface area, and distribution of surface curvature, respectively. α, β and γ are the respective weights of these effects on the deformation energy. A study on the determination of these weights is presented in [34].

Note that the domain considered for component surface S_k in eq. (6) is not D_k but $\bar{D}_k$. $\bar{D}_k$ is D_k subtracting trimmed regions defined by features that overlay this component surfaces.

For NURBS surfaces, eqs. (6) leads to a quadratic equation with respect to the control points if homogeneous representation is used. The homogeneous representations of scaled component surface $\hat{S}_k = T_s \circ S_k(u,v)$ and the deformed component surface $\bar{S}_k(u,v)$ are

$$\sum_{i=0}^{m_k}\sum_{j=0}^{n_k}(w_{i,j}^k \hat{Q}_{ij}^k,\ w_{i,j}^k) N_{i,p_k}(u) N_{j,q_k}(v)$$

and

$$\sum_{i=0}^{m_k}\sum_{j=0}^{n_k}(w_{i,j}^k \bar{Q}_{ij}^k,\ w_{i,j}^k) N_{i,p_k}(u) N_{j,q_k}(v),$$

respectively. For simplicity of notations, we shall use $\hat{S}_k(u,v)$ and $\bar{S}_k(u,v)$ to represent their own homogeneous forms, i.e., $\hat{Q}_{i,j}^k$ and $\bar{Q}_{i,j}^k$ are homogeneous control points of the following forms:

$$\hat{Q}_{i,j}^k = (w_{i,j}^k \hat{Q}_{ij}^k,\ w_{i,j}^k),\ \bar{Q}_{i,j}^k = (w_{i,j}^k \bar{Q}_{ij}^k,\ w_{i,j}^k).$$

$\hat{S}_k(u,v)$ and $\bar{S}_k(u,v)$ can be written as linear equations with respect to their control points as follows:

$$\hat{S}_k(u,v) = \sum_{l=0}^{\Delta_k} \hat{Q}_l^k N_l^k(u,v); \tag{9}$$

$$\bar{S}_k(u,v) = \sum_{l=0}^{\Delta_k} \bar{Q}_l^k N_l^k(u,v), \tag{10}$$

where

$$\Delta_k \equiv (m_k+1)\times(n_k+1) - 1, \tag{11}$$
$$(k = 1, ..., M)$$
$$\hat{Q}_l^k = \hat{Q}_{i,j}^k;\ \bar{Q}_l^k = \bar{Q}_{i,j}^k, \tag{12}$$
$$N_l^k(u,v) = N_{i,p_k}(u) N_{j,q_k}(v), \tag{13}$$

with

$$i = l - \lfloor l/(m_k+1) \rfloor \times (m_k+1),$$
$$j = \lfloor l/(m_k+1) \rfloor.$$

By substituting eqs. (9) and (10) into eq. (5) and eqs. (6), we get the following expression by simple algebra.

$$E(\bar{Q} - \hat{Q}) = \frac{1}{2}[\bar{Q} - \hat{Q}]^{\top} A [\bar{Q} - \hat{Q}] \tag{14}$$

where

$$\hat{Q} = [\hat{Q}^0, \hat{Q}^1, ..., \hat{Q}^M]$$
$$with\ \ \hat{Q}^k = [\hat{Q}_0^k, \hat{Q}_1^k, ..., \hat{Q}_{\Delta_k}^k] \tag{15}$$
$$\bar{Q} = [\bar{Q}^0, \bar{Q}^1, ..., \bar{Q}^M]$$
$$with\ \ \bar{Q}^k = [\bar{Q}_0^k, \bar{Q}_1^k, ..., \bar{Q}_{\Delta_k}^k] \tag{16}$$

and A is a $(\sum_{k=1}^{M}\Delta_k + 1) \times (\sum_{k=1}^{M}\Delta_k + 1)$ matrix.

The quadratic equation $E(\bar{Q} - \hat{Q})$ has a minimum at the point where $A(\bar{Q} - \hat{Q}) = 0$. Hence, if the optimization process is performed with no constraints, the resulting object $\bar{Q}$ will be exactly the same as the scaled object $\hat{Q}$.

7. Establishing constraints

Three types of constraints, *external boundary constraint*, *internal boundary constraint* and *feature constraint*, are used to control the deformation.

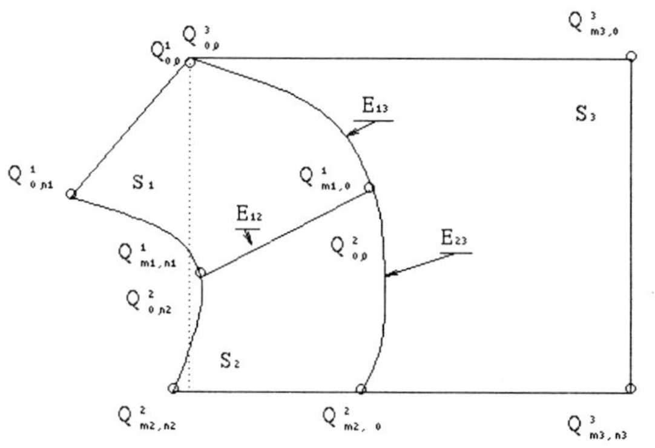

Figure 3. A multi-Surface Object.

7.1. External boundary constraints

The *external boundary constraint* is to prevent the boundary curves of the scaled object $\hat{O}$ from being changed during the deformation process. It also prevents the boundary smoothness condition between the given object and adjacent objects from being changed during the deformation process. Otherwise, the deformed object might not fit adjacent objects well. These requirements can be achieved by holding a banded area of the scaled object along its boundary unchanged during the deformation process. The width of the banded area depends on the smoothness condition.

The problem can be dealt with in two different cases. First, if an external boundary edge of the object is part of the parameter space boundary of some component surface, then a banded area along this external boundary edge can be fixed simply by fixing a band of boundary control points of that component surface. The width of the control point band depends on the boundary smoothness condition. The establishment of the constraint equations in this case is similar to the approach used in [33].

For the second case, we need to recall the definitions of two terms from [34]. The *envelop* of a trimming curve is the smallest union of domain patches in the parameter space of the component surface that contains the trimming curve. The set of control points needed to define the image of a trimming curve's envelop is called a *control envelop*.

If the external boundary edge is not part of the parameter space boundary of some component surface, then it must be a trimming curve of a component surface. In this case, one simply use the control envelop of this trimming curve in the corresponding component surface as the constraint condition for this external boundary edge. Note that, in this case, an extra subdivision process might be needed in the subdivision step to ensure non-overlapping of the control envelops of different external boundary edges and features. The subdivision process should be performed based on the degrees of the component surfaces. The establishment of the constraint equations in this case is similar to the approach used in [34].

7.2. Internal boundary constraints

The *internal boundary constraint* is only to prevent the internal boundary smoothness condition between adjacent component surfaces from being changed during the deformation process. Internal boundary curves between adjacent component surface can be changed.

Consider the example shown in Figure 3. There exists three internal boundary curves. Namely, E_{12} between component surfaces S_1 and S_2, E_{13} between component surfaces S_1 and S_3, and E_{23} between S_2 and S_3. These curves can be changed during the deformation process. But the smoothness condition between the corresponding component surfaces must be maintained.

If C^0 continuity is required, the resulting component surfaces $\bar{S}_k$ $(k = 1, 2, 3)$ of the object $\bar{O}$ should obey the following constrains.

$$\bar{S}_1(1, v_i^1) = \bar{S}_2(0, v_i^2), \quad i = 0, \ldots, n_{b1}; \tag{17}$$

$$\bar{S}_1(u_i^1, 0) = \bar{S}_3(u_i^3, v_i^3), \quad i = 0, \ldots, n_{b2}; \tag{18}$$

$$\bar{S}_2(u_j^2, 0) = \bar{S}_3(u_j^3, v_j^3), \quad j = 0, \ldots, n_{b3}; \tag{19}$$

where

$$n_{b1} >= max(n_1, n_2) \tag{20}$$

$$n_{b2} + n_{b3} >= max(m_3, (m_1 + m_2)) \tag{21}$$

$$n_{b2} >= m_1, n_{b3} >= m_2 \tag{22}$$

and $(1, v_i^1)$ and $(0, v_i^2)$ are parameters on the boundary edge E_{12}, corresponding to component surfaces S_1 and S_2, respectively, $(u_i^1, 0)$ and (u_i^3, v_i^3) are parameters on the edge E_{13}, corresponding to component surfaces S_1 and S_3, respectively, and $(u_j^2, 0)$ and (u_j^3, v_j^3) are parameters on the edge E_{23}, corresponding to component surfaces S_2 and S_3, respectively.

In Figure 3, if component surface $S_1(u, v)$ and component surface $S_2(u, v)$ in v direction have the same number of control points and knot distribution, we can simply set $n_{b1} = n_1 (= n_2)$ and $v_i^1 = v_i^2 = \sigma_{i+p_1}^1$ $(i = 0, \ldots, n_{b1})$.

If $S_1(u, v)$ and $S_2(u, v)$ do not have the same control points and knots distribution in v direction, one should determine v_i^1 first by setting $v_i^1 = 1.0 * i/(n_{b1}+1)$ $(i = 0, \ldots, n_{b1})$, and then evaluate v_i^2 using the fact that $S_2(v_i^2, 0) = S_1(v_i^1, 1)$.

If G^1 continuity on the internal boundary edges is required, the following equations are the constraints:

$$\bar{S}_u^1(1, v_i^1) = \bar{S}_v^2(0, v_i^2) \quad (23)$$

$$i = 0, \ldots, n_{b1}$$

$$\bar{S}_v^1(u_i^1, 0) = \bar{S}_u^3(u_i^3, v_i^3)cos\alpha_i + \bar{S}_u^3(u_i^3, v_i^3)sin\alpha_i \quad (24)$$

$$i = 0, \ldots, n_{b2}$$

$$\bar{S}_v^2(u_j^1, 0) = \bar{S}_u^3(u_j^3, v_j^3)cos\alpha_j + \bar{S}_u^3(u_j^3, v_j^3)sin\alpha_j \quad (25)$$

$$j = 0, \ldots, n_{b3}$$

where $\bar{S}_u^k$ $(k = 1, 2, 3)$ are the first derivatives of component surfaces $\bar{S}_k$ in u direction, $\bar{S}_v^k$ $(k = 1, 2, 3)$ are the first derivatives of component surfaces $\bar{S}_k$ in v direction, and α_i and α_j are the angles between the normal direction of the boundary curve (feature) E_{13} and E_{23} and direction u in the domain at the points (u_i^3, v_i^3) and (u_j^3, v_j^3), respectively.

7.3. Feature constraint

The *feature constraint* is to ensure that, after deformation, the relocated features of the given object O become features of the deformed object $\bar{O}$. This means that, after deformation, vertices of the relocated features should lie on the deformed object.

More specifically, if feature $O(C_l)$ overlays the component surface S_k and $P_{i,k}^l$, $i = 0, 1, ..., n_{l,k}$, are vertices of $O(C_l)$ on $\bar{S}_k$ then, after deformation, there must exist $(u_{i,k}^l, v_{i,k}^l)$, $i = 0, 1, ..., n_{l,k}$, in the domain D_k of $\bar{S}_k$ so that the values of $\bar{S}_k$ at these points are equal to the corresponding vertices of the relocated feature, i.e.,

$$R_l(P_{i,k}^l) = \bar{S}_k(u_{i,k}^l, v_{i,k}^l)$$

$$= \sum_{j=0}^{\Delta_k} \bar{Q}_j^k N_j^k(u_{i,k}^l, v_{i,k}^l) \quad (26)$$

where Δ_k, $\bar{Q}_j^k$ and N_j^k are defined in (11), (12), and (13). $\bar{Q}_j^k$ and N_j^k are control points and B-spline basis functions of the deformed component surface $\bar{S}_k$, respectively.

All the three type of constraints can be combined into a linear system which can be represented in a matrix form as follows:

$$\bar{B}\bar{Q} = \bar{\mathbf{b}} \quad (27)$$

where $\bar{Q}$ is defined in (16).

8. Implementation & test results

The optimization process of the quadratic system (14) with linear constraints (27) can be solved using the Lagrange Multiplier method. The presented approach has been implemented in Java on an SGI platform.

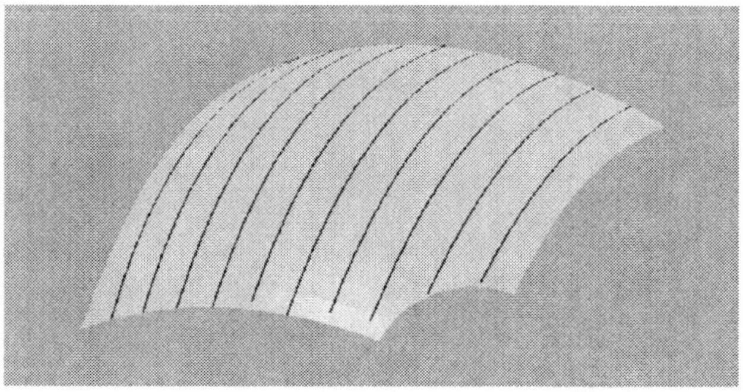

Figure 4. A single-component-surface-object with a corner feature before constraint scaling

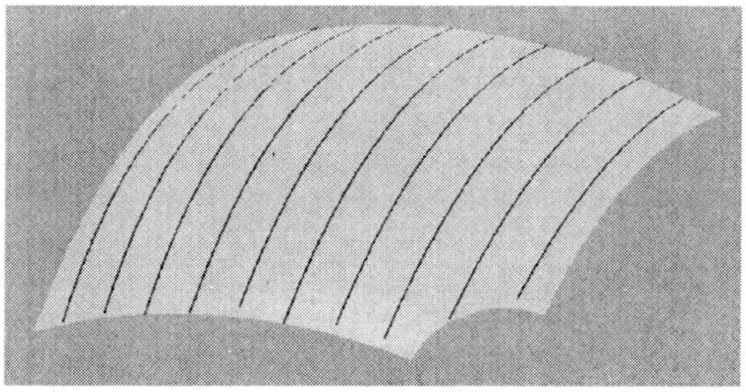

Figure 5. A single-component-surface-object with a corner feature after constraint scaling

Three test cases are presented here. These include a single-component-surface object with a corner feature (Figure 4), a two-component-surface object with a boundary feature (Figure 6), and an 18-component-surface object with all three kinds of features (Figure 8). All the three cases have same scaling factors, they are $s_x = 1.2$, $s_y = 1.15$ and $s_z = 1.1$. Figure 4,Figure 6 and Figure 8 are the object before the constained scaling process, while Figure 5, Figure 7 and Figure 9 are the resulting object of the constrained scaling process. It is easy to see that the features remain the same in both shape and dimension after the constrained scaling process in all three cases. The result of the last case is especially satisfactory considering the shape complexity of the object around the feature.

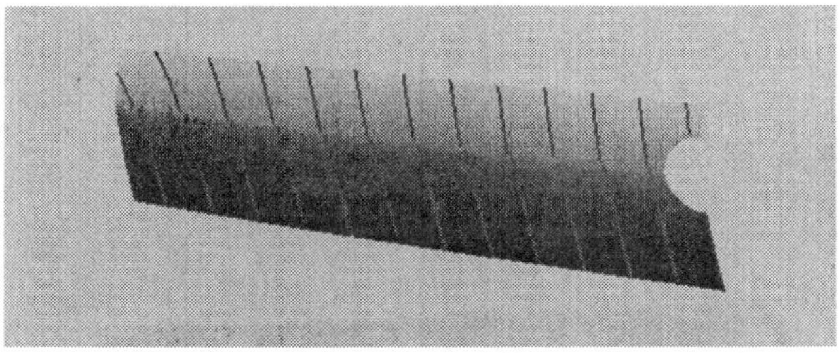

Figure 6. A two-component-surface-object with a boundary feature before constraint scaling

Figure 7. A two-component-surface-object with a boundary feature after constraint scaling

9. Conclusion

An attach-and-deform based constrained scaling method for multi-surface objects is presented. The presented method includes a previous version [33] for a single trimmed NURBS surface as a special case. The new method can be used for most free-form objects as long as the components surfaces of the object are trimmed NURBS surfaces and the features are trimming curves of the object defined in the domains of the component surfaces in NURBS form.

The new method can not tolerate scaling factors bigger than 1.5. The safe scaling range provided by this method is 1 to 1.3. This is enough for most of the reshaping tasks in the automotive industry where scaling factors used in the reshaping process are usually smaller than 1.3. However, it would be of interest to compare the performance of the presented approach with an fix-and-stretch based approach for multi-surface objects. The development of such a system will be one of the future works to be performed in the near future.

Acknowledgements. The financial support of Ford Motor Company through a URP grant and Honda Motor Company through two HIG grants, and the consulting support of Drs. Paul Stewart and Yifan Chen of the Ford Research Lab and Dr. Shane Chang of the Honda R&D North America are deeply appreciated. We also thank Shirish Pande for programming support on the generation of shaped images for this paper.

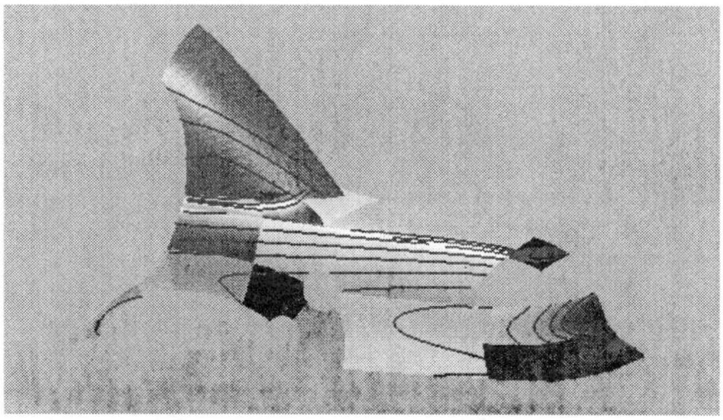

Figure 8. An automobile body piece with three features before constraint scaling

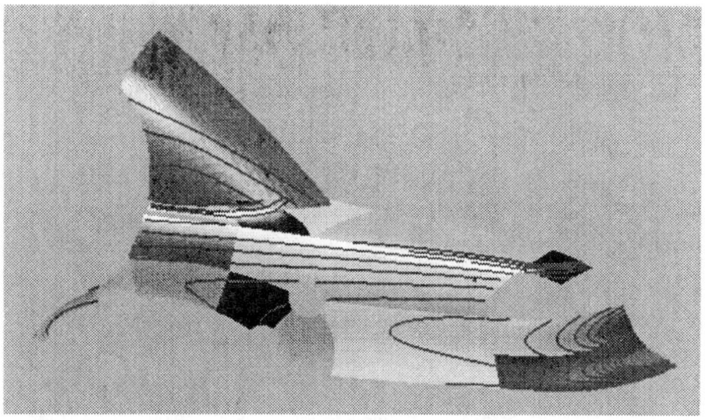

Figure 9. An automobile body piece with three features after constraint scaling

References

[1] Adhikary, N., and Gurumoorthy, B., Smooth surface interpolation with multiple patches, Proceedings of DETC'99, 1999 ASME Design Engineering Technical Conferences, Sep. 12-15, 1999, Las Vegas, Nevada, USA.

[2] Armit, A. P., Curve and Surface design using mul-

tipatch and multiobject design systems, *Computer-Aided Design* 25, 1993, 251-261.

[3] Bechmann, D., Space Deformation Models Survey, *Computer & Graphics*, 18, 4, 571-586.

[4] Beier, K-P, and Chen, Y., The highlight-line algorithm for real-time surface-quality assessment, *Computer-Aided Design* 26, 1994, 268-278.

[5] Beier, K-P, and Chen, Y., The highlight band, a simplified reflection model for interactive smoothness evaluation, in: Spapidis, N., ed., *Designing Fair Curves and Surfaces*, SIAM, 1994.

[6] Borrel, P. and Bechmann, D., Deformation of N-dimentional Objects, *International Journal of Computational Geometry & Applications*, 1(4), 1991, 427-453.

[7] Borrel, P. and Rappoport, A., Simple Constrained Deformations for Geometric Modeling and Interactive Design, *ACM Tansactions on Graphics*, 13(2), 1994, 137-155.

[8] Cavendish, J. C. and Marin, S. P., A procedural Feature-Based Approach for Designing Functional Surfaces, In *Topics in Surface Modeling*, Hans Hagen, editor, SIAM 1992, 145-168.

[9] Celniker, G. and Gossard, D., Deformable curve and surface finite elements for free form shape design, *Computer Graphics*, 25(4), 1991, 257-266.

[10] Celniker, G. and Welch, W., Linear constraints for deformable B-spline surfaces, In *Proceedings of the Symposium on Interactive 3D Graphics*, ACM. New York, 1992, 165-170.

[11] Chen, Y., Highlight lines for surface quality control and manipulation, Ph.D. Dissertation, Department of Naval Architecture and Marine Engineering, The University of Michigan, 1993.

[12] Chen, Y., Beier, K-P, and Papageorgiou, D., Direct highlight line modification on nurbs surface, *Computer Aided Geometric Design* 14, 1997, 583-601.

[13] Coquillart, S., Extended free form deformation: A Sculpting tool for 3D geometric modeling, *Computer Graphics*, 24(4), 1990, 187-193.

[14] Courant, R. and Hilbert, D., Methods of Mathematical Physics, Vol. I, Interscience, London, 1953.

[15] Feng, J., Ma, L. and Peng, Q., A new free-form deformation through the control of frametic surfaces, *Computer & Graphics*, 20(4), 1996, 531-539.

[16] Griessmair, J. and Parray, R., Deformation of Solids with Trivariate B-spline, *Proceedings of Eurographics'89*, 1989, 137-148.

[17] Hsu, W. H., Hughes, J. F. and Kaufman, H., Direct manipulation of free-form deformations, *Computer Graphics*, 26(2), 1992, 177-184.

[18] Kalra, P., et al., Simulation of facial muscle actions based on rational free-form deformations, *Computer Graphics Forum*, 2(3), 1992, C59-C69.

[19] Kumar, V., Burns, D., Dutta, D., and Hoffmann, C., A framework for object modeling, *Computer-Aided Design*, 31, 1999, 541-556.

[20] Lamousin, H. J. and Waggenspack, W. N., NURBS-based free-form deformations, *IEEE Transactions on Computer Graphics & Applications*, 14(6), 1994, 59-65.

[21] Lazarus, F., Coquillart S., and Jancene P., Axial deformations: an intuitive deformation technique, *Computer-Aided Design* 26, 1994, 607-613.

[22] Mansfield, E. H., The Bending and Stretching of Plates, Machillan, New York, 1964.

[23] Platt, J. and Barr, A., Constraints method for flexible models, *Computer Graphics*, 22(4), 1988, 279-288.

[24] Sedbergerg, T. W. and Parry, S. R., Free-form deformation of solid geometric models, *Computer Graphics*, 20(4), 151-160.

[25] Sarraga, R. F., Recent Methods for Surface Shape Optimization, *Computer Aided Geometric Design*, 15(1998), 417-436

[26] Terzopoulos, D., Multilevel Computational Processes for Visual Surface Reconstruction, *Computer Vision, Graphics, and Image processing*, Vol. 24, 1983, 52-96.

[27] Terzopoulos, D., Platt, J., Barr, A. and Fleischer, K., Elastically deformable models, *Computer Graphics*, 21(4), 1987, 205-214.

[28] Terzopoulos, D. and Fleischer, K., Deformable models, *Visual Computing*, 4(6), 1988, 306-331.

[29] Terzopoulos, D. and Qin, H., Dynamic NURBS with Geometric Constraints for interactive Sculpting, *ACM Transactions on Graphics*, 13(2), 1994, 103-136.

[30] Ugural, A. C., Ferster, S. K., Advanced Strength and Applied Elasticity, America Elsevier Publishing Co., New York, 1975.

[31] Wang, X., Cheng, F. and Barsky, B. A., Energy and B-spline interproximation, *Computer-Aided Design*, 29(7), 1997, 485-496.

[32] Welch, W. and Witkin, A., Variational surface modeling, *Computer Graphics*, 26(2), 1992, 157-166.

[33] Zhang, P., Zhang, C. and Cheng, F., Constrainted Shape Scaling of Trimmed NURBS Surfaces, Proc. 1999 ASME Design Theory and Methodology Conference, Las Vegas, Nevada, 1999.

[34] Zhang, C.,Zhang, P. and Cheng, F., Constrainted Shape Scaling of Trimmed NURBS Surfaces II, submitted.

[35] Zheng, J. M., Chan, K. W., and Gibson, I., Surface Feature Constraint Deformation for Free-form Surface and Interactive Design, Fifth Symposium on Solid Modeling, Ann Arbore, MI, 1999.

B-spline Free-Form Deformation of Polygonal Objects through Fast Functional Composition

Jieqing Feng Qunsheng Peng
State Key Lab. of CAD&CG, Zhejiang University
Hangzhou, 310027, P.R.China
Email: {jqfeng; peng}@cad.zju.edu.cn

Abstract

Free-form deformation, abbreviated to FFD, plays an important role in both computer animation and geometric modeling. When polygonal objects are deformed by the traditional FFD scheme, however aliasing will appear since the deformation only acts on the sample points of the objects. To eliminate this problem, a new method of accurate B-spline FFD was proposed by authors [13], which is based on functional composition. Unfortunately it is a time consuming process because generalized de Casteljau algorithm is adopted to solve the functional composition. In this paper, we propose a fast accurate B-spline FFD method. The proposed method makes use of the polynomial interpolation and symbolic computation to solve the functional composition. Both theoretical analysis and implementation results show that the proposed method runs faster than the original one [13].

Key words *Free-form deformation, B-spline, polynomial interpolation, symbolic computation*

1. Introduction

In geometric modeling and computer animation, modification of geometric shape is an important research topic. Space deformation methods can fill this role well. From a mathematical point of view, they can be regarded as performing a transformation $\mathbf{F} : \mathbf{R}^3 \rightarrow \mathbf{R}^3$ [2], which are more efficient and intuitive for modifying geometric shape than conventional methods such as pulling points, conducting 3D transformations,*etc*.

Barr [1] first proposed the global and local deformation concept, by which transformation $\mathbf{F}$ is represented as a matrix function of space position. Although the method is independent of the geometric representation of object, it is not intuitive enough. Free-form Deformation (FFD), proposed by Sederberg and Parry [21], supplies a general deformation framework. The object to be deformed is first embedded into an intermediate space. When the user deforms the embedding space, the deformation is passed automatically to the object inside. In [21] the intermediate space is the parametric space of a trivariate tensor product Bézier volume. Many extensions to the original mapping function were then proposed, such as B-spline FFD [14], Extended FFD [7], Rational FFD [16], NURBS FFD [17], continuous FFD [3], *etc*. Other extensions include animated FFD [8], direct manipulation FFD [15], arbitrary topological lattice FFD [19], Dirichlet FFD [20], which are application of FFD or generalized version of FFD. However they follow the same principle as above.

Unlike the above FFD methods, which are based 3D deformation tools (parametric volume), surface deformation [12], axial deformation [18, 6, 22] and constraint-based space deformation [5] adopt respectively parametric surfaces, parametric curve, points and their displacements as deformation tools whose dimensions are 2D, 1D and 0D. Bechmann gave a detailed survey of space deformation methods with different dimensional tools [4].

It should be pointed out that all of the above space deformation methods are applied to the object space and an *ideal* deformation result can be attained only if the deformation function is applied to *every* point of objects. Unfortunately it is impossible. In fact, only the positions of representative vertices or control points of the object are modified. This limitation is usually overlooked as for an object described by parametric surface, the deformed object will still keep smooth. For a polygonal object, however, if vertices are not sufficiently sampled, the resultant object will deviate greatly from the expected result. In the worst case, the deformed object remains unchanged, or the topology of object is changed. Obviously sampling is an important problem for practical use of FFD methods.

Little research work addressed this problem. In paper [14] a test-error approach was proposed for B-spline FFD.

0-7695-0562-7/00 $10.00 © 2000 IEEE

Specially, it examines the middle point of each triangle mesh edge and evaluates the error between true deformation and linear interpolation. If the error is beyond a given threshold, the triangle is further subdivided at the midpoints. However, such a solution may fail at singular points. An alternative way is to uniformly re-tessellate the polygonal object, then compare the current deformation result with the previous one and determine whether the object need to be tessellated further. The final result relies on user's perception. The authors have also proposed an accurate B-spline FFD, where the deformation of polygonal objects is recorded by a set of triangular Bézier patches [13]. The method is based on the Bernstein polynomials composition and generalized de Casteljau algorithm. Though the accurate FFD solves the sampling problem of polygonal objects deformation, its computational burden is heavy since generalized de Casteljau algorithm is a time-consuming procedure.

In this paper, we propose a more efficient accurate B-spline FFD method based on fast functional composition. The polygonal objects are first subdivided and re-triangulated according to knot vectors of the B-spline volume as in [13]. Then the B-spline volume is converted into piecewise continuous Bézier volumes. The accurate result is finally obtained through fast functional composition between a triangle and corresponding Bézier volume. The speed-up is achieved at the functional composition step. We adopt polynomial interpolation rather than the generalized de Casteljau algorithm to expand Bernstein polynomial composition. In fact, the interpolation is a procedure of solving a system of linear equations. By choosing sampling points in the parametric domain in a specified manner, the matrix of linear system can be diagonally dominant. Further more, the inversion of the matrix can be computed in advance and only once.

The rest of this paper is organized as follows. Section 2 sets notations, and introduces shifting operators. Section 3 gives brief introduction of original accurate FFD in [13]. In Section 4, we describe the fast functional composition algorithm to compute the control points of triangular Bézier patch. We also give theoretical analysis of both the algorithm's complexity and its numerical accuracy. Finally, we show implementation results in Section 5 and draw our conclusions in Section 6.

2. Preliminary

Bernstein polynomials can be concisely expressed using shifting operators. Let $\mathbf{r}(u,v,w)$ be a Bézier volume:

$$\mathbf{r}(u,v,w) = \sum_{i=0}^{n_u}\sum_{j=0}^{n_v}\sum_{k=0}^{n_w} \mathbf{r}_{ijk} B_{i,n_u}(u) B_{j,n_v}(v) B_{k,n_w}(w) \tag{1}$$

where $\mathbf{r}_{ijk} \in \mathbf{R}^3$, $(u,v,w) \in [0,1]^3$, and $B_{i,n_u}(u)$, $B_{j,n_v}(v)$ and $B_{k,n_w}(w)$ are Bernstein polynomials. Define operators E_u, E_v, E_w, I by:

$$E_u\mathbf{r}_{ijk} = \mathbf{r}_{i+1,j,k} \quad E_v\mathbf{r}_{ijk} = \mathbf{r}_{i,j+1,k}$$
$$E_w\mathbf{r}_{ijk} = \mathbf{r}_{i,j,k+1} \quad I\mathbf{r}_{ijk} = \mathbf{r}_{ijk}$$

Based on above definitions, (1) can be expressed via binomial expansion as:

$$\begin{aligned}\mathbf{r} \;=\; & [(1-u)I + uE_u]^{n_u}[(1-v)I + vE_v]^{n_v} \\ & [(1-w)I + wE_w]^{n_w}\mathbf{r}_{000}\end{aligned} \tag{2}$$

Let $\mathbf{c}(u)$ be a Bézier curve. It can be expressed in terms of shifting operators as:

$$\mathbf{c}(u) = [(1-u)I + uE]^n \mathbf{c}_0 \quad u \in [0,1] \tag{3}$$

We know that a portion of the curve $\mathbf{c}(u)$ restricted on interval $[u_0, u_1] \subseteq [0,1]$ is still a Bézier curve. The generalized de Casteljau algorithm can also be described using shifting operators. Let interval $[u_0, u_1]$ be parameterized as:

$$u(t) = (1-t)u_0 + tu_1 \quad t \in [0,1]$$

Substituting this into (3) gives:

$$\begin{aligned}\mathbf{c}(u(t)) &= [(1-u(t))I + u(t)E]^n\,\mathbf{c}_0 \\ &= [((1-u_0)I + u_0E)(1-t) \\ &\quad +((1-u_1)I + u_1E)t]^n\,\mathbf{c}_0 \\ &= \sum_{i=0}^{n} \tilde{\mathbf{c}}_i B_{i,n}(t)\end{aligned}$$

where the control points of sub-curve $\mathbf{c}(u(t))$ are $\tilde{\mathbf{c}}_i = ((1-u_0)I + u_0E)^{n-i}((1-u_1)I + u_1E)^i\mathbf{c}_0$. When the generalized de Casteljau algorithm is adopted to compute $\tilde{\mathbf{c}}_i$, the constants u_0 and u_1 are used $(n-i)$ and i times respectively. This is a kernel step of the algorithm for computing control points of result triangular Bézier patch in [13].

Next, we define the initial B-spline volume for the FFD. Let $[X_{min}, X_{max}]\times[Y_{min}, Y_{max}]\times[Z_{min}, Z_{max}]$ be the bounding box of the object to be deformed. Then the user specifies, or the system automatically generates, knot vectors $\mathbf{u}$ as follows:

$$\mathbf{u} = \{u_0, \cdots, u_{k_u}, u_{k_u+1}, \cdots, u_{n_u}, \cdots, u_{n_u+k_u}\}$$

where k_u is the u-degree of the B-spline volume, n_u is the number of its control points along the u-direction, and $u_0 = \cdots = u_{k_u} = X_{min}$, $u_{n_u} = \cdots = u_{n_u+k_u} = X_{max}$. The knot vectors $\mathbf{v}$ and $\mathbf{w}$ are defined similarly. The B-spline volume $\mathbf{P}(u,v,w)$ of degree k_u, k_v, k_w determined by $\mathbf{u}, \mathbf{v}, \mathbf{w}$ is:

$$\mathbf{P}(u,v,w) = \sum_{i=0}^{n_u}\sum_{j=0}^{n_v}\sum_{k=0}^{n_w} \mathbf{P}_{ijk} N_{i,k_u}(u) N_{j,k_v}(v) N_{k,k_w}(w) \tag{4}$$

An advantage of adopting the bounding box as the parametric domain of B-spline volume is that the point coordinates in object space can be used as local coordinates corresponding to $\mathbf{P}(u, v, w)$, saving computation time. The B-spline volume is then converted to piecewise continuous Bézier volumes by implementing a knot insertion algorithm [11]. In total, $(n_u - k_u) \cdot (n_v - k_v) \cdot (n_w - k_w)$ Bézier volumes are generated, which are defined in $[u_i, u_{i+1}] \times [v_j, v_{j+1}] \times [w_k, w_{k+1}]$ and $k_u \leq i < n_u$, $k_v \leq j < n_v$, $k_w \leq k < n_w$.

3. Accurate FFD using generalized de Casteljau algorithm

3.1. Subdivision and re-triangulation of polygonal objects

The kernel of the algorithm proposed in [13] is the functional composition of a Bézier volume and a triangular mesh. Therefore it is necessary to subdivide the polygonal objects based on the knot vector of B-spline volume so that each triangular mesh produced after subdivision lies within a parametric space of Bézier volume. According to the definition of knot vectors of B-spline volume, the parametric space is identical with the object space. For clarity, we rewrite knot vectors $\mathbf{u}, \mathbf{v}, \mathbf{w}$ as $\mathbf{x}, \mathbf{y}, \mathbf{z}$ and $x_i = u_i$, $y_j = v_j$, $z_k = v_k$ for all possible i, j, k. Assuming planes $x = x_i$ $(i = i_0, \cdots, i_1)$ cross the triangle $\mathbf{P}_0\mathbf{P}_1\mathbf{P}_2$, our purpose is to subdivide and re-triangulate the triangle according to the planes $x = x_i$. We first compute the intersection between each plane and triangle. Then we trace the triangle vertices and intersection points to form loops. Finally, we triangulate loops with more than four edges. These three steps are then repeated for the y and z planes successively. The triangles produced have the same outside normal with those of corresponding original ones. For further details of the subdivision algorithm, readers can refer to the paper [13].

3.2. Functional composition via shifting operators

For a line segment or a triangular mesh which is defined in the parametric space of a Bézier volume, its image after deformation is controlled by the control vertices of the corresponding Bézier volume. To derive the accurate deformation results, we perform functional compositions between their respective Bézier expression and the Bézier volume. The functional composition for Bernstein polynomials has been studied by DeRose *et al.* [9, 10]. Unlike their methods, we explore this problem using shifting operators to make the procedure of functional composition clear and intuitive. We first give the composition result between line segment and Bézier volume.

Let $\mathbf{P}_0\mathbf{P}_1$ be an edge of a triangle. Assume the local coordinates of $\mathbf{P}_0$ and $\mathbf{P}_1$ within the parametric space of Bézier volume be (u_0, v_0, w_0) and (u_1, v_1, w_1) respectively. Then the line segment $\mathbf{P}_0\mathbf{P}_1$ can be linearly parametrized in interval $[0, 1]$ as follows:

$$\mathbf{P}(t) = (u(t), v(t), w(t)) = (1-t)\mathbf{P}_0 + t\mathbf{P}_1 \qquad (5)$$

By substituting (5) into (2), we can get:

$$\begin{aligned} \mathbf{C}(t) &= \mathbf{r}(u(t), v(t), w(t)) \\ &= [(1-u(t))I + u(t)E_u]^{n_u} \\ &\quad [(1-v(t))I + v(t)E_v]^{n_v} \\ &\quad [(1-w(t))I + w(t)E_w]^{n_w}\,\mathbf{r}_{000} \\ &= [(1-t)U_0 + tU_1]^{n_u}\,[(1-t)V_0 + tV_1]^{n_v} \\ &\quad [(1-t)W_0 + tW_1]^{n_w}\,\mathbf{r}_{000} \\ &= \sum_{s=0}^{N} \mathbf{C}_s B_{s,N}(t) \end{aligned}$$

where

$$\left\{ \begin{array}{lcl} U_s &=& [(1-u_s)I + u_s E_u] \\ V_s &=& [(1-v_s)I + v_s E_v] \\ W_s &=& [(1-w_s)I + w_s E_w] \end{array} \right. \qquad s = 0, 1$$

and $N = n_x + n_y + n_z$. Obviously, the composition result $\mathbf{C}(t)$ is a Bézier curve. The control point $\mathbf{C}_s$ of resultant Bézier curve $\mathbf{C}(t)$ is:

$$\mathbf{C}_s = \sum_{i+j+k=s} \begin{array}{l} (\omega^{(C)}_{ijkuvw} U_0^{n_u - i} U_1^i V_0^{n_v - j} V_1^j \\ W_0^{n_w - k} W_1^k \mathbf{c}_{000}) \end{array} \qquad (6)$$

where $\omega^{(C)}_{ijkuvw} = \left(C_i^{n_u} C_j^{n_v} C_k^{n_w}\right) / C_s^N$. And it can be shown easily that $\sum_{i+j+k=s} \omega^{(C)}_{ijkuvw} \equiv 1$. The control points $\mathbf{C}_s$ can be evaluated by the generalized de Casteljau algorithm which is described in [13].

Let $\mathbf{P}_0\mathbf{P}_1\mathbf{P}_2$ be a triangle, whose local coordinates within the Bézier volume are (u_0, v_0, w_0), (u_1, v_1, w_1), (u_2, v_2, w_2) respectively. This triangle can be parameterized by barycentric coordinates as follows:

$$\begin{aligned} \mathbf{P}(x, y, z) &= (u(x,y,z), v(x,y,z), w(x,y,z)) \\ &= \mathbf{P}_0 x + \mathbf{P}_1 y + \mathbf{P}_2 z \end{aligned} \qquad (7)$$

where $x, y, z \geq 0$ and $x + y + z = 1$. By substituting (7) into (2), we can derive the accurate deformation result of a triangle, i.e.:

$$\begin{aligned} \mathbf{R}(x, y, z) &= \mathbf{r}(u(x,y,z), v(x,y,z), w(x,y,z)) \\ &= [(1-u(x,y,z))I + u(x,y,z)E_u]^{n_u} \\ &\quad [(1-v(x,y,z))I + v(x,y,z)E_v]^{n_v} \\ &\quad [(1-w(x,y,z))I + w(x,y,z)E_w]^{n_w} \end{aligned}$$

$$
\begin{aligned}
& \mathbf{r}_{000} \\
= & [Au_0 x + Au_1 y + Au_2 z]^{n_u} \\
& [Av_0 x + Av_1 y + Av_2 z]^{n_v} \\
& [Aw_0 x + Aw_1 y + Aw_2 z]^{n_w} \mathbf{r}_{000} \\
= & \sum_{i+j+k=N} \mathbf{R}_{ijk} B^N_{ijk}(x, y, z)
\end{aligned}
$$

Where

$$
\left\{ \begin{array}{lcl} Au_s & = & [(1-u_s)I + u_s E_u] \\ Av_s & = & [(1-v_s)I + v_s E_v] \\ Aw_s & = & [(1-w_s)I + w_s E_w] \end{array} \right. \qquad s = 0, 1
$$

$N = n_u + n_v + n_w$. The control point $\mathbf{R}_{ijk}$ of the resultant triangular Bézier patch $\mathbf{R}(x, y, z)$ is:

$$
\begin{aligned}
\mathbf{R}_{ijk} = & \sum_{S(i,j,k,u,v,w)} \Big(\omega^{(R)}_{ijkuvw} Au_0^{i_u} Au_1^{j_u} Au_2^{k_u} \\
& Av_0^{i_v} Av_1^{j_v} Av_2^{k_v} Aw_0^{i_w} Aw_1^{j_w} Aw_2^{k_w} \mathbf{r}_{000} \Big)
\end{aligned} \tag{8}
$$

where $\omega^{(R)}_{ijkuvw} = \left(C^{n_u}_{i_u, j_u, k_u} C^{n_v}_{i_v, j_v, k_v} C^{n_w}_{i_w, j_w, k_w} \right) / C^N_{i,j,k}$, the expression $S(i, j, k, u, v, w)$ means:

$$
\left\{ \begin{array}{l} i_u + i_v + i_w = i \\ j_u + j_v + j_w = j \\ k_u + k_v + k_w = k \\ i_u + j_u + k_u = n_u \\ i_v + j_v + k_v = n_v \\ i_w + j_w + k_w = n_w \end{array} \right.
$$

Obviously the coefficients $\omega^{(R)}_{ijkuvw}$ in equation (8) satisfy: $\sum_{S(i,j,k,u,v,w)} \omega^{(R)}_{ijkuvw} \equiv 1$. The control points $\mathbf{R}_{ijk}$ of the Bézier triangular patch can be evaluated by the generalized de Casteljau algorithm [13].

3.3. Computational complexities of generalized de Casteljau algorithms

Eqs. (6) and (8) show that the control points of resultant curve and patch are the linear convex combinations of control points of Bézier volume. Like the standard de Casteljau algorithm, the procedures described in [13] are numerically stable, thus the accumulated errors are not spread. Nevertheless the computational cost is very high. To compute a point on a Bézier volume with degrees n_u, n_v, n_w along three directions u, v, w successively using standard de Casteljau algorithm, the numbers of floating multiplication($\otimes$), addition($\oplus$) and subtraction($\ominus$) are:

$$
\begin{aligned}
V(n_u, n_v, n_w) = & \tfrac{3}{2} \Big[n_u(n_u+1)(n_v+1)(n_w+1) \\
& + n_v(n_v+1)(n_w+1) \\
& + n_w(n_w+1) \Big] (2 \otimes + \oplus + \ominus)
\end{aligned} \tag{9}
$$

The computational costs of the procedures `Compute_AsIJK` and `Compute_Auvwijk` in [13] are identical with $V(n_u, n_v, n_w)$ except for loops control.

The computational cost $Crv^{(deC)}(n_u, n_v, n_w)$ for evaluating all control points of Bézier curve $\mathbf{C}(t)$ in (6) is:

$$
\begin{aligned}
Crv^{(deC)}(n_u, n_v, n_w) = & (n_u+1)(n_v+1) \\
& (n_w+1) V(n_u, n_v, n_w)
\end{aligned} \tag{10}
$$

Here we assume the combinatorial number is pre-computed and saved in a look-up table. Its computational cost is omitted. Similarly, to evaluate all the control points of the triangular Bézier patch $\mathbf{R}(u, v, w)$ in (8), the floating computation cost $Pth^{(deC)}(n_u, n_v, n_w)$ is:

$$
\begin{aligned}
Pth^{(deC)}(n_u, n_v, n_w) = & \tfrac{1}{8} \Big[(n_u+1)(n_u+2) \\
& (n_v+1)(n_v+2)(n_w+1) \\
& (n_w+2) \Big] V(n_u, n_v, n_w)
\end{aligned} \tag{11}
$$

4. Solving functional composition by polynomial interpolation

According to the theory of numerical analysis, a polynomial of degree n can be reconstructed precisely by interpolating $(n+1)$ different sampling points in its domain. In FFD, the deformation results of a line and a triangle are Bézier curve and patch, which are Bernstein polynomials of 1 and 2 variables respectively. Thus we can obtain the control points of these curve and patch by polynomial interpolation. Nevertheless the conditional number of interpolation matrix will increase fast when the degree of curve or patch increases, the traditional numerical algorithm will introduce large error. To avoid this problem, we adopt the symbolic computation algorithm to compute the inverse matrix. Apparently it is free of numerical error. Further more the matrix can be pre-computed and saved for further use.

4.1. Interpolation of Bézier curve

Let $\{t_i\}_{i=0}^N$ be the $(N+1)$ sampling points on a Bézier curve $\mathbf{C}(t)$ which is the accurate deformation result of a line segment. The sample points $\mathbf{C}(t_i)$ can be calculated by traditional FFD algorithm. Then we can set up a linear equations system as follows:

$$
\mathbf{B}^{(1,N)} \begin{pmatrix} \mathbf{C}_0 \\ \mathbf{C}_1 \\ \vdots \\ \mathbf{C}_N \end{pmatrix} = \begin{pmatrix} \mathbf{C}(t_0) \\ \mathbf{C}(t_1) \\ \vdots \\ \mathbf{C}(t_N) \end{pmatrix} \tag{12}
$$

where matrix $\mathbf{B}^{(1,N)}$ is

$$\mathbf{B}^{(1,N)} = \begin{pmatrix} B_{0,N}(t_0) & B_{1,N}(t_0) & \cdots & B_{N,N}(t_0) \\ B_{0,N}(t_1) & B_{1,N}(t_1) & \cdots & B_{N,N}(t_1) \\ \vdots & \vdots & \ddots & \vdots \\ B_{0,N}(t_N) & B_{1,N}(t_N) & \cdots & B_{N,N}(t_N) \end{pmatrix}$$

$B_{i,N}(t)$ is Bernstein polynomial of degree N. The matrix $\mathbf{B}^{(1,N)}$ is non-degenerate if no two sampling points among $\{t_i\}_{i=0}^N$ are identical. Although the sampling points can be specified freely by user. A standard sampling method is suggested here. We take $\{\frac{i}{N}\}_{i=0}^N$ as the sample points. For such sampling points, the corresponding matrix $\mathbf{B}^{(1,N)}$ is central symmetry because of the following equality:

$$B_{i,n}\left(\frac{j}{n}\right) = B_{n-i,n}\left(\frac{n-j}{n}\right)$$

Since $B_{i,N}(t)$ arrives its maximum at $t = \frac{i}{N}$, the transpose of matrix $\mathbf{B}^{(1,N)}$ is diagonally dominant. This also ensures that the matrix $\mathbf{B}^{(1,N)}$ is invertible. The control points of the Bézier curve $\mathbf{C}(t)$ can be obtained as follows:

$$\begin{pmatrix} \mathbf{C}_0 \\ \mathbf{C}_1 \\ \vdots \\ \mathbf{C}_N \end{pmatrix} = \left(\mathbf{B}^{(1,N)}\right)^{-1} \begin{pmatrix} \mathbf{C}(t_0) \\ \mathbf{C}(t_1) \\ \vdots \\ \mathbf{C}(t_N) \end{pmatrix} \quad (13)$$

Remember in mind that matrix $\left(\mathbf{B}^{(1,N)}\right)^{-1}$ need to be computed only once. They can be pre-computed and saved for further use. In practice, the degree of a B-spline is often less than $3 \times 3 \times 3$. The pre-computed matrices with $N \leq 9$ are enough for practical use. Obviously, each element of matrix $\mathbf{B}^{(1,N)}$ is not great than 1. With the increase of degree N, most elements become smaller and approach to 0. Though the matrix is diagonally dominant, it may still cause large numerical error during computing inverse. The deficiency of matrix $\mathbf{B}^{(1,N)}$ can be described by its conditional number (see Table 1).

N	Conditional Number of matrix $\mathbf{B}^{(1,N)}$	Conditional Number of matrix $\mathbf{B}^{(2,N)}$
2	2.318	2.682
3	5.054	5.761
4	11.802	13.399
5	28.642	32.546
6	71.147	80.427
7	179.452	202.549
8	457.521	514.904
9	1175.77	1321.722

Table 1. Examples of conditional numbers

Symbolic computation can eliminate this problem. By this way, the elements in the matrix are regarded as symbols, rather than floating points with numerical error. Thus evaluation of inverse matrix of $\mathbf{B}^{(1,N)}$ through symbolic computation is free of numerical error. Though the cost of symbolic computation is high, the inverse matrix of $\mathbf{B}^{(1,N)}$ need only be computed for $N \leq 9$, and they can be computed in advance.

As described above, the computation consists of calculating $(N+1)$ sampling points within the Bézier volume and a multiplication between matrix and vector. The total number of floating computation to compute all of control points of the curve is:

$$\begin{aligned} Crv^{(int)}(n_u, n_v, n_w) &= [(N+1)V(n_u, n_v, n_w)] + \\ &\quad [(N+1)^2 \otimes + N(N+1)\oplus] \end{aligned} \quad (14)$$

Where $N = n_u + n_v + n_w$. The first squarely bracketed term is for evaluating the $(N+1)$ sample points through traditional FFD method. The second term is for matrix multiplication.

4.2. Interpolation of triangular Bézier patch

For the case of triangular Bézier patch, we first arrange Bernstein base functions $B_{i,j,N-i-j}(u, v, 1-u-v)$, which are defined on the 2D simplex, in alphabetical order of their subscripts (i, j) as follows:

$$\begin{matrix} B_{0,0,N} & B_{0,1,N-1} & B_{0,2,N-2} & \cdots & B_{0,N,0} \\ & B_{1,0,N-1} & B_{1,1,N-2} & \cdots & B_{1,N-1,0} \\ & \cdots & \cdots & \ddots & \cdots \\ & & B_{N-2,0,2} & B_{N-2,1,1} & B_{N-2,2,0} \\ & & & B_{N-1,0,1} & B_{N-1,1,0} \\ & & & & B_{N,0,0} \end{matrix}$$

Let them be re-noted as $\left\{\tilde{B}_i(\mathbf{u})\right\}_{i=0}^{\tilde{N}}$, where $\tilde{N} = \frac{(N+1)(N+2)}{2}$ and $\mathbf{u} = (u, v, 1-u-v)$. $\tilde{N}$ sampling points are needed to interpolate a triangular Bézier patch. Let the interpolation matrix be noted as $\mathbf{B}^{(2,N)}$ The sufficient condition for choosing such $\tilde{N}$ sampling points, which make the interpolation valid, is the research topic of multivariate splines. It will not be discussed in this paper. We suggest a standard sampling here. The proposed sampling points are just triples of subscripts of base function divided by N, *i.e.*, $\left\{\frac{i}{N}, \frac{j}{N}, \frac{N-i-j}{N}\right\}_{i+j=0}^{N}$. It is an valid sampling scheme since the matrix formed $\mathbf{B}^{(2,N)}$ are also diagonally dominated as $\mathbf{B}^{(1,N)}$. With the increase of degree N, the conditional number of matrix $\mathbf{B}^{(2,N)}$ also increase fast. Some conditional numbers are listed in the table 1. Symbolic computation is employed again to evaluate the inverse matrix of $\mathbf{B}^{(2,N)}$.

Since the inverse matrix of $\mathbf{B}^{(2,N)}$ can be pre-computed, similar with the curve case, the computation cost also consists of two parts, calculating the sampling points and matrix multiplication. The floating computation involved in evaluating all control points of deformation triangular Bézier patch is:

$$Pth^{(int)}(n_u, n_v, n_w) = \left[\tilde{N}V(n_u, n_v, n_w)\right] + \left[\tilde{N}^2 \otimes + \tilde{N}(\tilde{N}-1)\oplus\right] \quad (15)$$

Where $\tilde{N}$ is the number of the control points of resultant triangular Bézier patch.

5. Discussion and implementation

The proposed subdivision and re-triangulation method is suitable for the object, whose representation contains no topological information. The polygonal objects used currently are usually represented by means of vertex list and face list. Thus the proposed method is valid for such case. Our subdivision and re-triangulation process may result very small triangles. Obviously it is not necessary to perform the composition computation for these triangles since they are less affected sample problem. Thus during our implementation, we check the size of each triangle, as measured by maximal edge length. When this is less than a given threshold, only three vertices are deformed as in the original FFD. The resulting triangular mesh can be regarded as a triangular Bézier patch of degree 1. For such case, cracks may arise during the rendering. Some tricks should be adopted to remove it. For a triangle parallel to a coordinate plane in the parametric surface, namely $u = u_0$, the composition can be simplified along the u direction so that the degree of patch is $k_v + k_w$, rather than $k_u + k_v + k_w$.

The proposed interpolation in this paper is also numerical stable since no numerical error is introduced in evaluating the inverse matrix. Some examples of theoretical comparisons between proposed method and method in [13] for some examples are listed in Table 2.

We have implemented the proposed algorithm on an SGI Octane workstation. In each example from Figure. 1 to Figure. 3, Part(a)(up-left) displays an object to be deformed, represented by triangular meshes. Part (b) (up-right) shows the result of traditional FFD. The B-spline control lattice is shown by points and dash lines. Part (c) (bottom-left) shows the object after subdivision and re-triangulation. The parametric space of B-spline volume is described by dashed lines. When the B-spline volume is identical with a Bézier volume, no subdivision is performed. Part (d) (bottom-right) is the accurate FFD result, where each mesh is a triangular Bézier surface patch. Time comparisons of practical implementation are listed in Table 3.

$Crv^{(deC)}(2,2,2) = 2106 \otimes +1053 \oplus +1053\ominus$
$Crv^{(int)}(2,2,2) = 595 \otimes +315 \oplus +273\ominus$
$Crv^{(deC)}(3,3,3) = 16128 \otimes +8064 \oplus +8064\ominus$
$Crv^{(int)}(3,3,3) = 2620 \otimes +1350 \oplus +1260\ominus$
$Pth^{(deC)}(2,2,2) = 16848 \otimes +8424 \oplus +8424\ominus$
$Pth^{(int)}(2,2,2) = 2968 \otimes +1848 \oplus +1092\ominus$
$Pth^{(deC)}(3,3,3) = 252000 \otimes +126000 \oplus +126000\ominus$
$Pth^{(int)}(3,3,3) = 16885 \otimes +9900 \oplus +6930\ominus$

Table 2. Examples of comparisons of two methods in theory

Fig	$Pth^{(deC)}/Pth^{(int)}$	$Crv^{(deC)}/crv^{(int)}$
1	$1.25s/0.24s = 5.21$	$0.57s/0.16s = 3.56$
2	$1.37s/0.39s = 3.51$	$0.84s/0.30s = 2.80$
3	$57.43s/7.35s = 7.81$	$19.84s/4.07s = 3.56$

Table 3. Implementation results

6. Conclusion

In this paper, an accelerated B-spline FFD approach is presented by means of fast functional composition. Rather than calculating the deformed object by generalized de Casteljau algorithm, we adopt interpolation method to obtain all the control points of the deformed Bézier curve and triangular Bézier patch. Computation comparisons between the proposed method and the method in [13] are demonstrated that the proposed method can be easily integrated into existing geometric modeling and computer animation system. Further research should be focused on the accurate free-form deformation of curved object.

Acknowledgements

The authors would like to thank Professor Dominique Bechmann from University of Louis Pasteur in France and Professor Wenping Wang from the University of Hong Kong, who gave us valuable suggestions during our implementation. The work is supported by National Natural Science Foundation of China (Grant No. 69903008). The authors also acknowledge the K.C. Wong Foundation who gives us the financial support to attend the GMP'2000 in Hong Kong.

References

[1] A. Barr, "Global and Local Deformation of Solid Primitives", *Computer Graphics (Siggraph'84)*, 18(3):21–30, 1984.

[2] D. Bechmann, "Space Deformation Survey", *Computers&Graphics*, 18(4):571–586, 1994.

[3] D.Bechmann, Y.Bertrand and S.Thery, "Continuous Free-Form Deformation", *COMPUGRAPHICS'96 and a special issue of Computer Networks and ISDN systems*, Elsevier Science, Paris, Dec. 1996.

[4] D. Bechmann,"Multidimensional Free-Form Deformation Dools", In de Sousa A and Hopgood B Eds. *State of the art report (Eurographics'98)*, pp.102–110, Blackwell Publisher, Lisbon, Portugal, 1998.

[5] P. Borrel, D. Bechmann, "Deformation of N-dimensional Objects", *Symposium on Solid Modeling Foundations and CADCAM Applications*, ACM Press, Austin, Texas, June 1991. Also published in *International Journal of Computational Geometry & Applications*, 1(4):137–155, 1991.

[6] Y. Chang, A. Rockwood, "A Generalized de Casteljau Approach to 3D Free-Form Deformation", *Siggraph'97*, pp. 187–196, 1997.

[7] S. Coquillart,"Extended Free-Form Deformation: A Sculpturing Tool for 3D Geometric Modeling", *Computer Graphics (Siggraph'90)* 24(4):187–193, 1990.

[8] S. Coquillart, P. Jancene, "Animated Free-Form Deformations: An Interactive Animation Technique", *Computer Graphics(Siggraph'91)*, 25(4):23–26, 1991.

[9] T. DeRose, "Compositing Bézier simplex", *ACM Trans. on Graphics*, 7(3):198–221, 1988.

[10] T. DeRose, R. Goldman, H. Hagen, ans S. Mann, "Functional Composition Algorithms via Blossoming", *ACM Trans. on Graphics*, 12(3):113–135, 1993.

[11] G. Farin, *Curves and Surfaces for Computer-Aided Geometric Design*, 2nd Edition, Academic Press, New York, 1990.

[12] J. Feng, L. Ma and, Q.Peng, "A New Free-Form Deformation through the Control of Parametric Surfaces", *Computers & Graphics*, 20(4):531–539. 1996.

[13] J. Feng, P. Heng, and T. Wong, "Accurate B-spline Free-Form Deformation of Plygonal Objects". *Journal of Graphics Tools*, 3(3):11-27, 1999.

[14] J. Griessmair, W. Purgathofer, "Deformation of Solids with trivariate B-splines", *Eurographics'89*, pp. 137–148, 1989.

[15] W. Hsu, J. Hughes, and H. Kaufmann, "Direct Manipulation on Free-Form Deformation", *Computer Graphics(Siggraph'92)*, 26(2):177–184, 1992.

[16] P. Karla, A. Mangli, N.Thalmann, and D.Thalmann, "Simulation of Facial Muscle Actions Based on Rational Free-Form Deformation",*Computer Graphics Forum (Eurographics'92)*, 2(3):59–69, 1992.

[17] H. Lamousin, W. Waggenspack, "NURBS-based Free-form Deformation", *IEEE Computer Graphics and Applications*, 14(9):59–65, 1994.

[18] F. Lazarus, S. Coquillart, and P. Jancene, "Axial Deformation: An Intuitive Deformation Technique", *Computer-Aided Design*, 26(8):607–613, 1994.

[19] R. MacCracken, K. Joy, "Free-Form Deformation with Lattice of Arbitrary Topology", *Siggraph'96*, pp. 181–183, 1996.

[20] L. Moccozet, N. Magnenat-Thalmann, "Dirichlet Free-Form Deformations and Their Application to Hand Simulation", *Computer Animation'97*, IEEE Computer Society, pp. 93–102, 1997.

[21] T. Sederberg, S. Parry, "Free-Form Deformation of Solid Geometric Models", *Computer Graphics (Siggraph'86)*, 20(4):537–541, 1986.

[22] K. Singh, E. Fiume, "Wires: A Geometric Deformation Technique", *Siggraph'98*, pp. 404-413 , 1998.

Author Index

Notes

Revised 9 November 1999